T0173012

advances in commutative ring theory

PURE AND APPLIED MATHEMATICS

A Program of Monographs, Textbooks, and Lecture Notes

LECTURE NOTES IN PURE AND APPLIED MATHEMATICS

61. *O. A. Nielson*, Direct Integral Theory
62. *J. E. Smith et al.,* Ordered Groups
63. *J. Cronin,* Mathematics of Cell Electrophysiology
64. *J. W. Brewer,* Power Series Over Commutative Rings
65. *P. K. Kamthan and M. Gupta,* Sequence Spaces and Series
66. *T. G. McLaughlin,* Regressive Sets and the Theory of Isols
67. *T. L. Herdman et al.,* Integral and Functional Differential Equations
68. *R. Draper,* Commutative Algebra
69. *W. G. McKay and J. Patera,* Tables of Dimensions, Indices, and Branching Rules for Representations of Simple Lie Algebras
70. *R. L. Devaney and Z. H. Nitecki,* Classical Mechanics and Dynamical Systems
71. *J. Van Geel,* Places and Valuations in Noncommutative Ring Theory
72. *C. Faith,* Injective Modules and Injective Quotient Rings
73. *A. Fiacco,* Mathematical Programming with Data Perturbations I
74. *P. Schultz et al.,* Algebraic Structures and Applications
75. *L Bican et al.,* Rings, Modules, and Preradicals
76. *D. C. Kay and M. Breen,* Convexity and Related Combinatorial Geometry
77. *P. Fletcher and W. F. Lindgren,* Quasi-Uniform Spaces
78. *C.-C. Yang,* Factorization Theory of Meromorphic Functions
79. *O. Taussky,* Ternary Quadratic Forms and Norms
80. *S. P. Singh and J. H. Burry,* Nonlinear Analysis and Applications
81. *K. B. Hannsgen et al.,* Volterra and Functional Differential Equations
82. *N. L. Johnson et al.,* Finite Geometries
83. *G. I. Zapata,* Functional Analysis, Holomorphy, and Approximation Theory
84. *S. Greco and G. Valla,* Commutative Algebra
85. *A. V. Fiacco,* Mathematical Programming with Data Perturbations II
86. *J.-B. Hiriart-Urruty et al.,* Optimization
87. *A. Figa Talamanca and M. A. Picardello,* Harmonic Analysis on Free Groups
88. *M. Harada,* Factor Categories with Applications to Direct Decomposition of Modules
89. *V. I. Istrăţescu,* Strict Convexity and Complex Strict Convexity
90. *V. Lakshmikantham,* Trends in Theory and Practice of Nonlinear Differential Equations
91. *H. L. Manocha and J. B. Srivastava,* Algebra and Its Applications
92. *D. V. Chudnovsky and G. V. Chudnovsky,* Classical and Quantum Models and Arithmetic Problems
93. *J. W. Longley,* Least Squares Computations Using Orthogonalization Methods
94. *L. P. de Alcantara,* Mathematical Logic and Formal Systems
95. *C. E. Aull,* Rings of Continuous Functions
96. *R. Chuaqui,* Analysis, Geometry, and Probability
97. *L. Fuchs and L. Salce,* Modules Over Valuation Domains
98. *P. Fischer and W. R. Smith,* Chaos, Fractals, and Dynamics
99. *W. B. Powell and C. Tsinakis,* Ordered Algebraic Structures
100. *G. M. Rassias and T. M. Rassias,* Differential Geometry, Calculus of Variations, and Their Applications
101. *R.-E. Hoffmann and K. H. Hofmann,* Continuous Lattices and Their Applications
102. *J. H. Lightbourne III and S. M. Rankin III,* Physical Mathematics and Nonlinear Partial Differential Equations
103. *C. A. Baker and L. M. Batten,* Finite Geometrics
104. *J. W. Brewer et al.,* Linear Systems Over Commutative Rings
105. *C. McCrory and T. Shifrin,* Geometry and Topology
106. *D. W. Kueke et al.,* Mathematical Logic and Theoretical Computer Science
107. *B.-L. Lin and S. Simons,* Nonlinear and Convex Analysis
108. *S. J. Lee,* Operator Methods for Optimal Control Problems
109. *V. Lakshmikantham,* Nonlinear Analysis and Applications
110. *S. F. McCormick,* Multigrid Methods
111. *M. C. Tangora,* Computers in Algebra
112. *D. V. Chudnovsky and G. V. Chudnovsky,* Search Theory
113. *D. V. Chudnovsky and R. D. Jenks,* Computer Algebra
114. *M. C. Tangora,* Computers in Geometry and Topology
115. *P. Nelson et al.,* Transport Theory, Invariant Imbedding, and Integral Equations
116. *P. Clément et al.,* Semigroup Theory and Applications
117. *J. Vinuesa,* Orthogonal Polynomials and Their Applications
118. *C. M. Dafermos et al.,* Differential Equations
119. *E. O. Roxin,* Modern Optimal Control
120. *J. C. Díaz,* Mathematics for Large Scale Computing
121. *P. S. Milojevič* Nonlinear Functional Analysis
122. *C. Sadosky,* Analysis and Partial Differential Equations

123. *R. M. Shortt,* General Topology and Applications
124. *R. Wong,* Asymptotic and Computational Analysis
125. *D. V. Chudnovsky and R. D. Jenks,* Computers in Mathematics
126. *W. D. Wallis et al.,* Combinatorial Designs and Applications
127. *S. Elaydi,* Differential Equations
128. *G. Chen et al.,* Distributed Parameter Control Systems
129. *W. N. Everitt,* Inequalities
130. *H. G. Kaper and M. Garbey,* Asymptotic Analysis and the Numerical Solution of Partial Differential Equations
131. *O. Arino et al.,* Mathematical Population Dynamics
132. *S. Coen,* Geometry and Complex Variables
133. *J. A. Goldstein et al.,* Differential Equations with Applications in Biology, Physics, and Engineering
134. *S. J. Andima et al.,* General Topology and Applications
135. *P Clément et al.,* Semigroup Theory and Evolution Equations
136. *K. Jarosz,* Function Spaces
137. *J. M. Bayod et al.,* p-adic Functional Analysis
138. *G. A. Anastassiou,* Approximation Theory
139. *R. S. Rees,* Graphs, Matrices, and Designs
140. *G. Abrams et al.,* Methods in Module Theory
141. *G. L. Mullen and P. J.-S. Shiue,* Finite Fields, Coding Theory, and Advances in Communications and Computing
142. *M. C. Joshi and A. V. Balakrishnan,* Mathematical Theory of Control
143. *G. Komatsu and Y. Sakane,* Complex Geometry
144. *I. J. Bakelman,* Geometric Analysis and Nonlinear Partial Differential Equations
145. *T. Mabuchi and S. Mukai,* Einstein Metrics and Yang–Mills Connections
146. *L. Fuchs and R. Göbel,* Abelian Groups
147. *A. D. Pollington and W. Moran,* Number Theory with an Emphasis on the Markoff Spectrum
148. *G. Dore et al.,* Differential Equations in Banach Spaces
149. *T. West,* Continuum Theory and Dynamical Systems
150. *K. D. Bierstedt et al.,* Functional Analysis
151. *K. G. Fischer et al.,* Computational Algebra
152. *K. D. Elworthy et al.,* Differential Equations, Dynamical Systems, and Control Science
153. *P.-J. Cahen, et al.,* Commutative Ring Theory
154. *S. C. Cooper and W. J. Thron,* Continued Fractions and Orthogonal Functions
155. *P. Clément and G. Lumer,* Evolution Equations, Control Theory, and Biomathematics
156. *M. Gyllenberg and L. Persson,* Analysis, Algebra, and Computers in Mathematical Research
157. *W. O. Bray et al.,* Fourier Analysis
158. *J. Bergen and S. Montgomery,* Advances in Hopf Algebras
159. *A. R. Magid,* Rings, Extensions, and Cohomology
160. *N. H. Pavel,* Optimal Control of Differential Equations
161. *M. Ikawa,* Spectral and Scattering Theory
162. *X. Liu and D. Siegel,* Comparison Methods and Stability Theory
163. *J.-P. Zolésio,* Boundary Control and Variation
164. *M. Křížek et al.,* Finite Element Methods
165. *G. Da Prato and L. Tubaro,* Control of Partial Differential Equations
166. *E. Ballico,* Projective Geometry with Applications
167. *M. Costabel et al.,* Boundary Value Problems and Integral Equations in Nonsmooth Domains
168. *G. Ferreyra, G. R. Goldstein, and F. Neubrander,* Evolution Equations
169. *S. Huggett,* Twistor Theory
170. *H. Cook et al.,* Continua
171. *D. F. Anderson and D. E. Dobbs,* Zero-Dimensional Commutative Rings
172. *K. Jarosz,* Function Spaces
173. *V. Ancona et al.,* Complex Analysis and Geometry
174. *E. Casas,* Control of Partial Differential Equations and Applications
175. *N. Kalton et al.,* Interaction Between Functional Analysis, Harmonic Analysis, and Probability
176. *Z. Deng et al.,* Differential Equations and Control Theory
177. *P. Marcellini et al.* Partial Differential Equations and Applications
178. *A. Kartsatos,* Theory and Applications of Nonlinear Operators of Accretive and Monotone Type
179. *M. Maruyama,* Moduli of Vector Bundles
180. *A. Ursini and P. Aglianò,* Logic and Algebra
181. *X. H. Cao et al.,* Rings, Groups, and Algebras
182. *D. Arnold and R. M. Rangaswamy,* Abelian Groups and Modules
183. *S. R. Chakravarthy and A. S. Alfa,* Matrix-Analytic Methods in Stochastic Models
184. *J. E. Andersen et al.,* Geometry and Physics
185. *P.-J. Cahen et al.,* Commutative Ring Theory

Additional Volumes in Preparation

advances in commutative ring theory

proceedings of the third international conference on commutative ring theory in Fez, Morocco

edited by

David E. Dobbs
University of Tennessee
Knoxville, Tennessee

Marco Fontana
Università degli Studi Roma Tre
Rome, Italy

Salah-Eddine Kabbaj
King Fahd University of
Petroleum and Minerals
Dhahran, Saudi Arabia

CRC Press
Taylor & Francis Group
Boca Raton London New York

CRC Press is an imprint of the
Taylor & Francis Group, an **informa** business

CRC Press
Taylor & Francis Group
6000 Broken Sound Parkway NW, Suite 300
Boca Raton, FL 33487-2742
First issued in hardback 2017

First issued in hardback 2018

© 1999 by Taylor & Francis Group, LLC
CRC Press is an imprint of Taylor & Francis Group, an Informa business

No claim to original U.S. Government works

ISBN 13: 978-1-138-40217-1 (hbk)
ISBN 13: 978-0-8247-7147-8 (pbk)

**Visit the Taylor & Francis Web site at
http://www.taylorandfrancis.com**

**and the CRC Press Web site at
http://www.crcpress.com**

Library of Congress Cataloging-in-Publication Data

International Conference on Commutative Ring Theory (3rd: Fés, Morocco)
    Advances in commutative ring theory: proceedings of the Third International Conference on Commutative Ring Theory in Fez, Morocco / edited by David E. Dobbs, Marco Fontana, Salah–Eddine Kabbaj.
    p.   cm.— (Lecture notes in pure and applied mathematics; 205)
    Includes bibliographical references and index.
    ISBN 0-8247-7147-8 (alk. paper)
    1. Commutative rings—Congresses. I. Dobbs, David E. II. Fontana, Marco. III. Kabbaj, Salah–Eddine. IV. Title. V. Series: Lecture notes in pure and applied mathematics; v. 205.
QA251.3.I58 1999
512'.4—dc21   99-18800
                                                                                      CIP

Preface

This volume comprises the proceedings of the third International Conference in Commutative Ring Theory, held at the University of Fez, Fez, Morocco. All the articles in this volume were subject to a strict refereeing process. Most of the papers reflect the authors' contribution to the conference.

As was the case with the first and second conferences, the purpose of this conference is to present recent progress and new trends in commutative ring theory. The papers appearing in this volume adhere to this goal. While 26 of them are completely original and provide new results on some problems of current interest, the six remaining (survey) papers present useful reports on some basic topics in commutative algebra.

It is our hope that the volume will be a valuable source for specialists in algebra and related fields. Further, by providing extensive references and a detailed index, it should help young researchers to find valuable access to the literature.

Topics treated include group rings, n-generator property, π-domains, factorization and irreducibility criteria, root closure, class group, $A + X\,B[X]$ domains, pseudo-valuation rings, spectral topology, Hermite rings, semi-Steinitz rings, Krull dimension, integer-valued polynomials, Skolem properties, normsets, coherence, Kaplansky ideal transform, polynomial closure, pullbacks, polynomial functions, Koszul algebras, primary decomposition, polynomial rings, Prüfer domains, Noetherian domains, completions, trace properties, plane cubic curves, monoid rings, GCD domains, seminormality, and the Krull–Schmidt property.

We are grateful to the participants, who came from five continents to make the conference a success. No such event can take place without much hard work and financial support. We received funding from the Faculty of Sciences "Dhar Al-Mehraz"—Fez, the EST—Fez, and the International Mathematical Union (IMU/CDE); whose support we gratefully acknowledge. We wish to express our gratitude to the local committee, especially to Prof. Rachid Ameziane Hassani, Chairman of the Mathematics Department. We also are particularly grateful to Prof. Mohamed Saghi and Mr. Aziz Chad for their wise suggestions and advice during the organization of the conference. In addition, D. E. Dobbs thanks the University of Tennessee, Knoxville, for travel support. We also thank Maria Allegra, our editor at Marcel Dekker, Inc., for her encouragement and cooperation.

David E. Dobbs
Marco Fontana
Salah-Eddine Kabbaj

Contents

Contributors

Souad Ameziane Hassani, Department of Mathematics, FST Fez-Saïss, University of Fez, Fez, Morocco. E-mail: fst-saiss-fes@fesnet.net.ma

D. D. Anderson, Department of Mathematics, The University of Iowa, Iowa City, IA 52242, USA. E-mail: dan-anderson@uiowa.edu

David F. Anderson, Department of Mathematics, University of Tennessee, Knoxville, TN 37996-1300, USA. E-mail: anderson@novell.math.utk.edu

Ayman Badawi, Department of Mathematics and Computer Science, Birzeit University, P.O.Box 14, Birzeit, West Bank, Palestine, via Israel. E-mail: abring@math.birzeit.edu

Ezzeddine Bouacida, Department of Mathematics, Faculty of Sciences, Sfax, Tunisia

Abdelmalek Bouanane, Department of Mathematics, Faculty of Sciences, University of Tetouan, Tetouan, Morocco

Samir Bouchiba, Department of Mathematics, Faculty of Sciences, University of Meknes, Meknes, Morocco

J. Boulanger, Institut Universitaire de Formation des Maîtres, Amiens, France

Paul-Jean Cahen, Département de Mathématiques - Case 322, Faculté des Sciences de St. Jérôme, Université d'Aix-Marseille III, 13397 Marseille Cedex 13, France. E-mail: paul-jean.cahen@math.u-3mrs.fr

Jean-Luc Chabert, Faculté de Mathématiques et Informatique, Université de Picardie, Amiens, France. E-mail: jlchaber@worldnet.fr

Maria Contessa, Dipartimento de Matematica el Applicazioni, Università degli Studi de Palermo, 90123 Palermo, Italy

Jim Coykendall, Department of Mathematics, North Dakota State University, Fargo, North Dakota 58105-5075, USA

David E. Dobbs, Department of Mathematics, University of Tennessee, Knoxville, TN 37996-1300, USA. E-mail: dobbs@novell.math.utk.edu

Othman Echi, Department of Mathematics, Faculty of Sciences, Sfax, Tunisia

Said Elbaghdadi, Department of Mathematics, FST, Beni Mellal, Morocco. E-mail: baghdadi@fstbm.ac.ma

S. Evrard, Institut Universitaire de Formation des Maîtres, Amiens, France

Marco Fontana, Dipartimento di Matematica, Università degli Studi di Roma Tre, 00146 Roma, Italy. E-mail: fontana@matrm3.mat.uniroma3.it

Sophie Frisch, Institut für Mathematik, Technische Universität Graz, A-8010 Graz, Austria. E-mail: frisch@blah.math.tu-graz.ac.at

Ralf Fröberg, Matematiska Institutionen, Stockholms Universitet, 10691 Stockholm, Sweden. E-mail: ralff@matematik.su.se

Guang Fu, Department of Mathematics, Florida State University, Tallahassee, Florida 32306-4510, USA

Stefania Gabelli, Dipartimento di Matematica, Università di Roma "La Sapienza", 00185 Roma, Italy. E-mail: gabelli@mat.uniroma1.it

G. Gerboud, Institut Universitaire de Formation des Maîtres, Amiens, France

Robert Gilmer, Department of Mathematics, Florida State University, Tallahassee, Florida 32306-4510, USA

Florida Girolami, Dipartimento di Matematica, Università degli Studi di Roma Tre, 00146 Roma, Italy. E-mail: girolami@matrm3.mat.uniroma3.it

William Heinzer, Department of Mathematics, Purdue University, West Lafayette, IN 47907-1968, USA

Lahoucine Izelgue, Department of Mathematics, Faculty of Sciences "Semlalia", University of Marrakech, Marrakech, Morocco. E-mail: fssm.math@cybernet.net.ma

Laura E. Johnson, Department of Mathematics, University of Tennessee, Knoxville, TN 37996-1300, USA

Salah-Eddine Kabbaj, Department of Mathematics, Faculty of Sciences I, University of Fez, Fez, Morocco. Current address: Department of Mathematical Sciences, King Fahd University of Petroleum and Minerals (KFUPM), P.O.Box 849, Dhahran 31261, Saudi Arabia. E-mail: kabbaj@kfupm.edu.sa

Thomas G. Lucas, Department of Mathematics, University of North Carolina at Charlotte, Charlotte, NC 28223, USA. E-mail: tglucas@charweb.org

Najib Mahdou, Department of Mathematics, FST Fez-Saïss, University of Fez, Fez, Morocco. E-mail: fst-saiss-fes@fesnet.net.ma

Abdeslam Mimouni, Department of Mathematics, Faculty of Sciences I, University of Fez, Fez, Morocco

Joe L. Mott, Department of Mathematics, Florida State University, Tallahassee, Florida 32306-4510, USA

S. B. Mulay, Department of Mathematics, University of Tennessee, Knoxville, TN 37996-1300, USA. E-mail: mulay@math.utk.edu

Bernadette Mullins, Department of Mathematics and Statistics, Youngstown State University, Youngstown, Ohio 44555-3302, USA

Warren Nichols, Department of Mathematics, Florida State University, Tallahassee, Florida 32306-4510, USA

James S. Okon, Department of Mathematics, California State University, San Bernardino, CA 92407, USA. E-mail: jokon@wiley.csusb.edu

Jeanam Park, Department of Mathematics, Inha University, Inchon, 402-751, Korea

Gabriel Picavet, Laboratoire de Mathématiques Pures, Université Blaise Pascal, 63177 Aubière Cedex, France. E-mail: picavet@ucfma.univ-bpclermont.fr

Martine Picavet-L'Hermitte, Laboratoire de Mathématiques Pures, Université Blaise Pascal, 63177 Aubière Cedex, France. E-mail: picavet@ucfma.univ-bpclermont.fr

Christel Rotthaus, Department of Mathematics, Michigan State University, East Lansing, MI 48824-1027, USA

Ezzeddine Salhi, Department of Mathematics, Faculty of Sciences, Sfax, Tunisia

Mohamed Sobrani, Department of Mathematics, FST Fez-Saïss, University of Fez, Fez, Morocco. E-mail: fst-saiss-fes@fesnet.net.ma

Francesca Tartarone, Dipartimento di Matematica, Università di Roma "La Sapienza", 00185 Roma, Italy. E-mail: tartaron@marte.mat.uniroma1.it

Ayse A. Teymuroglu, Ziya-ur-Rahman cad, 11, Sokak N. 32/13, 06610 Gankaya Ankara, Turkey

J. Paul Vicknair, Department of Mathematics, California State University, San Bernardino, CA 92407, USA. E-mail: jvicknai@wiley.csusb.edu

Sylvia Wiegand, Department of Mathematics and Statistics, University of Nebraska, Lincoln, NE 68588-0323, USA. E-mail: swiegand@math.unl.edu

Roger Wiegand, Department of Mathematics and Statistics, University of Nebraska, Lincoln, NE 68588-0323, USA. E-mail: rwiegand@math.unl.edu

Group Rings $R[G]$ with 4-Generated Ideals When R is an Artinian Ring with the 2-Generator Property

SOUAD AMEZIANE HASSANI   Department of Mathematics, Faculty of Sciences Saïss, University of Fez, Fez, Morocco.

SALAH-EDDINE KABBAJ   Department of Mathematical Sciences, KFUPM, P.O.Box 849, Dhahran 31261, Saudi Arabia.

INTRODUCTION

For the convenience of the reader, let's recall the following facts. We have from the restriction on Krull dimension, $1 \geq dimR[G] = dimR + r$, where r denotes the torsion free rank of G. If $r = 0$, then G must be a finite group. If $r = 1$, then $G \cong Z \oplus H$, where H is a finite abelian group and Z the group of the integers. We will concentrate on the case in which R is Artinian and r = 0, that is, G is a finite abelian group. The cases $n = 2$ and $n = 3$ were considered in [15, Theorem 4.1] and [1], respectively. However, for $n \geq 4$, the problem of when $R[G]$ has the $n-$generator property remains open.

As the problem of determining when a group ring $R[G]$ has the $4-$generator property, when R is an Artinian principal ideal ring and G is a finite group is resolved in [2], in this paper, we consider the case where R is an Artinian ring with the 2-generator property.

Rings and groups are taken to be commutative and the groups are written additively. If p is a prime integer, then the $p-$sylow subgroup of the finite abelian group G will be denoted G_p. When I is an ideal of R, we shall use $\mu(I)$ to denote the number of generators

1

in a minimal basis for I. Finally, recall that in a local ring (R, m), if I is n−generated, then the n generators of I may be chosen from elements of a given set of generators of I (cf. [13, (5.3), p. 14]).

PROPOSITION 1 *Assume that* G *is a nontrivial finite 2−group,* (R, M) *is an Artinian local ring with the 2-generator property but* R *is not a principal ideal ring and that* $2 \in M$. *Then* $R[G]$ *has the 4−generator property if and only if* $G \cong Z/2^i Z$, *where*
   (1) $i \geq 1$ *if* M^2 *is a principal ideal and* $M^3 = 0$
   (2) $1 \leq i \leq 2$ *if* M^2 *is a principal ideal,* $M^3 \neq 0$ *and* $M^2 \subset (2)$.
   (3) $i = 1$ *otherwise.*

Proof.   ⇒] Assume that G is not a cyclic group and $R[G]$ has the 4−generator property. Then the homomorphic image $R[Z/2Z \oplus Z/2Z]$ does also. Hence N^2 is 4−generated where $N = (u, v, 1 - X^g, 1 - X^h)$, $M = (u, v)$ and $< g > \oplus < h >= Z/2Z \oplus Z/2Z$. Since $| < g > | = 2$ and $2 \in M$, then $N^2 = (u^2, v^2, uv, u(1 - X^g), v(1 - X^g), u(1 - X^h), v(1 - X^h), (1 - X^g)(1 - X^h))$.

It is easy to see that $(1 - X^g)(1 - X^h)$ is required as a generator of N^2. Since $M = (u, v)$ is not a principal ideal, it is also easy to verify that $u(1 - X^g), v(1 - X^g), u(1 - X^h)$ and $v(1 - X^h)$ are required as generators of N^2. Therefore N^2 needs more than four generators, a contradiction.

(1) Trivial.

(2) Since M^2 is a principal ideal, one can easily check that M^3 is a principal ideal too. Further, we may assume $M = (2, v)$ since $2 \in M \setminus M^2$. Suppose that $R[Z/8Z]$ has the 4-generator property and let $< g >= Z/8Z$, $M^2 = (\alpha)$, and $M^3 = (\mu)$. We have
   $N = (2, v, 1 - X^g)$;
   $N^2 = (\alpha, 2(1 - X^g), v(1 - X^g), (1 - X^g)^2)$;
   $N^3 = (\mu, \alpha(1 - X^g), 2(1 - X^g)^2, v(1 - X^g)^2, (1 - X^g)^3)$.
   Since $M^3 \neq 0$ and $| < g > | > 3$, it is clear that μ and $(1 - X^g)^3$ are required as generators of N^3.

If $\alpha(1 - X^g)$ is a redundant generator of N^3, then by passing to the homomorphic image $R/M^3[< g >]$ and by using [1, Lemma 1.4], we get $\alpha = 8\lambda$ for some $\lambda \in R/M^3$. It follows that $\alpha \in M^3$, whence $M^2 = M^3$, i. e., $M^2 = 0$, a contradiction.

If $2(1 - X^g)^2$ is redundant, then passing to the homomorphic image $R/(4, v)[< g >]$ yields $2(1 - X^g)^2 = \sum_{i=0}^{i=7} a_i X^{ig}(1 - X^g)^3$ where $a_i \in R/(4, v)$. After setting corresponding terms equal, we obtain a system of 8 linear equations in 8 unknowns. After resolving this system, we obtain $2 = 0$ in $R/(4, v)$, i. e., $M = (2, v) = (2^2, v) = (2^3, v) = \cdots = (v)$, since R is Artinian, a contradiction.

If $v(1 - X^g)^2$ is redundant, then passing to the homomorphic image $R/(2, v^2)[< g >]$, yields $v(1 - X^g)^2 \in (1 - X^g)^3 R/(2, v^2)[< g >]$, whence $v(1 - X^g)^7 \in (1 - X^g)^8 R/(2, v^2)[< g >] = 0$. Therefore $v \in (2, v^2)$ i. e., $M = (2, v) = (2)$, a contradiction. Consequently, N^3 is not 4−generated.

(3) We consider separately three subcases. *case*1 : Assume M^2 is not a principal ideal. It suffices to prove that $R[Z/4Z]$ does not have the 4−generator property.

Since M and M^2 are not principal ideals and $| < g > | > 3$, it is easily seen that $N^2 = (u^2, v^2, uv, u(1 - X^g), v(1 - X^g), (1 - X^g)^2)$ is not 4−generated where $M = (u, v)$ and $< g >= Z/4Z$.

*case*2: Assume M^2 is a principal ideal, $M^3 \neq 0$, and $2 \in M^2$. We claim that N^3 is not 4−generated in $R[Z/4Z]$, where $N = (u, v, 1 - X^g)$ and $< g >= Z/4Z$. Indeed, we have $N^2 = (\alpha, u(1 - X^g), v(1 - X^g), (1 - X^g)^2)$ and $N^3 = (\alpha u, \alpha v, \alpha(1 - X^g), u(1 - X^g)^2, v(1 - X^g)^2, (1 - X^g)^3)$, where $M^2 = (\alpha)$.

$| < g > | = 4$ implies that $(1 - X^g)^3$ is required as a generator of N^3. If $u(1 - X^g)^2$ is redundant, then passing to the homomorphic image $R/(u^2, v)[< g >]$ yields $u(1 - X^g)^2 \in (1 - X^g)^3 R/(u^2, v)[< g >]$, whence $u(1 - X^g)^3 \in (1 - X^g)^4 R/(u^2, v)[< g >] \subset 2R/(u^2, v)[< g >]$. Since $2 \in M^2$ and $R/(u^2, v)[< g >]$ is a free $(R/(u^2, v))$-module, then $u \in (u^2, v)$, a contradiction. Likewise for $v(1 - X^g)^2$.

If $\alpha(1 - X^g)$ is a redundant generator of N^3, then passing to the homomorphic image $R/M^3[< g >]$ yields $\alpha(1 - X^g) \in (1 - X^g)^2 R/M^3[< g >]$. By [1, Lemma 1.4] $\alpha = 4\lambda$, for some $\lambda \in R/M^3$. It follows that $\alpha = 0$ in R/M^3, i.e., $M^2 = (\alpha) = 0$, a contradiction.

Since $M^3 \neq 0$, it is clear that N^3 needs more than four generators. Consequently, $R[Z/4Z]$ does not have the 4−generator property.

*Case*3: Assume M^2 is a principal ideal, $M^3 \neq 0$, $2 \in M \setminus M^2$, and $M^2 \not\subset (2)$. Clearly, M^3 is principal. Further, we may assume $M = (2, v)$, and hence $M^2 = (v^2)$. We claim that $R[Z/4Z]$ does not have the 4−generator property. Effectively,

Suppose $4 \notin M^4$. It follows from the assumption $M^2 \not\subset (2)$ that $4 \in M^3 \setminus M^4$, and hence $M^3 = (4)$.

In $R[Z/4Z]$, let $I = (4, v^2(1 - X^g), 2(1 - X^g), v(1 - X^g)^2, (1 - X^g)^3)$ where $< g >= Z/4Z$. Since $4 \neq 0$ and $| < g > | > 3$, it is easily checked that 4 and $(1 - X^g)^3$ are required as generators of I. Moreover, using techniques similar to ones used above, we prove that $v(1 - X^g)^2$ must appear in a party of 4 generators extracted from the original set of generators of I. If $v^2(1 - X^g)$ is redundant, then passing to the homomorphic image $R/(2)[< g >]$ yields $v^2(1-X^g) \in (1 - X^g)^2 R/(2)[< g >]$. By [1, Lemma 1.4], we have $v^2 = 0$ in $R/(2)[< g >]$, i. e., $v^2 \subset (2)$, a contradiction since $M^2 = (v^2) \not\subset (2)$. Therefore $I = (4, v^2(1-X^g), v(1-X^g)^2, (1-X^g)^3)$. Now $2(1-X^g) \in I$, then passing to the homomorphic image $R/(4,v)[< g >]$ yields $2(1-X^g) = \sum_{i=0}^{i=3} a_i X^{ig}(1-X^g)^3$, where $a_i \in R/(4,v)$. After setting corresponding terms equal, we obtain the following equations :

X^o	$a_o - a_1 + 3a_2 - 3a_3 = 2$
X^g	$-3a_o + a_1 - a_2 + 3a_3 = -2$
X^{2g}	$3a_o - 3a_1 + a_2 - a_3 = 0$
X^{3g}	$-a_o + 3a_1 - 3a_2 + a_3 = 0$

This yields $2 = 0$ in $R/(4,v)$, i. e., $M = (2,v) = (v)$, a contradiction. Consequently, I needs more than four generators.

Suppose $4 \in M^4$. Let $M^3 = (\mu)$, if $M^3 \not\subset (2)$, we consider $I = (2, \mu, v^2(1-X^g), v(1-X^g)^2, (1-X^g)^3)$. Since $2 \notin M^2$, $M^3 \not\subset (2)$ and $| < g > | > 3$, it is an easy matter to verify that 2, μ and $(1-X^g)^3$ are required as generators of I. Moreover, using arguments similar to ones used above, it is easy to check that $v^2(1-X^g)$ and $v(1-X^g)^2$ are required as generators of I. Thus I is not 4-generated.
If $M^3 \subset (2)$, then $\mu = 2\lambda$ where $\lambda \in M$ since $2 \in M \setminus M^2$. Therefore $\mu = 4\alpha_1 + 2\alpha_2 v$, where $\alpha_1, \alpha_2 \in R$.

Since $M^3 \neq 0$, $M^2 \not\subset (2)$, M^3 is a principal ideal and $4 \in M^4$, then $M^3 = (2v)$.

On the other hand, $M^3 = (v^3, 2v^2)$. Since R is an Artinian ring and $2v \in M^3$, then $M^3 = (v^3)$, whence there exists λ a unit in R such that $2v = \lambda v^3$. Let $I = (v^3, 2-\lambda v^2, v^2(1-X^g), v(1-X^g)^2, (1-X^g)^3)$. As before, one can easily check that $v(1 - X^g)^2$ and $(1 - X^g)^3$ are required as generators of I. If $2 - \lambda v^2$ is redundant, then passing to the homomorphic image $R[< g >]/((1 - X^g)) \simeq R$ yields $2 - \lambda v^2 \in$

(v^3), i. e., $2 - \lambda v^2 = \beta v^3$ where $\beta \in R$. Hence $v^2 \in (2)$, so that $M^2 = (v^2) \subset (2)$, a contradiction. If v^3 is redundant, then passing to the homomorphic image $R[< g >]/((1 - X^g)) \simeq R$, we obtain that $v^3 \in (2 - \lambda v^2)$, i. e., $v^3 = \beta(2 - \lambda v^2)$ where $\beta \in R$. Since $M^2 = (v^2) \not\subset (2)$, $M^3 \subset (2)$ and λ is a unit, then β is not a unit in R, whence

$$v^3 = (2\beta_1 + v\beta_2)(2 - \lambda v^2), \quad \text{where } \beta_1, \beta_2 \in R$$
$$= \beta_1(4 - \lambda 2v^2) + \beta_2(2v - \lambda v^3)$$
$$= \beta_1(4 - \lambda 2v^2)$$

$4 \in M^4$ and $2v \in M^3$, then $(4 - \lambda 2v^2) \in M^4$, whence $M^3 = (v^3) \subset M^4$, a contradiction since $M^3 \neq 0$. Finally, if $v^2(1 - X^g)$ is redundant, then $v^3(1 - X^g) \in (v^4, 2v - \lambda v^3, v^2(1 - X^g)^2, v(1 - X^g)^3) = (v^4, v^2(1 - X^g)^2, v(1 - X^g)^3)$. By passing to the homomorphic image $R/(v^4)[< g >]$, we obtain that $v^3(1 - X^g) \in ((1 - X^g)^2)R/(v^4)[< g >]$. By [1, lemma 1.4], we get $v^3 = 4\gamma$ where $\gamma \in R/(v^4)$. Since $4 \in M^4 = (v^4)$, then $v^3 \in (v^4)$, i. e., $M^3 = M^4$, a contradiction $(M^3 \neq 0)$. Consequently, I needs more than four generators. Thus, $R[Z/4Z]$ does not have the 4−generator property.

$\Leftarrow)$ Now, $R[G]$ is a local ring with maximal ideal $N = (u, v, 1 - X^g)$ where u v are the generators of M and g generates the cyclic group G.

Step 1.   We claim that N, N^2, N^3, and N^4 are 4−generated. Indeed,

(1) Assume $M^2 = (\alpha)$ is a principal ideal and $M^3 = 0$. Clearly,

$$N = (u, v, 1 - X^g) \;;$$
$$N^2 = (\alpha, u(1 - X^g), v(1 - X^g), (1 - X^g)^2) \;;$$
$$N^3 = (1 - X^g)N^2 \text{ and } ;$$
$$N^4 = (1 - X^g)^2 N^2.$$

(2) Assume M^2 is a principal ideal, $M^3 \neq 0$, $M^2 \subset (2)$, and $G = Z/2^i Z$ with $1 \leq i \leq 2$.

Since $M^2 = (\alpha) \subset (2)$, then

$$
\begin{aligned}
N &= (2, v, 1 - X^g) \ ; \\
N^2 &= (\alpha, 2(1 - X^g), v(1 - X^g), (1 - X^g)^2) \ ; \\
N^3 &= (\mu, \alpha(1 - X^g), v(1 - X^g)^2, (1 - X^g)^3) \ \text{ where } \ M^3 = (\mu) \ ; \\
N^4 &= (\alpha^2, \mu(1 - X^g), v(1 - X^g)^3, (1 - X^g)^4).
\end{aligned}
$$

(3)*Case1*   Assume M^2 is not a principal ideal and $G = Z/2Z$. Clearly,

$$
\begin{aligned}
N &= (u, v, 1 - X^g) \ ; \\
N^2 &= (a, b, u(1 - X^g), v(1 - X^g)) \ \text{ where } \ M^2 = (a, b) \ ; \\
N^3 &= (a', b', a(1 - X^g), b(1 - X^g)) \ \text{ where } \ M^3 = (a', b') \ ; \\
N^4 &= (a", b", a'(1 - X^g), b'(1 - X^g)) \ \text{ where } \ M^4 = (a", b").
\end{aligned}
$$

(3)*Case2*   Assume $M^2 = (\alpha)$ is a principal ideal, $M^3 \neq 0$, $2 \in M^2$, and $G = Z/2Z$. We verify that

$$
\begin{aligned}
N &= (u, v, 1 - X^g) \ ; \\
N^2 &= (\alpha, u(1 - X^g), v(1 - X^g)) \ ; \\
N^3 &= (\alpha u, \alpha v, \alpha(1 - X^g)) = \alpha N \ ; \\
N^4 &= \alpha N^2.
\end{aligned}
$$

(3) *Case3*   Assume $M^2 = (\alpha)$ is a principal ideal, $M^3 = (\mu) \neq 0$, $2 \in M \setminus M^2$, $M^2 \not\subset (2)$, and $G = Z/2Z$. We easily check that

$$
\begin{aligned}
N &= (2, v, 1 - X^g) \ ; \\
N^2 &= (\alpha, 2(1 - X^g), v(1 - X^g)) \ ; \\
N^3 &= (\mu, \alpha(1 - X^g)) \ ; \\
N^4 &= (\alpha^2, \mu(1 - X^g)).
\end{aligned}
$$

Step 2.   Let I be an ideal of $R[G]$, we claim that I is 4$-$generated. Indeed, (1) Assume M^2 is a principal ideal and $M^3 = 0$. Then $N^3 = (1 - X^g)N^2$, whence by [12, Lemma 2] $\mu(I) \leq \mu(I + N^2)$.

Since N^2 is 4−generated, we may assume $N^2 \subset I$. Let $x \in I \setminus N^2$.
Then $\mu\left(\dfrac{N}{(x)}\right) = \mu(N) - 1 = 3 - 1 = 2$, so that $\dfrac{N}{(x)} = (\overline{u}, \overline{v})$ or
$\dfrac{N}{(x)} = (\overline{u}, \overline{1 - X^g})$ or $\dfrac{N}{(x)} = (\overline{v}, \overline{1 - X^g})$, where $N = (u, v, 1 - X^g)$.

If $\dfrac{N}{(x)} = (\overline{u}, \overline{v})$, then $(N/(x))^2$ is 2−generated since $M^2 = (u^2, v^2, uv)$ is 2−generated. By [12, Theorem 1, 1⟸ 6], $R[G]/(x)$ has the 2−generator property. Hence I is 4−generated.

If $\dfrac{N}{(x)} = (\overline{u}, \overline{1 - X^g})$, then $\left(\dfrac{N}{(x)}\right)^2 = \dfrac{N^2 + (x)}{(x)} \subseteq \dfrac{I}{(x)}$.
We consider separately two cases :

Assume $\left(\dfrac{N}{(x)}\right)^2 \subset \dfrac{I}{(x)}$. Choose $z \in I$ such that
$\overline{z} \in \dfrac{I}{(x)} \setminus \left(\dfrac{N}{(x)}\right)^2$. We have

$$
\begin{aligned}
\mu\left(\frac{N}{(x, z)}\right) &= \mu\left(\frac{N/(x)}{(\overline{z})}\right) \\
&\leq \mu\left(\frac{N}{(x)}\right) - 1 \\
&\leq 2 - 1 = 1.
\end{aligned}
$$

Consequently, $\left(\dfrac{R[G]}{(x, z)}\right)$ is a principal ideal ring, so that $\left(\dfrac{I}{(x, z)}\right)$ is a principal ideal, whence I is 4−generated.

Assume $\left(\dfrac{N}{(x)}\right)^2 = \dfrac{N^2 + (x)}{(x)} = \dfrac{I}{(x)}$. Then $I = N^2 + (x)$, where
$N^2 = (\alpha, u(1 - X^g), v(1 - X^g), (1 - X^g)^2)$ and $M^2 = (\alpha)$. Since $x \in N$, $x = \lambda u + \mu v + \gamma(1 - X^g)$ for some $\lambda, \mu, \gamma \in R[G]$. Moreover, we may assume that γ is not a unit. Hence there exist $\lambda', \mu', \gamma' \in R[G]$ such that $x = \lambda' u + \mu' v + \gamma'(1 - X^g)^2$. Clearly, since $x \notin N^2$, λ' or μ', say λ' is a unit. Since $I = N^2 + (x)$ we may choose $x = u + \beta v$ for some $\beta \in R[G]$ then $x(1 - X^g) = u(1 - X^g) + \beta v(1 - X^g)$ therefore $I = (\alpha, v(1 - X^g), (1 - X^g)^2, x)$ which is 4−generated.

Likewise, for $\dfrac{N}{(x)} = (\overline{v}, \overline{1 - X^g})$.

From now on, (3) case 1, (3) case 2, and (3) case 3 refer to the three subcases considered in the proof of the "only if" assertion (3).

We first handle (2) and (3) case 1 simultaneous. We have $\mu(N^4) \leq 4$, then by [2, Lemma 4], $\mu(I) \leq \mu(I+N^3)$. Since N^3 is 4-generated, we may assume that $N^3 \subset I$.

Case I: Suppose there exists $x \in I \setminus N^2$, then $\mu\left(\dfrac{N}{(x)}\right) = \mu(N) - 1 = 3 - 1 = 2$, therefore $\dfrac{N}{(x)} = (\overline{u}, \overline{v})$ or $\dfrac{N}{(x)} = (\overline{u}, \overline{1 - X^g})$ or $\dfrac{N}{(x)} = (\overline{v}, \overline{1 - X^g})$, here $N = (u, v, 1 - X^g))$.

If $\dfrac{N}{(x)} = (\overline{u}, \overline{v})$, using arguments similar to ones used above, we can check that I is 4-generated.

If $\dfrac{N}{(x)} = (\overline{v}, \overline{1 - X^g})$, $\left(\dfrac{N}{(x)}\right)^3 = \dfrac{N^3 + (x)}{(x)} \subseteq \dfrac{I}{(x)}$. We consider separately two cases:

If $\left(\dfrac{N}{(x)}\right)^3 \subset \dfrac{I}{(x)}$, the proof is similar to that one given in the proof of [2, proposition 3] (see pages 8,9)

If $\left(\dfrac{N}{(x)}\right)^3 = \dfrac{N^3 + (x)}{(x)} = \dfrac{I}{(x)}$, then $I = N^3 + (x)$.

(2) $I = N^3 + (x) = (x, \mu, \alpha(1 - X^g), v(1 - X^g)^2, (1 - X^g)^3)$.

$x \in N = (2, v, 1 - X^g)$ then $x = 2\lambda + \beta v + \gamma(1 - X^g)$ for some $\lambda, \beta, \gamma \in R[G]$. Moreover, we may assume that γ is not a unit. Hence there exist $\lambda', \beta', \gamma' \in R[G]$, with λ' or β' is a unit such that $x = 2\lambda' + \beta'v + \gamma'(1 - X^g)^2$.

If β' is a unit, then $v \in (2, x, 1 - X^g)$. Therefore $v(1 - X^g)^2 \in (2(1 - X^g)^2, x(1 - X^g)^2, (1 - X^g)^3) \subset (4(1 - X^g), x, (1 - X^g)^3) \subset (\alpha(1 - X^g), x, (1 - X^g)^3)$ (see [2, page 6]. Consequently, $I = N^3 + (x) = (x, \mu, \alpha(1 - X^g), (1 - X^g)^3)$.

If β' is not a unit, then λ' is a unit because $x \notin N^2$. Now, $x(1-X^g) = 2\lambda'(1-X^g)+\beta'v(1 \pm X^g)^2+\gamma'(1-X^g)^3$ then $2(1-X^g) \in I$. Since $M^2 = (\alpha) \subset (2)$, $I = (x, \mu, 2(1-X^g), v(1-X^g)^2, (1-X^g)^3)$. Finally, since λ' is a unit, $I = (x, \mu, v(1 - X^g)^2, (1 - X^g)^3)$.

(3) *case 1*: M^2 is not a principal ideal and $< g >= Z/2Z$. We are in the situation where $\left(\dfrac{N}{(x)}\right)^3 = \dfrac{I}{(x)}$. We have

$$\frac{N}{(x)} = \left(\overline{v}, \overline{1 - X^g}\right) \text{ and } ;$$

$$\left(\frac{N}{(x)}\right)^2 = \left(\overline{v^2}, \overline{v(1 - X^g)}, \overline{(1 - X^g)^2}\right);$$

$$\left(\frac{N}{(x)}\right)^3 = \left(\overline{v^3}, \overline{v^2(1 - X^g)}, \overline{v(1 - X^g)^2}, \overline{(1 - X^g)^3}\right);$$

$$= \left(\overline{v^3}, \overline{v^2(1 - X^g)}, \overline{2v(1 - X^g)}, \overline{4(1 - X^g)}\right);$$

$$= \left(\overline{v^3}, \overline{a(1 - X^g)}, \overline{b(1 - X^g)}\right) \text{ where } M^2 = (a, b).$$

Thus, $\left(\dfrac{N}{(x)}\right)^3$ is 3-generated, and hence so is $\dfrac{N}{(x)}$. It follows that I is 4−generated.

The argument is similar if $\dfrac{N}{(x)} = \left(\overline{u}, \overline{1 - X^g}\right)$.

<u>Case II</u>: $(N^3 \subset) I \subseteq N^2$. In this case, we claim that there exists $x \in I \setminus N^3$ such that $\mu\left(\left(\dfrac{N}{(x)}\right)^3\right) \leq 3$. Indeed,

(2) We have

$$N = (2, v, 1 - X^g) ;$$
$$N^2 = (\alpha, 2(1 - X^g), v(1 - X^g), (1 - X^g)^2) ;$$
$$N^3 = (\mu, \alpha(1 - X^g), v(1 - X^g)^2, (1 - X^g)^3).$$

Let $x \in I \setminus N^3$, $x = a_x\alpha + b_x 2(1 - X^g) + c_x v(1 - X^g) + d_x(1 - X^g)^2$ for some $a_x, b_x, c_x, d_x \in R[G]$, where at least one of a_x, b_x, c_x, d_x is a unit.

If a_x is a unit, then $\overline{\alpha} \in \left(\overline{2(1 - X^g)}, \overline{v(1 - X^g)}, \overline{(1 - X^g)^2}\right)$, whence $\overline{\mu} \in \left(\overline{\alpha(1 - X^g)}, \overline{2(1 - X^g)^2}, \overline{v(1 - X^g)^2}\right) \subseteq$

$\left(\overline{\alpha(1 - X^g)}, \overline{(1 - X^g)^3}, \overline{v(1 - X^g)^2}\right)$. So that $\left(\dfrac{N}{(x)}\right)^3 = \dfrac{N^3 + (x)}{(x)}$

$= \left(\overline{\alpha(1 - X^g)}, \overline{(1 - X^g)^3}, \overline{v(1 - X^g)^2}\right).$

If c_x is a unit, then $\overline{v(1-X^g)} \in \left(\overline{2(1-X^g)}, \overline{\alpha}, \overline{(1-X^g)^2}\right)$, whence

$\overline{v(1-X^g)^2} \in \left(\overline{2(1-X^g)^2}, \overline{\alpha(1-X^g)}, \overline{(1-X^g)^3}\right) \subseteq$

$\left(\overline{\alpha(1-X^g)}, \overline{(1-X^g)^3}\right)$. Therefore $\left(\dfrac{N}{(x)}\right)^3 = \dfrac{N^3 + (x)}{(x)} =$

$\left(\overline{\mu}, \overline{\alpha(1-X^g)}, \overline{(1-X^g)^3}\right)$.

If d_x is a unit, then $\overline{(1-X^g)^2} \in \left(\overline{2(1-X^g)}, \overline{\alpha}, \overline{v(1-X^g)}\right)$, whence

$\overline{v(1-X^g)^2} \in \left(\overline{\mu}, \overline{\alpha(1-X^g)}\right)$. Hence

$$\left(\frac{N}{(x)}\right)^3 = \frac{N^3 + (x)}{(x)} = \left(\overline{\mu}, \overline{\alpha(1-X^g)}, \overline{(1-X^g)^3}\right).$$

Otherwise, for each $x \in I \setminus N^3$, a_x, c_x, and d_x are not units, Necessarily, b_x is a unit. It follows that $2(1-X^g) \in I \setminus N^3$.

$$\left(\frac{N}{(2(1-X^g))}\right)^3 = \frac{N^3 + (2(1-X^g))}{(2(1-X^g))}$$
$$= \frac{(\mu, \alpha(1-X^g), v(1-X^g)^2, (1-X^g)^3, 2(1-X^g))}{(2(1-X^g))}.$$

Since $M^2 = (\alpha) \subset (2)$, then

$$\left(\frac{N}{(2(1-X^g))}\right)^3 = \left(\overline{\mu}, \overline{v(1-X^g)^2}, \overline{(1-X^g)^3}\right).$$

(3) _case1_: We have

$$\begin{aligned}
N &= (u, v, 1-X^g); \\
N^2 &= (a, b, u(1-X^g), v(1-X^g)) \text{ where } M^2 = (a, b); \\
N^3 &= (a', b', a(1-X^g), b(1-X^g)) \text{ where } M^3 = (a', b').
\end{aligned}$$

Let $x \in I \setminus N^3$. Clearly, $x = a_x a + b_x b + c_x u(1-X^g) + d_x v(1-X^g)$ for some $a_x, b_x, c_x, d_x \in R[G]$, where at least one of a_x, b_x, c_x, d_x is a unit. In each case, one may verify that $\mu\left(\left(\dfrac{N}{(x)}\right)^3\right) \leq 3$ (Assume $a \in \{u^2, uv\}$ and $b = v^2$).

We get

$$\mu\left(\frac{I}{(x)}\right) \le \mu\left(\frac{I}{(x)} + \left(\frac{N}{(x)}\right)^2\right) = \mu\left(\frac{I+N^2}{(x)}\right) \quad \text{by [2, Lemma 4]} \; ;$$

$$= \mu\left(\frac{N^2}{(x)}\right) \quad \text{since } I \subseteq N^2 \; ;$$

$$\le 3 \quad \text{since } x \in N^2 \setminus N^3 \text{ and } N^2 \text{ is 4-generated} .$$

Consequently, I is 4−generated.

(3) <u>*cases* 2 *and* 3</u>: We have $\mu(N^3) \le 3$, then by [2, Lemma 4], $\mu(I) \le \mu(I + N^2)$.
Since N^2 is 4−generated, we can assume that $N^2 \subset I$. We ape the proof of (1) (page 7) to reach the desired conclusion when $\left(\frac{N}{(x)}\right)^2 \subset \frac{I}{(x)}$. Otherwise, $I = N^2 + (x)$ is 4−generated because in (3) cases 2 and 3, N^2 is 3−generated. $\diamondsuit$

PROPOSITION 2 *Assume G is a non trivial finite 3−group, (R, M) is an Artinian local ring with the 2-generator property but R is not a principal ideal ring, and that $3 \in M$. Then $R[G]$ has the 4−generator property if and only if*
(a) G is a cyclic group.
(b_1) When M^2 is a principal ideal and $M^3 \ne 0$ then
  (α_1) If $3 \in M^2$, then $G \cong Z/3Z$ and M^3 is a principal ideal.
  (α_2) If $3 \in M \setminus M^2$, then $G \cong Z/3^i Z$ with $1 \le i \le 2$, moreover, if $9 \in M^3$ then $G \cong Z/3Z$.

(b_2) When M^2 is not a principal ideal, then $3 \notin M^2$, $G \cong Z/3Z$, moreover, if $M^3 \ne 0$ and $M^2 \not\subset (3)$ then M^3 is a principal ideal and
  (θ_1) If $9 \in M^2 \setminus M^3$ then $M^3 \subset (9)$.
  (θ_2) If $9 \in M^3$ then $M^3 = 3M^2$.

Proof. $\Rightarrow$] (a) Assume that G is not a cyclic group and $R[G]$ has the 4−generator property. Necessarily, the homomorphic image $R[Z/pZ \oplus Z/pZ]$ does also, when p $= 3$. Then N^2 is 4−generated, where $N = (u, v, 1 - X^g, 1 - X^h)$, $M = (u, v)$, and $< g > \oplus < h > = Z/pZ \oplus Z/pZ$.

$N^2 = (u^2, v^2, uv, u(1-X^g), v(1-X^g), u(1-X^h), v(1-X^h), (1-X^g)(1-X^h), (1-X^g)^2, (1-X^h)^2)$.

Since $|<g>| = 3$, via [1, Lemma 1.4], it is easy to verify that N^2 needs more than four generators. Thus $G = Z/p^m Z$, with $m \geq 1$.

(b_1) Assume $M^2 = (\alpha)$ is a principal ideal and $M^3 \neq 0$.

(α_1) Suppose $p = 3 \in M^2$. If $G = Z/p^m Z$ with $m > 1$, we claim that N^3 is not 4−generated in $R[Z/p^m Z]$ where $N = (u, v, 1 - X^g)$, $M = (u, v)$, and $<g> = Z/p^m Z$.
We have $N^2 = (\alpha, u(1-X^g), v(1-X^g), (1-X^g)^2)$ and $N^3 = (\alpha u, \alpha v, \alpha(1-X^g), u(1-X^g)^2, v(1-X^g)^2, (1-X^g)^3)$.
By [1, Lemma 1.7], $u(1-X^g)^2$ and $v(1-X^g)^2$ are required as generators of N^3.
Since $|<g>| > 3$ it is clear that $(1-X^g)^3$ is required as generator of N^3.

If $\alpha(1-X^g)$ is a redundant generator of N^3, then passing to the hommorphic image $R/M^3[<g>]$, yields $\alpha(1-X^g) \in (1-X^g)^2 R/M^3[<g>]$. By [1, Lemma 1.4] $\alpha = \lambda p^m$ for some $\lambda \in R/M^3$. It follows that $\alpha = 0$ in R/M^3. That is, $M^2 = (\alpha) = 0$, a contradiction. Further, αu or αv is required as a generator of N^3, since $M^3 \neq 0$.

Now, suppose M^3 is not a principal ideal and $G = Z/3Z$.

By [1, Lemma 1.7] and the fact that M^3 is not a principal ideal, we can easily check that $\alpha u, \alpha v, u(1-X^g)^2 and v(1-X^g)^2$ are required as generators of N^3, then, if N^3 is 4-generated, necessarily, $N^3 = (\alpha u, \alpha v, u(1-X^g)^2, v(1-X^g)^2)$. Further $|<g>| = 3$ and $3 \in M^2$, $\alpha(1-X^g) \notin (\alpha u, \alpha v, u(1-X^g)^2, v(1-X^g)^2)$. Then N^3 needs more than four generators.

(α_2) Suppose $p = 3 \in M \setminus M^2$. Let's show that N^3 is not 4−generated in $R[Z/p^3 Z]$.
We have:

$N = (p, v, 1-X^g)$ and $N^3 = (p\alpha, v\alpha, \alpha(1-X^g), p(1-X^g)^2, v(1-X^g)^2, (1-X^g)^3)$.

Since $|<g>| = p^3 > 3$ and $M = (p, v)$ is not a principal ideal, by [1, Lemma 1.4 and Lemma 1.7], $\alpha(1-X^g), p(1-X^g)^2, v(1-X^g)^2$, and $(1-X^g)^3$ are required as generators of N^3. Furthermore, since $M^3 \neq 0$, it is clear that N^3 needs more than four generators. It follows that $G = Z/p^i Z$ with $1 \leq i \leq 2$, as desired.

Suppose in addition that $p^2 = 9 \in M^3$. Using the arguments similar to ones used above, it is easy to verify that $p(1-X^g)^2, v(1-X^g)^2$, and $(1 - X^g)^3$ must appear in a party of four generators extracted from the original set of generators of N^3. Furthermore, if $\alpha(1 - X^g)$ is redundant, then $\alpha(1 - X^g) \in (p\alpha, \alpha v, p(1 - X^g)^2, v(1 - X^g)^2, (1 - X^g)^3)$, whence passing to the homomorphic image $R/M^3[< g >]$, we get $\alpha(1 - X^g) \in (1 - X^g)^2 R/M^3[< g >]$. By [1, Lemma 1.4], $\alpha = \lambda p^2 = 0$ for some $\lambda \in R/M^3$, a contradiction ($M^3 \neq 0$). Thus, $G = Z/pZ$, as desired.

(b_2) Assume M^2 is not a principal ideal. one may easily show that N^2 is not 4–generated neither if $< g >= Z/9Z$ nor if $3 \in M^2$ and $< g >= Z/3Z$. Necessarily, $3 \in M \setminus M^2$ and $< g >= Z/3Z$.

Set $p = 3$. Assume in addition $M^3 \neq 0$ and $M^2 \not\subset (p)$. we claim that M^3 is a principal ideal. Deny. Let $N = (p, v, 1 - X^g)$ and $< g >= Z/pZ$. Clearly, $N^3 = (a', b', a(1 - X^g), b(1 - X^g), p(1 - X^g)^2, v(1 - X^g)^2, (1 - X^g)^3)$, where $M^2 = (a, b)$ and $M^3 = (a', b') = (p^3, p^2 v, pv^2, v^3)$. Further, $M^2 = (a, b) = (v^2, p^2, pv)$, since $M^2 \not\subset (p)$, we can take $a = v^2$ and $b \in \{p^2, pv\}$. Then $N^3 = (a', b', (1 - X^g)^3, v^2(1 - X^g), v(1 - X^g)^2)$.

Since M^3 is not a principal ideal, by [1, Lemma 1.4 and Lemma 1.7], a',b', $v^2(1 - X^g)$, and $v(1 - X^g)^2$ are required as generators of N^3. Since N^3 is 4-generated, then $(1 - X^g)^3 = -3X^g(1 - X^g) \in (a', b', v^2(1 - X^g), v(1 - X^g)^2)$ (Here p = 3). By passing to the homomorphic image $R/(v)[< g >]$, we obtain that $3 \in (27, v)$. It follows that $M = (3, v) = (v)$ since R is Artinian, a contradiction. Consequently, $M^3 = (\mu)$ is a principal ideal.

Let $I = (v^3, b, v^2(1 - X^g), v(1 - X^g)^2, (1 - X^g)^3)$.

Since $| < g > | > 3$ and $b \notin M^3$ (R Artinian and M^2 not principal), it is clear that b and $(1 - X^g)^3$ are required as generators of I.

If $v(1 - X^g)^2$ is redundant, then by passing to the homomorphic image $R/M^2[< g >]$, and by using [1, Lemma 1.7], we get $v = \lambda p$ for some $\lambda \in R/M^2$. Hence, $v \in (p, v^2)$. Therefore $M = (p, v) = (p, v^2) = \cdots = (p)$, a contradiction.

If $v^2(1 - X^g)$ is redundant, then passing to the homomorphic image $R/(v^3, b)[< g >]$ yields $v^2(1 - X^g) \in (1 - X^g)^2 R/(v^3, b)[< g >]$. By [1, Lemma 1.4], we have $v^2 = \lambda p$ for some $\lambda \in R/(v^3, b)$. Similarly, $(p, v^2) = (p, v^3) = \cdots = (p)$, whence $M^2 = (p^2, v^2, pv) \subset (p)$, a contradiction.

Thus $I = (b, v^2(1 - X^g), v(1 - X^g)^2, (1 - X^g)^3)$. Further, $v^3 \in I$, by passing to the homomorphic image $R[< g >]/(1 - X^g) \cong R$, we obtain that $v^3 \in (b)$.

(θ_1) If $p^2 \in M^2 \setminus M^3$, we may assume $b = p^2$.
$M^3 = (p^3, p^2v, pv^2, v^3)$. Since $pv \in (p^2, v^2)$ then $pv^2 \in (v^3, p^2v)$. Therefore $M^3 \subset (p^2)$, as desired.

(θ_2) If $p^2 \in M^3$, we may assume $b = pv$. We have $M^2 = (v^2, pv) = vM$ so that $M^3 = v^2M = (v^3, pv^2)$. Since $v^3 \in (b) = (pv)$ and $pv \notin M^3$, then $v^3 \in (p^2v, pv^2)$. Therefore $M^3 = (pv^2, p^2v) = p(v^2, pv) = pM^2$, as desired.

$\Leftarrow$)   Now, $R[G]$ is a local ring with maximal ideal $N = (u, v, 1 - X^g)$ where u and v are the generators of M and g is a generator of the cyclic group G.

Step 1.   We claim that N, N^2, N^3, and N^4 are 4−generated. Indeed,

(b_1) Assume $M^2 = (\alpha)$ is a principal ideal. If $M^3 = 0$, then the proof is straightforward (see the case $p = 2$).
In the sequel, we suppose $M^3 \neq 0$.

α_1) Assume $3 \in M^2$, $G = Z/3Z$, and $M^3 = (\mu)$ is a principal ideal. We easily check that

$$
\begin{aligned}
N &= (u, v, 1 - X^g) \; ; \\
N^2 &= (\alpha, u(1 - X^g), v(1 - X^g), (1 - X^g)^2) \; ; \\
N^3 &= (\mu, \alpha(1 - X^g), u(1 - X^g)^2, v(1 - X^g)^2) \; ; \\
N^4 &= (\alpha^2, \mu(1 - X^g), \alpha(1 - X^g)^2).
\end{aligned}
$$

(α_2) Assume $p = 3 \in M \setminus M^2$ and $G \cong Z/p^iZ$, $1 \leq i \leq 2$. Since M^2 is a principal ideal, it is easy to verify that $M^3 = (\mu)$ is a principal ideal.

Suppose $p^2 = 9 \in M^2 \setminus M^3$. Clearly $M^2 = (p^2)$.

$$
\begin{aligned}
1 &= (1 - X^g + X^g)^{p^2} \\
&= \sum_{i=o}^{i=p^2} \binom{p^2}{i}(1 - X^g)^i X^{(p^2-i)g} \\
&= 1 + p^2(1 - X^g)X^{(p^2-1)g} + \frac{p^2(p^2-1)}{2}(1 - X^g)^2 X^{(p^2-2)g} \\
&\quad + (1 - X^g)^3 \left(\sum_{i=3}^{i=p^2} \binom{p^2}{i}(1 - X^g)^{(i-3)} X^{(p^2-i)g} \right).
\end{aligned}
$$

Then $p^2(1 - X^g) \in (p^2(1 - X^g)^2, (1 - X^g)^3) \subset ((1 - X^g)^3)$. Therefore,

$$
\begin{aligned}
N &= (p, v, 1 - X^g) \ ; \\
N^2 &= (p^2, p(1 - X^g), v(1 - X^g), (1 - X^g)^2) \ ; \\
N^3 &= (\mu, p(1 - X^g)^2, v(1 - X^g)^2, (1 - X^g)^3) \ ; \\
N^4 &= (p^4, \mu(1 - X^g), p(1 - X^g)^3, v(1 - X^g)^3, (1 - X^g)^4).
\end{aligned}
$$

We have $M^3 = (\mu) \subset M^2 = (p^2)$, whence $\mu \in (p^3, p^2 v)$ since $p^2 \notin M^3$. Therefore $\mu(1 - X^g) \in (p^3(1 - X^g), p^2 v(1 - X^g)) \subset (p(1 - X^g)^3, v(1 - X^g)^3)$. It results that
$$N^4 = (p^4, p(1 - X^g)^3, v(1 - X^g)^3, (1 - X^g)^4)$$

Suppose $9 \in M^3$ and $G = Z/3Z$. Clearly,

$$
\begin{aligned}
N &= (3, v, 1 - X^g) \ ; \\
N^2 &= (\alpha, v(1 - X^g), (1 - X^g)^2) \ ; \\
N^3 &= (\mu, \alpha(1 - X^g), v(1 - X^g)^2, 3(1 - X^g)) \ ; \\
N^4 &= (\alpha^2, \mu(1 - X^g), \alpha(1 - X^g)^2, 3(1 - X^g)^2).
\end{aligned}
$$

(b_2) Set $p = 3$. Assume $M^2 = (a, b)$ is not a principal ideal, $p \notin M^2$, and $G = Z/pZ$. Clearly,

$$
\begin{aligned}
N &= (p, v, 1 - X^g) \ ; \\
N^2 &= (a, b, v(1 - X^g), (1 - X^g)^2).
\end{aligned}
$$

If $M^3 = 0$, then

$$
\begin{aligned}
N^3 &= (a(1 - X^g), b(1 - X^g), v(1 - X^g)^2, (1 - X^g)^3) = (1 - X^g)N^2; \\
N^4 &= (1 - X^g)^2 N^2 \quad (\text{Recall that } p(1 - X^g) \in (1 - X^g)^3).
\end{aligned}
$$

In the sequel, we suppose $M^3 \neq 0$.

If $M^2 \subset (p)$, then $M^2 = (p^2, pv)$ $(p \notin M^2)$, whence $N^3 = (p^3, p^2 v, v(1 - X^g)^2, (1 - X^g)^3)$ and $N^4 = (p^4, p^3 v, v(1 - X^g)^3, (1 - X^g)^4)$, since $p(1 - X^g) \in ((1 - X^g)^2)$.

Now, assume $M^2 \not\subset (p)$ and $M^3 = (\mu)$ is a principal ideal. We may assume $a = v^2$ and $b \in \{pv, p^2\}$.

It is easily seen that $N^3 = (\mu, v^2(1 - X^g), v(1 - X^g)^2, (1 - X^g)^3)$. It remains to show that N^4 is 4-generated.

If $p^2 \in M^3$ and $M^3 = pM^2$, then $M^4 = p^2 M^2 \subset M^5$, whence $M^4 = 0$. Therefore $N^4 = (\mu(1 - X^g), v^2(1 - X^g)^2, v(1 - X^g)^3, (1 - X^g)^4)$.

If $p^2 \in M^2 \setminus M^3$ and $M^3 \subset (p^2)$, since M^3 is a principal ideal, it is easy to verify that $(M^4 = (\gamma))$ is a principal ideal. So that $N^4 = (\gamma, \mu(1 - X^g), v^2(1 - X^g)^2, v(1 - X^g)^3, (1 - X^g)^4)$.

Since $p(1 - X^g) \in ((1 - X^g)^2)$ and $\mu \in (p^3, p^2 v)$ $(M^3 = (\mu) \subset (p^2))$, then $\mu(1 - X^g) \in (v(1 - X^g)^3, (1 - X^g)^4)$. Therefore $N^4 = (\gamma, v^2(1 - X^g)^2, v(1 - X^g)^3, (1 - X^g)^4)$.

Step 2.   Let I be an ideal of $R[G]$, we claim that I is 4−generated. Indeed,

If $M^3 = 0$, then the proof is similar to the one given for p = 2.

If $M^3 \neq 0$, as in the proof of Proposition 1 (cases (2) and (3) *case1*), we may assume $N^3 \subset I$.

Case I: Suppose that there exists $x \in I \setminus N^2$. Via the proof of Proposition 1, it suffices to consider the case $I = N^3 + (x)$.

(b_1) Assume $M^2 = (\alpha)$ is a principal ideal and $M^3 \neq 0$.

(α_1) We got from step 1 that $N^3 = (\mu, \alpha(1 - X^g), u(1 - X^g)^2, v(1 - X^g)^2)$. Since $x \in N = (u, v, 1 - X^g)$, $x = \lambda u + \beta v + \gamma(1 - X^g)$ for some $\lambda, \beta, \gamma \in R[G]$, where λ or β or γ is a unit. If γ is a unit, $\frac{N}{(x)} = (\bar{u}, \bar{v})$. We conclude in the same way as in the case p = 2 step 2 page 7.

If γ is not a unit, necessarily, λ or β is a unit, say λ. Clearly, $u \in (x, v, 1 - X^g)$, then $u(1 - X^g)^2 \in (x, v(1 - X^g)^2, 3(1 - X^g)) \subset (x, v(1 - X^g)^2, \alpha(1 - X^g))$, since $|< g >| = 3$ and $3 \in M^2$. Therefore $I = (x, \mu, v(1 - X^g)^2, \alpha(1 - X^g))$.

(α_2) Assume $p = 3 \in M \setminus M^2$ and $p^2 \in M^2 \setminus M^3$. We got from step1 that $N^3 = (\mu, p(1 - X^g)^2, v(1 - X^g)^2, (1 - X^g)^3)$. Since $x \in N = (p, v, 1 - X^g)$, $x = \lambda p + \beta v + \gamma(1 - X^g)$, where λ or β or γ is a unit $(x \notin N^2)$. In each case, it is easy to verify that $I = N^3 + (x)$ is 4−generated.

Now, assume $9 \in M^3$ and $G = Z/3Z$. If $M^2 \subset (3)$, we are done via [1, Proposition 2.1]. Let's suppose $M^2 \not\subset (3)$. We have $N^3 = (\mu, v^2(1 - X^g), v(1 - X^g)^2, 3(1 - X^g))$. Since $x \in N \setminus N^2$, $x = 3\lambda + \beta v + \gamma(1 - X^g)$ for some $\lambda, \beta, \gamma \in R[G]$, with λ or β or γ is a unit. The cases in which β or γ is a unit are straightforward.

We assume then that β and γ are not units. Then $x = 3\lambda' + \beta'v^2 + \mu'v(1-X^g) + \gamma'(1-X^g)^2$ for some $\lambda', \beta', \mu', \gamma' \in R[G]$. Clearly, λ' is a unit. Therefore $x(1-X^g) = 3\lambda'(1-X^g) + \beta'v^2(1-X^g) + \mu'v(1-X^g)^2 + \gamma'(1-X^g)^3$.

If μ' or β' is a unit, it is easy to see that $I = (x, \mu, v^2(1-X^g), 3(1-X^g))$ or $I = (x, \mu, v(1-X^g)^2, 3(1-X^g))$.

If μ' and β' are not units, since $I = N^3 + (x)$, we can take $x = 3\lambda' + \gamma'(1-X^g)^2$. Furthermore, if γ' is not a unit, we may take $x = 3$, whence $I = (3, \mu, v^2(1-X^g), v(1-X^g)^2)$. If γ' is a unit, then $(1-X^g)^2 \in (3, x)$, hence $v(1-X^g)^2 \in (3v, x) \subset (\mu, x)$ since $3v \in M^3 = (\mu)$ (Recall M^2 is a principal ideal and $M^2 \not\subset (3)$). Thus, $I = (x, \mu, v^2(1-X^g), 3(1-X^g))$.

(b_2) Assume M^2 is not a principal ideal, $p = 3 \notin M^2$ and $G = Z/pZ$.

If $M^2 \subset (p)$, we have $N^3 = (p^3, p^2v, v(1-X^g)^2, (1-X^g)^3)$. Since, $p(1-X^g) \in ((1-X^g)^3)$, we easily show that $I = N^3 + (x)$ is 4-generated.

If $M^2 \not\subset (p)$, we have $N^3 = (\mu, v^2(1-X^g), v(1-X^g)^2, (1-X^g)^3)$. Similarly, $x = \lambda p + \beta v + \gamma(1-X^g)$ for some $\lambda, \beta, \gamma \in R[G]$, with λ or β or γ is a unit. We can assume that β and γ are not units, hence, there exist $\lambda', \beta', \gamma', \delta' \in R[G]$ such that $x = \lambda'p + \beta'v^2 + \gamma'v(1-X^g) + \delta'(1-X^g)^2$, where λ' is a unit.

If β' or γ' is a unit, it is easy to verify that $I = (x) + N^3$ is 4-generated.

If β' and γ' are not units, since $I = (x) + N^3$, we can suppose that $x = \lambda'p + \delta'(1-X^g)^2$, whence $p^2 \in (x, p(1-X^g)^2) \subset (x, (1-X^g)^3)$ and $pv \in (x, v(1-X^g)^2)$. Furtheremore, under the present hypotheses, one may check that $M^3 = (\mu) \subset (b)$ where $b \in \{p^2, pv\}$. Hence, $\mu \in (x, v(1-X^g)^2, (1-X^g)^3)$. Therefore $I = (x, v^2(1-X^g), v(1-X^g)^2, (1-X^g)^3)$.

Case II Suppose $(N^3 \subset)I \subseteq N^2$. Using step1 and arguments similar to ones used above, we show that there exists $x \in I \setminus N^3$ such that $\mu\left(\left(\dfrac{N}{(x)}\right)^3\right) \leq 3$.

Actually, it remains to handle the following case : Assume M^2 is not principal, $p = 3 \notin M^2$, $< g > = Z/pZ$, $M^3 = (\mu)$ is a nonzero principal ideal, and $M^2 \not\subset (p)$. we got by step1 that $N^2 = (v^2, b, v(1-X^g), (1-X^g)^2)$, and $N^3 = (\mu, v^2(1-X^g), v(1-X^g)^2, (1-$

$X^g)^3)$, where $b \in \{p^2, pv\}$. Let $x \in I \setminus N^3$, $x = a_x v^2 + b_x b + c_x v(1 -$ $X^g) + d_x(1 - X^g)^2$ for some $a_x, b_x, c_x, d_x \in R[G]$, with a_x or b_x or c_x or d_x is a unit.

If a_x or c_x or d_x is a unit, easily we check that $\mu\left(\left(\dfrac{N}{(x)}\right)^3\right) \leq 3$.

Otherwise, since b_x is a unit, $x \notin N^3$, and hence $b \in (x) + N^3$. Therefore $N^3 + (x) = N^3 + (b)$. Since $b \notin N^3$ and $M^3 = (\mu) \subset (b)$, then $\mu\left(\dfrac{N}{(b)}\right)^3 = \mu\left(\dfrac{N^3 + (b)}{(b)}\right) \leq 3$.

By the same proof for $p = 2$, we claim that I is 4−generated. $\Diamond$

PROPOSITION 3 *Assume that G is a non trivial finite p−group, (R, M) is an Artinian local ring with the 2-generator property but R is not a principal ideal ring and that $p \in M$. Then $R[G]$ has the 4−generator property if and only if*

    (a) G is a cyclic group.

    (b_1) When M^2 is a principal ideal and $M^3 \neq 0$ then

        (α_1) If $p \in M^2$, then $G \cong Z/pZ$, $p \notin M^3$, and M^3 is a principal ideal.

        (α_2) If $p \in M \setminus M^2$, then $G \cong Z/p^i Z$ with $1 \leq i \leq 2$, moreover, if $p^2 \in M^3$ then $G \cong Z/pZ$ and either $M^2 \subset (p)$ or $M^3 \subset (p)$.

    (b_2) When M^2 is not a principal ideal, then $p \notin M^2$, $G \cong Z/pZ$, moreover, if $M^3 \neq 0$ and $M^2 \not\subset (p)$ then M^3 is a principal ideal and

        (θ_1) If $p^2 \in M^2 \setminus M^3$ then $M^3 \subset (p^2)$.

        (θ_2) If $p^2 \in M^3$ then $M^3 = pM^2$.

Proof of Proposition 3. It is almost similar to the proof of Proposition 2. Here the main fact is that $|< g >| = p > 3$. The remaining two cases are : (α_1) and (α_2) when $p^2 \in M^3$.

$\Rightarrow]$ (α_1) Assume $p \in M^2$. If M^3 is not a principal ideal or $p \in M^3$ or $G = Z/p^m Z$ with $m > 1$, by the same proof given for Proposition 2 (α_1) we verify that N^3 is not 4−generated in $R[Z/p^m Z]$ where $N = (u, v, 1 - X^g)$, $M = (u, v)$, and $< g >= Z/p^m Z$.

(α_2) Assume $p \in M \setminus M^2$ and $p^2 \in M^3$. Necessarily, $< g >= Z/pZ$. Let's suppose $M^2 \not\subset (p)$ and $M^3 \not\subset (p)$. Let $I = (p, \mu(1 - X^g), \alpha(1 - X^g)^2, v(1 - X^g)^3, (1 - X^g)^4)$.

Since $p \in M \setminus M^2$ and $|<g>| > 4$, then p and $(1 - X^g)^4$ are required as generators of I.

If $v(1 - X^g)^3$ is redundant then by passing to the homomorphic image $R/(\alpha, p)[<g>]$, we obtain that $v(1 - X^g)^3 \in (1 - X^g)^4 R/(\alpha, p)[<g>]$, and whence $v(1 - X^g)^{p-1} = 0$ in $R/(\alpha, p)[<g>]$. Therefore $M = (p, v) = (p, v^2) = \cdots\cdots = (p)$, a contradiction.

If $\mu(1 - X^g)$ is a redundant generator then by passing to the homomorphic image $R/(p)[<g>]$, we obtain that $\mu(1 - X^g) \in (1 - X^g)^2 R/(p)[<g>]$. By [1, Lemma 1.4], we get $\mu = \lambda p$ for some $\lambda \in R/(p)$. Hence $M^3 = (\mu) \subset (p)$, a contradiction.

If $\alpha(1 - X^g)^2$ is redundant, then by passing to the homomorphic image $R/(p, \mu)[<g>]$, we obtain that $\alpha(1 - X^g)^2 \in (1 - X^g)^3 R/(p, \mu)[<g>]$. By [1, Lemma 1.7], we get $\alpha = \lambda p$ for some $\lambda \in R/(p, \mu)$, whence $v^2 \in (p, v^3)$. Hence $(p, v^2) = (p, v^3) = \cdots = (p)$, so that $M^2 = (v^2, p^2, pv) \subset (p)$, a contradiction.

Consequently, I is not 4$-$generated.

$\Leftarrow$)   Now, we know that $R[G]$ is a local ring with maximal ideal $N = (u, v, 1 - X^g)$, where u and v are the generators of M and g is a generator of the cyclic group G.

Step : 1.   We claim that N, N^2, N^3, and N^4 are 4$-$generated. Indeed,

α_1) Assume $p \in M^2$, $G = Z/pZ$, $p \notin M^3$, and M^3 is a principal ideal. Necessarily, $M^2 = (p)$.

Since $p(1 - X^g) \in ((1 - X^g)^3)$, we get

$$
\begin{aligned}
N &= (u, v, 1 - X^g) \ ; \\
N^2 &= (p, u(1 - X^g), v(1 - X^g), (1 - X^g)^2) \ ; \\
N^3 &= (\mu, u(1 - X^g)^2, v(1 - X^g)^2, (1 - X^g)^3) \ ; \\
N^4 &= (p^2, u(1 - X^g)^3, v(1 - X^g)^3, (1 - X^g)^4).
\end{aligned}
$$

(α_2) Assume $p \in M \setminus M^2$, $p^2 \in M^3$, $G \cong Z/pZ$, and either $M^2 \subset (p)$ or $M^3 \subset (p)$. We have

$$
\begin{aligned}
N &= (p, v, 1 - X^g) \ ; \\
N^2 &= (\alpha, v(1 - X^g), (1 - X^g)^2) \text{ where } M^2 = (\alpha) \ ; \\
N^3 &= (\mu, \alpha(1 - X^g), v(1 - X^g)^2, (1 - X^g)^3) \text{ where } M^3 = (\mu) \ ; \\
N^4 &= (\alpha^2, \mu(1 - X^g), \alpha(1 - X^g)^2, v(1 - X^g)^3, (1 - X^g)^4).
\end{aligned}
$$

If $M^2 \subset (p)$ then $\alpha(1 - X^g) \in (p^2(1 - X^g), pv(1 - X^g)) \subset (v(1 - X^g)^3, (1 - X^g)^4)$, whence $N^4 = (\alpha^2, \mu(1-X^g), v(1-X^g)^3, (1-X^g)^4)$.

If $M^3 \subset (p)$. Since $p \in M \setminus M^2$, $\mu \in (p^2, pv)$. Further, $p^2(1 - X^g) \in ((1-X^g)^4)$ and $pv(1-X^g) \in (v(1-X^g)^3)$. Then $\mu(1-X^g) \in (v(1 - X^g)^3, (1 - X^g)^4)$. Therefore $N^4 = (\alpha^2, \alpha(1 - X^g)^2, v(1 - X^g)^3, (1 - X^g)^4)$.

Step : 2   Let I be an ideal of $R[G]$, we claim that I is 4−generated. As in Proposition 1, we may assume that $N^3 \subset I$.

Case I: Suppose that there exists $x \in I \setminus N^2$. As above, it suffices to consider the case $I = N^3 + (x)$.

(α_1) By step1, it is easily seen that $I = N^3 + (x)$ is 4−generated.

(α_2) Assume $p \in M \setminus M^2$, $p^2 \in M^3$, $G \cong Z/pZ$, and either $M^2 \subset (p)$ or $M^3 \subset (p)$. By step1, $N^3 = (\mu, \alpha(1 - X^g), v(1 - X^g)^2, (1 - X^g)^3)$. Since $x \in N \setminus N^2$ then $x = \lambda p + \beta v + \gamma(1 - X^g)$ for some $\lambda, \beta, \gamma \in R[G]$, with λ or β or γ is a unit. We can assume that β and γ are not units. Therefore $p \in (x, v, 1 - X^g)$.

If $M^2 \subset (p)$, $\alpha \in (p^2, pv)$. Hence $\alpha(1-X^g) \in (p^2(1 - X^g), pv(1 - X^g)) \subset ((1 - X^g)^3)$. So that $I = (x, \mu, v(1 - X^g)^2, (1 - X^g)^3)$.

If $M^3 \subset (p)$, $x = \lambda'p + \beta'\alpha + \gamma'v(1 - X^g) + \delta'(1 - X^g)^2$ for some $\lambda', \beta', \gamma', \delta' \in R[G]$. Clearly λ' is a unit ($x \notin N^2$).

If β' or γ' is a unit, we verify that I is 4−generated.

Otherwise, since $I = N^3 + (x)$, we can suppose that $x = \lambda'p + \delta'(1 - X^g)^2$. By hypothesis, $M^3 = (\mu) \subset (p)$. Then $\mu = \theta p$ for some $\theta \in M$ ($p \notin M^2$), hence $x\theta = \lambda'\mu + \delta'\theta(1 - X^g)^2$. Therefore $\mu \in (x, v(1-X^g)^2, p(1 - X^g)^2) \subset (x, v(1-X^g)^2, (1-X^g)^3)$. Consequently, $I = (x, \alpha(1 - X^g), v(1 - X^g)^2, (1 - X^g)^3)$.

Case II: Suppose $(N^3 \subseteq)I \subset N^2$. The proof is the same as in Proposition 2. $\Diamond$

THEOREM. *Let R be an Artinian ring with the 2−generator property and let G be a finite abelian group. Then $R[G]$ has the 4−generator property if and only if $R = R_1 \oplus R_2 \oplus \cdots \oplus R_s$ where, for each j, (R_j, M_j) is a local Artinian ring with the 2−generator property subject to :*

(I) *Assume R_j is a field of characteristic $p \neq 0$.*

(α) when $p = 2$, then G_p is a homomorphic image of $Z/2Z \oplus Z/2Z \oplus Z/2^i Z$ or $Z/4Z \oplus Z/2^i Z$ where $i > 0$

(β) *when* $p = 3$, *then* G_p *is a homomorphic image of* $Z/3Z \oplus Z/3^i Z$ *where* $i > 0$

(γ) *when* $p > 3$, *then* G_p *is a cyclic group.*

(II) *Assume* (R_j, M_j) *is a principal ideal ring which is not a field, and* p *a prime integer such that* p *divides* $Ord(G)$ *and* $p \in M_j$, *then*

(α) *Assume* $p = 2$,

A) (i) $G_p \cong Z/2Z \oplus Z/2^i Z$ *with* $i > 1$

  (ii) *when* $M_j^2 \neq 0$, *then* $G_p \cong Z/2Z \oplus Z/2Z$.

B) (i) G_p *is a cyclic group*

  (ii) *When* $M_j^4 \neq 0$, *then*

    (a) $G_p \cong Z/2^i Z$, *where* $1 < i < 2$, *if* $2 \in M_j^2$

    (b) $G_p \cong Z/2^i Z$, *where* $1 < i < 3$, *if* $2 \in M_j \setminus M_j^2$.

(β) *Assume* $p = 3$,

A) $G_p \cong Z/3Z \oplus Z/3Z$, $3 \in M_j \setminus M_j^2$ *and* $M_j^2 = 0$.

B) (i) G_p *is a cyclic group*

  (ii) *When* $M_j^4 \neq 0$, *then*

    (a) $G_p \cong Z/3Z$, *if* $3 \in M_j^2$

    (b) $G_p \cong Z/3^i Z$, *where* $1 < i < 3$, *if* $3 \in M_j \setminus M_j^2$.

(γ) *Assume* $p > 3$,

  (i) G_p *is a cyclic group*

  (ii) *If* $M_j^4 \neq 0$, *then* $p \notin M_j^4$ *and*

    (a) $G_p \cong Z/pZ$, *if* $p \in M_j^2$

    (b) $G_p \cong Z/p^i Z$, *where* $1 < i < 3$, *if* $p \in M_j \setminus M_j^2$.

(III) *Assume* (R_j, M_j) *has the* $2-$*generator property but is not a principal ideal ring and* p *a prime integer such that* p *divides* $Ord(G)$ *and* $p \in M_j$, *then*

(α) *Assume* $p = 2$,

  $G_p \cong Z/2^i Z$,

  (1) $i \geq 1$ *if* M_j^2 *is a principal ideal and* $M_j^3 = 0$.

  (2) $1 \leq i \leq 2$ *if* M_j^2 *is a principal ideal,* $M_j^3 \neq 0$, *and* $M^2 \subset (2)$.

  (3) $i = 1$ *otherwise.*

(β) *Assume* $p = 3$,

  (a) G_p *is a cyclic group*

  (b_1) *When* M_j^2 *is a principal ideal and* $M_j^3 \neq 0$ *then*

    (α_1) *If* $3 \in M_j^2$, *then* $G_p \cong Z/3Z$ *and* M_j^3 *is a principal ideal.*

    (α_2) *If* $3 \in M_j \setminus M_j^2$, *then* $G_p \cong Z/3^i Z$ *with* $1 \leq i \leq 2$, *moreover, if* $9 \in M_j^3$

      then $G_p \cong Z/3Z$.

(b_2) *When M_j^2 is not a principal ideal, then $3 \notin M_j^2$, $G_p \cong$*
$Z/3Z$, moreover, if

$M_j^3 \neq 0$ *and* $M_j^2 \not\subset (3)$ *then M_j^3 is a principal ideal and*

 (θ_1) *If $9 \in M_j^2 \setminus M_j^3$ then $M_j^3 \subset (9)$.*

 (θ_2) *If $9 \in M_j^3$ then $M_j^3 = 3M_j^2$.*

(γ) *Assume $p > 3$,*

 (a) *G_p is a cyclic group*

 (b) (b_1) *When M_j^2 is a principal ideal and $M_j^3 \neq 0$ then*

  (α_1) *If $p \in M_j^2$, then $G_p \cong Z/pZ$, $p \notin M_j^3$, and M_j^3 is a*
principal ideal.

  (α_2) *If $p \in M_j \setminus M_j^2$ then $G_p \cong Z/p^iZ$ with $1 \leq i \leq 2$,*
moreover, if $p^2 \in M^3$,

   then $G_p \cong Z/pZ$ and either $M_j^2 \subset (p)$ or $M_j^3 \subset (p)$

 (b_2) *When M_j^2 is not a principal ideal, then $p \notin M_j^2$, $G_p \cong$*
Z/pZ, moreover, if

$M_j^3 \neq 0$ *and* $M_j^2 \not\subset (p)$ *then M_j^3 is a principal ideal and*

 (θ_1) *If $p^2 \in M_j^2 \setminus M_j^3$ then $M_j^3 \subset (p^2)$.*

 (θ_2) *If $p^2 \in M_j^3$ then $M_j^3 = pM_j^2$.*

Proof. We appeal to [2, Theorem], Propositions 1, 2, and 3, and
similar techniques used in the proof of [1, Theorem]. $\Diamond$

REFERENCES

[1] S. Ameziane Hassani, M. Fontana and S. Kabbaj. Group rings
    $R[G]$ with 3−generated ideals when R is an Artinian, Communi-
    cations in Algebra 24(4) (1996) 1253-1280.

[2] S. Ameziane Hassani, and S. Kabbaj. Group rings $R[G]$ with
    4−generated ideals when R is an Artinian principal ideal ring,
    Lecture Notes of Pure and Applied Mathematics, Marcel Dekker,
    New York, 185 (1997) 1-14.

[3] J. T. Arnold and R. Gilmer. The dimension theory of commuta-
    tive semigroup rings, Houston J. Math. 2 (1976) 299-313.

[4] J. T. Arnold and R. Matsuda. The n−generator property for
    semigroup rings, Houston J. Math. 12 (1986) 345-356.

[5] I. S. Cohen. Commutative rings with restricted minimum condi-
    tion, Duke Math. J. 17 (1950) 27-42.

[6] R. Gilmer. <u>Commutative Semigroup Rings</u>, University of Chicago
    Press, Chicago, 1984.

[7] R. Gilmer. Multiplicative Ideal Theory, Marcel Dekker, New York, 1972.

[8] N. Jacobson. Basic Algebra. Freeman, 1985 and 1989.

[9] I. Kaplansky. Commutative Rings, University of Chicago Press, Chicago, 1974.

[10] R. Matsuda. Torsion free abelian semigroup rings V, Bull. Fac. Sci. Ibaraki Univ. 11 (1979) 1-37.

[11] R. Matsuda. n−Generator property of a polynomial ring, Bull. Fac. Sci., Ibaraki Univ., Series A Math. 16 (1984) 17-23.

[12] K. R. McLean. Local rings with bounded ideals, Journal of Algebra 74 (1982) 328-332.

[13] M. Nagata. Local Rings. Interscience, New York, 1962.

[14] J. Okon, D. Rush and P. Vicknair. Semigroup rings with two-generated ideals, J. London Math. Soc. 45 (1992) 417-432.

[15] J. Okon and P. Vicknair. Group rings with n−generated ideals, Comm. Algebra 20 (1) (1992) 189-217.

[16] J. D. Sally. Number of Generators of Ideals in Local Rings, Lecture Notes in Pure and Applied Mathematics 35, Marcel Dekker, New York, 1978.

[17] A. Shalev. On the number of generators of ideals in local rings, Advances in Math. 59 (1986) 82-94.

[18] A. Shalev. Dimension subgroups, nilpotency indices and the number of generators of ideals in p−group algebras, J. Algebra 129 (1990) 412-438.

π-Domains without Identity

D. D. ANDERSON   Department of Mathematics, The University of Iowa, Iowa
City, IA 52242

ABSTRACT

Let D be a commutative integral domain without identity. We show that every
principal ideal of D is a product of prime ideals if and only if $D[1]$ is a quasilocal
UFD with D as its unique maximal ideal.

By a commutative ring we mean a nonzero commutative ring that does not necessarily have an identity. An integral domain is a commutative ring without proper
zero divisors. A commutative ring (integral domain) R is called a π-*ring* (π-*domain*)
if every principal ideal of R is a product of prime ideals. In a series of papers, S.
Mori [8, 9, 10, 11, 12] studied π-rings. In [11] he showed that a commutative ring
with identity is a π-ring if and only if it is a finite direct product of π-domains
and SPIR's (principal ideal rings with a single prime ideal and that prime ideal is
nilpotent). A very readable account can be found in [6, Section 46]. Examples of
π-domains with identity include UFD's and Dedekind domains. Many characterizations of π-domains with identity are collected in [1, Theorem 3.1]. For example,
for an integral D with identity the following conditions are equivalent: (1) D is a
π-domain, (2) every invertible ideal of D is a product of invertible prime ideals,
(3) every nonzero prime ideal of D contains an invertible prime ideal, (4) D is a
Krull domain with each height-one prime ideal invertible, (5) D is a locally factorial
Krull domain, and (6) $D(X) = \{\frac{f}{g} \mid f, g \in D[X],$ the coefficients of g generate $D\}$
is a UFD.

Less well known is Mori's characterization of π-rings without identity. In [12],
Mori showed that a π-ring without identity is either (1) an integral domain, (2) a

ring $R = (p)$ where every ideal of R including (0) is a power of R, (3) $K \times R$, where K is a field and R is a ring as in (2), or (4) $K \times D$, where K is a field and $D = (p)$ is a π-domain where every nonzero ideal of D is a power of D. He showed that in a π-domain without identity every principal ideal is actually a product of principal prime ideals (here the whole ring itself is considered to be a prime ideal) and that an integral domain D without identity is a π-domain if and only if (1) every nonzero element of D is a product of irreducible elements and (2) each irreducible element of D generates a principal prime ideal. Mori also remarked that $2\mathbb{Z}_{(2)} = \{\frac{m}{n} \mid m, n \in \mathbb{Z}, m \text{ is even and } n \text{ is odd}\}$ is an example of a π-domain without identity in which the factorization of an element into irreducible elements is not unique but for each irreducible element p of $2\mathbb{Z}_{(2)}$, $(p) = 2\mathbb{Z}_{(2)}$. In his review of Mori's paper, O.F.G. Schilling [13] states that in a π-domain D without identity, $D = (p)$ for each irreducible element p of D. Mori does *not* make this assertion which we shall later see is not correct.

Before continuing, a few remarks concerning integral domains without identity are in order. Let D be an integral domain without identity. Then $K = D_{D-\{0\}}$ is a field. Let $D^* = D[1] = \{d + n1 \mid d \in D, n \in \mathbb{Z}\}$, where 1 is the identity of K. Here D^* is an integral domain with identity and D is a proper ideal of D^*. More generally, it is easily seen ([2, Lemma 1]) that a subset A of D is an ideal of D if and only if it is an ideal of D^*. For $d \in D$, the principal ideal of D generated by d is $(d) = \{rd + nd \mid r \in D, n \in \mathbb{Z}\} = dD^*$. Recall that a fractional ideal A of D is invertible if there exists a fractional ideal B with $AB = D^*$. Finally, we consider D to be a prime ideal of D.

We next show that if (D, M) is a quasilocal UFD (with identity) and maximal ideal M such that $D = M[1]$, then M is a π-domain without identity. Thus if $D = \mathbb{Z}_{(p)}$ or $\mathbb{Z}_p[X_1, \ldots, X_n]$, $p > 0$ prime, $M = p\mathbb{Z}_{(p)}$ or $(X_1, \ldots, X_n)$ is a π-domain without identity. The purpose of this paper is to conversely show that if D is a π-domain without identity, then $D^* = D[1]$ is a quasilocal UFD with D as its maximal ideal.

THEOREM 1. *Let (D, M) be a quasilocal domain with maximal ideal M such that $M^* = M[1] = D$. Then M is a π-domain if and only if D is a UFD. In this case, every principal ideal of M is uniquely (up to order of factors) a product of principal prime ideals of M.*

Proof. ($\Leftarrow$) Let $0 \neq d \in M$, so $d = p_1 \cdots p_n$, where $p_1, \ldots, p_n$ are principal primes of D. Since $M[1] = D$, for $a \in M$, $aD = (a)$, the principal ideal of M generated by a. Hence $(d) = (p_1) \cdots (p_n)$ and each (p_i) is a prime ideal of D and hence of M.

($\Rightarrow$) Let $0 \neq d \in M$, so $(d) = P_1 \cdots P_n$, where P_i is a prime ideal of the ring M, possibly M itself. Since $M[1] = D$, each P_i is an ideal of D contained in M. Moreover, each P_i is actually a prime ideal of D. For let $d_1 d_2 \in P_i$, where $d_1, d_2 \in D$. If $d_1, d_2 \in M$, then d_1 or $d_2 \in P_i$ since P_i is a prime ideal of M. So assume, say, $d_2 \notin M$. Then d_2 is a unit in D, so $d_2^{-1} = m + n1$, where $m \in M$ and $n \in \mathbb{Z}$. Then $d_1 = d_1 d_2 d_2^{-1} = d_1 d_2 (m + n1) = d_1 d_2 m + n d_1 d_2 \in P_i$. So in D, each proper principal ideal is a product of prime ideals each necessarily principal since D is quasilocal. So D is a UFD.

Moreover, the above proof shows that the representation in M for (d) as a product of (principal) prime ideals is the same as in D, and hence is unique up to order. $\square$

If we take $D = \mathbb{Z}_{(p)}[\![X_1, \ldots, X_n]\!]$, where $p > 0$ is prime and $n \geq 2$, then $(X_1, \ldots, X_n)$ is a π-domain without identity, $X_1 \in (X_1, \ldots, X_n)$ is irreducible, but $(X_1, \ldots, X_n) \neq (X_1)$. This shows that Schilling's statement is not correct.

Our proof of the converse of Theorem 1 requires a number of lemmas. Besides the work of Mori, π-domains without identity were also investigated by C. Wood [14]. The only result concerning π-domains from Mori and Wood that we need is that each prime ideal occurring in a factorization of a nonzero principal ideal is a minimal principal prime ideal (Lemma 4). Since neither the papers of Mori nor Wood's dissertation may be easily accessible to many readers, we feel it is worthwhile to give a complete proof of this result. Lemma 2 is taken directly from [14]. While Lemma 3 follows from results in [12] and [14], its proof is different.

LEMMA 2. *Let D be a π-domain.*

(1) *Then each minimal prime ideal P of D is invertible. If D does not have an identity, then P is actually principal.*

(2) *If $P \subsetneq D$ is an invertible prime ideal, then P is a minimal prime ideal.*

Proof. (1) Let $0 \neq a \in P$, so $P_1 \cdots P_n = (a) \subseteq P$, where each P_i is a prime ideal of D. Hence some $P_i \subseteq P$ and so $P_i = P$. Since P is a factor of a principal ideal, it is invertible. Suppose that D doesn't have an identity. Then $PD \subsetneq P$ (for $PD = P$ gives $PD = PD^*$, and hence $D = D^*$ since P is invertible). Let $p \in P - PD$. Write $(p) = P_1 \cdots P_n$, a product of prime ideals. So some $P_i \subseteq P$, say P_1, and hence $P = P_1$ since P is minimal. So $(p) = PP_2 \cdots P_n$. Since $p \notin PD$, $n = 1$ and so $(p) = P$.

(2) It is easily checked that D_P is a quasilocal π-domain with identity and P_P is the unique maximal ideal of D_P. So D_P is a UFD and P_P is a principal ideal of D_P. Thus $\text{ht } P = \text{ht } P_P = 1$. $\square$

LEMMA 3. *Let D be a π-domain without identity. If D is invertible, then D is principal and every nonzero ideal of D is a power of D.*

Proof. Since D is invertible, $D^2 \subsetneq D$. Let $p \in D - D^2$. Since (p) is a product of prime ideals, (p) actually must be prime. Suppose $(p) \subsetneq D$. Now since D is invertible, $(p) = BD$ for some ideal B of D^*. Then $(p)^2 = ((p) B) D$, where $(p) B$ is an ideal of D. Now $(p)^2$ is (p)-primary and $D \not\subseteq (p)$, so $(p) B \subseteq (p)^2$. (We may see that $(p)^2$ is (p)-primary as follows. Let $xy \in (p)^2$, where $x \notin (p)$; so $y \in (p)$. Since (p) is a minimal prime ideal (Lemma 2), either $(y) = (p)$ or $(y) = (p) P_2 \cdots P_n$, where $P_2, \ldots, P_n$ are prime ideals and $n \geq 2$. Now if $(y) = (p)$, $(x)(p) \subseteq (p)^2$, and hence cancelling (p) gives $(x) \subseteq (p)$, a contradiction. If $(y) = (p) P_2 \cdots P_n$, then $(x)(p) P_2 \cdots P_n \subseteq (p)^2$ gives $(x) P_2 \cdots P_n \subseteq (p)$, and hence some $P_i \subseteq (p)$, so $y \in (p)^2$.) Cancelling (p) gives $B \subseteq (p)$, and hence $(p) = BD \subseteq (p) D \subseteq D^2$, a contradiction. So $D = (p)$.

Suppose there is a prime ideal $(0) \neq P \subsetneq D = (p)$. Then by Lemma 2, there is a principal prime ideal (q) with $(0) \neq (q) \subsetneq (p)$. Now since (p) is invertible, there is an ideal B of D^* with $(p) B = (q)$, and hence $(q)^2 = ((q) B)(p)$, where $(q) B$ is an ideal of D. Since $(q)^2$ is (q)-primary, $(q) B \subseteq (q)^2$, and hence cancelling (q) gives $B \subseteq (q)$. So $(q) = (p) B \subseteq (p) (q)$ and hence $D^* \subseteq (p)$, a contradiction. So $D = (p)$ is the only nonzero prime ideal of D. So every nonzero principal ideal, and hence every nonzero ideal of D is a power of D. $\square$

Gilmer [3, Theorem 4] has shown that an integral domain D without identity has the property that every nonzero ideal of D is a power of D if and only if D^* is a discrete valuation ring with D as its unique maximal ideal. This is of course a special case of our main result, which is independent of Gilmer's result. Gilmer also characterizes the discrete valuation domains $(V, (\pi))$ for which $V = (\pi)^*$.

LEMMA 4. ([12], [14]) *Let D be a π-domain without identity. Then each invertible prime ideal of D is minimal and principal. Thus each principal ideal of D is a product of principal prime ideals. Moreover, this factorization of a principal ideal into a product of (principal) prime ideals is unique up to order of the factors.*

Proof. Let P be an invertible prime ideal of D. If $P \subsetneq D$, P is minimal principal by Lemma 2, while if $P = D$, P is minimal and principal by Lemma 3.

We next show the uniqueness of the factorization. Let $0 \neq d \in D$ and suppose $(d) = (p_1) \cdots (p_n) = (q_1) \cdots (q_m)$, where each $(p_i), (q_j)$ is a principal prime ideal. Observe that each $(p_i), (q_j)$ is a minimal prime ideal of D. Now $(p_1) \cdots (p_n) \subseteq (q_1)$, so some $(p_i) \subseteq (q_1)$, say $(p_1) \subseteq (q_1)$, and hence $(p_1) = (q_1)$. Cancelling (p_1) gives $(p_2) \cdots (p_n) = (q_2) \cdots (q_m)$. By induction $n = m$ and after rearranging, if necessary, $(p_i) = (q_i)$ for $i = 2, \ldots, n$. $\square$

The uniqueness of factorization of (d) into prime ideals in Lemma 4 is actually a special case of [5, Theorem 2].

LEMMA 5. *Let D be a π-domain without identity. Then for $d \in D$, $1 + d$ is a unit in D^*.*

Proof. We can assume $d \neq 0$. Write $(d) = (p_1) \cdots (p_n)$ as a product of principal primes. Now $0 \neq (1 + d) d \in D$, so write $((1 + d) d) = (q_1) \cdots (q_m)$ as a product of principal primes. Now $(q_1) \cdots (q_m) = ((1 + d) d) \subseteq (d) = (p_1) \cdots (p_n)$, so say $(q_1) \subseteq (p_1)$ and hence $(q_1) = (p_1)$. Now either $n = 1$ or $n > 1$ and we can cancel $(p_1) = (q_1)$. Continuing, we either get $n = m$ and $((1 + d) d) = (d)$ or $m > n$ and $((1 + d) d) = (d) (q_{n+1}) \cdots (q_m)$. In the first case, $(1 + d) dD^* = dD^*$, and hence $1 + d$ is a unit in D^*. In the second case, $(1 + d) dD^* = dq_{n+1} \cdots q_m D^*$, and hence $(1 + d) D^* = q_{n+1} \cdots q_m D^*$. Then $1 + d \in (q_{n+1} \cdots q_m) \subseteq D$ and hence $1 \in D$, a contradiction. $\square$

LEMMA 6. *Let D be a π-domain without identity. Then there exists at most one prime number p such that $p1$ is not a unit in D^*. And if $p1$ is not a unit in D^*, then $p1 \in D$. Hence for $n \in \mathbb{Z}$, if $n1 \notin D$, $n1$ is a unit in D^*.*

Proof. We may assume char $D = 0$. Let $p > 0$ be a prime number. Let (q) be a nonzero principal prime ideal of D. Now $(0) \neq (pq) \subseteq (q)$, so either $(pq) = (q)$ or $(pq) = (q) (q_2) \cdots (q_n)$, where $(q_2), \ldots, (q_n)$ are other principal prime ideals of D. In the first case, $(pq) = (q)$ gives $p1D^* = D^*$ and hence $p1$ is a unit in D^*. In the second case, $p1qD^* = qq_2 \cdots q_n D^*$ gives $p1 \in p1D^* = q_2 \cdots q_n D^* \subseteq D$. Suppose we have distinct primes $p_1, p_2 \in \mathbb{Z}$ with $p_11, p_21 \in D$. Then $n_1 p_1 + n_2 p_2 = 1$ in $\mathbb{Z}$ gives $1 = n_1 p_1 1 + n_2 p_2 1 \in D$, a contradiction. So there exists at most one prime $p > 0$ of $\mathbb{Z}$ with $p1$ not a unit in D^*. $\square$

LEMMA 7. *Let D be a π-domain without identity. Then D is the unique maximal ideal of D^*.*

Proof. It suffices to show that if $x \in D^* - D$, then x *is* a unit in D^*. Now since $x \notin D$, $x = n1 + d$, where $n \in \mathbb{Z}$ with $n1 \notin D$ and $d \in D$. Since $n1 \notin D$, $n1$ is a unit in D^* by Lemma 6. Now $(n1)^{-1} x = 1 + (n1)^{-1} d$ is a unit in D^* by Lemma 5. Hence x is a unit in D^*. $\square$

We are now ready for the main result of this paper.

THEOREM 8. *Let D be an integral domain without identity. Then D is a π-domain if and only if $D^* = D[1]$ is a quasilocal UFD with D as its maximal ideal.*

Proof. ($\Leftarrow$) Theorem 1. ($\Rightarrow$) Suppose that D is a π-domain without identity. By Lemma 7, D^* is a quasilocal domain with maximal ideal D. By Theorem 1, D^* is a UFD. $\square$

COROLLARY 9. ([3] and [7]) *Let D be an integral domain without identity. Suppose that each ideal of D generated by two elements is a product of prime ideals. Then D^* is a discrete valuation domain with maximal ideal D. Conversely, if D^* is a discrete valuation domain with maximal ideal D, then every nonzero ideal of D is a power of D.*

Proof. ($\Rightarrow$) Clearly D is a π-domain. Hence D^* is a quasilocal UFD with maximal ideal D. Since every ideal of D generated by two elements is a product of prime ideals of D, the proof of Theorem 1 shows that every ideal of D^* generated by two elements is a product of prime ideals. By [7] or [6, Theorem 46.12] D^* is a discrete valuation domain.
($\Leftarrow$) Theorem 1. $\square$

Let D be an integral domain with identity. It is well known that D is a π-domain (UFD) if and only if every nonzero prime ideal of D contains an invertible (nonzero principal) prime ideal ([1], [6, Theorem 46.1]). Thus it seems reasonable to conjecture that if D is an integral domain without identity, then D is a π-domain if and only if each nonzero prime ideal contains a nonzero principal prime ideal. However, this need not be the case as shown by an example of Gilmer [4, Example 5.3]. Let $D = X\mathbb{Z}[X]$. Then $D^* = \mathbb{Z} + X\mathbb{Z}[X] = \mathbb{Z}[X]$ is not quasilocal, so D is not a π-domain (alternatively, observe that $(4X)$ is not a product of prime ideals), but every prime ideal of D is principal.

As mentioned earlier, Mori remarked that while $2\mathbb{Z}_{(2)}$ is a π-domain, elements of $2\mathbb{Z}_{(2)}$ do not have unique factorization into irreducible elements. This is actually typical of integral domains without identity. Let D be an integral domain without identity. Since D has no units, two elements are associates if and only if they are equal. Thus to say that D has unique factorization into irreducible elements would mean that each nonzero element of D is a product of irreducible elements and that this factorization is unique up to order of factors. Let us call an integral domain without identity that satisfies these two conditions a *strong UFD*. The next proposition shows that a strong UFD must be trivial.

PROPOSITION 10. *Let D be an integral domain containing a nonzero element. Then D cannot be a strong UFD.*

Proof. Suppose that D is a strong UFD that contains a nonzero irreducible element m. Then $m(1 + m) = m + m^2$ is a nonzero nonunit of D, for $m + m^2 = 0$ gives $m = m(-m)$ which contradicts that m is irreducible. So write $m + m^2 = a_1 \cdots a_s,$

where each a_i is irreducible. Likewise, $m(1+m)^2 = m + 2m^2 + m^3$ is nonzero and $m(1+m)^2 = b_1 \cdots b_t$, where each b_j is irreducible. Then $(m+m^2)^2 = (m(1+m))^2 = m\left(m(1+m)^2\right) = m\left(m + 2m^2 + m^3\right)$ gives $a_1^2 \cdots a_s^2 = mb_1 \cdots b_t$ as two factorizations into irreducibles. Hence, some a_i must be m, say $a_1 = m$. Then $m + m^2 = m(a_2 \cdots a_s)$ and $s \geq 2$ since $m + m^2 = m$ gives $m^2 = 0$. But then $m = m(a_2 \cdots a_s - m)$ again contradicts the hypothesis that m is irreducible. $\square$

REFERENCES

[1] D.D. Anderson, *Globalization of some local properties in Krull domains*, Proc. Amer. Math. Soc. **85** (1982), 141–145.

[2] R.W. Gilmer, *Commutative rings containing at most two prime ideals*, Michigan Math. J. **10** (1963), 263–268.

[3] R.W. Gilmer, *On a classical theorem of Noether in ideal theory*, Pacific J. Math. **13** (1963), 579–583.

[4] R. Gilmer, *Commutative rings in which each prime ideal is principal*, Math. Ann. **183** (1969), 151–158.

[5] R. Gilmer, *On factorization into prime ideals*, Comment. Math. Helv. **47** (1972), 70–74.

[6] R. Gilmer, *Multiplicative Ideal Theory*, Queen's Papers in Pure and Applied Mathematics, vol. 90, Queen's University, Kingston, Ontario, 1992.

[7] K. Levitz, *A characterization of general Z.P.I.-rings* II, Pacific J. Math. **42** (1972), 147–151.

[8] S. Mori, *Über die Produktzerlegung der Hauptideale*, J. Sci. Hiroshima Univ. Ser. A **8** (1938), 7–13.

[9] S. Mori, *Über die Produktzerlegung der Hauptideale*, II, J. Sci. Hiroshima Univ. Ser. A **9** (1939), 145–155.

[10] S. Mori, *Allgemeine Z.P.I.-Ringe*, J. Sci. Hiroshima Univ. Ser. A **10** (1940), 117–136.

[11] S. Mori, *Über die Produktzerlegung der Hauptideale*, III, J. Sci. Hiroshima Univ. Ser. A. **10** (1940), 85–94.

[12] S. Mori, *Über die Produktzerlegung der Hauptideale*, IV, J. Sci. Hiroshima Univ. Ser. A. **11** (1941), 7–14.

[13] O.F.G. Schilling, Review in *Mathematical Reviews* of [12], Math. Rev. **2** (1941), 121.

[14] C.A. Wood, *On General Z.P.I.-Rings*, Dissertation, Florida State University, Tallahassee, Florida, December 1967.

Extensions of Unique Factorization: A Survey

D. D. ANDERSON   Department of Mathematics, The University of Iowa, Iowa City, IA 52242

I. INTRODUCTION

Unique factorization domains (a.k.a. UFD's or factorial domains) have a central place in mathematics. Some of the first theorems a student of abstract algebra encounters are that certain familiar rings are UFD's. The Fundamental Theorem of Arithmetic states that the ring of integers $\mathbb{Z}$ is a UFD. The well-known implications Euclidean $\Rightarrow$ PID $\Rightarrow$ UFD yield that for a field K, $K[X]$ is a UFD. Ideal theory owes its start to the fact that while a ring of algebraic integers need not be a UFD, unique factorization can be restored via factorization into prime ideals. At a more advanced level, one of the first triumphs of homological algebra was that a regular local ring is a UFD [29]. Perhaps the first name that comes to mind when thinking about UFD's is Pierre Samuel. He has done fundamental work on Euclidean rings [76], UFD's, and power series rings over UFD's [72]. The reader is referred to his survey article [75] and monographs [73], [74]. His collected works [77], which contain a number of papers related to unique factorization, have recently been published in the Queen's Papers in Pure and Applied Mathematics series. Moreover, several of his students have done work dealing with unique factorization and its extensions to commutative rings with zero divisors and to modules.

Let D be an integral domain with identity. A nonzero nonunit $d \in D$ is said to be *irreducible* or an *atom* if $d = ab$ $(a, b \in D)$ implies a or b is a unit. And D is *atomic* if each nonzero nonunit of D is a finite product of atoms. Then D is a *UFD* if (1) D is atomic and (2) if $a_1 \cdots a_n = b_1 \cdots b_m$ where a_i, b_j are atoms, then $n = m$ and after reordering, if necessary, a_i and b_i are associates. As with any important definition, a number of different characterizations of UFD's are known. The following theorem lists several of the more important ones.

THEOREM 1.1. *For an integral domain D, the following conditions are equivalent.*

(1) *D is a UFD (as defined in the previous paragraph).*

(2) *For each nonzero nonunit $a \in D$, $a = p_1 \cdots p_n$, where each p_i is a principal prime (i.e., (p_i) is a prime ideal).*

(3) *Every proper principal ideal of D is a product of principal prime ideals.*

(4) *For each nonzero nonunit $a \in D$, $a = up_1^{a_1} \cdots p_n^{a_n}$ where u is a unit, each $a_i \geq 1$, and the p_i's are principal primes with $[p_i, p_j] = 1$ for $i \neq j$.*

(5) *Every nonzero prime ideal of D contains a nonzero principal prime ideal.*

(6) *D is a Krull domain with divisor class group $\mathrm{Cl}(D) = 0$.*

The purpose of this article is to survey some of the possible extensions of unique factorization when we weaken any of the conditions (1)–(6) of Theorem 1.1 and/or when we allow D to be replaced by a ring with zero divisors (possibly without identity) or by a module. This survey article is divided into six sections including the introduction. The second section covers some extensions of unique factorization to integral domains obtained by weakening any of conditions (1)–(6) of Theorem 1.1. We cover atomic domains with "good factorization properties" such as half-factorial domains (HFD's), bounded factorization domains (BFD's), and finite factorization domains (FFD's) and "non-atomic unique factorization" where elements are factored as products of elements other than atoms or primes such as primary elements or (completely) primal elements. We also mention π-domains (i.e., domains in which every principal ideal is a product of prime ideals) and Krull domains D with conditions on $\mathrm{Cl}(D)$ weaker than $\mathrm{Cl}(D) = 0$. The third section surveys unique factorization rings with zero divisors. We cover the two different unique factorization rings defined by Bouvier and Galovich and by Fletcher. A third type of unique factorization ring which unifies the above two types is given. We also briefly survey some of the generalizations of unique factorization given in Section II for domains to rings with zero divisors. The fourth section surveys the (unique) factorization of regular elements (i.e., non-zero-divisors) in a commutative ring R with zero divisors. Of particular interest is the relationship between regular elements of R having unique factorization into atoms or primes and R being a Krull ring with $\mathrm{Cl}(R) = 0$. The fifth section considers extensions of unique factorization to integral domains or rings without identity. It is shown (Theorem 5.1) that a commutative ring R has "absolutely unique factorization" into irreducibles if and only if $R^2 = 0$. We also consider π-rings without identity and a recent alternate approach to factorization in rings without identity by Ağargün and Fletcher. The final section covers (unique) factorization in modules. Work by A.-M. Nicolas and the author and S. Valdes-Leon is surveyed.

This paper is an elaboration of the talk given at the Fez workshop. Throughout all rings are commutative and have an identity unless explicitly stated otherwise. We will follow standard notation and definitions as found in [50], [56], or [58]. For Krull domains, the reader is referred to [44]. We use R to denote a commutative ring, but reserve D for an integral domain. For a ring R, $R^* = R - \{0\}$ and $U(R)$ denotes the group of units of R. For $a \in R$, $(a) = \{ra + na \mid r \in R, n \in \mathbb{Z}\}$, the principal ideal of R generated by a. So if R has an identity, $(a) = Ra$. The GCD of two elements a and b is denoted by $[a, b]$. If A is a nonzero ideal of R, $A_v = \left(A^{-1}\right)^{-1}$. We use $\mathrm{Cl}(R)$ to denote the divisor class group of a Krull domain or Krull ring R.

Finally, as with any survey article, the topics covered are subjective. No claim is made for exhaustiveness. I have been interested in and worked with extensions of unique factorization for over twenty years. The topics I will survey represent areas that I have worked in or am particularly interested in. I have included an extensive (but not exhaustive) bibliography of nearly one hundred articles. For a survey of factorization in integral domains the reader is referred to the conference proceedings [5], which contain several excellent survey articles ([28], [35], [46], [54], [68]).

II. EXTENSIONS OF UNIQUE FACTORIZATION IN INTEGRAL DOMAINS

A. Atomic Domains with "Good" Factorization Properties. Throughout this section D denotes an integral domain with identity. We begin by considering atomic integral domains with "good" factorization properties. Perhaps the first such weaker factorization property to be considered is that of being a half-factorial domain.

An atomic integral domain D is a *half-factorial domain* (HFD) if w $a_1 \cdots a_n = b_1 \cdots b_m$ where a_i, b_j are irreducible, then $n = m$. This property was first considered by Carlitz [34], who showed that a ring of algebraic integers D is an HFD $\Leftrightarrow$ $|\text{Cl}(D)| \leq 2$. HFD's were first formally defined and systematically studied by Zaks [86], [87]. Most of the investigation of HFD's has been for Dedekind domains or more generally for Krull domains. Generalizing Carlitz's result, a Krull domain D with $|\text{Cl}(D)| \leq 2$ is an HFD and if each class of $\text{Cl}(D)$ contains a height-one prime ideal (w is the case for rings of algebraic integers or polynomial rings) the converse is true [87]. However, in general a Krull (or even Dedekind) HFD D need not have $|\text{Cl}(D)| \leq 2$. Recently, Coykendall [41] has shown that if $D[X]$ is an HFD, then D must be integrally closed. Hence for D Noetherian, $D[X]$ is an HFD $\Leftrightarrow$ D is a Krull domain with $|\text{Cl}(D[X])| \leq 2$. For various generalizations of the HFD property, the reader is referred to [36], [37].

If D is not an HFD, it is of interest to know the possible different lengths of factorizations into irreducibles. One tool for this is the elasticity. The *elasticity of* D is $\rho(D) = \sup\left\{\frac{n}{m} \mid a_1 \cdots a_n = b_1 \cdots b_m, a_i, b_j \text{ irreducible}\right\}$. Thus D is an HFD $\Leftrightarrow \rho(D) = 1$. Elasticity, especially for Krull domains, has been extensively studied. For a survey of elasticity containing an extensive bibliography, the reader is referred to [28]. For other surveys of lengths of factorizations, see [35], [46], and [54].

Another generalization of unique factorization is to assume that only factorizations of a certain length are unique. This theme is investigated in [12] and [13]. Let $n \geq 1$. Then an atomic integral domain D is defined to be *n-factorial* (*quasi-n-factorial*) if $a_1 \cdots a_n = b_1 \cdots b_m$ ($a_1 \cdots a_n = b_1 \cdots b_n$) where a_i, b_j are irreducible $\Rightarrow n = m$ and after reordering, if necessary, a_i and b_i are associates. And D is said to be *square-factorial* if $u^2 = vw$ for u, v, w irreducible $\Rightarrow u$ and v are associates. In [12] it is shown that if D is a non-maximal order in an algebraic number field, then D is *not* square-factorial (earlier, Halter-Koch [52] showed that a maximal order D is factorial $\Leftrightarrow$ it is square-factorial) and the polynomial ring $D[X]$ is 2-factorial $\Leftrightarrow$ D is factorial. It is still an open question whether D being n-factorial for some $n \geq 2$ forces D to be factorial. However, examples show that an n-factorial monoid need not be factorial.

We have already defined an atomic domain, HFD, and UFD. Let D be an integral domain. Then D satisfies the *ascending chain condition on principal ideals* (ACCP) if each ascending chain of principal ideals of D is finite. If D satisfies ACCP, then D is atomic, but not conversely [51]. A (necessarily atomic) domain D is a *bounded factorization domain* (BFD) if for each nonzero nonunit $a \in D$ there exists a natural number $N(a)$ so that if $a = a_1 \cdots a_n$ where each a_i is a nonunit, then $n \leq N(a)$. An integral domain D is a *finite factorization domain* (FFD) if each nonzero nonunit element of D has only finitely many nonassociate divisors. More generally, a not necessarily atomic domain D is an *idf-domain* if each nonzero nonunit of D has only finitely many nonassociate irreducible factors. Now D is an FFD $\Leftrightarrow$ D is an atomic idf-domain. For a detailed discussion of these different factorization properties, the reader is referred to [8], [10]. FFD's are also studied in [20]. The simplest FFD's are those with only finitely many nonassociate atoms. Such domains are called Cohen–Kaplansky domains and are studied in [3], [17]. The following diagram from [8] gives the various implications for the above defined integral domains. Examples given in [8] show that these are the only possible implications.

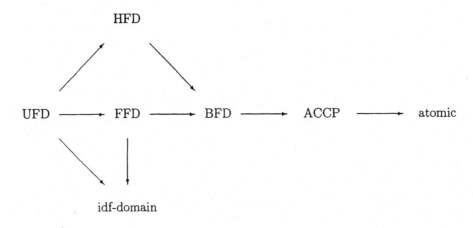

FIGURE 1.

B. **Non-atomic Unique Factorization.** Now UFD's can also be characterized by the property that every nonzero nonunit is a product of principal prime elements or equivalently that every nonzero nonunit x can be written in the form $x = u p_1^{a_1} \cdots p_n^{a_n}$ where u is a unit, $p_1, \ldots, p_n$ are nonassociate primes and each $a_i \geq 1$. Each of the $p_i^{a_i}$, in addition to being a power of a prime, has a number of other properties, each of which is subject to generalization. The goal of this subsection is to survey various generalizations of (unique) factorization into prime powers in integral domains that need not be atomic. This follows the thesis of M. Zafrullah that the $p_i^{a_i}$'s are the "building blocks" in a UFD. The author would like to thank M. Zafrullah for a number of discussions of this thesis. Much of this subsection comes from [19].

Now a primary element (i.e., (x) is a primary ideal) generalizes the notion of a prime power p^n, p a prime element. Integral domains with the property that every nonzero nonunit is a product of primary elements are called *weakly factorial*, see [15], [14]. Suppose that D is weakly factorial. By combining primary elements with the same radical, each nonzero nonunit x of D has a unique factorization into primary elements in the sense that if $x = q_1 \cdots q_n = q_1' \cdots q_m'$, where each q_i, q_i' is primary with $\sqrt{(q_i)} \neq \sqrt{(q_j)}$ ($\sqrt{(q_i')} \neq \sqrt{(q_j')}$) for $i \neq j$, then $n = m$ and after reordering, if necessary, q_i and q_i' are associates. Two results from [15] are: (1) a one-dimensional Noetherian domain D is weakly factorial $\Leftrightarrow$ Pic $(D) = 0$, and (2) a

Krull domain is weakly factorial $\Leftrightarrow$ it is factorial. In [60] it is shown that if D is a Noetherian weakly factorial domain, then the integral closure $\bar{D}$ of D is factorial. Also, an atomic domain with almost all atoms prime is weakly factorial [9]. Thus a Cohen–Kaplansky domain is weakly factorial.

Another property of a prime power p^n is that if $x, y | p^n$, then $x | y$ or $y | x$. With this in mind, following P.M. Cohn, Zafrullah [81] defined a nonzero nonunit $h \in D$ to be *rigid* if $x, y | h \Rightarrow x | y$ or $y | x$. Evidently an integral domain D is *rigid* (i.e., every nonzero nonunit of D is rigid) if and only if D is a valuation domain. And D is said to be *semi-rigid* if each nonzero nonunit of D is a product of rigid elements. Semi-rigid GCD domains are investigated in [81], [82]. While a primary element h need not be rigid, if (h) is P-primary, then P is a maximal t-ideal and hence h is contained in a unique maximal t-ideal. (Recall that an ideal A is a *t-ideal* if $a_1, \ldots, a_n \in A \Rightarrow (a_1, \ldots, a_n)_v = \left((a_1, \ldots, a_n)^{-1} \right)^{-1} \subseteq A$.) Let P be a prime ideal of D. An ideal A is said to be *P-pure* [18] if $A_P \cap D = A$. Certainly if (h) is P-primary, then (h) is P-pure. In [19] we defined a nonzero nonunit $h \in D$ to be *t-pure* if (h) is P-pure for some maximal t-ideal P, or equivalently, if h is contained in a unique maximal t-ideal. Let us call an integral domain D *t-pure* (resp., *semi-t-pure*) if every nonzero nonunit of D is t-pure (resp., a product of t-pure elements). Although not given a name there, a number of characterizations of semi-t-pure domains were given in [18, Corollary 4.4]. For example, D is semi-t-pure $\Leftrightarrow D = \bigcap_{P \in t\text{-Max}(D)} D_P$ has finite character, for distinct $P, Q \in t\text{-Max}(D)$, the set of maximal t-ideals of D, $P \cap Q$ does not contain a nonzero prime ideal, and the t-class group $\text{Cl}_t(D) = 0$.

For a nonzero nonunit h, put $P(h) = \{x \in D \mid (x, h)_v \neq D\}$. So $P(h) = \bigcup \{M \in t\text{-Max}(D) \mid h \in M\}$. Thus $P(h)$ is an ideal if and only if h is t-pure. Suppose h is contained in a unique maximal t-ideal P and suppose $x, y | h$. If x and y are both nonunits, then $x, y \in P$ and hence $(x, y)_v \neq D$. Thus if $x, y | h$ and $(x, y)_v = D$, then x or y is a unit. We define a nonzero nonunit $h \in D$ to be *homogeneous* (resp., *strongly homogeneous*) if $x, y | h$ and $(x, y)_v = D$ (resp., $[x, y] = 1$) implies x or y is a unit. We say that D is *(strongly) homogeneous* if every nonzero nonunit of D is (strongly) homogeneous and that D is *semi-(strongly-)homogeneous* if each nonzero nonunit of D is a product of (strongly-)homogeneous elements. Clearly a rigid element is strongly homogeneous and a strongly homogeneous element is homogeneous. Thus a (semi-)rigid domain is a (semi-)strongly-homogeneous domain and a (semi-)strongly-homogeneous domain is a (semi-)homogeneous domain.

To study factorization in a non-atomic setting, P.M. Cohn [39] defined a nonzero nonunit $h \in D$ to be *primal* if $h | xy \Rightarrow h = h_1 h_2$ where $h_1 | x$ and $h_2 | y$. Thus an atom is primal if and only if it is prime. While a nonunit factor of a prime, primary, irreducible, rigid, t-pure, strongly homogeneous or homogeneous element has the same property, a factor of a primal element need *not* be primal. Cohn defined a nonzero nonunit h to be *completely primal* if each nonunit factor of h is primal. The product of two completely primal elements is again completely primal. Cohn defined an integral domain D to be a *Schreier domain* if D is integrally closed and every nonzero nonunit of D is (completely) primal. Later, Zafrullah [85] defined an integral domain to be *pre-Schreier* if every nonzero nonunit is (completely) primal. For results on completely primal elements, also see [27].

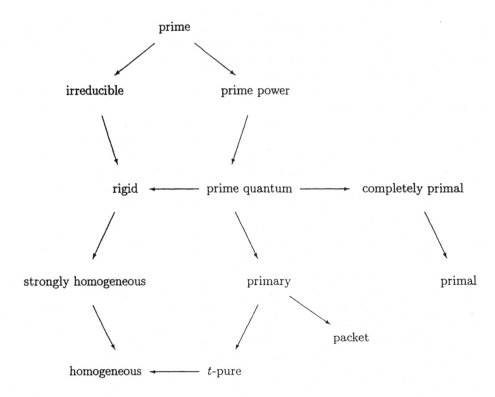

In [11] a nonzero nonunit q was defined to be a *prime quantum* if q is completely primal, each power of q is rigid, and for every nonunit $r|q$, there exists a natural number n with $q|r^n$. A prime power p^n is a prime quantum and a prime quantum q is $P(q)$-primary. An integral domain D was defined to be a *generalized unique factorization domain* (GUFD) if every nonzero nonunit of D is a product of prime quanta. Clearly a UFD is a GUFD and a GUFD is weakly factorial and a GCD domain.

Zafrullah [83] considered yet another generalization of a prime power. He defined a nonzero nonunit $x \in D$ to be a *packet* if $\sqrt{(x)}$ is prime, that is, there is a unique minimal prime over (x). He studied GCD domains, called *unique representation domains*, with the property that every nonzero nonunit is a product of packets.

The various generalizations of a prime element are indicated in the diagram above (Figure 2). If $0 \neq h \in D$ is a nonunit completely primal element, then the following are equivalent ([19, Theorem 2.3]): (1) h is homogeneous, (2) h is strongly homogeneous, (3) h is t-pure. For a GCD domain D (where every nonzero nonunit is completely primal) we can add: (4) h is rigid.

Corresponding to Figure 2, we have Figure 3 below showing the various generalizations of UFD's. Again, the only implications are the obvious ones.

When considering factorization into elements from Figure 2, only factorization into primes is necessarily unique. However, it is easily seen that factorization into prime powers is unique once powers of associated primes are combined. And as

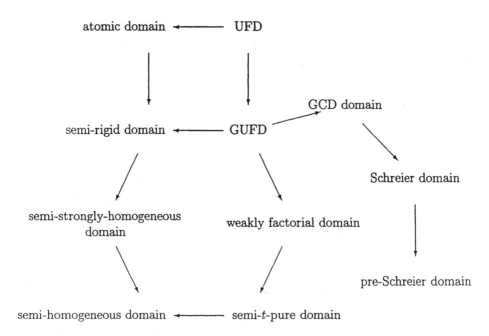

FIGURE 3.

previously mentioned, a similar result is also true for factorization into primary elements. We can do the same thing for t-pure elements [19, Theorem 3.1]. If x and y are P-pure where P is a maximal t-ideal, then xy is again P-pure. So if $0 \neq x \in D$ is a product of t-pure elements, then x is a product of v-coprime t-pure elements. Suppose $x = x_1 \cdots x_n = y_1 \cdots y_m$ where each x_i, y_i is t-pure and x_i, x_j (resp., y_i, y_j) are v-coprime (i.e., $(x_i, x_j)_v = D$) for $i \neq j$. Then $n = m$ and after reordering, if necessary, x_i and y_i are associates. Thus if $x \in D$ is a finite product of completely primal homogeneous elements, then x is uniquely expressible as a finite product of mutually coprime completely primal elements. So we get the promised unique factorization in semi-t-pure domains ([19, Theorem 3.4]) and in pre-Schreier semi-homogeneous domains ([19, Corollary 3.7]).

Kaplansky [56, Theorem 5] proved that D is a UFD if and only if each nonzero prime ideal of D contains a nonzero principal prime. For each of the properties (∗) of an element given in Figure 2 we can ask whether every nonzero nonunit of an integral domain D is a product of elements having property (∗) if and only if each nonzero prime ideal of D contains a (nonzero) element with property (∗). See [19] for details. Further Kaplansky-like theorems are investigated in [26].

Another generalization of UFD's we have not touched upon are GCD domains. Of course a GCD domain is a UFD ⇔ it is atomic. For a number of generalizations of GCD domains and their relationship to Gauss's lemma, see [23]

C. Krull Domains D with Conditions on $\mathrm{Cl}(D)$. An integral domain D is a π-*domain* if every proper principal ideal of D is a product of prime ideals. Thus a UFD or Dedekind domain is a π-domain. S. Mori introduced and studied π-domains in [63, 64, 65, 66, 67]. A number of characterizations of π-domains are known; for example, see [4]. The following theorem lists some of these characterizations.

THEOREM 2.1. *For an integral domain D, the following conditions are equivalent.*

(1) *D is a π-domain.*

(2) *D is a Krull domain and D_M is a UFD for each maximal ideal M of D.*

(3) *Each nonzero prime ideal of D contains an invertible prime ideal.*

(4) *$D(X) = \left\{ \frac{f}{g} \mid f, g \in D[X], \text{ the coefficients of } g \text{ generate } D \right\}$ is a UFD.*

(5) *D is a Krull domain with $\mathrm{Cl}(D) = \mathrm{Pic}(D)$.*

Another generalization of a UFD, that is, a Krull domain with $\mathrm{Cl}(D) = 0$, is a Krull domain with $\mathrm{Cl}(D)$ a torsion abelian group. Such domains, called *almost factorial domains*, were introduced by Storch [79]; also see [44]. One characterization of almost factorial domains is that they are Krull domains with the property that some power of each nonzero nonunit is a product of primary elements. For a generalization of almost factorial domains, see [84].

An integral domain D is *locally factorial* if $D\left[\frac{1}{f}\right]$ is a UFD for each nonunit $f \in D^*$. If D is a locally factorial Krull domain, then $\mathrm{Cl}(D) \approx \mathbb{Z}^n \oplus \mathbb{Z}/m\mathbb{Z}$. Conversely, for any abelian group $G \approx \mathbb{Z}^n \oplus \mathbb{Z}/m\mathbb{Z}$, there is a locally factorial Dedekind domain D with $\mathrm{Cl}(D) \approx G$. For these results and others concerning locally factorial domains, the reader is referred to [7]. It appears to be an open question whether a quasilocal locally factorial Krull domain must be a UFD. Also, all known examples of locally factorial Krull domains which are not factorial are Dedekind domains.

III. UNIQUE FACTORIZATION RINGS WITH ZERO DIVISORS

In this section and the next section we consider extensions of unique factorization to commutative rings with zero divisors Throughout this section R will denote a commutative ring with identity. There are basically two approaches to generalizing results about integral domains to commutative rings with zero divisors. The first is to directly extend a definition concerning elements or ideals of an integral domain to all the elements or ideals of a commutative ring. For example, a Prüfer domain generalizes to an arithmetical ring. This is the approach taken in this section. The second approach is to extend a definition concerning nonzero elements or ideals to just the regular elements or regular ideals of a commutative ring. For example, Prüfer domains have Prüfer rings as a generalization since Prüfer rings are defined to be commutative rings in which every finitely generated regular ideal is invertible. See [58] and [21] for results on Prüfer domains that typify this approach. In the next section we take the second approach. Two excellent references for commutative rings with zero divisors are [55] and [62],

The notion of a UFD may be extended to commutative rings with zero divisors in a number of ways depending on how "irreducible" and "associate" are defined. Perhaps the first paper to consider unique factorization rings was [30] which considered unique factorization in $\mathbb{Z}/m\mathbb{Z}$. Let R be a commutative ring with identity. Bouvier [33] defined a nonzero nonunit $r \in R$ to be "irreducible" if (r) is a maximal element of the set of proper principal ideals of R and defined two elements $a, b \in R$ to be "associates", denoted $a \sim b$, if $a|b$ and $b|a$, or equivalently if $(a) = (b)$. He then defined R to be a "unique factorization ring" if (1) each nonzero nonunit of R

is a product of "irreducible" elements and (2) if $0 \neq a_1 \cdots a_n = b_1 \cdots b_m$ where each a_i, b_j is "irreducible", then $n = m$ and after reordering, if necessary, $a_i \sim b_i$. He proved that R is a "unique factorization ring" if and only if R is a UFD, a special principal ideal ring (SPIR), or a quasilocal ring (R, M) with $M^2 = 0$. (Recall that an SPIR is a principal ideal ring with a unique prime ideal and that prime ideal is nilpotent.) Galovich [45] defined a nonzero nonunit $r \in R$ to be "irreducible" if $r = ab$ implies a or b is a unit and defined two elements $a, b \in R$ to be "associates", denoted $a \approx b$, if $a = ub$ for some unit $u \in R$. He then defined R to be a "unique factorization ring" if (1) each nonzero nonunit of R is a product of "irreducible" elements and (2) if $0 \neq a_1 \cdots a_n = b_1 \cdots b_m$ where each a_i, b_j is "irreducible", then $n = m$ and after reordering, if necessary, $a_i \approx b_i$. He also proved that R is a "unique factorization ring" if and only if R is a UFD, SPIR, or quasilocal ring (R, M) with $M^2 = 0$. We use the term *BG-UFR (Bouvier-Galovich unique factorization ring)* for the (equivalent) unique factorization rings introduced by Bouvier and Galovich. We state these two characterizations as the first theorem of this section.

THEOREM 3.1. *For a commutative ring R the following conditions are equivalent.*

(1) *R is a unique factorization ring as defined by Bouvier.*
(2) *R is a unique factorization ring as defined by Galovich.*
(3) *R is either a UFD, SPIR, or a quasilocal ring (R, M) with $M^2 = 0$.*

Fletcher [42] extended the notion of UFD to commutative rings with zero divisors in yet another way. He defined two elements $a, b \in R$ to be "associates" if $a \sim b$ and defined a nonzero nonunit $a \in R$ to be "irreducible" if $a = a_1 \cdots a_n$ implies some $a_i \sim a$ (or equivalently, $a = bc \Rightarrow a \sim b$ or $a \sim c$ [24, Theorem 2.6]). For $a \in R$, let $U(a) = \{r \in R \mid r(a) = (a)\}$. He defined a U-*decomposition* of an element: $a = (a_1 \cdots a_k)(b_1 \cdots b_n)$ where each a_i, b_j is "irreducible", $a_i \in U(b_1 \cdots b_n)$ for each $i = 1, \ldots, k$ and $b_j \notin U(b_1 \cdots \hat{b}_j \cdots b_n)$ for $j = 1, \ldots, n$. It is easily seen that if $a \in R$ is a product of "irreducibles", then a has a U-decomposition. He defined R to be a "unique factorization ring" (which we will call a *Fletcher unique factorization ring*) if (1) every nonunit of R has a U-decomposition, and (2) if $(a_1 \cdots a_k)(b_1 \cdots b_n) = (a'_1 \cdots a'_{k'})(b'_1 \cdots b'_{n'})$ are two U-decompositions of a (possibly zero) nonunit element, then $n = n'$, and after reordering, if necessary, $b_i \sim b'_i$. He then proved [43] that R is a (Fletcher) "unique factorization ring" if and only if R is a finite direct product of UFD's and SPIR's. Equivalently, R is a Fletcher UFR if and only if every nonunit of R is a product of principal primes. For an alternative treatment of Fletcher UFR's, see [16]. So we have the following theorem. In this theorem and in the remainder of this section we use "irreducible" in the sense of Fletcher.

THEOREM 3.2. *For a commutative ring R the following conditions are equivalent.*

(1) *R is a Fletcher UFR.*
(2) *Every nonunit element of R is a product of irreducible elements and each irreducible element is prime.*
(3) *Every nonunit of R is a product of principal primes.*
(4) *Every proper principal ideal of R is a product of principal prime ideals.*
(5) *R is a finite direct product of UFD's and SPIR's.*
(6) *Every height-zero prime ideal of R is principal and every prime ideal of height > 0 contains a principal prime ideal of height > 0.*

Recall that R is a *π-ring* if each proper principal ideal of R is a product of prime ideal. S. Mori [63, 64, 65, 66, 67] showed that R is a π-ring $\Leftrightarrow$ R is a finite direct product of π-domains and SPIR's. (Also, see [50] for a more modern treatment of π-rings.) Thus R is a Fletcher UFD $\Leftrightarrow$ R is a π-ring with every invertible ideal principal.

We next give another definition [1] of "unique factorization ring" which includes both BG-UFR's and Fletcher UFR's. We define R to be a *weak UFR* if (1) R is *atomic*, that is, each nonzero nonunit of R is a product of irreducible elements (in the sense used by Fletcher) and (2) any two factorizations of a nonzero nonunit are *weakly homomorphic*, that is, if $0 \neq a_1 \cdots a_n = b_1 \cdots b_m$ where each a_i, b_j is irreducible, then for each $i \in \{1, \ldots, n\}$, there exists $j \in \{1, \ldots, m\}$ such that $a_i | b_j$ and for each $i \in \{1, \ldots, m\}$ there exists $j \in \{1, \ldots, n\}$ such that $b_i | a_j$. We also define a nonzero nonunit $p \in R$ to be *weakly prime* if $p | ab \neq 0$ implies $p | a$ or $p | b$.

In [1] it is shown that R is a Fletcher UFR $\Leftrightarrow$ (1) R is atomic and (2) any two factorizations of a nonunit element (possibly 0) $a \in R$ into irreducibles are weakly homomorphic. Thus a Fletcher UFR is a weak UFR. It is clear that a quasilocal ring (R, M) with $M^2 = 0$ is a weak UFR. Thus a BG-UFR is a weak UFR. So if R is a finite direct product of UFD's and SPIR's or a quasilocal ring with $M^2 = 0$, R is a weak UFR. The converse is also true. The main result of [1] is the following characterization of weak UFR's.

THEOREM 3.3. *For a commutative ring R the following conditions are equivalent.*

    (1) *R is a weak UFR.*

    (2) *Every nonzero nonunit of R is a product of weak primes.*

    (3) *R is atomic and each nonzero irreducible of R is weakly prime.*

    (4) *R is either a BG-UFR or a Fletcher UFR, that is, R is either a finite direct product of UFD's and SPIR's or a quasilocal ring (R, M) with $M^2 = 0$.*

We have seen that different authors have defined "associate" and "irreducible" in different ways. A detailed study of the notions "associate" and "irreducible" has been carried out in [24]. The point of view taken there was to define various notions of irreducible elements via different notions of associates. Briefly, if $\backsimeq$ was an "associate" relation on R, then a nonunit $a \in R$ was said to be "$\backsimeq$-irreducible" if $a = bc \Rightarrow a \backsimeq b$ or $a \backsimeq c$. In the case of integral domains, all these notions agree with the usual ones.

Let R be a commutative ring with identity and group of units $U(R)$. Let $a, b \in R$. Then a and b are *associates*, denoted by $a \sim b$, if $a | b$ and $b | a$, that is, $aR = bR$. (This is the sense used by Bouvier and Fletcher.) If $a = ub$ for some $u \in U(R)$, we say that a and b are *strong associates*, denoted by $a \approx b$. (This is the sense used by Galovich.) Finally, we say that a and b are *very strong associates*, denoted by $a \cong b$, if (i) $a \sim b$ and (ii) either $a = b = 0$ or $a \neq 0$ and $a = rb \Rightarrow r \in U(R)$. Clearly, $a \cong b \Rightarrow a \approx b \Rightarrow a \sim b$, but examples given in [24] show that none of these implications can be reversed. Of course, all three notions of "associate" coincide if R is an integral domain, or more generally, if a is a regular element.

We next define four different forms of "irreducible elements". Let a be a nonunit of R. Then a is *irreducible* (respectively, *strongly irreducible*, *very strongly irreducible*) if $a = bc \Rightarrow a \sim b$ or $a \sim c$ (respectively, $a \approx b$ or $a \approx c$, $a \cong b$ or $a \cong c$). It can be shown [24] that a nonzero $a \in R$ is strongly irreducible $\Leftrightarrow$ $a = bc \Rightarrow b$ or $c \in U(R)$. For their "irreducible" elements, Fletcher used irreducible elements while Galovich used very strongly irreducible elements. We defined a nonunit $a \in R$ to be m-*irreducible* if (a) is maximal in the set of principal ideals of R excluding R itself. This is the type of "irreducible" used by Bouvier. Now for $a \in R$, a very strongly irreducible $\Rightarrow a$ is m-irreducible $\Rightarrow a$ is strongly irreducible $\Rightarrow a$ is irreducible (for the first implication we need $a \neq 0$). Again, examples given in [24] show that none of these implications can be reversed. All four forms of "irreducible" agree if R is an integral domain or quasilocal, or more generally, if R is *présimplifiable* (that is, for $x, y \in R$, $x = xy \Rightarrow x = 0$ or $y \in U(R)$). Note that we do allow $a = 0$. In this case, 0 is (very strongly) irreducible if and only if R is an integral domain.

Each form of irreducibility leads to a form of atomicity. We say that a commutative ring R is *atomic*, respectively, *strongly atomic, very strongly atomic, m-atomic, p-atomic*, if every nonzero, nonunit element of R is a finite product of irreducible, respectively, strongly irreducible, very strongly irreducible, m-irreducible, or principal prime elements. As expected, if R satisfies ACCP, R is atomic.

In a natural way the notions of HFD, BFD, FFD, and idf-domain can be extended to commutative rings with zero divisors. A commutative ring R is called a *bounded factorization ring* (BFR) if for each nonzero nonunit $a \in R$ there exists a natural number $N(a)$ so that for any factorization $a = a_1 \cdots a_n$ of a where each a_i is a nonunit we have $n \leq N(a)$. As in the domain case, if R is a BFR, then R satisfies ACCP. Note that a BFR is présimplifiable and hence each irreducible is very strongly irreducible. Thus a BFR is very strongly atomic. For Noetherian rings there is a satisfactory characterization of BFR's ([24, Theorem 3.9]). Here the following are equivalent: (1) R is a BFR, (2) R is présimplifiable, (3) $\bigcap_{n=1}^{\infty} I^n = 0$ for each proper (principal) ideal I of R.

To illustrate the differences between integral domains and rings, we note that the notion of an FFD can be generalized in at least three different ways. Let R be a commutative ring with identity. Then R is called a *finite factorization ring* (FFR) if every nonzero nonunit of R has only a finite number of factorizations up to order and associates; R is called a *weak finite factorization ring* (WFFR) if every nonzero nonunit of R has only a finite number of nonassociate divisors; and R is called an *atomic idf-ring* if R is atomic and each nonzero element of R has at most a finite number of nonassociate irreducible divisors. Clearly, FFR $\Rightarrow$ WFFR $\Rightarrow$ atomic idf-ring. However, $R = \mathbb{Z}_2 \times \mathbb{Z}_2$ is a WFFR, but $(0, 1) = (0, 1)^n$ is an infinite set of nonassociate factorizations of $(0, 1)$, so R is not an FFR. Next let $R = \mathbb{Z}_{(2)} \times \mathbb{Z}_{(2)}$. Then R has only four nonassociate irreducibles: $(0, 1)$, $(1, 2)$, $(1, 0)$, $(2, 1)$, so R is an atomic idf-ring, but $(1, 0) = (1, 0)(1, 2)^n$ for each $n \geq 1$. Thus R is not a WFFR. Note that neither $\mathbb{Z}_2 \times \mathbb{Z}_2$ nor $\mathbb{Z}_{(2)} \times \mathbb{Z}_{(2)}$ is a BFR or is présimplifiable. It is easily seen [24, Proposition 6.6] that a WFFR R or atomic idf-ring R is an FFR $\Leftrightarrow R$ is a BFR $\Leftrightarrow R$ is présimplifiable. Since an FFR is présimplifiable, all the forms of associate and irreducible coincide. However, in the definition of WFFR's and atomic idf-rings, we could get variations of these definitions by replacing "nonassociate" by "non-strongly-associate" or "non-very-strongly-associate". In the definition of an atomic idf-ring we could replace the atomic hypothesis by strongly atomic, very strongly atomic, etc. For R an integral domain, R is BFD $\Leftrightarrow R[X]$ is a BFD. For a commutative ring R the question of when $R[X]$ or $R[\![X]\!]$ is a BFR or one of the forms of an FFR remains open.

For other results on factorization in commutative rings with zero divisors, the reader is referred to the work of A. Bouvier, for example [31, 32, 33].

IV. (UNIQUE) FACTORIZATION OF REGULAR ELEMENTS

In this section we take the second approach to studying factorization in commutative rings with zero divisors: we are only concerned with the factorization of regular elements. Our treatment closely follows [25, Section 5]. Throughout this section R denotes a commutative ring with identity, $Z(R)$ the set of zero divisors of R, and $U(R)$ the group of units of R. Now reg $(R) = R - Z(R)$, the set of regular elements of R, is a cancellative monoid. Recall that an ideal is regular if it contains a regular element.

One of the earliest papers to use this approach was [78] where it was shown that if a commutative ring R has the ascending chain condition on regular principal ideals and if each pair of regular elements has a gcd, then the elements of reg(R) have unique factorization into irreducible elements up to order and associates. Also see [22] where a lattice-theoretic approach is used to study the regular ideals of a general commutative ring.

One simplification in dealing with only the regular elements of a ring R is that the three associate relations $\sim$, $\approx$, and $\cong$ all agree for regular elements. Hence for a regular nonunit $a \in R$, the notions of irreducible, strongly irreducible, very strongly irreducible, and m-irreducible all coincide, so we will simply use the term irreducible. Likewise, the various forms of atomicity when restricted to regular elements also coincide. We will say that R is r-*atomic* if every regular, nonunit element of R is a product of irreducible elements. Finally, we say that R satisfies r-*ACCP* if every ascending chain of regular principal ideals stabilizes. Clearly if R satisfies r-ACCP, then R is r-atomic, but the converse is false.

There are natural generalizations of integral domains satisfying certain factorization properties to commutative rings with zero divisors whose regular elements satisfy the corresponding factorization properties. We define R to be *factorial* if (1) R is r-atomic and (2) if $a_1 \cdots a_n = b_1 \cdots b_m$ where each a_i, b_j is a regular irreducible element of R, then $n = m$ and after reordering a_i and b_i are associates. The ring R is an r-*half-factorial ring* (r-HFR) if (1) R is r-atomic and (2) if $a_1 \cdots a_n = b_1 \cdots b_m$ where each a_i, b_j is a regular irreducible element of R, then $n = m$. The *regular elasticity of* R is r-$\rho(R) = \sup\{\frac{m}{n} \mid a_1 \cdots a_n = b_1 \cdots b_m$ where a_i, b_j are regular irreducibles of $R\}$. Clearly R is an r-HFR if and only if r-$\rho(R) = 1$. A ring R is defined to be an r-*bounded factorization ring* (r-BFR) if for each regular nonunit $a \in R$, there exists a natural number $N(a)$ so that if $a = a_1 \cdots a_n$ where each a_i is a nonunit, then $n \leq N(a)$. We define R to be an r-*finite factorization ring* (r-FFR) if one of the following three equivalent conditions holds: (1) every regular element of R has only a finite number of factorizations up to order and associates, (2) every regular element of R has only finitely many nonassociate factors, (3) R is r-atomic and each regular element of R has at most a finite number of nonassociate irreducible divisors. (That these three conditions are equivalent is identical to the proof of the integral domain case.) It is clear that factorial $\Rightarrow$ r-HFR $\Rightarrow$ r-BFR $\Rightarrow$ r-ACCP and that factorial $\Rightarrow$ r-FFR $\Rightarrow$ r-BFR. Moreover, none of these implications can be reversed.

Many of the definitions and results on factorization in integral domains have natural analogs in cancellative monoids. Let H be a commutative cancellative monoid that is written multiplicatively. We denote the units of H by $H^{\times}$ and the quotient group of H by $\langle H \rangle$. The notions $a|b$, associate, irreducible, and prime have obvious definitions in H. (Since H is a cancellative monoid, we don't need to worry about the various forms of associate and irreducible.) An ideal A of H is a nonempty subset A of H with the property that $a \in A$ and $h \in H$ implies that $ah \in A$. Of course, for us here the most important example of a monoid will be $\text{reg}(R)$, the set of regular elements of a commutative ring R under multiplication. For $a, b \in \text{reg}(R)$, $a|b$ in R (equivalently, $Rb \subseteq Ra$) $\Leftrightarrow a|b$ in $\text{reg}(R)$ (equivalently, $\text{reg}(R)b \subseteq \text{reg}(R)a$), $a \sim b$ in $R \Leftrightarrow a \sim b$ in $\text{reg}(R)$, $a \in U(R) \Leftrightarrow a \in \text{reg}(R)^{\times}$, and a is irreducible as an element of $R \Leftrightarrow a$ is irreducible as an element of $\text{reg}(R)$. If p is a regular prime of R, then certainly p is prime as an element of $\text{reg}(R)$. However, as we will later see, an element $a \in \text{reg}(R)$ that is prime as an element of $\text{reg}(R)$ need not be a prime element of R. This leads to certain pathologies: see Examples 4.2 and 4.4.

The various definitions for integral domains satisfying certain factorization properties also carry over to monoids. Let H be a commutative cancellative monoid. Then H is *atomic* if every nonunit of H may be written as a finite product of irreducible elements. The monoid H is *factorial* if H is atomic and for atoms $a_i, b_j \in H$ with $a_1 \cdots a_n = b_1 \cdots b_m$, we have $n = m$ and after reordering $a_i \sim b_i$. It is well known that H is factorial $\Leftrightarrow H \approx G \times (\bigoplus \mathbb{N}_0)$ where G is a group and $\mathbb{N}_0 = \{0, 1, 2, \ldots\}$ under addition $\Leftrightarrow$ every nonunit element of H is a product of prime elements. We say that H is a *half-factorial* (HF-) *monoid* if H is atomic and for atoms $a_i, b_j \in H$ with $a_1 \cdots a_n = b_1 \cdots b_m$ we have $n = m$. H is defined to be a *bounded factorization* (BF-) *monoid* if for each $a \in H$, the lengths of factorizations are bounded and H is defined to be a *finite factorization* (FF-) *monoid* if H is atomic and each nonunit $a \in H$ possesses only finitely many different factorizations (into irreducibles) up to order and associates. For each property α defined above, a commutative ring R is r-α if and only if the monoid $\text{reg}(R)$ satisfies α. Thus one way to study factorization properties of an integral domain or of the regular elements of a commutative ring R is to study the corresponding factorization properties in monoids. This approach has been used by F. Halter-Koch, A. Geroldinger, and others. For a survey of this approach, see [35], [46], and [54]. Also, see [25, Section 5] where this approach is used to study Krull rings and their regular elasticity.

The notion of a Krull domain has been generalized to both Krull rings and Krull monoids. Kennedy [57] defined a commutative ring R to be a Krull ring if $R = \bigcap_{\alpha} (V_{\alpha}, P_{\alpha})$ where each (V_{α}, P_{α}) is a rank one discrete valuation pair with total quotient ring $T(R)$, each P_{α} is a regular prime ideal, and if v_{α} is the associated valuation of (V_{α}, P_{α}), then for each $a \in \text{reg}(R)$, $v_{\alpha}(a) = 0$ for almost all α. While Kennedy required $R \neq T(R)$, we let a total quotient ring be a Krull ring with empty defining family. Equivalently, R is a Krull ring if and only if R is completely integrally closed and R has ACC on integral regular v-ideals (a regular ideal A is a v-ideal if $\left(A^{-1}\right)^{-1} = A$) [57], [61]. For results on Krull rings, also see [55] and [71]. Chouinard [38] defined a cancellative monoid S to be a Krull monoid if there exists a family $(v_i)_{i \in I}$ of discrete valuations on $\langle S \rangle$ (that is, each $v_i \colon \langle S \rangle \to \mathbb{Z}$ is a group homomorphism) such that $S = \bigcap V_i$ where $V_i = \{x \in \langle S \rangle \mid v_i(x) \geq 0\}$ and for every $x \in S$, the set $\{i \in I \mid v_i(x) > 0\}$ is finite. So a group is a Krull monoid with $I = \varnothing$. He showed that S is a Krull monoid if and only if S is completely integrally closed (that is, if $x \in \langle S \rangle$ with $rx^n \in S$ for all $n \geq 1$ where $r \in S$, then $x \in S$) and S has ACC on integral v-ideals (an ideal A of S is a v-ideal

$\Leftrightarrow \left(A^{-1}\right)^{-1} = A \Leftrightarrow A = \bigcap \{Sx \mid A \subseteq Sx \text{ where } x \in \langle S \rangle \}$). For results on Krull monoids, also see [49].

Now an integral domain R is a Krull domain if and only if $(R - \{0\}, \cdot)$ is a Krull monoid. Unfortunately, the theorem we would like, namely that R is a Krull ring if and only if $\text{reg}(R)$ is a Krull monoid, is not true. While a Krull ring R has $\text{reg}(R)$ a Krull monoid, $\text{reg}(R)$ can be a Krull monoid, even factorial, without R being a Krull ring. However, if R is a Marot ring and $\text{reg}(R)$ is a Krull monoid, then R is a Krull ring. Recall that a ring R is a Marot ring if every regular ideal of R is generated by regular elements. A Noetherian ring, or more generally a ring with $Z(R)$ a finite union of prime ideals, is a Marot ring. If R is a Marot ring and A is a regular fractional ideal of R, then $\left(A^{-1}\right)^{-1} = \bigcap \{Rx \mid A \subseteq Rx\}$. In fact, in Theorems 4.1, 4.3, and 4.5, the Marot hypothesis can be replaced by the condition that $\left(A^{-1}\right)^{-1} = \bigcap \{Rx \mid A \subseteq Rx\}$ for each regular fractional ideal A of R (which was called Property (D) in [16, corrigendum]).

THEOREM 4.1. ([25], [53]) *Let R be a commutative ring and let $\text{reg}(R)$ be the multiplicative monoid of regular elements of R.*

(1) *If R is a Krull ring, then $\text{reg}(R)$ is a Krull monoid.*

(2) *If R is a Marot ring and $\text{reg}(R)$ is a Krull monoid, then R is a Krull ring.*

EXAMPLE 4.2. ([25]) (A ring R with $\text{reg}(R)$ a Krull monoid, even factorial, but nevertheless R is not a Krull ring.) Let $D = K[X,Y]$, K a field and $A = \bigoplus \{D/M \mid M \in \max(D), Y \notin M\}$. Put $D_2 = K[Y, X^2, XY, X^3]$ and $R_2 = D_2 \oplus A$ (idealization). So R_2 is the ring of [22, Example 4.3]. Since $\left(X^2, 0\right) \in R_2$, but $(X, 0) \notin R_2$, R_2 is not (completely) integrally closed and hence R_2 is not a Krull ring. However, $\text{reg}(R_2) = \{(\alpha Y^m, a) \mid \alpha \in K - \{0\}, m \geq 0, a \in A\}$. Since $(\alpha Y^m, a) \sim (Y^m, 0)$ [22, page 407, second paragraph], $\text{reg}(R_2) \cong U(R_2) \times \mathbb{N}_0$ is a Krull monoid, even factorial. Note that $(Y, 0)$ is a prime element of $\text{reg}(R)$, but $(Y, 0) R_2 = Y D_2 \oplus A$ is not a prime ideal of R_2 since $\left(X^2, 0\right) \cdot (XY, 0) = \left(X^3 Y, 0\right) \in (Y, 0) R_2$ but $\left(X^2, 0\right) \notin (Y, 0) R_2$ and $(XY, 0) \notin (Y, 0) R_2$.

If R is a Krull ring, then the regular fractional v-ideals $D(R)$ form a group under the v-product $A * B = (AB)_v$ with subgroup $\text{Princ}(R)$ of regular fractional principal ideals. The quotient group $\text{Cl}(R)$ is called the *divisor class group* of R. In a similar fashion, we define $D(H)$, $\text{Princ}(H)$, and the *divisor class group* $\text{Cl}(H)$ for a Krull monoid H. We next state that for R a Krull ring, $\text{Cl}(\text{reg}(R))$ is naturally isomorphic to a subgroup of $\text{Cl}(R)$; moreover, if further R is a Marot ring, then $\text{Cl}(\text{reg}(R))$ and $\text{Cl}(R)$ are naturally isomorphic. But we give an example where $\text{Cl}(\text{reg}(R)) = 0$, but $\text{Cl}(R) \approx \mathbb{Z}$.

THEOREM 4.3. ([25]) *Let R be a Krull ring. Then the map $\psi \colon D(\text{reg}(R)) \to D(R)$ given by $\psi(A) = (RA)_v$ is a group monomorphism with $\psi(\text{Princ}(\text{reg}(R))) = \text{Princ}(R)$. Moreover, if $\psi(A)$ is principal, so is A. Thus the induced map $\bar{\psi} \colon \text{Cl}(\text{reg}(R)) \to \text{Cl}(R)$ given by $\bar{\psi}([A]) = [(RA)_v]$ is a group monomorphism. If further R is a Marot ring, then ψ and $\bar{\psi}$ are both surjective.*

Let A be a v-ideal of $\text{reg}(R)$. If $(RA)_v$ is a prime ideal of R, then A is a prime ideal of $\text{reg}(R)$. If further R is a Marot ring, the converse is true.

EXAMPLE 4.4. ([25]) (A Krull ring R with $\text{Cl}(\text{reg}(R)) \subsetneq \text{Cl}(R)$.) Let D be a Dedekind domain with maximal ideal M that is *not* principal, but some power of

M is principal, say $m > 1$ is the least positive integer with $M^m = (t)$ principal. Let $A = \bigoplus \{D/Q \mid Q \neq M$ is a maximal ideal of $D\}$ and take $R = D \oplus A$ to be the idealization of D and A. Then $\{t^n R\}_{n=0}^{\infty}$ is the set of regular principal ideals of R. So $\text{reg}(R) \approx U(R) \times (\mathbb{N}_0, +)$ and $\text{Cl}(\text{reg}(R)) = 0$. Let $P = M \oplus A$, so P is the unique regular prime ideal of R and $P^m = tR$, so P is invertible but not principal. Note that P is not an intersection of principal fractional ideals of R. So $\text{Cl}(R) \neq 0$; in fact, $\text{Cl}(R) = \langle [P] \rangle \approx \mathbb{Z}$. For details, see [22, Example 3.6]. Also, note that $t \, \text{reg}(R)$ is a prime ideal of $\text{reg}(R)$, but tR is not a prime ideal of R.

An integral domain R is factorial if and only if R is a Krull domain with $\text{Cl}(R) = 0$. However, Example 4.2 shows that a factorial ring need not be a Krull ring. And Example 4.4 gives an example of a Krull ring R that is factorial but with $\text{Cl}(R) \neq 0$.

THEOREM 4.5. ([25]) *If R is a Krull ring with $\text{Cl}(R) = 0$, then R is factorial. If R is a Marot ring, then R is factorial $\Leftrightarrow$ R is a Krull ring with $\text{Cl}(R) = 0$.*

If R is a weak UFR, then R is a Marot Krull ring with $\text{Cl}(R) = 0$ and hence R is factorial. However, even a Marot Krull ring R with $\text{Cl}(R) = 0$ need not be a weak UFR. For as remarked by Matsuda [61, 7.11], if D is a UFD and T is a total quotient ring that is not a field, SPIR, or quasilocal ring (T, M) with $M^2 = 0$, then $R = D \times T$ is a Marot Krull ring with $\text{Cl}(R) = \text{Cl}(D) \oplus \text{Cl}(T) = 0$, but R is not a weak UFR.

We end this section with the following theorem ([16, Theorem 3.6], but see the theorem in the corrigendum).

THEOREM 4.6. *Let R be a commutative ring with identity. Consider the following seven conditions on R.*

(1) *R is factorial.*
(2) *R is a Krull ring with $\text{Cl}(R) = 0$.*
(3) *R is a Krull ring and every maximal v-ideal is principal.*
(4) *R satisfies r-ACCP and the intersection of two regular principal ideals is principal.*
(5) *R is a Krull ring and the intersection of two regular principal ideals is principal.*
(6) *Every regular prime ideal of R contains a regular principal prime ideal.*
(7) *Every regular element (principal ideal) of R is a product of regular principal prime elements (ideals).*

Then we have the following implications:

$$(2) \Longleftrightarrow (3) \Longleftrightarrow (6) \Longleftrightarrow (7) \Longrightarrow (5) \Longrightarrow (4) \Longrightarrow (1).$$

However, $(5) \nRightarrow (7)$.

If R is a Marot ring, or more generally a ring in which $A_v = \bigcap \{Rx \mid A \subseteq Rx\}$ for each regular ideal A of R, then conditions (1)–(7) are all equivalent.

V. UNIQUE FACTORIZATION IN RINGS WITHOUT IDENTITY

In this section we discuss extensions of unique factorization to commutative rings without identity.

Let R be a commutative ring without identity. Since R has no units, the strongest forms of "associate" and "irreducible" are as follows. Two elements $a, b \in R$ are "associates" precisely when $a = b$ and a is "irreducible" if a cannot be written as

$a = bc$ where $b, c \in R$. We say that $a \in R$ is *totally irreducible* if $a \neq bc$ for any $b, c \in R$. Let us define R to be a *strong UFR* if (1) every nonzero element of R is a product of totally irreducible elements and (2) if $0 \neq a_1 \cdots a_s = b_1 \cdots b_t$ where each a_i, b_j is totally irreducible, then $s = t$, and after reordering, if necessary, $a_i = b_i$ for $i = 1, \ldots, s$. The following result generalizes [6, Proposition].

THEOREM 5.1. *Let R be a commutative ring. Then R is a strong UFR $\Leftrightarrow R^2 = 0$.*

Proof. ($\Leftarrow$) Clear. ($\Rightarrow$) We may suppose $R \neq 0$. Let $m \in R$ be totally irreducible. Suppose $m^2 \neq 0$. Then $m(1 + m) = m + m^2 \neq 0$ for $m + m^2 = 0$ gives $m = m(-m)$ which contradicts that m is totally irreducible. (While R is not assumed to have an identity, we will use $1 + m$ in the formal product $m(1 + m)$.) A similar argument shows that $0 \neq m(1 + m)^2$, so we can write $m(1 + m) = a_1 \cdots a_s$ and $m(1 + m)^2 = b_1 \cdots b_t$ where each a_i, b_j is totally irreducible. From the uniqueness of factorizations we get $(m + m^2)^2 \neq 0$ (for $(m + m^2)^2 = 0$ gives $m^2 = m^2(-2m - m^2)$). Hence $a_1^2 \cdots a_s^2 = (m(1 + m))^2 = m\left(m(1 + m)^2\right) = mb_1 \cdots b_t$. So we get, say, $a_1 = m$. So $m(1 + m) = a_1 \cdots a_s = ma_2 \cdots a_s$ where $s \geq 2$ since $m^2 \neq 0$. Thus $m = m(a_2 \cdots a_s - m)$ which contradicts that m is totally irreducible. Thus $m^2 = 0$.

Let n be another totally irreducible element. Suppose $mn \neq 0$. Then $mn = m(m + n)$ since $m^2 = 0$. Thus $m + n = n$ and hence $m = 0$, a contradiction. So for any two totally irreducible elements $m, n \in R$, $mn = 0$. Thus $R^2 = 0$.

We next consider π-rings without identity. In [67], Mori showed that a π-ring without identity is either (1) an integral domain, (2) a ring $R = (p)$ where every ideal of R including 0 is a power of R, (3) $K \times R$ where K is a field and R is a ring as in (2), or (4) $K \times D$ where K is a field and $D = (p)$ is a π-domain where every nonzero ideal of D is a power of D. He showed that in a π-domain without identity every principal ideal is actually a product of principal prime ideals (here the whole ring itself is considered to be a prime ideal) and that an integral domain without identity is a π-domain if and only if (1) every nonzero element of D is a product of irreducible elements and (2) each irreducible element of D generates a principal prime ideal. For an alternative treatment of π-domains without identity, see [80] and [6, Proposition]. Mori remarked that $2\mathbb{Z}_{(2)}$ is an example of a π-domain without identity. Here $\mathbb{Z}_{(2)}$ is a local UFD with $\mathbb{Z}_{(2)} = \mathbb{Z} + 2\mathbb{Z}_{(2)}$. More generally, the following example gives an easy way to construct π-domains without identity.

EXAMPLE 5.2. ([6, Theorem 1]) Let (D, M) be a quasilocal UFD with $D = M[1] = \{M + n1 \mid m \in M, n \in \mathbb{Z}\}$. Then M is a π-domain without identity. Since $M[1] = D$, for $a \in M$, $aD = (a)$, the principal ideal of M generated by a. Let $0 \neq d \in M$, so $d = p_1 \cdots p_n$ where p_i is a principal prime of D. Hence $(d) = (p_1) \cdots (p_n)$ where each (p_i) is a prime ideal of D and hence of M.

The main result of [6] is the converse of the preceding example which generalizes [47].

THEOREM 5.3. *Let D be an integral domain without identity. Then D is a π-domain if and only if $D^* = D[1]$ is a quasilocal UFD with D as its maximal ideal.*

Let D be an integral domain without identity. Then D is a UFD if and only if every nonzero prime ideal of D contains a nonzero principal prime ideal. Thus it seems reasonable to conjecture that if D is an integral domain without identity, then

D is a π-domain if and only if each nonzero prime ideal of D contains a nonzero principal prime ideal. However, this need not be the case as shown by the following example of Gilmer [48, Example 5.3].

EXAMPLE 5.4. Let $D = XZ[X]$. Then every prime ideal of D is principal. However, by Theorem 5.3, D is not a π-domain since $Z + XZ[X] = Z[X]$ is not quasilocal. (Alternatively, observe that $(4X)$ is not a product of prime ideals.)

We next give a characterization of π-domains without identity in terms of totally irreducible elements. Let D be an integral domain without identity. Recall that $0 \neq a \in D$ is totally irreducible if $a \neq bc$ for $b, c \in D$. However, here we define $a, b \in D$ to be associates, denoted $a \sim b$, if $(a) = (b)$.

THEOREM 5.5. *Let D be an integral domain without identity. Then the following two conditions are equivalent.*

(1) *D is a π-domain.*
(2) (a) *Each nonzero $d \in D$ is a product of totally irreducible elements.*
    (b) *If $0 \neq d_1 \cdots d_n = d'_1 \cdots d'_{n'}$, where each d_i, d'_i is totally irreducible, then $n = n'$, and after reordering, if necessary, $d_i \sim d'_i$.*

Moreover, when D is a π-domain without identity, $0 \neq d \in D$ is totally irreducible if and only if (d) is a prime ideal of D.

Proof. Suppose that D is a π-domain without identity. Then by Theorem 5.3, $D[1]$ is a quasilocal UFD with maximal ideal D. Thus $0 \neq d \in D$ is totally irreducible $\Leftrightarrow d$ is a nonzero principal prime of $D[1] \Leftrightarrow (d)$ is a prime ideal of D.

$(1) \Rightarrow (2)$. Let $0 \neq d$. Let $d = d_1 \cdots d_n$ where d_i is a principal prime of $D[1]$. Then each d_i is totally irreducible. If $0 \neq d = d_1 \cdots d_n = d'_1 \cdots d'_{n'}$ where each d_i, d'_i is totally irreducible, then in $D[1]$ each d_i is prime and $(d_1) \cdots (d_n) = (d'_1) \cdots (d'_{n'})$. Since $D[1]$ is a UFD, $n = n'$ and after reordering, if necessary, $(d_i) = (d'_i)$.

$(2) \Rightarrow (1)$. Let $0 \neq d \in D$ be irreducible. We show that (d) is prime. Let $x, y \in D$ with $xy \in (d)$, say $xy = rd + nd$ where $r \in D$ and $n \in \mathbb{Z}$. Write $x = x_1 \cdots x_s$, $y = y_1 \cdots y_t$ where each x_i, y_i is totally irreducible. Now $x_1^2 \cdots x_s y_1 \cdots y_t = x_1(rd + nd) = d(rx_1 + nx_1)$. Factoring $rx_1 + nx_1$ into totally irreducible elements and applying (2b) gives that either $(d) = (x_i)$ or $(d) = (y_i)$ for some i. Thus x or $y \in (d)$. So (d) is prime. Now let $0 \neq x \in D$. Then $x = d_1 \cdots d_n$ where d_i is totally irreducible and hence (d_i) is a prime ideal. Thus $(x) = (d_1 \ldots d_n) = (d_1) \cdots (d_n)$ is a product of prime ideals. Hence D is a π-domain.

An alternate approach to unique factorization in rings without identity has been given by Ağargün and Fletcher [2]. Let R be a commutative ring not necessarily with identity, let R' be a commutative ring with identity, and let $\theta \colon R \to R'$ be a ring monomorphism such that $\theta(R)$ is an ideal of R'. Then R is a *unique factorization ring with respect to θ* if (1) every element $\theta(a)$ is a product of "neo-irreducible" elements of R' where $p \in R'$ is "neo-irreducible" $\Leftrightarrow$ for every $y \in R'$, $yp = ya_1 \cdots a_r \Rightarrow ypR' = ya_iR'$ for some i and (2) if $\theta(a) = (p'_1 \cdots p'_k)(p_1 \cdots p_n) = (q'_1 \cdots q'_l)(q_1 \cdots q_m)$ are two U-decompositions into neo-irreducibles in R', then $m = n$ and after reordering, if necessary, $rp_iR' = rq_iR'$ for each i, and $r \in \theta(R)$. It is then shown that a Fletcher UFR is a unique factorization ring with respect to 1_R and that for $0 \neq p \in \mathbb{Z}$, $p\mathbb{Z}$ is a unique factorization ring with respect to the inclusion map $p\mathbb{Z} \hookrightarrow \mathbb{Z} \Leftrightarrow p$ is prime.

VI. FACTORIZATION IN MODULES

Throughout this section R will be a commutative ring with identity and M a nonzero R-module. Again, we reserve D for an integral domain. We begin this section with factorial modules, the extension of factorial domains to torsion-free modules.

A.-M. Nicolas [69] first defined and studied factorial modules. Let D be an integral domain with quotient field K and M a torsion-free D-module. She defined $0 \neq \xi \in M$ to be "irreducible" if $\xi = a\eta$ ($a \in D$, $\eta \in M$) $\Rightarrow a \in U(D)$ and ξ to be "primitive" if $a\eta = b\xi \Rightarrow \eta \in D\xi$, or equivalently, $K\xi \cap M = D\xi$. Now "primitive" $\Rightarrow$ "irreducible", but not conversely. Let us say M is *atomic* if for each $0 \neq m \in M$, $m = a_1 \cdots a_s \eta$ where each a_i is irreducible in D and η is irreducible in M. She defined two factorizations in M, $0 \neq a_1 \cdots a_n \xi = b_1 \cdots b_m \eta$, to be *equivalent* if $n = m$ and after reordering, if necessary, $a_i \sim b_i$ (and hence $\xi \sim \eta$, i.e., $D\xi = D\eta$, since M is torsion-free). Finally, she defined M to be *factorial* if (1) M is atomic, (2) p irreducible in $D \Rightarrow p$ is prime, and (3) ξ irreducible in $M \Rightarrow \xi$ is primitive. She proved the following theorem.

THEOREM 6.1. ([69, Théorème 2.3]) *Let D be an integral domain and M a torsion-free D-module. Then M is factorial $\Leftrightarrow$ D and M are atomic and any two factorizations of a nonzero element of M into irreducibles are equivalent.*

Theorem 6.1 shows that a factorial module does have unique factorization into irreducibles. Let us define M to be a *unique factorization module* (UFM) if (1) M is atomic and (2) any two factorizations of $0 \neq m \in M$ into irreducibles are equivalent. Thus Theorem 6.1 says M factorial $\Rightarrow M$ is a UFM. This raises the natural question of whether M is a UFM $\Rightarrow M$ is factorial. An equivalent question is whether M a UFM $\Rightarrow D$ is a UFD. If D is atomic, this is easy to show. For let a be a nonzero nonunit of D and suppose $a = a_1 \cdots a_n = b_1 \cdots b_m$ where each a_i, b_j is irreducible. Let $\xi \in M$ be irreducible. Then $a_1 \cdots a_n \xi = a\xi = b_1 \cdots b_n \xi$, so since M is a UFM, $n = m$ and after reordering, if necessary, $a_i \sim b_i$. Thus D is a UFD. Nicolas showed that for D atomic, M is a UFM $\Leftrightarrow M$ is factorial, but was unable to show that if M is a UFM, D must be atomic. The author and S. Valdes-Leon [25, Theorem 4.4] proved that M a UFM $\Rightarrow M$ is factorial. (Note that in (5) of Theorem 4.4, "R" should be "M".)

THEOREM 6.2. *Let D be an integral domain and M a torsion-free D-module. Then D is factorial $\Leftrightarrow M$ is a UFM. Thus if M is a UFM, D must be a UFD.*

Factorial modules have also been considered by D. Costa [40] and Chin-Pi Lu [59]. For example, Costa showed that the symmetric algebra $S_D(M)$ is a UFD if and only if D is a UFD and $S_D(M)$ is a factorial D-module. Lu proved a number of results concerning UFM's. However, some care is needed when reading [59] because Lu mistakenly assumed that Nicolas showed that if M is a UFM, then D is necessarily a UFD (see [59, page 127, line 8]). However, thanks to Theorem 6.2, the results of [59] are of course all correct. Further results on factorization in torsion-free modules are given in Nicolas [70] and [25, Section 4].

In a manner similar for rings done in Section III, we can use different associate relations on M to define different notions of irreducible element of M. However, we prefer not to use "irreducible" to avoid confusion with elements of R, and hence instead use "primitive". The rest of the material from this section comes from

[25, Section 2]. Let R be a commutative ring with identity and M a nonzero R-module. Let $m, n \in M$. Then m and n are *associates*, denoted by $m \sim n$, if $Rm = Rn$; m and n are *strong associates*, denoted by $m \approx n$, if $m = un$ for some $u \in U(R)$; and m and n are *very strong associates*, denoted by $m \cong n$, if (i) $m \sim n$ and (ii) either $m = n = 0$ or $m \neq 0$ and $m = rn \Rightarrow r \in U(R)$. We define $0 \neq m \in M$ to be *primitive* (respectively, *strongly primitive, very strongly primitive*) if $m = an \Rightarrow m \sim n$ (respectively, $m \approx n$, $m \cong n$). Finally, we define $0 \neq m \in M$ to be *superprimitive* if $bm = an \Rightarrow a | b$. We have m superprimitive $\Rightarrow m$ is very strongly primitive $\Rightarrow m$ is strongly primitive $\Rightarrow m$ is primitive, but none of these implications can be reversed. If m is superprimitive, then $\mathrm{ann}(m) = 0$. In fact, it is easily checked that $m \in M$ is superprimitive $\Leftrightarrow \mathrm{ann}(m) = 0$ and m is "pure" in the sense that $rM \cap Rm = rRm$ for each $r \in R$. Note that our definitions do not agree with those of Nicolas or Lu. Nicolas's "irreducible" is our very strongly primitive and Nicolas's "primitive" is our superprimitive (for the case where M is torsion-free).

In a natural way, we can define *bounded factorization module* (BFM), *half-factorial module* (HFM) and *unique factorization module* (UFM) for an R-module M with respect to each type of "irreducible" in R, "primitive" in M and associate relation on R and M. See [25, Definition 2.7] for details. (Under certain circumstances, these factorization properties force M to be *présimplifiable*, i.e., for $a \in R$ and $m \in M$, $am = m \Rightarrow a \in U(R)$ or $m = 0$, and hence the three forms of primitive all agree.) The following theorem ([25, Theorem 2.8]) is a sampling of results from [25].

THEOREM 6.3. *Let R be a commutative ring and M a nonzero R-module.*

(1) *If M is a faithful BFM, then R is a BFR.*

(2) *If M is an HFM and each primitive $m \in M$ has $\mathrm{ann}(m) = 0$, then R is an HFR and M is présimplifiable.*

(3) *If M is a UFM and each primitive $m \in M$ has $\mathrm{ann}(m) = 0$, then R is a BG-UFR and M is présimplifiable.*

It is reasonable to attempt to characterize unique factorization modules over a BG-UFR R. Recall that a commutative ring R is a BG-UFR if R is a UFD, SPIR, or quasilocal ring (R, M) with $M^2 = 0$. Our last theorem characterizes unique factorization modules over the last two classes of BG-UFR's.

THEOREM 6.4. ([25, Theorem 2.9])

(1) *Let $(R, (p))$ be an SPIR, but not a field, and let M be a nonzero R-module. Then M is a UFM $\Leftrightarrow pM = 0$ or M is cyclic.*

(2) *Let $(R, \mathcal{M})$ be a quasilocal ring with $\mathcal{M}^2 = 0$, but not a field, and let M be a nonzero R-module. Then M is a UFM $\Leftrightarrow \mathcal{M}M = 0$ or M is cyclic.*

One way to study factorization in an R-module M is to embed M into a ring containing isomorphic copies of R and M that preserves the R-module structure of M. Two natural such rings are the symmetric algebra $S_R(M)$ and the idealization $R(M)$ of R and M. Let $S_R(R) = \bigoplus_{n \geq 0} S_R^n(M)$ where $S_R^0(M) = R$ and $S_R^1(M) = M$. In a sense $S_R(M)$ is the "freest" commutative R-algebra generated by R and M. Now $R(M) = R \oplus M$ where $(r, m)(r', m') = (rr', rm' + r'm)$, $R \approx \{(r, 0) \mid r \in R\}$, $M \approx \{(0, m) \mid m \in M\}$ and $(0 \oplus M)^2 = 0$. So $R(M)$ is the R-algebra generated by R and M with the "most relations". Note that $R(M) \approx S_R(M) / \bigoplus_{n \geq 2} S_R^n(M)$. See [25, Section 3] for this approach.

REFERENCES

1. A.G. Ağargün, D.D. Anderson, and S. Valdes-Leon, *Unique factorization rings with zero divisors*, Comm. Algebra, to appear.

2. A.G. Ağargün and C.R. Fletcher, *Unique factorisation for commutative rings without identity*, Turkish J. Math., to appear.

3. D.D. Anderson, *Some finiteness conditions on a commutative ring*, Houston J. Math. 4 (1978), 289–299.

4. D.D. Anderson, *Globalization of some local properties in Krull domains*, Proc. Amer. Math. Soc. 85 (1982), 141–145.

5. D.D. Anderson (ed.), *Factorization in Integral Domains*, Lecture Notes in Pure and Appl. Math., vol. 189, Marcel Dekker, Inc., New York, 1997.

6. D.D. Anderson, *π-domains without identity*, these proceedings.

7. D.D. Anderson and D.F. Anderson, *Locally factorial integral domains*, J. Algebra 90 (1984), 265–283.

8. D.D. Anderson, D.F. Anderson, and M. Zafrullah, *Factorization in integral domains*, J. Pure Appl. Algebra 69 (1990), 1–19.

9. D.D Anderson, D.F. Anderson, and M. Zafrullah, *Atomic domains in which almost all atoms are prime*, Comm. Algebra 20 (1992), 1447–1462.

10. D.D Anderson, D.F. Anderson, and M. Zafrullah, *Factorization in integral domains, II*, J. Algebra 152 (1992), 78–93.

11. D.D Anderson, D.F. Anderson, and M. Zafrullah, *A generalization of unique factorization*, Boll. Un. Mat. Ital. A (7) 9 (1995), 401–413.

12. D.D. Anderson, S.T. Chapman, F. Halter-Koch, and M. Zafrullah, *Criteria for unique factorization in integral domains*, J. Pure Appl. Algebra, to appear.

13. D.D. Anderson, S.T. Chapman, and D.J. Kwak, *Integral domains with highly nonunique factorization*, Results in Mathematics, to appear.

14. D.D. Anderson and L.A. Mahaney, *Commutative rings in which every ideal is a product of primary ideals*, J. Algebra 106 (1987), 528–535.

15. D.D. Anderson and L.A. Mahaney, *On primary factorizations*, J. Pure Appl. Algebra 54 (1988), 141–154.

16. D.D. Anderson and R. Markanda, *Unique factorization rings with zero divisors*, Houston J. Math. 11 (1985), 15–30, corrigendum 423–426.

17. D.D. Anderson and J.L. Mott, *Cohen-Kaplansky domains: Integral domains with a finite number of irreducible elements*, J. Algebra 148 (1992), 17–41.

18. D.D. Anderson, J.L. Mott, and M. Zafrullah, *Finite character representations for integral domains*, Boll. Un. Mat. Ital. B (7) 6 (1992), 613–630.

19. D.D. Anderson, J.L. Mott, and M. Zafrullah, *Unique factorization in non-atomic integral domains*, Boll. Un. Mat. Ital., to appear.

20. D.D. Anderson and B. Mullins, *Finite factorization domains*, Proc. Amer. Math. Soc. 124 (1996), 389–396.

21. D.D. Anderson and J. Pascual, *Characterizing Prüfer rings via their regular ideals*, Comm. Algebra 15 (1987), 1287–1295.

22. D.D. Anderson and J. Pascual, *Regular ideals in commutative rings, sublattices of regular ideals, and Prüfer rings*, J. Algebra 111 (1987), 404–426.

23. D.D. Anderson and R. Quintero, *Some generalizatons of GCD-domains*, Factorization in Integral Domains (D.D. Anderson, ed.), Lecture Notes in Pure and Appl. Math., vol. 189, Marcel Dekker, Inc., New York, 1997, pp. 189–195.

24. D.D. Anderson and S. Valdes-Leon, *Factorization in commutative rings with zero divisors*, Rocky Mountain J. Math. **26** (1996), 439–480.

25. D.D. Anderson and S. Valdes-Leon, *Factorization in commutative rings with zero divisors*, II, Factorization in Integral Domains (D.D. Anderson, ed.), Lecture Notes in Pure and Appl. Math., vol. 189, Marcel Dekker, New York, 1997, pp. 197–219.

26. D.D. Anderson and M. Zafrullah, *On a theorem of Kaplansky*, Boll. Un. Mat. Ital. A (7) **8** (1994), 397–402.

27. D.D. Anderson and M. Zafrullah, *P.M. Cohn's completely primal elements*, Zero-dimensional Commutative Rings (D.F. Anderson and D.E. Dobbs, eds.), Lecture Notes in Pure and Appl. Math., vol. 171, Marcel Dekker, Inc., New York, 1995, pp. 115–123.

28. D.F. Anderson, *Elasticity of factorizations in integral domains: A survey*, Factorization in Integral Domains (D.D. Anderson, ed.), Lecture Notes in Pure and Appl. Math., vol. 189, Marcel Dekker, Inc., New York, 1997, pp. 1–29.

29. M. Auslander and D.A. Buchsbaum, *Unique factorization in regular local rings*, Proc. Nat. Acad. Sci. U.S.A. **45** (1959), 733–734.

30. M. Billis, *Unique factorization in the integers modulo n*, Amer. Math. Monthly **75** (1968), 527.

31. A. Bouvier, *Anneaux de Gauss*, C. R. Acad. Sci. Paris Ser. A **273** (1971), 443–445.

32. A. Bouvier, *Anneaux présimplifiables et anneaux atomiques*, C. R. Acad. Sci. Paris Ser. A **272** (1971), 992–994.

33. A. Bouvier, *Structure des anneaux à factorisation unique*, Publ. Dép. Math. (Lyon) **11** (1974), 39–49.

34. L. Carlitz, *A characterization of algebraic number fields with class number two*, Proc. Amer. Math. Soc. **11** (1960), 391–392.

35. S.T. Chapman and A. Geroldinger, *Krull domains and monoids, their sets of lengths, and associated combinatorial problems*, Factorization in Integral Domains (D.D. Anderson, ed.), Lecture Notes in Pure and Appl. Math., vol. 189, Marcel Dekker, Inc., New York, 1997, pp. 73–112.

36. S.T. Chapman and W.W. Smith, *On the HFD, CHFD, and k-HFD properties in Dedekind domains*, Comm. Algebra **20** (1992), 1955–1987.

37. S.T. Chapman and W.W. Smith, *On the k-HFD property in Dedekind domains with small class group*, Mathematika **39** (1992), 330–340.

38. L.G. Chouinard II, *Krull semigroups and divisor class groups*, Canad. J. Math. **33** (1981), 1459–1468.

39. P.M. Cohn, *Bezout rings and their subrings*, Math. Proc. Cambridge Philos. Soc. **64** (1968), 251–264.

40. D.L. Costa, *Unique factorization in modules and symmetric algebras*, Trans. Amer. Math. Soc. **224** (1976), 267–280.

41. J. Coykendall, *A characterization of polynomial rings with the half-factorial property*, Factorization in Integral Domains (D.D. Anderson, ed.), Lecture Notes in Pure and Appl. Math., vol. 189, Marcel Dekker, Inc., New York, 1997, pp. 291–294.

42. C.R. Fletcher, *Unique factorization rings*, Math. Proc. Cambridge Philos. Soc. **65** (1969), 579–583.

43. C.R. Fletcher, *The structure of unique factorization rings*, Math. Proc. Cambridge Philos. Soc. **67** (1970), 535–540.

44. R. Fossum, *The Divisor Class Group of a Krull Domain*, Springer-Verlag, New York, 1973.

45. S. Galovich, *Unique factorization rings with zero divisors*, Math. Mag. **51** (1978), 276–283.

46. A. Geroldinger, *The catenary degree and tameness of factorizations in weakly Krull domains*, Factorization in Integral Domains (D.D. Anderson, ed.), Lecture Notes in Pure and Appl. Math., vol. 189, Marcel Dekker, Inc., New York, 1997, pp. 113–153.

47. R. Gilmer, *On a classical theorem of Noether in ideal theory*, Pacific J. Math. **13** (1963), 579–583.

48. R. Gilmer, *Commutative rings in which each prime ideal is principal*, Math. Ann. **183** (1969), 151–158.

49. R. Gilmer, *Commutative Semigroup Rings*, Chicago Lectures in Mathematics, University of Chicago Press, Chicago, Ill., 1984.

50. R. Gilmer, *Multiplicative Ideal Theory*, Queen's Papers in Pure and Appl. Math., vol. 90, Queen's University, Kingston, Ontario, 1992.

51. A. Grams, *Atomic rings and the ascending chain condition for principal ideals*, Math. Proc. Cambridge Philos. Soc. **75** (1974), 321–329.

52. F. Halter-Koch, *On the factorization of algebraic integers into irreducibles*, Topics in classical number theory, Colloq. Math. Soc. János Bolyai, vol. 34, North-Holland, Amsterdam-New York, 1984, pp. 699–707.

53. F. Halter-Koch, *A characterization of Krull rings with zero divisors*, Arch. Math. (Brno) **29** (1993), 119–122.

54. F. Halter-Koch, *Finitely generated monoids, finitely primary monoids, and factorization properties of integral domains*, Factorization in Integral Domains (D.D. Anderson, ed.), Lecture Notes in Pure and Appl. Math., vol. 189, Marcel Dekker, Inc., New York, 1997, pp. 31–72.

55. J.A. Huckaba, *Commutative Rings with Zero Divisors*, Monographs and Textbooks in Pure and Applied Mathematics, vol. 117, Marcel Dekker, Inc., New York, 1988.

56. I. Kaplansky, *Commutative Rings*, revised ed., Polygonal Publishing House, Washington, N.J., 1994.

57. R.E. Kennedy, *Krull rings*, Pacific J. Math. **89** (1980), 131–136.

58. M.D. Larsen and P.J. McCarthy, *Multiplicative Theory of Ideals*, Academic Press, New York, 1971.

59. Chin-Pi Lu, *Factorial modules*, Rocky Mountain J. Math. **7** (1977), 125–139.

60. M.B. Martin and M. Zafrullah, *t-linked overrings of Noetherian weakly factorial domains*, Proc. Amer. Math. Soc. **15** (1992), 601–604.

61. R. Matsuda, *On Kennedy's problems*, Comment. Math. Univ. St. Paul. **31** (1982), 143–145.

62. R. Matsuda, *Generalizations of multiplicative ideal theory to commutative rings with zerodivisors*, Bull. Fac. Sci. Ibaraki Univ. Ser. A **17** (1985), 49–101.

63. S. Mori, *Über die Produktzerlegung der Hauptideale*, J. Sci. Hiroshima Univ. Ser. A **8** (1938), 7–13.

64. S. Mori, *Über die Produktzerlegung der Hauptideale, II*, J. Sci. Hiroshima Univ. Ser. A **9** (1939), 145–155.

65. S. Mori, *Allgemeine Z.P.I.-Ringe*, J. Sci. Hirosima Univ. Ser. A **10** (1940), 117–136.
66. S. Mori, *Über die Produktzerlegung der Hauptideale, III*, J. Sci. Hirosima Univ. Ser. A. **10** (1940), 85–94.
67. S. Mori, *Über die Produktzerlegung der Hauptideale, IV*, J. Sci. Hirosima Univ. Ser. A. **11** (1941), 7–14.
68. J.L. Mott, *The theory of divisibility*, Factorization in Integral Domains (D.D. Anderson, ed.), Lecture Notes in Pure and Appl. Math., vol. 189, Marcel Dekker, Inc., New York, 1997, pp. 155–187.
69. A.-M. Nicolas, *Modules factoriels*, Bull. Sci. Math. (2) **95** (1971), 33–52.
70. A.-M. Nicolas, *Extensions factorielles et modules factorables*, Bull. Sci. Math. (2) **98** (1974), 117–143.
71. D. Portelli and W. Spangher, *Krull rings with zero divisors*, Comm. Algebra **11** (1983), 1817–1851.
72. P. Samuel, *On unique factorization domains*, Illinois J. Math. **5** (1961), 1–17.
73. P. Samuel, *Anneaux Factoriels*, Sociedade de Matemática de São Paulo, 1963.
74. P. Samuel, *Lectures on Unique Factorization Domains*, Tata Institute for Fundamental Research Lectures on Mathematics, no. 30, Tata Institute for Fundamental Research, Bombay, 1964.
75. P. Samuel, *Unique factorization*, Amer. Math. Monthly **75** (1968), 945–952.
76. P. Samuel, *About Euclidean rings*, J. Algebra **19** (1971), 282–301.
77. P. Samuel, *Collected papers of Pierre Samuel* (P. Ribenboim, ed.), Queen's Papers in Pure and Appl. Math., vols. 99–100, Queen's University, Kingston, Ontario, 1995.
78. Th. Skolem, *Eine Bemerkung über gewisse Ringe mit Anwendung auf die Produktzerlegung von Polynomen*, Norsk Mat. Tidsskr. **21** (1939), 99–107.
79. U. Storch, *Fastfaktorielle Ringe*, Schriftenreihe Math. Inst. Univ. Münster **36** (1967).
80. C.A. Wood, *On General Z.P.I.-Rings*, Ph.D. thesis, Florida State University, Tallahassee, Florida, December 1967.
81. M. Zafrullah, *Semirigid GCD domains*, Manuscripta Math. **17** (1975), 55–66.
82. M. Zafrullah, *Rigid elements in GCD domains*, J. Natur. Sci. Math. **17** (1977), 7–14.
83. M. Zafrullah, *Unique representation domains*, J. Natur. Sci. Math. **18** (1978), 19–29.
84. M. Zafrullah, *A general theory of almost factoriality*, Manuscripta Math. **51** (1985), 29–62.
85. M. Zafrullah, *On a property of pre-Schreier domains*, Comm. Algebra **15** (1987), 1895–1920.
86. A. Zaks, *Half factorial domains*, Bull. Amer. Math. Soc. **82** (1976), 721–723.
87. A. Zaks, *Half-factorial-domains*, Israel J. Math. **37** (1980), 281–302.

Root Closure in Commutative Rings: A Survey

DAVID F. ANDERSON   Department of Mathematics, The University of Tennessee, Knoxville, TN 37996-1300

1   INTRODUCTION

All rings considered in this survey are commutative with identity. Let A be a subring of a ring B (with the same 1) and let $n \geq 1$ be an integer. We say that A is *n-root closed* in B if $b^n \in A$ with $b \in B$ implies $b \in A$, and that A is *n-root closed* if it is n-root closed in its total quotient ring $T(A)$. If A is n-root closed (in B) for all $n \geq 1$, we say that A is *root closed* (in B). For S a nonempty subset of $\mathbb{N}$, we say that A is *S-root closed* in B if $b^n \in A$ for some $n \in S$ and $b \in B$ implies $b \in A$, and that A is *S-root closed* if it is S-root closed in $T(A)$. Also, recall that A is *seminormal* in B if $b^2, b^3 \in A$ with $b \in B$ implies $b \in A$; equivalently, if $b^n \in A$ for all sufficiently large n with $b \in B$ implies $b \in A$. In case $B = T(A)$ and the above conditions hold, we say that A is *seminormal*. If A is n-root closed (in B) for some $n \geq 2$, then A is seminormal (in B); but the converse is false.

Let $\mathcal{C}(A,B) = \{n \in \mathbb{N} \,|\, A \text{ is } n\text{-root closed in } B \,\}$. Since A is trivially 1-root closed in B, and A is mn-root closed in B if and only if A is both m-root closed and n-root closed in B, $\mathcal{C}(A,B)$ is a multiplicative submonoid of $\mathbb{N}$ generated by primes. We write $\mathcal{C}(A) = \mathcal{C}(A, T(A))$. Thus A is root closed if and only if $\mathcal{C}(A) = \mathbb{N}$, and A is S-root closed in B if and only if $S \subset \mathcal{C}(A,B)$. We extend this notation by defining $\mathcal{C}_0(A,B) = \mathcal{C}(A,B) \cup \{0\}$ if A is seminormal in B, and $\mathcal{C}_0(A,B) = \mathcal{C}(A,B)$ otherwise. Similarly, we define $\mathcal{C}_0(A) = \mathcal{C}_0(A, T(A))$. Thus $\mathcal{C}_0(A,B)$ is a multiplicative submonoid of $\mathbb{Z}_+$, and either $\mathcal{C}_0(A,B) = \{1\}$ (i.e., A is not seminormal in B), $\mathcal{C}_0(A,B) = \{0,1\}$ (i.e., A is seminormal in B, but not n-root closed in B for any $n \geq 2$), or $\mathcal{C}_0(A,B)$ is infinite and $\{0,1\} \subset \mathcal{C}_0(A,B)$ (i.e., A is n-root closed in B for some $n \geq 2$).

Much of the literature on n-root closedness has focused on domains and connections with seminormality. We will attempt to state results in as general a setting as possible, using the $\mathcal{C}_0(A,B)$ notation. Most proofs will be omitted, but we give extensive literature references. We also include a few new results and extend several earlier results stated only for integral domains to rings with zero divisors. Results concerning $\mathcal{C}_0(A)$ are usually simpler in the domain setting because the total quotient ring construction need not behave well under extensions.

In general, the following connections hold between closedness properties, but even for one-dimensional quasilocal domains none of the implications is reversible (see Section 7).

Completely integrally closed $\implies$ integrally closed $\implies$ root closed $\implies$ n-root closed for some $n \geq 2$ $\implies$ seminormal.

In the second section, we include basic results about the behavior of root closedness for directed unions, intersections, pullbacks, localizations, and other ring-theoretic constructions. The third section studies root closedness for polynomial rings, monoid rings, and graded rings. The more difficult case for power series rings is discussed in the fourth section. The fifth section considers the S-root closure of a ring and related properties. The sixth section investigates the question of realizing a multiplicative submonoid of $\mathbb{N}$ generated by primes as $\mathcal{C}(A)$ for an integral domain A. In the final section, we give several examples and mention several related closedness properties, their history, and their relationship to n-root closedness.

Let $\mathbb{N}$, $\mathbb{Z}$, $\mathbb{Q}$, $\mathbb{R}$, and $\mathbb{C}$ denote the sets of positive integers, integers, rational numbers, real numbers, and complex numbers, respectively. Also, $\mathbb{Z}_+$, $\mathbb{Q}_+$, and $\mathbb{R}_+$ will denote the sets of nonnegative integers, nonnegative rational numbers, and nonnegative real numbers, respectively. Throughout, A will always be a commutative ring with 1 and $\mathbf{X} = \{X_\alpha\}$ a family of indeterminates. For an integral domain A, we denote its integral closure by A' and its quotient field by $qf(A)$. For general terminology, see [27] for rings and [28] for monoids and monoid rings. For an earlier survey article on root closedness, see [33].

2   BASIC RESULTS

In this section, we collect basic results about root closedness and seminormality. Many of these results are either well known or easily proved (most are both). We first give a unifying closedness notion which was introduced in [10]. Let S be a nonempty subset of $\mathbb{N}$ and let A be a subring of a ring B. We define A to be *S-closed* in B if $b^n \in A$ for all $n \in S$ with $b \in B$ implies $b \in A$. We also define A to be *S-closed* if A is S-closed in $T(A)$. Thus A is n-root closed in B if and only if A is $\{n\}$-closed in B, and A is seminormal in B if and only if A is $\{2,3\}$-closed in B. Clearly A is S-closed (resp., S-root closed) in B if and only if A is $[S]$-closed (resp., $\langle S \rangle^*$-root closed) in B, where $[S]$ (resp., $\langle S \rangle^*$) is the additive subsemigroup of $\mathbb{N}$ generated by S (resp., multiplicative submonoid of $\mathbb{N}$ generated by the prime divisors of each $s \in S$). Recall that an additive subsemigroup S of $\mathbb{N}$ is finitely generated; if $S = [n_1, \dots, n_t]$, we write $gcd\, S = gcd\,(n_1, \dots, n_t)$. Our first result shows that this more general concept really leads to nothing new.

PROPOSITION 2.1 ([10, Theorem 3.2])   Let S be a proper additive subsemigroup of $\mathbb{N}$ with $gcd\, S = d$ and let A be a subring of a ring B.

(a) Let $d = 1$. Then A is S-closed in B if and only if A is seminormal in B.

(b) Let $d > 1$. Then A is S-closed in B if and only if A is d-root closed in B. $\square$

The next two results follow directly from the definitions. One may easily construct examples to show that each of the inclusions may be proper. Recall that $0 \in C_0(A, B)$ if and only if A is seminormal in B, and for $n \geq 1$, that $n \in C_0(A, B)$ if and only if A is n-root closed in B. As usual, $\text{nil}(A)$ is the ideal of nilpotent elements of A.

PROPOSITION 2.2 Let A and B be subrings of a ring C with $A \subset B \subset C$. Then

(a) $C_0(A, B) \cap C_0(B, C) \subset C_0(A, C)$ and $C_0(A, C) \subset C_0(A, B)$. Hence, if $C_0(A, B) \subset C_0(B, C)$ (e.g., if B is root closed in C), then $C_0(A, C) = C_0(A, B)$.

(b) Suppose that $T(A) = T(B)$. Then $C_0(A, B) \cap C_0(B) \subset C_0(A)$ and $C_0(A) \subset C_0(A, B)$. Hence, if $C_0(A, B) \subset C_0(B)$ (e.g., if B is root closed), then $C_0(A) = C_0(A, B)$.

(c) If A is seminormal in B, then the conductor $(A : B) = \{x \in B \,|\, xB \subset A\}$ is a radical ideal of B, and thus $\text{nil}(A) = \text{nil}(B)$. Hence, if $\text{nil}(A) \neq \text{nil}(B)$, then $C_0(A, B) = \{1\}$. $\square$

PROPOSITION 2.3 Let $\{A_\alpha\}$ and $\{B_\alpha\}$ be families of subrings of a ring B with each $\{A_\alpha\} \subset \{B_\alpha\}$. Then

(a) $\cap C_0(A_\alpha, B_\alpha) \subset C_0(\cap A_\alpha, \cap B_\alpha)$. In particular, $\cap C_0(A_\alpha, B) \subset C_0(\cap A_\alpha, B)$. Moreover, $\cap C_0(A_\alpha) \subset C_0(\cap A_\alpha)$ if each A_α is an integral domain.

(b) $C_0(\Pi A_\alpha, \Pi B_\alpha) = \cap C_0(A_\alpha, B_\alpha)$.

(c) Suppose that $\{A_\alpha\}$ and $\{B_\alpha\}$ are directed families. Then $\cap C_0(A_\alpha, B_\alpha) \subset C_0(\cup A_\alpha, \cup B_\alpha)$. In particular, $\cap C_0(A_\alpha, B) \subset C_0(\cup A_\alpha, B)$. Moreover, $\cap C_0(A_\alpha) \subset C_0(\cup A_\alpha)$ if each A_α is an integral domain. $\square$

The next result gives an easy method for constructing seminormal or n-root closed rings.

PROPOSITION 2.4 ([15, Theorems 1.7 and 1.8]) Let A be a subring of a ring B with common ideal I. Then $C_0(A, B) = C_0(A/I, B/I)$. $\square$

Special cases of the above include: if B is an integral domain and I is nonzero (so A and B have the same quotient field), then $C_0(A) \subset C_0(A/I, B/I)$ and $C_0(B) \cap C_0(A/I, B/I) \subset C_0(A)$, with $C_0(A) = C_0(A/I, B/I)$ if B is root closed. This shows that even for integral domains the relative case arises naturally. We give the pullback version below; special cases for the $D + M$ construction are given in [10, Lemma 2.1], [14, Lemma 2.1(g)], and [17, Lemma 2.15].

COROLLARY 2.5 Let T be an integral domain with nonzero maximal ideal M, residue field $K = T/M, \phi : T \longrightarrow K$ the natural projection, D a subring of K, and $R = \phi^{-1}(D)$. Then $C_0(R, T) = C_0(D, K), C_0(R) \subset C_0(D, K)$, and $C_0(T) \cap C_0(D, K) \subset C_0(R)$. In particular, $C_0(R) = C_0(D, K)$ if T is root closed. $\square$

We next consider localizations; again, the inclusions may all be proper.

PROPOSITION 2.6 Let A be a subring of a ring B and let S be a multiplicative set of A. Then $\mathcal{C}_0(A, B) \subset \mathcal{C}_0(A_S, B_S)$. In particular, $\mathcal{C}_0(A) \subset \mathcal{C}_0(A_S)$ if A is an integral domain.   □

Our next result gives a special case where $\mathcal{C}_0(A) = \mathcal{C}_0(A_S)$ when A is an integral domain.

PROPOSITION 2.7 ([7, Propositions 4.1 and 4.2]) Let A be an integral domain with integral closure A' and let S be a multiplicative set of A generated by primes. Then $A = A_S \cap A'$, and hence $\mathcal{C}_0(A) = \mathcal{C}_0(A_S)$.   □

Finally, we consider to what extent n-root closedness is a local property. For integral domains, things behave nicely.

PROPOSITION 2.8 Let A be a subring of an integral domain B. Then $\mathcal{C}_0(\cap A_{P_\alpha}, \cap B_{P_\alpha}) = \cap \mathcal{C}_0(A_{P_\alpha}, B_{P_\alpha})$ and $\mathcal{C}_0(\cap A_{P_\alpha}) = \cap \mathcal{C}_0(A_{P_\alpha})$, where $\{P_\alpha\}$ is any nonempty set of prime ideals of A. In particular, $\mathcal{C}_0(A, B) = \cap \mathcal{C}_0(A_{P_\alpha}, B_{P_\alpha})$ and $\mathcal{C}_0(A) = \cap \mathcal{C}_0(A_{P_\alpha})$ if $\cap A_{P_\alpha} = A$
Proof: The "$\subset$" inclusion follows from Proposition 2.6 since $(\cap A_{P_\alpha})_{P_\beta} = A_{P_\beta}$ and $(\cap B_{P_\alpha})_{P_\beta} = B_{P_\beta}$ for each prime ideal P_β. The reverse inclusion follows from Proposition 2.3(a).   □

A locally n-root closed (resp., seminormal) ring A is n-root closed (resp., seminormal), i.e., $\cap \mathcal{C}_0(A_M) \subset \mathcal{C}_0(A)$ (this follows from the proof of [36, Theorem 2.1]); but the localization of an n-root closed (resp., seminormal) ring need not be n-root closed (resp., seminormal). In fact, the localization of an integrally closed ring need not be seminormal (cf. [36, p. 2406]). However, we do have the following positive result (the seminormal part is from [47] and the n-root closed part from [36]). Recall that a (commutative) ring A is *von Neumann regular* if for each $a \in A$, there is a $b \in A$ with $a^2 b = a$ (equivalently, A is reduced and zero-dimensional).

PROPOSITION 2.9 ([36, Theorem 2.1], [47, Corollary 5]) Let A be a ring and let $\{M_\alpha\}$ be the set of maximal (prime) ideals of A. Then
    (a) $\cap \mathcal{C}_0(A_{M_\alpha}) \subset \mathcal{C}_0(A)$.
    (b) If A is a (reduced) ring with $T(A)$ von Neumann regular, then $\mathcal{C}_0(A) = \cap \mathcal{C}_0(A_{M_\alpha})$.   □

Other related results concerning root closedness and localization may be found in [36]. We close with

PROPOSITION 2.10 Let A be an integral domain with quotient field K and let $b \in K$. Suppose that for each prime ideal P of A, we have $b^{n_P} \in A_P$ for some integer $n_p \geq 1$. Then $b^n \in A$ for some integer $n \geq 1$.
Proof: Let $S = \{s \in A \mid sb^{ns} \in A$ for some integer $n_s \geq 1\}$. By hypothesis, $S \not\subset P$ for any prime ideal P of R. Hence $(S) = A$; so $(s_1, \ldots, s_t) = A$ for some $s_1, \ldots, s_t \in S$. For each s_i, suppose that $s_i b^{n_i} \in A$. Let $n = n_1 \cdots n_t$ and $m_i = n/n_i$ for each $1 \leq i \leq t$. Then $s_i^{m_i} b^n \in A$ for each $1 \leq i \leq t$. Also,

$(s_1^{m_1}, \ldots, s_t^{m_t}) = A$. Thus $1 = r_1 s_1^{m_1} + \cdots + r_t s_t^{m_t}$ for some $r_1, \ldots, r_t \in A$, and hence $b^n = r_1 s_1^{m_1} b^n + \cdots + r_t s_t^{m_t} b^n \in A$. $\qquad\square$

3   POLYNOMIAL RINGS, MONOID RINGS, AND GRADED RINGS

Let A be a subring of a ring B and let $\mathbf{X} = \{X_\alpha\}$ be a family of indeterminates. Clearly $\mathcal{C}_0(A[\mathbf{X}], B[\mathbf{X}]) \subset \mathcal{C}_0(A, B)$ and $\mathcal{C}_0(A[\mathbf{X}]) \subset \mathcal{C}_0(A)$. That the reverse inclusions also hold are the next two results. These results are due to Brewer, Costa, and McCrimmon [23] for $\mathbf{X}$ finite; the general case follows directly from Proposition 2.3(c) (see Theorem 3.4 for a generalization of Corollary 3.2).

THEOREM 3.1 ([23, Theorem 1 and Proposition 1])   Let A be a subring of a ring B and let $\mathbf{X}$ be a family of indeterminates. Then $\mathcal{C}_0(A[\mathbf{X}], B[\mathbf{X}]) = \mathcal{C}_0(A, B)$. $\qquad\square$

COROLLARY 3.2 ([23, Theorem 2])   Let A be a reduced ring whose total quotient ring is von Neumann regular and let $\mathbf{X}$ be a family of indeterminates. Then $\mathcal{C}_0(A[\mathbf{X}]) = \mathcal{C}_0(A)$. In particular, this holds if A is an integral domain.
Proof: First note that $T(A)[\mathbf{X}]$ is integrally closed since $T(A)$ is von Neumann regular [1, Theorem and Corollary 3.4] (cf. [28, Theorem 18.7]). Thus $\mathcal{C}_0(A[\mathbf{X}]) = \mathcal{C}_0(A[\mathbf{X}], T(A)[\mathbf{X}])$ by Proposition 2.2(b). Hence $\mathcal{C}_0(A[\mathbf{X}]) = \mathcal{C}_0(A, T(A)) = \mathcal{C}_0(A)$. $\qquad\square$

Theorem 3.1 easily extends to monoid rings. Recall that a commutative cancellative monoid S (written additively) is *torsionfree* if $nx = ny$ for $x, y \in S$ and $n \geq 1$ implies $x = y$ (equivalently, its quotient group $G = \langle S \rangle \, (= \{a - b \,|\, a, b \in S\})$ is torsionfree). For S a submonoid of a torsionfree commutative cancellative monoid T, we say that S is *n-root closed* (resp., *seminormal*) in T if $nt \in S$ (resp., $2t, 3t \in S$) with $t \in T$ implies $t \in S$. We write $\mathcal{C}(S, T) = \{n \in \mathbb{N} \,|\, S$ is n-root closed in $T\}$ and $\mathcal{C}(S) = \mathcal{C}(S, \langle S \rangle)$; each is a multiplicative submonoid of $\mathbb{N}$ generated by primes. Similarly, we define $\mathcal{C}_0(S, T)$. For any ring A and torsionfree commutative cancellative monoid S, the monoid ring of S over A is $A[S] = \{\Sigma a_s X^s \,|\, a_s \in A, s \in S\}$ with $X^s X^t = X^{s+t}$.

THEOREM 3.3   Let A be a subring of a ring B and let S be a torsionfree commutative cancellative monoid. Then $\mathcal{C}_0(A[S], B[S]) = \mathcal{C}_0(A, B)$.
Proof: Clearly $\mathcal{C}_0(A[S], B[S]) \subset \mathcal{C}_0(A, B)$. First note that $\mathcal{C}_0(A, B) = \mathcal{C}_0(A[\mathbf{X}], B[\mathbf{X}]) \subset \mathcal{C}_0(A[\mathbf{X}^*], B[\mathbf{X}^*]) \subset \mathcal{C}_0(A, B)$ by Proposition 2.6, where $\mathbf{X} = \{X_\alpha\}$ is any family of indeterminates and $\mathbf{X}^* = \{X_\alpha, X_\alpha^{-1}\}$. Thus $\mathcal{C}_0(A[F], B[F]) = \mathcal{C}_0(A, B)$ for any free abelian group F. Since any torsionfree abelian group G is the directed union of its (finitely generated) free subgroups, we have $\mathcal{C}_0(A[G], B[G]) = \mathcal{C}_0(A, B)$ by Proposition 2.3(c). Let $G = \langle S \rangle$. Then $\mathcal{C}_0(A, B) = \mathcal{C}_0(A[G], B[G]) \cap \mathcal{C}_0(B[S], B[S]) \subset \mathcal{C}_0(A[S], B[S])$ by Proposition 2.3(a) since $A[G] \cap B[S] = A[S]$. Thus $\mathcal{C}_0(A[S], B[S]) = \mathcal{C}_0(A, B)$. $\qquad\square$

We next consider the more subtle case for $\mathcal{C}_0(A)$ when A has zero divisors. We have already observed that $\mathcal{C}_0(A[\mathbf{X}]) \subset \mathcal{C}_0(A)$. However, the reverse containment need not hold. Let S be any nonzero torsionfree commutative cancellative monoid. If A is not reduced, then there is a $0 \neq a \in A$ with $a^2 = 0$. Then for any $0 \neq s \in S, b = a/(1 + X^s) \in T(A[S])$, $b^n = 0$ in $A[S]$ for all $n \geq 2$, but $b \notin A[S]$ (cf. Proposition 2.2(c)). Thus, if A is not reduced, then $C_0(A[S]) = \{1\}$. However, we need not have $C(A[X]) = C(A)$ even if A is reduced. The difficulty is that $T(A)[X]$ need not be integrally closed; in [23, Example 1], an example is given of a reduced ring A with $A = T(A)$, so $C_0(A) = \mathbb{Z}_+$, but $C_0(A[X]) = \{1\}$. The special case considered in Corollary 3.2 holds since $B[X]$ is integrally closed when B is von Neumann regular.

The determination of $C_0(A[X])$ for an arbitrary (reduced) ring A is due to Lucas [35] and is based on his work on the integral closure of $A[X]$ (cf. [34], [36], and [37]). Let $Q_0(A)$ be the so-called ring of finite fractions of A (cf. [34] and [35]); $Q_0(A)$ is a subring of $Q(R)$, the complete ring of fractions of A (cf. [32]).

THEOREM 3.4 ([35, Corollaries 5 and 6]) Let A be a reduced ring and let $\mathbf{X}$ be a nonempty family of indeterminates. Then $C_0(A[\mathbf{X}]) = C_0(A, Q_0(A))$, where $Q_0(A)$ is the ring of finite fractions of A. In particular, $C_0(A[\mathbf{X}]) = C_0(A, T(A[\mathbf{X}]))$.
Proof: The "in particular" statement is observed for a single indeterminate in the Introduction of [36]. It follows since the integral closure of $A[X]$ in $T(A[X]))$ is $B[X]$, where B is the integral closure of A in $Q_0(A)$ [35, Theorem 3]. The result also extends to any family of indeterminates. $\qquad\square$

Next we consider graded versions. Let $B = \oplus B_\alpha$ be a ring graded by a torsion-free commutative cancellative monoid S. A subring A of B is a *graded subring* if $A = \oplus A_\alpha$ with each $A_\alpha \subset B_\alpha$. We say that a graded subring A of B is *homogeneously n-root closed* (resp., *homogeneously seminormal*) in B if $b \in B$ homogeneous with $b^n \in A$ (resp., $b^2, b^3 \in A$) implies $b \in A$; and we define $C(A, B)_h = \{n \in \mathbb{N} \mid A \text{ is homogeneously } n\text{-root closed in } B\}$ and $C(A)_h = C(A, B)_h$, where B is the homogeneous quotient ring of A. Clearly $C(A, B)_h$ is a multiplicative submonoid of $C(A, B)$ generated by primes. In the obvious way, we define $C_0(A, B)_h$ and $C_0(A)_h$. Our next result for $\mathbb{Z}_+$-graded rings is from [46]; other special cases are in [4] and [9].

THEOREM 3.5 ([46, Lemma 1.2 and Corollary 1.3]) Let A be a graded subring of a graded ring $B = \oplus B_\alpha$. Then $C_0(A, B) = C_0(A, B)_h$. $\qquad\square$

COROLLARY 3.6 ([10, Theorem 2.4], [4, Theorem 6.1]) Let $A = \oplus A_\alpha$ be a graded integral domain. Then $C_0(A) = C_0(A)_h$.
Proof: This follows from Proposition 2.2(b) since the homogeneous quotient field of A is integrally closed [3, Proposition 2.1]. $\qquad\square$

We now generalize Theorem 3.3 by also letting the monoids vary. The monoid ring $A[S]$ is S-graded with $\deg aX^s = s$ for $0 \neq a \in A$ and $s \in S$. If S is a submonoid of T, then $A[S]$ is a graded subring of $A[T]$.

COROLLARY 3.7 Let A be a ring and let S be a submonoid of a torsionfree commutative cancellative monoid T.
   (a) If A is reduced, then $C_0(A[S], A[T]) = C_0(S, T)$.
   (b) If A is not reduced and S is a proper submonoid of T, then $C_0(A[S], A[T]) =$

$\{1\}$.

Proof: Part (b) follows from Proposition 2.2(c). For part (a), we show that $\mathcal{C}_0(A[S], A[T])_h = \mathcal{C}_0(S, T)$. The result then follows from Theorem 3.5. The "$\subset$" inclusion is clear, even if A is not reduced. For the reverse inclusion, let $n \in \mathcal{C}_0(S, T)$ with $n \geq 2$. If $(aX^t)^n = a^n X^{nt} \in A[S]$ for some $0 \neq a \in A$ and $t \in T$, then $nt \in S$ since $a^n \neq 0$. Thus $t \in S$, so $aX^t \in A[S]$. The proof of the seminormal case is similar. $\square$

COROLLARY 3.8 Let A be a ring and let S be a nonzero torsionfree commutative cancellative monoid. If A is reduced, then $\mathcal{C}_0(A, A[S]) = \mathbb{Z}_+$. If A is not reduced, then $\mathcal{C}_0(A, A[S]) = \{1\}$. $\square$

Combining Theorem 3.3 and Corollary 3.7, we obtain our next result (it is also a direct consequence of Theorem 3.5).

THEOREM 3.9 Let A be a reduced subring of a ring B and let S be a submonoid of a torsionfree commutative cancellative monoid T. Then $\mathcal{C}_0(A[S], B[T]) = \mathcal{C}_0(A, B) \cap \mathcal{C}_0(S, T)$.

Proof: Since $A[S] \subset A[T] \subset B[T]$ and $A[S] \subset B[S] \subset B[T]$, we have $\mathcal{C}_0(A[S], B[T]) \subset \mathcal{C}_0(A[S], A[T]) \cap \mathcal{C}_0(A[S], B[S]) = \mathcal{C}_0(S, T) \cap \mathcal{C}_0(A, B)$ by Proposition 2.2(a), Corollary 3.7, and Theorem 3.3. Also, Proposition 2.2(a), together with Corollary 3.7 and Theorem 3.3, yields $\mathcal{C}_0(S, T) \cap \mathcal{C}_0(A, B) = \mathcal{C}_0(A[S], A[T]) \cap \mathcal{C}_0(A[T], B[T]) \subset \mathcal{C}_0(A[S], B[T])$. Thus $\mathcal{C}_0(A[S], B[T]) = \mathcal{C}_0(S, T) \cap \mathcal{C}_0(A, B)$. $\square$

COROLLARY 3.10 ([10, Corollary 2.5], [4, Corollary 6.2]) Let A be an integral domain and let S be a torsionfree commutative cancellative monoid. Then $\mathcal{C}_0(A[S]) = \mathcal{C}_0(A) \cap \mathcal{C}_0(S)$.

Proof: Let K be the quotient field of A and let G be the quotient group of S. Then $A[S] \subset A[G] \subset K[G]$ and $K[G]$ is integrally closed. By Proposition 2.2(b), $\mathcal{C}_0(A[S]) = \mathcal{C}_0(A[S], K[G]) = \mathcal{C}_0(A, K) \cap \mathcal{C}_0(S, G) = \mathcal{C}_0(A) \cap \mathcal{C}_0(S)$. $\square$

COROLLARY 3.11 Let A be a ring and let S be a nonzero torsionfree commutative cancellative monoid. If A is reduced, then $\mathcal{C}_0(A[S]) = \mathcal{C}_0(A, Q_0(A)) \cap \mathcal{C}_0(S) = \mathcal{C}_0(A, T(A[S])) \cap \mathcal{C}_0(S)$. If A is not reduced, then $\mathcal{C}_0(A[S]) = \{1\}$.

Proof: Theorem 3.4 generalizes to monoid rings using the ideas in [37, Section 3]. $\square$

Theorem 3.9 and Corollary 3.11 will be generalized in Section 5. We end this section by noting that Theorem 3.3 does not extend to arbitrary abelian groups. For example, let $A = K = \mathbb{Z}/2\mathbb{Z}, B = K[Y]$, and $G = \mathbb{Z}/2\mathbb{Z} = \{\bar{0}, \bar{1}\}$. Then $\mathcal{C}_0(A, B) = \mathbb{Z}_+$ by Corollary 3.8, but $2 \notin \mathcal{C}_0(A[G], B[G])$ since $(Y + YX^{\bar{1}})^2 = 0$. For related results on root closedness in $R(X)$ and $R\langle X \rangle$, see [6] and [19].

4   POWER SERIES RINGS

As in the polynomial ring case, it is clear that $\mathcal{C}_0(A[[\mathbf{X}]], B[[\mathbf{X}]]) \subset \mathcal{C}_0(A, B)$ and $\mathcal{C}_0(A[[\mathbf{X}]]) \subset \mathcal{C}_0(A)$ (here $A[[\mathbf{X}]] = \cup\{A[[\mathbf{Y}]] \mid \mathbf{Y} \subset \mathbf{X}$ is finite $\}$, cf. [27, (1.1)]). In general, the two converses fail. Indeed, $\mathbb{Z}[[X]]$ is completely integrally closed;

however, $\mathbb{Z}[[X]]$ is not n-root closed in $\mathbb{Q}[[X]]$ for any $n \geq 2$ even though $\mathbb{Z}$ is root closed [53, Example 1]. Also, for A a valuation domain with $dim A \geq 2$, $A[[X]]$ is never root closed (Corollary 4.10). In this section, we investigate to what extent the converses hold. Even if A is integrally closed, $A[[X]]$ need not be (root closed) integrally closed (cf. [39], [48], and Corollary 4.10). Further difficulties arise since if A has quotient field K, then $K[[X]]$ need not be in the quotient field of $A[[X]]$ (cf. [21]).

Root closedness in $A[[X]]$ was first studied by Watkins [53], who concentrated on the case where A is von Neumann regular. Most of the results in this section are from [16]; see [19] and [13] for other results on root closedness in power series rings.

We first consider the seminormal case; the main result is due to Brewer and Nichols [24]. The special case when B is an extension of a von Neumann regular ring A is given in [53, Theorem 1], and sufficient conditions for $A[[X]]$ to be seminormal when A is von Neumann regular are given in [53, Theorem 2]. Also, [19, Theorem 2] handles the case when A is a domain with $A = qf(A) \cap B$.

THEOREM 4.1 ([24, Theorem]) Let A be a subring of a ring B and let **X** be a family of indeterminates. Then $A[[\mathbf{X}]]$ is seminormal in $B[[\mathbf{X}]]$ if and only if A is seminormal in B. If A is a reduced ring with only finitely many minimal primes (for example, an integral domain or a Noetherian reduced ring), then $A[[\mathbf{X}]]$ is seminormal if and only if A is seminormal. $\qquad\square$

For A a subring of a ring B and $n \in \mathbb{N}$, the following key property was introduced in [16] in order to characterize when $A[[\mathbf{X}]]$ is n-root closed in $B[[\mathbf{X}]]$.

$$\mathcal{P}_n(A, B): \quad \text{If} \quad b \in B, a \in A, \quad \text{and} \quad nab \in A, \quad \text{then} \quad nab^2 \in A.$$

Clearly, if $m \mid n$, then property $\mathcal{P}_m(A, B)$ implies property $\mathcal{P}_n(A, B)$. In particular, property $\mathcal{P}_1(A, B)$ implies property $\mathcal{P}_n(A, B)$ for all $n \in \mathbb{N}$. See [16] for more details about this property. We next determine when $A[[\mathbf{X}]]$ is n-root closed in $B[[\mathbf{X}]]$. Note that if $char A = p > 0$, then property $\mathcal{P}_p(A, B)$ holds and clearly $A[[\mathbf{X}]]$ is p-root closed in $B[[\mathbf{X}]]$ if and only if A is p-root closed in B.

THEOREM 4.2 ([16, Theorem 1.18]) Let A be a subring of a ring B, let **X** be a nonempty family of indeterminates, and let $n \geq 2$. Then the following three conditions are equivalent.

1. $A[[\mathbf{X}]]$ is n-root closed in $B[[\mathbf{X}]]$.
2. A is n-root closed in B, and
   - If n has at least two distinct prime factors, then property $\mathcal{P}_1(A, B)$ holds.
   - If n is a power of a prime p, then property $\mathcal{P}_p(A, B)$ holds.
3. A is n-root closed in B and property $\mathcal{P}_p(A, B)$ holds for each prime factor p of n. $\qquad\square$

COROLLARY 4.3 ([16, Corollary 1.22 and Theorem 1.23]) Let A be a subring of a ring B and let **X** be a nonempty family of indeterminates. Then the following three conditions are equivalent.

1. $C(A[[\mathbf{X}]], B[[\mathbf{X}]]) = C(A, B)$.
2. • If $C(A, B)$ contains at least two distinct primes, then property $\mathcal{P}_1(A, B)$ holds.
   • If $C(A, B)$ contains exactly one prime p, then property $\mathcal{P}_p(A, B)$ holds.
3. Property $\mathcal{P}_p(A, B)$ holds for each prime p in $C(A, B)$.

In particular, $A[[\mathbf{X}]]$ is root closed in $B[[\mathbf{X}]]$ if and only if A is root closed in B and property $\mathcal{P}_1(A, B)$ holds. $\square$

A ring A is said to be *p-injective* if each principal ideal of A is an annihilator of some subset of A (cf. [30] and [38]). A reduced ring is p-injective if and only if it is zero-dimensional, that is, von Neumann regular (however, a zero-dimensional ring need not be p-injective, and a p-injective ring may have infinite Krull dimension). The property $\mathcal{P}_1(A, B)$ is satisfied for all ring extensions B of A if and only if A is p-injective [16, Proposition 1.14]. Hence the next corollary follows directly from the previous corollary. The case when A is von Neumann regular has been observed in [53, Theorem 1].

COROLLARY 4.4 ([16, Theorem 1.27]) Let A be a p-injective ring and let $\mathbf{X}$ be a family of indeterminates. Then $C(A[[\mathbf{X}]], B[[\mathbf{X}]]) = C(A, B)$ for any ring extension B of A. $\square$

The next result can be used to reduce the study of root closedness for power series over a pair of integral domains $A \subset B$ to the case where B is an overring of A. This result extends [19, Theorem 2].

PROPOSITION 4.5 ([16, Theorem 1.29]) Let A be an integral domain with quotient field K which is a subring of a ring B, suppose there is a ring containing both B and K, and let $\mathbf{X}$ be a nonempty family of indeterminates. Then $C(A[[\mathbf{X}]]), B[[\mathbf{X}]]) = C(A, B) \cap C(A[[\mathbf{X}]], (K \cap B)[[\mathbf{X}]])$. $\square$

Many other results are given in [16]. For example, if A and B have a common maximal ideal M, then $C(A[[\mathbf{X}]], B[[\mathbf{X}]]) = C(A, B) = C(A/M, B/M)$ [16, Proposition 1.30] (cf. Proposition 2.4).

We next consider the case for $C(A[[\mathbf{X}]])$ when A is an integral domain with quotient field K. Unlike the polynomial ring case, this does not follow trivially from the general case since $K[[X]]$ is not usually contained in the quotient field of $A[[X]]$. In fact, if A is not a field, then $C(A[[\mathbf{X}]], K[[\mathbf{X}]])$ can contain at most one prime p (cf. [16, Proposition 1.28]). As in [16], we define property $\mathcal{P}_n(A)$ as property $\mathcal{P}_n(A, A^*)$, where A^* is the complete integral closure of A. Unlike property $\mathcal{P}_n(A, B)$, property $\mathcal{P}_n(A)$ is *not* a local property. Property $\mathcal{P}_n(A)$ is used to characterize when $A[[\mathbf{X}]]$ is n-root closed since $C(A[[\mathbf{X}]]) = C(A[[\mathbf{X}]], A^*[[\mathbf{X}]])$ by [16, Proposition 2.3]. These results do not extend to rings with zero divisors (cf. [53, Example 6] and Theorem 6.4). Some sufficient conditions for $A[[\mathbf{X}]]$ to be n-root closed when A is von Neumann regular are given in [53, Theorem 2].

THEOREM 4.6 ([16, Theorem 2.4]) Let A be an integral domain, let $\mathbf{X}$ be a nonempty family of indeterminates, and let $n \geq 2$. Then the following three conditions are equivalent.

1. $A[[\mathbf{X}]]$ is n-root closed.
2. A is n-root closed, and
   - If n has at least two distinct prime factors, then property $\mathcal{P}_1(A)$ holds.
   - If n is a power of a prime p, then property $\mathcal{P}_p(A)$ holds.
3. A is n-root closed and property $\mathcal{P}_p(A)$ holds for each prime factor p of n. $\square$

COROLLARY 4.7 ([16, Theorem 2.5])   Let A be an integral domain and let $\mathbf{X}$ be a nonempty family of indeterminates. Then the following three conditions are equivalent.

1. $\mathcal{C}(A[[\mathbf{X}]]) = \mathcal{C}(A)$.
2. • If $\mathcal{C}(A)$ contains at least two distinct primes, then property $\mathcal{P}_1(A)$ holds.
   • If $\mathcal{C}(A)$ contains exactly one prime p, then property $\mathcal{P}_p(A)$ holds.
3. Property $\mathcal{P}_p(A)$ holds for each prime p in $\mathcal{C}(A)$.

In particular, $A[[\mathbf{X}]]$ is root closed if and only if A is root closed and property $\mathcal{P}_1(A)$ holds. $\square$

We next consider several special classes of integral domains. Recall that an integral domain A is a *Mori domain* if it satisfies ACC on integral divisorial ideals. In particular, Noetherian domains and Krull domains are Mori domains. The next theorem follows directly from Corollary 4.7 since a seminormal Mori domain satisfies property $\mathcal{P}_1(A)$ [45, Proposition 1.4].

THEOREM 4.8 ([16, Theorem 2.13])   Let A be a Mori domain (for example, a Noetherian domain) and let $\mathbf{X}$ be a family of indeterminates. Then $\mathcal{C}(A[[\mathbf{X}]]) = \mathcal{C}(A)$. $\square$

In many cases, $A[[X]]$ is root closed only in the trivial case when A is completely integrally closed (recall that $A[[X]]$ is completely integrally closed if and only if A is completely integrally closed). It will be convenient to consider $\mathcal{C}(A, B)^* = \mathcal{C}(A, B) - \{1\}$. Note that $\mathcal{C}(A, B)^*$ generates the ideal A if and only if either A is m-root closed in B for some $m \geq 2$ invertible in A, or A is both p- and q-root closed in B for distinct primes p and q (if $char\,A = p > 0$, this means A is q-root closed for some prime $q \neq p$). Our next result follows from Corollary 4.7 since a Prüfer domain A satisfies property $\mathcal{P}_1(A)$ if and only if it is completely integrally closed [16, Proposition 2.15].

PROPOSITION 4.9 ([16, Theorem 2.16])   Let A be a Prüfer domain and let $\mathbf{X}$ be a nonempty family of indeterminates. Then the following conditions are equivalent.

1. $A[[\mathbf{X}]]$ is root closed.
2. $\mathcal{C}(A[[\mathbf{X}]])^*$ generates the ideal $A[[X]]$.
3. A is completely integrally closed. $\square$

The next corollary follows from Proposition 4.9 since a valuation domain A is completely integrally closed if and only if $dim\,A \leq 1$ (note that being completely integrally closed is *not* a local property). The first four conditions are also equivalent for A a GCD-domain or a QR-domain [16, Proposition 2.18].

COROLLARY 4.10 ([16, Corollary 2.17]) Let A be a valuation domain and let $\mathbf{X}$ be a nonempty family of indeterminates. Then the following conditions are equivalent.

1. $A[[\mathbf{X}]]$ is root closed.
2. $\mathcal{C}(A[[\mathbf{X}]])^*$ generates the ideal $A[[\mathbf{X}]]$.
3. A is completely integrally closed.
4. A is Archimedean (i.e., $\cap\, a^n A = 0$ for any nonunit $a \in A$).
5. $\dim A \leq 1$. $\qquad\qquad\qquad\qquad\qquad\qquad\qquad\qquad\qquad\qquad\qquad\qquad$ □

If $\mathcal{C}(A[[\mathbf{X}]], B[[\mathbf{X}]]) \neq \mathcal{C}(A, B)$ (resp., $\mathcal{C}(A[[\mathbf{X}]]) \neq \mathcal{C}(A)$), then $\mathcal{C}(A, B)$ (resp., $\mathcal{C}(A[[\mathbf{X}]])$) can contain at most one prime [16, Remark 1.26] (resp., [16, Remark 2.14]), and both possibilities (one given prime or no primes) may occur. Examples are given in [16, Examples 2.22 and 2.23] of a valuation domain A with $\mathcal{C}(A[[\mathbf{X}]]) = \{1\}$ (just let $\mathbb{Q} \subset A$ and have $\dim A \geq 2$, then use Corollary 4.10), and for a given prime p, a two-dimensional valuation domain A with $char\, A = 0$ and $\mathcal{C}(A[[\mathbf{X}]]) = \langle p \rangle$. However, for any set of primes $\mathcal{P}$, there is a one-dimensional quasilocal domain A with $\mathcal{C}(A) = \mathcal{C}(A[[\mathbf{X}]]) = \langle \mathcal{P} \rangle$ [13, Theorem 3.3] (see Example 7.1(d)).

These results have been extended to generalized power series rings by Zhongkui [55]. (See [44] for background material on generalized power series rings). We briefly recall a few definitions. Let $(S, \leq)$ be a strictly ordered monoid (i.e., $s + t < s + t'$ implies $t < t'$), let A be a commutative ring, and let $R = [[A^{S,\leq}]]$ be the set of all maps $f : S \longrightarrow A$ such that every strictly decreasing sequence of elements of $supp\,(f) = \{s \in S \,|\, f(s) \neq 0\}$ is finite and each subset of pairwise order-incomparable elements of $supp\,(f)$ is finite (using the terminology of Ribenboim [44], one would say that $supp\,(f)$ is Artinian and narrow). Then R is a commutative ring with 1 under pointwise addition and convolution. If $S = \mathbb{N}$ with the usual order, then $[[A^{S,\leq}]] \cong A[[X]]$, and if S is given the trivial order, then $[[A^{S,\leq}]] = A[S]$. The generalized power series ring analog of Theorem 4.1 is given in [44, (6.9)]. Note that a torsionfree commutative cancellative ordered monoid is strictly ordered.

THEOREM 4.11 ([55, Theorem 2.3]) Let $(S, \leq)$ be a torsionfree commutative cancellative (strictly) ordered monoid, and let A be a subring of a ring B and $n \geq 2$. If A is n-root closed in B and property $\mathcal{P}_p(A, B)$ holds for each prime factor p of n, then the generalized power series ring $[[A^{S,\leq}]]$ is n-root closed in $[[B^{S,\leq}]]$. $\qquad$ □

COROLLARY 4.12 ([55, Corollary 2.4]) Let A be a subring of a ring B and let $n \geq 2$. Then the following three conditions are equivalent.

1. For each torsionfree commutative cancellative (strictly) ordered monoid $(S, \leq)$, the generalized power series ring $[[A^{S,\leq}]]$ is n-root closed in $[[B^{S,\leq}]]$.
2. $A[[X]]$ is n-root closed in $B[[X]]$.
3. A is n-root closed in B and property $\mathcal{P}_p(A, B)$ holds for each prime factor p of n. $\qquad\qquad\qquad\qquad\qquad\qquad\qquad\qquad\qquad\qquad\qquad\qquad\qquad\qquad$ □

COROLLARY 4.13 ([55, Corollary 2.5]) Let A be a subring of a ring B. Then the following conditions are equivalent.

1. $\mathcal{C}([[A^{S,\leq}]], [[B^{S,\leq}]]) = \mathcal{C}(A, B)$ for each torsionfree commutative cancellative (strictly) ordered monoid $(S, \leq)$.
2. Property $\mathcal{P}_p(A, B)$ holds for each prime p in $\mathcal{C}(A, B)$. $\qquad\qquad\qquad$ □

5   THE ROOT CLOSURE AND RELATED PROPERTIES

Let A be a subring of a ring B. For any nonempty subset S of $\mathbb{N}$, we define the *S-root closure* of A in B, denoted by $\mathcal{R}^S(A,B)$, to be the smallest subring of B containing A which is S-root closed in B. For $B = T(A)$, we write $\mathcal{R}^S(A) = \mathcal{R}^S(A, T(A))$. Clearly $\mathcal{R}^S(A,B)$ is the intersection of all the S-root closed subrings of B containing A; we next show that $\mathcal{R}^S(A,B)$ is also a union (usually, $\{b \in B \mid b^n \in A$ for some $n \in S\}$ is *not* a subring of B).

For $0 \le m < \infty$, define $\mathcal{R}^S_m(A,B)$ inductively as follows: $\mathcal{R}^S_0(A,B) = A$; and for $m > 0, \mathcal{R}^S_m(A,B)$ is the subring of B generated by $\mathcal{R}^S_{m-1}(A,B)$ and the elements $b \in B$ such that $b^n \in \mathcal{R}^S_{m-1}(A,B)$ for some $n \in S$. Then for $0 < m < \infty, \mathcal{R}^S_m(A,B) = \mathcal{R}^S_1(\mathcal{R}^S_{m-1}(A,B),B)$. Clearly $\mathcal{R}^S_m(A,B) \subset \mathcal{R}^T_m(A,B)$ if $S \subset T$, and $\mathcal{R}^S_m(A,B) = \mathcal{R}^{\langle S \rangle^*}_m(A,B)$, where $\langle S \rangle^*$ is the multiplicative submonoid of $\mathbb{N}$ generated by the prime divisors of elements of S. If $S = \mathbb{N}$, we write $\mathcal{R}^S_m(A,B) = \mathcal{R}_m(A,B)$ and $\mathcal{R}^S(A,B) = \mathcal{R}(A,B)$.

PROPOSITION 5.1 ([15, Proposition 1.1]) Let A be a subring of a ring B and let S be a nonempty subset of $\mathbb{N}$. Then $\mathcal{R}^S(A,B) = \cup \mathcal{R}^S_m(A,B)$. $\square$

We say that the S-root closure of A in B is *obtainable in n-steps* if $\mathcal{R}^S(A,B) = \mathcal{R}^S_n(A,B)$. Examples are given in [15] to show that (a) there is a quasilocal one-dimensional seminormal domain A such that $\mathcal{R}(A)$ is not obtainable in a finite number of steps, and (b) for each $n \ge 1$, there is a quasilocal (resp., affine) one-dimensional seminormal domain A such that $\mathcal{R}(A)$ is obtainable in exactly n steps. It was also observed in [15, Theorem 2.2] that if A is a Noetherian domain and $\mathcal{R}(A)$ is a finitely generated A-module, then $\mathcal{R}^S(A)$ is obtainable in finitely many steps; in particular, this holds if A is affine. However, it was conjectured that a Noetherian domain may need infinitely many steps. This conjecture was established by Roitman [46].

THEOREM 5.2 ([46, Theorem 2.11]) Let S be a nonempty subset of $\mathbb{N}$. For each positive integer d, there is a d-dimensional Noetherian domain A such that $\mathcal{R}^S(A)$ is not obtainable in finitely many steps. $\square$

The next result generalizes Theorem 3.9 and Corollary 3.11 (cf. [46, page 421] and [37, Section 3]).

PROPOSITION 5.3 Let S be a nonempty subset of $\mathbb{N}$.

   (a) Let A be a subring of a ring B and let T be a submonoid of a torsionfree commutative cancellative monoid W. Then $\mathcal{R}^S(A[T], B[W]) = \mathcal{R}^S(A,B)[T^*]$, where $T^* = \{w \in W \mid nw \in T$ for some $n \in S\}$.

   (b) Let A be a reduced ring and let T be a torsionfree commutative cancellative monoid with $T^* = \{t \in \langle T \rangle \mid nt \in T$ for some $n \in S\}$. Then $\mathcal{R}^S(A[T]) = \mathcal{R}^S(A, Q_o(A))[T^*]$. $\square$

The following result from [10] is a consequence of [26] and [14, Proposition 2.11]. Related results are also given in [12].

PROPOSITION 5.4 ([10, Proposition 3.5]) The following conditions are equivalent for an integral domain A.

1. Each subring of A is integrally closed.
2. Each subring of A is root closed.
3. Each subring R of A is n_R-root closed for some integer $n_R > 1$.
4. Each subring of A is seminormal.
5. Either A is isomorphic to an overring of $\mathbb{Z}$ or A is an algebraic extension of a finite field. $\qquad\square$

Let A be a subring of an integral domain B. We next consider when each subring of B containing A is n-root closed for some $n \geq 2$ (the seminormal case was investigated in [14]). By [2, Corollary 2.9] (cf. [14], [31, Theorem 1.2]), if each ring between A and B is seminormal, then either A is a field or B is an overring of A. The overring case is characterized in our next result.

PROPOSITION 5.5 ([10, Theorem 3.3]) Let A be an integral domain with quotient field K. Then every overring of A is n-root closed for some fixed integer $n > 1$ if and only if each integral overring of A is n-root closed and A', the integral closure of A, is a Prüfer domain. $\qquad\square$

6 REALIZING $\mathcal{C}(A)$

We have already observed that $\mathcal{C}(A)$ is a multiplicative submonoid of $\mathbb{N}$ generated by primes. It is natural to ask if conversely each such submonoid of $\mathbb{N}$ can be realized as $\mathcal{C}(A)$ for some integral domain A. This was first proved in [10] using a monoid domain construction (cf. Corollary 3.10).

THEOREM 6.1 ([10, Theorem 2.7]) Let S be a multiplicative submonoid of $\mathbb{N}$ generated by primes. Then there is an integral domain A such that $\mathcal{C}(A) = S$. $\qquad\square$

The domain A constructed in Theorem 6.1 is generally quite large. In [11], an elementary construction was used to show that if the monoid S is generated by odd primes, then there is a one-dimensional seminormal (affine or local) Noetherian domain A with $\mathcal{C}(A) = S$. Specifically, for K a field and $A = K[X^2 - X, X^3 - X^2], \mathcal{C}(A) = \{n \geq 1 \mid 1$ is the only nth root of unity in $K \}$. In [13, Theorem 3.3], a $D + M$ construction (cf. Corollary 2.5) was used to reduce the problem to finding fields $k \subset K$ with $\mathcal{C}(k, K) = S$, and thus show that the domain A in Theorem 6.1 can be one-dimensional quasilocal, and Noetherian in certain cases (see Example 7.1(d)). That the domain in Theorem 6.1 can always be chosen to be Noetherian was demonstrated by Roitman [46] using the same construction as in Proposition 5.2.

THEOREM 6.2 ([46, Proposition 2.9 and Theorem 3.11]) For each positive integer d, there is a d-dimensional Noetherian domain A such that for any multiplicative submonoid S of $\mathbb{N}$ generated by primes, $\mathcal{R}^S(A)$ is Noetherian and $\mathcal{C}(\mathcal{R}^S(A)) = S$. $\qquad\square$

Theorem 6.1 may be generalized in the following way. For an ideal I of a ring A, define I to be *n-root closed* if $b^n \in I$ with $b \in T(A)$ implies $b \in I$; and set $\mathcal{C}(I) = \{n \in \mathbb{N} \mid I \text{ is } n\text{-root closed}\}$. Then $\mathcal{C}(I)$ is a multiplicative submonoid of $\mathbb{N}$ generated by primes, and $\mathcal{C}(A) \subset \mathcal{C}(I)$ if I is a radical ideal of A. The construction in [10] can be localized at a suitable maximal ideal M to give $\mathcal{C}(A_M) = \mathcal{C}(M_M) = S$. For the $D + M$ example in [13], $\mathcal{C}(A) = S \subset \mathcal{C}(M) = \mathbb{N}$. In [17], the above two-mentioned constructions were combined.

THEOREM 6.3 ([17, Theorem 2.16]) Let $S \subset T$ be multiplicative submonoids of $\mathbb{N}$ generated by primes. Then there is a quasilocal integral domain A with maximal ideal M such that $S = \mathcal{C}(A) \subset \mathcal{C}(M) = T$. $\square$

We close with a realization theorem for nondomains (cf. Corollary 4.7).

THEOREM 6.4 ([13, Theorem 3.5]) Let S be a multiplicative submonoid of $\mathbb{N}$ generated by primes. Then there is a von Neumann regular (and hence root closed) ring A with $\mathcal{C}(A[[X]]) = S$. $\square$

7   EXAMPLES AND RELATED CLOSEDNESS CONDITIONS

We first give several examples.

EXAMPLES 7.1 (a) (A is seminormal, but not n-root closed for any $n \geq 2$ [10, Example 2.3].) Let $A = \mathbb{R} + X\mathbb{C}[[X]]$. Then A is a one-dimensional local Noetherian domain with $\mathcal{C}_0(A) = \mathcal{C}_0(\mathbb{R}, \mathbb{C}) = \{0, 1\}$ by Corollary 2.5.

(b) (A is root closed, but not integrally closed.) Let L be the algebraic closure of $\mathbb{Q}$ and let k be the subfield of L consisting of all elements β of L such that the minimal polynomial for β over $\mathbb{Q}$ is solvable by radicals over $\mathbb{Q}$. Choose $\alpha \in L$, but not in k, and let $F = k(\alpha)$. Then $A = k + XF[[X]]$ is root closed by Corollary 2.5. However, A is not integrally closed since α is integral over A, but not in A. Finally, A is a one-dimensional local Noetherian domain. This example from [10, Example 2.2] is a modification of [27, Exercise 6, page 184]. (A similar example is given in [25, Remark 2.7(c)].) For a simpler, but not quasilocal, example $\mathbb{Z}[\sqrt{d}]$ is root closed, but not integrally closed, if and only if $d \equiv 1 \pmod 8$ [18, Proposition] (also, see [51]).

(c) (A is integrally closed, but not completely integrally closed.) Let $A = \mathbb{Q} + X\mathbb{Q}(t)[[X]]$, where t is an indeterminate. Then A is a one-dimensional quasilocal domain which is integrally closed, but not completely integrally closed. Such a domain cannot be Noetherian.

(d) (For any set $\mathcal{P}$ of primes, a one-dimensional quasilocal integral domain A with $\mathcal{C}(A) = \langle \mathcal{P} \rangle$.) Let $S = \langle \mathcal{P} \rangle$. Define inductively an increasing sequence of subfields of $\mathbb{R}$ by $K_0 = \mathbb{Q}$ and $K_{n+1} = K_n(\{x \in \mathbb{R} \mid x^p \in K_n \text{ for some } p \in \mathcal{P}\})$. Let $K_S = \cup K_n$. In particular, $K_S = \mathbb{Q}$ if $\mathcal{P}$ is empty. Then K_S is n-root closed in $\mathbb{R}$ if and only if $n \in S$ [13, Lemma 3.2]. Let T be an indeterminate and $A = K_S + T\mathbb{R}[[T]]$. Then $\mathcal{C}(A) = \mathcal{C}(A[[X]]) = S$ for any family of indeterminates $\mathbf{X}$ by Corollary 2.5 and [16, Proposition 1.30]. This construction is from [13]. $\square$

Root closedness was first studied as a weaker condition than being integrally closed. In fact, in some cases a root closed ring is necessarily integrally closed (cf. [10, Corollary 2.6], [18], [19], and [23, Theorem 3]). The term "root closed" seems to have been introduced by Sheldon in his work [49] on how changing $A[[X]]$ changes its quotient field. Brewer, Costa, and McCrimmon [23] were the first to study root closedness in its own right, where in particular, they showed that $C_0(A[X], B[X]) = C_0(A, B)$ and $C_0(A[X]) = C_0(A)$ (when $T(A)$ is von Neumann regular). This investigation was continued by the author in [10], where the $C(A)$ notation was intoduced, and by Angermüller in [18] and [19].

Much of the early interest in root closedness was centered on its connections with seminormality. Seminormality was mainly of interest because for an integral domain or reduced Noetherian ring A, $Pic(A) = Pic(A[X])$ if and only if A is seminormal. This result is due to Gilmer and Heitmann [29] and extends earlier work of Traverso [52]. The above result does not hold for an arbitrary reduced ring. In [50], Swan gave an "internal" definition of seminormality which agrees with our definition whenever $T(A)$ is a product of fields (but not in general). He defined a (reduced) commutative ring A to be *seminormal* if whenever $b, c \in A$ satisfy $b^3 = c^2$, there is an $a \in A$ with $a^2 = b$ and $a^3 = c$ (Costa has observed that such an A is necessarily reduced). Then $Pic(A) = Pic(A[X_1, \ldots, X_n])$ for (some) all $n \geq 1$ if and only if $A/nil(A)$ is seminormal [50, Theorem 6.1]. These results have been extended to graded domains in [9] and [4]. Brewer and Nichols [24, Theorem] showed that $A[[\mathbf{X}]]$ is seminormal (Swan's definition) if and only if A is seminormal (cf. Theorem 4.1).

Several other closedness properties have been investigated with respect to the property that $Pic(A) = Pic(A[X, X^{-1}])$, such a ring A is called *quasinormal*. Clearly a quasinormal ring is seminormal. In [11], it was shown that quasinormality neither implies nor is implied by root closed. Let A be a subring of an integral domain B. Then A is *u-closed* (resp., *t-closed*) in B if $t^2 - t, t^3 - t^2 \in A$ with $t \in B$ (resp., $t^2 - at, t^3 - at^2 \in A$ for some $a \in A$) implies $t \in A$. Clearly t-closed implies u-closed. These properties have been studied in [31], [40], [41], [42], and [43]. In [41], the u-closed notion and the result that if $dim A = 1$, then A is quasinormal if and only if A is seminormal and u-closed [41, Corollary 1.14] are attributed to Asanuma. In [54], Weibel computed $Pic(A[X, X^{-1}])$ for any commutative ring A and showed that Asanuma's result does not extend to rings with $dim A \geq 2$ (also, see [20]).

REFERENCES

1. T. Akiba, Integrally-closedness of polynomial rings, Japan. J. Math. 6(1980), 67-75.
2. T. Akiba, Remarks on flat and relatively seminormal pairs, Kobe J. Math. 6(1989), 217-222.
3. D. D. Anderson and D. F. Anderson, Divisibility properties of graded domains, Canad. J. Math. 34(1982), 196-215.
4. D. D. Anderson and D. F. Anderson, Divisorial ideals and invertible ideals in a graded integral domain, J. Algebra 76(1982), 549-569.

5. D. D. Anderson and D. F. Anderson, Multiplicatively closed subsets of fields, Houston J. Math. 13(1987), 1-11.

6. D. D. Anderson, D. F. Anderson, and R. Markanda, The rings $R(X)$ and $R\langle X\rangle$, J. Algebra 95(1985), 96-115.

7. D. D. Anderson, D. F. Anderson, and M. Zafrullah, Factorization in integral domains, II, J. Algebra 152(1992), 78-93.

8. D. D. Anderson, D. F. Anderson, and M. Zafrullah, Rings between $D[X]$ and $K[X]$, Houston J. Math. 17(1991), 109-129.

9. D. F. Anderson, Seminormal graded rings, J. Pure Appl. Algebra 21(1981), 1-7.

10. D. F. Anderson, Root closure in integral domains, J. Algebra 79(1982), 51-59.

11. D. F. Anderson, Root closure in integral domains, II, Glasgow Math. J. 31(1989), 127-130.

12. D. F. Anderson and D. E. Dobbs, Fields in which seminormality implies normality, Houston J. Math. 16(1990), 231-247.

13. D. F. Anderson and D. E. Dobbs, Root closure in integral domains, III, Canad. Math. Bull. 41(1998), 3-9.

14. D. F. Anderson, D. E. Dobbs, and J. A. Huckaba, On seminormal overrings, Comm. Algebra 10(1982), 1421-1448.

15. D. F. Anderson, D. E. Dobbs, and M. Roitman, Root closure in commutative rings, Ann. Sci. Math. Univ. Clermont II, Sér. Math. 26(1990), 1-11.

16. D. F. Anderson, D. E. Dobbs, and M. Roitman, When is a power series ring n-root closed?, J. Pure Appl. Algebra 114(1997), 111-131.

17. D. F. Anderson and J. Park, Rooty and root closed domains, these proceedings.

18. G. Angermüller, On the root and integral closure of noetherian domains of dimension one, J. Algebra 83(1983), 437-441.

19. G. Angermüller, Root closure, J. Algebra 90(1984), 189-197.

20. T. Asanuma, $Pic(R[X, X^{-1}])$ for R a one-dimensional reduced noetherian ring, J. Pure Appl. Algebra 71(1991), 111-128.

21. J. W. Brewer, *Power Series over Commutative Rings*, Lecture Notes in Pure and Applied Mathematics, Marcel Dekker 64(1981).

22. J. W. Brewer and D. L. Costa, Seminormality and projective modules over polynomial rings, J. Algebra 58(1979), 208-216.

23. J. W. Brewer, D. L. Costa, and K. McCrimmon, Seminormality and root closure in polynomial rings and algebraic curves, J. Algebra 58(1979), 217-226.

24. J. W. Brewer and W. D. Nichols, Seminormality in power series rings, J. Algebra 82(1983), 282-284.

25. D. E. Dobbs, Divided rings and going down, Pacific J. Math. 67(1976), 353-363

26. R. Gilmer, Rings with integrally closed subrings, Math. Japon. 16(1971), 9-11.

27. R. Gilmer, *Multiplicative Ideal Theory*, Marcel Dekker, New York, 1972.

28. R. Gilmer, *Commutative Semigroup Rings*, Chicago Lectures in Mathematics, University of Chicago Press, Chicago, 1984.

29. R. Gilmer and R. C. Heitman, On Pic($R[X]$) for R seminormal, J. Pure Appl. Algebra 16(1980), 251-257.

30. M. Ikeda and T. Nakayama, On some characteristic properties of quasi-Frobenius and regular rings, Proc. Amer. Math Soc. 5(1954), 15-19.

31. M. Kanemitsu and R. Matsuda, Note on seminormal overrings, Houston J. Math. 22(1966), 217-224.

32. J. Lambek, *Rings and Modules*, Chelsea, New York, 1986.

33. D. Läseke, Wurzelabgeschlossenheit von Integritätsringen, Diplomarbeit, Institut für Mathematik der Universität Hannover, 1995.

34. T. G. Lucas, Characterizing when $R[X]$ is integrally closed, Proc. Amer. Math. Soc. 105(1989), 861-867.

35. T. G. Lucas, Charcacterizing when $R[X]$ is integrally closed, II, J. Pure Appl. Algebra 61(1989), 49-52.

36. T. G. Lucas, Root closure and $R[X]$, Comm. Algebra 17(1989), 2395-2414.

37. T. G. Lucas, The complete integral closure of R[X], Trans. Amer. Math. Soc. 330(1992), 757-768.

38. R. Yue Chi Ming, On (von Neumann) regular rings, Proc. Edinburgh Math. Soc. (2) 19(1974-1975), 89-91.

39. J. Ohm, Some counterexamples related to integral closure in $D[[X]]$, Trans. Amer. Math. Soc. 122(1966), 321-333.

40. N. Onoda, T. Sugatani, and K. Yoshida, Local quasinormality and closedness type criteria, Houston J. Math. 11(1985), 247-256.

41. N. Onoda and K. Yoshida, Remarks on quasinormal rings, J. Pure Appl. Algebra 33(1984), 59-67.

42. G. Picavet and M. Picavet-L'Hermite, Anneaux t-clos, Comm. Algebra 23(1995), 2643-2677.

43. M. Picavet-L'Hermite, t-closed pairs, in: Commutative Ring Theory, Lecture Notes in Pure and Applied Mathematics, Marcel Dekker 185(1997), 401-415.

44. P. Ribenboim, Special properties of generalized power series, J. Algbera 173(1995), 566-586.

45. M. Roitman, On the complete integral closure of a Mori domain, J. Pure Appl. Algebra 66(1990), 55-79.

46. M. Roitman, On root closure in Noetherian domains, in: Factorization in Integral Domains, Lecture Notes in Pure and Applied Mathematics, Marcel Dekker 189(1997), 417-428.

47. D. E. Rush, Seminormality, J. Algebra 67(1980), 377-384.

48. A. Seidenberg, Derivations and integral closure, Pacific J. Math. 16(1966), 167-173.

49. P. B. Sheldon, How changing $D[[X]]$ changes its quotient field, Trans. Amer. Math. Soc. 159(1971), 223-244.

50. R. G. Swan, On seminormality, J. Algebra 67(1980), 210-229.

51. H. Tanimoto, Root closedness of $\mathbb{Z}[\sqrt[n]{m}]$, Comm. Algebra 17(1989), 2101-2108.

52. C. Traverso, Seminormality and Picard group, Ann. Scuola Norm. Sup. Pisa 24(1970), 585-595.

53. J. J. Watkins, Root and integral closure for $R[[X]]$, J. Algebra 75(1982), 43-58.

54. C. Weibel, Pic is a contracted functor, Invent. Math. 103(1991), 351-377.

55. L. Zhongkui, On n-root closedness of generalized power series rings over pairs of rings, to appear in J. Pure Appl. Algebra.

On the Class Group of $A + XB[X]$ Domains

DAVID F. ANDERSON   Department of Mathematics, University of Tennessee, Knoxville, Tennessee 37996-1300

SAID EL BAGHDADI   Department of Mathematics, FST, Beni Mellal, Morocco

SALAH-EDDINE KABBAJ (*)   Department of Mathematical Sciences, KFUPM, P.O.Box 849, Dhahran 31261, Saudi Arabia

INTRODUCTION

All the rings considered in this paper are integral domains, and all modules and ring homomorphisms are unital. In this paper, we deal with the class group (see definition below) of $A + XB[X]$ domains. That is, let $A \subset B$ be an extension of integral domains. Then $A + XB[X]$ is a subring of the polynomial ring $B[X]$. This construction has been studied by many authors and has proven to be useful in constructing interesting examples and counterexamples, see for instance [2], [3], [5], [9], and [10].

If D is an integral domain, two well-known results on polynomial rings are that $Pic(D[X]) = Pic(D)$ if and only if D is seminormal, and $Cl(D[X]) = Cl(D)$ if and only if D is integrally closed (cf. [14] and [12], respectively). In [2], it is shown that $Pic(A + XB[X]) = Pic(A)$ if and only if B is seminormal. The purpose of this work is to study the question of when $Cl(A+XB[X]) = Cl(A)$, paying particular attention to the case where B is integrally closed. Namely, Theorem 4.4 establishes

---

(*) Supported by KFUPM

that if B is integrally closed and a flat overring of A, then $Cl(A + XB[X])$ is canonically isomorphic to $Cl(A)$. We also show that if B is integrally closed, then this isomorphism holds in the cases $qf(A) \subset B$ or $B = A[\mathbf{Y}]$, where $\mathbf{Y}$ is a set of indeterminates. Theorem 4.10 allows us to construct explicit examples showing that this canonical isomorphism does not hold in general even if B is integrally closed.

In this paper, $A \subset B$ is an extension of integral domains and K is the quotient field of B. Let D be an integral domain with quotient field $qf(D) = k$. By an ideal of D we mean an integral ideal of D. Given a nonzero fractional ideal I of D, we define $I^{-1} = \{x \in k \mid xI \subset D\}$ and $I_v = (I^{-1})^{-1}$. We say that I is divisorial or a v-ideal if $I_v = I$; while I is v-finite if $I = J_v$ for some finitely generated fractional ideal J of D. For I a nonzero fractional ideal of D, we define $I_t = \cup\{J_v \mid J \subset I \text{ finitely generated}\}$. Then I is a t-ideal if $I_t = I$. The mappings $I \mapsto I_v$ and $I \mapsto I_t$ are particular star-operations on fractional ideals of D, see [13, Sections 32 and 34] for a general theory. As in [7] and [8], we define the class group of D, $Cl(D)$, to be the group of t-invertible (fractional) t-ideals of D modulo the subgroup of principal ideals of D. If D is a Krull domain, then $Cl(D)$ is the usual divisor class group of D, see [11]. In this case, $Cl(D) = 0$ if and only if D is factorial.

This paper consists of four sections in addition to the introduction. In Sections 1 and 2, we state basic results on divisorial ideals and t-invertibility in $A + XB[X]$ domains. Section 3 establishes necessary and sufficient conditions for a v-invertible v-ideal or a t-invertible t-ideal of $A + XB[X]$ to be extended from A (see definition below). In Section 4, we give the proofs of the theorems mentioned above.

1. DIVISORIAL IDEALS IN $A + XB[X]$

Let $R = A + XB[X]$. In what follows, we consider the natural grading on R, that is, $R = \bigoplus_{n \geq 0} R_n$, where $R_0 = A$ and $R_n = X^n B$ for $n \geq 1$. An element (resp., an ideal) of R is said to be homogeneous if it is homogeneous with respect to this grading. If f is a polynomial over an integral domain A, we denote by A_f the content of f.

LEMMA 1.1   Let $R = A + XB[X]$ with B integrally closed. Let I be a homogeneous divisorial ideal of R, J the ideal of B generated by the coefficients of all polynomials of I, n the least integer k such that $aX^k \in I$ for some nonzero $a \in B$, and $W \subset J$ the A-module generated by all $a \in B$ such that $aX^n \in I$. Then J is a divisorial ideal of B and $I = X^n W + X^{n+1} J[X]$.

Proof. Since I is homogeneous, it is easy to see that $I \subset X^n W + X^{n+1} J[X]$. Conversely, since I is divisorial and $X^n W \subset I$, it suffices to show that $X^{n+1} J_v[X] \subset \frac{f}{g} R$ for each $f, g \in R$ such that $I \subset \frac{f}{g} R$. Choose a nonzero $a \in W$. Let $f, g \in R$ such that $I \subset \frac{f}{g} R$. Since $aX^n \in I$, we have $\frac{f}{g} = \frac{aX^n}{r}$ for some $r \in R$. Hence we can assume $f = aX^n$. Let $0 \neq h \in I$. Then $h \in \frac{aX^n}{g} R$, and hence $gh \in aX^n R \subset aB[X]$ and $A_{gh} \subset aB$. Since B is integrally closed, by [15, Lemme 1, Sect.2], $(A_g A_h)_v = (A_{gh})_v$. Hence $A_g A_h \subset aB$, so that $gA_h[X] \subset aB[X]$. Then $gJ[X] \subset aB[X]$, and by taking the v-closure, we get $gJ_v[X] \subset aB[X]$ [12, Lemma 1.6]. Since $XB[X] \subset R$, then $X^{n+1} J_v[X] \subset \frac{aX^n}{g} R$; whence $X^{n+1} J_v[X] \subset I$, as desired.

LEMMA 1.2 Let $R = A + XB[X]$ with B integrally closed. Then for each divisorial ideal I of R, there exist $u \in K[X, X^{-1}]$ and J a homogeneous divisorial ideal of R such that $I = uJ$.

Proof. Let S be the (multiplicatively closed) set of nonzero homogeneous elements of R. We have $R_S = K[X, X^{-1}]$. Let I be a divisorial ideal of R. Then $IR_S = fR_S$ for some $f \in B[X]$ such that $f(0) \neq 0$. Hence $I \subset fK[X]$. Since B is integrally closed, by [15, Section 2, Lemme 1], $fK[X] \cap B[X] = fA_f^{-1}[X]$; hence $I \subset fA_f^{-1}[X]$. Let $0 \neq b \in A_f$ and set $J = bXf^{-1}I$. Clearly, J is a divisorial ideal of R. We next show that J is homogeneous. Let $f, g \in R$ such that $J \subset \frac{f}{g} R$. Since $J \cap S \neq \emptyset$, and as in the proof of Lemma 1.1, we can assume that $f = aX^m$ for some nonzero element a of B and some integer $m \geq 0$. Now let $0 \neq h \in J$. Then $h \in \frac{aX^m}{g} R$, and hence $gh \in aX^m R \subset aB[X]$. Then $A_g A_h \subset aB$ by an argument similar to that in the proof of Lemma 1.1. On the other hand, $gh \in aX^m R$ implies that $g = X^r g_1$ and $h = X^s h_1$ with $r + s \geq m$ and $g_1(0)h_1(0) \neq 0$. We have $A_g = A_{g_1}$ and $A_h = A_{h_1}$, so $A_{g_1} A_{h_1} \subset aB$, and thus $g_1 A_{h_1}[X] \subset aB[X]$. Therefore $X^{s+1} A_{h_1}[X] \subset J$. It follows that the homogeneous components of h are in J, which proves that J is homogeneous.

REMARK 1.3 Lemma 1.2 can be generalized to N-graded domains by using other techniques. Let $R = \bigoplus_{n \geq 0} R_n$ be an N-graded integral domain and S be the (multiplicatively closed) set of nonzero homogeneous elements of R. Then R_S is a Z-graded domain. In [1], the authors define a graded domain R to be almost normal if it is integrally closed with respect to nonzero homogeneous elements of R_S of nonzero degree. They showed [1, Corollary 3.8] that the following statements are equivalent:

(i) R is almost normal.

(ii) For each v-ideal I of R, $I = uJ$ for some $u \in R_S$ and some homogeneous v-ideal J of R.

If $R = A + XB[X]$ is graded in the natural way, it is not difficult to show that R is almost normal if and only if B is integrally closed.

By Lemma 1.1 and Remark 1.3, we have the following theorem.

THEOREM 1.4   Let $R = A + XB[X]$. The following statements are equivalent.
(1) B is integrally closed.
(2) For each v–ideal I of R, $I = u(W + XJ[X])$ for some $u \in K[X, X^{-1}]$, J a v–ideal of B, and $W \subset J$ a nonzero A-module.

2. v–INVERTIBLE IDEALS AND t–INVERTIBLE IDEALS IN $A + XB[X]$

LEMMA 2.1   Let $R = A + XB[X]$. Let F_1 (resp., F_2) be a nonzero fractional ideal of A (resp., B) such that $F_1 \subset F_2$. Then $F_1 + XF_2[X]$ is a fractional ideal of R, and we have $(F_1 + XF_2[X])^{-1} = F_1^{-1} \cap F_2^{-1} + XF_2^{-1}[X]$.

Proof.   It is obvious that $I = F_1 + XF_2[X]$ is a fractional ideal of R. Now since $F_1 \subset I$, if $u \in I^{-1}$, then $u \in K[X]$. Thus if $u \in K[X]$, then $u \in I^{-1}$ if and only if $u(0)F_1 \subset A$ and $uF_2[X] \subset B[X]$. Hence $u \in I^{-1}$ if and only if $u \in F_1^{-1} \cap F_2^{-1} + XF_2^{-1}[X]$.

LEMMA 2.2   Let $R = A + XB[X]$. Then $XB[X]$ and $B[X]$ are divisorial ideals of R.

Proof.   Let $C(A, B) = \{x \in A \mid xB \subset A\}$. It is easy to see that $[R : B[X]] = C(A, B) + XB[X]$. If $C(A, B) = 0$, then $(B[X])^{-1} = XB[X]$; hence $(B[X])_v = B[X]$. If $C(A, B) \neq 0$, by Lemma 2.1, $(B[X])_v = (C(A, B)^{-1} \cap B) + XB[X] = B[X]$. Hence $B[X]$ and $XB[X]$ are divisorial ideals of R.

THEOREM 2.3   Let $R = A + XB[X]$ with B integrally closed. If I is a fractional v–invertible v–ideal, then $I = u(J_1 + XJ_2[X])$ for some $u \in qf(R)$, J_2 a v–invertible v–ideal of B, and $J_1 \subset J_2$ a nonzero ideal of A.

Proof.   By Theorem 1.4, we can assume that $I = W + XJ[X]$ for some v–ideal J of B and $W \subset J$ a nonzero A-module. First, we show that there exists nonzero $c \in K$ such that $cW \subset A$ and $cJ \subset B$. Let $a \in W$ be a nonzero element. Then one can easily show that $aI^{-1} \subset R$ satisfies the hypothesis of Lemma 1.1. Thus there exist an integer m, J' a divisorial ideal of B, and $W' \subset J'$ a nonzero A-module such that

$$aI^{-1} = X^m W' + X^{m+1} J'[X].$$

Since I is v−invertible,

$$aR = X^m \big((W + XJ[X])(W' + XJ'[X]) \big)_v.$$

On the other hand, we have $(W + XJ[X])(W' + XJ'[X]) \subset B[X]$. Further, since $B[X]$ is divisorial, $aR \subset X^m B[X]$. Hence $m = 0$ and

$$aR = \big((W + XJ[X])(W' + XJ'[X]) \big)_v.$$

Thus $a^{-1}WW' \subset A$ and $a^{-1}JJ' \subset B$. Let $c \in a^{-1}W'$ be a nonzero element. Then $J_1 = cW \subset A$ and $J_2 = cJ \subset B$. Hence there exist $J_2 \subset B$ a divisorial ideal of B and $J_1 \subset J_2$ a nonzero ideal of A such that $I = u(J_1 + XJ_2[X])$ for some $u \in qf(R)$. It remains to show that J_2 is v−invertible. By Lemma 2.1, we have

$$I^{-1} = u^{-1}(J_1^{-1} \cap J_2^{-1} + XJ_2^{-1}[X]).$$

Hence, $II^{-1} \subset J_1(J_1^{-1} \cap J_2^{-1}) + XJ_2J_2^{-1}[X] \subset R$, and since I is v−invertible, we have

$$\big(J_1(J_1^{-1} \cap J_2^{-1}) + XJ_2J_2^{-1}[X] \big)^{-1} = R.$$

By applying Lemma 2.1, we conclude that $(J_2J_2^{-1})^{-1} = B$. Hence J_2 is a v−invertible v−ideal of B.

COROLLARY 2.4 Let $R = A + XB[X]$ with B integrally closed. If I is a fractional t−invertible t−ideal, then $I = u(J_1 + XJ_2[X])$ for some $u \in qf(R)$, J_2 a t−invertible t−ideal of B and $J_1 \subset J_2$ a nonzero ideal of A.

Proof. It remains to show that J_2 and J_2^{-1}, from Theorem 2.3, are v−finite. Since I is a t−invertible t−ideal, then $J_1 + XJ_2[X] = (f_1, \ldots, f_n)_v$ for some $f_1, \ldots, f_n \in R$. Thus there exists $F_1 \subset J_1$ (resp., $F_2 \subset J_2$) a finitely generated ideal of A (resp., B) such that $F_1 \subset F_2$ and $f_1, \ldots, f_n \in F_1 + XF_2[X]$. Hence $J_1 + XJ_2[X] = (F_1 + XF_2[X])_v$. Applying Lemma 2.1 yields $J_2^{-1} = F_2^{-1}$; hence J_2 is v−finite. Similarly, one shows that J_2^{-1} is also v−finite by using the fact that I^{-1} is v−finite.

3. t−INVERTIBLE IDEALS OF $A + XB[X]$ EXTENDED FROM A

A fractional ideal I of $R = A + XB[X]$ is said to be extended from A if $I = uJR$ for some $u \in qf(R)$ and some ideal J of A.

LEMMA 3.1 Let $R = A + XB[X]$ with B integrally closed and I be a fractional divisorial ideal of R. Then the following statements are equivalent.

(1) There exist $u \in qf(R)$ and W a nonzero A-module ($\subset B$) such that $I = uWR$.
(2) $IB[X]$ is a divisorial ideal of $B[X]$.

Proof. (1) $\Rightarrow$ (2). We can assume that $I = XWR$; hence $I = XW + X^2WB[X]$. By Lemma 1.1, WB is a divisorial ideal of B; so $IB[X] = XWB[X]$ is divisorial in $B[X]$.
(2) $\Rightarrow$ (1) By Theorem 1.4, $I = u(W + XJ[X])$, where $u \in K[X, X^{-1}]$, J is a divisorial ideal of B and $W \subset J$ is a nonzero A-module. Let $I_1 = u^{-1}IB[X]$. Then we have $I_1 = WB + XJ[X]$. Applying Lemma 2.1 to I_1 in the case $A = B$, we get $(I_1)_v = J[X]$. Further, since I_1 is divisorial in $B[X]$, then $J = WB$. Therefore $I = uWR$.

REMARK 3.2 If B is integrally closed, then divisorial ideals of R are not always of the form uWR, where $u \in qf(R)$ and $W \subset B$ is a nonzero A-module. For, let $A = Z$ and $B = Z[i]$. Let's consider the ideal $I = 2Z + (1 + i)XZ[i][X]$ of $R = Z + XZ[i][X]$. By applying Lemma 2.1, one can easily show that I is a divisorial ideal. Notice that I is also a t−invertible t−ideal (see Remark 4.15 and Example 4.16). Now assume $I = uWR$. Then $u \in Q(i)$, so $uW = 2Z$ and $uWZ[i] = (1 + i)Z[i]$. Hence $2Z[i] = (1 + i)Z[i]$, a contradiction.

LEMMA 3.3 Let $R = A + XB[X]$. Let I be a divisorial ideal of R of the form $I = J_1 + XJ_2[X]$, where J_2 is an ideal of B and $J_1 \subset J_2$ is a nonzero ideal of A. Then the following statements are equivalent.
(1) I is extended from A.
(2) $J_2 = J_1B$.
(3) $IB[X]$ is divisorial in $B[X]$.

Proof. (1) $\Rightarrow$ (2) We assume that $I = uJR$ for some $u \in qf(R)$ and some ideal J of A. Since $I \cap A \neq 0$, $u \in qf(A)$. It follows that $J_1 = uJ$ and $J_2 = uJB$. Hence $J_2 = J_1B$.
(2) $\Rightarrow$ (1) Clear.
(2) $\Leftrightarrow$ (3) Notice that by Lemma 2.1, J_2 is necessarily a divisorial ideal of B. We have $IB[X] = J_1B + XJ_2[X]$, and applying Lemma 2.1 in the case where $A = B$ yields $(IB[X])_v = J_2[X]$. Therefore $IB[X]$ is divisorial if and only if $J_2 = J_1B$.

THEOREM 3.4 Let $R = A + XB[X]$ with B integrally closed. Let I be a fractional v−invertible v−ideal of R. Then the following statements are equivalent.
(1) I is extended from A.
(2) $IB[X]$ is a divisorial ideal of $B[X]$.

Proof. It follows from Theorem 2.3 and Lemma 3.3.

REMARK 3.5 The implication (2) $\Rightarrow$ (1) in Theorem 3.4 is not true in general if I is a v−ideal which is not v−invertible. To see this, let A and B be such that

$C(A, B) = 0$ (see Lemma 2.2), and consider the fractional ideal $I = B[X]$ of R. By Lemma 2.2, I is a divisorial ideal of R, but it is not $v-$invertible in R since $I^{-1} = XB[X]$ and $(II^{-1})_v = XB[X]$. Note that $IB[X] = B[X]$ is divisorial in $B[X]$. If $I = uJR$ for some $u \in qf(R)$ and some ideal J of A, then $u \in K$. Hence $B = uJ$, and thus $u^{-1}B = J \subset A$. Hence $u^{-1} \in C(A, B)$, a contradiction. Note that in this case, the implication $(2) \Rightarrow (1)$ in Lemma 3.1 is true, namely $W = B$.

LEMMA 3.6   Let $R = A + XB[X]$. Then R is a flat A-module if and only if B is a flat A-module.

Proof.   Just note that $R = A \oplus \bigoplus_{n \geq 1} X^n B$.

LEMMA 3.7   Let $S \subset T$ be an extension of integral domains such that T is a flat S-module. If I is a finitely generated ideal of S, then $(IT)^{-1} = I^{-1}T$

Proof.   See for instance [6, Alg. Comm., Chap.1].

LEMMA 3.8   Let $R = A + XB[X]$ and J be an ideal of A.
(1) If $(JR)_v = R$, then $J_v = A$.
(2) If $(JR)_t = R$, then $J_t = A$.

Proof.   (1) Assume $(JR)_v = R$. Let $u \in qf(A)$ such that $J \subset uA$. Then $JR \subset uR$, and hence $R = (JR)_v \subset uR$. Thus $1 \in uA$ and $J_v = A$. (2) is a consequence of (1) since $(JR)_t = \cup\{(FR)_v \mid F \subset J$ finitely generated $\}$.

PROPOSITION 3.9   Let $R = A + XB[X]$ such that B is a flat A-module. Let J be an ideal of A. Then the following statements are equivalent.
(1) J is a $t-$invertible $t-$ideal of A.
(2) JR is a $t-$invertible $t-$ideal of R.

Proof.   If B is a flat A-module, then R is a flat A-module by Lemma 3.6, and by [4, Prop. 2.2], we have $(1) \Rightarrow (2)$.
$(2) \Rightarrow (1)$ Assume that $I = JR$ is a $t-$invertible $t-$ideal of R. Then $J = I \cap A$ is a $t-$ideal. To see this, let $F \subset J$ be a finitely generated ideal of A. By using the formula $(F_v R)_v = (FR)_v$ which is a consequence of Lemma 3.7 (see [4, Prop. 2.2], we conclude that $F_v \subset J$. On the other hand, since JR is $t-$invertible, there exists $J_1 \subset J$ a finitely generated ideal of A such that $JR = (J_1 R)_v$. Thus $(JJ_1^{-1}R)_t = ((JR)(JR)^{-1})_t = R$, and by Lemma 3.8, $(JJ_1^{-1})_t = A$. Hence J is a $t-$invertible $t-$ideal of A.

THEOREM 3.10   Let $R = A + XB[X]$ such that B is integrally closed and a flat A-module. Let I be a fractional $t-$invertible $t-$ideal of R. Then the following statements are equivalent.

(1) $I = uJR$ for some $u \in qf(R)$ and some t−invertible t−ideal J of A.

(2) $IB[X]$ is a divisorial ideal of $B[X]$

Proof. $(1) \Rightarrow (2)$ is a particular case of $(1) \Rightarrow (2)$ in Lemma 3.1. Since t−invertible t−ideals are v−invertible v−ideals, $(2) \Rightarrow (1)$ is a consequence of $(2) \Rightarrow (1)$ of Theorem 3.4 and Proposition 3.9.

4. THE CLASS GROUP OF $A + XB[X]$

LEMMA 4.1 Let S be an integral domain and T an overring of S. Then the following statements are equivalent.

(1) T is a flat S-module.

(2) For each maximal ideal M of T, $T_M = S_{M \cap S}$.

Proof. See [11, Lemma 6.5].

LEMMA 4.2 Let $R = A + XB[X]$ such that B is a flat A-module. Then, $B[X]$ is a flat R-module if and only if B is an overring of A.

Proof. We will use Lemma 4.1. First suppose that $B[X]$ is a flat R-module and let M be a maximal ideal of $B[X]$ such that $XB[X] \subset M$; then $B[X]_M = R_{M \cap R}$. Let $x \in B$. Then $x \in R_{M \cap R}$, and hence $x = \frac{f}{g}$ for some $f, g \in R$ with $g \notin M$. Since $XB[X] \subset M$, $g(0) \neq 0$, so that $x = \frac{f(0)}{g(0)} \in qf(A)$, hence $B \subset qf(A)$.

Conversely, assume that B is an overring of A and let M be a maximal ideal of $B[X]$. We will show that $B[X]_M = R_{M \cap R}$. If $X \in M$, then $M = m + XB[X]$ for some maximal ideal m of B, and we have $M \cap R = (m \cap A) + XB[X]$. Since B is a flat A-module, by Lemma 4.1, $B_m = A_{m \cap A}$, and one can easily verify that $B[X]_M = R_{M \cap R}$. Now if $X \notin M$, let $u \in B[X]_M$. Then $u = \frac{f}{g}$ for some $f, g \in B[X]$ with $g \notin M$; thus $u = \frac{Xf}{Xg} \in R_{M \cap R}$. Hence $B[X]_M \subset R_{M \cap R}$ and $B[X]_M = R_{M \cap R}$.

LEMMA 4.3 Let $R = A + XB[X]$ such that B is a flat A-module. Then the canonical map $\varphi : Cl(A) \to Cl(R)$, $[J] \mapsto [JR]$ is well-defined and it is an injective homomorphism.

Proof. Since B is a flat A-module, by Lemma 3.6, R is a flat A-module. Hence by [4, Prop. 2.2], φ is well-defined and it is a homomorphism. φ is injective since R is a faithfully flat A-module.

THEOREM 4.4 Let $R = A + XB[X]$ such that B is integrally closed and a flat overring of A. Then $Cl(A + XB[X]) \cong Cl(A)$.

Proof. It suffices to show that the canonical homomorphism φ in Lemma 4.3 is surjective. Let I be a t−invertible t−ideal of R. Since B is a flat overring of A, by Lemma 4.2, $B[X]$ is a flat R-module, and thus $IB[X]$ is a t-invertible t-ideal of $B[X]$ [4, Prop. 2.2]. The surjectivity of φ now follows from Theorem 3.10.

COROLLARY 4.5 Let S be a multiplicatively closed subset of A. If A is integrally closed, then $Cl(A + XA_S[X]) \cong Cl(A)$.

For $A = B$, we have the following corollary [12, Theorem 3.6]

COROLLARY 4.6 If A is integrally closed, then $Cl(A[X]) \cong Cl(A)$.

THEOREM 4.7 Let $R = A + XB[X]$. If B is integrally closed and $qf(A) \subset B$, then $Cl(A + XB[X]) \cong Cl(A)$

Proof. Since $qf(A) \subset B$, B is a flat A-module, and hence by Lemma 4.3, it suffices to show that the canonical homomorphism $\varphi : Cl(A) \to Cl(R)$ is surjective. Let I be a t-invertible t-ideal of R. By Corollary 2.4, $I = u(J_1 + XJ_2[X])$, where $u \in qf(R)$, J_2 is a t-invertible t-ideal of B, and $J_1 \subset J_2$ is a nonzero ideal of A. Since $qf(A) \subset B$, $J_2 = B$, and hence $I = uJ_1R$ is extended from A. By Proposition 3.9, J_1 is a t-invertible t-ideal of A, as desired.

COROLLARY 4.8 If B is integrally closed and A is a field, then $Cl(A+XB[X]) = 0$.

THEOREM 4.9 Let $Y = \{Y_i\}_i$ be a set of indeterminates. If A is integrally closed, then $Cl(A + XA[Y][X]) \cong Cl(A)$.

Proof. $B = A[Y]$ is a flat A-module. By Lemma 4.3, it suffices to show that the canonical homomorphism φ is surjective. Let I be a t-invertible t-ideal of R. By Corollary 2.4, we can assume that $I = J_1 + XJ_2[X]$ for some t-invertible t-ideal J_2 of B and $J_1 \subset J_2$ a nonzero ideal of A. Let $J = J_2 \cap A$; $J \neq 0$ since $J_1 \neq 0$. By [12, Lemma 3.3 and Prop. 3.2], J is a t-invertible t-ideal of A and $J_2 = J[Y]$. Hence $I = J_1 + XJ[Y][X]$. By applying Lemma 2.1 and using the fact that $J[Y]^{-1} = J^{-1}[Y]$ ([12, Lemma 1.6]), we obtain

$$\begin{aligned} I^{-1} &= J_1^{-1} \cap J[Y]^{-1} + XJ[Y]^{-1}[X] \\ &= J_1^{-1} \cap J^{-1}[Y] + XJ^{-1}[Y][X] \\ &= J^{-1} + XJ^{-1}[Y][X], \end{aligned}$$

so that $I = (I^{-1})^{-1} = JR$. Hence φ is surjective.

From the above results, the following natural question arises: Assume that B is integrally closed and a flat A-module. Does the canonical isomorphism $Cl(A + XB[X]) \cong Cl(A)$ always hold? The answer is negative, in general, as is shown by the following examples.

THEOREM 4.10   Let A be an integral domain and α an element of some extension domain of A such that $\alpha \notin qf(A)$, $\alpha^2 \in A$, and α is not a unit in $B = A[\alpha]$. Let $R = A + XB[X]$ and $I = \alpha^2 A + \alpha X B[X]$. Then

(1) I is a divisorial ideal of R.

(2) I is a t–invertible t–ideal of R.

(3) I is not extended from A.

Proof.   Since $\alpha^2 \in A$ and $\alpha \notin qf(A)$, then $B = A + A\alpha$ and it is a free A-module with basis $\{1, \alpha\}$.

(1) By applying Lemma 2.1 to the ideal I, we get

$$I^{-1} = \alpha^{-2} A \cap \alpha^{-1} B + \alpha^{-1} X B[X].$$

On the other hand, $A \cap \alpha B = A \cap (A\alpha + A\alpha^2) = A\alpha^2$. Thus $\alpha^{-2} A \cap \alpha^{-1} B = A$, and hence $I^{-1} = A + \alpha^{-1} X B[X]$. Thus

$$\begin{aligned}
I_v &= A \cap \alpha B + \alpha X B[X] \\
&= \alpha^2 A + \alpha X B[X] \\
&= I.
\end{aligned}$$

(2) It suffices to show that I and I^{-1} are v–finite and I is v–invertible. First we show that $I = (\alpha^2, \alpha X)$. It is obvious that $(\alpha^2, \alpha X) \subset I$. For the reverse inclusion, let $f \in I$. We have $f = \alpha^2 a + \alpha X g(X)$ for some $a \in A$ and some $g \in B[X]$. Then g has the form $g = b + c\alpha + X h(X)$, where $b, c \in A$ and $h \in B[X]$. Thus $f = \alpha^2 (a + cX) + \alpha X (b + X h(X))$; hence $f \in (\alpha^2, \alpha X)$. In (1), we have shown that $I^{-1} = A + \alpha^{-1} X B[X]$. Hence $I^{-1} = \alpha^{-2} I = (1, \alpha^{-1} X)$. It remains to show that I is v–invertible. We have

$$I I^{-1} = (\alpha^2, \alpha X)(1, \alpha^{-1} X) = (\alpha^2, \alpha X, X^2).$$

Let $u \in qf(R)$ such that $(\alpha^2, \alpha X, X^2) \subset uR$. Since $\alpha^2 \in uR$, we can assume that $u = \frac{\alpha^2}{f}$ for some $f \in R$. Since $X^2 \in \frac{\alpha^2}{f} R$, then $X^2 f = \alpha^2 g$ for some $g \in R$; so g has the form $g = X^2 h$ for some $h \in B[X]$. Thus $f = \alpha^2 h$; so that $\alpha^2 h(0) = f(0) \in A$. Hence $h(0) \in A$ and $h \in R$. Thus $1 = uh \in uR$ and $(I I^{-1})_v = R$.

(3) If $I = uJR$ for some $u \in qf(R)$ and some ideal J of A, then $u \in qf(A)$. Hence $\alpha^2 A = uJ$ and $\alpha B = uJB$; so $B = \alpha B$, a contradiction.

EXAMPLE 4.11   To construct simple examples illustrating Theorem 4.10, let's consider $A = Z$ and let $d \in Z$, $d \neq -1$ and not a square. Then $\alpha = \sqrt{d}$ satisfies the conditions of Theorem 4.10. The cases where $d \equiv 2, 3 \pmod{4}$ and square-free give examples with B integrally closed.

We have the following corollary of Theorem 4.10.

COROLLARY 4.12 The canonical homomorphism $Cl(A) \rightarrow Cl(A + XB[X])$ is not surjective in general even if B is integrally closed.

EXAMPLE 4.13 Let k be a field and $A = k[[S^2]]$, ($\alpha = S$ an indeterminate); so $B = A[S] = k[[S]]$. Let $R = A + XB[X]$. Then $Cl(A) = 0$ and $Cl(R) = Z/2Z$. To see this, let I be a t–invertible t–ideal of R. Since A and B are DVRs, and by applying Corollary 2.4, we can assume that $I = S^{2n}A + S^m XB[X]$. By dividing out suitable powers of S^2, we may assume that $m = 0$ or $m = 1$. For $I = S^{2n}A + XB[X]$; $I^{-1} = A + XB[X] = R$, and hence $I = I_v = R$ and $n = 0$. For $I = S^{2n}A + SXB[X]$; $I^{-1} = A + S^{-1}XB[X]$, and hence $I = I_v = S^2A + SXB[X]$ and $n = 1$. By Theorem 4.10, the only nonzero class in $Cl(R)$ is $[I]$, where $I = S^2A + SXB[X]$, and its order is two. Thus $Cl(R) = Z/2Z$.

We can say more about the surjectivity of the canonical homomorphism $Cl(A) \rightarrow Cl(A + XB[X])$; this is a consequence of Theorem 4.10.

COROLLARY 4.14 Let G be an abelian group. Then there exists an extension $A \subset B$ of integral domains such that B is integrally closed, $Cl(A) = G$, and the canonical homomorphism $Cl(A) \rightarrow Cl(A + XB[X])$ is not surjective.

Proof. By Claborn's theorem there exists a Dedekind domain D such that $Cl(D) = G$. Let S be an indeterminate, and let $A = D[S^2]$, $B = D[S]$, and $R = A + XB[X] = D[S^2, SX, X]$. We have $Cl(A) = Cl(D) = G$. By Theorem 4.10, the natural homomorphism $Cl(A) \rightarrow Cl(R)$ is not surjective. Also, note that in this case, $Pic(R) = Pic(A) = Pic(D)$.

REMARK 4.15 By modifying the hypothesis "α not a unit in B" in Theorem 4.10 to "$a = 1 - \alpha^2$ not a unit in B" and considering the ideal $I = aA + (1 + \alpha)XB[X]$, one can show, using similar arguments, that $I = (a, (1 + \alpha)X)$ and it is a t–invertible t–ideal of R. In this case, if we take $A = Z$ and $\alpha = i$, we have the simple example $R = Z + XZ[i][X]$. See the example below for the class group of this ring.

EXAMPLE 4.16 Let $R = Z + XZ[i][X]$. Then $Cl(R)$ is the direct sum of $Z/2Z$ and a countably infinite number of copies of Z. The $Z/2Z$ summand corresponds to 2, the prime in Z that splits in $Z[i]$ with two associate prime factors, and each of the Z summands corresponds to a positive prime p in Z that splits in $Z[i]$ with two nonassociate prime factors. Let $2 = ab$ in $Z[i]$; a and b are associates. Then for $I = 2Z + aXZ[i][X]$, we have $[I] = -[I] = [2Z + bXZ[i][X]]$ is nonzero. For $p \neq 2$, a positive prime in Z that splits as $p = ab$ in $Z[i]$, we have $-[pZ + aXZ[i][X]] = [pZ + bXZ[i][X]]$ and each has infinite order in $Cl(R)$. These statements and those below follow from the form of a divisorial ideal in R and the formula for I^{-1} for such a divisorial ideal.

We next show that the above classes of ideals generate $Cl(R)$. Such a divisorial ideal I has the form $nZ + aXZ[i][X]$ with $n = ab$. Using the above comments, one can show that any prime divisor p of n in Z that does not split in $Z[i]$ must also divide a. If $p = cd$ splits in $Z[i]$ and p^k exactly divides n, then one can show that either c^k or d^k is exactly the prime power that divides a. Thus $[nZ + aXZ[i][X]] = \sum k_j[p_jZ + c_jXZ[i][X]]$, where $\{p_j = c_jd_j\}$ is the set of positive primes in Z that split in $Z[i]$. We show that the above classes are independent. Assume $\sum k_j[p_jZ + c_jXZ[i][X]] = 0$ in $Cl(R)$. We may assume that each $k_j \geq 0$ (replace c_j by d_j, or conversely, if needed). Thus the corresponding $nZ + aXZ[i][X]$ is principal. One then uses the above comments to show that each k_j is 0 (or for the prime 2, that k_j is even).

REFERENCES

[1] D.D. Anderson and D.F. Anderson. Divisorial ideals and invertible ideals in a graded integral domain, J. Algebra 76 (1982), 549-569.

[2] D.D. Anderson, D.F. Anderson, and M. Zafrullah. Rings between $D[X]$ and $K[X]$, Houston J. Math. 17 (1991), 109-129.

[3] D.F. Anderson and D. Nour EL Abidine. Factorization in integral domains, III, J. Pure Appl. Algebra. To appear.

[4] D.F. Anderson and A. Ryckaert. The class group of $D+M$, J. Pure Appl. Algebra 52 (1988), 199-212.

[5] V. Barruci, L. Izelgue, and S. Kabbaj. Some factorization properties of $A+XB[X]$ domains, Lecture Notes in Pure and Applied Mathematics, Marcel Dekker, New York, 185 (1997), 69-78.

[6] N. Bourbaki. Algèbre commutative, Chap. 1 to 4, Masson, 1985.

[7] A. Bouvier. Le groupe des classes d'un anneau intègre, 107ème congrés des sociétés savantes, Brest, 1982, IV 85-92.

[8] A. Bouvier and M. Zafrullah. On some class groups of an integral domain, Bull. Soc. Math. Greece 29 (1988), 45-59.

[9] M. Fontana, L. Izelgue, and S. Kabbaj. Krull and valuative dimension of the rings of the form $A + XB[X]$, Lecture Notes in Pure and Applied Mathematics, Marcel Dekker, New York, 153 (1993), 111-130.

[10] M. Fontana, L. Izelgue, and S. Kabbaj. Quelques propriétés des chaines d' idéaux premiers dans les anneaux $A + XB[X]$, Comm. Algebra 22 (1994), 9-24.

[11] R. Fossum. The divisor class group of a Krull domain, Springer, 1973.

[12] S. Gabelli. On divisorial ideals in polynomial rings over Mori domains, Comm. Algebra 15 (1987), 2349-2370.

[13] R. Gilmer. Multiplicative ideal theory, Marcel Dekker, New York, 1972.

[14] R. Gilmer and R.C. Heitmann. On Pic(R[X]) for R seminormal, J. Pure Appl. Algebra 16 (1980), 251-257.

[15] J. Querre. Idéaux divisoriels d'un anneau de polynôme, J. Algebra 64 (1980), 270-284.

Rooty and Root Closed Domains

DAVID F. ANDERSON Department of Mathematics, The University of Tennessee, Knoxville, Tennessee 37996-1300.

JEANAM PARK Department of Mathematics, Inha University, Inchon, Korea 402-751

0. INTRODUCTION

In this paper, we study various ring-theoretic properties related to root closure in integral domains. In particular, we continue the investigation of multiplicatively closed subsets of fields begun in [AA] and of rooty rings as introduced by Sato and Sugatani in [SS]. In the next several paragraphs, we include a rather long list of relevant definitions.

Let R be an integral domain with quotient field K. Then R is said to be *root closed* if whenever $x^n \in R$ for some $x \in K$ and positive integer n, then $x \in R$. For a positive integer n, we say that R is *n-root closed* if whenever $x^n \in R$ for some $x \in K$, then $x \in R$. Thus R is root closed if and only if it is n-root closed for each positive integer n. More generally, a subring A of a ring B is *n-root closed in B* if $b \in A$ whenever $b^n \in A$ for $b \in B$. References on root closure include [A1], [A2], [AD], [ADR], [BCM], and [R].

We say that an ideal I of R is a *strongly radical ideal* of R if whenever $x \in K$ satisfies $x^n \in I$ for some positive integer n, then $x \in I$ [AA]. An ideal I of R is a *boldly radical ideal* if whenever $x \in K$ satisfies $x^n \in I$ for all sufficiently large integers n, then $x \in I$ [SS]. A prime ideal P of R is said to be *strongly prime* if whenever $xy \in P$ for some $x, y \in K$, then either $x \in P$ or $y \in P$ [HH]. As in [SS], we say that R is a *rooty ring* if each radical ideal of R is strongly radical. We say that R is *seminormal* if whenever $x^2, x^3 \in R$ for some $x \in K$, then $x \in R$ (equivalently, if R contains each $x \in K$ such that $x^n \in R$ for all sufficiently large integers n). We say that R is a *pseudo-valuation domain* (PVD) if each prime ideal of R is strongly prime [HH]. General references for seminormal rings are [BCM] and [GH].

A nonempty subset $S \subseteq K$ is called *power closed* in case $x^n \in S$ for all $x \in S$

and $n \geq 1$ [AA]. We define a nonempty subset $S \subseteq K$ to be *eventually power closed* if for each $x \in S$ there is an $N \in \mathbb{N}$ such that $x^n \in S$ for all $n \geq N$, and define S to be *sequentially closed* if, for each $x \in S$ and $N \in \mathbb{N}$, $x^n \in S$ for some $n \geq N$. Thus S is sequentially closed if and only if for each $x \in S$ there is a strictly increasing sequence $(a_n)_{n=1}^{\infty}$ in $\mathbb{N}$ such that $x^{a_n} \in S$ for all $n \geq 1$. Hence multiplicatively closed $\Rightarrow$ power closed $\Rightarrow$ eventually power closed $\Rightarrow$ sequentially closed. Let I be an ideal of R. Then we have $R - I$ is power closed iff $R - I$ is eventually power closed iff $R - I$ is sequentially closed.

Analogous to the well-known fact that an ideal I of R is prime iff $R - I$ is multiplicatively closed, it is easy to see that I is a radical ideal of R iff $R - I$ is power closed, and that I is strongly radical iff $K - I$ is power closed, and I is boldly radical iff $K - I$ is sequentially closed (see Theorem 1.1 and Theorem 1.2). Also, we have I is strongly prime iff $K - I$ is multiplicatively closed.

Counting the introduction, this paper is divided into three sections. In section 1, we review some of material from [AA] and [SS], and we show that an integral domain R with quotient field K is rooty (resp., seminormal) if and only if $(K - R) \cup U(R)$ is power (resp., sequentially) closed. This continues the investigation of multiplicatively closed subsets of fields begun in [AA].

Theorem 2.1 shows that if a rooty domain R is not quasilocal, then R is root closed. For this reason, we shall usually assume that R is a quasilocal domain in section 2. In fact, every quasilocal rooty domain R may be realized as a pullback square of the following type:

$$
\begin{array}{ccc}
R & \longrightarrow & T \\
\downarrow & & \downarrow{\scriptstyle \pi} \\
k & \longrightarrow & T/M
\end{array}
$$

where T is a root closed integral domain, M a nonzero radical ideal of T such that M is contained in the Jacobson radical of T, and k is a field contained in T/M.

For any quasilocal domain $R = (R, M)$, we define $\mathcal{C}(R) = \{n \in \mathbb{N} \mid R$ is n-root closed $\}$ as in [A1], and define $\mathcal{C}(M) = \{n \in \mathbb{N} \mid M$ is n-root closed$\}$. Then $\mathcal{C}(R) \subseteq \mathcal{C}(M)$ and each is a multiplicative submonoid of $\mathbb{N}$ generated by primes. We show that for any multiplicative submonoids $S \subseteq T$ of $\mathbb{N}$ generated by primes, there is a quasilocal domain (R, M) with $\mathcal{C}(R) = S \subseteq T = \mathcal{C}(M)$ (Theorem 2.16). This theorem combines and generalizes earlier work in [A1], [A2], [AD], and [R] about realizing a multiplicative submonoid of $\mathbb{N}$ generated by primes as $\mathcal{C}(R)$ for a suitable integral domain R.

Throughout, R will denote an integral domain with integral closure $\overline{R}$ and quotient field K. As usual, an overring of R is a subring of K containing R. Let K^* be the multiplicative group $K - \{0\}$ and $U(R)$ the group of units of R. The quotient group $G(R) = K^*/U(R)$ is called the *group of divisibility* of R. If I, J are ideals of R, we define $I : J = \{x \in K \mid xJ \subseteq I\}$, and $I^{-1} = R : I$. In particular, $I : I$ is the largest overring of R in which I is an ideal. Also, $\mathbb{N}$, $\mathbb{Z}$, $\mathbb{Q}$, $\mathbb{R}$, and $\mathbb{C}$ will denote, respectively, the sets of positive integers, integers, rational numbers, real numbers, and complex numbers. A general reference is [G].

1. STRONGLY RADICAL, BOLDLY RADICAL IDEALS, AND MULTIPLICATIVE SUBSETS OF K

In this section, we first give several equivalent conditions for strongly radical and boldly radical ideals. For completeness, we include the following characterization of strongly radical ideals from [AA, Proposition 1.4.]. Recall that the root closure $\widehat{R}$ of R (in K) is the smallest root closed overring of R (i.e., the intersection of all the root closed subrings of K that contain R).

THEOREM 1.1. *Let I be an ideal of an integral domain R with quotient field K. Then the following statements are equivalent.*

(1) *I is strongly radical.*
(2) *$K - I$ is power closed.*
(3) *I is a radical ideal of $I : I$ and $I : I$ is root closed.*
(4) *I is a radical ideal in the root closure $\widehat{R}$ of R.*
(5) *I is a radical ideal in some root closed overring of R.*

Proof. The equivalence of (1), (2), (3), and (5) are from [AA, Proposition 1.4]. (4) $\Rightarrow$ (5) is clear. (3) $\Rightarrow$ (4). First, observe that $\widehat{R} \subseteq I : I$ since $I : I$ is root closed and $\widehat{R}$ is the smallest root closed overring of R. Thus I is also a radical ideal of $\widehat{R}$. $\square$

Note that an ideal I of R is boldly radical if and only if $x \in I$ whenever $x^2, x^3 \in I$ for some $x \in K$. Recall that the seminormalization R^+ of R (in K) is the smallest seminormal overring of R (i.e., the intersection of all the seminormal subrings of K that contain R)(cf. [ADF, Lemma 2.4(a)]).

THEOREM 1.2. *Let I be an ideal of an integral domain R with quotient field K. Then the following statements are equivalent.*

(1) *I is boldly radical.*
(2) *$K - I$ is sequentially closed.*
(3) *I is a radical ideal of $I : I$ and $I : I$ is seminormal.*
(4) *I is a radical ideal in the seminormalization R^+ of R.*
(5) *I is a radical ideal in some seminormal overring of R.*

Proof. (1) $\Rightarrow$ (2). Let $x \in K - I$. Since I is boldly radical, for any $N \in \mathbb{N}$, there exists $a_N \geq N$ such that $x^{a_N} \in K - I$. Thus we can find a strictly increasing sequence $(a_n)_{n=1}^{\infty}$ with $x^{a_n} \in K - I$ for all $n \geq 1$. Hence $K - I$ is sequentially closed. (2) $\Rightarrow$ (3). Let $x \in I : I$. Suppose that $x^k \in I$ for some k, but $x \notin I$. Since $K - I$ is sequentially closed, there is a strictly increasing sequence $(a_n)_{n=1}^{\infty}$ such that $x^{a_n} \notin I$ for all $n \geq 1$. Choose m such that $a_m \geq k$. Then $x^{a_m - k} \in I : I$, and so $x^{a_m} = x^{a_m - k} \cdot x^k \in I$, a contradiction. Hence I is a radical ideal of $I : I$. Suppose now that there exists $x \in K$ such that $x^2, x^3 \in I : I$, but $x \notin I : I$. Then there exists $y \in I$ with $xy \notin I$. Since $K - I$ is sequentially closed, there is a strictly increasing sequence $(a_n)_{n=1}^{\infty}$ such that $(xy)^{a_n} \notin I$ for all $n \geq 1$. However, $x^n \in I : I$ for all $n \geq 2$ and $y \in I \Rightarrow (xy)^n = x^n y^n \in I$ for all $n \geq 2$, a contradiction. Thus $I : I$ is seminormal. (3) $\Rightarrow$ (4). Note that $R^+ \subseteq I : I$ since $I : I$ is seminormal. Thus I is also a radical ideal of R^+. (4) $\Rightarrow$ (5) is clear. (5) $\Rightarrow$ (1). Suppose that

I is a radical ideal in a seminormal overring T of R. Let $x \in K$ such that x^2, x^3 $\in I \subseteq T$. Thus $x \in T$ since T is seminormal. Hence $x \in I$ since I is a radical ideal of T. Thus I is boldly radical. $\square$

Theorem 1.1 (resp., Theorem 1.2) shows that an ideal I of a domain R is a radical ideal in some root closed (resp., seminormal) overring of R if and only if I is strongly radical (resp., boldly radical). It is easily verified that I is a radical ideal in some valuation overring of R if and only if I is strongly prime. We next determine when I is a radical ideal in some integrally closed overring of R. We define I to be *strongly integrally closed* if whenever $x^n + a_1 x^{n-1} + \cdots + a_{n-1}x + a_n \in I$ for $x \in K$ and $a_i \in I$, then $x \in I$. Note that strongly prime $\Rightarrow$ strongly integrally closed $\Rightarrow$ strongly radical $\Rightarrow$ boldly radical $\Rightarrow$ radical. Also, we define $K - I$ to be *polynomially closed with respect to I* if $x \in K - I$ implies $x^n + a_1 x^{n-1} + \cdots + a_{n-1}x + a_n \in K - I$ for any $a_i \in I$.

THEOREM 1.3. *Let I be an ideal of an integral domain R with quotient field K. Then the following statements are equivalent.*

 (1) *I is strongly integrally closed.*
 (2) *$K - I$ is polynomially closed with respect to I.*
 (3) *I is a radical ideal of $I : I$ and $I : I$ is integrally closed.*
 (4) *I is a radical ideal in the integral closure $\overline{R}$ of R.*
 (5) *I is a radical ideal in some integrally closed overring of R.*

Proof. (1) $\Leftrightarrow$ (2) follows easily from the definitions. (1) $\Rightarrow$ (3). Let $x \in K$ be integral over $I : I$; so $x^n + a_1 x^{n-1} + \cdots + a_{n-1}x + a_n = 0$ for some $a_j \in I : I$. Let $i \in I$. Then $(ix)^n + (ia_1)(ix)^{n-1} + \cdots + (i^{n-1}a_{n-1})(ix) + (i^n a_n) = 0 \in I$ with each $i^m a_m \in I$. Hence $ix \in I$ since I is strongly integrally closed. Thus $x \in I : I$; so $I : I$ is integrally closed. Clearly I is a radical ideal of $I : I$. (3) $\Rightarrow$ (4) follows easily since $\overline{R} \subseteq I : I$ when $I : I$ is integrally closed. (4) $\Rightarrow$ (5) is clear. (5) $\Rightarrow$ (1). Suppose that I is a radical ideal in an integrally closed overring T of R. Let $x^n + a_1 x^{n-1} + \cdots + a_{n-1}x + a_n \in I \subseteq T$ for some $x \in K$ and $a_i \in I$. Then $x \in T$ since T is integrally closed. Hence $x^n \in I$, and thus $x \in I$ since I is a radical ideal of T. Hence I is strongly integrally closed. $\square$

REMARK 1.4. The integral closure of an ideal I of an integral domain R with quotient field K is defined to be $\bar{I} = \{x \in K \mid x^n + a_1 x^{n-1} + \cdots + a_{n-1}x + a_n = 0 \text{ for some } a_j \in I^j\}$, and I is said to be *integrally closed* if $\bar{I} = I$. One can easily verify that I is strongly integrally closed if and only if I is integrally closed and I is a radical ideal in $I : I$.

The first two parts of our next result are from [SS, Remark 1.2]. The third part follows directly from the definitions.

LEMMA 1.5. *Let I, J be ideals of an integral domain R with $I \subseteq J$.*

 (1) *If J is boldly radical and I is radical, then I is boldly radical.*
 (2) *If J is strongly radical and I is radical, then I is strongly radical.*

(3) *If J is strongly integrally closed and I is radical, then I is strongly integrally closed.* $\square$

In [S], P. Samuel showed that a subring R of a field K is a valuation domain with quotient field K if and only if $K - R$ is multiplicatively closed. In [AA], it was observed that R is root closed if and only if $K - R$ is power closed. Among other things, we next show that R is seminormal if and only if $K - R$ is sequentially closed. (The equivalence of (1) and (5) is [SS, Theorem 1.3]; however, we give a somewhat easier proof.)

THEOREM 1.6. *The following statements are equivalent for an integral domain R with quotient field K.*

(1) R *is seminormal.*
(2) *For each $x \in \overline{R} - R$, $(R : R[x])$ is a radical ideal of $R[x]$.*
(3) $K - R$ *is sequentially closed.*
(4) R *is itself a boldly radical ideal of R.*
(5) *Every prime ideal of R is boldly radical.*
(6) *Every maximal ideal of R is boldly radical.*
(7) $K - P$ *is sequentially closed for each prime ideal P of R.*
(8) $K - M$ *is sequentially closed for each maximal ideal M of R.*
(9) $(K - R) \cup U(R)$ *is sequentially closed.*

Proof. (1) $\Leftrightarrow$ (5) $\Leftrightarrow$ (6) [SS, Remark 1.2, Theorem 1.3]. ((6) $\Rightarrow$ (5) follows from Lemma 1.5(1).) (1) $\Leftrightarrow$ (2) [GH, Theorem 1.1]. (1) $\Leftrightarrow$ (4), (3) $\Leftrightarrow$ (4), (5) $\Leftrightarrow$ (7), and (6) $\Leftrightarrow$ (8) all follow from Theorem 1.2. It only remains to prove (3) $\Leftrightarrow$ (9). (3) $\Rightarrow$ (9). Suppose that $K - R$ is sequentially closed. Since $U(R)$ is a group, $(K - R) \cup U(R)$ is also sequentially closed. (9) $\Rightarrow$ (3). Suppose that $S = (K - R) \cup U(R)$ is sequentially closed. If $K - R$ is not sequentially closed, then there is an $x \in K - R$ and an integer N such that $x^n \in R$ for all $n \geq N$. Hence $x^m \in U(R)$ for some $m \geq N$ since S is sequentially closed. Thus $x = x^{m+1}(x^m)^{-1} \in R$, a contradiction. Hence $K - R$ is sequentially closed. (An easier proof of (5) $\Rightarrow$ (1). Let $x^2, x^3 \in R$. If $x^2 \in U(R)$, then $x = x^3(x^2)^{-1} \in R$; otherwise, $x^2 \in P$ for some prime ideal P of R. Then $(x^3)^2 = (x^2)^3 \in P$; so also $x^3 \in P$. Thus $x \in P \subset R$ since P is boldly radical. Hence R is seminormal.) $\square$

REMARK 1.7. In [SS, Theorem 1.4], it was observed that a Noetherian domain R is seminormal if and only if each height one prime ideal of R is boldly radical. Our proof of (5) $\Rightarrow$ (1) of Theorem 1.6 yields the following generalization: An integral domain R is seminormal if and only if each nonunit of R is contained in a boldly radical ideal of R.

Our next theorem is the analog of Theorem 1.6 for rooty domains.

THEOREM 1.8. *The following statements are equivalent for an integral domain R with quotient field K.*

(1) R *is a rooty domain (i.e., every radical ideal of R is strongly radical).*
(2) *Every prime ideal of R is strongly radical.*

(3) *Every maximal ideal of R is strongly radical.*
(4) $K - P$ *is power closed for each prime ideal P of R.*
(5) $K - M$ *is power closed for each maximal ideal M of R.*
(6) $(K - R) \cup U(R)$ *is power closed.*
(7) $(K - R) \cup U(R)$ *is eventually power closed.*

Proof. (1) $\Leftrightarrow$ (2) $\Leftrightarrow$ (3) follows from Lemma 1.5(2), or [SS, Remark 1.2.3]. (2) $\Leftrightarrow$ (4) and (3) $\Leftrightarrow$ (5) follow from Theorem 1.1. Cleary (6) $\Rightarrow$ (7). We give the proofs for (5) $\Leftrightarrow$ (6) and (7) $\Rightarrow$ (6). (5) $\Rightarrow$ (6). Suppose that $K - M$ is power closed for each maximal ideal M of R. Then $\bigcap \{K - M \mid M$ is a maximal ideal of $R\} = (K - R) \cup U(R)$ is also power closed. (6) $\Rightarrow$ (5). Suppose that $(K - R) \cup U(R)$ is power closed. Let M be a maximal ideal of R and let $x \in K - M$. By assumption, we may assume that x is a nonunit in $R - M$. Then it is clear that for any $n \geq 1$, $x^n \in R - M$ and is a nonunit. Hence $K - M$ is power closed. (7) $\Rightarrow$ (6). Suppose that $(K - R) \cup U(R)$ is eventually power closed. Since $R - U(R)$ is power closed, $(K - R) \cup U(R)$ is also power closed. $\square$

An integral domain R with quotient field K is a pseudo-valuation domain (PVD) if and only if $(K - R) \cup U(R)$ is multiplicatively closed [AA, Theorem 1.2]. Thus valuation (resp., root closed) domains may be distinguished from pseudo-valuation (resp., rooty) domains by whether $K - R$ or $(K - R) \cup U(R)$ is multiplicatively (resp., power) closed. However, by Theorem 1.6, $K - R$ is sequentially closed $\Leftrightarrow$ $(K - R) \cup U(R)$ is sequentially closed $\Leftrightarrow$ R is seminormal. Also, observe that if S is a nonempty subset of K, then S is multiplicatively closed $\Rightarrow$ S is power closed $\Rightarrow$ S is sequentially closed; and if $K - R$ is multiplicatively (resp., power, sequentially) closed, then $(K - R) \cup U(R)$ is also multiplicatively (resp., power, sequentially) closed. From these and earlier observations, we easily obtain the well-known results in the next remark.

REMARK 1.9. Let R be an integral domain with quotient field K, and let S be a nonempty subset of K. Then:

(1) (a) Letting $S = K - R$, we have that R a valuation domain $\Rightarrow$ R root closed $\Rightarrow$ R seminormal.
    (b) Letting $S = (K - R) \cup U(R)$, we have that R a PVD $\Rightarrow$ R rooty $\Rightarrow$ R seminormal.
(2) R root closed $\Rightarrow$ R rooty $\Rightarrow$ R seminormal.

2. ROOTY DOMAINS

In this section, we investigate when a rooty domain R is root closed. If a rooty domain R is not quasilocal, then R is root closed [SS, Proposition 1.7]. This result, with a different proof, is Theorem 2.1. We will thus be mainly interested in quasilocal rooty domains.

THEOREM 2.1. *([SS, Proposition 1.7]) Let R be a rooty domain with quotient field K. If R is not quasilocal, then R is root closed.*

Proof. Let M and N be distinct maximal ideals of R. Let $a \in M$ and $b \in N$ such that $1 = a + b$. Suppose that $x^n \in R$ for some $x \in K$. Then $(xa)^n = x^n a^n \in M$ and $(xb)^n = x^n b^n \in N$. Hence $xa \in M$ and $xb \in N$ since M and N are each strongly radical. Then $x = xa + xb \in M + N = R$. Hence R is root closed. □

REMARK 2.2. Let I and J be comaximal ideals of R (i.e., $I + J = R$). One can then easily verify that $(I : I) \cap (J : J) = R$. Hence, if R has comaximal strongly radical (resp., boldly radical, strongly integrally closed) ideals, then R is root closed (resp., seminormal, integrally closed) by Theorem 1.1 (resp., Theorem 1.2, Theorem 1.3).

THEOREM 2.3. *The following statements are equivalent for an integral domain R with quotient field K.*

(1) *R is root closed.*
(2) *Every prime ideal of R is strongly radical and $G(R)$ is torsionfree.*
(3) *$K - R$ is power closed.*
(4) *$K - R$ is eventually power closed.*

Proof. The proofs for (3) $\Rightarrow$ (4) and (1) $\Leftrightarrow$ (3) are by definition. We give the proofs for (1) $\Leftrightarrow$ (2) and (4) $\Rightarrow$ (3). (1) $\Rightarrow$ (2). Let R be root closed and let P be a prime ideal of R. Suppose that $x^n \in P$ for some $n \in \mathbb{N}$ and $x \in K$. Since R is root closed, $x \in R$ and so $x \in P$. Thus P is strongly radical. Similarly, if $xU(R)$ has finite order in $G(R) = K^* / U(R)$, then $x^n \in U(R)$ for some $n \geq 1$, and so $x \in R$. Hence $x \in U(R)$. Thus $G(R)$ is torsionfree. (2) $\Rightarrow$ (1). Suppose that $x^n \in R$ for some $x \in K$. If $x^n \notin U(R)$, then there is a prime ideal P such that $x^n \in P$. Since P is strongly radical, $x \in P$. On the other hand, if $x^n \in U(R)$, then $x \in U(R)$ since $G(R)$ is torsionfree. Thus R is root closed. (4) $\Rightarrow$ (3). Suppose that $K - R$ is eventually power closed. Let $x \in K - R$. Then there is a positive integer $N \geq 2$ such that $x^n \in K - R$ for all $n \geq N$. If $x^i \in R$ for some $1 \leq i \leq N - 1$, then $x^{iN} \in R$, a contradiction. Hence $K - R$ is power closed. □

COROLLARY 2.4. *An integral domain R is root closed if and only if R is rooty and $G(R)$ is torsionfree.* □

Let (R, M) be a quasilocal rooty domain. Then $M : M$ is root closed by Theorem 1.1. Suppose that R is not root closed. Then $M : M \neq R$, and hence M is a nonprincipal ideal of R. Since M is not invertible, necessarily $MM^{-1} = M$. Hence, we have $M : M = M^{-1}$ and M is divisorial (cf. [A3, Proposition 2.1]). Note that $M^{-1} = M : M \subseteq \tilde{R}$, the complete integral closure of R. (Recall that the *complete integral closure* of an integral domain R (in its quotient field K) is $\tilde{R} = \{x \in K \mid rx^n \in K \text{ for some nonzero } r \in R \text{ and all } n \geq 1\}$.) Our next example shows that we may have $M : M$ a proper overring of R even when R is root closed.

EXAMPLE 2.5. ($R = (R, M)$ is root closed (and so rooty by Corollary 2.4), but $M : M \neq R$.) Let $V = \mathbb{Q}(t)[[X]] = \mathbb{Q}(t) + M$, where $M = X\mathbb{Q}(t)[[X]]$. Then $R = \mathbb{Q} + M$ is a one-dimensional quasilocal domain with $M : M = V$. Moreover, R is root closed since $\mathbb{Q}$ is root closed in $\mathbb{Q}(t)$ [A1, Lemma 2.1].

Let (R, M) be a quasilocal rooty domain with quotient field K. Then by Theorem 1.8(5) or Theorem 1.1, $(K - R) \cup U(R) = K - M$ is power closed. So, if R is not root closed, then there exists $x \in K - R$ such that $x^n \in U(R)$ for some $n \geq 2$. Thus, we set $\overline{U(R)} = \{x \in K^* \mid x^n \in U(R) \text{ for some } n \geq 1\}$. Then $\overline{U(R)}$ is a multiplicative subgroup of K^* containing $U(R)$. Moreover, the factor group $\overline{U(R)}/U(R)$ is the torsion subgroup of $G(R) = K^*/U(R)$, and thus, $\overline{U(R)}/U(R)$ measures how far the quasilocal rooty domain R is from being root closed.

We next show how quasilocal rooty domains may be realized as pullbacks.

Let (R, M) be a quasilocal rooty domain. Then M is a radical ideal of the root closed domain $M : M$ by Theorem 1.1. Since R is quasilocal, $1 - ax \in U(R) \subseteq U(M : M)$ for all $a \in M$ and all $x \in M : M$. Thus $M \subseteq J(M : M)$, where $J(M : M)$ is the Jacobson radical of $M : M$. Then we have the following pullback square, where π is the canonical surjection.

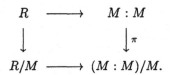

Our next theorem determines when such an R is root closed.

THEOREM 2.6. *Let (R, M) be a quasilocal rooty domain with quotient field K. Then the following statements are equivalent.*

    (1) *R is root closed.*
    (2) *$U(M : M)/U(R)$ is torsionfree.*
    (3) *$U(M : M/M)/U(R/M)$ is torsionfree.*

Proof. (1) $\Rightarrow$ (2). Recall from Theorem 2.3 that $G(R)$ is torsionfree since R is root closed. Thus $U(M : M)/U(R)$ is a subgroup of $G(R) = K^*/U(R)$, and hence is also torsionfree. (2) $\Rightarrow$ (1). Suppose that $x^n \in R$ for some $x \in K$. Then $x \in M : M$ since $M : M$ is root closed. If $x^n \in U(R)$, then $xU(R)$ has finite order, and so $x \in U(R) \subset R$ since $U(M : M)/U(R)$ is torsionfree. If $x^n \notin U(R)$, then $x^n \in M$, and so $x \in M \subset R$ since M is strongly radical. Thus R is root closed. The proof of (1) $\Leftrightarrow$ (3) is similar, and hence will be left to the reader. $\square$

Conversely, we next show that every such pullback square yields a quasilocal rooty domain.

THEOREM 2.7. *Let T be a root closed integral domain with quotient field K, M a nonzero radical ideal of T such that $M \subseteq J(T)$, where $J(T)$ is the Jacobson radical of T, and k a field contained in T/M. Let $\pi : T \to T/M$ be the canonical surjection and $R = \pi^{-1}(k)$ be the pullback arising from*

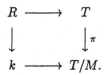

Then R is a quasilocal rooty domain with maximal ideal M and quotient field K.

Proof. Since $R/M \cong k$, M is a maximal ideal of R. Let $a \in M \subseteq J(T)$. Then $1 - ab \in U(T)$ for all $b \in R$. But, since M is also an ideal of T, we have $1 - ab \in U(R)$ for all $b \in R$. Thus $M \subseteq J(R)$, and hence M is the unique maximal ideal of R. Since M is a strongly radical ideal of T, M is also a strongly radical ideal of R. Hence R is a quasilocal rooty domain. Note that K is also the quotient field of R since R and T have M as a common nonzero ideal. $\square$

Probably the most important case of Theorem 2.7 is when T is quasilocal with maximal ideal M. In this case, R is root closed if and only if R/M is root closed in T/M (i.e., $(T/M)^*/(R/M)^*$ is torsionfree) ([ADR, Theorem 1.7]). Also, in this case without assuming that T is root closed, it is clear that R is rooty if and only if T is rooty.

Let $R = \pi^{-1}(k) \subseteq T$ be as in Theorem 2.7. Consider the exact sequence

$$0 \to \overline{U(R)}/U(R) \to U(T)/U(R) \to U(T)/\overline{U(R)} \to 0.$$

Note that $U(T)/\overline{U(R)}$ is torsionfree since $\overline{U(R)}/U(R)$ is the torsion subgroup of $U(T)/U(R) \subseteq K^*/U(R) = G(R)$. Thus, if $U(T)/\overline{U(R)}$ is finitely generated, then it is free. In this case,

$$U(T)/U(R) \cong \overline{U(R)}/U(R) \oplus U(T)/\overline{U(R)},$$

the direct sum of the torsion part and torsionfree part of $U(T)/U(R)$.

Since $M \subseteq J(T)$, π induces a surjective homomorphism $U(T) \to U(T/M)$, which in turn induces the isomorphism

$$U(T)/U(R) \cong U(T/M)/k^*.$$

As in Theorem 2.6, R is root closed if and only if $U(T/M)/k^*$ is torsionfree. Moreover, if $U(T)/U(R)$ is torsion, then $\overline{U(R)}/U(R) = U(T)/U(R) \cong U(T/M)/k^*$.

EXAMPLE 2.8. (A quasilocal rooty domain R such that $\overline{U(R)}/U(R)$ is a finite group.) (1) Let $k \subset K$ be algebraic number fields and let T be a discrete valuation domain with residue field K, for example, $T = K[[X]]$. Then we have a pullback square of the following type:

$$
\begin{array}{ccc}
R & \longrightarrow & T \\
\downarrow & & \downarrow{\scriptstyle\pi} \\
k & \longrightarrow & K
\end{array}
$$

By the above remarks, we have $U(T)/U(R) \cong K^*/k^*$. Hence $U(T)/U(R) \cong A \oplus B$, where A is free abelian of rank $\aleph_0$ and B is finite [K, Theorem 4.3.8]. Hence $\overline{U(R)}/U(R) \cong B$.

(2) Let $T = GF(p^{mn})[[X]]$. Then $R = GF(p^m) + XGF(p^{mn})[[X]]$ is a quasilocal one-dimensional rooty domain with $U(T)/U(R) = \overline{U(R)}/U(R)$ a cyclic group of order $(p^{mn} - 1)/(p^m - 1)$.

(3) Recall that by Brandis' Theorem [K, Theorem 4.3.11, page 144], K^*/k^* is finitely generated if and only if $K = k$ or K is a finite field. Thus, with the same notation as in Theorem 2.7 and $T/M = K$ a field, $\overline{U(R)}/U(R)$ is a nonzero finite group if and only if $k \subsetneq K$ are finite fields.

Since a non-quasilocal rooty domain is root closed, we restrict our investigation to quasilocal rooty domains. Let (R, M) be quasilocal with quotient field K. Then R is rooty if and only if M is strongly radical. For $n \in \mathbb{N}$, we define M to be *n-root closed* if whenever $x^n \in M$ for some $x \in K$, then $x \in M$. Thus R is rooty if and only if M is n-root closed for each $n \geq 1$, if and only if $(K - R) \cup U(R) = K - M$ is power closed. Moreover, if M is n-root closed for some integer $n \geq 2$, then M is boldly radical, and hence R is seminormal. In [A1], for any integral domain R the first author defined $\mathcal{C}(R) = \{n \in \mathbb{N} \mid R \text{ is } n\text{-root closed}\}$. We thus define $\mathcal{C}(M) = \{n \in \mathbb{N} \mid M \text{ is } n\text{-root closed}\}$. It is easy to see that $\mathcal{C}(M)$, like $\mathcal{C}(R)$, is a multiplicative submonoid of $\mathbb{N}$ generated by primes, since M is mn-root closed if and only if M is both m- and n-root closed. In general, we have $\mathcal{C}(R) \subseteq \mathcal{C}(M)$.

More generally, for I any ideal of an integral domain R with quotient field K, one can define $\mathcal{C}(I) = \{n \in \mathbb{N} \mid x^n \in I, x \in K \Rightarrow x \in I\}$. Our next proposition generalizes Lemma 1.5(2).

PROPOSITION 2.9. *Let R be an integral domain with quotient field K. If $I \subseteq J$ are ideals and I is a radical ideal of R, then $\mathcal{C}(J) \subseteq \mathcal{C}(I)$.*

Proof. Suppose that $n \in \mathcal{C}(J)$ and $x^n \in I$ for some $x \in K$. Then $x^n \in I \subseteq J$. Thus $x \in J \subseteq R$, and so $x \in I$ since I is a radical ideal of R. Hence $n \in \mathcal{C}(I)$. $\square$

Our next example shows that it is necessary to assume that I is a radical ideal in Proposition 2.9.

EXAMPLE 2.10. *Let K be a field, $R = K[[X]]$, $J = (X)$, and $I = (X^2)$. Then $\mathcal{C}(J) = \mathbb{N}$ and $\mathcal{C}(I) = \{1\}$.*

THEOREM 2.11. *Let (R, M) be a quasilocal domain with quotient field K. Then $\mathcal{C}(R) \subseteq \mathcal{C}(M)$. If $G(R)$ is torsionfree, then $\mathcal{C}(R) = \mathcal{C}(M)$.*

Proof. It is clear that $\mathcal{C}(R) \subseteq \mathcal{C}(M)$. Now assume that $G(R)$ is torsionfree. Suppose we have $n \in \mathcal{C}(M)$, but $n \notin \mathcal{C}(R)$. Then there exists $x \in K$ such that $x^n \in R$, but $x \notin R$. Since $n \in \mathcal{C}(M)$, we conclude that $x^n \in R - M$, so $x^n \in U(R)$. But since $xU(R)$ has finite order in $G(R)$, we obtain that $x \in U(R) \subset R$, a contradiction. Hence $\mathcal{C}(R) = \mathcal{C}(M)$. $\square$

COROLLARY 2.12. *Let (R, M) be a quasilocal integral domain. Then R is root closed if and only if R is rooty and $C(R) = C(M)$.* $\square$

Let (R, M) be a quasilocal domain. Our next example shows that the equality $C(R) = C(M)$ does not imply that $G(R)$ is torsionfree.

EXAMPLE 2.13. Let $T = \mathbb{C}[[X]] = \mathbb{C} + M$, where $M = X\mathbb{C}[[X]]$. Let $A = \mathbb{R} + M$, a one-dimensional local Noetherian domain with maximal ideal M. Note that A is not n-root closed for any $n \geq 2$ since $\mathbb{C}$ contains n-th roots of unity not in $\mathbb{R}$. However, M is n-root closed for all $n \in \mathbb{N}$ since T is integrally closed; so A is rooty (i.e., $\{1\} = C(A) \subsetneqq C(M) = \mathbb{N}$). Next, let $B = \mathbb{R} + \mathbb{R}X + X^2\mathbb{C}[[X]]$ and $D = \mathbb{C} + X^2\mathbb{C}[[X]]$. Each is a one-dimensional local Noetherian domain with maximal ideal $N = \mathbb{R}X + X^2\mathbb{C}[[X]]$ and $P = X^2\mathbb{C}[[X]]$, respectively. Then $C(B) = C(N) = \{1\}$ and $C(D) = C(P) = \{1\}$. However, $G(B) = \mathbb{Z} \oplus \mathbb{C}^*/\mathbb{R}^* \oplus (\mathbb{C}, +)$ is not torsionfree, while $G(D) = \mathbb{Z} \oplus (\mathbb{C}, +)$ is torsionfree.

In [A1, Theorem 2.7], the first author showed that any multiplicative submonoid of $\mathbb{N}$ generated by primes may be realized as $C(R)$ for some integral domain R, i.e., if $\mathcal{P}$ is any set of primes, then there is an integral domain R such that $C(R)$ is generated by $\mathcal{P}$ (also, cf. [A2, Theorem 6] and [R, Proposition 2.9 and Theorem 2.11]). Recall that if R is an integral domain, then $C(R) \subseteq C(R_M)$ for each maximal ideal M of R (i.e., if R is n-root closed, then R_M is also n-root closed).

We next show that the construction in the proof of [A1, Theorem 2.7] localizes.

THEOREM 2.14. *Let S be a multiplicative submonoid of $\mathbb{N}$ generated by primes. Then there is a quasilocal domain (A, N) such that $C(A) = C(N) = S$.*

Proof. Let S be generated by some set $\mathcal{P}$ of primes. If $S = \mathbb{N}$, then we may take A to be any quasilocal root closed domain. Otherwise as in [A1, Theorem 2.7], let $Q = \{p_i \mid i \in I\}$ be the set consisting of those primes p_i which are not in $\mathcal{P}$. Let $\{X_i, Y_i \mid i \in I\}$ be a set of indeterminates over a field F. Then let $R = F[\{X_i^{p_i} \mid i \in I\}, \{X_i^m Y_j^n \mid m \geq 0, n \geq 1, \text{and } i, j \in I\}] = F + M$, where $M = (\{X_i^{p_i}\}, \{X_i^m X_j^n \mid m \geq 0, n \geq 1\})$, a maximal ideal of R. Then $C(R) \subseteq C(R_M) \subseteq C(M_M)$, and $C(R) = S$ by [A1, Theorem 2.7]. We now show that also $C(M_M) = S$. Let $p_i \in Q$, so $X_i^{p_i} \in R$, but $X_i \notin R$. It suffices to show that $p_i \notin C(M_M)$, and hence $C(R) = C(R_M) = C(M_M)$. Now, $X_i^{p_i} \in M \Rightarrow X_i^{p_i} \in M_M$. Suppose that $X_i \in M_M$. Then $X_i = m/s$ for some $m \in M$, $s \in R - M$; so $s = \alpha + f$ with $0 \neq \alpha \in F$ and $f \in M$. Then $sX_i = \alpha X_i + X_i f \in M \subset R$. Thus $\alpha X_i \in R$ since R is generated by monomials over F, and hence $X_i \in R$, a contradiction. Let $A = R_M$ and $N = M_M$. Then $C(A) = C(N) = S$. $\square$

In [AD, Theorem 3.3], it was shown that for any multiplicative submonoid S of $\mathbb{N}$ generated by primes, there is a one-dimensional quasilocal domain (R, M) with $C(R) = S$. It is easily verified that $C(M) = \mathbb{N}$ for the domain (R, M) constructed in the proof of [AD, Theorem 3.3], so this R is rooty. We next generalize these two results about realizing S as $C(R)$ for a suitable domain R. But first we need a lemma.

LEMMA 2.15. *Let $T = K + M$ be a quasilocal domain with nonzero maximal ideal M and K a field, and let $R = k + M$, where k is a subfield of K. If $C(T) = C(M)$, then $C(R) = C(k, K) \cap C(M)$, where $C(k, K) = \{n \in \mathbb{N} \mid k$ is n-root closed in $K\}$.*

Proof. Cleary $C(R) \subseteq C(k, K) \cap C(M)$. Conversely, let $n \in C(k, K) \cap C(M)$. Suppose that $x^n \in R$ for some x in the quotient field of R. Then $x = \alpha + m \in T$ ($\alpha \in K$, $m \in M$) since $n \in C(M) = C(T)$. Thus $\alpha^n \in k$, so $\alpha \in k$ since $n \in C(k, K)$. Hence $x \in R$, so $n \in C(R)$. Thus $C(R) = C(k, K) \cap C(M)$. $\square$

THEOREM 2.16. *For any multiplicative submonoids $S \subseteq T$ of $\mathbb{N}$ generated by primes, there is a quasilocal domain (B, I) with $C(B) = S \subseteq T = C(I)$.*

Proof. Let R and M be as in the proof of Theorem 2.14 with $F = \mathbb{R}$ and $C(R) = T$. Then $R_M = \mathbb{R} + M_M$ with $C(R_M) = C(M_M) = T$. Let $B = k + I$, where $I = M_M$ and $k = K_S$ is the subfield of $\mathbb{R}$ defined as in [AD, Lemma 3.2]. (Let S be generated by a set $\mathcal{P}$ of positive primes. Define inductively an increasing sequence of subfields of $\mathbb{R}$ by $K_0 = \mathbb{Q}$ and $K_{n+1} = K_n(\{x \in \mathbb{R} \mid x^p \in K_n$ for some $p \in \mathcal{P}\})$. Let $K_S = \bigcup_{n=0}^{\infty} K_n$. Thus $K_S = \mathbb{Q}$ if $\mathcal{P}$ is the empty set.) Then K_S is p-root closed in $\mathbb{R}$ if and only if $p \in S$ [AD, Lemma 3.2], so $C(K_S, \mathbb{R}) = S$. Thus $C(B) = C(k, \mathbb{R}) \cap C(I) = S \cap T = S$ by Lemma 2.15. $\square$

EXAMPLE 2.17. (A quasilocal rooty domain R which is neither 2- nor 3-root closed, but is n-root closed for each n not divisible by 2 or 3.) Follow the notation in Example 2.8(2). Select $p = 5, n = 2, m = 1$. Then we have a quasilocal rooty domain $R = GF(5) + XGF(5^2)[[X]]$ such that $\overline{U(R)}/U(R)$ is the cyclic group of order 6. Then R is the desired domain. $\square$

ACKNOWLEDGMENTS

This research was conducted while the second author visited The University of Tennessee at Knoxville. He gratefully acknowledges support received under The Inha University Research Foundation and the hospitality of The University of Tennessee at Knoxville.

REFERENCES

[AA] D.D. Anderson and D.F. Anderson, *Multiplicatively closed subsets of fields*, Houston J. Math. **13** (1987), 1-11.

[A1] D.F. Anderson, *Root closure in integral domains*, J. Algebra **79** (1982), 51-59.

[A2] D.F. Anderson, *Root closure in integral domains, II*, Glasgow Math. J. **31** (1989), 127-130.

[A3] D.F. Anderson, *When the dual of an ideal is a ring*, Houston J. Math. **9** (1983), 325-332.

[AD] D.F. Anderson and D.E. Dobbs, *Root closure in integral domains, III*, Canad. Math. Bull., to appear.

[ADF] D.F. Anderson, D.E. Dobbs and Marco Fontana, *On treed Nagata rings*, J. Pure Appl. Algebra **61** (1989), 107-122.

[ADR] D.F. Anderson, D.E. Dobbs and Moshe Roitman, *Root closure in commutative rings*, Ann. Sci. Univ. Clermont II, Sér.Math. **26** (1990), 1-11.

[BCM] J.W. Brewer, D.L. Costa and K. McCrimmon, *Seminormality and root closure in polynomial rings and algebraic curves*, J. Algebra **58** (1979), 217-226.

[G] R. Gilmer, Multiplicative Ideal Theory, Marcel Dekker, New York, 1972.

[GH] R. Gilmer and R. Heitmann, *On Pic(R[x]) for R seminormal*, J. Pure Appl. Algebra **16** (1980), 251-258.

[HH] J.R. Hedstrom and E.G. Houston, *Pseudo-valuation domains*, Pacific J. Math. **75** (1978), 137-147.

[K] G. Karpilovsky, Unit Groups of Classical Rings, Oxford Univ. Press, Oxford, 1988.

[R] M. Roitman, *On root closure in Noetherian domains*, in: Factorization in Integral Domains, Lecture Note in Pure and Applied Mathematics, Marcel Dekker **189** (1997), 417-428.

[S] P. Samuel, *La notion de place dans un anneau*, Bull. Soc. Math. France **85** (1957), 123-133.

[SS] J. Sato and T. Sugatani, *On the radical ideals of seminormal rings*, Comm. Algebra **18** (1990), 441-451.

On Φ-Pseudo-Valuation Rings

AYMAN BADAWI Department of Mathematics and Computer
Science, Birzeit University, P.O. Box 14, Birzeit
WestBank, Palestine, via Israel.

1   INTRODUCTION

Throughout this paper, all rings are commutative with
identity and if R is a ring, then Z(R) denotes the set of
zerodivisors of R and Nil(R) denotes the set of nilpotent
elements of R. Our main purpose is to provide another
generalization of pseudo-valuation domains (as introduced in
[10]) to the context of arbitrary rings (with Z(R)
possibly nonzero). Recall from [10] that an integral domain
R with quotient field K is called a pseudo-valuation domain
(PVD) in case each prime ideal P of R is strongly prime (or a
strong prime), in the sense that xy ∈ P, x ∈ K, y ∈ K implies
that either x ∈ P or y ∈ P. Anderson, Dobbs, and the author
in [7] generalized the study of pseudo-valuation domains to
the context of arbitrary rings. Recall from [7] that a prime
ideal P of a ring R is said to be strongly prime (or a strong
prime) if aP and bR are comparable for all a,b ∈ R. If R is
an integral domain this is equivalent to the original
definition of strongly prime as introduced by Hedstrom and
Houston in [10] (cf. [1, Proposition 3.1], [2 Proposition
4.2], and [5, Proposition3]). If each prime ideal of R is
strongly prime, then R is called a pseudo-valuation ring
(PVR).

2   RESULTS

First, recall from [6] and [8] that a prime ideal of R
is called divided if it is comparable to every principal
ideal of R; equivalently, if it is comparable to every ideal
of R. If every prime ideal of R is divided, then R is called
a divided ring.

In the following proposition, we show that if a ring R
admits a strongly prime ideal, then Nil(R) is a strongly
prime ideal and thus Nil(R) is a divided prime. This result
justifies our focus in studying pseudo-valuation rings to be
restricted to rings R where Nil(R) is a divided prime.

PROPOSITION 0   Let P be a strongly prime ideal of a ring R.
Then the prime ideals of R contained in P are strongly prime
and are linearly ordered. In particular, Nil(R) is strongly
prime and therefore it is a divided prime.
Proof: Let Q be a prime ideal of R contained in P. By
applying the same argument as in the proof of [7, Theorem 2],
we conclude that Q is strongly prime. By [7, Lemma 1], P is
comparable to every prime ideal of R and the prime ideals of
R contained in P are linearly ordered. Hence, Nil(R) is
prime and therefore it is strongly prime and divided.

Now we state our definition of ϕ-pseudo-valuation
rings.

DEFINITION Let R be a ring such that Nil(R) is a divided
prime, let S be the set of nonzerodivisors of R, let $T = R_S$ be
the total quotient ring of R, and let $K = R_{Nil(R)}$. Define ϕ :
$T \longrightarrow K$ by $\phi(a/b) = a/b$ for every $a \in R$ and $b \in S$. Then ϕ
is a ring homomrphism from T into K, and ϕ restricted to R is
also a ring homomorphism from R into K given by $\phi(x) = x/1$
for every $x \in R$. Also, observe that $\phi(R)$ is a subring of K
with identity. A prime ideal Q of $\phi(R)$ is called K-strongly
prime if $xy \in Q$, $x \in K$, $y \in K$ implies that either $x \in Q$ or
$y \in Q$. If each prime ideal of $\phi(R)$ is K-strongly prime, then
$\phi(R)$ is called a K-pseudo-valuation ring (K-PVR). A prime
ideal P of R is called ϕ-strongly prime, if $\phi(P)$ is a K-
strongly prime ideal of $\phi(R)$. If each prime ideal of R is
ϕ-strongly prime then R is called a ϕ-pseudo-valuation ring
(ϕ-PVR). Observe that Q is a prime ideal of $\phi(R)$ if and only
if $Q = \phi(P)$ for some prime ideal P of R, and R is a ϕ-PVR if
and only if $\phi(R)$ is a K-PVR.

Throughout this section, R denotes a commutative ring with identity such that Nil(R) is a divided prime. Given a ring R, let K =$R_{Nil(R)}$ and T = R_s, where S is the set of nonzerodivisors of R.

Observe that an integral domain R is a PVD if and only if it is a ϕ-PVR. In fact, in Corollary 7, we show that a PVR (in the sense of [7]) is always a ϕ-PVR. Also, observe that a quasilocal zero-dimensional ring is a ϕ-PVR. The following is an example of a zero-dimensional ϕ-PVR that is not a PVR.

EXAMPLE 1 ([7, Remark 15]) Let K be a field, X,Y, and Z be indeterminates, and R = K[X,Y,Z] / (X^2,Y^2,Z^2) = K[x,y,z]. Then R is quasilocal zero-dimensional with maximal ideal Nil(R) = (X,Y,Z) / (X^2,Y^2,Z^2) = (x,y,z); hence R is a ϕ-PVR. However, R is not a PVR since xz $\notin$ yR and y $\notin$ xNil(R).

PROPOSITION 2   For a ring R, we have the following :
(1) Ker (ϕ) is contained in Nil(R).
(2) ϕ(R) is an integral domain if and only if for every nonzero w $\in$ Nil(R) there exists a z $\in$ Z(R)\Nil(R) such that zw = 0 in R.
Proof: (1). Let x $\in$ Ker(ϕ). Then x = a/b for some a $\in$ R and b $\in$ S such that ϕ(a/b) = a/b = 0/1 in K. Hence, za = 0 in R for some z $\in$ Z(R)\Nil(R). Thus, a $\in$ Nil(R) since Nil(R) is prime. Hence, x = a/b = w $\in$ Nil(R) since b $\in$ S and Nil(R) is divided. (2). Suppose that ϕ(R) is an integral domain. Since R/Ker(ϕ) $\approx$ ϕ(R) and Ker(ϕ) $\subset$ Nil(R), we have Ker(ϕ) = Nil(R), and the claim is now clear. Conversely, since for every nonzero w $\in$ Nil(R) there is a z $\in$ Z(R)\Nil(R) such that zw = 0 in R, we have Ker(ϕ) = Nil(R). Since Nil(R) is prime and R/Nil $\approx$ ϕ(R), ϕ(R) is an integral domain.

PROPOSITION 3 For a ring R, we have the following:
(1). Nil(T) = Nil(R) and Nil(K) = Nil(ϕ(R))= ϕ(Nil(R)).
(2). Let x $\in$ Nil(K) and write x=a/b for some a $\in$ R and b $\in$ R\Nil(R). Then a $\in$ Nil(R) and x = a/b = w/1 in K for some w $\in$ Nil(R).

(3). Let $x \in K$ and write $x = a/b$ for some $a \in R$ and
$b \in R \backslash \text{Nil}(R)$. If $a/b = i/1$ in K for some $i \in R$, then $b|a$ in
R; in particular, $a = (i+w)b$ in R for some $w \in \text{Nil}(R)$, and
therefore a is contained in every prime ideal of R which
contains i.
(4). Let $x \in R$ and $y \in R \backslash \text{Nil}(R)$. If $x/1 = y/1$ in K, then x
$= uy$ in R for some unit u of R; in particular, $(x) = (y)$ in
R.
Proof: **(1).** Note that $\text{Nil}(T) = \text{Nil}(R)$ since $\text{Nil}(R)$ is a
divided prime ideal of R. For the second equality, we only
need show that $\text{Nil}(K) \subset \text{Nil}(\phi(R))$. Let $x \in \text{Nil}(K)$ and write
$x = a/b$ for some $a \in R$ and $b \in R \backslash \text{Nil}(R)$. Since $\text{Nil}(R)$ is
prime, it follows that $a \in \text{Nil}(R)$. Since $\text{Nil}(R)$ is a divided
prime and $a \in \text{Nil}(R)$ and $b \in R \backslash \text{Nil}(R)$, $x = a/b = w/1$ for some
$w \in \text{Nil}(R)$. Thus, $x \in \text{Nil}(\phi(R))$. **(2).** Clear by the proof of
(1). **(3).** Since $a/b = i/1$ in K, $z(a-bi) = 0$ in R for some
$z \in R \backslash \text{Nil}(R)$. Thus, $a-bi = c \in \text{Nil}(R)$ since $\text{Nil}(R)$ is prime.
Since $b \in R \backslash \text{Nil}(R)$ and $\text{Nil}(R)$ is a divided prime, $c = wb$ for
some $w \in \text{Nil}(R)$. Hence, $a-bi = c = wb$. Thus, $a = (i+w)b$.
(4). Since $x/1 = y/1$ in K, $z(x-y) = 0$ in R for some
$z \in R \backslash \text{Nil}(R)$. Thus, $x-y = w \in \text{Nil}(R)$. Once again, since y
$\in R \backslash \text{Nil}(R)$, $w = dy$ for some $d \in \text{Nil}(R)$. Hence, $x-y = w = dy$.
Thus, $x = (1+d)y$. Since $1+d$ is a unit of R, the claim is
clear.

In light of the above proposition, observe that K is
quasilocal, zero-dimensional, and a K-PVR with maximal ideal
$\text{Nil}(\phi(R))$. In general, let A be a divided ring and I be an
ideal of A, and let $R = A/I$. Then K is a K-PVR with maximal
ideal $\text{Nil}(\phi(\text{Rad}(I)/I))$, where $\text{Rad}(I)$ is the radical ideal of
I in A.

The following result is an analogue of [10, Corollary
1.3] and [7, Lemma 1], also see [4, Proposition 1].

PROPOSITION 4   Let P be a ϕ-strongly prime ideal of R. Then
P (resp., $\phi(P)$) is a divided prime. In particular, if R is a
ϕ-PVR, then R (resp., $\phi(R)$) is a divided ring and hence is
quasilocal.
Proof: Deny. Then for some ideal I of R, there is an $i \in$
$I \backslash P$ and a $p \in P \backslash I$. Since $\text{Nil}(R) \subset P$, $i \in R \backslash \text{Nil}(R)$. Hence,

(p/i)(i/1) = p/1 ∈ φ(P). Since i/1 ∉ φ(P) by Proposition 3(4), p/i ∈ φ(P). Hence, i|p in R by Proposition 3(3). Thus, p ∈ I which is a contradiction.

The following result is an analogue of [10, Theorem 1.4], [2, Proposition 4.8], [4, Propositon2], and [7, Theorem 2].

PROPOSITION 5 1. Let P be a φ-strongly prime ideal of R and suppose that Q is a prime ideal of R contained in P. Then Q is φ-strongly prime. In particular, R is a φ-PVR if and only if some maximal ideal of R is φ-strongly prime.
2. Let P be a K-strongly prime ideal of φ(R). If Q is a prime ideal of φ(R) contained in P, then Q is K-strongly prime. In particular, φ(R) is a K-PVR if and only if some maximal ideal of φ(R) is K-strongly prime.
Proof: (1). Suppose that $xy \in \phi(Q)$ for some $x \in K$ and $y \in K$. If $xy \in Nil(\phi(R))$, then either $x \in Nil(\phi(R)) \subset \phi(Q)$ or $y \in Nil(\phi(R) \subset \phi(Q)$ since K is a K-PVR with maximal ideal $Nil(\phi(R))$. Hence, we may assume that $xy \notin Nil(\phi(R))$ and $x \in K \backslash \phi(R)$. Since $xy \in \phi(P)$ and $x \in K \backslash \phi(R)$, we must have $y \in \phi(P)$. Since $x(y^2/xy) = y \in \phi(P)$ and $x \in K \backslash \phi(R)$, we must have $y^2/xy = p/1 \in \phi(P)$ for some $p \in P$. Thus, $y^2 = (xy)(p/1)$ in K. Since $xy \in \phi(Q)$, $y^2 \in \phi(Q)$. Thus, $y \in \phi(Q)$. (2). Since every prime ideal of φ(R) is of the form φ(G) for some prime ideal G of R, the claim is clear.

The following lemma is an analogue of [10, Proposition 1.2]. Since the proof is exactly the same as in [10], we leave the proof to the reader.

LEMMA 6 A prime ideal P of R is φ-strongly prime if and only if $x^{-1}\phi(P) \subset \phi(P)$ for every $x \in K \backslash \phi(R)$.

COROLLARY 7 (1). A prime ideal P of R is φ-strongly prime if and only if for every $a,b \in R \backslash Nil(R)$, either a|b in R or aP ⊂ bP.
(2). A ring R is a φ-PVR if and only if for every $a,b \in R \backslash Nil(R)$, either a|b in R or b|ac in R for every nonunit c of R.
(3). If R is a PVR, then R is a φ-PVR.

Proof: (1). Suppose that P is φ-strongly prime and
a,b ∈ R\Nil(R) such that a/b in R. Then b/a ∈ K\φ(R) by
Proposition 3(3). Let p ∈ P. Then (a/b)(p/1) = q/1 in K for
some q ∈ P by Lemma 6. Thus, ap = (q+w)b in R for some
w ∈ Nil(R) by Proposition 3(3). Hence, ap ∈ bP in R. Thus,
aP ⊂ bP in R. Conversely, suppose that for every
a,b ∈ R\ Nil(R) either a|b or aP ⊂ bP. Let x ∈ K\φ(R). Then
x = b/a for some a,b ∈ R\Nil(R) (observe that b ∉ Nil(R)
since Nil(R) is divided). Hence, a/b in R by Proposition
3(3). Thus, aP ⊂ bP in R. Hence, (a/b) φ(P) ⊂ φ(P). Thus,
P is φ-strongly prime by Lemma 6. (2). If R is a φ-PVR with
maximal ideal M, then the claim is clear by (1). Conversely,
since for every a,b ∈ R either a|b^n or b|a^m for some n,m ≥ 1,
the prime ideals of R are linearly ordered by [5, Theorem 1].
Hence R is quasilocal with maximal ideal M. Once again, the
claim is clear by (1). (3). This is clear by [7, Theorem 5].

REMARK 8   It was shown in [7, Theorem 5] that a ring R is a
PVR if and only for every a,b ∈ R, either a|b or b|ac for
every nonunit c of R. Thus, Corollary 7(2) gives a clear
difference between a PVR and a φ-PVR.

The first part of the following proposition follows
easily since the prime ideals of a divided ring R are
linearly ordered and Z(R) is a union of prime ideals of R.

PROPOSITION 9   Let R be a divided ring. Then
(1). Z(R) is a prime ideal of R.
(2). If x ∈ T\R, then x^{-1} ∈ T.
Proof: (2). Let x = a/b ∈ T\R for some a ∈ R and b ∈ S. Then
a ∈ S since R is divided. Hence, x^{-1} = b/a ∈ T.

Given an ideal I of R, then I:I = {x∈T : xI⊂I} and
φ(I) : φ(I) = {x∈K : xφ(I) ⊂ φ(I)}

PROPOSITION 10   Let R be a quasilocal ring with maximal
ideal M. Then
(1). R is a φ-PVR if and only if M:M is a φ-PVR with maximal
ideal M.
(2). φ(R) is a K-PVR if and only if φ(M) : φ(M) is a K-PVR
with maximal ideal φ(M).

Proof: **(1)**. Suppose that R is a φ-PVR. Let x ∈ M:M\R. Then φ(x) ∈ K\φ(R) by Proposition 3(3). Since x is a unit of T by Proposition 9(2), φ(x⁻¹)φ(M) = φ(x)⁻¹φ(M) ⊂ φ(M) by Lemma 6. Thus, x⁻¹ ∈ M:M. Thus, x is a unit of M:M. Hence, M is the maximal ideal of M:M. Thus, M:M is a φ-PVR since φ(M) is K-strongly prime. The converse is clear. **(2)**. This follows by a similar argument to that in (1).

Recall that a ring B is called an overring of R (resp., φ(R)) if R ⊂ B ⊂ T (resp., φ(R) ⊂ B ⊂ K).

PROPOSITION 11 Suppose that R is a φ-PVR with maximal ideal M.
(1). If B is an overring of φ(R) which contains an element of the form 1/s for some nonunit s ∈ R\Nil(R), then x⁻¹ ∈ B for every x ∈ K\φ(R). Furthermore, B is a K-PVR.
(2). If B is an overring of R which contains an element of the form 1/s for some nonunit s ∈ S, then x⁻¹ ∈ B for every x ∈ T\R. Furthermore B is a φ-PVR.
Proof: **(1)**. Suppose that B is an overring of φ(R) which contains an element of the form 1/s for some nonunit s ∈ R\Nil(R). Let x ∈ K\φ(R). Then x⁻¹(s/1) ∈ φ(M) ⊂ φ(R) by Lemma 6. Hence, x⁻¹ = (x⁻¹s)/s ∈ B since s⁻¹ ∈ B. Now, let N be a maximal ideal of B and xy ∈ N for some x,y ∈ K with x ∈ K\φ(R). Then y = x⁻¹(xy) ∈ N since x⁻¹ ∈ B. Thus, N is K-strongly prime. Hence, B is a K-PVR. **(2)**. Suppose that B is an overring of R which contains an element of the form 1/s for some nonunit s ∈ S. Then 1/s ∈ φ(B). Hence, φ(B) is a K-PVR by (1) and therefore B is a φ-PVR. Let x = a/b ∈ T\R for some a ∈ R and b ∈ S. Then x⁻¹ = b/a ∈ T by Proposition 9(2). Since b/a in R, a|sb in R by Corollary 7(2). Hence, sb = ga in R for some g ∈ R. Thus, x⁻¹ = b/a = g/s ∈ B since s⁻¹ ∈ B.

COROLLARY 12 Let R be a φ-PVR with maximal ideal M. Then
(1). For every prime ideal P of R, P:P is a φ-PVR.
(2). For every prime ideal P of φ(R), P:P is a K-PVR.
(3). For every prime ideal P of φ(R), φ(R)ₚ is a K-PVR.
Proof: **(1)**. If P is maximal, then the claim follows by Proposition 10. Hence, assume that P is nonmaximal. Since P is divided, P:P either contains an element of the form 1/s

for some nonunit s ∈ S, and in this case P:P is a φ-PVR by
Proposition 11; or P:P does not contain such an element, and
in this case it is a φ-PVR since it equals R. (2). This
follows by a similar argument to that in (1). (3). Once
again, if P is maximal, then $\phi(R)_P = \phi(R)$ is a K-PVR. If P is
nonmaximal, then $\phi(R)_P$ contains an element of the form 1/s
for some nonunit s ∈ R\Nil(R) and therefore it is a K-PVR by
Proposition 11.

Recall that a ring B is called a chained ring if the
principal ideals of B are linearly ordered.

PROPOSITION 13   Let R be a φ-PVR and let B be an overring of
R (resp., φ(R)) which contains an element of the form 1/s for
some nonunit s ∈ S (resp., s ∈ R\Nil(R)).   Then B is a
chained ring if and only if for every a,b ∈ Nil(R)
(resp.,Nil(φ(R)) either a|b in B or b|a in B.
Proof: We only need prove the converse. Suppose that B is an
overring of φ(R).   Let x,y ∈ B such that neither
x ∈ Nil(φ(R)) nor y ∈ Nil (φ(R)) and x∤y in B.   Then
d = $x^{-1}y$ ∈ K\φ(R). Hence, $d^{-1} = xy^{-1}$ ∈ B by Proposition 11.
Thus, x = $(xy^{-1})y$ in B.   Next, suppose that B is an overring
of R.   Let x,y ∈ B such that neither x ∈ Nil(R) nor
y ∈ Nil(R) and y∤x in B.   Since each d ∈ B\R is a unit of B
by Proposition 11, we may assume that x,y ∈ R.   Since y∤x in
B, y∤x in R, and therefore x|ys in R by Corollary 7(2).
Hence, ys = cx for some c ∈ R.   Hence, y = (c/s)x.   Thus,
x|y in B since c/s ∈ B.

Given a ring R, then R' denotes the integral closure of
R in T, and φ(R)' denotes the integral closure of φ(R) in K.
The following result is an analogue of [7, Lemma 17 and
Theorem 19].

PROPOSITION 14   Let R be a φ-PVR with maximal ideal M.   Then
(1). R' ⊂ M:M and R' is a φ-PVR with maximal ideal M.
(2). φ(R)' ⊂ φ(M):φ(M) and φ(R)' is a K-PVR with maximal
ideal φ(M).
Proof: (1). Let x ∈ R'\R.   Then x^{-1}∉R.   For, if x^{-1} ∈ R, then
x = 1/s for some nonunit s ∈ S which is impossible by [12,

Theorem 15]. Since $x^{-1} \notin R$, $\phi(x^{-1}) \notin \phi(R)$ by Proposition 3(3), and hence $\phi(x)\phi(M) \subset \phi(M)$ by Lemma 6. Thus, $xM \subset M$. Hence, $x \in M:M$ and M is a prime ideal of R' (observe that if $zw \in M$ for some $z,w \in T$, then either $z \in M$ or $w \in M$ since M is ϕ-strongly prime). Since $R \subset R'$ satisfies the INC condition by [12, Theorem 47], M is the maximal ideal of R. Hence, R' is a ϕ-PVR. (2). Apply a similar argument as in (1).

Our final result is an analogue of [11, Proposition 2.7], [9, Proposition 4.2], and [7, Theorem 21].

PROPOSITION 15   Let R be a ϕ-PVR with maximal ideal M. Then (1). Every overring of R is a ϕ-PVR if and only if R' = M:M. (2). Every overring of $\phi(R)$ is a K-PVR if and only if $\phi(R)' = \phi(M):\phi(M)$.
Proof: (1). Let C be an overring of R that does not contain an element of the form 1/s for some nonunit $s \in S$. Then observe that $C \subset M:M$, and use a similar argument as in [7, Theorem 21]. (2). Once again, let C be an overring of $\phi(R)$ that does not contain an element of the form 1/s for some nonunit $s \in R \backslash Nil(R)$. Then observe that $C \subset \phi(M):\phi(M)$, and use a similar argument as in the proof of [7, Theorem 21].

ACKNOWLEDGMENT

I am very grateful to the referee for his many corrections.

REFERENCES

[1] D.F. Anderson, Comparability of ideals and valuation overrings, *Houston J. Math.* 5(1979), 451-463.
[2] D.F. Anderson, When the dual of an ideal is a ring, *Houston J. Math.* 9(1983), 451-463.
[3] D.F. Anderson and D.E. Dobbs, Pairs of rings with the same prime ideals, *Canad. J. Math.* 32 (1980), 362-384.
[4] A. Badawi, A visit to valuation and pseudo-valuation domains, *in Commutative Ring Theory*, Lecture Notes in Pure and Appl. Math., Vol. 171 (1995), 155-161, Marcel Dekker, Inc., New York/Basel.

[5] A. Badawi, On domains which have prime ideals that are linearly ordered, *Comm. Algebra* 23 (1995), 4365-4373.

[6] A. Badawi, On divided commutative rings, to appear in *Comm.Algebra*.

[7] A. Badawi, D.F. Anderson, D.E. Dobbs, Pseudo-valuation rings, *Proceedings of the Second International Conference on Commutative Rings*, Lecture Notes in Pure and Appl. Math., Vol. 185(1997), 57-67, Marcel Dekker, Inc., New York/Basel.

[8] D.E. Dobbs, Divided rings and going-down, *Pacific J. Math.*, 67(1976), 353-363.

[9] D.E. Dobbs, Coherence, ascent of going down, and pseudo-valuation domains, *Houston J. Math.* 4(1978), 551-567.

[10] J.R. Hedstrom and E.G. Houston, Pseudo-valuation domains, *Pacific J. Math.* 75(1978), 137-147.

[11] J. R. Hedstrom and E.G. Houston, Pseudo-valuation domains, II, *Houston J. Math.* 4(1978), 199-207.

[12] I. Kaplansky, Commutative Rings, rev. ed., Univ. Chicago Press, Chicago, 1974.

Foliations, Spectral Topology, and Special Morphisms

EZZEDDINE BOUACIDA, OTHMAN ECHI and EZZEDDINE SALHI

Department of Mathematics, Faculty of Sciences of Sfax, 3038 Sfax, Tunisia.

Abstract. *Let $\mathcal{F}$ be a codimension-one foliation of class $C^r (r \geq 0)$, transversally oriented on a closed connected manifold M. Set $X = M/\overline{\mathcal{F}}$ the space of leaves classes of $\mathcal{F}$ (the class of a leaf F is the union of the leaves G such that $\overline{F} = \overline{G}$). Suppose that $\mathcal{F}$ has a well defined height. Let X_0 be the union of open subsets of X homemorphic to $\mathbb{R}$ or S^1. In a recent paper we have shown that $X \setminus X_0$ is a spectral space. Let $Z = M/\mathcal{F}$ be the space of leaves of $\mathcal{F}$ and Z_0 be the union of open subsets of Z homeomorphic to $\mathbb{R}$ or S^1. In this paper we introduce the notion of special morphism used to prove that $Z \setminus Z_0$ is a quasi-spectral space.*

AMS Subject Classification: 54 A10, 54 B35, 57 R30.

INTRODUCTION

Let M be a closed connected manifold, $\mathcal{F}$ a foliation defined on M, transversally oriented, of class $C^r (r \geq 0)$ and of codimension 1. Let F be a leaf, the class $Cl(F)$, of F is the union of all leaves G of $\mathcal{F}$ such that $\overline{F} = \overline{G}$

[15]. We define the equivalence relation $\overline{\mathcal{F}}$ on M by : $x\overline{\mathcal{F}}y$ if and only if, $\overline{F}_x = \overline{F}_y$, where F_x and F_y are the leaves of $\mathcal{F}$ containing respectively x and y. The quotient space, denoted by $M/\overline{\mathcal{F}}$, is called the *space of leaves classes*, it is a T_0-space. However, the *space of leaves* $M/\mathcal{F}$ is not in general a T_0-space. We remark that if every leaf is proper then $M/\mathcal{F} = M/\overline{\mathcal{F}}$.

In this work, we compare the quotient topology of $M/\overline{\mathcal{F}}$ with a spectral topology.

Recall that a topological space Y is said to be a *spectral space* [13] if, it satisfies the following properties:

i) Y is a T_0-space (that is, for every pair of distinct points x and y of Y, there exists a neighborhood V_x of x which does not contain y or there exists a neighborhood V_y of y which does not contain x).

ii) Y is quasi-compact and has a basis of quasi-compact open subsets.

iii) Each irreducible closed subset of Y has a generic point (that is, if A is a closed subset of Y such that the intersection of two nonempty open subsets of A is nonempty, then there exists an element a of A, called a generic point, such that $\overline{\{a\}} = A$).

iv) The intersection of two quasi-compact open subsets is quasi-compact.

The space Y is said to be *quasi-spectral* [2] if it satisfies the above properties ii), iii) and iv).

Every spectral space is homeomorphic to the spectrum of a unitary commutative ring equipped with the Zariski topology [13].

Throughout this paper, we denote by $X = M/\overline{\mathcal{F}}$, X_0 the union of open subsets of X homeorphic to $\mathbb{R}$ or to S^1 and p the canonical projection of M onto X. The set X is ordered by the ordering defined by: $p(x) \leq p(y)$ if and only if, $\overline{F}_x \subseteq \overline{F}_y$.

A point a of X is said to be of height 0 if it is a minimal point of X. By transfinite induction, we say that a has a height the ordinal number α, if every point $b < a$ has a height equal to an ordinal number $\beta < \alpha$, and for each ordinal number $\beta < \alpha$, there exists an element b of X of height β

such that $b < a$ [3]. The level of a leaf (defined in [17]) corresponds to the height of its class in X. The foliation $\mathcal{F}$ is said to be of well defined height if every leaf has a level, the supremum of the levels is called the height of the foliation $\mathcal{F}$. The foliation $\mathcal{F}$ has a well defined height if and only if, for each leaf F and each saturated open subset U of M containing F, the intersection $\overline{F} \cap U$ contains a minimal set of the restriction of $\mathcal{F}$ to U [16] (this is the case if the foliation $\mathcal{F}$ is of class C^r, $(r \geq 2)$ [6]).

The main result of [4] consists to prove that, if the foliation $\mathcal{F}$ has a well defined height, then $X \setminus X_0$ is a spectral space and its height is countable. This result fails to be true if the height of $\mathcal{F}$ is not well defined. Moreover, we can construct some connected spectral spaces which are not homeorphic to a space $X \setminus X_0$.

In this work, when M is a closed manifold and $\mathcal{F}$ is a codimension one foliation, of class $C^r(r \geq 0)$, tranversally oriented, defined on M, we denote by Z the space of leaves $M / \mathcal{F}$ and Z_0 the union of the open subsets of Z homeomorphic to $\mathbb{R}$ or to S^1. The notion of *Special Morphism* introduced in the second paragraph of this paper permits to link some properties of a topological space which is not necessarely T_0, with the analogous properties in a T_0-space. This notion of special morphism recalls the notion of quasi-homeomorphism introduced in [10]. We use these topological instruments to prove, without geometrical consideration, that if $\mathcal{F}$ has a well defined height, then $Z \setminus Z_0$ is quasi-spectral.

Throughout this paper "⊂" and "⊆" will denote respectively proper containment and containment with possible equality.

1. FOLIATIONS AND SPECTRAL TOPOLOGY

In this section, we review some results established in [4]

a) Subsets compact by saturation

In order to study the quasi-compact open subsets of X, we introduce the notion of subsets of M compact by saturation.

DEFINITION 1.1 A saturated subset S of M is said to be *compact by saturation* if each covering of S by saturated open subsets has a finite subcovering (that is, S is quasi-compact for the topology defined by the saturated open subsets of M).

REMARK 1.2 *Every open subset compact by saturation has finitely many connected components.*

LEMMA 1.3 *A saturated open subset U of M is compact by saturation if and only if, it is the saturation of a compact subset T of U.*

LEMMA 1.4 *Let U be a saturated open subset of M. Then the following statements are equivalent.*
a) U is compact by saturation.
b) U has finitely many connected components and the two following properties are satisfied:

   i) For each leaf L of $\delta^\varepsilon U(\varepsilon = \pm)$, L is attractant from the side ε.

   ii) For each leaf F contained in U, the intersection $\overline{F} \cap U$ contains a minimal set of $\mathcal{F}$ restricted to U.

REMARK 1.5 [17] *If the foliation $\mathcal{F}$ restricted to U has a well defined height, then the condition b-ii) of Lemma 1.4 is satisfied. In particular, if $\mathcal{F}$ is of class $C^r(r \geq 2)$ this condition holds for all saturated open subset U of M [6].*

COROLLARY 1.6 *Let $\mathcal{F}$ be a foliation on M of well defined height and U a saturated open subset of M with finitely many connected components. The open subset U is compact by saturation if and only if, every leaf L of $\delta^\varepsilon U$ is attractant from the side ε $(\varepsilon = \pm)$.*

COROLLARY 1.7 *If $\mathcal{F}$ is a foliation of well defined height, then the intersection of two open subsets compact by saturation is compact by saturation.*

REMARK 1.8 *There exists a foliation defined on the manifold $M = V_2 \times S^1$ (where V_2 is a closed connected surface of genus 2) and two open subsets of M compact by saturation such that their intersection is not compact by saturation.*

b) Foliations

Since M is a compact manifold, the space $X = M/\overline{\mathcal{F}}$ is quasi-compact. We prove that the space $X \setminus X_0$ has a basis of quasi-compact open subsets. Example 1.11 proves that, in general, it is not the case for X.

PROPOSITION 1.9 *The topological space X is a T_0-space.*

THEOREME 1.10 *The following statements are equivalent:*

i) X is a T_2-space (Hausdorff space)

ii) X is a T_1-space(accessible space)

iii) The foliation $\mathcal{F}$ is minimal (that is, the closure of every leaf of $\mathcal{F}$ is a minimal set).

EXAMPLE 1.11 *We consider the two dimensional torus T^2 equipped with a foliation $\mathcal{F}$ which has a unique periodic leaf. The space of leaves classes X has no basis of quasi-compact open subsets. X is not a Hausdorff-space(that is, is not T_2), hence it is not accessible (that is, is not T_1)(cf. Theorem 1.10).*

REMARK 1.12 *It follows, from Theorem 1.10, that if $\mathcal{F}$ is not minimal, then X is a T_0-space which is not accessible and that if $\mathcal{F}$ is minimal, then X is a Hausdorff space. This result fails to be true if M is not compact.*

PROPOSITION 1.13 *The topological space $X \setminus X_0$ has a basis of quasi-compact open subsets.*

To prove this proposition, we need the following:

LEMMA 1.14 *An open subset V of X is quasi-compact if and only if, the open subset $U = p^{-1}(V)$ is compact by saturation.*

PROPOSITION 1.15 *If the foliation $\mathcal{F}$ has a well defined height, then the intersection of two quasi-compact open subsets of $X \setminus X_0$ is quasi-compact.*

For the proof of this proposition, we state the following:

LEMMA 1.16 *If the foliation $\mathcal{F}$ has a well defined height, then the intersection of two quasi-compact open subsets of X is quasi-compact.*

This result does not hold if the height of $\mathcal{F}$ is not well defined (cf. Remark 1.8).

PROPOSITION 1.17 *Each irreducible closed subset B of X has a generic point.*

For the proof, we establish the two following lemmas:

LEMMA 1.18 *Let $a = p(F)$. Then a is not a generic point of the irreducible closed subset B of X if and only if, there exists a leaf G of $A = p^{-1}(B)$ with $F < G$ (that is, $\overline{F} \subset \overline{G}$).*

LEMMA 1.19 [17] *Let $\mathcal{T} = (F_i, i \in I)$ be a totally ordered set of leaves, then $\mathcal{T}$ has an upper bound G. Moreover, we have $\overline{G} = \overline{\bigcup_{i \in I} F_i}$.*

Since $X \setminus X_0$ is a closed subset of X, we obtain:

COROLLARY 1.20 *Each irreducible closed subset of $X \setminus X_0$ has a generic point.*

Finally, we conclude with the following:

THEOREM 1.21 *Let M be a closed connected manifold and $\mathcal{F}$ a foliation on M, of codimension 1, of class C^r ($r \geq 0$) and transversally oriented. We denote $X = M/\overline{\mathcal{F}}$ the space of leaves classes and X_0 the union of open subsets of X homeomorphic to $\mathbb{R}$ or to S^1. If $\mathcal{F}$ has a well defined height, then $X \setminus X_0$ is a spectral space. Moreover the space X has a countable height.*

Note that, the hypothesis that $\mathcal{F}$ has a well defined height is necessary (cf. Remark 1.8 and Lemma 1.14).

2. SPECIAL MORPHISMS AND FOLIATIONS

Let $\mathcal{F}$ be a codimension-one foliation of class $C^r(r \geq 0)$, transversally oriented on a closed connected manifold M, $Z = M/\mathcal{F}$ be the space of leaves of $\mathcal{F}$ and Z_0 be the union of open subsets of Z homeomorphic to $\mathbb{R}$ or S^1. To study the relationship betwen the spaces $Z = M/\mathcal{F}$ and $X = M/\overline{\mathcal{F}}$, we introduce the notion of *special morphism* as a continuous and surjective map $p : X \longrightarrow Y$ such that every open subset U of X is saturated under p (that is, $p^{-1}(p(U)) = U$). This notion is used to prove that $Z \setminus Z_0$ is quasi-spectral.

a) Special morphisms

First, we define the notion of saturated subset under a map.

DEFINITION 2.1 *Let X, Y be two sets and $p : X \longrightarrow Y$ be a map. The subset S of X is said to be saturated under the map p if $p^{-1}(p(S)) = S$ (equivalently, there exists a subset L of Y such that $S = p^{-1}(L)$).*

The following proposition gives some elementary properties of saturated subsets.

PROPOSITION 2.2 *Let $p : X \longrightarrow Y$ be a map.*

1) The subset S of X is saturated under p if and only if, for every subset

T of X, $p(S \cap T) = p(S) \cap p(T)$.

2) Let $(S_i, i \in I)$ be a collection of saturated subsets of X under p, then we have,

    a) $\bigcup_{i \in I} S_i$ *is saturated under p*

    b) $\bigcap_{i \in I} S_i$ *is saturated under p, and* $p(\bigcap_{i \in I} S_i) = \bigcap_{i \in I} p(S_i)$

3) If S is a saturated subset of X under p, then $X \setminus S$ is also saturated under p, and $p(X \setminus S) = Y \setminus p(S)$.

We obtain at once the following result.

PROPOSITION 2.3 *Let X, Y be two topological spaces and $p : X \longrightarrow Y$ be a surjective map. The following statements are equivalent.*

i) p is an open map and each open subset of X is saturated under p.

ii) p is a closed map and each closed subset of X is saturated under p.

Proof.

    (i) $\Rightarrow$ (ii) Let F be a closed subset of X, $X - F$ is saturated under p, so F is also saturated under p and $p(X - F) = Y - p(F)$ (cf. Proposition 2.2).

    On the other hand, p is an open map, this yields $p(X - F) = Y - p(F)$ is an open subset of Y. Therefore $p(F)$ is a closed subset of Y. The converse can be proved similarly. ∎

DEFINITION 2.4 Let X, Y be two topological spaces and $p : X \longrightarrow Y$ be a continuous surjective map. We shall say that p is a *special morphism* if p satisfies one of the conditions of the previous proposition.

EXAMPLE 2.5 *a) Let X be a set, Y be a topological space and $p : X \longrightarrow Y$ be a surjective map. Consider the inverse image topology on X determined by p (the open subsets of X are the $p^{-1}(V)$, where V is an open subset of Y). Then p is a special morphism. Conversely, if $p : X \longrightarrow Y$ is a special morphism, then the topology on X is the inverse image by p of that of Y.*

In particular, if $M/\mathcal{F}$ is the space of leaves classes and $p : M \longrightarrow M/\mathcal{F}$ is the canonical projection. If we equip M with the topology of saturated open subsets of M, then p is a special morphism.

b) Let X be a topological space. Consider the equivalence relation $\mathcal{R}$ defined on X by: $x\mathcal{R}y$ if and only if, $\overline{\{x\}} = \overline{\{y\}}$. Let $Y = X/\mathcal{R}$ be the quotient space of X by $\mathcal{R}$. The projection p of X onto $X/\mathcal{R}$ is a special morphism. Moreover $Y = X/\mathcal{R}$ equipped with the quotient topology is always a T_0-space.

REMARKS 2.6

1) The composition of two special morphisms is a special morphism.

2) If $p : X \longrightarrow Y$ is a special morphism, then every locally closed subset of X is saturated under p.

It is clear that a homeomorphism is a special morphism. For the converse, we have:

PROPOSITION 2.7 *Let $p : X \longrightarrow Y$ be a special morphism. If X is a T_0-space, then p is a homeomorphism.*

Proof.

Since p is continuous and open, it suffices to prove that p is injective. Let $x, y \in X$ such that $p(x) = p(y)$, then $\overline{\{p(x)\}} = \overline{\{p(y)\}}$. Since p is a closed continuous map, we have $\overline{\{p(x)\}} = p(\overline{\{x\}})$ and $\overline{\{p(y)\}} = p(\overline{\{y\}})$, hence $p(\overline{\{x\}}) = p(\overline{\{y\}})$. Using the fact that every closed subset of X is saturated under p, we have $p^{-1}(p(\overline{\{x\}})) = \overline{\{x\}}$ and $p^{-1}(p(\overline{\{y\}})) = \overline{\{y\}}$. Therefore $\overline{\{x\}} = \overline{\{y\}}$, and so, $x = y$ (since X is a T_0-space). ∎

Recall that a continuous map $f : X \longrightarrow Y$ is said to be a *quasi-homeomorphism* [10], if for each open subset U of X, there exists a unique open subset V of Y such that $U = f^{-1}(V)$ (equivalently, for each closed subset F of X, there exists a unique closed subset G of Y such that $F = f^{-1}(G)$).

A subset S of a topological space X is said to be *strongly dense* (in french: *très dense*), if for every nonempty locally closed subset Z of X, we have: $S \cap Z \neq \emptyset$. It is well known that, a continuous map $f : X \longrightarrow Y$ is a quasi-homeomorphism if and only if, the topology of X is the inverse image by f of the topology of Y and the subset $f(X)$ is strongly dense in Y [10].

It follows immediately from the definitions that we have the following.

PROPOSITION 2.8 *Every special morphism is a quasi-homeomorphism.*

PROPOSITION 2.9 *Let $f : X \longrightarrow Y$ be a continuous map, then the following statements are equivalent.*
(i) f is a quasi-homeomorphism.
(ii) The map $f_1 : X \longrightarrow f(X)$; $x \mapsto f(x)$ is a special morphism and $f(X)$ is strongly dense in X.

Proof.
(i) $\Rightarrow$ (ii) The map $f_1 : X \longrightarrow f(X)$ is a continuous surjective map. Let U be an open subset of X. There exists an open subset V of Y such that $U = f^{-1}(V)$. Thus U is saturated under f_1. On the other hand, we have $f_1(U) = f(f^{-1}(V)) = V \cap f(X)$. Hence $f(U)$ is an open subset of $f(X)$, and f_1 is an open map. Finally, since f is a quasi-homeomorphism, the subset $f(X)$ is strongly dense in Y [10].
(ii) $\Rightarrow$ (i) This follows immediately from Proposition 2.8 and the fact that the composition of two quasi-homeomorphisms is a quasi-homeomorphism.
■

PROPOSITION 2.10 *Let $f : X \longrightarrow Y$ be a continuous map, then the following statements are equivalent.*
(i) f is a special morphism.
(ii) f is an open quasi-homeomorphism.
(iii) f is a closed quasi-homeomorphism.

(iv) f is a surjective quasi-homeomorphism.

The proof needs the next lemma:

LEMMA 2.11 *Let X be a topological space and Y be a strongly dense subset of X. Then the following conditions are equivalent.*

(i) $X = Y$.

(ii) Y is a closed subset of X.

(iii) Y is an open subset of X.

(iv) Y is a locally closed subset of X.

Proof.

Obviously, we have the implications (i) $\Rightarrow$ (ii), (i) $\Rightarrow$ (iii), (i) $\Rightarrow$ (iv), (ii) $\Rightarrow$ (iv), (iii) $\Rightarrow$ (iv). To complete the proof, it suffices to prove that (iv) $\Rightarrow$ (i). For let U be an open subset of X and F a closed subset of X such that $Y = U \cap F$. If $U \subset X$, then $(X - U)$ is a nonempty locally closed subset of X. So that $(X - U) \cap Y \neq \emptyset$, which is a contradiction. It follows that $U = X$ and $Y = F$ is a closed subset of X such that $Y \cap (X - F) = \emptyset$. Since, in addition, Y is strongly dense in X, we obtain $F = X$. Therefore $Y = X$. ∎

Proof of 2.10.

Clearly, we have the implications (i) $\Rightarrow$ (ii), (i) $\Rightarrow$ (iii), (i) $\Rightarrow$ (iv). The implication (iv) $\Rightarrow$ (i) follows immediately from Proposition 2.9.

Now, if f satisfies the condition (ii) (resp. condition (iii)) then $f(X)$ is strongly dense in Y, Lemma 2.11 gives $f(X) = Y$. Hence f is a surjective quasi-homeomorphism, that is, f is a special morphism, completing the proof. ∎

REMARK 2.12 *A quasi-homeomorphism needs not be surjective or open. It suffices to consider a topological space X and a strongly dense subset Y of X such that $Y \neq X$, then the canonical injection $i : Y \longrightarrow X$ is a quasi-homeomorphism which is not surjective nor open.*

Let $p : X \longrightarrow Y$ be a special morphism. We consider the equivalence relation $\mathcal{R}$ defined on X by: for every $(a, b) \in X^2$, $a\mathcal{R}b$ if and only if, $p(a) = p(b)$, and denote by $q : X \longrightarrow X/\mathcal{R}$ the canonical surjection . The map $\varphi : X/\mathcal{R} \longrightarrow Y$, $q(a) \mapsto p(a)$ is a continuous bijection. Let V be an open subset of $X/\mathcal{R}$. There exists an open subset U of X such that $V = q(U)$, hence $\varphi(V) = \varphi(q(U)) = p(U)$ and $\varphi(V)$ is an open subset of Y (since p is an open map). Therefore φ is an open bijection (that is, φ is a homeomorphism) and q is a special morphism (as a composition of two special morphisms). The following is now immediate.

PROPOSITION 2.13 *Let $p : X \longrightarrow Y$ be a special morphism, then there exists an equivalence relation $\mathcal{R}$ on X and a homeomorphism $\varphi : X/\mathcal{R} \longrightarrow Y$, such that $p = \varphi \circ q$ (where q is the projection of X onto $X/\mathcal{R}$). Moreover, q is a special morphism.*

One of our aims in this section is to prove the following.

THEOREM 2.14 *Let $p : X \longrightarrow Y$ be a quasi-homeomorphism.*

1) If X is quasi-spectral, then Y is also quasi-spectral.

2) Suppose that Y is quasi-spectral. Then the following properties hold.

(i) X is quasi-compact and has a basis of quasi-compact open subsets.

(ii) The intersection of two quasi-compact open subsets of X is quasi-copact.

3) If Y is spectral, then the following statements are equivalent.

(i) X is quasi-spectral.

(ii) p is sujective.

To prove this theorem, we need the following lemma:

LEMMA 2.15 *Let $p : X \longrightarrow Y$ be a quasi-homeomorphism, and S be a subset of Y.*

1) If S is an open subset of Y, then the following statements are equivalent.

(i) S is quasi-compact in Y.

(ii) $p^{-1}(S)$ *is quasi-compact in* X.

2) If S *is a closed subset of* Y, *then the following statements are equivalent.*

(i) S *is irreducible in* Y.

(ii) $p^{-1}(S)$ *is irreducible in* X.

Proof.

1) Suppose that S is a quasi-compact open subset of Y. Let $(U_i, \ i \in I)$ be an open covering of $p^{-1}(S)$; $p^{-1}(S) = \bigcup_{i \in I} U_i$. For each $i \in I$, there exists an open subset V_i of Y such that $U_i = p^{-1}(V_i)$, then $p^{-1}(S) = p^{-1}(\bigcup_{i \in I} V_i$). Hence $S = \bigcup_{i \in I} V_i$ (since p is a quasi-homeomorphism). It follows from the fact that S is quasi-compact, that there exists a finite subset J of I such that $S = \bigcup_{i \in J} V_i$, which gives $p^{-1}(S) = \bigcup_{i \in J} U_i$. Thus $p^{-1}(S)$ is quasi-compact.

Conversely, suppose that $p^{-1}(S)$ is quasi-compact. Let $(V_i, \ i \in I)$ be an open covering of S; $S = \bigcup_{i \in I} V_i$. Then $p^{-1}(S) = \bigcup_{i \in I} p^{-1}(V_i)$. Thus there exists a finite subset J of I such that

$$p^{-1}(S) = \bigcup_{i \in J} p^{-1}(V_i) = p^{-1}(\bigcup_{i \in J} V_i)$$

Using the fact that p is a quasi-homeomorpism, we obtain $S = \bigcup_{i \in J} V_i$. This proves that S is quasi-compact.

2) Suppose that S is an irreducible closed subset of Y. Let F_1 and F_2 be two closed subsets of X such that $p^{-1}(S) = F_1 \cup F_2$. Since p is a quasi-homeomorphism, there exists two closed subsets G_1 and G_2 of Y such that $F_1 = p^{-1}(G_1)$ and $F_2 = p^{-1}(G_2)$. Hence $p^{-1}(S) = p^{-1}(G_1 \cup G_2)$, which gives $S = G_1 \cup G_2$ (since p is a quasi-homeomorphism). From the fact that S is an irreducible closed subset of Y, it follows that $S = G_1$ or $S = G_2$, this yields $p^{-1}(S) = F_1$ or $p^{-1}(S) = F_2$. Therefore $p^{-1}(S)$ is an irreducible closed subset of X.

Conversely, suppose that $p^{-1}(S)$ is an irreducible closed subset of X. Let G_1 and G_2 be two closed subsets of Y such that $S = G_1 \cup G_2$, then $p^{-1}(S) = p^{-1}(G_1) \cup p^{-1}(G_2)$. Since $p^{-1}(S)$ is irreducible, we have $p^{-1}(S) = p^{-1}(G_1)$ or $p^{-1}(S) = p^{-1}(G_2)$. This yields $S = G_1$ or $S = G_2$ (since p is a quasi-homeomorphism). Thus S is an irreducible closed subset of Y. ∎

Proof of 2.14.

1) Suppose that X is quasi-spectral

– Since $X = p^{-1}(Y)$, Y is quasi-compact (cf. Lemma 2.15).

Let $(U_i, i \in I)$ be a basis of quasi-compact open subsets of X. For each $i \in I$, there exists an open subset V_i of Y such that $U_i = p^{-1}(V_i)$. From Lemma 2.15, it follows that V_i is quasi-compact. Let V be an open subset of Y, $p^{-1}(V)$ is an open subset of X, hence there exists a subset J of I such that

$$p^{-1}(V) = \bigcup_{i \in J} U_i = p^{-1}(\bigcup_{i \in J} V_i)$$

Since p is a quasi-homeomorphism, we obtain $V = \bigcup_{i \in J} V_i$. Thus $(V_i, i \in I)$ is a basis of quasi-compact open subsets of Y.

– Let U, V be two quasi-compact open subsets of Y, then $p^{-1}(U)$ and $p^{-1}(V)$ are two quasi-compact open subsets of X (cf. Lemma 2.15). The subset $p^{-1}(U) \cap p^{-1}(V) = p^{-1}(U \cap V)$ is quasi-compact (since X is quasi-spectral). Hence $U \cap V$ is quasi-compact (cf. Lemma 2.15).

– Let F be an irreducible closed subset of Y, then $p^{-1}(F)$ is an irreducible closed subset of X (cf. Lemma 2.15), whence there exists $x \in p^{-1}(F)$ such that $p^{-1}(F) = \overline{\{x\}}$. Hence $p^{-1}(F) = \overline{\{x\}} \subseteq p^{-1}(\overline{\{p(x)\}}) \subseteq p^{-1}(F)$. It follows that $p^{-1}(\overline{\{p(x)\}}) = p^{-1}(F)$. Thus $F = \overline{\{p(x)\}}$ (since p is a quasi- homeomorphism). Therefore F has a generic point.

We conclude that Y is a quasi-spectral space.

2) Suppose that Y is quasi-spectral.

i) Using Lemma 2.15, we claim that X is quasi-compact.

Let (V_i, $i \in I$) be a basis of quasi-compact open subsets of Y. For each i we consider $U_i = p^{-1}(V_i)$. Let U be an open subset of X, there exists an open subset V of Y such that $U = p^{-1}(V)$. On the other hand, there exists a subset J of I such that $V = \bigcup_{i \in J} V_i$, hence $U = \bigcup_{i \in J} U_i$. Therefore ($U_i, i \in I$) is a basis of quasi-compact open subsets of X.

ii) Let U, V be two quasi-compact open subsets of X. From lemma 2.15 and the fact that p is a quasi-homeomorphism, there exist two quasi-compact open subsets U_1, V_1 of Y such that $U = p^{-1}(U_1)$ and $V = p^{-1}(V_1)$. Since $U_1 \cap V_1$ is quasi-compact and $U \cap V = p^{-1}(U_1 \cap V_1)$, the subset $U \cap V$ of X is quasi-compact (cf. Lemma 2.15).

3) Suppose, that Y is spectral

– Assume that X is quasi-spectral.

For each element y of Y, $p^{-1}(\overline{\{y\}})$ has a generic point x. We have the containments $\overline{\{x\}} \subseteq p^{-1}(\overline{\{p(x)\}}) \subseteq p^{-1}(\overline{\{y\}}) = \overline{\{x\}}$. Hence $p^{-1}(\overline{\{p(x)\}}) = p^{-1}(\overline{\{y\}})$.

It follows, from the fact that p is a quasi-homeomorphism, that $\overline{\{p(x)\}} = \overline{\{y\}}$. Therefore $p(x) = y$ (since Y is a T_0-space). This proves that p is a surjective map.

– Assume that p is surjective.

In this case p, is a special morphism (that is, p is a closed map and every closed subset of X is saturated under p) (cf. Proposition 2.10). Let F be an irreducible closed subset of X. Using Lemma 2.15 and the fact that p is a quasi-homeomorphism, there exists an irreducible closed subset G of Y, such that $F = p^{-1}(G)$. The subset G of Y has a generic point y. Since p is a surjective map, there exists an element x of X such that $y = p(x)$. Hence $F = p^{-1}(\overline{\{p(x)\}})$, since p is continuous and closed, we have $F = p^{-1}(\overline{\{p(x)\}}) = p^{-1}(p(\overline{\{x\}})) = \overline{\{x\}}$, completing the proof. ∎

It follows immediately from the above theorem that we have.

COROLLARY 2.16 *Let* $p : X \longrightarrow Y$ *be a special morphism, then the following conditions are equivalent.*

i) X *is quasi-spectral.*

ii) Y *is quasi-spectral.*

REMARKS 2.17

1) Let $p : X \longrightarrow Y$ *be a special morphism. If* Y *is a spectral space then* X *is a quasi-spectral space, however* X *needs not be a* T_0*-space. If, for instance,* X *is a finite topological space which is not a* T_0*-space, then the space* Y *constructed from Examples 2.5 b) is a finite* T_0*-space; hence it is a spectral space. The canonical projection p of* X *onto* Y *is a special morphism and* X *is not a spectral space (since it is not a* T_0*-space.)*

2) From Theorem 2.14, it follows that, if X *is a strongly dense subspace of a spectral space* Y *such that* $Y \neq X$*, then* X *is never a quasi-spectral space.*

b) Foliations

We derive here some properties of the space of leaves of a foliation $\mathcal{F}$ on a compact closed manifold M, where $\mathcal{F}$ is of codimension one and transversally oriented of class $C^r (r \geq 0)$.

We are primarily concerned with quotient topology.

PROPOSITION 2.18 *Let* M *be a topological space equipped with an equivalence relation* $\mathcal{F}$ *such that the projection q of* M *onto* $Z = M/\mathcal{F}$ *is an open map.*

Let $x \in M$*, we denote by* F_x *the equivalence class of x by* $\mathcal{F}$*. Consider the equivalence relation* $\overline{\mathcal{F}}$ *defined on* M *by:* $x\overline{\mathcal{F}}y$ *if and only if,* $\overline{F_x} = \overline{F_y}$*.*

Then the following properties hold.

1) The canonical projection $p : M \longrightarrow M/\overline{\mathcal{F}} = X$ *is an open map.*

2) Consider, on the space X*, the ordering defined by:* $p(x) \leq p(y)$ *if and* *only if, $\overline{F_x} \subseteq \overline{F_y}$. Then the quotient topology on $X = M/\overline{\mathcal{F}}$ is compatible with the inverse order of this ordering.*

3) Consider the map $\varphi : Z = M/\mathcal{F} \longrightarrow X = M/\overline{\mathcal{F}}$ such that $\varphi \circ q = p$ (that is, $\varphi(F_x) = Cl(F_x) = \{y \in M/\overline{F_x} = \overline{F_y}\} = p(x))$. Then φ is a special morphism.

4) Let $\tilde{Z}$ be the quotient space of $Z = M/\mathcal{F}$ by the equivalence relation $\sim$ defined by: $F_x \sim F_y$ if and only if, $\overline{\{F_x\}} = \overline{\{F_y\}}$ (see Example 2.5).

Then the map $\psi : \tilde{Z} \longrightarrow X = M/\overline{\mathcal{F}}; \tilde{F}_x \mapsto Cl(F_x)$ is a homemorphism (where $\tilde{F}_x$ is the equivalence class of F_x by the equivalence relation $\sim$).

Proof.

1) Let U be an open subset of M, we show that $q^{-1}(q(U)) = p^{-1}(p(U))$.

Indeed, it is obvious to see that $q^{-1}(q(U)) \subseteq p^{-1}(p(U))$. Conversely, if $x \in p^{-1}(p(U))$, then there exits $y \in U$ such that $\overline{F_x} = \overline{F_y}$.

Since $y \in \overline{F_x}$, then $U \cap F_x$ is nonempty; hence there exists $z \in F_x$ such that $z \in U$. Thus $F_z = F_x$ and $x \in q^{-1}(q(U))$. This proves that $q^{-1}(q(U)) = p^{-1}(p(U))$.

Now, since q is an open continuous map, $q^{-1}(q(U))$ is an open subset of M. Hence $p^{-1}(p(U)) = q^{-1}(q(U))$ is an open subset of M. It follows that $p(U)$ is an open subset of X. Therefore the map p is open.

2) The goal is to prove that $\overline{\{p(x)\}} =] \downarrow, p(x)] = \{z \in X/z \le p(x)\}$, for every $x \in M$.

– Let $z = p(y) \in \overline{\{p(x)\}}$ and U be an open subset of M containing y, then $p(U)$ is an open subset of X containing $p(y)$ (p is open). Hence $p(x) \in p(U)$. There exists $t \in U$ such that $\overline{F_x} = \overline{F_t}$, so $U \cap F_x$ is nonempty. Thus $y \in \overline{F_x}$, $\overline{F_y} \subseteq \overline{F_x}$ and finally $z = p(y) \le p(x)$.

– Conversely, let $z = p(y) \le p(x)$ and V be an open subset of X containing $p(y)$. There exists an open subset U of M saturated under p such that $V = p(U)$. Since $z = p(y) \in p(U)$, and U is saturated under p, we have $y \in U$.

On the other hand, we have $\overline{F_y} \subseteq \overline{F_x}$, this yields $F_x \cap U$ is a nonempty subset of X. Hence $p(x) \in p(U) = V$, which proves that $z \in \overline{\{p(x)\}}$.

3) It is clear that φ is a continuous and surjective map, it remains to prove that every open subset of $M/\mathcal{F}$ is saturated under φ.

Let V be an open subset of $Z = M/\mathcal{F}$. There exists an open subset U of M saturated under q such that $V = q(U)$. We have

$$\varphi^{-1}(\varphi(V)) = \varphi^{-1}(\varphi(q(U))) = \varphi^{-1}(p(U))$$

Since q is surjective, we have

$$\varphi^{-1}(p(U)) = q(q^{-1}(\varphi^{-1}(p(U)))) = q(p^{-1}(p(U)))$$

As mentioned in 1), $p^{-1}(p(U)) = q^{-1}(q(U))$; which gives

$$\varphi^{-1}(\varphi(V)) = \varphi^{-1}(p(U)) = q(U) = V,$$

and V is saturated under φ. Therefore φ is a special morphism.

4) We prove now that $\psi : \tilde{Z} \longrightarrow X = M/\overline{\mathcal{F}}$; $\tilde{F}_x \mapsto Cl(F_x)$ is a homeomorphism.

– ψ is well defined:

If $\tilde{F}_x = \tilde{F}_y$, then $\overline{\{q(x)\}} = \overline{\{q(y)\}}$ (in $M/\mathcal{F}$), we shall prove that $Cl(F_x) = Cl(F_y)$; that is $\overline{F_x} = \overline{F_y}$. It suffices to prove that $x \in \overline{F_y}$. For this, let O be an open subset of M containing x. Suppose that $O \cap F_y$ is empty, then $q(O \cap F_y) = q(O) \cap q(F_y)$ is also empty (F_y is saturated under q). Hence $q(O)$ is an open subset of $M/\mathcal{F}$ containing $q(x)$ and not containing $q(y)$, which contradicts the fact that $q(x) \in \overline{\{q(y)\}}$.

– It is clear that ψ is a bijective continuous map:

– Finally, we prove that ψ is an open map.

Let $\tilde{q} : M/\mathcal{F} \longrightarrow \tilde{Z} = (M/\mathcal{F})/\sim$ be the projection of $M/\mathcal{F}$ onto $\tilde{Z}$, then we have $\psi \circ \tilde{q} = \varphi$.

Let $\tilde{U}$ be an open subset of $\tilde{Z}$; $\tilde{U} = \tilde{q}(V)$, where V is an open subset of $M/\mathcal{F}$. There exists an open subset O of M such that $V = q(O)$. Hence

$$\psi(\tilde{U}) = \psi(\tilde{q}(q(O))) = \varphi(q(O)) = p(O)$$

is an open subset of $M/\overline{\mathcal{F}}$ (since p is open). ■

It is worth noting that the next result extends Proposition 3.9 obtained in [4], besides, we have used geometrical properties of the foliation in proving that proposition. While our proof here is rather simple and elementary.

THEOREM 2.19 *Let E be a compact metric space equipped with an open equivalence relation $\mathcal{R}$. Then each irreducible closed subset of $E/\mathcal{R}$ has a generic point.*

The proof of this result will rely essentially on the following lemma.

LEMMA 2.20 *Let E be a topological space equipped with an open equivalence relation $\mathcal{R}$, $p : E \longrightarrow E/\mathcal{R}$ be the projection of E onto $E/\mathcal{R}$, and K be a saturated subset of E. Then the induced topology on $p(K)$ is the quotient topology on K by $\mathcal{R}$.*
Moreover the projection $p_1 : K \longrightarrow p(K)$ is an open map.

Proof.

 – Let V be an open subset of the subspace $p(K)$ of $E/\mathcal{R}$, then $V = p(K) \cap p(U)$, where U is a saturated open subset of E, and we have

$$p^{-1}(V) = p^{-1}(p(K)) \cap p^{-1}(p(U)) = K \cap U$$

which is an open subset of K.

 – Conversely, let V be an open subset of $p(K)$ for the quotient topology on K, hence $p_1^{-1}(V) = p^{-1}(V)$ is an open subset of K and there exists an open subset U of E such that $p^{-1}(V) = K \cap U$. Since K is saturated under p, we have

$$V = p(p^{-1}(V)) = p(K) \cap p(U)$$

On the other hand, p is open, so $p(U)$ is an open subset of $E/\mathcal{R}$. Therefore V is an open subset of $p(K)$ as a subspace of $E/\mathcal{R}$.

 – We prove now that $p_1 : K \longrightarrow p(K) = K/\mathcal{R}$ is an open map : Let U_1 be an open subset of K, there exists an open subset U of E such that

$U_1 = K \cap U$. Since K is saturated under p, we have $p(U_1) = p(K) \cap p(U)$ (cf. Proposition 2.2). Hence $p(U_1)$ is an open subset of the subspace $p(K)$ of $E/\mathcal{R}$. It follows that $p(U_1)$ is an open subset of $K/\mathcal{R}$. ■

Proof of Theorem 2.19.

Let F be an irreducible closed subset of $E/\mathcal{R}$. Set $K = p^{-1}(F)$, K is a closed subset of the compact metric space E; hence K is a compact metric suspace of E. According to the previous lemma, the topology of the subspace $p(K) = F$ concides with the quotient topology on $K/\mathcal{R}$, and the projection $p_1 : K \longrightarrow K/\mathcal{R}$ of K onto $K/\mathcal{R}$ is an open map. Finally, it follows from [5, Exercice 7, page 171] that F has a generic point. ■

As an immediate concequence of Proposition 2.18, we have:

COROLLARY 2.21 *Let M be a closed connected manifold and $\mathcal{F}$ be a codimension-one foliation defined on M, such that $\mathcal{F}$ is of class $C^r(r \geq 0)$ and transversally oriented. Set $Z = M/\mathcal{F}$ the space of leaves of $\mathcal{F}$ and $X = M/\overline{\mathcal{F}}$ the space of laeves classes of $\mathcal{F}$. Then the map $\varphi : Z \longrightarrow X$; $F_x \mapsto Cl(F_x)$ is a special morphism.*

Now, we state one of the principal results of this section.

THEOREM 2.22 *Let M be a connected closed manifold, $\mathcal{F}$ be a codimension-one foliation defined on M, which is transversally oriented and of class $C^r(r \geq 0)$. Suppose that $\mathcal{F}$ has a well defined height (in the sens of [17]). Let Z be the space of leaves of $\mathcal{F}$, and Z_0 be the union of open subsets of Z homemorphic to $\mathbb{R}$ or S^1, then $Z \setminus Z_0$ is a quasi-spectral space.*

Proof.

We denote $X = M/\overline{\mathcal{F}}$ the space of leaves classes of $\mathcal{F}$, X_0 the union of open subsets of X which are homeomorphic to $\mathbb{R}$ or to S^1, then $X_0 = p(\mathcal{U}_0)$, $Z_0 = q(\mathcal{U}_0)$, where $\mathcal{U}_0$ is the union of all stable and proper leaves of $\mathcal{F}$, $p : M \longrightarrow M/\overline{\mathcal{F}}$ and $q : M \longrightarrow Z = M/\mathcal{F}$ are the canonical surjections.

From Corollary 2.21, the map $\varphi : Z \longrightarrow X$; $F \mapsto Cl(F)$ is a special morphism . The map $\varphi_1 : Z \setminus Z_0 \longrightarrow X \setminus X_0$; $F \mapsto Cl(F)$ is also a special morphism. Applying [4, Théoreme 3.15] and Corollary 2.16, it follows that $Z \setminus Z_0$ is a quasi-spectral space. ∎

As a consequence of [4, Théoreme 3.15] and Corollary 2.16, we obtain the following.

THEOREM 2.23 *Under the same hypothesis as in (2.22), let $\mathcal{S}$ be the topology on M defined by the saturated open subsets. Let $\mathcal{U}_0$ be the union of stable and proper leaves of M. Then $M \setminus \mathcal{U}_0$ is a quasi-spectral subspace of $(M, \mathcal{S})$.*

Proof.

One may check easily that $p_0 : (M \setminus \mathcal{U}_0, \mathcal{S}) \longrightarrow X \setminus X_0$; $x \mapsto Cl(F_x)$ is a special morphism. Combining [4, Théoreme 3.15] and Corollary 2.16, we obtain the result. ∎

ACKNOWLEDGEMENT

We would like to thank the DGRST (E03/C15 and E05/C15) for his support. Also, we are grateful to the referee for his comments.

REFERENCES

[1] K. Belaïd, C. Bahri and O. Echi. On Spectral Binary Relation. *Lect. Not. Pure. Appl. Math. Dekker.* **185**(1997), 79-88.

[2] E. Bouacida, O. Echi, E. Salhi. Topologies associées à une relation binaire et relation binaire spectrale. *Boll. Un. Mat. Ital.* **(7)10-B** (1996), 417-439.

[3] E. Bouacida, O. Echi and E. Salhi. Nonfinite Heights. *Lect. Not. Pure. Appl. Math. Dekker.* **185**(1997), 113-130.

[4] E. Bouacida, O. Echi et E. Salhi. Feuilletage et Topologie Spectrale.

(*Submitted to publication*).

[5] N. Bourbaki. Algèbre Commutative Chapitre 1 à 4. *Masson Paris* (1985).

[6] J. Cantwell, L. Conlon. Analytic foliations and the theory of levels. *Math. Ann.* **265** (1983), 253-26.

[7] P. R. Dippolito. Codimension one foliations of closed manifolds. *Ann. of Math.* **107** (1978), 403-453.

[8] M. Fontana, P. Maroscia. Sur les anneaux de Goldman. *Boll. Un. Mat. Ital.* **(5) 13-B** (1976), 743-759.

[9] C. Godbillon. Feuilletage, étude géométrique. *Birkhäuser-Verlag* (1991)

[10] A. Grothendieck, J. Dieudonné. Eléments de géometrie algébrique. *Springer- Verlag* (1972).

[11] G. Hector. Sur un théorème de structure des feuilletages de codimension un. Thèse d'Etat. *Publication de l'IRMA* (1972). Strasbourg. France.

[12] G. Hector. Architecture des feuilletages de classe C^2. *Astérisque* **107-108**(1983) 243-258.

[13] M. Hochster. Prime ideal stucture in commutative rings. *Trans. Amer. Math. Soc.* **142** (1969), 43-60.

[14] E. Salhi. Problème de structure dans les feuilletages de codimension un de classe C^0. Thèse d'Etat. *Publication de l'IRMA* (1984). Strasbourg. France.

[15] E. Salhi. Sur les ensembles minimaux locaux. *C. R. A. S. Paris.* **295** (1982) , 691-692.

[16] E. Salhi. Sur un théorème de structure des feuilletages de codimension un. *C. R. A. S. Paris.* **300** (1985), 635-638.

[17] E. Salhi. Niveau des feuilles. *C. R. A. S. Paris.* **301** (1985), 219-222.

Hermite and Weakly Semi-Steinitz Properties in Pullbacks

ABDELMALEK BOUANANE Département de Mathématiques, Faculté des Sciences, Université Abdelmalek Essaâdi, BP. 2121, Tetouan, Morocco

NAJIB MAHDOU Département de Mathématiques et Informatique, Faculté des Sciences et Techniques Fès-Saïss, Université de Fès, Fès, Morocco

1 INTRODUCTION

All rings considered in this paper are commutative with unit. We say that a ring A is semi-Steinitz if every finite linearly independent subset of a finitely generated free R-module F can be extended to a basis of F by adjoining elements of a given basis of F. In [18], B. Nashier and W. Nichols showed that a ring A is semi - Steinitz if and only if it is a local ring and every finitely generated proper ideal of A has non-zero annihilator.

A projective A-module P is said to be stably free if there exists a finitely generated free A-module G such that $P \oplus G$ is free. We say that a ring A is a Hermite ring if every finitely generated stably free A-module is free. Local rings and semi -local rings are Hermite [16, I.4.7, p.26]. It is shown in [16, Corollary I.4.5, p.26] that A is Hermite if and only if for every $a_1, \ldots, a_n$ in A, such that $\sum_{i=1}^{n} a_i A = A$, the row $[a_1, \ldots, a_n]$ can be completed to an invertible square matrix. It is then not difficult to see that a ring A is Hermite if and only if it is so modulo its Jacobson radical.

We say that a ring A is weakly semi-Steinitz if every finite linearly independent subset of a finitely generated free A-module F can be extended to a basis of F. In [18], B. Nashier and W. Nichols showed that a ring A is weakly semi-Steinitz if and only if it is a Hermite ring and every finitely generated proper ideal of A has non-zero annihilator.

The goal of this work is to exhibit a new class of non local Hermite rings and a new class of weakly semi-Steinitz rings which are not semi-Steinitz rings. In section 2, we develop several results on transfer of Hermite properties, and these are used in section 3 to describe a new class of non local Hermite rings and weakly semi-Steinitz rings which are not semi-Steinitz.

2 HERMITE AND WEAKLY SEMI-STEINITZ PROPERTIES IN PULLBACKS

A ring A is Hermite if for every finitely generated projective A-module P and for every finitely generated free A-module G such that $P \oplus G$ is a free A-module, P is a free A-module. By using induction on $rg(G)$, we easily show that A is Hermite if and only if for every finitely generated projective A-module P such that $P \oplus A$ is a free A-module, P is a free A-module. For any ring A, we denote by $A^\star$ the set of all non-zero elements of A.

THEOREM 2.1 *Let B be a ring of the form: $B = L + I$, where L is a subring of B and I an ideal of B such that $L \cap I = 0$, and let $A = D + I$, where D is a subring of L. The following statements hold:*

1) *A is a Hermite ring $\Longrightarrow D$ is a Hermite ring.*

2) *If B is a Hermite ring, then:*

A is a Hermite ring $\Longleftrightarrow D$ is a Hermite ring.

3) *If $L = S^{-1}D$ for some multiplicatively closed subset S of D, and D is a valuation domain or a principal ideal domain, then:*

A is a Hermite ring $\Longleftrightarrow B$ is a Hermite ring.

Proof. **1)** Let P be a finitely generated and projective D-module and m a positive integer such that: $P \oplus Da = D^m$, where $a \in D^m$ is such that $Da \cong D$ as D-modules.

Our aim is to show that P is a free D-module.

If we tensor the equality $P \oplus Da = D^m$ of D-modules by A, we have:

$(P \otimes_D A) \oplus (Da \otimes_D A) \cong A^m$ and then we have $P \otimes_D A \cong A^{m-1}$ (isomorphism of A-modules) since $Da \otimes_D A \cong D \otimes_D A \cong A$ (isomorphisms of A-modules) and A is a Hermite ring.

We may assume that $P \otimes_D A = PA(\subseteq A^m)$ since P is a direct summand of D^m. Let $S_1 = \{e_1, \ldots, e_{m-1}\}$ be a basis of the A-module $PA = P(D+I) = P+IP$ and $e_i = d_i + m_i$, where $d_i \in P$ and $m_i \in I^m$. We have: $P + IP = PA = \sum_{i=1}^{m-1} Ae_i = \sum_{i=1}^{m-1} (D+I)(d_i + m_i) = \sum_{i=1}^{m-1} Dd_i + \sum_{i=1}^{m-1} (Dm_i + Id_i + Im_i)$. Hence, $P = \sum_{i=1}^{m-1} Dd_i$ since $D^m \cap I^m = 0$, and then $S_2 := \{d_1, \ldots, d_{m-1}\}$ generates the D-module P. Therefore, we have: $PA = \sum_{i=1}^{m-1} Ad_i$ and then S_2 is a basis of the A-module PA by [19, Cor.3.7, p.60] since $Card(S_2) = m-1$ and $PA \cong A^{m-1}$ (isomorphism of A-modules). Hence, S_2 is a linearly independent subset over A so over D and then S_2 is a basis of the D-module P and D is a Hermite ring.

2) If A is a Hermite ring, then so is D by 1). Conversely, assume that D is a Hermite ring. Let P be a finitely generated and projective A-module and m a positive integer such that: $P \oplus Aa = A^m$, where $a \in A^m$ is such that $Aa \cong A$ as A-modules.

Our aim is to show that P is a free A-module.

If we tensor the equality $P \oplus Aa = A^m$ of A-modules by B and A/I respectively, we get the isomorphisms:

$$(P \otimes_A B) \oplus (Aa \otimes_A B) \cong B^m \quad and \quad (P \otimes_A A/I) \oplus (Aa \otimes_A A/I) \cong (A/I)^m$$

of B and A/I-modules respectively. Therefore, we have the isomorphisms:

$$P \otimes_A B \cong B^{m-1} \quad and \quad P \otimes_A A/I \cong (A/I)^{m-1}$$

of B and A/I-modules respectively since $Aa \otimes_A B \cong A \otimes_A B \cong B$ (isomorphisms of B-modules), $Aa \otimes_A A/I \cong A \otimes_A A/I \cong A/I$ (isomorphisms of A/I-modules), and B and A/I are Hermite rings.

We may assume that $P \otimes_A B = PB(\subseteq B^m)$ since P is a direct summand of A^m. Let $S_1 = \{e_1, \ldots, e_{m-1}\}$ be a basis of the B-module PB. On the other hand, there exists a subsets $S_2 = \{x_1, \ldots, x_{m-1}\}$ of the A-module P such that $P = \sum_{i=1}^{m-1} Ax_i + IP$, since $(A/I)^{m-1} \cong P \otimes_A A/I \cong P/IP$ as A/I-modules. We have $x_i \in P \subseteq BP = \sum_{j=1}^{m-1} Be_j = \sum_{j=1}^{m-1} (L+I)e_j = \sum_{j=1}^{m-1} Le_j + \sum_{j=1}^{m-1} Ie_j = \sum_{j=1}^{m-1} Le_j + I \sum_{j=1}^{m-1} Be_j = \sum_{j=1}^{m-1} Le_j + IBP = \sum_{j=1}^{m-1} Le_j + IP$, and then we may assume that $x_i \in P \cap (\sum_{j=1}^{m-1} Le_j)$ since $IP \subseteq P$ and $P = \sum_{i=1}^{m-1} Ax_i + IP$.

Let $H = \sum_{i=1}^{m-1} Ax_i$ be the A-submodule of P generated by S_2. We have $P = H + IP$. Our aim is to show that $IP \subseteq H$.

We have $\sum_{i=1}^{m-1} Lx_i \subseteq \sum_{i=1}^{m-1} Le_i$ since $x_j \in \sum_{i=1}^{m-1} Le_i$ for each j. Our aim is to show that $e_j \in \sum_{i=1}^{m-1} Lx_i$ for each j and this suffices to show that $\sum_{i=1}^{m-1} Le_i = \sum_{i=1}^{m-1} Lx_i$. We have

$$\sum_{i=1}^{m-1} Le_i + IP = BP = B(H + IP) = BH + IBP = BH + IP = \sum_{i=1}^{m-1} Bx_i + IP =$$

$$\sum_{i=1}^{m-1}(L+I)x_i + IP = (\sum_{i=1}^{m-1} Lx_i) + (\sum_{i=1}^{m-1} Ix_i + IP) = \sum_{i=1}^{m-1} Lx_i + IP \text{ since } \sum_{i=1}^{m-1} Ix_i \subseteq IP$$

(since $x_i \in P$ for each i). Hence, $e_j \in \sum_{i=1}^{m-1} Lx_i + IP$ and then $e_j = a_j + m_j$, where

$a_j \in \sum_{i=1}^{m-1} Lx_i \subseteq \sum_{i=1}^{m-1} Le_i$ and $m_j \in IP$. Therefore, $e_j = a_j \in \sum_{i=1}^{m-1} Lx_i$ since

$m_j = e_j - a_j \in (\sum_{i=1}^{m-1} Le_i) \cap IP = (\sum_{i=1}^{m-1} Le_i) \cap (\sum_{i=1}^{m-1} Ie_i) = 0$ (since $L \cap I = 0$ and S_1 is a basis of the B-module BP).

Therefore, $\sum_{i=1}^{m-1} Le_i = \sum_{i=1}^{m-1} Lx_i$ and then we have $\sum_{i=1}^{m-1} Be_i = \sum_{i=1}^{m-1} Bx_i$. Hence,

$$IP = I(BP) = I(\sum_{i=1}^{m-1} Be_i) = I(\sum_{i=1}^{m-1} Bx_i) = \sum_{i=1}^{m-1} Ix_i = I(\sum_{i=1}^{m-1} Ax_i) = IH \subseteq H.$$

We have $P = H + IP = H = \sum_{i=1}^{m-1} Ax_i$ and then S_2 generates the A-module P. Therefore, we have $BP = \sum_{i=1}^{m-1} Bx_i$ and then S_2 is a basis of the B-module BP by [19,Cor.3.7,p.60] since $Card(S_2) = m - 1$ and $BP \cong B^{m-1}$ (isomorphism of B-modules). Hence, we have S_2 is a linearly independent subset over B so over A and then S_2 is a basis of the A-module P and A is a Hermite ring.

3) If B is a Hermite ring, then so is A by (Theorem 2.1, 2) since a valuation domain and a principal ideal domain are a Hermite rings. Conversely, assume that A is a Hermite ring. Let P_1 be a finitely generated and projectif B-module and let m a positive integer such that: $P_1 \oplus Ba = B^m$ where $a \in B^m$ such that $Ba \cong B$ as B-modules.

Our aim is to show that P_1 is a free B-module. We may assume that $P_1 = \sum_{i=1}^{m} Be_i$, where $e_i \in A^m$ and $a \in A^m$ since $B = S^{-1}A$.

Let $P = \sum_{i=1}^{m} Ae_i \subseteq A^m$ and let $H = P \oplus Aa \subseteq A^m$. Since S is a multiplicatively closed subset of a domain D, we have $BH = S^{-1}A.H = S^{-1}H = S^{-1}(P + Aa) = S^{-1}P + S^{-1}A.a = P_1 + Ba = B^m$. Therefore, we have $I^m = IB^m = IBH = IH \subseteq H \subseteq A^m = D^m + I^m$, so $H = G + I^m$, where $G = D^m \cap H$ is a finitely generated D-submodule of D^m since $G \cong H/I^m$ and H is a finitely generated A-module. Hence, $G \cong D^n$ (isomorphism of D-modules) for some positive integer n since D is a valuation domain or a principal ideal domain. Our aim is to show that $n = m$. We have $(S^{-1}D)^m + I^m = B^m = HB = (G + I^m)(S^{-1}D + I) = GS^{-1}D + (GI + I^m S^{-1}D + II^m)$, so $S^{-1}G = GS^{-1}D = (S^{-1}D)^m$. Hence $(S^{-1}D)^m = S^{-1}G = G \otimes_D S^{-1}D \cong D^n \otimes_D S^{-1}D \cong (S^{-1}D)^n$ (isomorphisms of $S^{-1}D$-modules) and then $n = m$.

Our aim is to show that $H \cong A^m$ as A-modules. We have $A^m \cong D^m \otimes_D A \cong G \otimes_A A = GA$ (since $G \cong D^m$, isomorphism of D-modules) $= G(D + I) = GD + GI = G + GI$, and $GI = G(S^{-1}D)I = (GS^{-1}D)I = (S^{-1}D)^m I = I^m$; hence $A^m \cong G + GI = G + I^m = H = P \oplus Aa \cong P \oplus A$ (isomorphism of A-modules). Therefore, $P \cong A^{m-1}$ as A-modules since A is a Hermite ring and then $P_1 = P \otimes_A B \cong B^{m-1}$ as B-modules and this completes the proof of Theorem (2.1).

Let $T = K + M$ be an integral domain, where K is a field and M is a maximal ideal of T. Let $R = D + M$ be the classical pullback construction, where D is a subring of the field K.

COROLLARY 2.2 *Let $T = K + M$ be any domain and let $R = D + M$ where D is a subring of the field K and M is a maximal ideal of T. The following statements hold:*

1) *R is a Hermite ring $\Longrightarrow D$ is a Hermite ring.*
2) *If T is a Hermite ring, then:*
   R is a Hermite ring $\Longleftrightarrow D$ is a Hermite ring.
3) *If D is a principal ideal domain or a valuation domain, and $qf(D) = K$, then:*
   R is a Hermite ring $\Longleftrightarrow T$ is a Hermite ring.

In [8] D. Costa, J.L. Mott and M. Zafrullah introduced the rings of the type $D^{(S)} = D + X(S^{-1}D)[X]$, where D is an integral domain and S is a multiplicatively closed subset of D.

COROLLARY 2.3 *Let $D^{(S)} = D + X(S^{-1}D)[X]$, where D is an integral domain and S is a multiplicatively closed subset of D. The following statements hold:*

1) $D^{(S)}$ *is a Hermite ring* $\Longrightarrow$ D *is a Hermite ring.*

2) *If* $S^{-1}D$ *is a principal ideal domain, then:*
$D^{(S)}$ *is a Hermite ring* $\Longleftrightarrow$ D *is a Hermite ring.*

3) *If* D *is a principal ideal domain, or a valuation domain; then:*
$D^{(S)}$ *is a Hermite ring* $\Longleftrightarrow$ $(S^{-1}D)[X]$ *is a Hermite ring.*

Proof. If $S^{-1}D$ is a principal ideal domain, then $(S^{-1}D)[X]$ is a Hermite ring by [16]. Therefore, the pullback:

$$
\begin{array}{ccc}
D^{(S)}(= D + X(S^{-1}D)[X]) & \longleftrightarrow & (S^{-1}D)[X] \\
\downarrow & & \downarrow \\
D(= D^{(S)}/X(S^{-1}D)[X]) & \longrightarrow & S^{-1}D(= (S^{-1}D)[X]/X(S^{-1}D)[X])
\end{array}
$$

and Theorem (2.1) allow us to complete the proof of Corollary (2.3).

REMARK 2.4 *In Theorem (2.1,3), we don't have in general that if A is a Hermite ring (and then D is a Hermite ring by Theorem (2.1,1)), then B is a Hermite ring (see Example (3.3)).*

The next result may be of general interest in studying rings considered in Theorem (2.1).

PROPOSITION 2.5 *Let A be a subring of a ring B and let I be a common ideal of A and B. If $J(A)$ and $J(B)$ are respectively the Jacobson radicals of A and B, then $I \cap J(B) \subseteq J(A)$. In particular, if $I \subseteq J(B)$ then $I \subseteq J(A)$.*

Proof. Let $x \in I \cap J(B)$ and $y \in A(\subseteq B)$. Our aim is to show that $1 + xy$ is an invertible element in A. Now, $1 + xy$ is an invertible element in B since $x \in J(B)$. Let $u \in B$ such that $1 = u(1 + xy) = u + uxy$. Hence, we have $u = 1 - uxy \in A$ since $uxy \in IB(= I)$ (since $x \in I$) and then $1 + xy$ is an invertible element in A for each $y \in A$ and this completes the proof of Proposition (2.5).

We say that a ring R satisfies (CH) if every finitely generated proper ideal has a non-zero annihilator. We next consider (CH) properties in pullbacks.

Let $A \to B$ be an injective ring homomorphism and I is a proper ideal of A such that $IB = I$. We don't have in general that if B and A/I satisfy (CH), then A satisfies (CH) (see Example (3.4)). Moreover, if $B = S^{-1}A$ for some multiplicatively closed subset of A, we have:

PROPOSITION 2.6 *Let A be a ring and let S be a multiplicatively closed subset of A consisting of regular elements. If $A \neq S^{-1}A$ and if I is a common ideal of A and $S^{-1}A$, then A/I does not satisfy (CH).*

Proof. I is a common ideal of A and $S^{-1}A$ means that for every $a \in I$ and for every $s \in S$ there exists $b \in I$ and $t \in S$ such that $a/s = bt/t$ which is equivalent to $a = sb$. Thus $I = sI$ for every $s \in S$.

Now, suppose, on the contrary, that A/I satisfies (CH). Let s be any element of S and let $J = (s)$ be the ideal of A generated by s. we have $I = sI \subseteq J$. If $\bar{d} = d + I \in A/I$ annihilates J/I, then $dJ \subseteq I$, so $ds \in I = Is$; since s is regular we have then $d \in I$, so that $\bar{d} = 0$. Thus the principal ideal J/I of A/I has no non-zero annihilator. Therefore $J/I = A/I$ and then $J = A$. Thus every $s \in S$ is invertible in A and then $A = S^{-1}A$. A contradiction. Therefore A/I does not satisfy (CH).

3 EXAMPLES AND COUNTER -EXAMPLES

In this section we exhibit a new class of examples of non local Hermite rings (Examples (3.1) and (3.2)), and then, by [2, Prop.5, p.138], we exhibit a new class of examples of weakly semi-Steinitz rings which are not semi-Steinitz rings. Next, we exhibit an example of an injective ring homomorphism $A(= D+I) \to B(L+I)$ (such that I is a common ideal of A and B and $L \cap I = 0$) such that A (and then D by Theorem (2.1,1)) is a Hermite ring and B is not a Hermite ring (Example (3.3)). Also, we give an example of an injective ring homomorphism $A \to B$ and I a common ideal of A and B such that B and A/I satisfy (CH), but A does not satisfy (CH) (Example (3.4)). Finally, we give a counter -example showing that the converse of properties: "Any localisation $S^{-1}A$ of a ring A satisfying the property (CH) such that $A \to S^{-1}A$ is an injective ring homomorphism, is a ring satisfying the property (CH)", is false (Example (3.5)).

EXAMPLE 3.1 *Let D be any principal ideal domain, K a field containing D as a subring, and X an indeterminate over K.*
  1) *Let $D_1 := D[X]$. The ring D_1 is a Hermite ring but not a local ring.*
  2) *Let $D_2 := D_1 + YK[X]_{(X)}[Y](= D[X] + YK[X]_{(X)}[Y])$, where Y is an*

*indeterminate over the principal ideal domain $K[X]_{(X)}$. Then $D_2 \to K[X]_{(X)}[Y]$
is an injective ring homomorphism and:*

a) *$D_2(= D[X] + YK[X]_{(X)}[Y])$ is a Hermite ring by Theorem (2.1,2) since:
$D_2 \to K[X]_{(X)}[Y]$ is an injective ring homomorphism, $K[X]_{(X)}[Y]$ is a Hermite
ring, and $YK[X]_{(X)}[Y]$ is a common ideal of D_2 and $K[X]_{(X)}[Y]$, and we have
$D_2/(YK[X]_{(X)}[Y])(\cong D_1 = D[X])$ is a Hermite ring.*

b) *D_2 is not a local ring by [6, Prop.3] since $D_2/(YK[X]_{(X)}[Y])(\cong D[X])$ is
not a local ring.*

c) *The trivial extension ring of D_2 by L_2 : $A = D_2 \times L_2$, where $L_2 = \bigoplus_{m \in max(D_2)} D_2/m$, is a weakly semi-Steinitz ring by [2, Prop.5, p.138] since D_2
is a Hermite ring. But A is not semi-Steinitz since $max(A) = max(D_2 \times L_2) =
\{M \times L_2/M \in Max(D_2)\}$ (by [2, Prop.5, p.138]) contains more than one element
by b).*

EXAMPLE 3.2 *Let D be a Hermite ring which is not local. Let $D_1 = D + XK_1[X]$
be a subring of the Hermite ring $K_1[X]$, where K_1 is any field containing D as a
subring and X is an indeterminate over K_1. Then:*

1) *$D_1 := D + XK_1[X]$ is a Hermite ring by Theorem (2.1,2) since $K_1[X]$ is a
Hermite ring, $XK_1[X]$ is a common ideal of D_1 and $K_1[X]$, and $D_1/XK_1[X](\cong D)$
is a Hermite ring.*

2) *$D_1(= D + XK_1[X])$ is not a local ring by [6, Prop.3] since $D_1/XK_1[X](\cong D)$
is not a local ring.*

3) *The trivial extension ring of D_1 by L_1 : $A = D_1 \times L_1$, where $L_1 = \bigoplus_{m \in max(D_1)} D_1/m$, is a weakly semi-Steinitz ring by [2, Prop.5, p.138] since D_1
is a Hermite ring. But A is not semi-Steinitz since $max(A) = max(D_1 \times L_1) =
\{M \times L_1/M \in Max(D_1)\}$ (by [2, Prop.5, p.138]) contains more than one element
by 2).*

*Using the same argument as above, we have also $D_2 := D_1 + YK_2[Y] = D +
XK_1[X] + YK_2[Y]$ is a Hermite ring which is not local, where K_2 is any field
containing $K_1[X]$ as a subring and Y is an indeterminate over K_2. Also we have:
$D_3 := D_2 + ZK_2[Z] = D + XK_1[X] + YK_2[Y] + ZK_3[Z]$ is a Hermite ring which
is not local, where K_3 is any field containing $K_2[X]$ as a subring and Z be an
indeterminate over K_3. Therefore, an argument as above shows that the trivial
extension ring of D_2 by L_2 : $D_2 \times L_2$ (where $L_2 = \bigoplus_{m \in max(D_2)} D_2/m$) and the
trivial extension ring of D_3 by L_3 : $D_3 \times L_3$ (where $L_3 = \bigoplus_{m \in max(D_3)} D_3/m$) are
weakly semi-Steinitz rings but not semi-Steinitz rings.*

EXAMPLE 3.3 *Let R be the field of real numbers, $L = R[X, Y, Z]/(X^2+Y^2+Z^2-1)$, where X, Y, and Z are indeterminates over R, $K = qf(L)$, and $I = TK[T]$ where T is an indeterminate over K. Let $A = R+I$, $B = L+I$, and $C = K+I(= K[T])$. We have $A \subseteq B \subseteq C$ and I is a common ideal of $A, B,$ and C. The ring B is not a Hermite ring by Theorem (2.1,1) since L is not a Hermite ring. On the other hand, the ring A is a Hermite ring by Theorem (2.1,2) since C and R are Hermite rings. Therefore, $A(= R + I) \rightarrow B(= L + I)$ is an injective ring homomorphism such that A and R are Hermite rings, but B is not a Hermite ring.*

EXAMPLE 3.4 *Let D be a local domain which is not a field, M its maximal ideal, and $L = \bigoplus_{m\in max(D)} D/m = D/M$. Let $A = D \times 0, B = D \times L,$ and $I = M \times 0$. Then $A \rightarrow B$ is an injective ring homomorphism and I is a common ideal of A and B. The ring $B(= D \times L)$ satisfies (CH), the ring $A/I(= (D \times 0)/(M \times 0)) \cong D/M =field)$ satisfies (CH), but the ring A does not satisfy (CH) since $A \cong D$ (isomorphism of rings) and D is a domain which is not a field.*

We know that any localisation $S^{-1}A$ of any ring A satisfying the property (CH) such that $A \rightarrow S^{-1}A$ is an injective ring homomorphism, is a ring satisfying the property (CH). The converse is false. The following example is an injective ring homomorphism $A \rightarrow B(= S^{-1}A)$ and a common ideal I of A and B, where B is a weakly semi-Steinitz ring and A is a Hermite ring which does not satisfy property (CH).

EXAMPLE 3.5 *Let D be a Hermite domain which is not a field, $K = qf(D)$, $T = K + M$ be any local domain where M is its maximal ideal, and $R = D + M$. Let $B := T \times L$ the trivial extension ring of T by L, where $L = \bigoplus_{m\in max(T)} T/m = T/M = K$ (since (T, M) is a local domain), and $A := R \times L$ be a subring of B. We have that $B = S^{-1}A$, where $S = D^\star \times 0$ is a multiplicatively closed subset of A, and $I(= 0 \times L = 0 \times K)$ is a proper ideal of A such that $IB = I$. We have:*

1) $B(= T \times L = T \times K)$ is a weakly semi-Steinitz ring by [2, Prop.5, p.138] since T is a Hermite ring (since T is a local domain).

2) A is a Hermite ring: Since D and T are a Hermite rings, then R is a Hermite ring by Theorem (2.1,2). Therefore, A is a Hermite ring since $A/I(\cong R)$ is a Hermite ring and $I(= 0 \times L) \subseteq J(A)$, where $J(A)$ is the Jacobson radical of A.

3) A does not satisfy (CH): Since D is not a field, let d be a non invertible element of $D^\star$. To show that A does not satisfy (CH), it suffices to show that the *element $(d, 0)$ of A is not an invertible element of A and has a zero annihilator in A.*

If $(d, 0)$ is an invertible element of A, then there exist $(x, y) \in A(= R \times K)$ such that: $(1, 0) = (d, 0)(x, y) = (dx, dy)$. Hence, $dx = 1$ and then d is an invertible element of R and then an invertible element of D, a contradiction. Therefore, $(d, 0)$ is not an invertible element of A.

If $(d, 0)$ has a non-zero annihilator in A, there exists a non-zero element (x, y) of $A(= R \times K)$ such that: $(0, 0) = (d, 0)(x, y) = (dx, dy)$. Hence we have $xd = 0$ in R and $dy = 0$ in $L(= T/M = K)$. Therefore, $xd = 0$ in R implies that $x = 0$ since R is a domain and $d \neq 0$; on the other hand, $dy = 0$ in $L(= T/M = K)$ implies that $y = 0$ since d is an invertible element of K; a contradiction since $(x, y) \neq (0, 0)$. So, $(d, 0)$ has a zero annihilator in A.

REFERENCES

1. H. Bass, "Finistic dimension and a homological characterization of semi primary rings", Trans. Amer. Math. Soc. 95 (1960) 466-488.

2. A. Bouanane and F. Kourki, "On weakly semi-Steinitz rings", Lecture notes in Pure and Applied Mathematics, Marcel Dekker, 185 (1997) 131-139.

3. N. Bourbaki, "Algèbre Commutative", Chapitre 7, Hermann, Paris, (1965).

4. N. Bourbaki, "Algèbre Commutative", Chapitres 1-4, Masson, Paris, (1985).

5. J. Brewer and E. Rutter, "D+M constructions with general overrings", Michigan Math. J. 23 (1976) 33-42.

6. P.-J. Cahen, "Couple d'anneaux partageant un ideal", Arch. Math. 51 (1988) 505-514.

7. B.S. Chwe and J. Neggers, "On the extension of linearly independent subsets of free modules to basis", Proc. Amer. Math. Soc. 24 (1970) 466-470.

8. B.S. Chwe and J. Neggers, "Local rings with left vanishing radical", J. London Math. Soc. 4 (1971) 374-378.

9. D.L. Costa, "Parameterizing families of non-noetherian rings", Comm. Algebra 22 (1994) 3997-4011.

10. D.L. Costa, J.L. Mott, and M. Zafrullah, "The construction $D + X D_S[X]$", J. Algebra 53 (1978) 423-439.

11. D.E. Dobbs, S. Kabbaj, and N. Mahdou, "n-coherent rings and modules", Lecture Notes in Pure and Applied Mathematics, Marcel Dekker, 185 (1997) 269-282.

12. C. Faith, "Annihilator ideals, associated primes, and Kasch-McCoy commutative rings" Comm. Algebra 19 (7) (1991) 1867-1892.

13. S. Glaz, "Commutative Coherent Rings", Lecture Notes in Mathematics, 1371, Springer-Verlag, Berlin, (1989).

14. J.A. Huckaba, "Commutative Rings With Zero Divizors", Marcel Dekker, New York - Basel, (1988).

15. I. Kaplansky, "Commutative Rings", Univ. Chicago Press, Chicago, (1974).

16. T.Y. Lam, "Serre's conjecture", Lecture Notes in Mathematics, Springer-Verlag, Berlin and New York (1978).

17. H. Lenzing, " A homological characterization of Steinitz rings", Proc. Amer. Math. Soc. 29 (1971) 269-271.

18. B. Nashier and W. Nichols, " On Steinitz properties", Arch. Math. 57 (1991) 247-253.

19. J.J. Rotman, "An Introduction to Homological Algebra", Academic Press, New York, (1979).

20. W.V. Vasconcelos, " Annihilators of modules with a finite free resolution", Proc. Amer. Math. Soc. 29 (1971) 440-442.

21. W.V. Vasconcelos, " Conductor, projectivity and injectivity", Pacific. J. Math. 46 (1973) 603-608.

The Dimension of Tensor Products of Commutative Algebras Over a Zero-Dimensional Ring

Samir Bouchiba   Université de Meknès, Meknès, Morocco

Florida Girolami[1]   Università degli Studi di Roma Tre, Roma, Italy

Salah-Eddine Kabbaj[2]   KFUPM, Dhahran, Saudi Arabia

0. Introduction

All the rings and algebras considered in this paper will be commutative, with identity elements and ring-homomorphisms will be unital. If A is a ring, then $\dim A$ will denote the (Krull) dimension of A, that is, the supremum of lengths of chains of prime ideals of A. An integral domain D is said to have valuative dimension n (in short, $\dim_v D = n$) if each valuation overring of D has dimension at most n and there exists a valuation overring of D of dimension n. If no such integer n exists, then D is said to have infinite valuative dimension (see [G]). For reader's convenience, recall that for any ring A, $\dim_v A = \sup\{\dim_v(A/P) \mid P \in \mathrm{Spec}(A)\}$, and that a finite-dimensional domain D is a Jaffard domain if $\dim D = \dim_v D$. As the class of Jaffard domains is not stable under localization, an integral domain D is defined to be a locally Jaffard domain if D_P is a Jaffard domain for each prime ideal P of D (see [ABDFK]). Analogous definitions are given in [C] for a finite-dimensional ring.

---

[1]Partially supported by MURST (60% Fund).
[2]Partially supported by KFUPM.

In [S] Sharp proved that if K and L are two extension fields of a field k, then

$$\dim(K \otimes_k L) = \min(\mathrm{t.\,d.}(K : k),\ \mathrm{t.\,d.}(L : k)).$$

This result provided a natural starting point to explore dimensions of tensor products of somewhat general k-algebras and it was concretized by Wadsworth in [W], where the result of Sharp was extended to AF-domains (for "altitude formula"), that is, integral domains A such that

$$\mathrm{ht}\, P + \mathrm{t.\,d.}(A/P : k) = \mathrm{t.\,d.}(A : k)$$

for all prime ideals P of A. He showed that if A_1 and A_2 are AF-domains, then

$$\dim(A_1 \otimes_k A_2) = \min(\dim A_1 + \mathrm{t.\,d.}(A_2 : k),\ \dim A_2 + \mathrm{t.\,d.}(A_1 : k)).$$

He also stated a formula for $\dim(A \otimes_k R)$ which holds for an AF-domain A, with no restriction on R. At this point, it is worthwhile to recall that an AF-domain is a (locally) Jaffard domain [Gi].

In [BGK1] we were concerned with AF-rings. A k-algebra A, where k is a field, is said to be an AF-ring provided $\mathrm{ht}\, P + \mathrm{t.\,d.}(A/P : k) = \mathrm{t.\,d.}(A_P : k)$, for all prime ideals P of A (for non domains, $\mathrm{t.\,d.}(A) = \sup\{\mathrm{t.\,d.}(A/P : k) \mid P \in \mathrm{Spec}\,(A)\}$). A tensor product of AF-domains is perhaps the most natural example of an AF-ring. We then developed quite general results for AF-rings, showing that the results do not extend trivially from integral domains to rings with zero-divisors.

The purpose of this note is to extend all the known results on the dimension of tensor products of k-algebras to the general case where k is any zero-dimensional ring, denoted by R. The most remarkable outcome is perhaps that $\dim(A_1 \otimes_R A_2)$, where A_1 and A_2 are two R-algebras, depends on a subtle relation which intertwines their two R-module structures. Such phenomenon does not hold in the field case since any k-algebra ($\neq \mathbf{0}$) contains k. Thus, our investigation relies on a mild new assumption that keeps under control most results involving the ideal structures of $A_1 \otimes_R A_2$.

In the first section we extend, in a natural way, the definition of the transcendence degree over a field as well as some basic Wadsworth's results to the zero-dimensional ring case. In the same regard, Section 2 and 3 establish adequate analogues of all the results stated in [BGK1] for AF-rings. Some examples throw more light on the new phenomenon (cited above).

1. Background and preliminaries on tensor products of algebras over a zero-dimensional ring

Throughout this paper R denotes a zero-dimensional ring. We denote by (A, λ_A) an R-algebra A and its associated ring homomorphism $\lambda_A : R \longrightarrow A$; we denote by λ_A^* the associated spectral map $\mathrm{Spec}(A) \longrightarrow \mathrm{Spec}(R)$.

If P is a prime ideal of A, $\lambda_A^{-1}(P)$ is a maximal ideal of R; so we can consider the transcendence degree of the integral domain A/P over the field $R/\lambda_A^{-1}(P)$; we put:

$$t(A :_{\lambda_A} R) = \sup\{\text{t.\,d.}\,(A/P : R/\lambda_A^{-1}(P)) \mid P \in \text{Spec}\,(A)\}$$
$$= \sup\{\text{t.\,d.}\,(A/P : R/\lambda_A^{-1}(P)) \mid P \in \text{Min}\,(A)\}$$

and we say that $t(A :_{\lambda_A} R)$ is the transcendence degree of the R-algebra A over R; we write $t(A : R)$ as an abbreviation for $t(A :_{\lambda_A} R)$, when there is no ambiguity. All along this note we consider only R-algebras (A, λ_A) such that $t(A :_{\lambda_A} R) < \infty$. This, of course, ensures that $\dim A < \infty$. If A is an integral domain, p_A denotes $\ker \lambda_A$.

First of all we note that the transcendence degree of an R-algebra A depends on its R-module structure, as it is shown by the next example:

Example 1. Let $R = k(X) \times k$ and $A = k(X)$, where k is a field. Let $\lambda_1 : R \longrightarrow A$ be the ring homomorphism defined by $\lambda_1(x, y) = x$, and $\lambda_2 : R \longrightarrow A$ be the ring homomorphism defined by $\lambda_2(x, y) = y$. We have $t(A :_{\lambda_1} R) = \text{t.\,d.}\,((k(X) : k(X)) = 0$ and $t(A :_{\lambda_2} R) = \text{t.\,d.}(k(X) : k) = 1$. Thus $t(A :_{\lambda_1} R) \neq t(A :_{\lambda_2} R)$.

We begin by giving a simple generalization of a well-known result [ZS] for algebras over a field.

Lemma 1.1. *Let (A, λ_A) be an R-algebra and $P \in \text{Spec}(A)$. Then*

$$\text{ht}\,P + t(A/P : R) \leq t(A_P : R).$$

Proof. Clearly, $t(A/P : R) = \text{t.\,d.}(A/P : R/\lambda_A^{-1}(P))$. If $P' \in \text{Spec}(A)$ and $P' \subseteq P$, then for the prime ideal P/P' of the $R/\lambda_A^{-1}(P)$-algebra A/P', by [ZS, p.10], we get

$$\text{ht}(P/P') + \text{t.\,d.}(A/P : R/\lambda_A^{-1}(P)) \leq \text{t.\,d.}(A/P' : R/\lambda_A^{-1}(P))$$
$$= \text{t.\,d.}(A_P/P'A_P : R/\lambda_A^{-1}(P')).$$

The result then follows. $\square$

The following elementary properties will be used frequently. These statements admit routine proofs.

Lemma 1.2. *Let (A, λ_A) be an R-algebra.*

(1) *If P is a prime ideal of A and $p = \lambda_A^{-1}(P)$, then*

$$\text{ht}\,P = \text{ht}(P/pA) \qquad \text{and} \qquad t(A_P : R) = t((A/pA)_{P/pA} : R).$$

(2) *If P is a prime ideal of A and $p = \lambda_A^{-1}(P)$, then for each $n \geq 1$*

$$\text{ht}\,P[X_1, \dots, X_n] = \text{ht}(P/pA)[X_1, \dots, X_n].$$

(3) *If A is a locally Jaffard ring, then A/pA is a locally Jaffard ring for each prime ideal p of R such that $pA \neq A$.*

Let (A_1, λ_1) and (A_2, λ_2) be R-algebras. For $i = 1, 2$, we denote by $\mu_i : A_i \to A_1 \otimes_R A_2$ the canonical A_i-algebra homomorphism. The R-algebra $A_1 \otimes_R A_2$, when not specifically indicated, has $\lambda_{A_1 \otimes_R A_2} = \mu_1 \circ \lambda_1 = \mu_2 \circ \lambda_2$ as its associated ring homomorphism. If $P_i \in \mathrm{Spec}(A_i)$, $i = 1, 2$, j_i denotes the inclusion of P_i into A_i, t_{P_i} denotes the transcendence degree of the local ring $A_{i_{P_i}}$ over R and $k(P_i)$ denotes the residue field of $A_{i_{P_i}}$. At last, we set $\Gamma(A_1, A_2) = \{(P_1, P_2) \mid P_1 \in \mathrm{Spec}(A_1),\ P_2 \in \mathrm{Spec}(A_2) \text{ and } \lambda_1^{-1}(P_1) = \lambda_2^{-1}(P_2)\}$.

A tensor product of R-algebras may be zero. We are interested in R-algebras (A_1, λ_1) and (A_2, λ_2) such that $A_1 \otimes_R A_2 \neq 0$, and say that such algebras are *tensorially compatible*. The next result provides some elementary and useful characterizations of tensorially compatible R-algebras. For a more general result, we refer the reader to [GD, Corollary 3.2.7.1].

Proposition 1.3. *Let (A_1, λ_1) and (A_2, λ_2) be R-algebras. The following conditions are equivalent:*

(1) *(A_1, λ_1) and (A_2, λ_2) are tensorially compatible.*
(2) *$\lambda_1^*(\mathrm{Spec}(A_1)) \cap \lambda_2^*(\mathrm{Spec}(A_2)) \neq \emptyset$.*
(3) *There exists a prime ideal P_1 of A_1 such that $\lambda_1^{-1}(P_1)A_2 \neq A_2$.*
(4) *There exists a prime ideal P_2 of A_2 such that $\lambda_2^{-1}(P_2)A_1 \neq A_1$.*
(5) *There exists a prime ideal p of R such that $pA_1 \neq A_1$ and $pA_2 \neq A_2$.*
(6) *$\ker \lambda_1 + \ker \lambda_2 \neq R$.*

Proof. (1) $\Longrightarrow$ (2). If (1) holds, then there exists a prime ideal Q of $A_1 \otimes_R A_2$; therefore $\mu_1^{-1}(Q) \in \mathrm{Spec}(A_1)$ and $\mu_2^{-1}(Q) \in \mathrm{Spec}(A_2)$ are such that $\lambda_1^{-1}(\mu_1^{-1}(Q)) = \lambda_2^{-1}(\mu_2^{-1}(Q))$, and hence, $\lambda_1^*(\mathrm{Spec}(A_1)) \cap \lambda_2^*(\mathrm{Spec}(A_2)) \neq \emptyset$.

(2) $\Longrightarrow$ (3). Let P_1 be a prime ideal of A_1 and P_2 a prime ideal of A_2 such that $\lambda_1^{-1}(P_1) = \lambda_2^{-1}(P_2)$; then $\lambda_1^{-1}(P_1)A_2 \subseteq P_2$ and so $\lambda_1^{-1}(P_1)A_2 \neq A_2$.

The implications (3) $\Longrightarrow$ (4), (4) $\Longrightarrow$ (5) and (5) $\Longrightarrow$ (6) are apparent.

Finally, assume (6). Since $\ker \lambda_1 + \ker \lambda_2 \neq R$, there exists a prime ideal p of R such that $\ker \lambda_1 + \ker \lambda_2 \subseteq p$; this ensures that there exist a prime ideal P_1 of A_1 and a prime ideal P_2 of A_2 such that $\lambda_1^{-1}(P_1) = \lambda_2^{-1}(P_2) = p$. Then

$$
\begin{aligned}
(A_1 \otimes_R A_2)/\left(\mathrm{Im}(j_1 \otimes id_{A_2}) + \mathrm{Im}(id_{A_1} \otimes j_2)\right) &\cong (A_1/P_1) \otimes_R (A_2/P_2) \\
&\cong (A_1/P_1) \otimes_{R/p} (A_2/P_2) \neq 0
\end{aligned}
$$

and so $A_1 \otimes_R A_2 \neq 0$. $\square$

By induction, we obtain the following:

Proposition 1.4. *Let* $(A_1, \lambda_1), \ldots, (A_n, \lambda_n)$ *be R-algebras. Then the following conditions are equivalent:*

(1) $A_1 \otimes_R \cdots \otimes_R A_n \neq 0$.
(2) $\lambda_1^*(\mathrm{Spec}(A_1)) \cap \lambda_2^*(\mathrm{Spec}(A_2)) \cap \cdots \cap \lambda_n^*(\mathrm{Spec}(A_n)) \neq \emptyset$.
(3) *There exists a prime ideal* p *of* R *such that* $pA_i \neq A_i$ *for every* $i = 1, 2, \ldots, n$.

The next result establishes an analogue to [W, Proposition 2.3].

Proposition 1.5. *Let* (A_1, λ_1) *and* (A_2, λ_2) *be R-algebras. Let* $P_1 \in \mathrm{Spec}(A_1)$ *and* $P_2 \in \mathrm{Spec}(A_2)$ *such that* $\lambda_1^{-1}(P_1) = \lambda_2^{-1}(P_2) = p$. *Let*

$$T = A_1 \otimes_R A_2 \quad and \quad \Omega = \{Q \in \mathrm{Spec}(A_1 \otimes_R A_2) \mid \mu_i^{-1}(Q) = P_i, \, i = 1, 2\}.$$

Then

(1) Ω *is lattice isomorphic to* $\mathrm{Spec}(T')$, *where* $T' = k(P_1) \otimes_{R/p} k(P_2)$.
(2) *A prime ideal* Q *of* Ω *is minimal in* Ω *if and only if* $\mathrm{t}(T/Q : R) = \mathrm{t}(A_1/P_1 : R) + \mathrm{t}(A_2/P_2 : R)$.
(3) *If* $Q_0 \in \mathrm{Spec}(T)$ *and* $\mu_i^{-1}(Q_0) \supseteq P_i$, $i = 1, 2$, *then there exists* $Q \in \Omega$ *such that* $Q \subseteq Q_0$.

Proof. (1) is an immediate consequence of [GD, Corollary 3.2.7.1(ii)]. The proof of (2) and (3) is quite similar to that of Wadsworth. $\square$

Proposition 1.5 allows us to extend partially some results of [W] to R-algebras.

Corollary 1.6. *Let* (A_1, λ_1) *and* (A_2, λ_2) *be tensorially compatible R-algebras. Let* $Q \in \mathrm{Spec}(A_1 \otimes_R A_2)$. *Then*

$$\mathrm{ht}\, Q \geq \mathrm{ht}(\mu_1^{-1}(Q)) + \mathrm{ht}(\mu_2^{-1}(Q)).$$

We omit the proof because of its similarity to that of [W, Corollary 2.5].

Corollary 1.7. *Let* (A_1, λ_1) *and* (A_2, λ_2) *be tensorially compatible R-algebras. Then*

$$\mathrm{t}(A_1 \otimes_R A_2 : R) = \sup\{\mathrm{t}(A_1/P_1 : R) + \mathrm{t}(A_2/P_2 : R) \mid (P_1, P_2) \in \Gamma(A_1, A_2)\}.$$

Consequently,

$$\mathrm{t}(A_1 \otimes_R A_2 : R) \leq \mathrm{t}(A_1 : R) + \mathrm{t}(A_2 : R).$$

Proof. Let $(P_1, P_2) \in \Gamma(A_1, A_2)$. Then by Proposition 1.5 there exists a prime ideal Q of $A_1 \otimes_R A_2$ such that $t\left((A_1 \otimes_R A_2)/Q : R\right) = t(A_1/P_1 : R) + t(A_2/P_2 : R)$. So

$\sup\{t(A_1/P_1 : R) + t(A_2/P_2 : R) \mid (P_1, P_2) \in \Gamma(A_1, A_2)\} \leq t(A_1 \otimes_R A_2 : R)$.

Conversely, let Q be a prime ideal of $A_1 \otimes_R A_2$; the prime ideals $P_1 = \mu_1^{-1}(Q)$ and $P_2 = \mu_2^{-1}(Q)$ are such that $\lambda_1^{-1}(P_1) = \lambda_2^{-1}(P_2) = p$; let $T = A_1 \otimes_R A_2$; then, by using [W, Corollary 2.4], we obtain:

$$
\begin{aligned}
t\,(T/Q : R) &= \text{t.d.}\,(T/Q : R/p) \\
&= \text{t.d.}\left(\frac{T/\left(\text{Im}(j_1 \otimes id_{A_2}) + \text{Im}(id_{A_1} \otimes j_2)\right)}{Q/\left(\text{Im}(j_1 \otimes id_{A_2}) + \text{Im}(id_{A_1} \otimes j_2)\right)} : R/p\right) \\
&\leq \text{t.d.}\left((A_1/P_1) \otimes_{R/p} (A_2/P_2)\right) : R/p) \\
&= \text{t.d.}(A_1/P_1 : R/p) + \text{t.d.}(A_2/P_2 : R/p) \\
&= t(A_1/P_1 : R) + t(A_2/P_2 : R),
\end{aligned}
$$

as desired. $\square$

Remark 1. Let (A_1, λ_1) and (A_2, λ_2) be tensorially compatible R-algebras. Clearly, $t(A_1 \otimes_R A_2 : R) = t(A_1 : R) + t(A_2 : R)$ if and only if there exist $P_1 \in \text{Spec}(A_1)$, and $P_2 \in \text{Spec}(A_2)$ such that $\lambda_1^{-1}(P_1) = \lambda_2^{-1}(P_2)$ and $t(A_1 : R) = t(A_1/P_1 : R)$, $t(A_2 : R) = t(A_2/P_2 : R)$. The second condition holds, for instance, if A_1 and A_2 are integral domains or if $\text{Spec}(R)$ is reduced to only one prime ideal. In general, the equality fails as it is shown in the next example. Moreover, when R is a field, we have $\dim(A_1 \otimes_R A_2) \geq \dim A_1 + \dim A_2$ [W, Corollary 2.5]. This is not, in general, true in the zero-dimensional case. The next example deals with these matters.

Example 2. There exist two tensorially compatible R-algebras (A_1, λ_1) and (A_2, λ_2) with

$$t(A_1 \otimes_R A_2 : R) < t(A_1 : R) + t(A_2 : R)$$

and

$$\dim(A_1 \otimes_R A_2 : R) < \dim A_1 + \dim A_2.$$

Let $R = \mathbb{R} \times \mathbb{R}$, $A_1 = \mathbb{R}$ and $A_2 = \mathbb{R} \times \mathbb{R}[X]$. Let $\lambda_1 : R \longrightarrow A_1$ be the ring homomorphism defined by $\lambda_1(x, y) = x$ and let $\lambda_2 : R \longrightarrow A_2$ be the ring homomorphism defined by $\lambda_2(x, y) = (x, y)$. Then

$$t(A_1 :_{\lambda_1} R) = t(\mathbb{R} :_{\lambda_1} R) = \text{t.d.}(\mathbb{R} : \mathbb{R}) = 0$$

and

$$t(A_2 :_{\lambda_2} R) = t(\mathbb{R} \times \mathbb{R}[X] :_{\lambda_2} R) = \sup\{\text{t.d.}(\mathbb{R} : \mathbb{R})\,,\, \text{t.d.}(\mathbb{R}[X] : \mathbb{R})\} = 1$$

and so

$$t(A_1 :_{\lambda_1} R) + t(A_2 :_{\lambda_2} R) = 1.$$

Moreover, by Corollary 1.7,

$t(A_1 \otimes_R A_2 : R) = \sup\{t(A_1/P_1 : R) + t(A_2 : P_2 : R) \mid (P_1, P_2) \in \Gamma(A_1, A_2)\} = t(A_1 :_{\lambda_1} R) + t(A_2/((0) \times I\!\!R[X]) : R) = \mathrm{t.\,d.}(I\!\!R : I\!\!R) + \mathrm{t.\,d.}(I\!\!R : I\!\!R) = 0.$

Further

$$\dim(A_1 \otimes_R A_2 : R) \leq t(A_1 \otimes_R A_2 : R) = 0 < \dim A_1 + \dim A_2 = 1. \quad \square$$

2. Tensor products of AF-rings

Definition 2.8. *An R-algebra (A, λ_A) is an AF-ring if for every $P \in \mathrm{Spec}(A)$*

$$\mathrm{ht}\, P + t(A/P : R) = t(A_P : R).$$

Remark 2. The AF-ring concept does not depend on the structure of algebra over R defined by the associated ring homomorphism.

Indeed, let A be a ring and let λ and $\overline{\lambda}$ be two ring homomorphisms defining two different structures of algebra over R on A. Let $P \in \mathrm{Spec}(A)$. Let $\pi : A \to A/P$ be the natural ring homomorphism. Let $p = \ker(\pi \circ \lambda) = \lambda^{-1}(P)$ and $q = \ker(\pi \circ \overline{\lambda}) = \overline{\lambda}^{-1}(P)$. We can view R/p and R/q as subfields of A/P. Let $k = R/p \cap R/q$. We have:

$$t(A/P :_\lambda R) = \mathrm{t.\,d.}(A/P : R/p) = \mathrm{t.\,d.}(A/P : k) - \mathrm{t.\,d.}(R/p : k)$$

and

$$t(A/P :_{\overline{\lambda}} R) = \mathrm{t.\,d.}(A/P : R/q) = \mathrm{t.\,d.}(A/P : k) - \mathrm{t.\,d.}(R/q : k).$$

On the other hand

$$
\begin{aligned}
t(A_P :_\lambda R) &= \sup\{\mathrm{t.\,d.}(A/Q : R/p) \mid Q \in \mathrm{Spec}(A) \text{ and } Q \subseteq P\} \\
&= \sup\{t(A/Q : k) \mid Q \in \mathrm{Spec}(A) \text{ and } Q \subseteq P\} - \mathrm{t.\,d.}(R/p : k)
\end{aligned}
$$

and

$$
\begin{aligned}
t(A_P :_{\overline{\lambda}} R) &= \sup\{\mathrm{t.\,d.}(A/Q : R/q) \mid Q \in \mathrm{Spec}(A) \text{ and } Q \subseteq P\} \\
&= \sup\{t(A/Q : k) \mid Q \in \mathrm{Spec}(A) \text{ and } Q \subseteq P\} - \mathrm{t.\,d.}(R/q : k).
\end{aligned}
$$

Therefore $t(A/P :_\lambda R) - \mathrm{t.\,d.}(A_P :_\lambda k) = t(A/P :_{\overline{\lambda}} R) - t(A_P :_{\overline{\lambda}} R)$. Consequently, (A, λ) is an AF-ring if and only if $(A, \overline{\lambda})$ is an AF-ring.

Let $\mathcal{R}$ be the class of R-algebras that are AF-rings. Since $R[X_1, \ldots, X_n]$ satisfies the first chain condition for prime ideals [G, Corollary 31.17], any finitely generated R-algebra or any integral extension of such an algebra is an AF-ring. Moreover the class $\mathcal{R}$ is stable under localization and direct product.

The next result presents some properties of the class $\mathcal{R}$, and our proof of Proposition 2.3 uses the following lemma.

Lemma 2.9. *Let (A, λ_A) be an R-algebra. Then A is an AF-ring if and only if A/pA is an AF-ring over the field R/p, for each prime ideal p of R such that $pA \neq A$.*

Proof. Let P be a prime ideal of A and let $p = \lambda_A^{-1}(P)$. According to Lemma 1.2, $\operatorname{ht} P = \operatorname{ht}(P/pA)$ and $\operatorname{t}(A_P : R) = \operatorname{t.d.}((A/pA)_{P/pA} : R/p)$.

Assume that A is an AF-ring and let p be a prime ideal of R such that $pA \neq A$. Let P be a prime ideal of A containing pA; then

$$
\begin{aligned}
\operatorname{ht}(P/pA) + \operatorname{t.d.}((A/pA)/(P/pA) : R/p) &= \operatorname{ht} P + \operatorname{t}(A/P : R) \\
&= \operatorname{t}(A_P : R) \\
&= \operatorname{t.d.}((A/pA)_{P/pA} : R/p).
\end{aligned}
$$

Conversely, let P be a prime ideal of A and let $p = \lambda_A^{-1}(P)$. Then $pA \neq A$; so by hypothesis A/pA is an AF-ring over R/p, hence

$$
\begin{aligned}
\operatorname{ht} P + \operatorname{t}(A/P : R) &= \operatorname{ht}(P/pA) + \operatorname{t.d.}(A/P : R/p) = \\
&= \operatorname{ht}(P/pA) + \operatorname{t.d.}((A/pA)/(P/pA) : R/p) \\
&= \operatorname{t.d.}((A/pA)_{P/pA} : R/p) = \operatorname{t}(A_P : R). \quad \square
\end{aligned}
$$

Proposition 2.10. *The class $\mathcal{R}$ satisfies the following properties:*

(1) *Let $(A_1, \lambda_1), \dots, (A_n, \lambda_n)$ be tensorially compatible R-algebras. If $A_1, \dots, A_n$ are AF-rings, then $A_1 \otimes_R \cdots \otimes_R A_n$ is an AF-ring.*

(2) *Let A be an AF-ring. Then the polynomial ring $A[X]$ is an AF-ring and for each prime ideal P of A, $\operatorname{ht} P = \operatorname{ht} P[X]$.*

(3) *An AF-ring A is a locally Jaffard ring.*

Proof. (1) By induction, it suffices to consider the case $n = 2$. Let (A_1, λ_1) and (A_2, λ_2) be tensorially compatible AF-rings. Let p be a prime ideal of R such that $p(A_1 \otimes_R A_2) \neq A_1 \otimes_R A_2$. By Lemma 2.2 A_1/pA_1 and A_2/pA_2 are AF-rings over the field R/p ; hence by [W, Proposition 3.1] $(A_1/pA_1) \otimes_{R/p} (A_2/pA_2)$ is an AF-ring over R/p, so that $(A_1 \otimes_R A_2)/p(A_1 \otimes_R A_2) \cong (A_1/pA_1) \otimes_{R/p} (A_2/pA_2)$ is an AF-ring over R/p. The proof is complete via Lemma 2.2.

(2) Since $A[X] \cong A \otimes_R R[X]$, the result follows from (1). Let P be any prime ideal of A; so

$$
\begin{aligned}
\operatorname{ht} P &\leq \operatorname{ht} PA[X] \\
&= \operatorname{t}\left(A[X]_{PA[X]} : R\right) - \operatorname{t}(A[X]/PA[X] : R) \\
&\leq \operatorname{t}(A_P[X] : R) - \operatorname{t}((A/P)[X] : R) \\
&= \operatorname{ht} P.
\end{aligned}
$$

(3) Let A be an AF-ring. By (2) we obtain that for any prime ideal P of A and for each positive integer n, ht $P = $ ht $P[X_1, \ldots, X_n]$. Hence, by [C, p.127], A is a locally Jaffard ring. $\square$

In the same regard, this section establishes adequate analogues of the main results stated in [BGK1] on the dimension of tensor products of AF-rings over a field. Let us consider for R-algebras the following functions (introduced in [W] for k-algebras) : let (A_1, λ_1) and (A_2, λ_2) be tensorially compatible R-algebras; let $P_1 \in \mathrm{Spec}(A_1)$ and $P_2 \in \mathrm{Spec}(A_2)$ such that $\lambda_1^{-1}(P_1) = \lambda_2^{-1}(P_2)$. Set

$$\delta(P_1, P_2) = \sup\{\mathrm{ht}\, Q \mid Q \in \mathrm{Spec}(A_1 \otimes_R A_2) \text{ and } \mu_i^{-1}(Q) = P_i, \ i = 1, 2\}.$$

One may easily check that

$$\dim(A_1 \otimes_R A_2) = \sup\{\delta(P_1, P_2) \mid (P_1, P_2) \in \Gamma(A_1, A_2)\},$$

and

$$\delta(P_1, P_2) = \delta(P_1/pA_1, P_2/pA_2),$$

where $p = \lambda_1^{-1}(P_1) = \lambda_2^{-1}(P_2)$.

Let (A, λ_A) be an R-algebra, $P \in \mathrm{Spec}(A)$ and d and s integers with $0 \leq d \leq s$. Set

$$\triangle(s, d, P) = \mathrm{ht}\, PA[X_1, \ldots, X_s] + \min(s, \ d + \mathrm{t}(A/P : R)),$$

$$D(s, d, A) = \sup\{\triangle(s, d, P) \mid P \in \mathrm{Spec}(A)\}.$$

Next we provide a formula for the dimension of the tensor product $A \otimes_R B$, where A is an AF-ring and B is any ring.

Theorem 2.11. *Let (A, λ_A) be an AF-ring and (B, λ_B) be any R-algebra such that $A \otimes_R B \neq \mathbf{0}$. Let $(P, I) \in \Gamma(A, B)$. Then*

$$\delta(P, I) = \Delta(\mathrm{t}_P, \mathrm{ht}\, P, I)$$

where $\mathrm{t}_P = \mathrm{t}(A_P : R)$, and consequently
$\dim(A \otimes_R B) = \sup\{D(\mathrm{t}_P, \mathrm{ht}\, P, B/pB) \mid P \in \mathrm{Spec}(A), \ p = \lambda_A^{-1}(P) \text{ and } pB \neq B\}$
$= \sup\{\mathrm{ht}\, I\,[X_1, \ldots, X_{\mathrm{t}_P}] + \min(\mathrm{t}_P, \ \mathrm{ht}\, P + \mathrm{t}(B/I : R)) \mid (P, I) \in \Gamma(A, B)\}.$

Proof. Let $P \in \mathrm{Spec}(A)$, $I \in \mathrm{Spec}(B)$ such that $\lambda_A^{-1}(P) = \lambda_B^{-1}(I) = p$. As noted previously, $\delta(P, I) = \delta(P/pA, I/pB)$; moreover, by Lemma 2.2, A/pA is an AF-ring over the field R/p; so we can apply Theorem 1.4 from [BGK1] to the (R/p)-algebras A/pA and B/pB, obtaining that

$$\delta(P/pA, I/pB) = \Delta\left(\mathrm{t}_P, \mathrm{ht}(P/pa), I/pB\right).$$

By Lemma 1.2, $\text{ht}(P/pa) = \text{ht}\,P$ and $t(A_P : R) = \text{t}\,((A/pA)_{P/pA} : R))$. Further, for any $n \geq 1$, $\text{ht}\,I[X_1,\ldots,X_n] = \text{ht}(I/pB)[X_1,\ldots,X_n]$. Hence

$$\delta(P/pA, I/pB) = \Delta(\text{t}_P, \text{ht}\,P, I).$$

It follows that $\delta(P, I) = \Delta(\text{t}_P, \text{ht}\,P, I)$, as asserted. Consequently, using the definitions of δ, Δ and D and the stated condition on $\delta(P, I)$, yields $\dim(A \otimes_R B) = \sup\{\delta(P, I) \mid (P, I) \in \Gamma(A, B)\} = \sup\{\Delta(\text{t}_P, \text{ht}\,P, I) \mid (P, I) \in \Gamma(A, B)\} = \sup\{D(\text{t}_P, \text{ht}\,P, B/pB) \mid P \in \text{Spec}(A),\ p = \lambda_A^{-1}(P)$ and $pB \neq B\} = \sup\{\text{ht}\,I\,[X_1,\ldots,X_{t_P}]+\min(\text{t}_P,\ \text{ht}\,P+\text{t}(B/I : R)) \mid (P, I) \in \Gamma(A, B)\}$ as we wished to show. $\square$

It is worthwhile to note that $\dim(A \otimes_R B)$ depends on the R-module structure of A and B. The next example illustrates this fact:

Example 3. Let (A, λ_A) and (B, λ_B) be R-algebras such that A is an AF-ring and $A \otimes_R B \neq \mathbf{0}$. Let p be a prime ideal of R and let $\pi : R \longrightarrow R/p$ be the canonical ring homomorphism. Let $\lambda_1 : R \times R \times R \longrightarrow R/p \times A$ and $\lambda_2 : R \times R \times R \longrightarrow R/p \times B$ be the ring homomorphisms defined respectively by $\lambda_1(x, y, z) = (\pi(x), \lambda_A(y))$ and $\lambda_2(x, y, z) = (\pi(x), \lambda_B(z))$. It is an easy matter to verify that $\Gamma(R/p \times A, R/p \times B) = \{((0) \times A,\ (0) \times B)\}$. Hence via Theorem 2.4, it is easy to check that the dimension of the tensor product of $((R/p \times A), \lambda_1)$ and $((R/p \times B), \lambda_2)$ is zero. On the other hand, let $\lambda_2' : R \times R \times R \longrightarrow R/p \times B$ be the ring homomorphism defined by $\lambda_2'(x, y, z) = (\pi(x), \lambda_B(y))$; now by Theorem 2.4 we obtain that the dimension of the tensor product of $((R/p \times A), \lambda_1)$ and $((R/p \times B), \lambda_2')$ is equal to $\dim(A \otimes_R B)$. Thus, it suffices to choose A and B such that $\dim(A \otimes_R B) > 0$ (for instance, when R is a field and A, B are non trivial R-algebras). Therefore the two values are different according to the $(R \times R \times R)$-module structure of $R/p \times A$ and $R/p \times B$.

With the further assumption that A is an AF-domain, we obtain the following :

Corollary 2.12. *Let (A, λ_A) be an AF-domain and let (B, λ_B) be any R-algebra such that $A \otimes_R B \neq \mathbf{0}$. Then*

$$\dim(A \otimes_R B) = D\,(\text{t}(A : R), \dim A, B/p_A B)$$

where $p_A = \ker \lambda_A$. Furthermore, if B is an integral domain, then

$$\dim(A \otimes_R B) = D\,(\text{t}(A : R), \dim A, B)\,.$$

Proof. Since A is an integral domain, for any prime ideal P of A, $\lambda_A^{-1}(P) = p_A$ and $\text{t}(A_P : R) = \text{t}(A : R)$; so Theorem 2.4 implies that $\dim(A \otimes_R B) = \sup\{D(\text{t}(A : R), \text{ht}\,P, B/p_A B) \mid P \in \text{Spec}(A)\}$. Since $D(s, d, A)$ is a nondecreasing function of the second argument, then $\dim(A \otimes_R B) = D(\mathrm{t}(A : R), \dim A, B/p_A B)$, as asserted. $\square$

Next, we state a technical result that allows us to determine a necessary and sufficient condition under which the dimension of the tensor product of AF-rings over a zero-dimensional ring satisfies the formula of Wadsworth's Theorem 3.8.

Proposition 2.13. *Let* $(A_1, \lambda_1), \ldots, (A_n, \lambda_n)$ *be tensorially compatible AF-rings. Then*
$\dim(A_1 \otimes_R \cdots \otimes_R A_n) =$
$\sup\{\min(\mathrm{ht}\, M_1 + \mathrm{t}_{M_2} + \cdots + \mathrm{t}_{M_n},\, \mathrm{t}_{M_1} + \mathrm{ht}\, M_2 + \cdots + \mathrm{t}_{M_n},\, \cdots,\, \mathrm{t}_{M_1} + \ldots + \mathrm{t}_{M_{n-1}} + \mathrm{ht}\, M_n) \mid M_i \in \mathrm{Max}(A_i)$ *and* $\lambda_1^{-1}(M_1) = \lambda_2^{-1}(M_2) = \cdots = \lambda_n^{-1}(M_n)\}$.

Proof. It is deduced from the fact that $\dim(A_1 \otimes_R \cdots \otimes_R A_n) = \sup\{\dim((A_1/P_1) \otimes_R (A_2/P_2) \otimes_R \cdots \otimes_R (A_n/P_n)) \mid P_i \in \mathrm{Spec}(A_i)$ for $i = 1, \ldots, n$ and $\lambda_1^{-1}(P_1) = \lambda_2^{-1}(P_2) = \cdots = \lambda_n^{-1}(P_n)\} = \sup\{\dim((A_1/pA_1) \otimes_{R/p} (A_2/pA_2) \otimes_{R/p} \cdots \otimes_{R/p} (A_n/pA_n)) \mid p \in \mathrm{Spec}(R),$ and $pA_i \neq A_i$ for $i = 1, \ldots, n\}$; now we conclude via [BGK1, Lemma 1.6 and Remark 1.7]. $\square$

Theorem 2.14. *Let* $(A_1, \lambda_1), \ldots, (A_n, \lambda_n)$ *be tensorially compatible AF-rings with* $\mathrm{t}_i = \mathrm{t}(A_i : R)$ *and* $d_i = \dim A_i$. *Then* $\dim(A_1 \otimes_R \ldots \otimes_R A_n) = \mathrm{t}_1 + \cdots + \mathrm{t}_n - \max\{\mathrm{t}_i - d_i \mid 1 \leq i \leq n\}$ *if and only if there exist maximal ideals* $M_1, \ldots, M_n$ *belonging respectively to* $A_1, \ldots, A_n$ *such that* $\lambda_1^{-1}(M_1) = \cdots = \lambda_n^{-1}(M_n)$, *and there exists* $r \in \{1, \ldots, n\}$ *such that* $\mathrm{ht}\, M_r = d_r$ *and for any* $j \in \{1, \ldots, n\} - \{r\}$, $\mathrm{t}_{M_j} = \mathrm{t}_j$ *and* $\mathrm{t}(A_j/M_j : R) \leq \mathrm{t}(A_r/M_r : R)$.

Proof. It is deduced from the fact that $\dim(A_1 \otimes_R \cdots \otimes_R A_n) = \sup\{\dim((A_1/pA_1) \otimes_{R/p} (A_2/pA_2) \otimes_{R/p} \cdots \otimes_{R/p} (A_n/pA_n)) \mid p \in \mathrm{Spec}(R)$ and $pA_i \neq A_i$, for $i - 1, 2, \ldots, n\}$ and [BGK1, Theorem 1.8]. $\square$

Corollary 2.15. *Let* $(A_1, \lambda_1), \ldots, (A_n, \lambda_n)$ *be tensorially compatible AF-rings with* $\mathrm{t}_i = \mathrm{t}(A_i : R)$ *and* $d_i = \dim A_i$. *If one of the following conditions is satisfied:*

(1) *There exist maximal ideals* $M_1, \ldots, M_n$ *belonging respectively to* $A_1, \ldots, A_n$ *such that* $\lambda_1^{-1}(M_1) = \cdots = \lambda_n^{-1}(M_n)$ *and* $\mathrm{ht}\, M_i = d_i$, $\mathrm{t}_{M_i} = \mathrm{t}_i$ *for* $i = 1, 2, \ldots, n$.

(2) *If* $M_1, \ldots, M_n$ *are maximal ideals belonging respectively to* $A_1, \ldots, A_n$ *such that* $\lambda_1^{-1}(M_1) = \cdots = \lambda_n^{-1}(M_n)$, *then* $\mathrm{t}_{M_i} = \mathrm{t}_i$ *for* $i = 1, \ldots, n$.

(3) *If $P_1, \ldots, P_n$ are minimal prime ideals belonging respectively to $A_1, \ldots, A_n$ such that $\lambda_1^{-1}(P_1) = \cdots = \lambda_n^{-1}(P_n)$, then $\mathrm{t}(A_i/P_i : R) = t_i$, for $i = 1, \ldots, n$.*

(4) *$A_1, \ldots, A_n$ are equicodimensional.*

then

$$\dim(A_1 \otimes_R \cdots \otimes_R A_n) = t_1 + \cdots + t_n - \max\{t_i - d_i \mid 1 \le i \le n\}.$$

The proofs of (1), (2), (3) and (4) are similar to those of [BGK1, Corollaries 1.10, 1.11, 1.13, and 1.14], respectively.

Corollary 2.16. *Let $(A_1, \lambda_1), \ldots, (A_n, \lambda_n)$ be tensorially compatible AF-domains with $t_i = \mathrm{t}(A_i : R)$ and $d_i = \dim(A_i)$. Then*

$$\dim(A_1 \otimes_R \cdots \otimes_R A_n) = t_1 + \cdots + t_n - \max\{t_i - d_i \mid 1 \le i \le n\}.$$

Proof. Since $A_1 \otimes_R \cdots \otimes_R A_n \neq 0$, by Proposition 1.4 we have $p_{A_1} = p_{A_2} = \cdots = p_{A_n} = p$; then $A_1 \otimes_R \cdots \otimes_R A_n \cong A_1 \otimes_{R/p} \cdots \otimes_{R/p} A_n$. The result follows from [W, Theorem 3.8]. $\square$

Now we consider the special case in which $(A_1, \lambda_1) = (A_2, \lambda_2)$.

Corollary 2.17. *Let (A, λ_A) be an AF-ring. Then $\dim(A \otimes_R A) = \dim A + \mathrm{t}(A : R)$ if and only if there exist maximal ideals M and N in A such that $\lambda_A^{-1}(M) = \lambda_A^{-1}(N)$, $\mathrm{ht}\, M = \dim A$, $\mathrm{t}(A_N : R) = \mathrm{t}(A : R)$ and $\mathrm{t}(A/N : R) \le \mathrm{t}(A/M : R)$.*

3. The valuative dimension of tensor products and Jaffard rings

[BGK1, Theorem 2.1] establishes that if A is an AF-ring over a field k and B is a locally Jaffard ring, then $A \otimes_k B$ is a locally Jaffard ring. We next extend this result to AF-rings over a zero-dimensional ring.

Theorem 3.18. *Let (A, λ_A) be an AF-ring and (B, λ_B) a locally Jaffard ring such that $A \otimes_R B \neq 0$. Then $A \otimes_R B$ is a locally Jaffard ring.*

Proof. It is sufficient to prove that for each prime ideal Q of $A \otimes_R B$ and for each nonnegative integer n, $\mathrm{ht}\, Q[X_1, \ldots, X_n] = \mathrm{ht}\, Q$ (see [ABDFK] and [C]). Let $P = \mu_A^{-1}(Q)$, $I = \mu_B^{-1}(Q)$ and $p = \lambda_{A \otimes_R B}^{-1}(Q)$; according to Lemma 2.2, A/pA is an AF-ring over the field R/p; moreover, by Lemma 1.2 B/pB is a locally Jaffard ring; so we can apply Theorem 2.1 of [BGK1] to the (R/p)-algebras A/pA and B/pB obtaining that $(A/pA) \otimes_{R/p} (B/pB)$ is a locally Jaffard ring. Since $(A/pA) \otimes_{R/p} B/pB \cong (A \otimes_R B)/p(A \otimes_R B)$, then for each nonnegative integer n, it results that $\mathrm{ht}((Q/p(A \otimes_R B))[X_1, \ldots, X_n]) =$

ht $(Q/p(A \otimes_R B))$; so according to Lemma 1.2, ht $Q = $ ht $Q[X_1, \ldots, X_n]$, as desired. $\square$

Remark 3. Let (A, λ_A) be an AF-ring and (B, λ_B) any R-algebra such that $A \otimes_R B \neq 0$. Let $Q \in \mathrm{Spec}(A \otimes_R B)$, $P = \mu_A^{-1}(Q)$ and $I = \mu_B^{-1}(Q)$. We obtain from [BGK 1, Lemma 2.2] the following result:

$$\mathrm{ht}\, Q + \mathrm{t}\,((A \otimes_R B)/Q : R) = \mathrm{t}_P + \mathrm{ht}\, I[X_1, \ldots, X_{t_p}] + \mathrm{t}(B/I : R).$$

Let us recall that the valuative dimension of tensor products of algebras over a field does not seem to be effectively computable in general. However, [Gi, Proposition 3.1] states that provided A_1 and A_2 are two algebras over a field k, then

$$\dim_v(A_1 \otimes_k A_2) \leq \min \left(\dim_v A_1 + \mathrm{t.\,d.}(A_2 : k),\ \mathrm{t.\,d.}(A_1 : k) + \dim_v A_2 \right).$$

The next result establishes the analogue of this result for the zero-dimensional case.

Proposition 3.19. *Let (A_1, λ_1) and (A_2, λ_2) be tensorially compatible R-algebras. Then*

$$\dim_v(A_1 \otimes_R A_2) \leq \min \left(\dim_v A_1 + \mathrm{t}(A_2 : R),\ \mathrm{t}(A_1 : R) + \dim_v A_2 \right).$$

Proof. Let Q be any prime ideal of $A_1 \otimes_R A_2$; let $P_1 = \mu_1^{-1}(Q)$, $P_2 = \mu_2^{-1}(Q)$ and $p = \lambda_1^{-1}(P_1) = \lambda_2^{-1}(P_2)$. Let $T = A_1 \otimes_R A_2$. Then

$$\dim_v (T/Q) \leq \dim_v (T/ (\mathrm{Im}(j_1 \otimes id_{A_2}) + \mathrm{Im}(id_{A_1} \otimes j_2))).$$

Moreover, using the canonical isomorphism

$$T/(\mathrm{Im}(j_1 \otimes id_{A_2}) + \mathrm{Im}(id_{A_1} \otimes j_2)) \cong (A_1/P_1) \otimes_{R/p} (A_2/P_2)$$

and [Gi, Proposition 3.1], yields

$$\begin{aligned}
\dim_v (T/Q) &\leq \dim_v ((A_1/P_1) \otimes_{R/p} (A_2/P_2)) \\
&\leq \min (\dim_v A_1/P_1 + \mathrm{t}(A_2/P_2 : R),\ \dim_v A_2/P_2 + \mathrm{t}(A_1/P_1 : R)) \\
&\leq \min (\dim_v A_1 + \mathrm{t}(A_2 : R),\ \dim_v A_2 + \mathrm{t}(A_1 : R)).\ \square
\end{aligned}$$

The next result handles the case where one of two R-algebras is an AF-ring.

Proposition 3.20. *Let (A, λ_A) and (B, λ_B) be tensorially compatible R-algebras and A an AF-ring. Then, for any $r \geq \dim_v B - 1$,*
$\dim_v(A \otimes_R B) =$
$\sup\{D(\mathrm{t}_P + r, \mathrm{ht}\, P + r, B/pB) \mid P \in \mathrm{Spec}(A),\ p = \lambda_A^{-1}(P) \text{ and } pB \neq B\} -$
$r =$
$\sup\{\mathrm{ht}I\,[X_1, \ldots, X_r] + \min (\mathrm{t}_P,\ \mathrm{ht}\, P + \mathrm{t}(B/I : R)) \mid (P, I) \in \Gamma(A, B)\}.$

Proof. Let $r \geq \dim_v B - 1$. Then, by [C, Proposition 1, ii)], $B[X_1, ..., X_r]$ is a locally Jaffard ring. So, according to Theorem 3.1, $A \otimes_R B[X_1, ..., X_r]$ is a locally Jaffard ring and hence a Jaffard ring. Therefore, by Corollary 2.5, $\dim_v(A \otimes_R B[X_1, ..., X_r]) = \dim(A \otimes_R B[X_1, ..., X_r]) = \sup\{D(t_P, htP, (B/pB)[X_1, ..., X_r]) \mid P \in \mathrm{Spec}(A)$ with $\lambda_A^{-1}(P) = p$ and $pB \neq B\}$. Hence, according to [BGK1, Lemma 2.3],
$\dim_v(A \otimes_R B) = \sup\{D(t_P + r, \ htP + r, \ B/pB) \mid P \in \mathrm{Spec}(A)$ with $\lambda_A^{-1}(P) = p$ and $pB \neq B\} - r = \sup\{htI[X_1, ..., \ X_r] + \min(t_P, \ htP + t(B/I : R)) \mid (P, I) \in \Gamma(A, B)\}$. □

We conclude this section with two results on AF-domains.

Corollary 3.21. *Let (A, λ_A) be an AF-domain and B any R-algebra such that $A \otimes_R B \neq 0$. Then for any $r \geq \dim_v B - 1$*
$\dim_v(A \otimes_R B) = D(t+r, \ d+r, \ B/p_A B) - r = \sup\{ht \, Q[X_1, ... X_r] + \min(t, \ d + t(B/I : R)) \mid I \in \mathrm{Spec}(B)$ and $\lambda_B^{-1}(I) = p_A\}$,
where $t = t(A : R)$ *and* $d = \dim A$.

Corollary 3.22. *Let (A, λ_A) and (B, λ_B) be R-algebras such that A is an AF-domain and $A \otimes_R B \neq 0$. If $\dim_v B \leq t(A : R) + 1$, then $A \otimes_R B$ is a Jaffard ring.*

Remark 4. We thank the Referee for the following observation. Let A_{red} be the reduced ring associated to a ring A. Then $t(A : R) = t(A_{red} : R_{red})$ for any R-algebra (A, λ_A); moreover, if (A_1, λ_1) and (A_2, λ_2) are R-algebras, then $(A_1 \otimes_R A_2)_{red} = ((A_1)_{red} \otimes_{R_{red}} (A_2)_{red})_{red}$ [GD, Corollary 4.5.12]. One may therefore assume that R is absolutely flat and (A_1, λ_1), (A_2, λ_2) are reduced R-algebras.

REFERENCES

[ABDFK] D.F. Anderson, A. Bouvier, D.E. Dobbs, M. Fontana and S. Kabbaj, On Jaffard domains, Expo. Math. 6 (1988), 145-175.

[B]   N. Bourbaki, Algèbre, Ch.3, Paris, Hermann, 1958.

[BGK1] S. Bouchiba, F. Girolami and S. Kabbaj, The dimension of tensor products of AF-rings, Lect. Notes Pure Appl. Math., M. Dekker, 185 (1997), 141-154.

[BGK2] S. Bouchiba, F. Girolami and S. Kabbaj, The dimension of tensor products of k-algebras arising from pullbacks, J. Pure Appl. Algebra (to appear).

[BK]   A. Bouvier and S. Kabbaj, Examples of Jaffard domains, J. Pure Appl. Algebra 54 (1988), 155-165.

[BMRH]   J.W. Brewer, P.R. Montgomery, E.A. Rutter and W.J. Heinzer, Krull dimension of polynomial rings, Lect. Notes in Math., Springer Berlin-New York, 311 (1972), 26-45.

[C]   P.-J. Cahen, Construction B,I,D et anneaux localement ou résiduellement de Jaffard, Arch. Math., 54 (1990), 125-141.

[G]   R. Gilmer, Multiplicative ideal theory, M. Dekker, New York, 1972.

[GD]   A. Grothendieck and J.A. Dieudonné, Eléments de Géométrie Algébrique I, Die Grundlehren der mathematischen Wissenschaften 166, Springer, Berlin, 1971.

[Gi]   F. Girolami, AF-rings and locally Jaffard rings, Lect. Notes Pure Appl. Math. 153 (1994), M. Dekker, 151-161.

[J]   P. Jaffard, Théorie de la dimension dans les anneaux de polynômes, Mém. Sc. Math., 146 (1960), Gauthier-Villars, Paris.

[M]   H. Matsumura, Commutative ring theory, Cambridge University Press, Cambridge, 1989.

[N]   M. Nagata, Local rings, Interscience, New York, 1962.

[S]   R.Y. Sharp, The dimension of the tensor product of two field extensions, Bull. London Math. Soc. 9 (1977), 42-48.

[SV]   R.Y. Sharp, P. Vamos, The dimension of the tensor product of a finite number of field extensions, J. Pure Appl. Algebra, 10 (1977), 249-252.

[V]   P. Vamos, On the minimal prime ideals of a tensor product of two fields, Math. Proc. Camb. Phil. Soc., 84 (1978), 25-35.

[W]   A.R. Wadsworth, The Krull dimension of tensor products of commutative algebras over a field, J. London Math. Soc., 19 (1979), 391-401.

[ZS]   O. Zariski, P. Samuel, Commutative Algebra, Vol. I, Van Nostrand, New York, 1960.

Current address: Samir BOUCHIBA, Faculté des Sciences de Meknès, Université Moulay Ismail, Meknès (50000), Morocco
E-mail address: bouchiba@caramail.com

Current address: Florida GIROLAMI, Dipartimento di Matematica, Università degli Studi Roma Tre, Largo S. Leonardo Murialdo 1, 00146 Roma, Italy
E-mail address: girolami@mat.uniroma3.it

Current address: Salah-Eddine KABBAJ, Department of Mathematical Sciences, KFUPM, P.O. Box 849, Dhahran 31261, Saudi Arabia
E-mail address: kabbaj@kfupm.edu.sa

The Characteristic Sequence of Integer-Valued Polynomials on a Subset

J. BOULANGER, Institut Universitaire de Formation des Maîtres d'Amiens, Amiens, France

J.-L. CHABERT, Faculté de Mathématiques et d'Informatique, Université de Picardie, Amiens, France

S. EVRARD, Institut Universitaire de Formation des Maîtres d'Amiens, Amiens, France

G. GERBOUD, Institut Universitaire de Formation des Maîtres d'Amiens, Amiens, France

Abstract Let V be a discrete valuation domain with quotient field K and let E be an infinite subset of V. The V-module
$$\text{Int}(E, V) = \{f \in K[X] \mid f(E) \subseteq V\}$$
of integer-valued polynomials on E is isomorphic to $\oplus_{k=0}^{\infty} I_k g_k$ where the g_k are monic polynomials in $V[X]$ and the I_k are the characteristic ideals of $\text{Int}(E, V)$. We compute here the valuation of these ideals I_k in the case where E is a homogeneous subset of V and we give explicit formulas in several particular cases.

1 Introduction

Two papers about integer-valued polynomials on an arbitrary subset E of a Dedekind domain D appeared recently. The first one by Bárbácioru [1] essentialy extends the results of Cahen [2] which concern the case where $E = D$. The second paper by Bhargava [3] is more general, but in some sense gives less precise results than [1]. The aim of the present paper is to use results of [3] to improve results of [1]. For notation, definitions and well known results we refer to [4].

Let D be a Dedekind domain and let E be an infinite subset of D. Let K denote the quotient field of D and $\text{Int}(E, D)$ denote the ring of *integer-valued polynomials* on E, that is:

$$\text{Int}(E, D) = \{f \in K[X] \mid f(E) \subseteq D\}.$$

We are interested in the D-module structure of $\text{Int}(E, D)$. Bárbácioru's main result is the following (in the case where the residue fields are finite): for each integer $n \geq 0$,

$$\text{Int}_n(E, D) = \{f \in \text{Int}(E, D) \mid \deg(f) \leq n\} = \bigoplus_{k=0}^{n} J_k^{(n)} f_k$$

where $J_0^{(n)}, \ldots, J_n^{(n)}$ are fractional ideals of D and $f_0, \ldots, f_n$ are monic polynomials in $D[X]$ [1, Theorem 1]. It follows from the proofs of [1] that the ideals $J_k^{(n)}$ depend a priori on the integer n. A natural question raised by this result is then: does the ideal $J_k^{(n)}$ actually depend on n? The answer is no because, if such fractional ideals $J_k^{(n)}$ exist, then they are necessarily the characteristic ideals I_k of $\text{Int}(E, D)$.

Recall that, for each integer $k \geq 0$, the *characteristic ideal* I_k of $\text{Int}(E, D)$ is the fractional ideal formed by 0 and the set of leading coefficients of polynomials in $\text{Int}(E, D)$ of degree $\leq k$ (see [4, §II.1]).

The fact that $J_k^{(n)} = I_k$ is easy to check by induction on k.

In fact, among several results of Bhargava [3], there is the following assertion (without any assumption on the residue fields) which clearly contains Bárbácioru's assertion:

$$\text{Int}(E, D) = \bigoplus_{k=0}^{\infty} I_k g_k$$

where the I_k are the characteristic ideals of $\text{Int}(E, D)$ and the g_k are monic polynomials in $D[X]$ [3, Theorems 12 and 13].

But, on the other hand, there is one interesting point in [1] that we do not find in [3]: an attempt to characterize, and actually compute, the fractional ideals I_k. In order to do that, Bárbácioru first notices that to study the D-module $\text{Int}_n(E, D)$, we may replace E by the set

$$E_n = \{x \in K \mid \forall f \in \text{Int}_n(E, D),\ f(x) \in D\}$$

since

$$\text{Int}_n(E, D) = \text{Int}_n(E_n, D).$$

Of course,

$$E_0 = K,$$

and, for $n \geq 1$,

$$E \subseteq E_n \subseteq D$$

because X belongs to $\mathrm{Int}_n(E, D)$.

Since E is infinite, the D-module $\mathrm{Int}_n(E, D)$ is finitely generated (see for instance [4, Proposition II.1.1]), and hence, 'by continuity' of the integer-valued polynomials which generate $\mathrm{Int}_n(E, D)$ (see for instance [4, Proposition III.2.1]), there is a nonzero ideal A of D such that

for each $x \in E_n$, $x + A = \{x + a \mid a \in A\} \subseteq E_n$.

Such a subset E_n is called by McQuillan [5, §2] a *homogeneous subset* of D with ideal A. Since the subsets E_n are homogeneous subsets, in almost all the proofs of [1], the subset E is supposed itself to be homogeneous.

We may notice that in the case where the Dedekind domain D has finite residue fields, a homogeneous subset is necessarily of the form

$$E = \bigcup_{i=1}^{r} b_i + A$$

where $b_1, \ldots, b_r$ are elements of D pairwise non-congruent modulo A (this is no more true with infinite residue fields).

Moreover, to study the characteristic ideals I_n we may localize because, for each ideal M of D, $(\mathrm{Int}(E, D))_M = \mathrm{Int}(E, D_M)$ [4, Proposition I.2.7], and hence, the fractional ideals $(I_n)_M$ are the characteristic ideals of $\mathrm{Int}(E, D_M)$. Then, D_M is a discrete valuation domain and, if E is the union of r cosets modulo M^l, that is

$$E = \bigcup_{i=1}^{r} b_i + M^l,$$

letting $s = |D/M^l|$, we find in [1, Theorem 4] that

$$(I_n)_M = M^{-S(n)} D_M \quad \text{with} \quad S(n) = l \sum_{\alpha \geq 0} \left[\frac{n}{r s^\alpha} \right].$$

This formula is correct for the classical case where $E = D$ provided one considers that it corresponds to the union of q cosets modulo M (where $q = |D/M|$), so that

$$S(n) = \sum_{\alpha \geq 0} \left[\frac{n}{q^{\alpha+1}} \right]$$

which is the formula given by Pólya [6] (see, for instance, [4, Corollary II.2.9]). In fact, the formula fails for $l > 1$. For example, if $D = \mathbf{Z}$, $M = 2\mathbf{Z}$, and $E = 4\mathbf{Z}$, then $l = 2$, $r = 1$, $s = 4$, and $S(2) = 2\sum_{\alpha \geq 0}\left[\frac{2}{4^\alpha}\right] = 4$, while $\frac{X(X-4)}{2^5} \in \mathrm{Int}(E,V)$. The correct value is indeed 5 (as follows from Proposition 3.3).

The aim of this paper is to give a correct formula. In order to do this, we use a general result of Bhargava [3] that we first recall.

2   Bhargava's revisited result

HYPOTHESIS. From now on, V denotes a discrete valuation domain and E an infinite subset of V. We denote by K the quotient field of V, v the valuation of K associated to V, M the maximal ideal of V, and t a generator of M.

DEFINITION [4, §IX.3]. The *characteristic sequence* of $\mathrm{Int}(E,V)$ is the sequence of positive integers $\{-v(I_n)\}_{\{n \in \mathbf{N}\}}$ where I_n denotes the characteristic ideal of $\mathrm{Int}(E,V)$ (that is, the fractional ideal formed by 0 and the leading coefficients of the elements of $\mathrm{Int}(E,V)$ with degree $\leq n$).

Similarly to [3], we set the following.

DEFINITION. A *v-ordered sequence of elements of E* is a (finite or infinite) sequence $\{a_n\}_{n \geq 0}$ of elements of E such that, for $n > 0$,

$$v\left(\prod_{k=0}^{n-1}(a_n - a_k)\right) = \inf_{a \in E} v\left(\prod_{k=0}^{n-1}(a - a_k)\right).$$

There always exist infinite v-ordered sequences of elements of E. Such sequences may be constructed inductively on n: choose any element a_0 in E, choose a_1 in E such that $v(a_1 - a_0) = \inf_{a \in E} v(a - a_0)$, and so on. We may notice that a V.W.D.W.O. sequence in V [4, Definition II.2.1] is a v-ordered sequence of E in the case where $E = V$ and V/M is finite.

From our point of view, the main result of Bhargava is the following [3, Theorem 1].

PROPOSITION 2.1 (Bhargava) *The sequence $\{w_E(n)\}_{\{n \in \mathbf{N}\}}$ defined by $w_E(n) = \sum_{k=0}^{n-1} v(a_n - a_k)$ where $\{a_n\}_{\{n \in \mathbf{N}\}}$ is a v-ordered sequence of E does not depend on the choice of the sequence $\{a_n\}$.*

This is an easy consequence of the fact that the sequence $w_E(n)$ is the characteristic sequence of $\text{Int}(E, V)$, since the characteristic sequence of $\text{Int}(E, V)$ only depends on $\text{Int}(E, V)$. For the sake of completness we give a straightforward proof of these assertions (shorter than Barghava's proof).

PROPOSITION 2.2 *Assume* $\{a_n\}_{\{n \in \mathbb{N}\}}$ *is a v-ordered sequence. Then the polynomials*

$$f_n(X) = \prod_{k=0}^{n-1} \frac{X - a_k}{a_n - a_k}$$

form a basis of the V-module $\text{Int}(E, V)$.

Proof. By construction, for each $a \in E$, $v(f_n(a)) \geq v(f_n(a_n))$, and hence, $f_n(E) \subseteq V$. Moreover, $f_n(a_0) = f_n(a_1) = \ldots = f_n(a_{n-1}) = 0$ and $f_n(a_n) = 1$. It then follows that the f_n, $n \in \mathbb{N}$, form a basis of $\text{Int}(E, V)$ (as in the classical case of $\text{Int}(\mathbb{Z})$ [4, Proposition I.1.1]).$\square$

COROLLARY 2.3 *Assume* $\{a_n\}_{\{n \in \mathbb{N}\}}$ *is a v-ordered sequence of elements of E. Then,*

1. $v(I_n) = -\sum_{k=0}^{n-1} v(a_n - a_k)$,

2. $w_E(n) = -v(I_n)$,

3. *if* $\{a_n\}_{\{n \in \mathbb{N}\}}$ *is a v-ordered sequence of elements of E, then the polynomials* $t^{-w_E(n)} \prod_{k=0}^{n-1} (X - a_k)$ *form a basis of* $\text{Int}(E, V)$.

Now, let us return to the particular case where E is supposed to be homogeneous.

3 Bárbácioru's corrected formula

We begin with some easy remarks concerning the function w_E. For a fixed subset E of V, $w_E(n)$ is an increasing function of n. More precisely, for all m and $n \in \mathbb{N}$,

$$w_E(m + n) \geq w_E(m) + w_E(n).$$

Moreover, if $E \subseteq F \subseteq V$, then, for each $n \in \mathbb{N}$,

$$0 \leq w_V(n) \leq w_F(n) \leq w_E(n).$$

We now consider translations and homotheties: for each $a \in V$, let $a + E = \{a + x \mid x \in E\}$ and, for each $l \in \mathbb{N}$, let $t^l E = \{t^l x \mid x \in E\}$.

PROPOSITION 3.1 *Let E be a subset of V.*

1. *For each $a \in V$, $w_{a+E}(n) = w_E(n)$.*

2. *For each $l \in \mathbb{N}$, $w_{t^l E}(n) = w_E(n) + ln$.*

Proof. Let τ (resp., σ) be the K-automorphism of $K(X)$ such that $\tau(X) = X - a$ (resp., $\sigma(X) = X/t^l$). Then $\tau(\text{Int}(E,V)) = \text{Int}(a + E, V)$ (resp., $\sigma(\text{Int}(E,V)) = \text{Int}(t^l E, V))$). Indeed, if $f(X)$ belongs to $\text{Int}(E,V)$, then $\tau(f(X)) = f(X - a)$ belongs to $\text{Int}(a + E, V)$ (resp., $\sigma(f(X)) = f(X/t^l)$ belongs to $\text{Int}(t^l E, V)$). If $\{f_n\}_{n \in \mathbb{N}}$ is a basis of $\text{Int}(E,V)$, then $\{\tau(f_n)\}$ (resp., $\{\sigma(f_n)\}$) is a basis of $\text{Int}(a + E, V)$ (resp., $\text{Int}(t^l E, V)$). If the leading coefficient of f_n is α, then the leading coefficient of $\tau(f_n)$ (resp., $\sigma(f_n)$) is α (resp., α/t^{ln}). $\square$

The case where $E = V$ is well known. We first recall a notation.

NOTATION [4, §II.2]. For each $q \in \mathbb{N}^*$ and each $n \in \mathbb{N}$, let

$$w_q(n) = \sum_{\alpha > 0} \left[\frac{n}{q^\alpha} \right]$$

where $[x]$ denotes the entire part of x. We extend this notation to the case where q is infinite with $w_\infty(n) = 0$ for each $n \in \mathbb{N}$.

PROPOSITION 3.2 [4, Corollaries I.3.7 and II.2.9]. *Let q be the cardinal of the residue field V/M (q is finite or infinite). Then, for each $n \in \mathbb{N}$, one has $w_V(n) = w_q(n)$.*

This leads us to a first formula.

PROPOSITION 3.3 *If $E = a + M^l$ with $a \in V$ and $l \in \mathbb{N}$, then*

$$w_E(n) = w_q(n) + ln.$$

Indeed, $E = a + t^l V$.

Here are two basic technical lemmas. The first one is quite general.

LEMMA 3.4 *Let* $\{a_k\}$ *be a v-ordered sequence of* E *and* $E_1 = (b + M^l) \cap E$ *be the intersection of* E *with some coset modulo* M^l. *Then the subsequence formed by the elements* a_k *in* E_1 *is a v-ordered sequence of* E_1.

Proof. Note that, even if the sequence $\{a_k\}$ is infinite, the subsequence formed by the elements a_k in E_1 may be finite. If this subsequence is empty, or contains only one element, there is nothing to prove. Suppose, by induction, that the first n elements $a_{k_0}, a_{k_1}, \ldots, a_{k_{n-1}}$ of this subsequence form a v-ordered sequence of E_1. If there is a next one, a_{k_n}, we prove that $a_{k_0}, a_{k_1}, \ldots, a_{k_n}$ is a v-ordered sequence of E_1. We set $N = k_n$. For each α in E_1, we have

$$\sum_{k=0}^{N-1} v(\alpha - a_k) = \sum_{k<N,\, a_k \in E_1} v(\alpha - a_k) + \sum_{k<N,\, a_k \notin E_1} v(\alpha - a_k).$$

If $a_k \notin E_1$, we have $v(b - a_k) < l$, while $v(\alpha - b) \geq l$, thus

$$\sum_{k<N,\, a_k \notin E_1} v(\alpha - a_k) = \sum_{k<N,\, a_k \notin E_1} v(b - a_k)$$

and this sum is independent of the choice of α in E_1.
By hypothesis, $\sum_{k=0}^{N-1} v(\alpha - a_k)$ is minimal for $\alpha = a_{k_n}$. Hence

$$\sum_{k<N,\, a_k \in E_1} v(\alpha - a_k) = \sum_{i=0}^{n-1} v(\alpha - a_{k_i})$$

is also minimal for this choice of α. $\square$

HYPOTHESIS. From now on, we assume that E is of the following form:

$$E = \cup_{i=1}^{r} b_i + M^l$$

where $r \in \mathbf{N}^*$, $l \in \mathbf{N}$ and $b_1, \ldots, b_r$ are pairwise non-congruent modulo M^l.

NOTATION. For $j \in \{1, \ldots, r\}$ and $\delta_1, \ldots, \delta_r \in \mathbf{N}$, let

$$w_E^j(\delta_1, \ldots, \delta_r) = w_q(\delta_j) + l\delta_j + \sum_{i \neq j} v(b_j - b_i)\delta_i.$$

LEMMA 3.5 *For each* $j \in \{1, \ldots, r\}$, *let* $\delta_j \in \mathbf{N}$ *and let* $a_{j,1}, \ldots, a_{j,\delta_j}$ *be* δ_j *elements of* $b_j + M^l$ *which form a v-ordered sequence of elements of* $b_j + M^l$. *Consider the polynomial*

$$g(X) = \prod_{j=1}^{r} \left(\prod_{k=1}^{\delta_j} (X - a_{j,k}) \right).$$

Then, for each j, one has:

$$\inf\left\{v(g(x)) \mid x \in b_j + M^l\right\} = w_E^j(\delta_1, \ldots, \delta_r).$$

Proof. Let $x \in b_j + M^l$. For each $i \neq j$ and each $k \in \{1, \ldots, \delta_i\}$, we have $v(x - a_{i,k}) = v(b_j - b_i)$, and hence

$$v(g(x)) = \sum_{k=1}^{\delta_j} v(x - a_{j,k}) + \sum_{i \neq j} v(b_j - b_i)\delta_i.$$

Clearly $\sum_{i \neq j} v(b_i - b_j)\delta_i$ does not depend on the choice of x in $b_j + M^l$. Since $\{a_{j,1}, \ldots, a_{j,\delta_j}\}$ is a v-ordered sequence of elements of $b_j + M^l$, it follows from Proposition 3.3 that the minimal value of $\sum_{k=1}^{\delta_j} v(x - a_{j,k})$ is $w_q(\delta_j) + l\delta_j$.□

THEOREM 3.6 *Let V be a discrete valuation domain. Denote by v the corresponding valuation, by M the maximal ideal of V, and by q the cardinal (finite or infinite) of the residue field V/M. Let E be a subset of V such that $E = \cup_{i=1}^r b_i + M^l$ where $r \in \mathbb{N}^*$, $l \in \mathbb{N}$ and $b_1, \ldots, b_r \in V$ are pairwise non-congruent modulo M^l. The characteristic sequence $\{w_E(n)\}_{\{n \in \mathbb{N}\}}$ of $\text{Int}(E, V)$ may be computed by means of the following formulas:*

$$w_E(n) = \max_{\delta_1 + \cdots + \delta_r = n} \left(\min_{1 \leq j \leq r} w_E^j(\delta_1, \ldots, \delta_r)\right) \qquad (\delta_1, \ldots, \delta_r \in \mathbb{N})$$

where

$$w_E^j(\delta_1, \ldots, \delta_r) = w_q(\delta_j) + l\delta_j + \sum_{i \neq j} v(b_i - b_j)\delta_i$$

and

$$w_q(\delta_j) = \sum_{\alpha > 0} \left[\frac{\delta_j}{q^\alpha}\right].$$

Proof. Let n be a fixed integer and let

$$\omega(n) = \max_{\delta_1 + \cdots + \delta_r = n} \left(\min_{1 \leq j \leq r} w_E^j(\delta_1, \cdots, \delta_r)\right).$$

Let $\{a_0, a_1, \ldots, a_n\}$ be a v-ordered sequence of elements of E. For each $j \in \{1, \ldots, r\}$, let δ_j be the number of elements a_k ($k < n$) which are in $b_j + M^l$. Let $g(X) = \prod_{k=0}^{n-1}(X - a_k)$. Lemma 3.4 says that, for each j, the finite subsequence formed by the a_k which lie in $b_j + M^l$ is a v-ordered sequence of elements of $b_j + M^l$. Thus, the hypothesis of Lemma 3.5 is satisfied. It follows that we have

$$\inf_{x \in E} \{v(g(x))\} = \min_{1 \le j \le r} w_E^j(\delta_1, \ldots, \delta_r).$$

By definition of a v-ordered sequence, we also have

$$v(g(a_n)) = \inf_{x \in E} \{v(g(x))\}.$$

Consequently,

$$w_E(n) = \sum_{k=0}^{n-1} v(a_n - a_k) = \min_{1 \le j \le r} w_E^j(\delta_1, \ldots, \delta_r).$$

Since $\delta_1 + \cdots + \delta_r = n$, we have in particular

$$\omega(n) \ge \min_{1 \le j \le r} w_E^j(\delta_1, \ldots, \delta_r)$$

and hence

$$\omega(n) \ge w_E(n).$$

Conversely, let now $d_1, \ldots, d_r \in \mathbf{N}$ be such that

$$d_1 + \ldots + d_r = n$$

and such that

$$\inf_{1 \le j \le r} w_E^j(d_1, \ldots, d_r) = \omega(n).$$

For each $j \in \{1, \ldots, r\}$, let $\{a_{j,1}, \ldots, a_{j,d_j}\}$ be a v-ordered sequence of elements of $b_j + M^l$. Then, let

$$g(X) - \prod_{j=1}^{r} \left(\prod_{k=1}^{d_j} (X - a_{j,k}) \right).$$

It follows from Lemma 3.5 that, for each $x \in E$, we have

$$v(g(x)) \ge \min_{1 \le j \le r} w_E^j(d_1, \ldots, d_r).$$

Thus, $\omega(n))^{-1} g(X)$ belongs to $\mathrm{Int}(E, V)$; and hence, by definition of I_n,

$$\omega(n) \le w_E(n) = -v(I_n).$$

Finally

$$\omega(n) = w_E(n).$$

☐

REMARKS

a) Theorem 3.6 shows that in order to compute $w_E(n)$, we do not really have to know any v-ordered sequence in E. In fact, we may forget the original question on integer-valued polynomials, we just have to know the integers q, l and $v(b_i - b_j)$. We may also notice that, for each n, $w_E(n)$ may be computed in finitely many steps since there are only finitely many $(\delta_1, \ldots, \delta_r) \in \mathbf{N}^r$ such that $\delta_1 + \cdots + \delta_r = n$.

b) On the other hand, for each n, the computation of $w_E(n)$ may help us to determine a v-ordered sequence of n elements. If $(\delta_1, \ldots, \delta_r) \in \mathbf{N}^r$ is such that $\delta_1 + \cdots + \delta_r = n$ and $\inf_{1 \leq j \leq r} w_E^j(\delta_1, \ldots, \delta_r) = w_E(n)$, then it suffices to consider, for each $j \in \{1, \ldots, r\}$, a v-ordered sequence of δ_j elements in $b_j + M^l$. Such sequences are easy to construct : if $\{a_k\}$ is a v-ordered sequence in V, then $\{b_j + a_k t^l\}$ is a v-ordered sequence in $b_j + M^l$. Moreover, v-ordered sequences in V are well known : if q is finite, see for instance [4, Proposition II.2.3], and if q is infinite, any sequence of elements of V which are pairwise non-congruent modulo M is a v-ordered sequence in V.

c) In the case where q is infinite, the problem becomes a classical linear programming problem. Let us consider the symmetric matrix

$$B = (\beta_{i,j}) \in \mathcal{M}_r(\mathbf{N})$$

defined by

$$\begin{cases} \beta_{ii} & = \quad l & \text{for each } i \\ \beta_{ij} & = \quad \beta_{ji} = v(b_j - b_i) & \text{for } i \neq j. \end{cases}$$

We have to determine the function w_E such that

$$w_E(n) = \max_{\delta_1 + \cdots + \delta_r = n} \left(\min_{1 \leq j \leq r} w_E^j(\delta_1, \ldots, \delta_r) \right)$$

with

$$W_E(\Delta) = \Delta B$$

where

$$\Delta = (\delta_1, \ldots, \delta_r) \in \mathbf{N}^r \quad \text{and} \quad W_E(\Delta) = \left(w_E^1(\Delta), \ldots, w_E^r(\Delta) \right) \in \mathbf{N}^r.$$

REFEREE'S REMARK The previous results may be slightly improved in the case where the residue field is infinite: Lemma 3.5 and Theorem 3.6, in particular, remain valid if E is of the form

$$E = \bigcup_{i=1}^{r} b_i + M^{l_i},$$

where the l_i are positive integers and the b_i are elements of V which are pairwise non-congruent modulo M^l with $l = \inf_{1 \leq i \leq r} l_i$. We just have to consider the new following functions

$$w_E^j(\delta_1, \ldots, \delta_r) = w_q(\delta_j) + l_j \delta_j + \sum_{i \neq j} v(b_i - b_j) \delta_i.$$

[Note that if the residue field is finite, we may always assume that all the l_i are equal.]

4 Some explicit formulas

There are some cases where the *maximin* which gives the value for $w_E(n)$ (see Theorem 3.6) may be described by an explicit formula. The first one is the case where $l = 1$, that is, the only case where Bárbácioru's formula is correct (see Section 1).

PROPOSITION 4.1 *If* $E = \cup_{j=1}^r b_j + M$ *where* $b_1, \ldots, b_r$ *are pairwise non-congruent modulo* M, *then*

$$w_E(n) = w_q \left(\left[\frac{n}{r} \right] \right) + \left[\frac{n}{r} \right] = \sum_{\alpha \geq 0} \left[\frac{n}{r q^\alpha} \right].$$

This is a particular case of Proposition 4.2 below where, for each $i \neq j$, $\beta_{ij} = v(b_j - b_i) = 0$. We already encountered an example of such a case in the literature: $\mathrm{Int}(E, V)$ where $V = \mathbf{Z}_{(p)}$, $E = \mathbf{Z} \setminus p\mathbf{Z}$ and p is a prime number. Then $\mathrm{Int}(E, V) = \mathrm{Int}(\overline{E}, V)$ where $\overline{E} = \mathbf{Z}_{(p)} \setminus p\mathbf{Z}_{(p)}$ [4, Theorem IV.1.15], and $\overline{E} = \left\{ 1 + \mathbf{Z}_{(p)} \right\} + \cdots + \left\{ (p-1) + \mathbf{Z}_{(p)} \right\}$ corresponds to $l = 1$, $q = p$, and $r = p - 1$. Hence,

$$w_E(n) = \sum_{\alpha \geq 0} \left[\frac{n}{(p-1)p^\alpha} \right] \qquad \text{[8, Lemma 4].}$$

There are at least two reasons which explain the difficulty in replacing the *maximin* by explicit formulas:

— the gaps of the function w_q are difficult to control unless the residue field is infinite (in this case, $w_q(n) \equiv 0$, see the last remark of the previous section),

— the weights of the cosets modulo M^l may be different unless all the $\beta_{i,j} = v(b_i - b_j)$ are equal.

A first case where the later difficulty is avoided is those where all the $\beta_{i,j}$ are equal to zero.

PROPOSITION 4.2 *If $E = \cup_{i=1}^r b_j + M^l$ where $l \in \mathbb{N}$ and $b_1, \ldots, b_r$ are pairwise non-congruent modulo M, then*

$$w_E(n) = w_q\left(\left[\frac{n}{r}\right]\right) + l\left[\frac{n}{r}\right].$$

Proof. Since $v(b_i - b_j) = 0$ for $i \neq j$, it follows from the definition of $w_E^j(\delta_1, \ldots, \delta_r)$ that we have

$$w_E^j(\delta_1, \ldots, \delta_r) = w_q(\delta_j) + l\delta_j.$$

Then

$$w_E^j(\delta_1, \ldots, \delta_r) = \varphi(\delta_j) \text{ where } \varphi(\delta) = w_q(\delta) + l\delta$$

is an increasing function of δ. Thus

$$\min_j w_E^j(\delta_1, \ldots, \delta_r) = \varphi(\delta) \text{ where } \delta = \inf_j \delta_j,$$

and

$$\max_{\delta_1 + \cdots + \delta_r = n}\left(\min_j\left(w_E^j(\delta_1, \ldots, \delta_r)\right)\right) = \max_{\delta_1 + \cdots + \delta_r = n} \varphi\left(\min_j \delta_j\right)$$

$$= \varphi\left(\max_{\delta_1 + \ldots + \delta_r = n}\left(\min_j \delta_j\right)\right).$$

Since

$$\max_{\delta_1 + \cdots + \delta_r = n}\left(\min_j \delta_j\right) = \left[\frac{n}{r}\right],$$

one has

$$w_E(n) = \varphi\left(\left[\frac{n}{r}\right]\right). \quad \square$$

Another case where things are relatively easy is those where $r = 2$ since in that case there is only one $\beta_{i,j}$ to consider ($\beta_{1,2} = \beta_{2,1}$).

PROPOSITION 4.3 *If* $E = \{b_1 + M^l\} \cup \{b_2 + M^l\}$ *with* $l \in \mathbf{N}^*$ *and* $v(b_1 - b_2) < l$, *then*

$$w_E(n) = w_q\left(\left[\frac{n}{2}\right]\right) + l\left[\frac{n}{2}\right] + v(b_1 - b_2)\left[\frac{n+1}{2}\right].$$

Proof. For $j = 1, 2$, one has

$$w_E^j(\delta_1, \delta_2) = w_q(\delta_j) + l\delta_j + h\delta_{3-j} \text{ where } h = \beta_{12} = \beta_{21}.$$

Thus, if $\delta_1 + \delta_2 = n$, then

$$w_E^j(\delta_1, \delta_2) = \psi(\delta_j)$$

where

$$\psi(\delta) = w_q(\delta) + l\delta + h(n - \delta) = w_q(\delta) + (l - h)\delta + hn$$

is an increasing function of δ. As in the previous proof:

$$\max_{\delta_1 + \delta_2 = n}\left(\min_j w_E^j(\delta_1, \delta_2)\right) = \psi\left(\max_{\delta_1 + \delta_2 = n}\left(\min_j \delta_j\right)\right) = \psi\left(\left[\frac{n}{2}\right]\right)$$

$$= w_q\left(\left[\frac{n}{2}\right]\right) + (l - h)\left[\frac{n}{2}\right] + hn$$

$$= w_q\left(\left[\frac{n}{2}\right]\right) + l\left[\frac{n}{2}\right] + h\left[\frac{n+1}{2}\right]. \square$$

In fact, both previous propositions are particular cases of the following where the $\beta_{i,j}$ are equal to each other.

PROPOSITION 4.4 *If* $E = \cup_{j=1}^r b_j + M^l$ *where* $\beta_{i,j} = v(b_j - b_i) = h$ *for each* $i \neq j$, $l \in \mathbf{N}$ *and* $0 \leq h < l$, *then*

$$w_E(n) = w_q\left(\left[\frac{n}{r}\right]\right) + (l - h)\left[\frac{n}{r}\right] + hn.$$

Proof. By hypothesis, for each $j \in \{1, \ldots, r\}$, one has $b_j - b_1 = t^h c_j$ where t is a generator of M and the elements $c_1, \ldots, c_r$ of V are pairwise non-congruent modulo M. Let $E_1 = E - b_1$, then $E_1 = t^h E_2$ where $E_2 = \cup_{j=1}^r c_j + M^{l-h}$. Then, Proposition 3.1 shows that

$$w_E(n) = w_{E_1}(n) = w_{E_2}(n) + hn,$$

and Proposition 4.2 shows that

$$w_{E_2}(n) = w_q\left(\left[\frac{n}{r}\right]\right) + (l - h)\left[\frac{n}{r}\right]. \square$$

References

[1] C. BÁRBÁCIORU, *Integer-Valued Polynomials on a Subset*, J. Number Theory **65** (1997), 40–47.

[2] P.-J. CAHEN, *Polynômes à valeurs entières*, Canad. J. Math. **24** (1972), 747–754.

[3] M. BHARGAVA, *P-orderings and polynomial functions on arbitrary substes of Dedekind rings*, J. reine angew. Math. **490** (1997), 101–127.

[4] P.-J. CAHEN AND J.-L. CHABERT, *Integer-Valued Polynomials*, Amer. Math. Soc. Surveys and Monographs, **48**, Providence, 1997.

[5] D.L. MCQUILLAN, *On ideals in Prüfer domains of polynomials*, Arch. Math. **45** (1985), 517–527.

[6] PÓLYA, *Ueber ganzwertige Polynome in algebraischen Zahlkörpern*, J. reine angew. Math. **149** (1919), 97–116.

[7] J.-L. CHABERT, S. CHAPMAN AND W. SMITH, *Algebraic Properties of the Ring of Integer-Valued Polynomials on Prime Numbers*, Comm. Algebra **25** (1997), 1945–1959.

[8] J.-L. CHABERT, *Une caractérisation des polynômes prenant des valeurs entières sur tous les nombres premiers*, Canad. Math. Bull. **39**, 402–407.

Skolem Properties and Integer-Valued Polynomials: A Survey

PAUL-JEAN CAHEN, Case Cr. A, Faculté des Sciences de Saint Jérôme, 13397 Marseille cedex 20, France, email: paul-jean.cahen@math.u-3mrs.fr

JEAN-LUC CHABERT, Faculté de Mathématiques et d'Informatique, 80039 Amiens Cedex 01, France, email: jlchaber@worldnet.fr

ABSTRACT. Let $\mathrm{Int}(D) = \{f \in K[X] \mid f(D) \subseteq D\}$ be the ring of integer-valued polynomials on a domain D with quotient field K. One says that $\mathrm{Int}(D)$ has the strong Skolem property if the finitely generated ideals of $\mathrm{Int}(D)$ are characterized by their values, that is, $\mathfrak{A} = \mathfrak{B}$ if and only if, for each $a \in D$, $\mathfrak{A}(a) = \{g(a) \mid g \in \mathfrak{A}\}$ is equal to $\mathfrak{B}(a) = \{g(a) \mid g \in \mathfrak{B}\}$. For example, it is well known that, if D is the ring of integers of a number field, then $\mathrm{Int}(D)$ has the strong Skolem property.

After a survey of the main known results, we show how these results may be extended to the ring $\mathrm{Int}(E, D) = \{f \in K[X] \mid f(E) \subseteq D\}$ of integer-valued polynomials on a subset E of D. In particular, if D is Noetherian, local, one-dimensional, and analytically irreducible, we show that the finitely generated ideals of $\mathrm{Int}(E, D)$ containing nonzero constants are characterized by their values if and only if the topological closure $\widehat{E}$ of E (in the topology defined by the maximal ideal) is compact.

INTRODUCTION

Let E be a set, D be a ring, $\mathcal{F}(E, D)$ be the ring of functions from E to D, and R be a subring of $\mathcal{F}(E, D)$. For instance, R may be the ring $\mathcal{C}(E, D)$ of continuous functions, if E and D are topological spaces. If D is a domain, we shall mainly be concerned by the ring

$$\mathrm{Int}(E, D) = \{f \in K[X] \mid f(E) \subseteq D\}$$

of integer-valued polynomials on some subset E of the quotient field K of D. More particularly, for $E = D$, we denote by $\mathrm{Int}(D)$ the ring of integer-valued polynomials on D.

Given an ideal $\mathfrak{A}$ of the subring R of $\mathcal{F}(E, D)$, we may, for each $a \in E$, consider the *set of values* of $\mathfrak{A}$, that is,

$$\mathfrak{A}(a) = \{g(a) \mid g \in \mathfrak{A}\}.$$

Note that, if R contains D (as it is the case, for the ring $\text{Int}(E, D)$ of integer-valued polynomials on E), then $\mathfrak{A}(a)$ is an ideal of D. The problem we address here is: to what extent are the ideals of R characterized by their values? For a given ideal $\mathfrak{A}$, our problem can be rephrased as follows: for $f \in R$, does the condition $f(a) \in \mathfrak{A}(a)$ for each $a \in E$ implies $f \in \mathfrak{A}$? We thus set the following definition.

Definition 1 (Skolem closure). Let E be a set, D be a ring, R be a subring of $\mathcal{F}(E, D)$, and $\mathfrak{A}$ be an an ideal of R. Then

$$\mathfrak{A}^* = \{f \in R \mid f(a) \in \mathfrak{A}(a) \text{ for each } a \in E\}$$

is called *the Skolem closure* of $\mathfrak{A}$. If $\mathfrak{A} = \mathfrak{A}^*$, then $\mathfrak{A}$ is said to be *Skolem closed*.

Note that $\mathfrak{A}^*$ is clearly an ideal of R. Note also that the Skolem closure does indeed behave like a closure [4, Proposition VII.1.6]: $\mathfrak{A}^*$ is the smallest Skolem closed ideal containing $\mathfrak{A}$; an intersection of Skolem closed ideals is Skolem closed; if $\mathfrak{A} \subseteq \mathfrak{B}$, then $\mathfrak{A}^* \subseteq \mathfrak{B}^*$.

In particular, if $\mathfrak{A}^* = R$, does it imply that $\mathfrak{A} = R$? (in other words, in the case where R contains D, does $\mathfrak{A}(a) = D$, for each $a \in D$, imply that $\mathfrak{A} = R$?). In the thirties, Thoralf Skolem [23] pointed out that this particular question has a negative answer if R is the ring of polynomials $\mathbb{Z}[X]$ (considered as a ring of functions from $\mathbb{Z}$ to itself): the ideal $\mathfrak{A} = (3, X^2 + 1)$ is such that $\mathfrak{A}(n) = \mathbb{Z}$ for all n, while $\mathfrak{A} \neq \mathbb{Z}[X]$. However, he proved the answer to be positive for the *finitely generated* ideals of the ring $R = \text{Int}(\mathbb{Z}) = \{f \in \mathbb{Q}[X] \mid f(\mathbb{Z}) \subseteq \mathbb{Z}\}$ (which is not Noetherian). We then say that $\text{Int}(\mathbb{Z})$ has the Skolem property. In fact, it was shown by D. Brizolis [1] in the seventies that $\text{Int}(\mathbb{Z})$ has a stronger property: the finitely generated ideals are characterized by their value ideals.

As in the case of $\text{Int}(\mathbb{Z})$, we shall often restrict ourselves to finitely generated ideals. For other reasons (developed below), we shall sometimes also restrict ourselves to *unitary* ideals, that is, ideals containing nonzero constant functions. We then set the following definitions (similar to [4, Definitions VII.1.1 & VII.2.3]):

Definitions 2 (Skolem properties). Let E be a set, D be a ring and R be a subring of $\mathcal{F}(E, D)$.
(i) R is said to satisfy the *Skolem property* (resp., the *almost Skolem property*) if, for each finitely generated ideal (resp., each finitely generated unitary ideal) $\mathfrak{A}$ of R, $\mathfrak{A}^* = R$ implies $\mathfrak{A} = R$.
(ii) R is said to satisfy the *strong Skolem property* (resp., the *almost strong Skolem property*) if each finitely generated ideal (resp., each finitely generated unitary ideal) of R is Skolem closed.
(iii) R is said to satisfy the *super Skolem property* (resp., the *almost super Skolem property*) if each ideal (resp., each unitary ideal) of R is Skolem closed.

The Skolem and strong Skolem properties are classical, the super Skolem property is a new notion. In fact, most papers consider only finitely generated ideals, since in the very classical case of the ring $R = \text{Int}(\mathbb{Z})$ (which satisfies the strong Skolem property), some non-finitely generated ideals are not Skolem closed (as, for instance, the maximal ideal $\mathfrak{M}_x = \{f \in \text{Int}(\mathbb{Z}) \mid f(x) \in p\widehat{\mathbb{Z}_p}\}$, where x belongs to $\widehat{\mathbb{Z}_p}$, the p-adic completion of $\mathbb{Z}$, but not to $\mathbb{Z}$).

Usually the context makes clear in which ring of functions the ring R is contained; let us note, however, that the Skolem properties are relative to it. Indeed, we shall in particular consider the case where R is contained in a ring $\mathcal{C}(E, D)$ of continuous functions (in some topology) and extend these functions to continuous functions from the completion $\widehat{E}$ of E to the completion $\widehat{D}$ of D. Then R may satisfy the super Skolem property as a subring of $\mathcal{F}(\widehat{E}, \widehat{D})$, but only the strong Skolem property as a subring of $\mathcal{F}(E, D)$; indeed, one considers the values at each $x \in \widehat{E}$, in the first case, and only at each $a \in E$, in the second one.

In the first section of this paper, we survey the main known results for the ring $R = \mathrm{Int}(D)$ of integer-valued polynomials on a domain D. This case has been extensively studied by several authors (Brizolis [1], [2], [3], Chabert [8], [9], [10], [11], McQuillan [16], [17], [18]). In particular, we recall the notion of d-ring (equivalent to the Skolem property restricted to non-unitary ideals), and we end this short survey with the study of the almost strong Skolem property in the case of a one-dimensional local Noetherian domain D with finite residue field. We recall that a sufficient condition is that D is *analytically irreducible* (that is, its completion is a domain). We announce also a new result: it is necessary that D is *unibranched*, that is, its integral closure is a local ring.

We then turn to the consideration of the ring $\mathrm{Int}(E, D)$, where E is a subset of the quotient field K of D. In a second section, we study various properties of E, beginning with a generalization of d-rings. In fact, contrary to the case of $\mathrm{Int}(D)$, where various properties are equivalent, we must introduce here several definitions: a d-set is such that each almost integer-valued rational function on E is a polynomial, an s-set is such that each unit-valued polynomial on E is a constant. We show that these notions are distinct. Contrary also to the case of integer-valued polynomials on the ring D, the Skolem closure of a finitely generated ideal $\mathfrak{A}$ of $\mathrm{Int}(E, D)$ is not determined by the ideal of values $\mathfrak{A}(a)$ for a in a cofinite subset of E. Thus we introduce various notions of coherence: we say E is *coherent*, if each almost integer-valued on E is in fact integer-valued, and that E is *strongly coherent*, if for each finitely generated ideal $\mathfrak{A}$ of $\mathrm{Int}(E, D)$, and each polynomial $f \in K[X]$, the condition $f(a) \in \mathfrak{A}(a)$ for almost each $a \in E$, implies $f \in \mathfrak{A}^*$. Again we show that these notions are distinct. Finally, we describe a large class of strongly coherent subsets.

In a third section, we link the Skolem properties to the notion of d-sets and coherence. We first show that, if $\mathrm{Int}(E, D)$ satisfies the almost Skolem property and if E is d-set, then $\mathrm{Int}(E, D)$ satisfies the Skolem property. Turning then to the almost Skolem property, we give an easy generalization of the results on $\mathrm{Int}(D)$. But mainly, we show in this section that $\mathrm{Int}(E, D)$ satisfies the strong Skolem property if and only if it satisfies the almost strong Skolem property and E is a strongly coherent d-set. In the case where $\mathrm{Int}(E, D)$ is a Prüfer domain (for instance if D is a Dedekind domain with finite residue fields), we may even conclude that $\mathrm{Int}(E, D)$ satisfies the strong Skolem property if and only if E is a coherent d-set.

In the last section, we finally turn to the almost strong Skolem property in the local case. As for $\mathrm{Int}(D)$, we consider the case where D is a one-dimensional local Noetherian domain, with maximal ideal $\mathfrak{m}$. We suppose that D is analytically irreducible, but we do not assume that the residue field is finite. Indeed, we show that $\mathrm{Int}(E, D)$ satisfies the almost strong Skolem property if and only if the topological closure $\widehat{E}$ of E (in the $\mathfrak{m}$-adic topology) is compact. We give a couple of examples.

1. THE RING $R = \mathrm{Int}(D)$

In this section we recall, without proofs, the known results concerning the ring $R = \mathrm{Int}(D)$ of integer-valued polynomials on a domain D with quotient field K. Most of these results are collected in Chapter VII of [4], to which we send the reader for references.

d-rings. Let us look in particular at the principal ideal (f) generated by a non-constant polynomial $f \in \mathrm{Int}(D)$.

— If $\mathrm{Int}(D)$ satisfies the Skolem property, the Skolem closure of (f) is distinct from $\mathrm{Int}(D)$, and hence, there exists a in D such that $f(a)$ is not a unit in D. We could say that D is an *s-ring* (*s* for solution): for each non-constant polynomial $f \in \mathrm{Int}(D)$, there exists a maximal ideal $\mathfrak{m}$ of D such that f has a root modulo $\mathfrak{m}$ [4, Proposition VII.2.3].

— If $\mathrm{Int}(D)$ satisfies the strong Skolem property, the ideal (f) is Skolem closed: if a polynomial $g \in \mathrm{Int}(D)$ is such that $g(a) \in f(a)D$ for each $a \in D$ (that is, the rational function g/f takes integer values at each a but possibly the zeros of f), then f divides g in $\mathrm{Int}(D)$ (in particular, g/f is a polynomial). We could say that D is a *d-ring* (*d* for divisibility): each rational function which is integer-valued for almost all $a \in D$ is in fact an integer-valued polynomial.

Obviously, a d-ring is an s-ring, but as it turns out these properties are equivalent [4, Proposition VII.2.3]. For the s-ring property, we may consider only polynomials with coefficients in D [4, Exercise VII.7]. Recall also that an almost integer-valued rational function (that is, a function which takes integer-values for each $a \in D$, but finitely many), is in fact integer-valued [4, Lemma VII.1.8]. Thus, we list below several equivalent statements for the definition of a d-ring.

Proposition-Definition 1.1. *We say that a domain D is a d-ring if it satisfies the following equivalent conditions:*

  (i) *each integer-valued rational function on D is a polynomial,*
  (ii) *each almost integer-valued rational function on D is an integer-valued polynomial,*
  (iii) *for each non-constant polynomial $f \in D[X]$ (resp., each non-constant polynomial $f \in \mathrm{Int}(D)$), there exists $a \in D$ such that $f(a)$ is not a unit of D,*
  (iv) *for each non-constant polynomial $f \in D[X]$ (resp., each non-constant polynomial $f \in \mathrm{Int}(D)$), the intersection of the maximal ideals $\mathfrak{m}$ of D for which f has a root modulo $\mathfrak{m}$ is (0).*

There are many examples of d-rings (for instance, see [14]): in particular, every domain which is a finitely generated $\mathbb{Z}$-algebra is a d-ring. However it is easy to note that this property is not stable under localization; in fact, a semi-local domain is never a d-ring (consider the polynomial $1 + mX$, where m is in the Jacobson radical of D).

As seen at the beginning of this section, D is a d-ring if and only if each principal ideal (f) of $\mathrm{Int}(D)$ is Skolem closed. In fact, we may then restrict our attention to unitary ideals (that is, to the almost Skolem properties), indeed we have the following [4, Proposition VII.2.14]:

Proposition 1.2. *Let D be a domain. Then* $\mathrm{Int}(D)$ *satisfies the Skolem property* (*resp., the strong Skolem property*) *if and only if*

a) *D is a d-ring,*

b) *$\mathrm{Int}(D)$ satisfies the almost Skolem property (resp., the almost strong Skolem property).*

Divisorial ideals and Prüfer domains. Recall that a *divisorial* ideal is an intersection of principal fractional ideals. For a *d*-ring D, each principal ideal, thus, each divisorial ideal of $\mathrm{Int}(D)$ is Skolem closed (since an intersection of Skolem closed ideals is Skolem closed). Restricting ourselves to unitary ideals, we have the following, without any hypothesis on D [4, Lemma VII.2.15]:

Lemma 1.3. *Let D be a domain. Each unitary divisorial ideal of* $\mathrm{Int}(D)$ *is Skolem closed.*

If $\mathrm{Int}(D)$ is a Prüfer domain, each finitely generated ideal is divisorial. Moreover, if D is a Noetherian domain, we know that $\mathrm{Int}(D)$ is a Prüfer domain if and only if D is a Dedekind domain with finite residue fields [4, Theorem VI.1.7]. Hence, we may summarize ourselves with the following.

Proposition 1.4. *Let D be a domain such that* $\mathrm{Int}(D)$ *is a Prüfer domain (for instance, a Dedekind domain with finite residue fields), then* $\mathrm{Int}(D)$ *satisfies the almost strong Skolem property. Moreover, the following assertions are equivalent:*

(i) *D is a d-ring,*

(ii) *$\mathrm{Int}(D)$ satisfies the Skolem property,*

(iii) *$\mathrm{Int}(D)$ satisfies the strong Skolem property.*

In particular, if D is the ring of integers of a number field, then $\mathrm{Int}(D)$ satisfies the strong Skolem property.

Using divisorial ideals, we arrived easily at a conclusion for the class of Dedekind domains. In fact, we can characterize the Noetherian domains D such that $\mathrm{Int}(D)$ satisfies the Skolem property, and give necessary or sufficient conditions for the strong Skolem property. We will see that D need not be integrally closed.

Almost Skolem property. We have the following characterization of the almost Skolem property in the Noetherian case [4, Proposition VII.4.5].

Proposition 1.5. *Assume D is Noetherian (but not a field). Then* $\mathrm{Int}(D)$ *satisfies the almost Skolem property if and only if*

a) *for each maximal ideal $\mathfrak{m}$ of D, either $D/\mathfrak{m}$ is algebraically closed, or $\mathfrak{m}$ is an height-one prime and $D/\mathfrak{m}$ is finite,*

b) *each nonzero prime ideal of D is an intersection of maximal ideals.*

Let us emphasize in particular that every one-dimensional Noetherian domain, with finite residue fields, satisfies the almost Skolem property. We would derive immediately a characterization of the Skolem property adding the condition that D is a *d*-ring.

Remarks 1.6. (1) If D is a *d*-ring then, in particular, the ideal (0) is an intersection of maximal ideals [Proposition-Definition 1.1]. Thus if $\mathrm{Int}(D)$ satisfies the Skolem property we derive that D is an Hilbert ring (that is, each prime ideal is an intersection of maximal ideals).

(2) Brizolis [1] pointed out a property which seems a priori stronger than the Skolem property, that he called the *Nullstellensatz property*: the Skolem closure of each finitely generated ideal $\mathfrak{A}$ of Int(D) is contained in the radical $\sqrt{\mathfrak{A}}$ of $\mathfrak{A}$. In fact, the Nullstellensatz and Skolem properties are equivalent in the Noetherian case [4, Proposition VII.4.5].

Almost strong Skolem property. For the almost strong Skolem property, we first have a necessary condition [4, Proposition VII.3.3].

Lemma 1.7. *Assume D is Noetherian (but not a field). If* Int(D) *satisfies the almost strong Skolem property, then D is one-dimensional with finite residue fields.*

Moreover, contrary to the Skolem properties, the almost strong Skolem property is local in the Noetherian case [4, Exercise VII.18]:

Lemma 1.8. *Assume D is Noetherian. Then* Int(D) *satisfies the almost strong Skolem property if and only if, for each maximal ideal $\mathfrak{m}$ of D,* Int($D_\mathfrak{m}$) *satisfies the almost strong Skolem property.*

Thus, we restrict ourselves to a one-dimensional Noetherian local domain D, with maximal ideal $\mathfrak{m}$, and we assume its residue field to be finite. Recall that D is said to be *analytically irreducible* if its completion $\widehat{D}$ in the $\mathfrak{m}$-adic topology is an integral domain, and that D is said to be *unibranched* if its integral closure D' is local (that is, a rank-one discrete valuation domain). It is known that an analytically irreducible domain is unibranched, while the converse does not hold in general. These conditions are linked to the almost strong Skolem property, and we may summarize what we know as follows:

Theorem 1.9. *Let D be a one-dimensional local Noetherian domain, with finite residue field.*

  (i) *If D is analytically irreducible, then* Int(D) *satisfies the almost strong Skolem property.*

  (ii) *If* Int(D) *satisfies the almost strong Skolem property, then D is unibranched.*

Let us give some comments on both assertions.

(i) We knew the first condition (D is analytically irreducible) to be sufficient [4, Theorem VII.3.8] (the result is given there in more generality). The proof is based on the following facts, using the compactness of $\widehat{D}$:

— A. The ring $\mathcal{C}(\widehat{D}, \widehat{D})$ satisfies the almost super Skolem property: the unitary ideals (not only the finitely generated unitary ideals) are characterized by their value ideals on $\widehat{D}$.

— B. Analogously to the classical Stone-Weierstrass theorem, Int(D) is dense in $\mathcal{C}(\widehat{D}, \widehat{D})$ for the uniform convergence topology [4, Theorem III.5.3]. It follows that, as a subring of $\mathcal{F}(\widehat{D}, \widehat{D})$, Int($D$) satisfies also the almost super Skolem property (in fact, for each unitary ideal $\mathfrak{A}$ of Int(D), if $f(x) \in \mathfrak{A}(x)\widehat{D}$ for each $x \in \widehat{D}$, then $f \in \mathfrak{A}$) [4, Theorem VII.3.7].

— C. For each finitely generated unitary ideal $\mathfrak{A}$ of Int(D), the ideals of values are locally constant on D: there exists a nonzero ideal $\mathfrak{b}$ of D such that $(a - b) \in \mathfrak{b}$ implies $\mathfrak{A}(a) = \mathfrak{A}(b)$ [4, Lemma VII.1.9]. It follows that $f(a) \in \mathfrak{A}(a)$ for each $a \in D$ implies that $f(x) \in \mathfrak{A}(x)\widehat{D}$ for each $x \in \widehat{D}$.

(ii) We proved recently that the second condition (D is unibranched) is necessary [a paper is under preparation]: if D is not unibranched, there exists a nonzero ideal $\mathfrak{a}$

in D, such that the ideal $\text{Int}(D, \mathfrak{a})$ (of integer-valued polynomials with values in $\mathfrak{a}$) is not finitely generated. However, it is the Skolem closure of the finitely generated ideal $\mathfrak{a}\text{Int}(D)$.

Question 1.10. If D is a one-dimensional local Noetherian domain, with finite residue field, we know that D is analytically irreducible if and only if $\text{Int}(D)$ satisfies the Stone-Weierstrass property (that is, is dense in $\mathcal{C}(\widehat{D}, \widehat{D})$). On the other hand, D is unbranched if and only if $\text{Int}(D)$ satisfies the following *interpolation property*: for each finite set $(a_1, \ldots, a_n)$ of distinct elements of D, and each corresponding set of "values" $(c_1, \ldots, c_n)$ in D, there exists $f \in \text{Int}(D)$ such that $f(a_i) = c_i$ for $1 \leq i \leq n$ [6, Theorem 3.1]. Is the almost strong Skolem property equivalent to one of these properties? In general, what relation is there between the almost strong Skolem property and the interpolation property?

Various generalizations. More generally, we may consider fractional ideals of $\text{Int}(D)$ (as in [5]), and see to what extent they are characterized by their values. If $\mathfrak{A}$ is a fractional ideal, there is a polynomial f such that $f\mathfrak{A}$ is an integral ideal. It follows that, if $\text{Int}(D)$ satisfies the super Skolem property (resp., the strong Skolem property), the fractional ideals (resp. the finitely generated fractional ideals) are characterized by their values: if $\mathfrak{A}$ is a fractional ideal (resp., a finitely generated fractional ideal), and φ a rational function such that $\varphi(a) \in \mathfrak{A}(a)$ for almost all $a \in \mathfrak{A}$, then $\varphi \in \mathfrak{A}$ (in particular, letting $\mathfrak{A} = \text{Int}(D)$, we recover the fact that D must be a d-ring).

We may also consider the ring of integer-valued polynomials in n indeterminates:

$$R = \text{Int}(D^n) = \{f \in K[X_1, \ldots, X_n] \mid f(D^n) \subseteq D\}.$$

For $n > 1$, the ring $\text{Int}(D^n)$ is never a Prüfer domain (even if D is a Dedekind domain with finite residue fields); nevertheless, we still have the following [4, Proposition XI.3.8]:

Proposition 1.11. *Let D be a (local, one-dimensional) analytically irreducible Noetherian domain with finite residue field. Then $\text{Int}(D^n)$ satisfies the almost strong Skolem property.*

We could finally consider the ring of integer-valued rational functions:

$$\text{Int}^R(D) = \{\varphi \in K(X) \mid \varphi(D) \subseteq D\}.$$

Even in the local case, although D is not a d-ring, we may have the strong Skolem property for the ring of rational functions. Indeed, we then have [4, Proposition X.3.8]:

Proposition 1.12. *Let D be a (local, one-dimensional) analytically irreducible Noetherian domain with finite residue field. Then $\text{Int}^R(D)$ satisfies the strong Skolem property.*

2. d-SETS, COHERENT AND HOMOGENEOUS SUBSETS

In this section, we let D be a domain with quotient field K, and we consider a subset E of K (which is not necessarily a subset of D). We consider the ring $\text{Int}(E, D)$ of *integer-valued polynomials on E*:

$$\text{Int}(E, D) = \{f \in K[X] \mid f(E) \subseteq D\}.$$

We study various properties of the subset E, related to the Skolem properties of $\text{Int}(E, D)$, which allow, in the next sections, to restrict ourselves to the almost Skolem properties.

s-sets and d-sets. As in the case of $\text{Int}(D)$, let us first examine the Skolem properties with respect to a principal ideal (f), where f is a non-constant polynomial in $\text{Int}(E, D)$.
— If $\text{Int}(E, D)$ satisfies the Skolem property, the Skolem closure of (f) is distinct from $\text{Int}(E, D)$, and hence, there exists a in D such that $f(a)$ is not a unit in D.
— If $\text{Int}(E, D)$ satisfies the strong Skolem property, the ideal (f) is Skolem closed: if a polynomial $g \in \text{Int}(E, D)$ is such that $g(a) \in f(a)D$ for each $a \in D$ (that is, g/f is an almost integer-valued rational function on E), then f divides g in $\text{Int}(E, D)$ (in particular, g/f is a polynomial).

We thus set the following definitions (the first one is in [12]).

Definitions 2.1. (i) We say that a subset E of K is a d-set (*with respect to D*) if each almost integer-valued rational function on E is a polynomial.
(ii) We say that E is an s-set (*with respect to D*) if each unit-valued polynomial on E is a constant.

If the context is clear, we may drop the reference to D. If E is a subset of D, we may also say that E is a d-subset (resp., an s-subset) of D.

If f is a unit-valued polynomial on E, then $(1/f)$ is an integer-valued rational function on E; also, if f is a unit-valued polynomial (resp., if φ is an integer-valued rational function) on some subset F of K, then a fortiori f is unit-valued (resp., φ is integer-valued) on a subset E of F. We thus have immediately the following properties.

Proposition 2.2. *Let E be a subset of K.*

  (i) *If E is a d-set, then E is an s-set.*
  (ii) *If E is a d-set (resp., an s-set), and if $E \subseteq F$, then F is also a d-set (resp., an s-set).*

In particular, if there is an s-subset in D, then D is an s-subset of itself and, in fact, a d-ring [Proposition-Definition 1.1]. We give below an example showing that, contrary to the case of $\text{Int}(D)$, the notions of d-set and of s-set are not equivalent [Example 2.4 (4)]. But first, we make some comments.

Remarks 2.3. (1) To say that E is not a d-set means that we can find an almost integer-valued rational function φ, which is not a polynomial. In fact, we may ask φ to be of the form $\varphi = d/g$, where $d \in D$ is a nonzero constant and $g \in D[X]$ is a non-constant polynomial with coefficients in D. Indeed, write $\varphi = f/g$, where f and g are polynomials of $D[X]$ which are coprime in $K[X]$. We then have $uf + vg = d$, where $u, v \in D[X]$, and $d \in D, d \neq 0$. Clearly, d/g is also almost integer-valued.
(2) To say that E is not an s-set means that we can find a non-constant polynomial f which is unit-valued on E. Necessarily $f \in \text{Int}(E, D)$, but that does not mean that we can find such a polynomial with coefficients in D. We could thus say that E is a *weak s-set*, if each polynomial $f \in D[X]$, which is unit-valued on E, is constant. Contrary to the case of $\text{Int}(D)$, a weak s-set need not be an s-set [Example 2.4 (5)].

Examples 2.4. (1) *Finite subsets.* A finite set $E = \{a_1, \ldots, a_r\}$ is never an s-set. Indeed, the polynomial $f = 1 + \prod_{i=1}^{r}(X - a_i)$ is unit-valued on E.

(2) *Subsets of $\mathbb{Z}$.* A subset of $\mathbb{Z}$ is a d-subset (resp., an s-subset) if and only if it is infinite [22, vol. II.8/II.93]. Indeed, the condition is necessary from the previous example. It is sufficient: if φ is an integer-valued rational function on an infinite subset of $\mathbb{Z}$, write $\varphi = f + p/q$, where f, p, q are polynomials with coefficients in $\mathbb{Q}$, and $\deg(p) < \deg(q)$. Let $d \in \mathbb{Z}, d \neq 0$, be such that $df \in \mathbb{Z}[X]$. Then, for each $x \in \mathbb{Z}, dp(x)/q(x) \in \mathbb{Z}$, and moreover converges to 0 as x goes to infinity. Necessarily, $p = 0$.

(3) *Finite group of units.* If the group of units of D is finite, every infinite set E is an s-set: if a polynomial is unit-valued on E, it takes finitely many values on an infinite set, thus it is a constant.

(4) *An s-set which is not a d-set.* Let $D = \mathbb{Z} + t\mathbb{Q}[t]$ be the ring formed by the polynomials with coefficients in $\mathbb{Q}$ and constant term in $\mathbb{Z}$, and $E = \mathbb{Z}$. Clearly, E is not a d-subset of D, since the rational function $\frac{t}{1+X^2}$ is integer-valued on $E = \mathbb{Z}$. On the other hand, the units of D are 1 and -1, and it follows from the previous example that E is an s-subset of D.

(5) *A subset E which is not an s-set, but such that each polynomial, with coefficients in D, which is unit-valued on E, is constant.* Let $D = \mathbb{Q}[t]$ be the ring of polynomials with coefficients in $\mathbb{Q}$, and $E = \{at \mid a \in \mathbb{Q}\}$ be the subset of D formed by the monomials of degree one. Clearly, E is not an s-subset, since the polynomial $\frac{X^2+t^2}{t^2}$ is unit-valued on E. Now, let $f \in D[X]$. We can write

$$f = h_0(t) + h_1(t)X + \ldots + h_n(t)X^n,$$

where each $h_i(t)$ is a polynomial with coefficients in $\mathbb{Q}$. Suppose that f is unit-valued: for each $a \in \mathbb{Q}, f(at)$ is a unit, that is, an element of $\mathbb{Q}$. Write

$$f(Xt) - f(0) = t[h_1(t)X + \ldots + h_n(t)t^{n-1}X^n].$$

Thus, $f(at) - f(0)$ is an element of $\mathbb{Q}$ which is divisible by t, and hence, is null for each $a \in \mathbb{Q}$. Since $\mathbb{Q}$ is infinite, the polynomial $f(Xt) - f(0)$ is identically null, whence so is each $h_i(t)$. Finally, $f = h_0(t)$ is a constant (that is, an element of D).

Coherent and strong coherent sets. Contrary to the case of $\mathrm{Int}(D)$, the Skolem closure of a finitely generated ideal $\mathfrak{A}$ of $\mathrm{Int}(E, D)$ is not necessarily determined by the ideal of values $\mathfrak{A}(a)$ for a in a cofinite subset of E; a polynomial f may even be such that $f(a) \in D$, for almost all $a \in E$, but not be integer-valued on E. This is obviously the case if E is finite, but here is a less trivial example: let $\mathbb{P}$ be the set of prime numbers; for each prime number p, the polynomial $\frac{(X-1)\cdots(X-p+1)}{p}$ is clearly integer-valued on each prime number but p. We then set the following definitions.

Definitions 2.5. (i) We say that a subset E of K is *coherent* (*with respect to* D) if each polynomial $f \in K[X]$ which is almost integer-valued on E is in fact integer-valued on E.

(ii) We say that E is *strongly coherent* (*with respect to* D) if, for each finitely generated ideal $\mathfrak{A}$ of $\mathrm{Int}(E, D)$, each polynomial $f \in K[X]$ such that $f(a) \in \mathfrak{A}(a)$ for almost each $a \in E$, belongs to the Skolem closure $\mathfrak{A}^*$ of $\mathfrak{A}$.

As for d-and s-sets, we often drop the reference to D. Note that a ring D is always a strongly coherent subset of itself [4, Lemma VII.1.8]. Note also that we restricted ourselves to finitely generated ideals of $\mathrm{Int}(E, D)$. Indeed, consider the maximal ideal $\mathfrak{M}_{p,n} = \{f \in \mathrm{Int}(\mathbb{Z}) \mid f(n) \in p\mathbb{Z}\}$ of $\mathrm{Int}(\mathbb{Z})$, where p is a prime number and n an integer: the constant polynomial $f = 1$ is such that $f(a) \in \mathfrak{M}_{p,n}(a)$, for each $a \neq n$, while $f(n) \notin \mathfrak{M}_{p,n}(n)$.

A strongly coherent set is coherent (consider the ideal $\mathfrak{A} = \mathrm{Int}(E, D)$). The following example shows that the converse does not hold.

Example 2.6. A coherent set which is not strongly coherent. Let $D = \mathbb{C}[t]$ be the ring of polynomials with coefficients in the complex field $\mathbb{C}$, and let $E = \mathbb{C}$. By a Vandermonde argument, it is easy to see that, for each infinite subset F of E, we have $\mathrm{Int}(F, D) = D[X]$ [4, Proposition I.3.1]. We obtain the following containments, thus, in fact, equalities

$$D[X] \subseteq \mathrm{Int}(D) \subseteq \mathrm{Int}(E, D) \subseteq \mathrm{Int}(F, D) \subseteq D[X].$$

In particular, E is coherent: each polynomial which is integer-valued on a cofinite subset F of E is, in fact, integer-valued on E. On the other hand, the polynomial X is clearly unit-valued on the complement of 0 on E, but not on E. Letting $\mathfrak{A} = (X)$ be the ideal generated by X, and $f = 1$, we then have $f(a) \in \mathfrak{A}(a)$ for all $a \in E$, but 0. And hence, E is not strongly coherent.

Remarks 2.7. (1) Extending [4, Definitions IV.1.2], we may say that a subset F of E is *polynomially dense* in E if $\mathrm{Int}(F, D) = \mathrm{Int}(E, D)$. To say that E is coherent thus means that each cofinite subset of E is polynomially dense in E. In fact, we could rather say that E is *polynomially coherent* in this case. We could then similarly say that F is *rationally dense* in E, if each rational function which is integer-valued on F is in fact integer-valued on E, and that E is *rationally coherent* if each cofinite subset of E is rationally dense in E. Example 2.6 shows that these two notions are distinct: the rational function $\frac{1}{X}$ is unit-valued on the complement of 0 on E.
(2) A strongly coherent set is even rationally coherent. This follows immediately from the fact that we can extend the property of the definition to rational functions. Suppose indeed that $\varphi(a) \in \mathfrak{A}(a)$ for almost each $a \in E$. Write $\varphi = f/g$, where f and g are integer-valued and coprime in $K[X]$, and set $\mathfrak{B} = g\mathfrak{A}$. If $\varphi(a) \in \mathfrak{A}(a)$, then $f(a) \in \mathfrak{B}(a)$ for almost each a. If E is strongly coherent, it follows that $f(a) \in \mathfrak{B}(a)$ for all a. In particular, if $g(a) = 0$, then $f(a) = 0$, but since f and g are coprime, this never happens: $g(a)$ never vanishes. In conclusion $\varphi(a) = f(a)/g(a)$ belongs to $\mathfrak{A}(a)$ for all a. We do not know if the notions of strong and rational coherence are distinct (for instance, we shall see that they are equivalent if $\mathrm{Int}(E, D)$ is a Prüfer domain [Proposition 3.10]).
(3) As in [12, Definition B], we could say that an element $a \in E$ is *(polynomially) isolated* in E if its complement is not polynomially dense in E (for instance, we saw above that each point of $\mathbb{P}$ is isolated in $\mathbb{P}$). Similarly we could say that $a \in E$ is *rationally isolated* in E if its complement is not rationally dense in E. A polynomially isolated point is rationally isolated; the converse does not hold (in Example 2.6 above, the point 0 is rationally, but not polynomially isolated). Of course a coherent (resp., a rationally coherent) set cannot have any isolated (resp., rationally isolated) point. We do not know if the converse holds, thus we end this paragraph with a question.

Question 2.8. Suppose that E is a subset of K without isolated points (resp., without rationally isolated points). Does this imply that E is coherent (resp., rationally coherent)?

We have a partial answer: if E is a subset of a completely integrally closed domain D, the following assertions are equivalent [12, Corollary 2.2 & Theorem 3.1], [4, Exercise VI.10].

— E is coherent,
— E has no isolated point,
— $\mathrm{Int}(E, D)$ is completely integrally closed.

Coherent d-sets. The notion of d-set or s-set on one-hand, and of coherence on the other, are clearly distinct. For instance, a ring D is always a strongly coherent subset of itself, but not necessarily an s-set (that is, a d-ring). On the other hand, the set $\mathbb{P}$ of prime numbers is infinite, and hence, a d-subset of $\mathbb{Z}$ [Example 2.4 (2)], but it is not coherent. We shall need the (immediate) following characterization of the cases where both properties hold.

Proposition 2.9. *A subset E of K is a a coherent d-set if and only if each almost integer-valued rational function on E is an integer-valued polynomial.*

Remark 2.10. We could say that a subset E of K is an *almost d-set* if each integer-valued rational function on E is a polynomial (and of course, such a polynomial is then integer-valued). If E is a d-set, each almost-integer valued rational function is a polynomial, but note that here, such a polynomial need not be integer-valued (unless E is coherent). Clearly, E is a coherent d-set if and only if it is a rationally coherent almost d-set. The following example (in fact, Example 2.6) shows that an almost d-set need not be a d-set (even if it is coherent).

Example 2.11. As in Example 2.6, let $D = \mathbb{C}[t]$ be the ring of polynomials with complex coefficients, and let $E = \mathbb{C}$. Then E is not a d-set: the rational function $\frac{1}{X}$ is almost integer-valued, yet not a polynomial. We have seen that E is coherent. Finally, we show that E is an almost d-set: each integer-valued rational function is a polynomial. From Remark 2.3 (1), it suffices to show that, if $d/g(X)$ is integer-valued, where d is a nonzero element of D, and $g(X)$ is a polynomial with coefficients in D, then $g(X)$ is a constant, that is, an element of D. Write $\frac{d}{g(X)} = \frac{d(t)}{g(t,X)}$, where $d(t)$ is a polynomial with coefficients in $\mathbb{C}$, and

$$g(t, X) = g_0(t) + g_1(t)X + \ldots + g_n(t)X^n,$$

each $g_i(t)$ being also a polynomial with coefficients in $\mathbb{C}$. To say that $d/g(X)$ is integer-valued means that, for each $a \in E = \mathbb{C}$, $d/g(a)$ belongs to $D = \mathbb{C}[t]$, that is,

$$d(t) = h_a(t)g(t, a) \text{ where } h_a(t) \in \mathbb{C}[t].$$

If, for some $\alpha \in \mathbb{C}$, the polynomial $g(\alpha, X)$ is not constant, it has a root in $\mathbb{C}$: there is $a \in \mathbb{C}$ such that $g(\alpha, a) = 0$, and hence, such that $d(\alpha) = 0$. Since $d(t)$ has only finitely many roots in $\mathbb{C}$, this implies that $g(\alpha, X)$ is a constant, for almost each $\alpha \in \mathbb{C}$. Since $g(\alpha, X) = g_0(\alpha) + g_1(\alpha)X + \ldots + g_n(\alpha)X^n$, it follows that $g_i(t) = 0$ for $i \geq 1$. Therefore $g(t, X) = g_0(t)$, that is, $g(t, X)$ is an element of $D = \mathbb{C}[t]$.

Homogeneous and weakly-homogeneous sets. We describe here a class of strongly coherent sets. It generalizes the *homogeneous* subsets (in a Dedekind domain) of D. McQuillan, which are union of cosets modulo some nonzero ideal [20].

Definitions 2.12. (i) We say that a subset E of K is *homogeneous (with respect to D)*, if there exists a nonzero ideal $\mathfrak{a}$ of D such that $(a + \mathfrak{a}) \subseteq E$ for each $a \in E$.
(ii) We say that E is *weakly-homogeneous (with respect to D)* if, for each nonzero ideal $\mathfrak{a}$ of D, and each $a \in E$, $(a + \mathfrak{a})$ contains at least one element of E distinct from a.

As usual, we shall often drop the reference to D. If E is homogeneous, and if α is a nonzero element of the ideal $\mathfrak{a}$, then clearly $(a + \alpha D) \subseteq E$. Thus we could define homogeneous sets using only principal ideals. The same holds for weakly-homogeneous sets.

Let us now introduce some topological ideas. We can consider the topology where the nonzero ideals form a basis of neighborhoods of 0, let us call it the *ideal topology*. It follows from [4, Lemma I.3.19] that each polynomial is uniformly continuous in this topology. (We could derive that a topologically dense subset F of E is polynomially dense; however the converse does not hold: a subset of $\mathbb{Z}$ is polynomially dense in $\mathbb{Z}$ if and only if it is dense in every p-adic topology, which does not imply that it is dense in the ideal topology [4, Remark IV.2.8].) To say that E is weakly homogeneous means that each element of E is an accumulation point of E in the ideal topology. In particular, for each nonzero ideal $\mathfrak{a}$ of D, and each $a \in E$, the intersection $(a + \mathfrak{a}) \cap E$ is infinite. We leave to the reader that a homogeneous set is weakly homogeneous. It is easy to see that the union of the intervals $[n!, n! + n]$ is weakly homogeneous in $\mathbb{Z}$ (since it is dense in $\mathbb{Z}$, in the ideal topology), but not homogeneous (since it does not contain any arithmetic progression).

We show now that a weakly homogeneous set is strongly coherent; this is very similar to the proof that D itself (which obviously is homogeneous) is strongly coherent [4, Lemma VII.1.8].

Proposition 2.13. *A weakly homogeneous set is strongly coherent.*

Proof. Let $\mathfrak{A}$ be a finitely generated ideal of $\mathrm{Int}(E, D)$, and f be a polynomial such that $f(a) \in \mathfrak{A}(a)$ for almost each $a \in E$. Let b be an element of E, we wish to show that $f(b) \in \mathfrak{A}(b)$. Denote by $g_1, \ldots, g_r$ a set of generators of $\mathfrak{A}$. The polynomials f and g_j have a common denominator d. Whatever $a \in E$, $f(a) \in \mathfrak{A}(a)$ if and only if $df(a) \in d\mathfrak{A}(a)$. With no loss of generality, we may thus assume the polynomials f and g_j to have their coefficients in D. Since E is weakly homogeneous, we may choose, for each nonzero ideal $\mathfrak{a}$, an element $a \in E$ of the form $a = b + \alpha$, where $\alpha \in \mathfrak{a}$, such that $f(a) \in \mathfrak{A}(a)$. We then have

- $f(b) = f(a) + \beta$, where $\beta \in \mathfrak{a}$, and
- $g_j(b) = g_j(a) + \beta_j$, where $\beta_j \in \mathfrak{a}$, for $1 \le j \le r$.

We then consider two cases:

— $\mathfrak{A}(b) = (0)$. Then $g_j(b) = 0$ for $1 \le j \le r$. For each choice of the ideal $\mathfrak{a}$, we have $g_j(a) \in \mathfrak{a}$, for $1 \le j \le r$. Hence $\mathfrak{A}(a) \subseteq \mathfrak{a}$. In particular $f(a) \in \mathfrak{a}$, and finally $f(b) \in \mathfrak{a}$. It follows that $f(b) = 0$ (since $f(b)$ belongs to every nonzero ideal of D).

— $\mathfrak{A}(b) \ne (0)$. Choose $\mathfrak{a} = \mathfrak{A}(b)$. We then have $g_j(a) \in \mathfrak{A}(b)$ for $1 \le j \le r$. Hence $\mathfrak{A}(a) \subseteq \mathfrak{A}(b)$. In particular $f(a) \in \mathfrak{A}(b)$, and finally $f(b) \in \mathfrak{A}(b)$. $\square$

Isomorphic subsets. We consider two elements $\alpha, \beta, \in K$, such that $\beta \ne 0$, and the K-isomorphism $\Psi_{\alpha,\beta} : K(X) \mapsto K(X)$ defined by $\Psi_{\alpha,\beta}(X) = \frac{X - \alpha}{\beta}$. For each subset E of K, we then denote by $E_{\alpha,\beta}$ the subset $E_{\alpha,\beta} = \alpha + \beta E$. With these notations, we obviously have the following isomorphisms.

Lemma 2.14. *The K-isomorphism $\Psi_{\alpha,\beta}$ induces an isomorphism from $\mathrm{Int}(E, D)$ onto $\mathrm{Int}(E_{\alpha,\beta}, D)$.*

We can consider that the subsets E and $E_{\alpha,\beta} = (\alpha + \beta E)$ are, in a sense, isomorphic, hence they share many properties:

Proposition 2.15. (i) E *is an s-set (resp., a d-set) if and only if $E_{\alpha,\beta}$ is an s-set (resp., a d-set).*

(ii) E *is coherent (resp., strongly coherent) if and only if $E_{\alpha,\beta}$ is coherent (resp., strongly coherent).*

(iii) E *is homogeneous (resp., weakly homogeneous) if and only if $E_{\alpha,\beta}$ is homogeneous (resp., weakly homogeneous).*

(iv) $\mathrm{Int}(E,D)$ *has some Skolem property if and only if $E_{\alpha,\beta}$ has the same Skolem property.*

Proof. Only the assertions dealing with ideals (that is, the strong coherence and the Skolem properties) deserve an explanation. For each ideal $\mathfrak{A}$ of $\mathrm{Int}(E,D)$, and each $a \in E$, $\mathfrak{A}(a) = \Psi_{\alpha,\beta}(\mathfrak{A})(\alpha + \beta a)$. It follows that $g(a) \in \mathfrak{A}(a)$ if and only if $\Psi_{\alpha,\beta}(g)(\alpha + \beta a) \in \Psi_{\alpha,\beta}(\mathfrak{A})(\alpha + \beta a)$; in other words, $\left(\Psi_{\alpha,\beta}(\mathfrak{A})\right)^* = \Psi_{\alpha,\beta}(\mathfrak{A}^*)$. $\square$

Remarks 2.16. (1) The morphism $\Psi_{\alpha,\beta}$ does not necessarily take polynomials with coefficients in D into polynomials with coefficients in D. Thus, for instance, the subset $E = \{at \mid a \in \mathbb{Q}\}$ of $D = \mathbb{Q}[t]$ is such that each unit-valued polynomial with coefficients in D is constant [Example 2.4 (5)], while the polynomial $X^2 + 1$ is unit-valued on $\mathbb{Q} = (1/t)E$.

(2) When studying the ring $\mathrm{Int}(E,D)$, one often assumes that E is a *fractional subset* of D, that is, a subset of K such that $dE \subseteq D$ for a nonzero element d of D (for instance, if D is integrally closed, this condition is necessary for $\mathrm{Int}(E,D)$ to contain nonzero constants [4, Corollary I.1.10]). It follows from the previous proposition, that we may often restrict ourselves to subsets of D (a fractional subset is isomorphic to a subset of D having similar properties).

We finally recover and generalize [12, Lemma 2.6].

Corollary 2.17. *If E is a homogeneous subset of a d-ring, then E is a d-set.*

Proof. By hypothesis, E contains a subset of the form $\alpha + \beta D$, which is isomorphic to D, and hence, which is a d-set. Then E itself is a d-set [Proposition 2.2]. $\square$

3. COHERENCE AND SKOLEM PROPERTIES

In this section, we consider again a subset E of K, and link the Skolem properties to the notion of d-sets and coherence. We begin with the Skolem property, then proceed to the strong Skolem property. But first, we examine the case where E is finite.

Finite sets. A finite set E is never an s-set [Example 2.4 (1)], thus $\mathrm{Int}(E,D)$ does not satisfy the Skolem property. Nevertheless, let us recall the following result of D. L. McQuillan [19].

Proposition 3.1. *Let E be a non-empty finite set, then $\mathrm{Int}(E,D)$ satisfies the almost super Skolem property.*

Proof. Let $E = \{a_1, \ldots, a_r\}$ and set $\psi = \prod_{i=1}^{r}(X - a_i)$. Consider a unitary ideal $\mathfrak{A}$ and a polynomial f in its Skolem closure $\mathfrak{A}^*$. Then $f(a_i) = g_i(a_i)$, where $g_i \in \mathfrak{A}$ for $1 \leq i \leq r$. Using Lagrange interpolation, we may write $f = \sum_{i=1}^{r} \varphi_i g_i + g$, where $\varphi_i = \prod_{j \neq i} \frac{X - a_j}{a_i - a_j}$, and thus, g vanishes on E. Since $\mathfrak{A}$ is unitary, it contains a nonzero constant a. Writing $g = a(g/a)$, it is clear that (g/a) vanishes also on E. Hence we have $(g/a) \in \mathrm{Int}(E,D)$, thus $g \in \mathfrak{A}$, and finally, $f \in \mathfrak{A}$. $\square$

Skolem property. In the previous section, considering a principal ideal (f), where f is a non-constant polynomial in $\text{Int}(E, D)$, we saw that, if $\text{Int}(E, D)$ satisfies the Skolem property, then E must be an s-subset. Clearly, $\text{Int}(E, D)$ must also satisfy the almost Skolem property. Contrary to the case of $\text{Int}(D)$ [Proposition 1.2], we have only a partial converse.

Proposition 3.2. *Let E be a subset of K.*

(i) *If $\text{Int}(E, D)$ satisfies the Skolem property, then E is an s-set.*

(ii) *If E is a d-set, and if $\text{Int}(E, D)$ satisfies the almost Skolem property, then $\text{Int}(E, D)$ satisfies the Skolem property.*

The first statement is immediate, and (ii) follows from the next lemma (which implies in particular that the Skolem closure of a non-unitary finitely generated ideal is a proper ideal of $\text{Int}(E, D)$).

Lemma 3.3. *Let E be a d-set of D. If $\mathfrak{A}$ is a finitely generated ideal of $\text{Int}(E, D)$, such that $\mathfrak{A} \subseteq \left(gK[X] \cap \text{Int}(E, D)\right)$, for some non-constant polynomial g, then the Skolem closure $\mathfrak{A}^*$ of $\mathfrak{A}$ is also such that $\mathfrak{A}^* \subseteq \left(gK[X] \cap \text{Int}(E, D)\right)$.*

Proof. Since $\mathfrak{A}$ is finitely generated, there is a nonzero element $d \in D$ such that $d\mathfrak{A} \subseteq g\text{Int}(E, D)$. Let $f \in \mathfrak{A}^*$: for each $a \in E$, $f(a) \in \mathfrak{A}(a)$, thus, $df(a) \in g(a)D$. The rational function df/g is then integer-valued at each a which is not a root of g. Since E is a d-set, it follows that g divides df in $K[X]$, that is, f belongs to the intersection $gK[X] \cap \text{Int}(E, D)$. $\square$

Remark 3.4. Assuming that $\text{Int}(E, D)$ satisfies the almost Skolem property, we do not know if it is necessary that E be a d-set for $\text{Int}(E, D)$ to satisfy the Skolem property, nor that if it is sufficient that E be an s-set. At least, we can show that it is sufficient that E be an almost d-set (each integer-valued rational function is a polynomial). Suppose indeed that $\mathfrak{A}$ is a finitely generated non-unitary ideal, then $\mathfrak{A} \subseteq \left(qK[X] \cap \text{Int}(E, D)\right)$, for some polynomial q which is irreducible in $K[X]$. There is a nonzero element $d \in D$ such that $d\mathfrak{A} \subseteq q\text{Int}(E, D)$. If $f \in \mathfrak{A}^*$, then $df(a) \in q(a)D$ for each $a \in E$.

— If q has no root in E, then df/q is an integer-valued rational function on E. Assuming that E is an almost d-set, it follows that df/q is a polynomial, that is, $f \in \left(qK[X] \cap \text{Int}(E, D)\right)$.

— If $q(a) = 0$, then $q = X - a$ (since q is irreducible in $K[X]$). On the other hand, $f(a) = 0$ (since $df(a) \in q(a)D$). Hence $q = X - a$ divides f in $K[X]$, and again $f \in \left(qK[X] \cap \text{Int}(E, D)\right)$.

Almost Skolem property. We just saw that, if E is a d-set, we may restrict ourselves to the almost Skolem property. Similarly to the case of $\text{Int}(D)$ [Proposition 1.5], we can conclude in the case of a one-dimensional Noetherian domain with finite residue fields:

Proposition 3.5. *Let D be a one-dimensional Noetherian domain with finite residue fields, and E be a fractional subset of D. Then $\text{Int}(E, D)$ satisfies the almost Skolem property.*

Proof. Replacing E by an isomorphic subset, one may in fact assume that E is a subset of D [Proposition 2.15]. We then know that the non-unitary maximal ideals of $\text{Int}(E, D)$ are of the form $\mathfrak{M}_{\mathfrak{m},\alpha} = \{f \in \text{Int}(E, D) \mid f(\alpha) \in \widehat{\mathfrak{m}D_{\mathfrak{m}}}\}$, where $\mathfrak{m}$ is a maximal ideal of D and α is an element of the topological closure of E in the

$\mathfrak{m}$-adic completion of D [4, Proposition V.2.2] . Let $\mathfrak{A}$ be a proper unitary finitely generated ideal of $\mathrm{Int}(E, D)$. Since $\mathfrak{A}$ is a proper unitary ideal, it is contained in some maximal ideal $\mathfrak{M}_{\mathfrak{m},\alpha}$. Since $\mathfrak{A}$ is finitely generated, there is some $a \in E$ close enough to α such that $\mathfrak{A} \subseteq \mathfrak{M}_{\mathfrak{m},a}$, that is, such that $\mathfrak{A}(a) \neq D$.

Corollary 3.6. *Let D be an order of a number field and E be a fractional subset of D. If E is a d-set, then $\mathrm{Int}(E, D)$ satisfies the Skolem property.*

The results of the last section will show that we may also obtain a positive conclusion, in the local case, even if the residue field of D is infinite, under some compactness condition for E (and other conditions).

Skolem closure of divisorial ideals. For the unitary divisorial ideals, Lemma 1.3 may be extended without difficulties to a subset.

Lemma 3.7. *Each unitary divisorial ideal of $\mathrm{Int}(E, D)$ is Skolem closed.*

Proof. Let $\mathfrak{A}$ be a unitary divisorial ideal of $\mathrm{Int}(E, D)$. We first show that $\mathfrak{A}$ is an intersection of fractional principal ideals of the form $(1/q)\mathrm{Int}(E, D)$ where $q \in K[X]$. Let $\varphi\mathrm{Int}(E, D)$ be a principal ideal containing $\mathfrak{A}$ and write $\varphi = p/q$ where p, q are relatively prime in $K[X]$. Since $\mathfrak{A}$ is unitary, there is a nonzero element a in $\mathfrak{A} \cap D$. Since $a \in \mathfrak{A}$, a fortiori $a \in \varphi\mathrm{Int}(E, D)$: we may write $aq = ph$, where $h \in \mathrm{Int}(E, D)$. Since p and q are relatively prime, it follows that p is a constant, as claimed.

Now let $f \in \mathfrak{A}^*$ and let $q \in K[X]$ be such that $\mathfrak{A} \subseteq (1/q)\mathrm{Int}(E, D)$. For each $a \in E$, one has $f(a) \in \mathfrak{A}(a)$, and thus, $f(a)q(a) \in D$, that is, $fq \in \mathrm{Int}(E, D)$. Therefore, $f \in (1/q)\mathrm{Int}(E, D)$, and finally, f belongs to the ideal $\mathfrak{A}$ which is the intersection of the principal ideals that contain it. $\square$

However, as noted in introduction of the notion of d-set, if the (non-unitary) principal ideals of $\mathrm{Int}(E, D)$ are Skolem closed, then E is a d-set. We show here that E must also be coherent. In fact, we have an equivalence.

Proposition 3.8. *Let E be a subset of K. The following assertions are equivalent:*

  (i) *E is a coherent d-set,*
  (ii) *the principal ideals of $\mathrm{Int}(E, D)$ are Skolem closed,*
  (iii) *the divisorial ideals of $\mathrm{Int}(E, D)$ are Skolem closed.*

Proof. (i) $\Rightarrow$ (iii) Let $\mathfrak{A}$ be a divisorial ideal of $\mathrm{Int}(E, D)$, that is, an intersection of fractional principal ideals. Consider $f \in \mathfrak{A}^*$. For each $a \in E$ we have $f(a) \in \mathfrak{A}(a)$. If $\varphi \in K(X)$ is such that the principal ideal $\varphi\mathrm{Int}(E, D)$ contains $\mathfrak{A}$, and if a is not a pole of φ, we then have $f(a) \in \varphi(a)D$. Hence, the rational function f/φ is almost integer-valued on E. Assuming that E is a coherent d-set, it follows that f/φ is in fact an integer-valued polynomial [Proposition 2.9], that is, $(f/\varphi) \in \mathrm{Int}(E, D)$. Hence $f \in \varphi\mathrm{Int}(E, D)$, and this holds for each principal ideal containing $\mathfrak{A}$.
(iii) $\Rightarrow$ (ii) Obvious.
(ii) $\Rightarrow$ (i) Let $\varphi = f/g$ be an almost integer-valued rational function, that is $\varphi(a) \in D$ for each $a \in E$, but possibly $\{a_1, \ldots, a_r\}$. Consider the polynomial $\psi = \prod_{i=1}^{r}(X - a_i)$. Then $(\psi f)(a) \in (\psi g)(a)D$ for each $a \in E$ (now, with no exception). Since the principal ideal $(\psi g)\mathrm{Int}(E, D)$ is Skolem closed, it follows that $\psi f = \psi g h$, where $h \in \mathrm{Int}(E, D)$. In conclusion, $\varphi = f/g = h$ is an integer-valued polynomial. $\square$

Strong coherence and strong Skolem property. Analogously to Proposition 1.2, we can now relate the strong and almost strong Skolem properties.

Proposition 3.9. *Let E be a subset of K. Then $\mathrm{Int}(E, D)$ satisfies the strong Skolem property if and only if the following two conditions are satisfied:*

(a) *E is a strongly coherent d-set with respect to D,*

(b) *$\mathrm{Int}(E, D)$ satisfies the almost strong Skolem property.*

Proof. — The conditions are necessary. It is immediate that $\mathrm{Int}(E, D)$ must satisfy the almost strong Skolem property. From Proposition 3.8, we know also that E is a coherent d-set, but prove that, in fact, E is strongly coherent. Let $\mathfrak{A}$ be a finitely generated ideal of $\mathrm{Int}(E, D)$, and $f \in \mathrm{Int}(E, D)$ be such that $f(a) \in \mathfrak{A}(a)$ for almost each $a \in E$, that is, each a but possibly $\{a_1, \ldots, a_r\}$. As in the proof of Proposition 3.8, consider the polynomial $\psi = \prod_{i=1}^{r}(X - a_i)$, then set $\mathfrak{B} = \psi\mathfrak{A}$. Thus $(\psi f)(a) \in \mathfrak{B}(a)$ for each $a \in E$ (now, with no exception). From the strong Skolem property, we have $\psi f \in \mathfrak{B}$, that is, $\psi f = \psi h$, where $h \in \mathfrak{A}$. Thus $f \in \mathfrak{A}$, and of course, $f(a) \in \mathfrak{A}(a)$ for all $a \in E$.

— The conditions are sufficient. Let $\mathfrak{A}$ be a finitely generated nonzero ideal of $\mathrm{Int}(E, D)$, and f be in the Skolem closure $\mathfrak{A}^*$ of $\mathfrak{A}$: for each $a \in E$, $f(a) \in \mathfrak{A}(a)$. As for $\mathrm{Int}(D)$, we may find a polynomial $g \in D[X]$, and a nonzero element $d \in D$, such that $d\mathfrak{A} = g\mathfrak{B}$, where $\mathfrak{B}$ is a finitely generated unitary ideal of $\mathrm{Int}(E, D)$ [4, Lemma VI.1.2]. We thus have $df(a) \in g(a)\mathfrak{B}(a)$ for each $a \in E$. In particular, $\varphi = df/g$ is an almost integer-valued rational function. Since E is a coherent d-set, φ is in fact an integer-valued polynomial [Proposition 2.9]. By hypothesis $\varphi(a) \in \mathfrak{B}(a)$ for each $a \in E$, but possibly the zeros of g. Since φ is a polynomial and since E is strongly coherent, it follows that $\varphi(a) \in \mathfrak{B}(a)$ for all $a \in E$. Since $\mathfrak{B}$ is a unitary ideal and since $\mathrm{Int}(E, D)$ satisfies the almost strong Skolem property, it follows that $\varphi \in \mathfrak{B}$. Finally, $f = (g/d)\varphi$ belongs to $\mathfrak{A} = (g/d)\mathfrak{B}$. $\square$

Prüfer domains. If $\mathrm{Int}(E, D)$ is a Prüfer domain, each finitely generated ideal is invertible, and a fortiori, divisorial. Hence it follows from Lemma 3.8 that it satisfies the strong Skolem property if and only if E is a coherent d-set. On the other hand, $\mathrm{Int}(E, D)$ satisfies always the almost strong Skolem property, and it follows from Proposition 3.9, that $\mathrm{Int}(E, D)$ satisfies the strong Skolem property if and only if E is a strongly coherent d-set. We thus obtain a partial generalization of Proposition 1.4, recovering also [12, Theorem 4.6]. This applies for instance to a fractional subset E of a Dedekind domain D with finite residue fields (we may replace E by a subset of D with the same properties [Proposition 2.15], then $\mathrm{Int}(E, D)$ is an overring of $\mathrm{Int}(D)$, which is a Prüfer domain).

Proposition 3.10. *Assume that $\mathrm{Int}(E, D)$ is a Prüfer domain (for instance, E is a fractional subset of a Dedekind domain with finite residue fields), then $\mathrm{Int}(E, D)$ satisfies the almost strong Skolem property. Moreover, the following assertions are equivalent:*

(i) *E is a coherent d-set,*

(ii) *E is a strongly coherent d-set,*

(iii) *$\mathrm{Int}(E, D)$ satisfies the strong Skolem property.*

However, if $\mathrm{Int}(E, D)$ satisfies the Skolem property, we know only that E is an s-set. We do not know if we can conclude that E is a strongly coherent d-set, we do not know if we can conclude that $\mathrm{Int}(E, D)$ satisfies the strong Skolem property (as in Proposition 1.4).

Example 3.11. Let E be a subset of $\mathbb{Z}$. Then E is a d-set if and only if it is infinite [Example 2.4 (2)]. On the other hand, since $\mathbb{Z}$ is completely integrally closed, E is coherent if and only if it has no polynomially isolated point [4, Exercise VI.10]. Since Int($\mathbb{Z}$) is a Prüfer domain, we conclude that Int($E, \mathbb{Z}$) satisfies the strong Skolem property, if and only if E is infinite without isolated points [12, Corollary 4.8].

4. ALMOST STRONG SKOLEM PROPERTY: THE LOCAL CASE

We do not know if the almost strong Skolem property for Int(E, D) implies that, for each maximal ideal $\mathfrak{m}$ of D, Int($E, D_\mathfrak{m}$) satisfies also this property (that is, whether Lemma 1.8 may be extended). Nevertheless, if Int($E, D_\mathfrak{m}$) satisfies the almost strong Skolem property for each maximal ideal $\mathfrak{m}$ of D, then clearly so does Int(E, D). Therefore, we restrict ourselves to the local case.

Notations and hypotheses. We let D be a one-dimensional local Noetherian domain with maximal ideal $\mathfrak{m}$, and E be a fractional subset of D. We denote by $\widehat{D}$ the completion of D, and by $\widehat{E}$ the topological closure of E (in the $\mathfrak{m}$-adic topology). We assume that D is analytically irreducible.

Assuming moreover that the residue field of D is finite, we have seen that Int(D) satisfies the almost strong Skolem property [Theorem 1.9]. This generalizes easily to a fractional subset E of D [4, Exercise VII.17]. The proof is in every respect similar to the case of Int(D), using the fact that $\widehat{E}$ is compact. However $\widehat{E}$ may be compact even if $D/\mathfrak{m}$ is infinite. This is clearly the case if E is finite (but we already know that Int(E, D) satisfies the almost super Skolem property in this case [Proposition 3.1]). Here are two less trivial examples:

Examples 4.1. 1) Let $k = \mathbb{F}_q$ be a finite field. The ring $D = \widehat{D} = k(y)[[t]]$ of power series with coefficients in $k(y)$ is not compact (its residue field is infinite); the subring $E = \widehat{E} = k[[t]]$ is compact (its residue field is finite).
2) Let $D = \widehat{D} = K[[t]]$ be the ring of power series with coefficients in an infinite field K. As above, D is not compact. Let $E = \{t^k \mid k \in \mathbb{N}^*\}$. Then $\widehat{E} = E \cup \{0\}$ is compact.

We shall prove that Int(E, D) satisfies the almost strong Skolem property, under the hypothesis that $\widehat{E}$ is compact (without supposing that the residue field of D is finite). We will most often assume that E is a subset of D (if E is a fractional subset, we may as well replace it by an isomorphic subset, of the form dE, which is contained in D). As for Int(D), the proof relies mainly on three points:

— A. The ring $\mathcal{C}(\widehat{E}, \widehat{D})$ of continuous functions from $\widehat{E}$ to $\widehat{D}$ satisfies the almost super Skolem property.
— B. Analogously to the classical Stone-Weierstrass theorem, the ring Int(E, D) is dense in the ring $\mathcal{C}(\widehat{E}, \widehat{D})$ for the uniform convergence topology [4, Corollary III.5.6].
— C. For each finitely generated unitary ideal $\mathfrak{A}$ of Int(E, D), the ideals of values are locally constant on E: there exists a nonzero ideal $\mathfrak{b}$ of D such that $(a - b) \in \mathfrak{b} \cap E$ implies $\mathfrak{A}(a) = \mathfrak{A}(b)$. It follows that $f(a) \in \mathfrak{A}(a)$ for each $a \in E$ implies that $f(x) \in \mathfrak{A}(x)\widehat{D}$ for each $x \in \widehat{E}$.

Assertion A: continuous functions. We first state and prove a very general result: the super Skolem property for the ring of continuous functions holds under the assumption that $\widehat{E}$ is compact.

Proposition 4.2. *Let X be a topological space and R be a one-dimensional local Noetherian domain, with maximal ideal* m. *Assume that X is compact and totally disconnected. Then $\mathcal{C}(X, R)$, the ring of continuous functions from X to R (when R is endowed with the* m-*adic topology), satisfies the almost super Skolem property.*

Proof. Let $\mathfrak{A}$ be a unitary ideal of $\mathcal{C}(X, R)$: there is a nonzero element a in $\mathfrak{A} \cap R$. Since R is a one-dimensional local Noetherian domain, there is an integer k such that $\mathfrak{m}^k \subseteq aR$. Consider $\varphi \in \mathfrak{A}^*$, we wish to prove that $\varphi \in \mathfrak{A}$. By hypothesis, for each $x \in X$, there is a function $\psi_x \in \mathfrak{A}$ such that $\varphi(x) = \psi_x(x)$. By continuity of the functions φ and ψ_x, there is a neighborhood U_x of x such that, for each $y \in U_x$, $\big(\varphi(y) - \psi_x(y)\big) \in \mathfrak{m}^k R$, and thus $\big(\varphi(y) - \psi_x(y)\big) \in aR$. Since X is totally disconnected, we may choose each U_x to be a clopen subset of X. Since X is compact, it may be covered by finitely many such $U_1, \ldots, U_s$, with corresponding functions $\psi_1, \ldots, \psi_s$, such that, for each $y \in U_j$, $\big(\varphi(y) - \psi_j(y)\big) \in aR$. Since these sets are clopen sets, we may even choose them to be pairwise disjoint. Let η_j be the characteristic function of U_j, then η_j is continuous. Set $\psi = \sum_{i=1}^s \psi_i \eta_i$, then $\psi \in \mathfrak{A}$. For each $y \in X$, there is j such that $y \in U_j$, and hence,

$$\varphi(y) - \psi(y) = \varphi(y) - \sum_{i=1}^s \psi_i(y)\eta_i(y) = \varphi(y) - \psi_j(y).$$

Therefore $\big(\varphi(y) - \psi(y)\big) \in aR$, that is, $(1/a)\big(\psi(y) - \varphi(y)\big) \in R$. In other words, the function $\theta = (1/a)(\psi - \varphi)$ belongs to $\mathcal{C}(X, R)$. Finally, since $\psi \in \mathfrak{A}$ and $a \in \mathfrak{A}$, we have $\varphi = \psi - a\theta \in \mathfrak{A}$. $\square$

For instance, $R = \widehat{D}$ may be the completion of an analytically irreducible domain D and $X = \widehat{E}$ be the topological closure of a subset E of D. Assuming that $\widehat{E}$ is compact, we conclude that the ring $\mathcal{C}(\widehat{E}, \widehat{D})$ of continuous functions from $\widehat{E}$ to $\widehat{D}$ satisfies the almost super Skolem property.

Assertion B: Stone-Weierstrass.

Proposition 4.3. *Let D be a one-dimensional local Noetherian analytically irreducible domain (with maximal ideal* m), *and E be a subset of D such that the completion $\widehat{E}$ of E (in the* m-*adic topology) is compact. Then* $\mathrm{Int}(E, D)$ *is dense in* $\mathcal{C}(\widehat{E}, \widehat{D})$ *for the uniform convergence topology.*

Proof. We first consider the case where D is a rank-one discrete valuation domain. We denote by K the quotient field of D and by $\widehat{K}$ its completion (in the topology given by the corresponding valuation, which clearly coincides with the m-adic topology on D). The polynomials with coefficients in $\widehat{K}$ can be considered as (uniformly) continuous functions from $\widehat{E}$ to $\widehat{K}$, and it follows from a classical generalization of the Stone-Weierstrass theorem, that $\widehat{K}[X]$ is dense in $\mathcal{C}(\widehat{E}, \widehat{K})$ for the uniform convergence topology ($\widehat{E}$ is compact and $\widehat{K}[X]$ "separates the points" of $\widehat{E}$: for $a \neq b$ in $\widehat{E}$, the polynomial $f = \frac{X-a}{b-a}$ is such that $f(a) = 0$, and $f(b) = 1$) [15], [4, Exercise III.20]. Clearly $K[X]$ is dense in $\widehat{K}[X]$, and hence, in $\mathcal{C}(\widehat{E}, \widehat{K})$. Finally, $\mathcal{C}(\widehat{E}, \widehat{D})$ is an open set in $\mathcal{C}(\widehat{E}, \widehat{K})$. Hence the intersection of $K[X]$ with this open set, that is, $\mathrm{Int}(E, D) = K[X] \cap \mathcal{C}(\widehat{E}, \widehat{D})$, is dense in $\mathcal{C}(\widehat{E}, \widehat{D})$.

Now, we consider the general case: D is a one-dimensional local Noetherian analytically irreducible domain. The integral closure D' of D is a rank-one discrete valuation domain, with maximal ideal $\mathfrak{m}'$, moreover D' is a finitely generated D-module [21, (32.2)]. Therefore, the $\mathfrak{m}'$-adic topology on D' induces the $\mathfrak{m}$-adic topology on D; in particular, there is an integer k such that $\mathfrak{m}'^k \subseteq \mathfrak{m}$. As in [4, Proposition III.2.4], it is not difficult to see that $\mathrm{Int}(E, D)$ is dense in $\mathcal{C}(\widehat{E}, \widehat{D})$, provided that, for each h and each n, the characteristic function of $\mathfrak{m}^h \cap E$ can be approximated modulo $\mathfrak{m}^n$ by an integer-valued polynomial (or equivalently the characteristic function of $\widehat{\mathfrak{m}}^h \cap \widehat{E}$ can be approximated modulo $\widehat{\mathfrak{m}}^n$ by an integer-valued polynomial). Clearly $\mathfrak{m}^h \cap E$ is a clopen set of E in the $\mathfrak{m}$-adic topology, hence also in the $\mathfrak{m}'$-adic topology (and $\widehat{\mathfrak{m}}^h \cap \widehat{E}$ is a clopen set of $\widehat{E}$ in the $\widehat{\mathfrak{m}}'$-adic topology). From the special case above of a valuation domain, there is a polynomial $f \in \mathrm{Int}(E, D')$ such that

$$
f(z) \equiv \begin{cases} 1 \pmod{\mathfrak{m}'^{kn}}, & \text{if } z \in \mathfrak{m}^h \cap E, \\[2mm] 0 \pmod{\mathfrak{m}'^{kn}}, & \text{if } z \in E, z \notin \mathfrak{m}^h. \end{cases}
$$

Since $\mathfrak{m}'^{kn} \subseteq \mathfrak{m}^n \subset D$, one has $f(E) \subseteq D$, and hence f is an approximation in $\mathrm{Int}(E, D)$ of the characteristic function of $\mathfrak{m}^h \cap E$ modulo $\mathfrak{m}^n$. $\square$

Compactness: a necessary condition. We obtained positive results under the assumption that $\widehat{E}$ is compact. It follows from the next two lemmas that this condition is necessary. The first one is a classical topological argument, we give it for the sake of completeness.

Lemma 4.4. *Let E be a subset of a local domain R, with maximal ideal $\mathfrak{m}$. Suppose that $\bigcap_n \mathfrak{m}^n = (0)$ and denote by $\widehat{E}$ the topological closure of E in the completion of R in the $\mathfrak{m}$-adic topology. Then $\widehat{E}$ is compact if and only if E meets only finitely many cosets of R modulo $\mathfrak{m}^n$ for each n.*

Proof. That the condition is necessary is clear. Conversely, since R is a metric space, to show that $\widehat{E}$ is compact amounts to show that, from a sequence $\{x_n\}$ in $\widehat{E}$, we may extract a converging subsequence. Infinitely many terms of the sequence $\{x_n\}$, forming a subset X_1 of $\widehat{E}$, are in the same coset modulo $\mathfrak{m}$ (since E meets only finitely much such cosets). Then infinitely many terms of X_1, forming a subset X_2 of X_1, are in the same coset modulo $\mathfrak{m}^2$. And so on. We thus define a decreasing sequence $\{X_n\}$ of subsets, the elements of X_n being in the same coset modulo $\mathfrak{m}^n$. Let x_{k_n} be the first term of X_n. The subsequence $\{x_{k_n}\}$ of $\{x_n\}$ is a Cauchy sequence in $\widehat{E}$, and hence, it converges. $\square$

Lemma 4.5. *Let D be a one-dimensional local Noetherian analytically irreducible domain, with maximal ideal $\mathfrak{m}$, and E be a subset of D. If $\mathrm{Int}(E, D)$ satisfies the almost strong Skolem property, then E meets only finitely many cosets of D modulo $\mathfrak{m}^n$ for each n.*

Proof. Assuming that E meets infinitely many cosets of D modulo some $\mathfrak{m}^k$, we shall prove that $\mathrm{Int}(E, D)$ does not satisfy the almost strong Skolem property.
— The integral closure D' of D is a discrete valuation domain with maximal ideal $\mathfrak{m}'$ and there is an integer r such that $\mathfrak{m}'^r \subseteq \mathfrak{m}$. Hence, E meets infinitely many cosets of D' modulo $\mathfrak{m}'^{rk}$.
— If n is the greatest integer such that E meets only finitely many cosets of D' modulo $\mathfrak{m}'^n$, there is a coset $a + \mathfrak{m}'^n$ such that $F = E \cap (a + \mathfrak{m}'^n)$ meets infinitely many cosets of D' modulo $\mathfrak{m}'^{n+1}$. Denoting by t a generator of the ideal $\mathfrak{m}'$, we may

replace E and F by the subsets $E' = -a + (1/t^n)E$ and $F' = -a + (1/t^n)F$, which have the same Skolem properties [Proposition 2.15]. Since E contains F, it follows that E' contains F'. But now, F' is a subset of D' which meets infinitely many cosets of D' modulo $\mathfrak{m}'$. From [4, Proposition I.3.1], we then have the containments

$$\text{Int}(E', D) \subseteq \text{Int}(F', D) \subseteq \text{Int}(F', D') \subseteq D'[X].$$

— Consider the ideal $\mathfrak{A}$ of $\text{Int}(E, D)$ generated by the maximal ideal $\mathfrak{m}$ of D and X^2. It is easy to see that X belongs to the Skolem closure $\mathfrak{A}^*$ of $\mathfrak{A}$. By way of contradiction, assume that $\text{Int}(E, D)$ satisfies the almost strong Skolem property. Then $X \in \mathfrak{A}$, and a fortiori, $X \in (\mathfrak{m}', X^2)D'[X]$. We reach a contradiction. $\square$

Assertion C and conclusion. Lastly, we need to prove assertion C: the value ideals of a finitely generated unitary ideal of $\text{Int}(E, D)$ are locally constant on E. This is very similar to the case of $\text{Int}(D)$ [4, Lemma VII.1.9].

Lemma 4.6. *Let D be a domain, E be a subset of D, and $\mathfrak{A}$ be a finitely generated unitary ideal of $\text{Int}(E, D)$. Then there exists a nonzero ideal $\mathfrak{b}$ of D such that $(a - b) \in \mathfrak{b} \cap E$ implies $\mathfrak{A}(a) = \mathfrak{A}(b)$.*

Proof. Let $f_1, \ldots, f_r$ be a system of generators of $\mathfrak{A}$, and d be a nonzero common denominator of their coefficients. For each a, b in D, and for each i, $(a - b)$ divides $\big(df_i(a) - df_i(b)\big)$ [4, Lemma I.3.19]. In particular, if $\mathfrak{a} = \mathfrak{A} \cap D$, and $\mathfrak{b} = d\mathfrak{a}$, then $(a - b) \in \mathfrak{b}$ implies $\big(f_i(a) - f_i(b)\big) \in \mathfrak{a}$. On the other hand, if a and b are in E, then $\mathfrak{A}(a)$ and $\mathfrak{A}(b)$ contain $\mathfrak{a}$. Therefore $(a - b) \in \mathfrak{b} \cap E$ implies

$$\mathfrak{A}(a) = \big(\mathfrak{a}, f_1(a), \ldots, f_r(a)\big) = \big(\mathfrak{a}, f_1(b), \ldots, f_r(b)\big) = \mathfrak{A}(b).$$

$\square$

Putting together all the results of this section, we can finally characterize the fractional subsets E of D such that $\text{Int}(E, D)$ satisfies the almost strong Skolem property.

Theorem 4.7. *Let D be a one-dimensional local Noetherian analytically irreducible domain (with maximal ideal $\mathfrak{m}$), and E be a fractional subset of D. Then $\text{Int}(E, D)$ satisfies the almost strong Skolem property if and only if the completion $\widehat{E}$ of E (in the $\mathfrak{m}$-adic topology) is compact.*

Proof. From Lemma 4.4 and Lemma 4.5, we know that the compactness of $\widehat{E}$ is necessary. Let us prove that it is sufficient. Let $\mathfrak{A} = (f_1, \ldots, f_r)$ be a finitely generated unitary ideal of $\text{Int}(E, D)$. Then $\mathfrak{A}$ contains a nonzero constant $a \in D$, and there is an integer k such that $\mathfrak{m}^k \subseteq aD$. Consider $f \in \mathfrak{A}^*$: for each $a \in D$, $f(a) \in \mathfrak{A}(a)$. From Lemma 4.6, it follows that, for each $x \in \widehat{E}$, $f(x) \in \mathfrak{A}(x)\widehat{D}$. From Proposition 4.2, we know that $\mathcal{C}(\widehat{E}, \widehat{D})$ satisfies the almost super Skolem property: f belongs to the ideal $\mathfrak{A}\mathcal{C}(\widehat{E}, \widehat{D})$ and we can write $f = \sum_{j=1}^{r} f_j \psi_j$, where each ψ_j belongs to $\mathcal{C}(\widehat{E}, \widehat{D})$. From the Stone-Weierstrass property [Proposition 4.3], each ψ_j can be approximated by an integer-valued polynomial $g_j \in \text{Int}(E, D)$, modulo $\mathfrak{m}^k \widehat{D}$: for each $x \in E$, $g_j(x) - \psi_j(x) \in \mathfrak{m}^k \widehat{D}$. Let $g = \sum_{j=1}^{r} f_j g_j$, then $g \in \mathfrak{A}$. The difference $(f - g)$ is a polynomial, and, for each $x \in E$, we have

$$(f - g)(x) = \sum_{j=1}^{r} f_j(x)(\psi_j(x) - g_j(x)) \in \mathfrak{m}^k \widehat{D}.$$

It follows that $(f - g)(x) \in aD$, that is, $(f - g) = ah$, where $h \in \mathrm{Int}(E, D)$. Finally, $f = g + ah$ belongs to the ideal $\mathfrak{A}$. $\square$

REFERENCES

[1] D. BRIZOLIS, *Hilbert rings of integral-valued polynomials*, Comm. Algebra **3** (1975), 1051–1081.

[2] _____, *Ideals in rings of integer-valued polynomials*, J. reine angew. Math. **285** (1976), 28–52.

[3] _____, *A theorem on ideals in Prüfer rings of integral-valued polynomials*, Comm. Algebra **7** (1979), 1065–1077.

[4] P.-J. CAHEN AND J.-L. CHABERT, *Integer-Valued Polynomials*, Amer. Math. Soc. Surveys and Monographs, **48**, Providence, 1997.

[5] P.-J. CAHEN, J.-L. CHABERT, E. HOUSTON AND T. G. LUCAS, *Skolem properties, value functions and divisorial ideals*. To appear *J. of Pure and Applied Algebra.*

[6] P.-J. CAHEN, J.-L. CHABERT, AND S. FRISCH, *Interpolation domains*, to appear.

[7] P.-J. CAHEN, F. GRAZZINI, AND Y. HAOUAT, *Intégrité du complété et théorème de Stone-Weierstrass*, Ann. Sci. Univ. Clermont, Ser. Math. **21** (1982), 47–58.

[8] J.-L. CHABERT, *Polynômes à valeurs entières et propriétés de Skolem*, J. reine angew. Math. **303/304** (1978), 366–378.

[9] _____, *Anneaux de Skolem*, Arch. Math. **32** (1979), 555–568.

[10] _____, *Le Nullstellensatz de Hilbert et les polynômes à valeurs entières*, Mh. Math. **95** (1983), 181–195.

[11] _____, *Idéaux de polynômes et idéaux de valeurs*, Manuscripta Math. **60** (1988), 277–298.

[12] J.-L. CHABERT, S. CHAPMAN AND W. SMITH, *The Skolem property in rings of integer-valued polynomials*, Proc. Amer. Math. Soc., to appear.

[13] J.-L. CHABERT AND A. LOPER, *Integer-valued Polynomials on a Subset and Prüfer domains*, to appear.

[14] H. GUNJI AND D.L. McQUILLAN, *On rings with a certain divisibility property*, Michigan Math. **22** (1975), 289–299.

[15] I. KAPLANSKY, *The Weierstrass theorem in fields with valuations*, Proc. Amer. Math. Soc. **1** (1950), 356–357.

[16] , _____, *On the coefficients and values of polynomials rings*, Arch. Math. **30** (1978), 8–13.

[17] D.L. McQUILLAN, *On ideals in Prüfer domains of polynomials*, Arch. Math. **45** (1985), 517–527.

[18] _____, *On Prüfer domains of polynomials*, J. reine angew. Math. **358** (1985), 162–178.

[19] _____, *Rings of integer-valued polynomials determined by finite sets*, Proc. Roy. Irish Acad. Sect. A **85** (1985), 177–184.

[20] _____, *On a theorem of R. Gilmer*, J. Number theory **39** (1991), 245–250.

[21] M. NAGATA, *Local Rings*, Interscience, New York, 1962.

[22] G. PÓLYA AND G. SZEGÖ, *Aufgaben und Lehrsätze aus der Analysis*, 1925; English transl. *Problems and Theorems in Analysis* I II, Springer-Verlag, Berlin, Heidelberg, New York, 1972.

[23] TH. SKOLEM, *Ein Satz über ganzwertige Polynome*, Norske Vid. Selsk. (Trondheim) **9** (1936), 111–113.

Multiplicative Groups of Fields

MARIA CONTESSA Dipartimento de Matematica el Applicazioni, University degli Studi de Palermo, Via Archirafi 34, Palermo, Italy, I-90123

JOE L. MOTT* Department of Mathematics, Florida State University, Tallahassee, Florida 32306-4510

WARREN NICHOLS† Department of Mathematics, Florida State University, Tallahassee, Florida 32306-4510

1. INTRODUCTION

In [10], L. Fuchs asked which abelian groups G can be the multiplicative group of a field. A necessary condition is that the torsion subgroup of G is locally cyclic [10, p. 296]. On the other hand, if the torsion subgroup of G is locally cyclic, P.M. Cohn [3] proved that G can be embedded as a subgroup of the multiplicative group of some field. Then Warren May [32] improved upon Cohn's Embedding Theorem by proving that there exists a field K and a group H such that the multiplicative group K^* of K is isomorphic to the direct product of G and H where H is either a free abelian group or the direct product of a free abelian group and a cyclic group of order two.

Still Fuch's problem, with this modification, remains open.

PROBLEM 1 Let G be an abelian group whose torsion group is locally cyclic. Determine necessary and sufficient conditions for G to be isomorphic to the multiplicative group of a field.

---

* This author thanks CNR (Italy) for support, and the Universities of Palermo, Catania, and Rome for hospitality during the preparation of this work.

† This author was partially supported by the National Science Foundation.

With the exception of R.M. Dicker's paper [7], most research on Fuchs' question have focused on one or both of the following problems:

PROBLEM 2 Starting with a class of fields, determine the structure of their multiplicative groups.

The multiplicative group structure is known for prime fields, algebraic number fields, algebraic function fields of one variable, algebraically closed fields, real closed fields, and local fields [11, p. 312–316] and [30, p. 92–136].

PROBLEM 3 Starting with an abelian group which satisfies certain properties, determine whether such a group can be (embedded in) the multiplicative group of some field.

For example, it is known that a torsion group T is isomorphic to the torsion part of K^* for some field K of characteristic 0 if and only if T is isomorphic to a subgroup of $\mathbb{Q}/Z$ with nontrivial 2-component [11, p. 313].

In [10, p. 299] and later in [11, p. 324] Fuchs suggested:

PROBLEM 4 Study the group K^*/k^* where K is an extension field of k.

Various aspects of the quotient group K^*/k^* have been studied by several authors [2,3,11,28,31,32,33,34,42]. Let us highlight some of their results.

In 1965, Brandis [2] showed that K^*/k^* is not finitely generated if $K \neq k$ and k is infinite. May's three seminal papers [32,33,34] show conditions when K^* is free modulo its torsion subgroup. Moreover, May shows in Theorems 1 and 2 of [33] that the problem of relating the group structure of K^* to that of k^* can be quite complicated even when K is a quadratic extension of k and k^* has the particularly simple structure of the direct product of Z_2 and a free abelian group A.

For a more detailed discussion of the papers by Brandis [2] and May [32,33,34] see [29, p. 439–533] or [30, p. 92–178].

Several authors have studied the torsion group $T(K^*/k^*)$ of all elements of finite order in K^*/k^* where K is a finite extension of k. For example, Gay and Velez [12] determine the structure of $T(K^*/k^*)$ in the case where $K = k(\alpha)$ if α is a root of an irreducible binomial $X^m - a \in k[X]$ such that the characteristic of k does not divide m. Then de Orozco and Velez [6] treated the case where K is an abelian extension of k. In [14] Greither and Harrison established conditions under which there is a one-to-one correspondence between subgroups of $T(K^*/k^*)$ and intermediate fields between K and k.

In this paper we obtain results pertinent to all four of the above problems. In Section 2, we study the torsion-free rank $r_0(K^*) = r_0(K^*/T(K^*))$ and make the simple observation that for any field K with prime subfield k_0, $r_0(K^*) = 0$ if char $K = q > 0$ and K is an algebraic extension of k_0; otherwise $r_0(K^*) =$ the maximum of $\aleph_0$ and the transcendence degree of K over k_0. In particular, $r_0(K^*)$ is either zero or infinity. Thus, an abelian group G for which $1 \leq r_0(G) < \infty$ cannot be isomorphic to the multiplicative group of a field.

In Section 3 we analyze $r_0(K^*/k^*)$ where K is an extension field of k. Familiar facts imply that if char $K = p > 0$ and K is either algebraic over a finite field or a purely inseparable field extension of k, then K^*/k^* is a torsion group. Moreover, in 1951, Kaplansky [28] showed these are the only instances for which a torsion quotient group K^*/k^* occurs. (Also see [30, p. 182].) Nagata [37] proved this in a slightly stronger form in 1968. Therefore, the case where the torsion-free rank of K^*/k^* is zero has been completely determined.

In all other cases, the torsion-free rank of K^*/k^* is infinite. The proof of this fact has an interesting history. Davis and Maroscia [4] first published the theorem in 1984. But in their paper they mention that L. Avramov also had proved the theorem. Apparantly, Avramov's proof was never published. In Section 3 we present a new proof less technical than the Davis/Maroscia proof.

We discovered our proof in 1990 independent of Davis/Maroscia, but after the second author presented it at an American Mathematical Society Meeting at the University of North Texas in November 1990, Roger Wiegand informed us of the existence of their paper. From Karpilovsky we learned about the existence of a preprint by J.D. Reid, which contained a proof. Recently Reid confirmed that he presented the theorem at an American Mathematical Society Meeting in Denver in January 1965 [40]. Although his paper was accepted with modifications for publication in the *Transactions*, Reid never resubmitted the modified manuscript. We do not know when Avramov discovered his proof, but at this point in time it appears that Reid has the priority of discovery.

The theorem of Reid, Avramov, Davis and Maroscia has been used in the study of $K^*/E_1^* E_2^* \cdots E_r^*$ where for each $i = 1, \ldots, r$, E_i is a proper intermediate field of the finite separable field extension K over k. This quotient group has been studied most notably by R. Wiegand and his students and colleagues [5,15,16,17,18,19,21,27,44]. If $r \leq 2$ then the n-torsion subgroup of the quotient group is finite for all integers $n > 0$, and its torsion-free rank is infinite (unless k is an algebraic extension of a finite field). For $r > 2$ examples are given in [5] to show that the group $K^*/E_1^* E_2^* \cdots E_r^*$ may be trivial, finite and nontrivial, infinite torsion, or of infinite torsion-free rank.

In Section 4, we study the multiplicative group structure of E^* where $K \subseteq E$ and E is a subfield of an algebraic closure $\bar{K}$ of K maximal with respect to the property that $T(E^*) = T(K^*)$. We show that E^* is p-divisible for each prime p such that $\zeta_p \notin K$ where ζ_p is a primitive p-th root of unity.

In Section 5, we apply the results of Section 4 to the special case where $T(K^*) = \{1\}$. In particular, we prove the following theorem:

THEOREM 5.5 Suppose that G is any torsion-free divisible group of infinite rank. Then G is the multiplicative group of some field.

We obtain a similar result in Section 6:

THEOREM 6.2 Let G be any torsion-free divisible group of infinite rank and let p be any prime integer. Then there is a field F of characteristic p such that F^* is isomorphic to the direct product $(Z_p)^* \times G$.

2. THE TORSION-FREE RANK OF K^*

In this section we investigate the torsion-free rank of K^* where K is a field; first we recall the definition.

DEFINITION If G is an abelian group, the cardinality of a maximal Z-linearly independent system is the torsion-free rank $r_0(G)$ of G. Alternatively, $r_0(G) = \dim_Q(G \otimes_Z Q)$ where $G \otimes_Z Q$ is considered as a vector space over Q.

We will need some basic facts about cardinality; good references to these facts are [30, p. 94-95, 106-107] and [13]. We especially need the following:

1. Suppose that D is a UFD with quotient field K and that $\{X_\lambda\}_{\lambda \in \Lambda}$ is a nonempty set of indeterminates over D. Let $\mathcal{I} = \{p_\alpha\}_{\alpha \in A}$ be a complete set of nonassociate irreducible elements of the domain $R = D[\{X_\lambda\}]$. Then $|\mathcal{I}| = |R|$.

2. Let F be an algebraically closed field, then
$$r_0(F^*) = \begin{cases} 0 & \text{if char } F = q > 0 \text{ and } F \text{ is algebraic over its prime subfield } F_0 \\ |F| & \text{otherwise.} \end{cases}$$

The first of these facts is listed in [13] and the second follows from [30, p. 106-107].

In general, if R is an integral domain with quotient field F, then we have the exact sequence of abelian groups:
$$\{1\} \to U(R) \to F^* \xrightarrow{\sigma} \mathcal{P}(R) \to \{1\} \tag{2.1}$$
where $U(R)$ denotes the group of units of R, $\mathcal{P}(R)$ is the group of principal fractional ideals of the form xR where $x \in F^*$, and $\sigma(x) = xR$ for each $x \in F^*$. If the sequence (2.1) splits, then $F^* \simeq U(R) \times \mathcal{P}(R)$. Specifically, if R is such that $\mathcal{P}(R)$ is a free abelian group (e.g., R is UFD, Dedekind [30, p. 93] or Krull [1, p. 480]) or if $U(R)$ is divisible, then the sequence (2.1) splits.

Suppose, in particular, that $R = k[B]$, where k is a field, B is a transcendence base over k, and $|B| \geq 1$. Then R is a UFD and $\mathcal{P}(R)$ is a free abelian group with basis the set $\mathcal{I}$ of nonassociated prime elements of R. Also, the quotient field $F = k(B)$ of R is such that $F^* \simeq k^* \times \mathcal{P}(R)$.

Then we see that
$$\begin{aligned} r_0(k(B)^*) &= r_0(k^*) + r_0(\mathcal{P}(R)) = r_0(k^*) + |\mathcal{I}| \\ &= r_0(k^*) + |R| = r_0(k^*) + |F| \\ &= r_0(k^*) + |k||B|\aleph_0 = \max\{|k|, |B|, \aleph_0\} \end{aligned} \tag{2.2}$$

since $r_0(k^*) \le |k|$.

Note that, if k is any finite field, then $k(B)^*$ is the direct product of a finite cyclic group and a free abelian group. Finally, if $k = Z_2$, then $k(B)^*$ is a free abelian group and hence is torsion-free.

THEOREM 2.1   Suppose K is a field extension of the field k. Let B be a transcendence base for $K\!:\!k$. Suppose $|B| \ge 1$. Then $r_0(K^*) = r_0(k(B)^*) = \max\{|k|, |B|, \aleph_0\}$. In particular, if k is the prime subfield of K, then $r_0(K^*) = \max\{|B|, \aleph_0\}$.

Proof: Let $F = k(B)$. Then K is algebraic over F. Let $\bar{F}$ be an algebraic closure of F containing K. Thus we have

$$|F| = |\overline{F}| = r_0(\overline{F}^*) \ge r_0(K^*) \ge r_0(F^*). \tag{2.3}$$

But we observed in equation (2.2) that $r_0(F^*) = |F| = \max\{|k|, |B|, \aleph_0\}$ so all inequalities in (2.3) became equalities.   ∎

COROLLARY 2.2   Suppose K is a field. Let k_0 be the prime subfield of K. Then

$$r_0(K^*) = \begin{cases} 0 & \text{if char } K = q > 0 \text{ and } K\!:\!k_0 \text{ is algebraic} \\ \max\{\mathrm{tr.\,deg}(K\!:\!k_0), \aleph_0\} & \text{otherwise.} \end{cases}$$

Proof: If char $K = q > 0$ and $K\!:\!k_0$ is algebraic, then (2) above gives $r_0(K^*) = 0$. If $k_0 = Q$ and $\mathrm{tr.\,deg}(K\!:\!k_0) = 0$, then $r_0(K^*) = \aleph_0 = \max\{\mathrm{tr.\,deg}(K\!:\!k_0), \aleph_0\}$. Otherwise let B be a transcendence base for K over k_0 where $|B| \ge 1$. Let $F = k_0(B)$ and let $\overline{F}$ be an algebraic closure of F containing K. Then by Theorem 2.1 and equation (2.3) we have the result.   ∎

COROLLARY 2.3   Suppose K is a field such that $K^* \ne T(K^*)$. Let k_0 be the prime subfield of K. Then

$$r_0(K^*) = \max\{\mathrm{tr.\,deg}(K\!:\!k_0), \aleph_0\}.$$

Proof: If $\mathrm{tr.\,deg}(K\!:\!k_0) = 0$, then since $K^* \ne T(K^*)$, char $K = 0$ [30, p. 107]. Therefore $r_0(K^*) = r_0(k_0^*) = \aleph_0$. If $\mathrm{tr.\,deg}(K\!:\!k_0) \ge 1$, then apply Theorem 2.1.   ∎

COROLLARY 2.4   Suppose that G is an abelian group such that $1 \le r_0(G) < \infty$, then G cannot be isomorphic to the multiplicative group of a field.

Proof: If we suppose the contrary, we contradict Corollary 2.2.   ∎

The next corollary is really a special case of a theorem of Brandis [2] or one of Isbell [23] but it also follows from Theorem 2.1.

COROLLARY 2.5   If K is a field for which K^* is finitely generated, then K is finite.

Proof: Since K^* is finitely generated $r_0(K^*) = r_0(K^*/T(K^*))$ is finite. Hence by Corollary 2.4, $r_0(K^*) = 0$. Hence $K^* = T(K^*)$ and K^* finitely generated implies K^* is finite.   ■

3.   THE TORSION-FREE RANK OF K^*/k^* AND A PROOF OF A THEOREM OF DAVIS AND MAROSCIA

Let K be a field extension of k with $K \neq k$. Our aim is to analyze the torsion-free rank of the factor group K^*/k^*. One result, due to May [33] and [30, p. 141] is important to our investigation and we state it as Theorem 3.1. We use Theorem 3.1 in the proof of Theorem 3.4.

THEOREM 3.1   (May) Let $k \subset K$ be algebraic number fields. Then $K^*/k^* \simeq A \times B$, where A is a free abelian group of rank $\aleph_0$ and B is finite.

In our analysis of K^*/k^* we give a new proof of the following theorem.

THEOREM 3.2   (Reid, Avramov, Davis and Maroscia) Let $k \subset K$ be an extension of fields. Then the following statements are equivalent:

(a) $r_0(K^*/k^*) < \infty$

(b) $r_0(K^*/k^*) = 0$

(c) K has prime characteristic $p > 0$ and either K is algebraic over its prime subfield Z_p or K is purely inseparable over k.

As Davis and Maroscia observe, it is clear that (c)→(b) →(a), so the goal is to show that (a)→(c). We note that as we mentioned in the introduction the weaker assertion (b)→(c) was established in 1951 by Kaplansky [28], and proven by Nagata in a slightly stronger form in 1968 [37].

Davis and Maroscia's proof rests upon a theorem of Nagata, Nakayama, and Tuzuku [38] on the splitting of valuations, whose proof requires much more technical machinery than we employ. Besides that, we believe the construction used in that proof may fail, unless the element c in paragraph two on page 60 is chosen more carefully. (Make sure that $c\alpha$ lies in every prime ideal of $I \cap R(\alpha)$ which contains $(c\alpha) \cap Z$.) The referee of this paper informed us that Jia carefully proved the result of [38] in [27], and in fact gets the following sharpened form: Let k be a field that is not algebraic over a finite field, and let K/k be a finite separable extension. Then there is a set V of pairwise inequivalent rational valuations (whose value group is contained in $\mathbb{Q}$) such that (1) V has finite character, (2) $|V| = |K|$, and (3) each $v \in V$ splits completely in K.

THEOREM 3.3   Let x be an indeterminate over a field k, and E a proper subfield of $K = k(x)$ which contains k. Then either

(1) char $K = p > 0$ and $E = k(x^{p^i})$ for some $i \geq 0$, or

(2) $r_0(K^*/E^*) = \infty$.

Proof: We may assume $k \neq E$, else (2) applies. By Luroth's theorem we have $E = k(p/q)$ for some $p, q \in k[x] \backslash \{0\}$. We may assume $\gcd(p, q) = 1$. Since $k(p/q) = k(q/p)$, we may assume $\deg p \geq \deg q$. If $\deg p = \deg q$, then $p = aq + r$ where $a \in k$ and $\deg r < \deg q$. Then $p/q = a + r/q$, so $k(p/q) = k(r/q) = k(q/r)$. Note that $\gcd(r, q) = 1$. Thus (after a change of notation) we may assume that $E = k(p/q)$, where $\deg p > \deg q$ and $\gcd(p, q) = 1$. Since $E \neq k(x)$, we have $\deg p \geq 2$. We may further assume that p, q are monic.

We have $K^* = k^* \times G$, where $G = \{f/g : f, g \in k[x]$ are monic$\}$. Moreover, G is free on the monic irreducibles of $k[x]$.

Write $p/q = v$. Then $E^* = k(v)^* = k^* \times H$, where $H = \{f(v)/g(v) : f, g \in k[x]$ are monic$\}$.

Suppose $f \in k[x]$ is monic. Write $f = x^n + a_{n-1}x^{n-1} + \ldots + a_0$. Let $\widetilde{f} = q^n f(v) = p^n + a_{n-1}p^{n-1}q + \ldots + a_0 q^n$. Since p is monic and $\deg p > \deg q$, $\widetilde{f}$ is monic. Since q is also monic, we have $f(v) \in G$. Thus $H \subseteq G$, and we have $K^*/E^* \simeq G/H$.

Suppose the monics f, g of $k[x]$ are relatively prime. Then $af + bg = 1$ for some $a, b \in k[x]$. So $a(v)f(v) + b(v)g(v) = 1$. Multiplying by a large power of q, we find that there exist $c, d \in k[x]$ and $r \geq 1$ with $c\widetilde{f} + d\widetilde{g} = q^r$. Thus any irreducible factor of $\widetilde{f}$ and $\widetilde{g}$ will divide q. But $\gcd(\widetilde{f}, q) = \gcd(p^n, q) = 1$. Thus $\widetilde{f}, \widetilde{g}$ are relatively prime.

Observe that the assignment $f \mapsto \widetilde{f}$ is injective. For suppose $\widetilde{f} = \widetilde{g}$. Then $(\deg f)(\deg p) = \deg \widetilde{f} = \deg \widetilde{g} = (\deg g)(\deg p)$, so $\deg f = \deg g$. Then $f(v) = q^{-\deg f} \widetilde{f} = q^{-\deg g} \widetilde{g} = g(v)$, so $f = g$ (since v is transcendental over k).

Let $f \in k[x]$ be a monic irreducible polynomial. We say that a monic irreducible ω of $k[x]$ is *associated* to f if ω divides $\widetilde{f}$. From the above, ω can be associated to at most one f. Since $\gcd(\widetilde{f}, q) = 1$, any ω associated to f must satisfy $\gcd(\omega, q) = 1$. Conversely, given $\omega \in k[x]$, a monic irreducible with $\gcd(\omega, q) = 1$, let λ be a root of ω in $\overline{k}$, and let $f = \text{Irr}(p(\lambda)/q(\lambda), k, x)$. Since $\widetilde{f}(\lambda) = q(\lambda)^{\deg f} f(p(\lambda)/q(\lambda)) = 0$, ω is associated to f.

We show in Lemma 2 below that, unless we are in situation (1) of our desired theorem, there exists an infinite set S of monic irreducibles f having more than one associated monic irreducible ω. For each $f \in S$, select an associated monic irreducible ω_f. We claim that $\{\omega_f H : f \in S\}$ is $\mathbb{Z}$-linearly independent in G/H.

If not, we can find distinct $f_1, \ldots, f_r$ in S and nonzero integers $n_1, \ldots, n_r$ such that $\omega_{f_1}^{n_1} \cdots \omega_{f_r}^{n_r} \in H$. Thus, there exist distinct monic irreducibles $g_1, \ldots, g_s$ in $k[x]$ and integers $m_1, \ldots, m_s$ with $\omega_{f_1}^{n_1} \cdots \omega_{f_r}^{n_r} = g_1(v)^{m_1} \ldots g_s(v)^{m_s}$. So, $\omega_{f_1}^{n_1} \cdots \omega_{f_r}^{n_r} = (\widetilde{g}_1)^{m_1} \ldots (\widetilde{g}_s)^{m_s} q^{-t}$, where $t = m_1(\deg g_1) + \ldots + m_s(\deg g_s)$. We view this equality using the basis of G consisting of the monic irreducibles. Recall that we have established that $\widetilde{g}_1, \ldots, \widetilde{g}_s, q$ are pairwise relatively prime.

Now ω_{f_1} must occur in the product $(\widetilde{g}_1)^{m_1} \ldots (\widetilde{g}_s)^{m_s} q^{-t}$. Since ω_{f_1} is not a factor of q, it must be a factor of (say) $\widetilde{g}_1$, with $m_1 \neq 0$. Then $\widetilde{g}_1$ and $\widetilde{f}_1$ have a common factor, so $g_1 = f_1$. Let ω be a monic irreducible other than ω_{f_1} which is associated to f_1. Then ω occurs in $(\widetilde{g}_1)^{m_1} \ldots (\widetilde{g}_s)^{m_s} q^{-t}$ but not in $\omega_{f_1}^{n_1} \cdots \omega_{f_r}^{n_r}$, a contradiction.

Thus, once we have established the lemmas below, we will have proved our theorem.

LEMMA 1   Let k be a field and let f and g be monic polynomials in $k[x]$ where $\deg f > \deg g$, and $\deg f = d \geq 2$. Assume $\bar{k}$ is an algebraic closure of k. Then either

(1) char $k = p > 0$, d is a power of p, $g = 1$, and $f = x^d + a$ for some $a \in k$, or

(2) there are at most d elements α of $\bar{k}$ such that $f + \alpha g$ has only one root in $\bar{k}$.

Proof: If there are fewer than 2 elements α of $\bar{k}$ such that $f + \alpha g$ has only one root in $\bar{k}$, then we are in situation (2). So suppose we have distinct $\alpha_1, \alpha_2, \ldots$ in $\bar{k}$ with

$$f + \alpha_i g = (x + \beta_i)^d, \qquad \beta_i \in \bar{k}, \qquad i = 1, 2, \ldots$$

Then $(\alpha_i - \alpha_1)g = (x + \beta_i)^d - (x + \beta_1)^d = \binom{d}{1}(\beta_i - \beta_1)x^{d-1} + \binom{d}{2}(\beta_i^2 - \beta_1^2)x^{d-2} + \cdots$.

Let us first suppose char $k = 0$. Let c_{d-2} be the coefficient of x^{d-2} in g. Since g is monic, we have, for $i > 1$,

$$c_{d-2} = \frac{\binom{d}{2}(\beta_i^2 - \beta_1^2)}{\binom{d}{1}(\beta_i - \beta_1)} = \frac{d-1}{2}(\beta_i + \beta_1).$$

Thus $\beta_i = \frac{2c_{d-2}}{(d-1)} - \beta_1$, for all $i > 1$. Thus α_1, α_2 are all that we can have.

Now assume that char $k = p > 0$. First suppose that d is a power of p, say $d = p^s$. Then $f + \alpha_i g = x^{p^s} + \beta_i^{p^s}$, for each i, so $(\alpha_2 - \alpha_1)g \in \bar{k}$. Since g is monic, we must have $g = 1$. Then $f = x^{p^s} + (\beta_i^{p^s} - \alpha_i)$, and we are in situation (1).

If d is not a power of p, then $d = p^s t$, where $\gcd(p, t) = 1$ and $t > 1$. We have

$$(x + \beta_i)^d = (x + \beta_i)^{p^s t} = (x^{p^s} + \beta_i^{p^s})^t = x^{p^s t} + t\beta_i^{p^s} x^{p^s(t-1)} + \ldots + \beta_i^d,$$

so $(\alpha_i - \alpha_1)g = t(\beta_i^{p^s} - \beta_1^{p^s})x^{p^s(t-1)} + \ldots + (\beta_i^d - \beta_1^d)$. Now $\beta_i^{p^s} - \beta_1^{p^s} = (\beta_i - \beta_1)^{p^s} \neq 0$ for $i > 1$. Thus $\deg g = p^s(t-1) > 0$. Since g is monic, the constant term c_0 of g is $\frac{\beta_i^d - \beta_1^d}{t(\beta_i^{p^s} - \beta_1^{p^s})}$ for $i > 1$. Thus, for all i, β_i is a root of $(x^d - \beta_1^d) - t(x^{p^s} - \beta_1^{p^s})c_0$. So there are at most d β_i's, and thus at most d α_i's. ∎

LEMMA 2 Let k be a field. Let $p, q \in k[x]$ be monic, with $\deg p > \deg q$, $\deg p \geq 2$, and $\gcd(p, q) = 1$. Suppose that it is not the case that

$$\text{char } k > 0, \quad q = 1, \quad p(x) = x^{(\text{char } k)^i} + a \quad \text{for some } i \geq 1, \quad a \in k.$$

Then there are infinitely many monic irreducible polynomials $f \in k[x]$ such that $\widetilde{f} = q^{\deg f} f(p/q)$ has at least two distinct monic irreducible factors.

Proof: First suppose that k is infinite. For $a \in k$ with $q(a) \neq 0$, write $v(a) = p(a)/q(a)$. Then $f = x - v(a)$ is monic irreducible, and $\widetilde{f} = p(x) - v(a)q(x)$ has factor $x - a$. If $\widetilde{f}$ has no other irreducible factor, then $p(x) - v(a)q(x) = (x - a)^d$, where $d = \deg p$. By Lemma 1, this can occur for at most d values $v(a)$. Now a value α of $v(a)$ is taken on by a root $a \in k$ of the polynomial $p(x) - \alpha q(x)$, so at most d a's take on the same value. Thus we can find infinitely many $f = x - v(a)$'s with $\widetilde{f}$ not of the forbidden form.

Now suppose that k is finite. Then $k[x]$ contains monic irreducibles of each degree. Thus there are infinitely many monic irreducibles ω in $k[x]$ satisfying: $\gcd(\omega, q) = 1$, $\deg \omega = n > 1$, $n \equiv 1 \pmod{d}$.

Given such an ω, let λ be a root of ω in $\overline{k}$, let $\mu = p(\lambda)/q(\lambda)$. Thus associate to ω the monic irreducible polynomial $f(x)$ for μ over k. Write $m = \deg f = [k[\mu] : k]$. Since $k \subseteq k[\mu] \subseteq k[\lambda]$ and $[k[\lambda] : k] = n$, we have $m | n$.

Suppose that ω is the only irreducible factor of $\widetilde{f}$. Then $\widetilde{f} = \omega^r$ for some r. We have $md = \deg \widetilde{f} = \deg \omega^r = nr$. Since $n \equiv 1 \pmod{d}$, we have $d | r$. Thus $1 = \frac{n}{m} \frac{r}{d}$ shows $n = m$ and $r = d$.

Since k is perfect, we have $f = (x - \alpha_1) \cdots (x - \alpha_n)$, where $\alpha_1, \ldots, \alpha_n \in \overline{k}$ are distinct. Then $\widetilde{f} = (p - \alpha_1 q) \cdots (p - \alpha_n q)$.

For $i \neq j$ we have $\gcd(p - \alpha_i q, p - \alpha_j q) = \gcd(p - \alpha_i q, (p - \alpha_j q) - (p - \alpha_i q)) = \gcd(p - \alpha_i q, (\alpha_j - \alpha_i)q) = \gcd(p, q) = 1$. Thus $p - \alpha_i q, p - \alpha_j q$ have no roots in common.

Now $n > d$, so, by Lemma 1, $p - \alpha_i q$ has more than one root for some i. Thus $\widetilde{f}$ has more than n roots. But ω^r has exactly n roots, so the assumption that $\widetilde{f} = \omega^r$ has led to a contradiction. Thus $\widetilde{f}$ has more than one irreducible factor.

Since only finitely many ω's are associated to any one f, our result follows. ∎

We can now give our proof of the implication (a)→(c) in Theorem 3.2.

THEOREM 3.4 Let K be an extension field of k which is not purely inseparable and for which K is not an algebraic extension of a finite field. Then $r_0(K^*/k^*) = \infty$.

Proof: Let k_0 be the prime subfield of k. If $u \in K \backslash k$ the key idea we will employ is to consider the factor group $k_0(u)^*/(k \cap k_0(u))^* \simeq k_0(u)^* k^*/k^*$ as a subgroup of K^*/k^*.

Case 1. If $k_0 = Q$ and u is transcendental over Q, then Theorem 3.3 shows

$$r_0(k_0(u)^*/(k \cap k_0(u))^*) = \infty.$$

If u is algebraic over Q, then May's Theorem 3.1 applies to give the same conclusion.

Case 2. Now suppose $k_0 = Z_p$ where char $k = p > 0$. If there exists $u \in K \backslash k$ such that u is transcendental over k, then u is transcendental over Z_p. Moreover, $k \cap Z_p(u) = Z_p$ since otherwise Luroth's Theorem implies u is algebraic over $k \cap Z_p(u)$ and therefore algebraic over k as well. Thus $r_0(k_0(u)^*/k \cap k_0(u)^*) = \infty$ by Theorem 3.3.

Thus, we may assume K is an algebraic extension of k. By hypothesis K is not a purely inseparable extension of k so there is $\alpha \in K \backslash k$ where α is separable over k. Also since K is not an algebraic extension of a finite field, there is a $t \in k$ which is transcendental over Z_p.

Note that α is not purely inseparable over $Z_p(\alpha) \cap k$. Thus, if α is transcendental over Z_p, Theorem 3.3 implies $r_0(Z_p(\alpha)^*/(k \cap Z_p(\alpha))^*) = \infty$.

Therefore, we assume α is algebraic over Z_p. Then $u = t + \alpha$ is transcendental over Z_p. Moreover, u is not purely inseparable over k, else α would be. So u is not purely inseparable over $Z_p(u) \cap k$. Theorem 3.3 implies $r_0(Z_p(u)^*/(k \cap Z_p(u))^*) = \infty$. ∎

REMARK   One conclusion from Section 2 is that $r_0(K^*)$ is either 0 or ∞ if K is a field. The conclusion also follows from the results of Section 3. If char $K = 0$, then $r_0(K^*) \geq r_0(Q^*) = \infty$; on the other hand, if char $K = p$, $r_0(K^*) = r_0(K^*/Z_p^*)$.

In his preprint and abstract J.D. Reid [40] also lists several other significant results as corollaries of his theorem. One is a special case of the Theorem 3.1 by Warren May. Another is a proof of a strengthened version of what now is known as Brandis's Theorem. A third result generalizes Kaplansky's Theorem. We include proofs for the last two observations; the proof of Theorem 3.6 is essentially Reid's.

THEOREM 3.5   (Brandis)   If K is a proper extension field of k such that K is infinite, then K^*/k^* is not finitely generated.

Proof: First note that if K^*/k^* is finitely generated then $r_0(K^*/k^*) < \infty$ and Theorem 3.2 implies that K^*/k^* is torsion and therefore finite. The theorem follows if we show that if K is infinite and $K \neq k$, then K^*/k^* is infinite. This is clear if k is finite. Assume k is infinite. Pick $c \in K \backslash k$. If s and t are distinct elements of k, then the cosets $k^*(c + s)$ and $k^*(c + t)$ are distinct. If not, then $c + s = u(c + t)$ for some $u \in k^*$ where $u \neq 1$. But then $c(1 - u) = ut - s$ or $c = \frac{ut - s}{1 - u} \in k$, a contradiction. ∎

In [28] Kaplansky used the equivalence of (b) and (c) of Theorem 3.2 and two theorems of Jacobson to prove that if D is a division ring with center k such that if D^*/k^* is torsion, then $D = k$ (and hence D is commutative). (See also

[30, p. 184].) The same ideas can be used to prove the following strengthening of Kaplansky's Theorem.

THEOREM 3.6 (Reid) If D is a division ring with center k with $D \neq k$, then D^*/k^* contains abelian subgroups of infinite torsion-free rank.

Proof: We prove the contrapositive. First observe that if D^*/k^* contains no subgroup of infinite torsion-free rank, then D is algebraic over k. For if D contains an element x transcendental over k, $k(x)^*/k^*$ has infinite torsion-free rank.

Next, since D algebraic over k, Jacobson's Proposition 2 in [25, p.180] or Lemma 5.16 in [30, p. 183] implies D contains an element $y \notin k$ separable over k. Let $K = k(y)$. Since k is not purely inseparable over k, Theorem 3.2 implies K and k are algebraic over the prime subfield Z_p. Therefore D is algebraic over Z_p. Jacobson's Theorem 2 of [25, p. 180] implies $D = k$ and proves the theorem. ∎

REMARKS We recall that Kaplansky's Theorem contains as special cases Wedderburn's Theorem that any finite division ring is commutative and Jacobson's generalization [25, p. 183] that asserts that any algebraic division algebra A over a finite field is commutative; for in such an algebra each non-zero element x has some power equal to 1 since the non-zero elements in the finite field $Z_p(x)$ is cyclic where Z_p is the prime subfield of A.

4.   MAXIMAL FIELDS DISJOINT FROM SETS OF ROOTS OF UNITY

One class of fields we want to investigate in this section are those fields K for which $T(K^*)$ is finite, that is, those fields which contain only finitely many roots of unity. Examples of such fields abound, for instance:

(1) $k(B)$ where k is a finite field and B is a transcendence base over k,

(2) any field finitely generated over its prime subfield [20], [37], and [41],

(3) any real closed field [30, p.107],

(4) any field with a discrete valuation with finite residue field (in particular, finite extensions of the rational field).

Risman [41] asserts that it is known that $T(K^*)$ is finite for fields K of type (4). Since a reference is lacking, we include a proof. We begin with the following result.

THEOREM 4.1   Let R be an integral domain and $I \neq R$ an ideal of R. Let $G = \{1 + x | x \in I$ and $(1 + x)^n = 1$ for some integer $n\}$. Then G is a group.

(i) If char $R = p > 0$, then $G = \{1\}$.

(ii) If char $R = 0$ and $\bigcap\limits_{n=1}^{\infty} I^n = (0)$, then G is finite.

Proof: First we assert: if G has an element of order $n > 1$, then $n \cdot 1 \in I$. This follows because if $x \in I$ is such that $1 + x \in G$ has order $n > 1$, then $x \neq 0$. From $(1 + x)^n = 1$ we get $x(n \cdot 1 + \binom{n}{2}x + \ldots + x^{n-1}) = 0$, and since $x \neq 0$, $n \cdot 1 = -(\binom{n}{2}x + \ldots + x^{n-1}) \in I$.

(i) Since char $R = p > 0$, the order of every element of G is relatively prime to p. If $\gcd(n, p) = 1$, then $n \cdot 1$ is a unit of R so that $n \cdot 1 \notin I$. Thus $G = \{1\}$.

(ii) We may assume, in this case, that $|G| > 1$. Note that $n \cdot 1 \in I$ and $m \cdot 1 \in I$ implies $d \cdot 1 \in I$ where $d = \gcd(m, n)$. Thus there exists a prime p such that the order of every element of G is a power of p. Since G is a subgroup of the multiplicative group of a field, G is infinite only if for each integer $r \geq 1$, G contains an element g_r of order p^r. Moreover, in this case, we conclude that $(g_{r+1})^p$ is a primitive p-th root of unity, so $g_r = (g_{r+1})^{pn_r}$ for some integer n_r. For each positive integer, write $g_r = 1 + x_r$ where $x_r \in I$. Then $(1 + x_r) = (1 + x_{r+1})^{pn_r}$ yields $x_r = x_{r+1}[p \cdot n_r \cdot 1 + \binom{pn_r}{2}x_{r+1} + \ldots + x_{r+1}^{pn_r-1}]$. Since G contains an element g_1 of order p, $p \cdot 1 \in I$. Thus, $x_r \in x_{r+1}I$ for all r. But then we see that $x_1 \in \bigcap\limits_{n=1}^{\infty} I^n = (0)$ so $g_1 = 1$ has order 1. This contradiction shows that G must be finite.   ∎

THEOREM 4.2 Let K be a field. If there exists a discrete valuation v on K with finite residue field, then $T(K^*)$ is finite.

Proof: Let R_v be the valuation ring of v where M_v is the maximal ideal of R_v. For any $x \in K^*$ either $x \in R_v$ or $x^{-1} \in R_v$. If $\omega \in T(K^*)$, then ω is a power of ω^{-1}. Thus, $\omega \in R_v$. Since $T(K^*) \subseteq R_v$, the ring homomorphism $\eta \colon R_v \to R_v/M_v$ induces a group homomorphism $f \colon T(K^*) \to (R_v/M_v)^*$. By hypothesis, Im f is finite. Moreover, $\ker f = G = \{1 + x \mid x \in M_v \text{ and } (1 + x)^n = 1 \text{ for some integer } n\}$ is finite by Theorem 4.1. Thus, $T(K^*)$ is finite.   ∎

REMARK Under the hypothesis of Theorem 4.2, the group G defined in Theorem 4.1 may not be the trivial group.

EXAMPLE 1 Let $K = Q$ and let $R_v = Z_S$ where $S = \{\text{odd integers}\}$. Then $G = \{\pm 1\}$.

EXAMPLE 2 Let $K = Q(i)$ and let $R_v = Z[i]_{(1+i)}$. Then $M_v = (1 + i)R_v$, $R_v/M_v = Z/2Z$, and $G = \{1, -1, i, -i\}$. Note $-i = 1 + (-1 - i) = 1 - (1 + i)$ and $i = 1 + (i - 1) = 1 + i(1 + i)$.

We have said that we are interested in fields K for which $|T(K^*)| < \infty$. Nevertheless, we want to formulate some results in a more general framework. We describe that context now. First let us fix the notation and terminology for the sequel. Let K be a field and assume $\overline{K}$ is a fixed algebraic closure of K. Let $S = \{$subfields L of $\overline{K} | K \subseteq L$ and $T(L^*) = T(K^*)\}$. Partial order S by inclusion, observe that S is inductive, and let E be a maximal element of S guaranteed by Zorn's Lemma. Thus E is maximal with respect to being disjoint from the set U of roots of unity in $\overline{K}\backslash K$. For any chain of fields $K \subseteq F \subseteq \overline{K}$ and a set S of $\overline{K}$, the fields M maximal with respect to the properties $F \subseteq M$ and M and S are disjoint have been studied by several authors [8,9,35,39,43].

We want to investigate the group structure of E^*. For this purpose, call the set $C(K) = \{$primes $p \,|\, $a primitive pth root of unity $\zeta_p \notin K\}$ the *co-spectrum* of K.

THEOREM 4.3 Let K be a field and let $p \in C(K)$. Let $a \in K\backslash K^p$ and let $F = K(\alpha)$ where $\alpha \in \overline{K}$ is such that $\alpha^p = a$. Then $T(F^*) = T(K^*)$.

Proof: We first observe that $X^p - a$ is irreducible in $K[X]$ by Abel's Theorem [29, p. 223] or [30, p. 146], so $[F:K] = p$.

Suppose that $T(F^*) \neq T(K^*)$. Then for some integer n, F contains a primitive nth root of unity ζ_n which is not in K.

First consider the case $p = \text{char } K$. Then $\gcd(p, n) = 1$ and ζ_n^p is also a primitive nth root of unity. But $F^p \subseteq K$ implies $\zeta_n^p \in K$ and then that $\zeta_n \in K$, a contradiction.

Now consider the case $p \neq \text{char } K$. Since $F = K(\zeta_n)$ is the splitting field of $X^n - 1$, F is a normal extension of K. Thus F contains a primitive pth root of unity ζ_p. Since $\zeta_p \notin K$, $F = K[\zeta_p]$ and we conclude $[F:K] \leq p - 1$, a contradiction. ∎

THEOREM 4.4 If E is maximal in S, then E^* and $G = E^*/T(E^*)$ are p-divisible for each prime p in the co-spectrum $C(K)$. In particular, if $t = |T(K^*)| < \infty$, then E^* and $G = E^*/T(E^*)$ are p-divisible for each prime p that does not divide t.

Proof: Let p be a prime such that $\zeta_p \notin K$. Then $\zeta_p \notin E$. Suppose that E^* is not p-divisible. Then for some $\theta \in E^*$, the polynomial $f(X) = X^p - \theta$ has no root in E^*. Therefore $\theta \in E\backslash E^p$. Let $\alpha \in \overline{K}$ be a root of $f(X)$ and consider $F = E(\alpha)$. By Theorem 4.3, $T(E^*) = T(F^*)$, contradicting the maximality of E. Thus, E^* is p-divisible and so is G. ∎

The next examples show the limitations for extending the conclusions of Theorem 4.4.

EXAMPLE 3 Let $K = \overline{Q} \cap \mathbf{R}$, where $\mathbf{R}$ is the real field. While it is known that K is real closed and $\overline{Q} = K(i)$ [24, p. 276-278], we give a proof without appealing to the full machinery of real closed fields. Clearly, for each $\alpha \in K$ where $\alpha > 0$, $\sqrt{\alpha} \in K$. Moreover, if $\beta \in \overline{Q}\backslash K$, then the conjugate $\overline{\beta}$ of β is in $\overline{Q}$. Therefore, the polynomial $f(X) = X^2 - (\beta + \overline{\beta})X + \beta\overline{\beta} \in K[X]$ has β as a root. Hence $[K[\beta]:K] = 2$

and $K[\beta] = K[\sqrt{\gamma}]$ where $\gamma \in K$. Since $\sqrt{\gamma} \notin K$, $\gamma < 0$, $i = \frac{\sqrt{\gamma}}{\sqrt{-\gamma}} \in K[\beta]$, and since $i \notin K$, $K[\beta] = K[i]$. Since β is an arbitrary element of $\overline{Q} \backslash K$, it follows that $\overline{Q} = K[i]$.

Now $T(K^*) = \{\pm 1\}$ and $[\overline{Q}:K] = 2$ implies that K is the maximal subfield of $\overline{Q}$ containing K such that $T(E^*) = T(K^*)$. Thus, K^* is q-divisible for each odd prime q by Theorem 4.3. Clearly K^* is not 2-divisible since $i \notin K$. Moreover, we claim that $G = K^*/\{\pm 1\}$ is divisible. To see this we need only show that G is 2-divisible. But if $\overline{\alpha}$ is the coset $\{\pm \alpha\}$ in G, then $X^2 = \overline{\alpha}$ has a solution in G since one of $\pm \alpha$ is positive and $\sqrt{|\alpha|} \in K$. ∎

Example 3 shows that $G = E^*/T(E^*)$ *may* be divisible if E is maximal in the set $\mathcal{S}$ of all fields $F \subseteq \overline{K}$ where $T(F^*) = T(K^*)$. The next example shows that G need not be divisible for every field K.

EXAMPLE 4 Let $K = Q(\{\zeta_n | \gcd(3, n) = 1\})$. We claim that $\zeta_3 \notin K$. In general, ζ_{n_1} and $\zeta_{n_2} \in Q(\zeta_{n_1 n_2})$, so if $\zeta_3 \in Q(\zeta_{n_1}, \zeta_{n_2}, \ldots, \zeta_{n_r})$ where $\gcd(n_i, 3) = 1$ for all i, then $\zeta_3 \in Q(\zeta_n)$ where $n = n_1 n_2 \cdots n_r$ and $\gcd(n, 3) = 1$. This implies that $Q(\zeta_3) \cap Q(\zeta_n) = Q(\zeta_3)$, contradicting the well known fact that $Q(\zeta_r) \cap Q(\zeta_s) = Q(\zeta_d)$ where $d = \gcd(r, s)$.

Expand K to a subfield $M = M_{\zeta_3}$ of $\overline{Q}$ maximal without ζ_3. We assert that $T(K^*) = T(M^*)$. If we assume the contrary, then $\zeta_n \in M$ for some integer n where $3|n$. But then M must contain $(\zeta_n)^{n/3}$, a primitive third root of unity. This contradiction proves the assertion and M is a maximal subfield of $\overline{Q}$ such that $T(K^*) = T(M^*)$.

Finally we assert that $M^*/T(M^*)$ is not 2-divisible. Quigley's Theorem ([39] or [8,9]) shows that $[\overline{Q}:M] = \infty$ and $\overline{Q} = M(a^{\frac{1}{p}\infty})$ where $p = [M(\zeta_3):M] = 2$. Thus, $x^2 = a$ has no solutions in M. If $M^*/T(M^*)$ is 2-divisible, then $x^2 = ta$ must have a solution $b \in M^*$ for some $1 \neq t \in T(M^*)$. Since t is a primitive nth root of unity for some n where $\gcd(n, 3) = 1$, $t = (\zeta_{2n})^2$ where ζ_{2n} is a primitive $2n$th root of unity. But then $\zeta_{2n} \in K$ because $\gcd(2n, 3) = 1$. Therefore $(\zeta_{2n}^{-1} b)^2 = b^2/t = \frac{ta}{t} = a$ gives the contradiction that $x^2 = a$ has a solution in M^*. ∎

The above examples still have not addressed the following question: Must the group $G = E^*/T(E^*)$ be divisible for those fields K where $T(K^*)$ is finite? This, too, has a negative answer.

EXAMPLE 5 Let K be the splitting field of $f(X)$ over Q where $f(X) = 144X^4 + 60X^2 + 5$. Then

(i) $T(K^*) = \{\pm 1\}$.

(ii) If $K \subseteq E$ is maximal in $\overline{Q}$ with respect to the property that $T(E^*) = T(K^*)$, then E^* and $G = E^*/\{\pm 1\}$ are not 2-divisible. In particular, $x^2 = -3$ has no solutions in G.

Solving $f(X) = 0$ for X^2, we have $X^2 = \frac{-60 \pm \sqrt{720}}{288} = \frac{-5 \pm \sqrt{5}}{24}$. Thus, the roots of $f(X)$ are $\pm u$ and $\pm v$ where $u = i\sqrt{\frac{5+\sqrt{5}}{24}}$ and $v = i\sqrt{\frac{5-\sqrt{5}}{24}}$. The roots α, β, γ of the resolvent cubic are of the form $x_1 x_2 + x_3 x_4$ where x_1, x_2, x_3, x_4 are the roots of $f(X)$. Therefore, $Q(\alpha, \beta, \gamma) = Q(u^2 + v^2, uv)$. Now u^2 and v^2 are the roots of $144y^2 + 60y + 5 = 0$ so $u^2 + v^2 = \frac{-60}{144}$ and $u^2 v^2 = \frac{5}{144}$. Thus $uv = \frac{\sqrt{5}}{12}$ or $uv = -\frac{\sqrt{5}}{12}$. Consequently, $Q(\alpha, \beta, \gamma) = Q(\sqrt{5})$ and $[Q(\alpha, \beta\gamma): Q] = 2$. Thus, $\mathrm{Gal}(K:Q)$ is either the dihedral group D_4 of order 8 or the cyclic group C_4 of order 4 [22, p. 273]. Since $v = \pm\frac{\sqrt{5}}{12u} \in Q(u)$, $K = Q(u)$ is the splitting field of $f(x)$. Thus, $\mathrm{Gal}(K:Q) = C_4$ and K contains the unique intermediate field $Q(\sqrt{5})$.

We next show that $T(K^*) = \{\pm 1\}$. If not, $\zeta_n \in K$ for some $n \neq 2$. But then $Q(\zeta_n) = K$ and $\deg \zeta_n = \phi(n) = 4$ implies $n = 5$. Since $\zeta_5 = \frac{-1+\sqrt{5}}{4} + i\sqrt{\frac{5+\sqrt{5}}{8}}$, $\zeta_5 \in K$ implies $z = i\sqrt{\frac{5+\sqrt{5}}{8}} \in K$ and $\frac{z}{u} = \sqrt{3} \in K$. But this would produce another intermediate field.

Finally, if $x^2 = \overline{3}$ has a solution in $E^*/\{\pm 1\}$, then either $\sqrt{-3} \in E$ or $\sqrt{3} \in E$. But $\sqrt{-3} \in E$ implies $\zeta_3 \in E^*$ and $\sqrt{3} \in E$ implies $\zeta_5 = \frac{-1+\sqrt{5}}{4} + \sqrt{3}u \in E^*$. Since $T(K^*) = \pm 1$, neither case can occur. ∎

THEOREM 4.5 Let E be maximal in $\mathcal{S}$. Let $\mathcal{F} = \{$subfields $L|K \subseteq L \subseteq E$ and L^* is p-divisible for each prime $p \in \mathcal{C}(K)$, the co-spectrum of $K\}$. Let $N_E = \cap\{L|L \in \mathcal{F}\}$. Then N_E^* is p-divisible.

Proof: Clearly, N_E is a subfield of E containing K. Suppose p is a prime such that $\zeta_p \notin K$. Let $\theta \in N_E^*$. Then $X^p = \theta$ has a solution $y_1 \in E$ since E is p-divisible. Moreover, if y_2 is another solution in E, then y_1/y_2 is a pth root of unity in E. But since $\zeta_p \notin K$, we conclude $y_1 = y_2$. Thus, $X^p - \theta$ has exactly one root α in E. Therefore, this same root α is in each subfield $L \in \mathcal{F}$. Therefore, $\alpha \in N_E^*$ and we conclude N_E^* is p-divisible. ∎

Let us describe an alternate way to obtain N_E. First some definitions. For an intermediate field F between K and E, $p \in \mathcal{C}(K)$ and $\alpha \in F$, let $\alpha^{\frac{1}{p}}$ be the unique root in E of the polynomial $X^p - \alpha$. Let $F^{\frac{1}{p}} = \{\alpha^{\frac{1}{p}}|\alpha \in F\}$. Define $\mathcal{D}(F)$ to be the field obtained by adjoining to F the sets $F^{\frac{1}{p}}$ for each $p \in \mathcal{C}(K)$. Then define $\mathcal{D}^2(F) = \mathcal{D}(\mathcal{D}(F))$ and, in general, $\mathcal{D}^n(F) = \mathcal{D}(\mathcal{D}^{n-1}(F))$ for $n > 1$. Finally, define $\overline{\mathcal{D}}(F) = \bigcup_{n=1}^{\infty} \mathcal{D}^n(F)$.

THEOREM 4.6 For the field N_E defined in Theorem 4.5, we have $N_E = \overline{\mathcal{D}}(K)$.

Proof: Note that $T(\overline{\mathcal{D}}(K)^*) = T(K^*)$ since $\overline{\mathcal{D}}(K) \subseteq E$. Moreover for each $p \in \mathcal{C}(K)$, $\overline{\mathcal{D}}(K)$ is p-divisible. This follows because if $\theta \in \overline{\mathcal{D}}(K)$, then $\theta \in \mathcal{D}^n(K)$ for some integer n. Hence $\theta^{\frac{1}{p}} \in \mathcal{D}^{n+1}(K)$ and therefore in $\overline{\mathcal{D}}(K)$. Clearly $\overline{\mathcal{D}}(K) \subseteq N_E$ and since N_E is the smallest subfield of E containing K with p-divisible multiplicative group we conclude $N_E = \overline{\mathcal{D}}(K)$.

The following example shows that E and N_E can be different.

EXAMPLE 6 Let $f(X) \in Q[X]$ where $\deg f = 5$ and the Galois group of f over Q is the symmetric group S_5. Let α be a real root of f and let $K = Q(\alpha)$. If $f(X) = (X - \alpha)g(X)$ where $g(X) \in K[X]$, then the splitting field N for $f(X)$ over Q can be obtained as a subfield of a radical extension L of K since Cardano's formulas show how to construct the roots of $g(X)$.

Let $E = \overline{Q} \cap \mathbb{R}$. We know that E is a subfield of $\overline{Q}$ maximal with respect to $T(E^*) = \{\pm 1\}$. Clearly $\alpha \in E$ since α is a real number.

The co-spectrum $C(Q) = \{\text{primes } p \neq 2\}$. Thus, $N_E = \cap\{L | L \in \mathcal{F}\}$ is such that N_E^* is p-divisible for all odd primes p.

We assert that $N_E \neq E$. If $N_E = E$ then $\alpha \in N_E$ and since $N_E = \overline{\mathcal{D}}(Q)$ we see that $K \subseteq F$ where F is a radical extension of Q. But then the splitting field N is contained the radical extension LF of Q. By [22, p. 304] we obtain the contradiction that the Galois group of $f(X)$ over Q must be solvable.   ∎

If, in Theorems 4.4 and 4.5, we assume $|T(K^*)| = 1$, then the conclusions can be strengthened.

THEOREM 4.7 If K is a field such that K^* is torsion-free, then any maximal element E of S has a torsion-free divisible multiplicative group. Moreover, E is perfect and if N_E is the intersection of all subfields L of E such that $K \subseteq L$ and L^* is torsion-free and divisible, then N_E^* is also divisible.

Proof: That E^* is divisible is immediate from Theorem 4.4. Next to show E is perfect all we need show is that $E = E^2$ since it is obvious that char $E = 2$. Suppose $e \in E^*$ and $\alpha \in \bar{K}$ is such that $\alpha^2 = e$. Since E^* is divisible, there exists $e_1 \in E$ such that $e = e_1^2 = \alpha^2$. But then $\alpha^2 - e_1^2 = (\alpha + e_1)(\alpha - e_1) = 0$ and char $E = 2$ imply that $\alpha = e_1 \in E$.   ∎

5. FIELDS WITH TORSION-FREE MULTIPLICATIVE GROUPS

Let us examine more closely the case where $|T(K^*)| = 1$. We start with a theorem mentioned in an earlier paper of the second author [36] and used above in the proof of Theorem 4.7.

THEOREM 5.1 The multiplicative group of a field K is torsion-free if and only if char $K = 2$ and the prime subfield Z_2 is relatively algebraically closed in K.

Proof: First, since $(-1)^2 = 1$ and K^* is torsion-free we see that $-1 = 1$ and that char $K = 2$. If $\alpha \in K^*$ is algebraic over Z_2, then $Z_2(\alpha)$ is a finite field so α satisfies $X^{2^m} - X$ for some positive integer m. But then α has finite order and we conclude $\alpha = 1 \in Z_2$.

Conversely, if $\alpha \in K^*$ has finite order m, α satisfies $X^m - 1 \in Z_2[X]$. But since Z_2 is relatively algebraically closed in K, $\alpha = 1 \in Z_2$ so that K^* is torsion-free. ∎

The next theorem and its corollary are basically restatements of Corollary 2.3 and Corollary 2.4.

THEOREM 5.2   Suppose K is a field such that K^* is torsion-free and $K \neq Z_2$. Then K^* (considered as an abelian group and hence as a vector space over Q) has torsion-free rank equal to $\max\{\aleph_0, \text{tr.}\deg(K : Z_2)\}$.

COROLLARY 5.3   No torsion-free group of finite rank is the multiplicative group of a field. In particular, the additive group of rational numbers is not isomorphic to the multiplicative group of a field.

COROLLARY 5.4   A free abelian group G is isomorphic to the multiplicative group of a field if and only if G has infinite rank.

Proof: If G is free of infinite rank τ, let B be a transcendence base over the field Z_2 where $|B| = \tau$. Then the exact sequence (2.1) shows that $F = Z_2(B)$ is such that F^* is free of rank τ. ∎

THEOREM 5.5   Let G be any torsion-free divisible group of infinite rank. Then there is a field F such that F^* is isomorphic to G.

Proof: Let τ be the rank of G. Let $K = Z_2(B)$ where B is an algebraically independent set of cardinality τ. Let E be a maximal element in S guaranteed by Theorem 4.7. Then $r_0(E^*) = \max\{\aleph_0, \text{tr.}\deg(E : Z_2)\}$ by Theorem 5.2. But since E is an algebraic extension of K, tr. $\deg(E : Z_2) = \tau$. Therefore, $r_0(E^*) = \tau$ and since E^* is torsion-free and divisible with the same rank as G, $E^* \simeq G$. ∎

COROLLARY 5.6   The additive group of real numbers is isomorphic to the multiplicative group of some field.

6. A METHOD OF CONSTRUCTING EXAMPLES.

Next we show that the direct product of a cyclic group of order $p - 1$ (for a prime p) and a torsion-free divisible group is the multiplicative group of a field. We use a method suggested by Stephen Comer to the second author many years ago for the special case $p = 2$.

Fix a prime p. Let p_i denote the i-th prime integer and let $n_i = p^{p_i} - 1 = |GF(p^{p_i})^*|$ where $GF(q)$ denotes the Galois field with q elements. The integers $n_i/(p-1)$ are pairwise relatively prime since $\gcd(x^m - 1, x^n - 1) = x^d - 1$ where

$d = \gcd(m, n)$. Moreover, if C_{n_i} denotes the cyclic group of order n_i isomorphic to $GF(p^{p_i})^*$, then $(C_{n_i})^{p-1}$, the group of $(p-1)$th powers of the elements of C_{n_i}, has order $n_i/(p-1)$ for each i.

THEOREM 6.1 If $\mathcal{F}$ is a nonprincipal ultrafilter on $\omega = \{0, 1, 2, 3, \ldots\}$, then the following conditions hold:

(i) The ultraproduct $K = \prod_{i \in \omega} GF(p^{p_i})/\mathcal{F}$ is a perfect field of characteristic p such that K^* is isomorphic to the ultraproduct $\prod_{i \in \omega} C_{n_i}/\mathcal{F}$.

(ii) $(K^*)^{p-1}$ is isomorphic to the ultraproduct $G = \prod_{i \in \omega}(C_{n_i})^{p-1}/\mathcal{F}$.

(iii) Moreover, G is a torsion-free divisible group.

(iv) $K^*/(K^*)^{p-1}$ is a cyclic group of order $p - 1$.

(v) $K^* \simeq (K^*)^{p-1} \times C_{p-1}$ so that $K^*/T(K^*)$ is a torsion-free divisible group and $T(K^*) = Z_p^*$.

Proof: Parts (i), (ii), and (iv) are standard consequences of Los' Principle [26, p. 2], which states that if $(R_\alpha)_{\alpha \in I}$ is a family of rings (resp. fields, modules, ...), and $\mathcal{F}$ is an ultrafilter on I, then a first-order sentence σ in the language of rings (resp. fields, modules, ...) holds for the ultraproduct $\prod_{\alpha \in I} R_\alpha/\mathcal{F}$ if and only if σ holds in R_α for almost all $\alpha \in I$. Conditions (iii) follows easily from the definition of ultraproduct. The crucial piece of information is that the elements $n_i/(p-1)$ are pairwise relatively prime. Since G is divisible, K^* is isomorphic to the product of G and $K^*/(K^*)^{p-1}$. So (v) also is easy.   ∎

THEOREM 6.2 Let G be a torsion-free divisible group of infinite rank and let p be an arbitrary prime integer. Then there is a field F of characteristic p such that F^* is isomorphic to $Z_p^* \times G$.

Proof: Let $\tau = r_0(G)$. Let K be a field of characteristic p guaranteed by Theorem 6.1. Apply Lowenheim-Skolem to get a field F of cardinality τ elementarily equivalent to K. Since $T(K^*) \cong Z_p^*$ is cyclic of order $p - 1$, K satisfies each γ_n, $n \in N$ where for $n \in N$, γ_n is the statement: "if $x^n = 1$, then $x^{p-1} = 1$." Also since $K^*/T(K^*)$ is divisible, K satisfies each δ_n, $n \in N$ where δ_n is the statement: "if $y \neq 0$, there exists x such that $x^n = y$ or $x^n = 2y$ or $x^n = 3y$ or ... or $x^n = (p-1)y$." Since F is elementarily equivalent to K, $T(F^*) \simeq Z_p^*$ and $F^*/T(F^*)$ is a divisible group. Since $|F| = \tau$, it follows that $|F^*/T(F^*)| = \tau$ and therefore $F^*/T(F^*)$ must have torsion-free rank τ. Thus, $F^*/T(E^*) \simeq G$.   ∎

References

1. N. Bourbaki, Elements of Mathematics: Commutative Algebra. Reading: Addison-Wesley, 1972.

2. R. Brandis, Über die multiplikativ struktur von korpererweiterungen. Math. Zeit. 87(1965), 71-73.

3. P.M. Cohn, Eine bemerkung über die multiplikativ gruppe eines köpern. Arch. Math. 13(1962), 344-348.

4. E. D. Davis and P. Maroscia, Affine curves on which every point is a set-theoretic complete intersection. J. Alg. 87(1984), 113-135.

5. J.L. Colliot-Thélène, R. Guralnick and R. Wiegand, Multiplicative groups of fields modulo products of subfields. J. Pure Appl. Algebra 106(1996), 233–262.

6. M.A. de Orozco and W.Y. Velez, The torsion group of a field defined by radicals. J. Number Th. 19(1984), 283-294.

7. R.M. Dicker, A set of independent axioms for a field and a condition for a group to be the multiplicative group of a field. Proc. Lond. Math. Soc. 18(1968), 114-124.

8. A.J. Engler and T.M. Viswanathan, Digging holes in algebraic closures à la Artin, I. Math. Ann. 265(1983), 263-271.

9. A.J. Engler and T.M. Viswanathan, Digging holes in algebraic closures à la Artin, II. Cont. Math. 8(1982), 351-360.

10. L. Fuchs, Abelian Groups. London: Pergammon Press, 1960.

11. L. Fuchs, Infinite Abelian Groups, Vol. II. New York: Academic Press, 1973.

12. D. Gay and W.Y. Velez, The torsion group of a radical extension. Pacific J. Math. 92(1981), 317-327.

13. R.W. Gilmer, Generating sets for a field as a ring extension of a subfield. Rocky Mountain J. of Math. 2(1972), 111-118.

14. C. Greither and D.K. Harrison, A Galois correspondence for radical extensions of fields. J. Pure Appl. Algebra 43(1986), 257-270.

15. R. Guralnick, D. Jaffe, W. Raskind and R. Wiegand, On the Picard group: torsion and the kernel induced by a faithfully flat map. J. Algebra 183(1996), 820–435.

16. R. Guralnick and R. Wiegand, Galois groups and the multiplicative structure of field extensions. Trans. Amer. Math. Soc. 331(1992), 563–584.

17. R. Guralnick and R. Wiegand, Picard groups, cancellation, and the multiplicative structure of fields. In: D. Anderson and D. Dobbs, eds. Zero-Dimensional Commutative Rings. New York: Marcel Dekker, 1995, pp. 65–79.

18. R. Guralnick and R. Wiegand, The Picard group of a certain pull-back domain: a non-commutative approach. In: P. Cahen, M. Fontana and S. Kabhaj, eds. Commutative Ring Theory. New York: Marcel Dekker, 1997, pp. 339–347.

19. W. Haboush, Multiplicative groups of Galois extensions. J. Algebra 165(1994), 122–137.

20. I.N. Herstein, C. Procesi, and M. Schacher, Algebraic valued functions on non-commutative rings. J. Algebra 36(1975), 128–150.

21. D. Holley and R. Wiegand. Torsion in quotients of the multiplicative group of a number field. In: R. Göbel, P. Hill and W. Liebert, eds. Abelian Group Theory and Related Topics. Obewolfach, 1993. Reprinted in Contemp. Math. 171(1994), 201-204.

22. T.W. Hungerford, Algebra. New York: Springer-Verlag, 1989.

23. J.R. Isbell, On the multiplicative semigroup of a commutative ring. Proc. Amer. Math. Soc. 10(1959), 908-909.

24. N. Jacobson, Lectures in Abstract Algebra, Vol. III. Princeton: D. Van Nostrand, 1964.

25. N. Jacobson, Structure of rings. Amer. Math. Soc., Providence, 1964.

26. C.U. Jensen and H. Lenzing. Model Theoretic Algebra. New York: Gordon and Breach Science Publishers, 1989.

27. B.-P. Jia, Splitting of rank one valuations. Comm. Algebra 19(1991), 777-794.

28. I. Kaplansky, A theorem on division rings. Canad. J. Math. 3(1951), 290-292.

29. G. Karpilovsky, Field Theory. New York: Mercel Dekker, 1988.

30. G. Karpilovsky, Unit Groups of Classical Rings. New York: Oxford University Press (Clarendon), 1988.

31. M. Kneser, Lineare Abhangigkeit von Wurzeln. Acta Arith. 26(1975), 307–308.

32. W. May, Multiplicative groups of fields. Proc. Lond. Math. Soc. 24(1972), 295-306.

33. W. May, Multiplicative groups under field extension. Can. J. Math. 31(1979), 436-440.

34. W. May, Fields with free multiplicative groups modulo torsion. Rocky Mountain J. 10(1980), 599-604.

35. P.J. McCarthy, Maximal fields disjoint from certain sets. Proc. Amer. Math. Soc. 18(1967), 347-351.

36. J.L. Mott, Groups of divisibility: A unifying concept for integral domains and partially ordered groups, Lattice-Ordered Groups, Kluwer Acedemic Publishers, (1989), 80-104.

37. M. Nagata, A type of integral extension. J. Math. Soc. Japan 20(1968), 266-267.

38. M. Nagata, T. Nakayama, and T. Tuzuku, On an existence lemma in valuation theory. Nagoya Math. J. 6(1957), 59-61.

39. F. Quigley, Maximal subfields of an algebraically closed field not containing a given element. Proc. Amer. Math. Soc. 13(1962), 562-566.

40. J.D. Reid, On the multiplicative structure of field extensions. (preprint) MR 12(1965), #619-68.

41. L.J. Risman, On the order and degree of solutions to pure equations. Proc. Amer. Math. Soc. 55(1976), 261-266.

42. P. Samuel, About Euclidean rings. J. Algebra 19(1971), 282–301.

43. W.R. Scott, Maximal fields disjoint from finite sets. Proc. Amer. Math. Soc. 19(1968), 1366-1368.

44. R. Wiegand, Picard groups of singular affine curves over a perfect field. Math. Zeit. 200(1989), 301–311.

Factorization in Antimatter Rings

JIM COYKENDALL Department of Mathematics, North Dakota State University, Fargo, North Dakota, 58105-5075

DAVID E. DOBBS Department of Mathematics, University of Tennessee, Knoxville, Tennessee, 37996-1300

BERNADETTE MULLINS Department of Mathematics and Statistics, Youngstown State University, Youngstown, Ohio, 44555-3302

1 INTRODUCTION

This study of factorization in antimatter rings builds on two established lines of research. In [4], we investigated the integral domains that have no atoms, called antimatter domains. That paper made no attempt, however, to characterize "antimatter" rings with zero divisors. On the other hand, several authors, including D.D. Anderson and S. Valdes-Leon ([1], [2]), have made an extensive study of factorization in commutative rings with zero divisors. Their work, however, focused on atomic rings and unique factorization. By contrast, this paper investigates the antimatter property in the context of commutative rings with zero divisors.

As noted in [1], the familiar concept of "associates" in an integral domain has three analogous definitions in the setting of a commutative ring with zero divisors. The three definitions of "associate" lead to three definitions of "atom" and, hence, to three types of atomic and antimatter rings. Our investigation begins in the quasilocal case where the antimatter property can be characterized in terms of the monoid of principal ideals of R (Proposition 8). It is shown that, in contrast to the integral domain case, the group of divisibility does not fully capture the antimatter property (Example 6).

We give a number of examples of antimatter rings and show that the product of any nonempty collection of antimatter rings is again an antimatter ring (Proposition 10). We exhibit a decomposition of the ring of integers modulo n as $A \times \prod U_i$, where each U_i has a unique (up to very strong associates) very strong atom and A is either zero or an antimatter ring (Proposition 12). We show, nevertheless, that no analogous decomposition theorem exists for total quotient rings in general (Example 15).

We also generalize the concept of a fragmented domain that was introduced by the second-named author [5] to the case of commutative rings with zero divisors and give examples of fragmented rings. Returning to the antimatter context, we prove that any commutative ring can be embedded in an antimatter ring (Theorem 18). Finally, it is shown that the direct limit of an ascending chain of antimatter rings need not preserve the antimatter property (Example 19).

Throughout this paper, R is a commutative ring with identity, $U(R)$ is the group of units of R, and T is the total quotient ring of R. Any unexplained notation is standard, as in Kaplansky [6].

This work was completed while the third-named author was on leave at the University of Tennessee, Knoxville. Thanks are expressed to UTK for its hospitality and Youngstown State University for reassigned time.

2 RESULTS

DEFINITION 1   Let R be a commutative ring and let $a, b \in R$.

1. a and b are associates if $Ra = Rb$, or equivalently, if $a|b$ and $b|a$.

2. a and b are strong associates if $a = bu$ for some $u \in U(R)$.

3. a and b are very strong associates if either (i) $a = b = 0$ or (ii) $a \neq 0$, a and b are associates, and $r \in U(R)$ for all $r \in R$ such that $a = rb$.

Clearly, $(3) \Rightarrow (2) \Rightarrow (1)$. The three types of "associate" are distinct in general [1], but in certain instances, they do coincide, for instance, in case R is quasilocal or an integral domain. More generally, R is said to be presimplifiable if for every $a, b \in R, a = ab \Rightarrow a = 0$ or $b \in U(R)$. If R is presimplifiable, then all three types of "associate" are equivalent on R [1].

DEFINITION 2   Let R be a commutative ring and let $a \in R$ be a nonunit. Then a is an atom (resp., strong atom; resp., very strong atom) if $a = bc$ with $b, c \in R \Rightarrow a$ is associated (resp., strongly associated; resp., very strongly associated) with either b or c.

Clearly, very strong atom $\Rightarrow$ strong atom $\Rightarrow$ atom, with equivalence of these concepts in presimplifiable rings. Moreover, in contrast to the customary usage for integral domains, it is easily seen that $0 \in R$ is an atom if and only if R is an integral domain. Also, the three types of "atoms" lead to three types of atomicity and, eventually, to three types of antimatter rings.

DEFINITION 3   Let R be a commutative ring.

1. R is atomic (resp., strongly atomic; resp., very strongly atomic) if every nonzero nonunit element of R is a finite product of atoms (resp., strong atoms; resp., very strong atoms).

2. R is antimatter (resp., highly antimatter; resp., very highly antimatter) if R does not contain any very strong atoms (resp., strong atoms; resp., atoms) except possibly zero.

Of course, R very strongly atomic $\Rightarrow$ R strongly atomic $\Rightarrow$ R atomic. Similarly, R very highly antimatter $\Rightarrow$ R highly antimatter $\Rightarrow$ R antimatter. It is also clear that if R is presimplifiable, then the various definitions above in (1) (resp., (2)) agree.

The following result, due to Anderson and Valdes-Leon, shows that the familiar concept of an atom in an integral domain is closely related to the concept of a very strong atom in a ring with zero divisors. For this reason, we will focus our attention on very strong atoms and, as a result, on antimatter rings.

PROPOSITION 4   [1, Theorem 2.5] Let a be a nonzero nonunit in a commutative ring R. Then a is a very strong atom if and only if $a = bc$ (with $b, c \in R$) implies either b or c is a unit of R.

We begin our "antimatter" investigations in the quasilocal case. Recall that in this setting, the three types of "atom" coincide. As in [6], we say that R is a valuation ring if for any $a, b \in R$, either a divides b or b divides a. The next proposition is a generalization of a result for integral domains [4, Proposition 2.1] and has essentially the same proof.

PROPOSITION 5   Let R be a commutative valuation ring. Then either R is an antimatter ring or R contains (up to very strong associates) exactly one very strong atom.

For valuation domains, the antimatter property can be characterized in terms of the value group. More generally, let D be an integral domain with quotient field K and group of divisibility $G(D) = K^*/U(D)$, partially ordered by $aU(D) \leq bU(D)$ if and only if $b = ra$ for some $r \in R$. Then D is an antimatter domain if and only if $G(D)$ has no minimal positive elements [4, Proposition 2.3 (b)]. The next example shows that this result does not extend to a commutative ring R even if R is a valuation ring.

We recall that the group of divisibility of R is $G(R) = U(T)/U(R)$, where T is the total quotient ring of R; $G(R)$ is partially ordered as in the case of domains. The difficulty in using $G(R)$ is that the units of T are exactly the regular elements of R, and hence $G(R)$ gives no information about the zero divisors of R. Thus R may have a very strong atom that is a zero divisor and hence is not detected by $G(R)$.

EXAMPLE 6   The valuation ring $R = \mathbf{Z}_4$ is not an antimatter ring, since it has a very strong atom, namely 2. However, 2 does not correspond to a minimal positive element of $G(R)$. In fact, since $\mathbf{Z}_4$ is its own total quotient ring, $G(R)$ is trivial.

Although the group of divisibility does not serve to characterize "antimatter" in the spirit of [4], we have an alternative way to characterize the antimatter property for the quasilocal case in terms of the monoid of principal ideals of R. For this, we need the following lemma.

LEMMA 7   Let a be a nonzero nonunit of a ring R and let $P^0(R)$ be the monoid of principal ideals of R, partially ordered by reverse inclusion; that is, $Ra \leq Rb$ if and only if $Rb \subseteq Ra$. Let $P^0(R)^+ = \{Ra \in P^0(R) : a \text{ is a nonunit of } R\}$. Then:

1. If a is a very strong atom of R, then Ra is a minimal element of $P^0(R)^+$.

2. If Ra is a minimal element of $P^0(R)^+$, then a is an atom of R.

Proof: 1. Let a be a very strong atom of R and suppose $Rb \leq Ra$ for some nonunit b of R. Then $Ra \subseteq Rb$ and hence $a = bc$ for some $c \in R$. Since a is a very strong atom, c must be a unit of R. Thus $Ra = Rb$ and hence Ra is a minimal element of $P^0(R)^+$, as claimed.

2. Let Ra be a minimal element of $P^0(R)^+$. To show that a is an atom, suppose $a = bc$ for some nonzero $b, c \in R$. Then $Ra \subseteq Rb$. Without loss of generality, b is a nonunit of R. Then $Rb \leq Ra$ in $P^0(R)^+$ and hence $Rb = Ra$ since Ra is minimal. But then a and b are associates, as required. $\square$

PROPOSITION 8   Let R be a quasilocal (more generally, presimplifiable) commutative ring. Then R is an antimatter ring if and only if $P^0(R)$ has no minimal positive elements.

Proof: In a quasilocal ring, an atom a is also a very strong atom. Thus the lemma shows that, in the quasilocal case, a is a very strong atom if and only if Ra is a minimal positive element of $P^0(R)$. $\square$

EXAMPLE 9   Unfortunately, the characterization of the antimatter property does not extend to the general case. For instance, $R = \mathbf{Z}_6$ is an antimatter ring although $2\mathbf{Z}$ is a minimal positive element of $P^0(R)$.

We pause to give some examples of antimatter rings (other than the antimatter domains in [4]). The following proposition will be useful.

PROPOSITION 10   Let $\{R_\alpha\}_{\alpha \in \Lambda}$ be a nonempty family of commutative rings and let $R = \prod R_\alpha$. Then R is an antimatter ring if and only if each R_α is an antimatter ring.

Proof: The proof is an immediate consequence of the following result [1]: a nonzero element $a = (a_\alpha)_{\alpha \in \Lambda}$ is a very strong atom of R if and only if each a_α is a unit of R_α except for one $\alpha_0 \in \Lambda$ and a_{α_0} is a nonzero very strong atom of R_{α_0}. $\square$

EXAMPLE 11   There exists an antimatter ring with zero divisors of every Krull dimension. For a proof, recall that in [4], one has several constructions of antimatter domains D_n of Krull dimension n, for any n, $0 \leq n \leq \infty$. Then, by Proposition 10 and Example 9, $R_n = D_n \times \mathbf{Z}_6$ is an antimatter ring of Krull dimension n.

We next show that the ring $\mathbf{Z}_n$ of integers modulo n can be decomposed into a product $A \times \prod U_i$ where each U_i has a unique (up to very strong associates) very strong atom and A is either 0 or an antimatter ring. As $\mathbf{Z}_n$ is its own total quotient ring, it is natural to ask whether a similar decomposition theorem holds for total quotient rings in general. Example 15 will give an example to the contrary.

PROPOSITION 12   Let $n \geq 2$ be an integer. Then there exists a decomposition of the ring of integers modulo n as $\mathbf{Z}_n = A \times \prod U_i$, where each U_i has a unique (up to very strong associates) very strong atom and A is either 0 or an antimatter ring.

Proof: Let $n = p_1 \cdots p_s q_1^{k_1} \cdots q_t^{k_t}$ where all of the p_i and q_j are pairwise distinct prime integers and the exponents $k_i \geq 2$ for all i. We will show that the q_i are (up to very strong associates) the only very strong atoms of $\mathbf{Z}_n$. After establishing this claim, it will follow that $\mathbf{Z}_n \cong A \times \prod U_i = \mathbf{Z}_{p_1 \cdots p_s} \times \prod \mathbf{Z}_{q_i^{k_i}}$ where $A = \mathbf{Z}_{p_1 \cdots p_s}$ is either 0 or a product of fields and hence an antimatter ring (by Proposition 10), and each $U_i = \mathbf{Z}_{q_i^{k_i}}$ has a unique (up to very strong associates) very strong atom, q_i.

First we show that the q_j are very strong atoms of $\mathbf{Z}_n$. Without loss of generality, $j = 1$. Suppose $q_1 \equiv ab \pmod{n}$ where a and b are nonunits of $\mathbf{Z}_n$. Then, in the ring of integers, $ab = q_1 + nx$ for some $x \in \mathbf{Z}$. Now in $\mathbf{Z}$, we have that q_1 divides n, hence q_1 must also divide ab and hence, without loss of generality, q_1 divides a. Thus $q_1 a' = a$ and $q_1 n' = n$ for some $a', n' \in \mathbf{Z}$. Then $a'b = 1 + n'x$ showing that n' and b are relatively prime integers. Now recall that b is not a unit of $\mathbf{Z}_n$. Since $U(\mathbf{Z}_n) = \{u \in \mathbf{Z}_n : n \text{ and any coset representative of } u \text{ are relatively prime integers}\}$, some p_i or q_j must divide b in $\mathbf{Z}$. But any such prime must also divide $n' = \frac{n}{q_1} = p_1 \cdots p_s q_1^{k_1-1} q_2^{k_2} \cdots q_t^{k_t}$. This contradicts the fact that b and n' are relatively prime integers.

Next we show that no p_i is a very strong atom of $\mathbf{Z}_n$. Without loss of generality, $i = 1$. Since, in the ring of integers, $\gcd(p_1, p_2 \cdots p_s q_1^{k_1} \cdots q_t^{k_t}) = 1$, $\exists x, y \in \mathbf{Z}$ so that $p_1 x + p_2 \cdots p_s q_1^{k_1} \cdots q_t^{k_t} y = 1$. Thus $p_1 = p_1(p_1 x + p_2 \cdots p_s q_1^{k_1} \cdots q_t^{k_t} y) = p_1^2 x + ny$. Thus $p_1 \equiv p_1^2 x \equiv (p_1)(p_1 x) \pmod{n}$ where neither factor is a unit in $\mathbf{Z}_n$.

Finally, suppose that d is any nonzero nonunit in $\mathbf{Z}_n$. Then, in the ring of integers, some p_i or q_j must divide d. If p_i divides d, then d cannot be a very strong atom since p_i is not. If q_j divides d, then $q_j d' = d$ for some $d' \in \mathbf{Z}$ and hence $q_j d' \equiv d \pmod{n}$. Now if d' is a nonunit in $\mathbf{Z}_n$, then this factorization shows that d is not a very strong atom. On the other hand, if d' is a unit in $\mathbf{Z}_n$, then this factorization shows that d is a very strong associate of q_j. Thus the q_j are the only very strong atoms of $\mathbf{Z}_n$ (up to very strong associates), as claimed. $\square$

COROLLARY 13   Let $n = p_1^{k_1} p_2^{k_2} \cdots p_m^{k_m} \geq 2$ be an integer, where the p_i are pairwise distinct prime integers and $k_i \geq 1$. Then:

1. $\mathbf{Z}_n$ is an antimatter ring if and only if n is square-free.

2. $\mathbf{Z}_n$ is very strongly atomic if and only if $k_i \geq 2$ for all i.

Proof: 1. If $k_i = 1$ for all i, then the proof of Proposition 12 shows that $\mathbf{Z}_n$ has no very strong atoms. Conversely, if some $k_i \geq 2$, then the proof of Proposition 12 shows that p_i is a very strong atom of $\mathbf{Z}_n$ and hence $\mathbf{Z}_n$ is not an antimatter ring.

2. Suppose $k_i \geq 2$ for all i. To see that $\mathbf{Z}_n$ is very strongly atomic, consider a nonzero nonunit $a \in \mathbf{Z}_n$. Since a and n are not relatively prime integers, we can

write $a = p_1^{s_1} p_2^{s_2} \cdots p_m^{s_m} x$ for some s_i (where $s_i \geq 0$ and not all the s_i are zero) and some integer x which is relatively prime to n. Thus $a \equiv p_1^{s_1} p_2^{s_2} \cdots p_m^{s_m} x \pmod{n}$ where the p_i are very strong atoms by the proof of Proposition 12 and x is a unit of $\mathbf{Z}_n$. Thus a has been expressed as a product in the desired form.

Conversely suppose that some $k_i = 1$. Then by the proof of Proposition 12, p_i is not a very strong atom of $\mathbf{Z}_n$. Furthermore, p_i cannot be written as a finite product of the p_j with $i \neq j$. Thus R is not very strongly atomic. $\square$

REMARK 14   1. For the sake of concreteness and future applications, Proposition 12 and Corollary 13 were stated for $\mathbf{Z}_n$. However, these results can be extended to hold for any ring of the form R/I where R is a principal ideal domain and I is a nonzero proper ideal of R. The proofs carry over in the obvious way.

2. The decomposition $\mathbf{Z}_n = A \times \prod U_i$, described in Proposition 12, is unique in the following sense. Suppose that $\mathbf{Z}_n = B \times \prod V_i$, where each V_i has a unique (up to very strong associates) very strong atom and B is either 0 or an antimatter ring; then $A \cong B$ and, after reordering, $U_i \cong V_i$ for each i.

EXAMPLE 15   There exists a total quotient ring R that does not decompose as $A \times \prod U_i$, where each U_i has a unique (up to very strong associates) very strong atom and A is either 0 or an antimatter ring. To this end, let $R = F[[X,Y]]/(X^2, XY)$, where F is any field of characteristic 2. Since R is a homomorphic image of the local ring $F[[X,Y]]$, the units of R are the cosets represented by the units of $F[[X,Y]]$. Then each nonunit of R is annihilated by the coset represented by X, and so is a zero divisor of R. Hence R is its own total quotient ring.

Write $R = F \oplus Fx \oplus \sum Fy^n$, where x (resp., y) is the coset represented by X (resp., Y). Suppose there is a decomposition $R = A \times \prod U_i$, where each U_i has a unique (up to very strong associates) very strong atom and A is either 0 or an antimatter ring. This cannot be a trivial product, for R is nonzero, R is not antimatter (since it has essentially distinct very strong atoms x and y) and R is not of the type U_i. Hence, R contains nontrivial idempotents such as $(1, 0, \ldots, 0)$. However, the above description of R shows that $e = e^2 \in R$ forces $e = 0, 1$. This contradiction completes the proof.

We now recall that the concept of a fragmented domain was introduced by the second-named author in [5]. This definition is easily extended to the case of rings with zero divisors, as follows.

DEFINITION 16   Let R be a commutative ring.

1. A nonzero nonunit $r \in R$ is said to fragment in R if there exists a nonzero nonunit s of R such that $r \in \cap_{k=1}^{\infty} R s^k$.

2. R is said to be a fragmented ring if every nonzero nonunit of R fragments in R.

It is easily seen that fragmented rings are antimatter rings, and thus provide an additional source of examples. In [4], it was shown that if n is a positive integer or ∞, then there exists an antimatter domain of Krull dimension n which is not fragmented. Since the second-named author [5] has provided numerous examples of infinite-dimensional fragmented domains R, we can use Proposition 10 to construct infinite-dimensional fragmented rings by considering $R \times \mathbf{F}_2$. We now give a family of examples of zero-dimensional fragmented rings (other than fields).

PROPOSITION 17 Let $n \geq 2$ be an integer. Then $\mathbf{Z}_n$ is fragmented $\Leftrightarrow$ $\mathbf{Z}_n$ is antimatter $\Leftrightarrow$ n is square-free.

Proof: In view of Corollary 13 and the above remarks, it is enough to show that if n is square-free, then $\mathbf{Z}_n$ is fragmented. Let $n = p_1 p_2 \cdots p_s$ where the p_i are distinct prime integers. Since each nonzero nonunit of $\mathbf{Z}_n$ is represented by an integral multiple of some p_i, it is enough to show that each p_i fragments in $\mathbf{Z}_n$. Without loss of generality, $i = 1$. In the ring of integers, $\gcd(p_1^{k-1}, p_2 \cdots p_s) = 1$ for each $k \geq 1$. Thus $\exists x, y \in \mathbf{Z}$ such that $p_1^{k-1}x + p_2 \cdots p_s y = 1$. Then $p_1 = p_1(p_1^{k-1}x + p_2 \cdots p_s y) = p_1^k x + ny$. Thus $p_1 \equiv p_1^k x \pmod{n}$. Hence, $p_1 \in \cap R p_1^k$, showing that p_1 fragments in R. $\square$

As in the case of Proposition 12 and Corollary 13, one easily checks that Proposition 17 holds in the setting of rings of the form R/I where R is a principal ideal domain and I is a nonzero proper ideal of R.

It was shown by the second-named author [5] that if R is a fragmented domain and P is a prime ideal contained in the Jacobson radical of R, then R/P is also a fragmented domain. This result extends easily to R/I, in case R is a fragmented ring with zero divisors and I is any ideal contained in the Jacobson radical of R. It is unknown whether an arbitrary homomorphic image of a fragmented ring is fragmented. We conjecture, however, that this result is true in the semi-quasilocal case. If so, it will follow that, in the semi-quasilocal case, a fragmented ring must be either zero- or infinite-dimensional, because it was shown in [3] that a fragmented domain must be either a field or infinite-dimensional.

We next turn our attention to showing that every ring can be embedded in an antimatter ring. Although there is an analogous result for integral domains [4, Theorem 2.13], the proof given below is different.

THEOREM 18 Every commutative ring can be embedded in an antimatter ring.

Proof: Let R be a commutative ring. We recursively define an ascending chain of rings. Let Λ_1 be an index set with $\Lambda_1 = R$. Let $A_1 = R[\{X_\alpha : \alpha \in \Lambda_1\}]$. Thus, we have adjoined to R a set of algebraically independent indeterminates where the cardinality of the set of indeterminates is equal to the cardinality of R. Now let I_1 be the ideal of A_1 generated by the set of polynomials $\{X_\alpha^2 - \alpha : \alpha \in \Lambda_1\}$. Let $R_1 = A_1/I_1$.

We claim that R is embedded in R_1 under the mapping $r \mapsto r + I_1$. To establish the claim, suppose $r \in R$ is in the kernel of this mapping. Then $r \in I_1$, and hence $r = \sum (f_i)(X_{\alpha_i}^2 - \alpha_i)$ for some $f_i \in A_1$. Since only finitely many of the indeterminates $X_{\alpha_1}, \ldots, X_{\alpha_n}$ appear in the description of r, it is enough to show that that the mapping $R \to R[\{X_{\alpha_1}, \ldots, X_{\alpha_n}\}]/(\{X_{\alpha_1}^2 - \alpha_1, \ldots, X_{\alpha_n}^2 - \alpha_n\})$ is an injection. For this, we induct on n. For the induction basis, suppose that $a \in R$ is in the kernel of $R \to R[X_{\alpha_1}]/(X_{\alpha_1}^2 - \alpha_1)$. Then $a = (f)(X_{\alpha_1}^2 - \alpha_1)$ for some $f \in R[X_{\alpha_1}]$. If a is nonzero, then the above factorization of a shows that $\deg(a) \geq 2$, a contradiction. This establishes the induction basis. Similarly, if we let $T = R[\{X_{\alpha_1}, \ldots, X_{\alpha_{n-1}}\}]/(\{X_{\alpha_1}^2 - \alpha_1, \ldots, X_{\alpha_{n-1}}^2 - \alpha_{n-1}\})$, then the mapping $T \to T[X_{\alpha_n}]/(X_{\alpha_n}^2 - \alpha_n)$ is also an injection. Thus, the mapping $R \to R[\{X_{\alpha_1}, \ldots, X_{\alpha_n}\}]/(\{X_{\alpha_1}^2 - \alpha_1, \ldots, X_{\alpha_n}^2 - \alpha_n\})$ is an injection, as desired.

We remark that if a were a very strong atom of R, then a could not be a very strong atom of R_1. To see this, observe that $X_a^2 - a = 0$ in R_1. Thus $a = X_a X_a$ in R_1. If a is a nonunit of R_1, then X_a must also be a nonunit in R_1, and hence this is a nontrivial factorization of a in R_1. Thus R_1 does not contain any very strong atoms a that belong to R. Of course, R_1 may contain other very strong atoms that are not elements of R.

In general, if $n \geq 2$, we let Λ_n be an index set with $\Lambda_n = R_{n-1}$. Then we let $A_n = R_{n-1}[\{X_\alpha : \alpha \in \Lambda_n\}]$ and let I_n be the ideal of A_n defined by $I_n = (\{X_\alpha^2 - \alpha : \alpha \in \Lambda_{n-1}\})$. Then we set $R_n = A_n/I_n$. As above, view $R_{n-1} \subseteq R_n$. Thus the R_i form an ascending chain of rings and we let $S = \cup_{i=1}^\infty R_i$. Then $R \subseteq S$.

Now we claim that S is an antimatter ring. To see this, let b be any nonzero nonunit in S. Then b must lie in R_i for some i. But then $b = X_b X_b$ in R_{i+1}. Of course, this is also a factorization of b in S. Furthermore, X_b must be a nonunit in S since otherwise, b would also be a unit of S. Thus b is not a very strong atom of S since $b = X_b X_b$ is a nontrivial factorization of b in S. Since b was arbitrary, it follows that S is an antimatter ring. $\square$

For certain classes of rings R, one need not resort to the methods of Theorem 18 and [4, Theorem 2.13] to find an antimatter ring containing R. For instance, if R has Jacobson radical 0, then R embeds in the ring $\prod\{R/M : M \text{ a maximal ideal}$ of $R\}$; this product is an antimatter ring by Proposition 10. In view of the different approaches that have been mentioned, we ask whether one can find a "smallest" antimatter ring containing a given ring R. Given Corollary 13 (1) regarding $\mathbf{Z}_n$, a natural first step would be to characterize the zero-dimensional antimatter rings or, more generally, to determine which total quotient rings are antimatter rings.

Given the role of directed union in the proof of Theorem 18, it seems natural to ask if the union of an ascending chain of antimatter rings is necessarily itself an antimatter ring. We conclude this paper by showing that this is not the case in general.

EXAMPLE 19 There exists an ascending chain $R_1 \subseteq R_2 \subseteq \cdots$ of antimatter rings (in fact, integral domains) such that $R = \cup_{i=1}^\infty R_i$ is not an antimatter ring.

Motivated by the constructions in [4], we begin with

$$R_0 = \mathbf{F}_2[\{X_1^{\alpha_1}, X_2^{\alpha_2}, \ldots, \frac{Y}{X_1^{\beta_1}}, \frac{Y}{X_1^{\beta_1} X_2^{\beta_2}}, \ldots : \alpha_i, \beta_i \in \mathbf{Q}^+\}],$$

where the X_i and Y are algebraically independent indeterminates over $\mathbf{F}_2$. For each n, let $P_n = (\{X_n^\alpha : \alpha \in \mathbf{Q}^+\})$ be the ideal of R_0 generated by the powers of X_n with positive rational exponents.

We first show that each P_n is a prime ideal of R_0. Let $f, g \in R_0$ with $fg \in P_n$. Then every term of fg is divisible by some positive rational power of X_n. By considering the least positive rational exponent of X_n among the terms of fg, we see that fg is divisible by X_n^α for some $\alpha \in \mathbf{Q}^+$. Suppose $f, g \notin P_n$. Then some term of f is not divisible by any positive rational power of X_n. Write $f = f_1 + f_2$ where f_1 is divisible by some positive rational power of X_n and no term of f_2 is divisible by any positive rational power of X_n. Similarly, write $g = g_1 + g_2$. Then $fg = (f_1 + f_2)(g_1 + g_2) = (f_1 g_1 + f_1 g_2 + f_2 g_1) + f_2 g_2$. Since $fg \in P_n$ and

$f_1g_1 + f_1g_2 + f_2g_1 \in P_n$, it follows that $f_2g_2 \in P_n$. Thus, X_n^β divides f_2g_2 for some $\beta \in \mathbf{Q}^+$. However, no term of f_2 or g_2 is divisible by any positive rational power of X_n, a contradiction.

Next we let $S_1 = R_0 \setminus \bigcup_{n=1}^{\infty} P_n$ be the set complement of the union of these prime ideals. Then S_1 is a multiplicatively closed set of R_0. Let $R_1 = (R_0)_{S_1}$. We next show that every nonzero nonunit h of R_1 can be written as a monomial times a unit of R_1. By multipling h by a unit of R_1, we may view h as lying in some P_i. Let n be the largest integer such that $h \in P_n$. Then h is divisible by some positive rational power of X_n. Of course, only finitely many X_i appear in the description of h. Suppose that at least one of the terms of h is not divisible by Y. Then, for some $k_i \in \mathbf{Q}^+$, we may write $h = X_1^{k_1} \cdots X_n^{k_n} h'$ where, for each i, h' is not divisible by any positive rational power of X_i. Then h' is not in any of the P_i, and hence h' is a unit of R_1. On the other hand, suppose that all terms of h are divisible by Y. Then, for some $k_0 \in \mathbf{Z}$ and $k_i \in \mathbf{Q}^+$, we may write $h = Y^{k_0} X_1^{k_1} \cdots X_m^{k_m} h''$, where h'' is a unit of R_1.

Now we claim that R_1 is an antimatter ring. To establish the claim, suppose that h is a nonzero nonunit of R_1. By multiplying h by a unit of R_1, we may view h as lying in some P_n as above. Then $h = X_n^\gamma h_0$ for some $\gamma \in \mathbf{Q}^+$ and some $h_0 \in R_0$. Then $h = (X_n^{\gamma/2})(X_n^{\gamma/2} h_0)$ is a nontrivial factorization of h in R_1, and hence h is not a (very strong) atom of R_1. Since h was arbitrary, R_1 is an antimatter ring.

In general, let $S_{i+1} = R_0 \setminus \bigcup_{n=i+1}^{\infty} P_n$. Then $S_{i+1} \supseteq S_i$ and S_{i+1} is a multiplicatively closed set of R_i. Let $R_{i+1} = (R_i)_{S_{i+1}}$. Then, by repeating the above argument, R_{i+1} is an antimatter ring, and so $R_1 \subseteq R_2 \subseteq \ldots$ forms an ascending chain of antimatter rings.

Now let $R = \bigcup_{i=1}^{\infty} R_i$. We will prove that R is not an antimatter ring by showing that Y is a (very strong) atom of R. Suppose $Y = fg$ for some $f, g \in R$. Then $f, g \in R_n$ for some $n \geq 1$. Suppose that f is not a (very strong) associate of Y in R_n. By the above proof, up to a unit of R_n, f must be a monomial of the form $X_{k_1}^{\alpha_1} \cdots X_{k_m}^{\alpha_m}$. Let k be the largest of the k_i. Clearly, X_i is a unit of R_{k+1}, for all $i \leq k$. Thus f is a unit in R_{k+1}, and hence a unit in R. Thus Y is a (very strong) atom of R, as claimed.

An alternative proof that $R = \cup R_i$ is not an antimatter ring can be given by observing that $R = F[Y]_{(Y)}$, where F is the field $\mathbf{F}_2(\{X_1^{\alpha_1}, X_2^{\alpha_2}, \ldots : \alpha_i \in \mathbf{Q}^+\})$. Since R is a DVR, Y is a prime element of R, and hence a (very strong) atom of R.

REFERENCES

1. D.D. Anderson and S. Valdes-Leon, Factorization in commutative rings with zero divisors, Rocky Mountain J. Math., 26 (1996), 439-480.

2. D.D. Anderson and S. Valdes-Leon, Factorization in commutative rings with zero divisors II, Lecture Notes in Pure and Applied Mathematics, vol. 189, Marcel Dekker, New York, 1997, 197-219.

3. J. Coykendall and D.E. Dobbs, Fragmented domains have infinite Krull dimension, submitted for publication.

4. J. Coykendall, D.E. Dobbs, and B. Mullins, On integral domains with no atoms, submitted for publication.

5. D.E. Dobbs, Fragmented integral domains, Port. Math., 43 (1985-1986), 463-473.

6. I. Kaplansky, Commutative Rings, Polygonal Publishing House, Washington, New Jersey, 1994.

Divisor Properties Inherited by Normsets of Rings of Integers

JIM COYKENDALL   Department of Mathematics, North Dakota State University, Fargo, ND 58105-5075.

AYSE A. TEYMUROGLU   Ziya-ur-Rahman cad, 11. Sokak No. 32/13, 06610 Gankaya Ankara, TURKEY.

1. INTRODUCTION

In this paper, there are two main goals. The first is to extend some of the known results on the interplay between factorization in Galois rings of algebraic integers and the inherited factorization in the *normset*; that is, the set of integral norms from the extension ring to the base ring.

The second goal is to introduce some new, computationally accessible results that characterize saturation properties (intuitively, saturation measures how close the normset is to containing its integral divisors) of the normset.

We begin here with some basic background material, notation, and terminology. Any unexplained material can be found in [8]. In this paper, $K \subseteq F$ will be algebraic number fields with respective rings of integers $T \subseteq R$. We recall the standard norm map from

R to T (denoted N_T^R or just N when no confusion exists) is given by

$$N(a) = \prod_\sigma \sigma(a)$$

where the product is taken over all distinct embeddings $\sigma : F \to \mathbf{C}$. In the Galois case, this is just the familiar "product of conjugates" under the Galois action.

More generally, the norm of a prime ideal in R is given by:

$$N(\mathcal{P}) = \wp^f$$

where $\wp$ is the (prime) ideal of R; $\wp := \mathcal{P} \bigcap R$, and f, the *degree* of $\mathcal{P}$, is the integer $f = [R/\mathcal{P} : T/\wp]$.

For a general ideal in $I \subseteq R$ we extend by linearity; that is

$$N(I) = N(\mathcal{P}_1)N(\mathcal{P}_2)...N(\mathcal{P}_n)$$

where $I = \mathcal{P}_1\mathcal{P}_2...\mathcal{P}_n$ is the (unique) prime ideal factorization of the ideal I.

There is a subtle distinction between the norm of an element $a \in R$ and the norm of the ideal generated by a. It is well-known that $(N(a)) = N(Ra)$; care must be taken when using an ideal-theoretic approach to computing norms of elements as there may be some ambiguity up to a unit in T.

It is well-known (and obvious from the above) that the norm map from R to T is a multiplicative map. At this juncture we digress into a bit of multiplicative set theory to set the groundwork for the theory of norms in general.

2. SOME MULTIPLICATIVE SET THEORY

This section lays the groundwork for our study of normset factorizations. Most of the material here is well-known, but for the sake of completeness we highlight some important concepts. Some good references are [6] and [7], to name a couple. (The interested reader is also encouraged to consult Factorization in Integral Domains (Lecture Notes in Pure and Applied Math, 189, Dekker) for many other references on factorization theory in monoids).

In this section S will be a multiplicative monoid; that is, a multiplicatively closed and cancellative commutative semigroup with unit element denoted by 1. We denote the group of units of S by G.

DEFINITION 2.1. We say that an element $s \in S \setminus G$ is irreducible if whenever $s = ut$ with $u, t \in S$, then either u or t is in G.

The notion of irreducibility allows us to consider various types of factorizations. In particular, we make the following definition.

DEFINITION 2.2. We say that an element $s \in S$ is atomic if s can be written in the form $s = \pi_1 \pi_2 ... \pi_n$ with π_i an irreducible element of S for $1 \leq i \leq n$. Additionally, we say the monoid S is atomic if every element of $S \setminus G$ is atomic.

The notion of atomicity is the first step toward factorization theory. Since the study of factorization (in integral domains) boils down to the study of the multiplicative monoid of non-zero elements of R (perhaps modulo the units of R) or the group of divisibility of R, almost all of the recent notions in modern factorization theory (e.g. bounded factorization domains (BFD's) and finite factorization domains (FFD's), just to name a couple of examples) can be couched in terms of the theory of factorization in multiplicative monoids ([7]). In this treatment of the theory, we will only concentrate on a couple of phenomena that can occur in an atomic monoid. Interested readers are encouraged to consult [1] for further reference on more general "factorization domains".

DEFINITION 2.3 We say that an atomic monoid M is a unique factorization monoid (UFM) if every element of $S \setminus G$ can be decomposed uniquely (up to units) into irreducible elements of S.

The obvious generalization of Definition 2.3 in the spirit of [2] and [10] is the following.

DEFINITION 2.4. We say the atomic monoid M is a half-factorial monoid (HFM) if given any two decompositions of $a \in S \setminus G$ into irreducible elements of S:

$$a = \alpha_1\alpha_2...\alpha_n = \beta_1\beta_2...\beta_m$$

we have $n = m$.

As is expected, the UFM condition implies HFM. Of course, the implication is not reversible (just consider the multiplicative monoid of nonzero elements of a non-UFD half-factorial domain). We now introduce a theorem (which can also be found in [6, Theorem 6.8]) that sheds some light on the structure of UFM's.

THEOREM 2.5. A multiplicative monoid S is a UFM if and only if S is isomorphic to $U \oplus (\bigoplus_{i \in I} \mathbf{Z}^+)$ where U is the unit group of S, $\mathbf{Z}^+$ denotes the non-negative integers, and I is some index set.

Given the structure of a UFM, one is tempted to consider free objects in the category of monoids since they are canonical examples of UFMs. For our purposes, however, we wish to generalize a bit because there are many examples of monoids with rather complicated unit structure that still have the UFM property. We begin with the definition of a free monoid.

DEFINITION 2.6. We say the monoid S is free if S is isomorphic to $\bigoplus_{j \in J} \mathbf{Z}^+$.

For our purposes there are many types of monoids that capture the essence of freeness if not the technical definition. With this in mind we make one more definition.

DEFINITION 2.7. We say the monoid S with unit group U is almost free if S/U is a free monoid.

This definition gives insight into the interplay between the notions of UFM and freeness. We record a useful observation.

THEOREM 2.8. The monoid S is a UFM if and only if S is almost free.

Proof: We appeal to the well-known (and straightforward) fact that S is a UFM if and only if S/U is a UFM. $\Diamond$

In the next section, we will specialize a bit more and explore the interplay between structure properties of rings of algebraic integers (in the Galois case for the most part) and structure properties of their normsets.

3. NORMSETS IN RINGS OF INTEGERS

In this section we apply some of the more general theory of the previous two sections to the study of factorization properties of normsets of rings of algebraic integers. Following the notation of the first section, we let N be the norm map from F to K and we let the set of norms from R to T be denoted by S.

In this section, we will generalize a unique factorization theorem from [3] to a slightly more general case. In [3], the main result was that (in the Galois case) R is a unique factorization domain if and only if the normset, S, of R is a UFM. A blanket assumption made was that T was a UFD. We will now extend the result to the case where T is not necessarily a UFD; [3, Theorem 2.7] will follow as an easy corollary.

To begin, we generalize the concept of normset introduced in [3], and declare that the normset of R is the set of *ideals* of the form $N((a))$ where (a) is a nonzero principal ideal of R.

THEOREM 3.1. If $K \subseteq F$ is a Galois extension with rings of integers $T \subseteq R$ respectively, then R is a UFD if and only if the monoid $\{N((a))|(a)$ is a principal ideal of $R\}$ is a UFM.

We remark here that the monoid mentioned above has trivial unit structure. When considering the monoid generated by *elements* of R there can be some difficulties caused by units in the normset (c.f. [3] and [4]). We also remark that the analog of [3, Theorem 2.7] is true for HFD's in the Galois case; that is, if R is a Galois ring of integers that is an HFD, then the normset S is an HFM ([5]).

Proof: ($\Longrightarrow$) Under the assumption that R is a UFD, we consider factorizations of an arbitrary norm $N((a)) \in S$ by considering the (unique) factorization of a into prime elements of R. In other words, given a factorization of a into primes of R:

$$a = \pi_1 \pi_2 ... \pi_n$$

we have a corresponding factorization in S:

$$N((a)) = N((\pi_1))N((\pi_2))...N((\pi_n)).$$

At this juncture, we note that $N((\pi))$ is irreducible in the monoid if and only if $\pi \in R$ is an irreducible (prime) element. Indeed, if $\pi \in R$ is reducible (say $\pi = \alpha\beta$ with α, β nonunits), then $N((\pi)) = N((\alpha))N((\beta))$. For the converse of this statement, we assume that $\pi \in R$ is prime and note that $N((\pi)) = \wp^f$ where $\wp$ is the prime ideal of T that (π) lies over and f is the degree. If $N((\pi))$ is not irreducible in S, then there exists an element $\gamma \in R$ with $N((\gamma)) = \wp^k$ with $k < f$. Therefore since F/K is Galois, we have the following ideal factorizations in R:

$$\wp^{kf}R = \prod_{\sigma \in G}(\sigma(\pi))^k = \prod_{\sigma \in G}(\sigma(\gamma))^f.$$

From this we get the factorization into elements

$$\prod_{\sigma \in G} \sigma(\pi)^k = u \prod_{\sigma \in G} \sigma(\gamma)^f$$

for some unit $u \in R$.

Since R is a UFD, this implies that (up to a unit) π is a conjugate of γ. Hence $N((\pi)) = N((\gamma))$ which is the desired contradiction.

With this statement in hand, we see that the factorization given for $N((a))$ above is indeed an irreducible factorization.

For uniqueness, we note that if we have two factorizations in S:

$$N((a)) = N((\pi_1))N((\pi_2))...N((\pi_k)) = N((\gamma_1))N((\gamma_2))...N((\gamma_r))$$

then the elements π_i, $1 \leq i \leq k$ and γ_j, $1 \leq j \leq r$ are necessarily prime elements of R by the previous argument. Since these elements are prime, we must have $k = r$ and that (without loss of generality) up to units and Galois conjugation $\pi_i = \gamma_i$ for $1 \leq i \leq k$. Therefore uniqueness follows inductively. This establishes the first direction.

($\Longleftarrow$) If we assume that R is not a UFD, then there must be a non-principal prime ideal $\mathcal{P}_1 \subseteq R$ of norm $N(\mathcal{P}_1) = \wp_1^{f_1}$, where $\wp_1$ is the prime of T that $\mathcal{P}_1$ lies over and f_1 is the degree. We now select a prime ideal $\mathcal{P}_2$ in the class $[\mathcal{P}_1]^{-1}$ of norm $\wp_2^{f_2}$. We will denote the orders of both $\mathcal{P}_1$ and $\mathcal{P}_2$ in the class group by $n > 1$.

It is easy to see that the right hand side of the following factorization is an irreducible factorization in S that cannot be reconciled with the factorization on the left:

$$(\wp_1^{f_1 n})(\wp_2^{f_2 n}) = (\wp_1^{f_1} \wp_2^{f_2})^n.$$

This establishes the theorem. $\diamondsuit$

COROLLARY 3.2. If $K \subseteq F$ is a Galois extension with rings of integers $T \subseteq R$ respectively, then R is a UFD if and only if S, the set of integral norms from R to T, is a UFM.

Proof: We denote $\{N((a))|(a)\text{is a principal ideal of } R\}$ by M, and we observe that

$$S/U \cong M$$

where U is the unit group of S. Hence M is (almost) free if and only if S is almost free. So Theorem 2.8 then gives that S is a UFM if and only if M is a UFM. This establishes the corollary. $\diamondsuit$

We now exhibit an example to show that the Galois assumption is needed for this result to hold.

EXAMPLE 3.3. Let R be the ring of integers of the field $\mathbf{Q}(\alpha)$, where α is a root of the polynomial $x^5 - x^3 + 1$. It is well-known that R is a UFD ([8]), but a simple computation shows that the rational prime 3 splits into two prime elements of R of norms 3^2 and 3^3 (and both of these norms must be irreducible since there is no element of norm 3). Therefore, in the normset of R we have the irreducible factorizations

$$(3^2)^3 = (3^3)^2$$

so despite the fact that R is a UFD, S is not a UFM (in fact it is not even an HFM). $\Diamond$

4. SATURATION PROPERTIES

In this section we take the investigation of saturation properties originally done in [4] a step further by exploring the differing types of saturation (saturation versus strong saturation) first discussed there.

The main results of [4] show that a normset of a Galois ring of algebraic integers, R, is saturated if and only if the Galois action on $\mathrm{Cl}(R)$ is trivial. In the quadratic case, this gives a complete characterization of saturation: namely, a ring of integers R has a saturated normset if and only if its class group is 2-elementary abelian. Care must be taken here, however, for the word "normset" here means the set of integral norms *modulo unit equivalence*. In other words, two norms are considered equivalent if they vary only by a unit in T. Put in terms of the second section, this means that we only want to consider the free monoid associated with our almost free monoid. The reason for this choice is that in the general case, there are examples that do not perform well in the strictest sense, but do not violate the spirit of the mathematical structure.

For the sake of completeness, we recall a couple of definitions first found in [4].

DEFINITION 4.1. We say the normset S is strongly saturated if the existence of $\alpha, \beta \in R$ with $N(\alpha)/N(\beta) \in T$ implies that there exists a $\gamma \in R$ such that $N(\gamma) = N(\beta)$ and $\gamma | \alpha$.

For the next definition we consider the modified normset alluded to in the introduction of this section. That is, we consider two norms equivalent if they vary by a unit in R. We refer to this normset associated with S as S_{eq}.

DEFINITION 4.2. We say that the normset S_{eq} is saturated if the existence of $\alpha, \beta \in R$ with $N(\alpha)/N(\beta) \in T$ implies that $N(\alpha)/N(\beta) \in S_{eq}$.

It is easy to see that the property "strong saturation" implies "saturation". In fact, it appears that from the defintions, saturation can be thought of as a property of the set S_{eq}, whereas the property strong saturation is a property that depends on the ring R. The thrust of the initial investigation of saturation properties previously performed concentrated exclusively on the saturation property. Saturation is the property that measures how non-trivial the Galois action is on the class group ([4]).

In this section, we will show how the properties of strong saturation and saturation are related, and we will reveal the somewhat surprising result that strong saturation depends less upon the parent ring than the intial definition would seem to imply. We now produce a (temporary) definition to assist us.

DEFINITION 4.3. We say that the normset S is *medium saturated* if the existence of $\alpha, \beta \in R$ with $N(\alpha)/N(\beta) \in T$ implies that $N(\alpha)/N(\beta) \in S$.

We remark that this definition is similar to Definition 4.2, but now we are considering the general set of integral norms. In the spirit of Section 2, we are now refusing to ignore the unit structure of the monoids.

We now introduce the following theorem which shows the relationship between the saturation properties.

THEOREM 4.4. Let S be a normset. Then S is strongly saturated if and only if S is medium saturated.

Proof: The implication that strong saturation implies medium saturation is clear. We will only show that medium saturation implies strong saturation.

Let $\alpha, \beta \in R$ with $N(\alpha)/N(\beta) = t \in T$. Since S is medium saturated, we must also have that $t \in S$. We now write the prime ideal factorizations of (α) and (β) respectively:

$$(\alpha) = \mathcal{P}_1 \mathcal{P}_2 ... \mathcal{P}_n$$

and

$$(\beta) = \mathcal{Q}_1 \mathcal{Q}_2 ... \mathcal{Q}_r.$$

So in terms of ideals we have that

$$\frac{N((\alpha))}{N((\beta))} = \frac{\prod_{\sigma \in G} \sigma(\mathcal{P}_1) \prod_{\sigma \in G} \sigma(\mathcal{P}_2) \cdots \prod_{\sigma \in G} \sigma(\mathcal{P}_n)}{\prod_{\sigma \in G} \sigma(\mathcal{Q}_1) \prod_{\sigma \in G} \sigma(\mathcal{Q}_2) \cdots \prod_{\sigma \in G} \sigma(\mathcal{Q}_r)} = (t).$$

In particular, the quotient of norms above is an integral ideal, so $n \geq r$, and we assume (without loss of generality) that $\mathcal{P}_i = \overline{\mathcal{Q}_i}$, where the overscore indicates a Galois conjugate, for $1 \leq i \leq r$.

With this notation we rewrite (α) as

$$(\alpha) = \overline{\mathcal{Q}_1} \cdots \overline{\mathcal{Q}_r} \mathcal{P}_{r+1} \cdots \mathcal{P}_n.$$

We now make the key observation in this theorem. Since S is medium saturated, the normset S (technically, modulo unit equivalence) is saturated. Applying [4, Theorem 3.8], we see that the Galois action on the class group of R must be trivial; that is, $[I] = [\sigma(I)]$ for any ideal $I \subseteq R$.

With this observation in hand, we notice that since the ideal $(\beta) = \mathcal{Q}_1 \mathcal{Q}_2 \cdots \mathcal{Q}_r$ is principal, then so is the ideal $(\gamma) = \overline{\mathcal{Q}_1} \cdots \overline{\mathcal{Q}_r}$. Therefore, the ideal $\mathcal{P}_{r+1} \mathcal{P}_{r+2} \cdots \mathcal{P}_n$ is also forced to be principal (and we will denote this ideal by (η)).

Immediately, we see that $N((\gamma)) = N((\beta))$ so that $N(\gamma) = uN(\beta)$ where u is a unit of T. In fact, since $u = \frac{N(\gamma)}{N(\beta)}$ we have that $u \in S$ (so it is the norm of some unit $v \in R$). By the choice of the ideal factorization of γ, we have that γ, and hence $\frac{\gamma}{v}$, must divide α, so we have produced an element of R with norm equal to the norm of β that divides α. Hence, S is strongly saturated. This completes the proof. $\Diamond$

At this juncture, we discard the terminology "medium saturated" since it is equivalent to strong saturation. An obvious question is whether the notion of strong saturation is equivalent to saturation (modulo unit equivalence). We now produce an example to show that this is not the case.

EXAMPLE 4.5. Let $R = \mathbf{Z}[\sqrt{34}]$ and let $T = \mathbf{Z}$. It is well-known that the class number of R is 2 and the fundamental unit has norm 1. [4, Theorem 3.7] says that the normset of R must be saturated (in fact this is true for any quadratic extension of $\mathbf{Q}$

with 2-elementary abelian class group). However, we note that there are elements of R (namely 3 and $5 - \sqrt{34}$) of norms 9 and -9 respectively. For strong saturation to occur, this implies that there must be an element of R of norm -1. But since the fundamental unit of R has positive norm, this cannot happen. $\Diamond$

5. SOME APPLICATIONS TO THE QUADRATIC CASE

In this final section, we record some useful applications of these saturation theorems to the case of (real) quadratic extensions of $\mathbf{Q}$. The imaginary case will be ignored for the most part because the different notions of saturation are equivalent in that setting (c.f.[4]).

THEOREM 5.1. Let R be a quadratic ring of integers with saturated normset S. Then S is strongly saturated if and only if either $-1 \in S$ or $-n^2 \notin S$ for any $n \in \mathbf{Z}$.

Proof: ($\Longrightarrow$) For this implication, we assume that -1 is not in S (a strongly saturated normset), and for some $n \in Z$, $-n^2 \in S$. Note that since S is the normset of a quadratic extension, every square integer is an element of S. As $-n^2, n^2$ are both elements of S, and S is strongly saturated, this means there must be an element of norm -1 in R, a contradiction.

($\Longleftarrow$) If $-1 \in S$, we consider two elements of $k, r \in S$ such that $k = rs$ with s an integer. Since S is a saturated normset, this implies that $\pm s \in S$ ([4]). So there is an $\alpha \in R$ with norm $\pm s$. If $N(\alpha) = s$ then we are done. On the other hand, if $N(\alpha) = -s$ then by assumption there is an element $\epsilon \in R$ such that $N(\epsilon) = -1$. So the element $\epsilon \alpha$ has norm s, and so by Theorem 4.4, S is strongly saturated.

If we assume that $-n^2 \notin S$ for any $n \in Z$, then as before we assume that we have two elements $k, r \in S$ such that $k = rs$ with s an integer. Since S is a saturated normset, again we have that $\pm s \in S$; we have to show that $s \in S$. If we assume that $-s \in S$, then $-k = r(-s) \in S$, which in turn implies that $-k^2 \in S$, a contradiction. Therefore, $s \in S$ which establishes the theorem. $\Diamond$

We note here that the previous theorem shows that for imaginary quadratic extensions, the various notions of saturation discussed here are equivalent. This was first noted in [4], but falls out here rather nicely. It is also worth noting that Theorem 5.1 shows that for any real quadratic field with fundamental unit of norm -1, the notions of saturation and strong saturation are equivalent.

As a final application, we produce an example to show that the various types of saturation (as alluded to in Theorem 5.1) occur in the quadratic case.

EXAMPLE 5.2. Consider the quadratic rings of integers $A=\mathbf{Z}[\sqrt{34}]$ and $B=\mathbf{Z}[\sqrt{15}]$. We have already noted that the normset of A is saturated, but not strongly saturated. It is well-known that the class number of both these rings is 2, and that they both have fundamental unit of positive norm. However, the quadratic form $x^2 - 15y^2$ (the norm form of the ring B) does not represent $-n^2$ for any $n \in \mathbf{Z}$ (this is easily seen via reduction mod 3). So B possesses a strongly saturated normset by Theorem 5.1.

REFERENCES

1. D. D. Anderson, D. F. Anderson, and M. Zafrullah, *Factorization in integral domains*, J. Pure Appl. Algebra **69** (1990), 1-19.

2. L. Carlitz, *A characterization of algebraic number fields with class number two*, Proc. Amer. Math. Soc. **11** (1960), 391-392.

3. J. Coykendall, *Normsets and determination of unique factorization in rings of algebraic integers*, Proc. Amer. Math. Soc. **124** (1996), 1727-1732.

4. J. Coykendall, *Properties of the normset relating to the class group*, Proc. Amer. Math. Soc. **124** (1996), 3587-3593.

5. J. Coykendall, *Elasticity properties preserved in the normset*, preprint.

6. R. Gilmer, "Commutative Semigroup Rings," The University

of Chicago Press, Chicago, 1984.

7. F. Halter-Koch, *Finiteness theorems for factorizations*, Semi-group Forum **44** (1992), 112-117.

8. W. Narkiewicz, "Elementary and Analytic Theory of Algebraic Numbers," Springer-Verlag/Polish Scientific Publishers, Warszawa, 1990.

9. M. Pohst and H. Zassenhaus, "Algorithmic Algebraic Number Theory," Cambridge University Press, Cambridge, 1993.

10. A. Zaks, *Half-factorial-domains*, Israel J. Math. **37** (1980), 281-302.

On the Probability That Eisenstein's Criterion Applies to an Arbitrary Irreducible Polynomial

DAVID E. DOBBS Department of Mathematics, University of Tennessee, Knoxville, Tennessee 37996-1300

LAURA E. JOHNSON Department of Mathematics, University of Tennessee, Knoxville, Tennessee 37996-1300

1 INTRODUCTION

A well-known algorithm of Kronecker (cf. [1, pages 126-127]) can be used to determine whether a given polynomial with integral coefficients is irreducible in $\mathbf{Q}[X]$. As it is often time-consuming to implement this algorithm, it has seemed desirable to find sufficient conditions for irreducibility. As described in [3], the development of such irreducibility criteria has a history exceeding 150 years and remains an active area of research today. Surely, the best-known and oft-cited irreducibility criterion is that named after Eisenstein (cf. [1, page 124]). In fact, in virtually any current textbook on modern algebra, the example given of an irreducible quintic polynomial $X^5 + aX + b \in \mathbf{Z}[X]$ with Galois group S_5 has its irreducibility established by an appeal to Eisenstein's Criterion. Because such examples may seem somewhat contrived, the question naturally arises as to what the probability is that Eisenstein's Criterion applies to a random polynomial $X^m + aX + b \in \mathbf{Z}[X]$, with $m \geq 2$. The purpose of this note is to give sharper meaning to this question and to show that, in a sense, this probability can be bounded independently of m, between 0.2 and 0.3.

Fix a positive integer $m \geq 2$. To avoid calculations which take the indeterminate form $\frac{\infty}{\infty}$, we approach the probability discussed above in the spirit of Niven and Zuckerman's approach [4, page 290] to the natural density of sets of positive integers. Specifically, for each positive integer n, let

$$a_n := |\{f = X^m + aX + b \in \mathbf{Z}[X] : |a| \leq n, |b| \leq n,$$

and Eisenstein's Criterion applies to f, in the sense that some prime $p \in \mathbf{Z}$ satisfies $p|a, p|b$, and $p^2 \nmid b\}|$;

$$b_n := |\{f = X^m + aX + b \in \mathbf{Z}[X] : |a| \leq n, |b| \leq n,$$

and f is irreducible in $\mathbf{Q}[X]\}|$; and

$$c_n := |\{f = X^m + aX + b \in \mathbf{Z}[X] : |a| \leq n, |b| \leq n\}|.$$

We interpret $\lim_{n \to \infty} \frac{a_n}{c_n}$ (resp., $\lim_{n \to \infty} \frac{a_n}{b_n}$), if it exists, as the probability that Eisenstein's Criterion applies to a random (resp., random irreducible) m-th degree polynomial $X^m + aX + b \in \mathbf{Z}[X]$.

Do these limits exist and are they independent of m? Our main result dispatches the $\underline{\lim}$ and the $\overline{\lim}$ of $\frac{a_n}{c_n}$ in general and the corresponding $\underline{\lim}$ and $\overline{\lim}$ of $\frac{a_n}{b_n}$ if $m = 2, 3$. Observe that $c_n = (2n + 1)^2 = 4n^2 + 4n + 1 \sim 4n^2$. (As usual, $s_n \sim t_n$ denotes that s_n is asymptotic to t_n, in the sense that $\lim_{n \to \infty} \frac{s_n}{t_n} = 1$.) The principal consequence of Section 2 is that if $m \leq 3$, then $b_n \sim 4n^2$. It then follows, for $m \leq 3$, that $b_n \sim c_n$, so that $\lim_{n \to \infty} \frac{a_n}{b_n} = \lim_{n \to \infty} \frac{a_n}{c_n}$, if these limits exist. In Section 3, we find the order of magnitude of a_n (which is independent of m). Our work culminates in the following result.

MAIN THEOREM *Fix a positive integer $m \geq 2$. Then*

$$0.2294 < \underline{\lim} \frac{a_n}{c_n} < \overline{\lim} \frac{a_n}{c_n} < 0.2784.$$

If $m \leq 3$, then $\underline{\lim} \frac{a_n}{b_n} = \underline{\lim} \frac{a_n}{c_n}$ and $\overline{\lim} \frac{a_n}{b_n} = \overline{\lim} \frac{a_n}{c_n}$.

The lower (resp., upper) bound for the $\underline{\lim}$ (resp., $\overline{\lim}$) is given as the sum of an infinite series T (resp., S) described below.

Section 4 considers analogues of the above context, in particular the corresponding probabilistic issues for arbitrary monic cubic integral polynomials. In addition to a proof that the series T converges, Section 5 contains some musings on the vagaries of raw data.

Throughout, we use notation with the meanings given above, in particular for m, n, a_n, b_n, c_n, and $\sim$. As usual, $[\ldots]$ denotes the greatest-integer (or "floor") function. Unexplained symbols such as $p, p_1, p_2, \ldots$ will be assumed to be prime natural numbers.

We thank Robert M. McConnel and Charles R. Collins for conversations about the integral test and rearrangements of series, respectively. Both authors were supported in part by the National Science Foundation through its Research Experiences for Undergraduates site at the University of Tennessee–Knoxville in 1998.

2 BOUNDING THE DENOMINATOR

We have seen that the denominator $c_n = (2n + 1)^2 \sim 4n^2$. This section is devoted to proving that if $2 \leq m \leq 3$, then the denominator $b_n \sim 4n^2$: see Proposition 2.1 for a more precise statement. As usual, τ denotes the number-of-integral-divisors function; i.e., for each non-zero integer k, $\tau(k) := |\{d \in \mathbf{Z} : 1 \leq d|k\}|$. It will also be useful to recall the following result of Dirichlet (cf.[2, Theorem 6.30]):

$$\sum_{b=1}^{n} \tau(b) = n \ln n + (2\gamma - 1)n + O(n^{\frac{1}{2}})$$

where γ denotes Euler's constant. In particular, $\sum_{b=1}^{n} \tau(b) \sim n \ln n$.

PROPOSITION 2.1 *If $m = 2$ or $m = 3$, then $b_n \geq 4n^2 + 2n - 4\sum_{b=1}^{n} \tau(b)$, and so $b_n \sim 4n^2$.*

Proof: We consider first the case $m = 2$. Let $f = X^2 + aX + b \in \mathbf{Z}[X]$. Suppose that f is *reducible* in $\mathbf{Q}[X]$ and that $b \neq 0$. Then f has a root $r \in \mathbf{Z}$. (Indeed, f has a root $r \in \mathbf{Q}$ since f has a linear factor in $\mathbf{Q}[X]$. Then $r \in \mathbf{Z}$ by the rational root test [1, page 115] or by its highbrow analogue, the fact that $\mathbf{Z}$ is integrally closed.) By the factor theorem [1, page 39], there exists $s \in \mathbf{Z}$ such that $f = (X - s)(X - r)$. Equating corresponding coefficients, we find that $a = -r - s$ and $b = sr$. Hence, $r \neq 0$ and $a = -r - br^{-1}$. It follows that if $-n \leq \beta \leq n$ with $0 \neq \beta \in \mathbf{Z}$, then each integral divisor r of β leads to at most one integral value of α (namely, $\alpha = -r - \beta r^{-1}$) such that $X^2 + \alpha X + \beta$ is reducible in $\mathbf{Q}[X]$. Of course, $\beta = 0$ leads to $X^2 + \alpha X + \beta = X(X + \alpha)$ being reducible, regardless of which $\alpha \in \mathbf{Z}$ is considered; restricting $|\alpha| \leq n$, we find $2n + 1$ such reducible polynomials of the form $X^2 + \alpha X$. Therefore, we have that

$$b_n \geq c_n - \sum_{1 \leq |\beta| \leq n} 2\tau(\beta) - (2n + 1) = 4n^2 + 2n - 4\sum_{b=1}^{n} \tau(b).$$

In case $m = 3$, the asserted lower bound for b_n is found similarly. Indeed, if $f = X^3 + aX + b \in \mathbf{Z}[X]$ is *reducible* in $\mathbf{Q}[X]$ and $b \neq 0$, then the factor theorem yields that $f = (X - r)(X^2 + \lambda X + \mu)$ for some $r, \lambda, \mu \in \mathbf{Z}$. Then, by equating corresponding coefficients, we find that $0 = \lambda - r, a = \mu - r\lambda$, and $b = -r\mu$. It follows that $\lambda = r, b = -r\mu$, and $a = -r^2 + \mu = -r^2 - br^{-1}$. Thus, if $0 \neq \beta \in \mathbf{Z}$, then each integral divisor r of β leads to at most one integer α (namely, $\alpha = -r^2 - \beta r^{-1}$) such that $X^3 + \alpha X + \beta$ is reducible in $\mathbf{Q}[X]$. Moreover, if $\alpha \in \mathbf{Z}$, then $X^3 + \alpha X = X(X^2 + \alpha)$ is reducible in $\mathbf{Q}[X]$. Therefore, the lower bound for b_n is again produced by the above-displayed expression.

To show that $b_n \sim 4n^2$, it follows from the "sandwich limit theorem" (cf. [5, Exercise 14, page 34]) that it suffices to find sequences d_n and c_n such that $d_n \leq b_n \leq c_n$ for each n and $d_n \sim c_n \sim 4n^2$. Now, it follows from the definitions in the Introduction that $b_n \leq c_n$, and we have seen that $c_n \sim 4n^2$. Put

$$d_n := 4n^2 + 2n - 4\sum_{b=1}^{n} \tau(b).$$

We have shown that $d_n \leq b_n$. Moreover, the result of Dirichlet recalled above,

$$d_n = 4n^2 + 2n + 4n\ln n - 4(2\gamma - 1)n + O(n^{\frac{1}{2}}).$$

It is now elementary to verify that $d_n \sim 4n^2$, completing the proof. $\square$

The sole impediment to removing the hypothesis $m \leq 3$ in the assertions regarding the $\underline{\lim}$ and the $\overline{\lim}$ of $\frac{a_n}{b_n}$ in the main theorem is the appearance of this hypothesis in Proposition 2.1. Therefore, we close this section by raising the question whether $b_n \sim 4n^2$ if $m \geq 4$.

3   DESCRIBING THE NUMERATOR

In this section, we determine asymptotic upper and lower bounds for the numerator a_n (see Proposition 3.3 (a) and 3.7 (a), respectively). This work combines with Proposition 2.1 to yield a proof of the main theorem stated in the Introduction.

Let $p_1 = 2, p_2 = 3, \ldots, p_\nu = N, \ldots$ enumerate the positive prime integers, listed in increasing order. We shall see that an asymptotic upper bound for a_n can be found by using the infinite series

$$S := \sum_{k=1}^{\infty}\left(\frac{1}{p_k^2} - \frac{1}{p_k^3}\right) := \sum_{p}\left(\frac{1}{p^2} - \frac{1}{p^3}\right).$$

We proceed to analyze S by means of its typical partial sum

$$S_\nu := \sum_{k=1}^{\nu}\left(\frac{1}{p_k^2} - \frac{1}{p_k^3}\right).$$

LEMMA 3.1 $S = \sum_{p}(\frac{1}{p^2} - \frac{1}{p^3}) \approx 0.28$, *rounded off to two decimal places.*

Proof: When S is approximated by the partial sum S_ν, the resulting (nonnegative) error is at most

$$\sum_{n=N+1}^{\infty}\left(\frac{1}{n^2} - \frac{1}{n^3}\right) \leq \int_{N}^{\infty}\left(\frac{1}{x^2} - \frac{1}{x^3}\right)dx = \frac{1}{N} - \frac{1}{2N^2}.$$

(The error estimate is provided by the integral test (cf. [5, Theorem II, page 408]) which can be applied here because $f(x) := x^{-2} - x^{-3}$ defines a positive decreasing function for $x > 1.5$.) We found it convenient to use $\nu = 168$, for which $N = p_\nu = p_{168} = 1,009$. A computer computation led to $S_{168} \approx 0.2773588$, while the above analysis gives an upper bound for the error of

$$\frac{1}{1009} - \frac{1}{2(1009)^2} \approx 0.000990589157444.$$

It follows that

$$0.2773588 \approx S_\nu < S \le S_\nu + 0.000990589157445 < 0.2783494$$

and so $S \approx 0.28$, rounded off to two decimal places, as asserted. $\square$

Recall that if e_n and f_n are sequences of real numbers, then $e_n = o(f_n)$ means that $\lim_{n\to\infty} \frac{e_n}{f_n} = 0$.

LEMMA 3.2 *(a) Fix a positive integer n. Let p be a positive prime integer such that $p \le n$. Then the number of ordered pairs $(a,b) \in \mathbf{Z} \times \mathbf{Z}$ such that $|a| \le n, |b| \le n, p|a, p|b$ and $p^2 \nmid b$ is*

$$2\left(2\left[\frac{n}{p}\right] + 1\right)\left(\left[\frac{n}{p}\right] - \left[\frac{n}{p^2}\right]\right).$$

(b) In describing a_n, if one ignores all the polynomials $X^m + aX + b$ which arise from a particular value of a (in particular, $a = 0$), the resulting approximation to a_n differs from a_n by an error which is $o(n^2)$.

(c) Fix positive integers $n \ge p$, with p prime. Then the error incurred by approximating $2(2[\frac{n}{p}])([\frac{n}{p}] - [\frac{n}{p^2}])$ by $4n^2(\frac{1}{p^2} - \frac{1}{p^3})$ is, in absolute value, less than $16(\frac{n}{p} + 1)$.

Proof: (a) There are $[\frac{n}{p}]$ positive values of a being counted, namely

$$1 \cdot p, 2 \cdot p, \ldots, \quad \text{and} \quad \left[\frac{n}{p}\right] p.$$

The additive inverses of the displayed numbers are the negative values of a being counted. As $a = 0$ also needs to be counted, we have $2[\frac{n}{p}] + 1$ choices for a in all. Since the choices of a and b are made independently, it remains to show that there are $2[\frac{n}{p}] - 2[\frac{n}{p^2}]$ choices for b in all. Note $b \ne 0$ since $p^2 \nmid b$; and that k is a possible value of b if and only if $-k$ is. Thus, it suffices to show that there are $[\frac{n}{p}] - [\frac{n}{p^2}]$ choices for positive values of b in all. This is evident, for the positive b's are obtained from the above-displayed list of $[\frac{n}{p}]$ numbers by deleting the following $[\frac{n}{p^2}]$ numbers:

$$1 \cdot p^2, 2 \cdot p^2, \ldots, \quad \text{and} \quad \left[\frac{n}{p^2}\right] p^2.$$

(b) The (nonnegative) error incurred by ignoring polynomials with a specified coefficient of X is bounded above by

$$|\{b \in \mathbf{Z} : |b| \le n\}| = 2n + 1 = o(n^2).$$

(c) Essentially because of the division algorithm in $\mathbf{Z}$, we can write

$$\frac{n}{p} = \left[\frac{n}{p}\right] + \epsilon \quad \text{and} \quad \frac{n}{p^2} = \left[\frac{n}{p^2}\right] + \delta$$

with $0 \le \epsilon < 1$ and $0 \le \delta < 1$. Then the absolute value of the error in question is

$$\left| 2 \left(2 \left[\frac{n}{p} \right] \right) \left(\left[\frac{n}{p} \right] - \left[\frac{n}{p^2} \right] \right) - 4n^2 \left(\frac{1}{p^2} - \frac{1}{p^3} \right) \right| =$$

$$\left| 2 \left(2\frac{n}{p} - 2\epsilon \right) \left(\frac{n}{p} - \frac{n}{p^2} - \epsilon + \delta \right) - 4n^2 \left(\frac{1}{p^2} - \frac{1}{p^3} \right) \right| =$$

$$\left| 4\frac{n}{p}(-\epsilon + \delta) - 4\epsilon \left(\frac{n}{p} - \frac{n}{p^2} - \epsilon + \delta \right) \right| <$$

$$4\frac{n}{p}(2) + 8\frac{n}{p} \left(1 - \frac{1}{p} \right) + 16 < 16 \left(\frac{n}{p} + 1 \right) . \quad \square$$

Our next result gives an upper bound for the $\overline{\lim}$ of the sequences $\frac{a_n}{b_n}$ and $\frac{a_n}{c_n}$ (assuming $m \le 3$ for the case of $\frac{a_n}{b_n}$). First, it is convenient to recall three results from number theory. As usual, $\pi(x)$ denotes $|\{p \in \mathbf{Z} : p$ is a positive prime number and $p \le x\}|$. The Prime Number Theorem asserts that $\pi(n) \sim \frac{n}{\ln n}$. Chebychev's weak version of the Prime Number Theorem [4, Theorem 8.1] asserts the existence of positive real numbers α and β such that

$$\alpha \frac{x}{\ln x} < \pi(x) < \beta \frac{x}{\ln x} \quad \text{for all real numbers} \quad x \ge 2.$$

As a consequence [4, Theorem 8.4], there exists a positive real number k such that

$$\sum_{2 < p \le x} \frac{1}{p} < k \ln \ln x \quad \text{for all real numbers} \quad x \ge 3.$$

PROPOSITION 3.3 (a) *For each* $n \ge 3, a_n < 4n^2 S + 16kn \ln \ln n + 16\beta \frac{n}{\ln n} + 10n + 1$.
  (b) $\overline{\lim} \frac{a_n}{c_n} \le S$, *and so* $\overline{\lim} \frac{a_n}{c_n} < 0.28$. *If* $m = 2$ *or* $m = 3$, *then* $\overline{\lim} \frac{a_n}{b_n} = \overline{\lim} \frac{a_n}{c_n} \le S$, *and so* $\overline{\lim} \frac{a_n}{b_n} < 0.28$.

Proof: (a) By the proof of Lemma 3.2 (a),

$$a_n = \left(\sum_{1 < p \le n} 2 \left(2 \left[\frac{n}{p} \right] \right) \left(\left[\frac{n}{p} \right] - \left[\frac{n}{p^2} \right] \right) \right) + h_n,$$

where $h_n := |\{f = X^m + b \in \mathbf{Z}[X] : |b| \le n$ and Eisenstein's Criterion applies to $f\}|$. By the proof of Lemma 3.2 (b), $0 \le h_n \le 2n + 1$. Thus, by Lemma 3.2 (c) and the results recalled above from [4], if $n \ge 3$, then

$$a_n \le \left(\sum_{1 < p \le n} \left(4n^2 \left(\frac{1}{p^2} - \frac{1}{p^3} \right) + 16 \left(\frac{n}{p} + 1 \right) \right) \right) + 2n + 1 <$$

$$4n^2 \sum_p \left(\frac{1}{p^2} - \frac{1}{p^3} \right) + 16n \sum_{1 < p \le n} \frac{1}{p} + 16\pi(n) + 2n + 1 <$$

$$4n^2 S + 16n \left(\frac{1}{2} + k \ln \ln n \right) + 16\beta \frac{n}{\ln n} + 2n + 1,$$

which easily simplifies to the asserted upper bound.

(b) Recall that $c_n \sim 4n^2$ and (from Proposition 2.1) that $b_n \sim 4n^2$ if $2 \leq m \leq 3$. Hence, $\overline{\lim} \frac{a_n}{c_n} = \overline{\lim} \frac{a_n}{4n^2} (= \overline{\lim} \frac{a_n}{b_n}$ if $2 \leq m \leq 3)$. By (a),

$$\overline{\lim} \frac{a_n}{4n^2} \leq \overline{\lim} \frac{4n^2 S + 16kn \ln \ln n + 16\beta \frac{n}{\ln n} + 10n + 1}{4n^2} =$$

$$\lim_{n \to \infty} \frac{4n^2 S + 16kn \ln \ln n + 16\beta \frac{n}{\ln n} + 10n + 1}{4n^2} = S.$$

The proof concludes by recalling from the proof of Lemma 3.1 that $S < 0.28$. $\square$

REMARK 3.4 In the proof of Proposition 3.3 (b), we used the fact that the sequences $\frac{a_n}{c_n}$ (and, if $m \leq 3$, $\frac{a_n}{b_n}$) and $\frac{a_n}{4n^2}$ have the same $\overline{\lim}$. Since the standard limit theorems of calculus do not carry over to $\overline{\lim}$ (cf. [5, Exercises 19 and 20, page 59]), we pause to record a justification. It uses not only the fact that $a_n \sim 4n^2$ but also the upper bound established in Proposition 3.3 (a). The pertinent detail is the following:

$$\left| \frac{a_n}{4n^2} - \frac{a_n}{c_n} \right| = a_n \left(\frac{1}{4n^2} - \frac{1}{c_n} \right) <$$

$$\left(\frac{4n^2 S + 16kn \ln \ln n + 16\beta \frac{n}{\ln n} + 10n + 1}{4n^2} \right) \left(\frac{c_n - 4n^2}{c_n} \right) =$$

$$\left(1 - \frac{4n^2}{c_n} \right) \left(S + \frac{(16kn \ln \ln n + 16\beta \frac{n}{\ln n} + 10n + 1)}{4n^2} \right).$$

The same reasoning shows that if α_n, β_n, and γ_n are sequences of real numbers such that $\beta_n \sim \gamma_n$ and $\frac{\alpha_n}{\gamma_n}$ is bounded, then $\overline{\lim} \frac{\alpha_n}{\beta_n} = \overline{\lim} \frac{\alpha_n}{\gamma_n}$. The analogous behavior for $\underline{\lim}$ can be established in the same way; we shall use it in the proof of Proposition 3.7 (b) without further mention. $\square$

We turn now to the more arduous task of developing an asymptotic lower bound for u_n. Once again, we work with the enumeration $p_1 = 2, p_2 = 3, \dots, p_\nu = N, \dots$ of the positive prime integers, listed in increasing order. We proceed to describe a_n as a sum $t_1^* + t_2^* + t_3^* + \cdots$.

In principle, the terms t_ν^* are not difficult to describe. We take t_1^* to be the number of polynomials $f = X^m + aX + b \in \mathbf{Z}[X]$ such that $|a| \leq n, |b| \leq n$, and f satisfies Eisenstein's criterion *because of* the prime $p_1 = 2$ (i.e. $2|a, 2|b$, and $4 \nmid b$). By Lemma 3.2 (a), $t_1^* = 2(2[\frac{n}{2}] + 1)([\frac{n}{2}] - [\frac{n}{4}])$.

In general, having found $t_1^*, t_2^*, \dots, t_{\nu-1}^*$, we say that t_ν^* counts the number of *new* (i.e., as yet uncounted) polynomials f which satisfy Eisenstein's Criterion because of the prime p_ν. To calculate t_ν^*, we shall use an elaborate inclusion/exclusion process. Given positive prime integers $p_{i_1} < \dots < p_{i_k}$ (with $p_{i_k} \leq n$), we define $g_{i_1 i_2 \dots i_k}$ to be the number of polynomials $f = X^m + aX + b \in \mathbf{Z}[X]$ such that $|a| \leq n, |b| \leq n$, and f satisfies Eisenstein's Criterion because of *each* of the primes $p_{i_1}, \dots, p_{i_k}$. By

Lemma 3.2 (a), $g_i = 2(2[\frac{n}{p_i}] + 1)([\frac{n}{p_i}] - [\frac{n}{p_i^2}])$. Of course, $t_1^* = g_1$. However, $t_2^* \neq g_2$. Indeed, by inclusion/exclusion considerations, we have

$$t_2^* = g_2 - g_{12}, \quad t_3^* = g_3 - g_{13} - g_{23} + g_{123},$$

$$t_4^* = g_4 - g_{14} - g_{24} - g_{34} + g_{124} + g_{134} + g_{234} - g_{1234}, \quad \text{etc.}$$

In general,

$$t_\nu^* = \sum_{\substack{1 \leq i_1 < \ldots < i_k < \nu \\ 0 \leq k \leq \nu - 1}} (-1)^k g_{i_1 \ldots i_k \nu}.$$

Moreover, $t_\nu^* \geq 0$, by the definition of t_ν^*. To say more about t_ν^*, we need to develop

a formula for $g_{i_1 \ldots i_k}$. As the case $k = 1$ was discussed above, we next consider g_{ij}, where $i < j$.

Of course, $g_{ij} = |\{f = X^m + aX + b \in \mathbf{Z}[X] : |a| \leq n, |b| \leq n, p_i|a, p_i|b, p_i^2 \nmid b, p_j|a, p_j|b, p_j^2 \nmid b\}| = r_{ij}k_{ij}$, where r_{ij} (resp., k_{ij}) is the number of choices for a (resp., b). As in the proof of Lemma 3.2 (a), $r_{ij} = 2[\frac{n}{p_i p_j}] + 1$, since p_i and p_j are relatively prime. Since $\mathbf{Z}$ is a unique factorization domain, we see by an inclusion/exclusion process that the number of choices for b is

$$k_{ij} = 2 \left(\left[\frac{n}{p_i p_j} \right] - \left[\frac{n}{p_i^2 p_j} \right] - \left[\frac{n}{p_i p_j^2} \right] + \left[\frac{n}{p_i^2 p_j^2} \right] \right).$$

We offer a proof (leaving the later, similar details for $g_{i_1 \ldots i_k}$ to the reader). It is enough to count the choices for positive b's, for k_{ij} is twice this number of choices. The term $[\frac{n}{p_i p_j}]$ arises by counting the set $\{1 \cdot p_i p_j, 2 \cdot p_i p_j, \ldots, [\frac{n}{p_i p_j}]p_i p_j\}$ of positive $b \leq n$ such that b is divisible by both p_i and p_j. Because of the condition $p_i^2 \nmid b$, we must exclude the set $\{1 \cdot p_i^2 p_j, 2 \cdot p_i^2 p_j, \ldots, [\frac{n}{p_i^2 p_j}]p_i^2 p_j\}$, thus accounting for the term $-[\frac{n}{p_i^2 p_j}]$. Similarly, the condition $p_j^2 \nmid b$ accounts for the term $-[\frac{n}{p_i p_j^2}]$. However, the set $\{1 \cdot p_i^2 p_j^2, 2 \cdot p_i^2 p_j^2, \ldots, [\frac{n}{p_i^2 p_j^2}]p_i^2 p_j^2\}$ has been excluded twice, and so the final term $[\frac{n}{p_i^2 p_j^2}]$ compensates for this overkill.

A formula for g_{ijk} can be derived similarly. As in the proof of Lemma 3.2 (a), the number of relevant a's is $2[\frac{n}{p_i p_j p_k}] + 1$. Then, by applying unique factorization and inclusion/exclusion, we may count the suitable b's, with the result that

$$g_{ijk} = 2 \left(2 \left[\frac{n}{p_i p_j p_k} \right] + 1 \right) \cdot \left\{ \left[\frac{n}{p_i p_j p_k} \right] - \left[\frac{n}{p_i^2 p_j p_k} \right] - \left[\frac{n}{p_i p_j^2 p_k} \right] - \left[\frac{n}{p_i p_j p_k^2} \right] + \right.$$

$$\left. \left[\frac{n}{p_i^2 p_j^2 p_k} \right] + \left[\frac{n}{p_i^2 p_j p_k^2} \right] + \left[\frac{n}{p_i p_j^2 p_k^2} \right] - \left[\frac{n}{p_i^2 p_j^2 p_k^2} \right] \right\}.$$

We trust that the reader can now discern the pattern in a general formula for $g_{i_1...i_k}$ (as well as its proof).

REMARK 3.5 Before developing a tractable approximation for $g_{i_1...i_k}$, we pause to illustrate the above material. Recall that

$$g_{ij} = 2\left(2\left[\frac{n}{p_i p_j}\right]+1\right)\left(\left[\frac{n}{p_i p_j}\right]-\left[\frac{n}{p_i^2 p_j}\right]-\left[\frac{n}{p_i p_j^2}\right]+\left[\frac{n}{p_i^2 p_j^2}\right]\right).$$

With $n = 200, i = 1$ and $j = 2$ (so that $p_i = 2$ and $p_j = 3$), we find that

$$g_{12} = 2\left(2\left[\frac{200}{6}\right]+1\right)\left(\left[\frac{200}{6}\right]-\left[\frac{200}{12}\right]-\left[\frac{200}{18}\right]+\left[\frac{200}{36}\right]\right) = 748.$$

Moreover, the number of positive b's being counted is $[\frac{200}{6}]-[\frac{200}{12}]-[\frac{200}{18}]+[\frac{200}{36}] = 11$. More tediously, we see that $\{b \in \mathbf{Z} : 1 \leq b \leq 200, 2|b, 4 \nmid b, 3|b, 9 \nmid b\} = \{6, 30, 42, 66, 78, 102, 114, 138, 150, 174, 186\}$, which is indeed a set of cardinality 11.
□

We have expressed a_n as a sum $\sum_{\nu} t_{\nu}^*$; each t_{ν}^* as a sum of terms of the form $\pm g_{i_1...i_k \nu}$; and each $g_{i_1...i_k}$ as a sum of terms of the form $\pm 2(2[\frac{n}{p_{i_1}...p_{i_k}}]+1)([\frac{n}{p_{i_1}^{\delta_{i_1}}...p_{i_k}^{\delta_{i_k}}}])$ with each $\delta_{i_j} \in \{1, 2\}$. Taking our cue from the proofs leading to the upper bounds in Proposition 3.3, we shall approximate each $g_{i_1...i_k}$ by deleting the " $+1$ " (arising from polynomials with $a = 0$) and by also deleting the greatest-integer function symbols. We let t_{ν}^{**} be the result of altering the above-displayed formula for t_{ν}^* by replacing each $g_{i_1...i_k \nu}$ with its canonical approximation. The key upshot of the above formulas is that

$$t_{\nu}^{**} = 4n^2 t_{\nu}$$

where t_{ν} is *independent* of n (provided only that $n \geq p_1^2 ... p_{\nu}^2$). We shall show in Proposition 3.7 (b) that, for each $\nu \geq 1$, the partial sum

$$T_{\nu} = \sum_{i=1}^{\nu} t_i$$

is a lower bound for $\underline{\lim}\frac{a_n}{c_n}$ (and, if $m \leq 3$, for $\underline{\lim}\frac{a_n}{b_n}$). First, we calculate T_{ν} for $\nu \leq 3$.

LEMMA 3.6 *(a) The approximation process described above replaces*

$$g_i \quad with \quad \frac{4n}{p_i}\left(\frac{n}{p_i}-\frac{n}{p_i^2}\right) = 4n^2\left(\frac{1}{p_i^2}-\frac{1}{p_i^3}\right);$$

$$g_{ij} \quad with \quad \frac{4n}{p_i p_j}\left(\frac{n}{p_i p_j}-\frac{n}{p_i^2 p_j}-\frac{n}{p_i p_j^2}+\frac{n}{p_i^2 p_j^2}\right) =$$

$$4n^2 \left(\frac{1}{p_i^2 p_j^2} - \frac{1}{p_i^3 p_j^2} - \frac{1}{p_i^2 p_j^3} + \frac{1}{p_i^3 p_j^3} \right);$$

$$g_{ijk} \quad \text{with} \quad 4n^2 \left\{ \frac{1}{p_i^2 p_j^2 p_k^2} - \frac{1}{p_i^3 p_j^2 p_k^2} - \frac{1}{p_i^2 p_j^3 p_k^2} - \frac{1}{p_i^2 p_j^2 p_k^3} + \right.$$

$$\left. \frac{1}{p_i^3 p_j^3 p_k^2} + \frac{1}{p_i^3 p_j^2 p_k^3} + \frac{1}{p_i^2 p_j^3 p_k^3} - \frac{1}{p_i^3 p_j^3 p_k^3} \right\}; \quad \text{etc.}$$

(b) $T_1 = t_1 = \frac{1}{8} = 0.125.$
(c) $t_2 = \frac{7}{108}$, and so $T_2 = \frac{41}{216} = 0.1898\ldots$
(d) $t_3 = \frac{7}{270}$, and so $T_3 = \frac{233}{1080} = 0.2157\ldots$
(e) $t_4 \approx 0.0137188208\ldots$, and so $T_4 = 0.2294\ldots$

Proof: (a) Apply the approximation process to the above formulas for g_i, g_{ij}, g_{ijk}, etc.

(b) Since $t_1^* = g_1$, it follows from (a) that

$$4n^2 t_1 = t_1^{**} = 4n^2 \left(\frac{1}{2^2} - \frac{1}{2^3} \right) = 4n^2 \left(\frac{1}{8} \right),$$

whence $t_1 = \frac{1}{8}$ and the remaining assertions are clear.

(c) Since $t_2^* = g_2 - g_{12}$, it follows from (a) that

$$4n^2 t_2 = t_2^{**} = 4n^2 \left(\frac{1}{3^2} - \frac{1}{3^3} \right) - 4n^2 \left(\frac{1}{2^2 3^2} - \frac{1}{2^3 3^2} - \frac{1}{2^2 3^3} + \frac{1}{2^3 3^3} \right) = 4n^2 \left(\frac{7}{108} \right),$$

whence $t_2 = \frac{7}{108}$. As $T_2 = t_1 + t_2$, the remaining assertions follow from (a).

(d), (e) Argue as above, using $t_3^* = g_3 - g_{13} - g_{23} + g_{123}$ and $t_4^* = g_4 - g_{14} - g_{24} - g_{34} + g_{124} + g_{134} + g_{234} - g_{1234}$. □

PROPOSITION 3.7 (a) *For each* $\nu \geq 1$, *there exist* $E_\nu, F_\nu \in \mathbf{Q}$ *and a positive integer* M *such that for all* $n \geq M$,

$$4n^2 T_\nu - E_\nu n - F_\nu \leq a_n.$$

(b) *Fix a positive integer* ν. *Then* $T_\nu \leq \underline{\lim} \frac{a_n}{c_n}$, *and so* $0.2294 < \underline{\lim} \frac{a_n}{c_n}$. *If* $m = 2$ *or* $m = 3$, *then* $T_\nu = \underline{\lim} \frac{a_n}{b_n} = \underline{\lim} \frac{a_n}{c_n}$, *and so* $0.2294 < \underline{\lim} \frac{a_n}{b_n}$.

Proof: (a) Fix ν, and consider all sufficiently large n (say, $n \geq p_\nu$). For the moment, we ignore the "+1" part of the formulas for the g's arising from polynomials with $a = 0$. Reasoning as in the proof of Lemma 3.2 (c), we see that for each $1 \leq i_1 < i_2 < \ldots < i_k < \nu$, the absolute value of the error incurred by estimating $g_{i_1 i_2 \ldots i_k \nu}$ by the above approximation process is at most $\gamma_{i_1 \ldots i_k \nu} n + \delta_{i_1 \ldots i_k \nu}$, where $\gamma_{i_1 \ldots i_k \nu}, \delta_{i_1 \ldots i_k \nu}$ are suitable positive real numbers which are independent of n. Applying the triangle inequality, we see from the formula for t_ν^* that the absolute value of the error incurred in estimating t_ν^* is thus at most $\gamma_\nu n + \delta_\nu$, for some positive constants γ_ν, δ_ν

which are independent of n. (For instance, take $\gamma_\nu := \sum_{\substack{1 \le i_1 < \ldots < i_k < \nu \\ 0 \le k \le \nu - 1}} (\gamma_{i_1 \ldots i_k \nu})$.)
Similarly, we then have that the absolute value of the resulting error in using $4n^2 T_\nu$
to approximate $\sum_{i=1}^{\nu} t_i^*$ is at most $e_\nu n + f_\nu + (2n + 1)$, where $e_\nu := \sum_{i=1}^{n} \gamma_i$ and
$f_\nu := \sum_{i=1}^{n} \delta_i$ are independent of n and the " $2n + 1$ " term bounds the number of
Eisensteinable polynomials with $a = 0$. Now, by choice of n, $\sum_{i=1}^{\nu} t_i^* \le a_n$ since
$p_\nu \le n$ and each $t_\lambda^* \ge 0$, and so

$$4n^2 T_\nu - \{e_\nu n + f_\nu + 2n + 1\} \le \sum_{i=1}^{\nu} t_i^* \le a_n.$$

With $E_\nu := e_\nu + 2, F_\nu := f_\nu + 1$ and $M := p_\nu$, the assertion follows.

(b) By (a), if $n \ge M$, then

$$4n^2 T_\nu - E_\nu n - F_\nu \le a_n,$$

and so

$$T_\nu - \frac{E_\nu n + F_\nu}{4n^2} \le \frac{a_n}{4n^2}.$$

Let $n \to \infty$. Hence, $T_\nu \le \lim \frac{a_n}{4n^2} = \lim \frac{a_n}{c_n} (= \lim \frac{a_n}{b_n})$, the equalities following by

reasoning as in Remark 3.4 with the aid of Proposition 3.3 (a). An application of
Lemma 3.6 (e) completes the proof. $\square$

The numerical estimate for the $\underline{\lim}$'s in Proposition 3.7 (b) could be enhanced by
routinely finding $T_5, T_6, \ldots$, for "large" ν by invoking the methodology of Lemma
3.6. However, such work will not settle the open question of whether, as we suspect,
$\lim_{n \to \infty} \frac{a_n}{c_n} (= \lim_{n \to \infty} \frac{a_n}{4n^2}; = \lim_{n \to \infty} \frac{a_n}{b_n}$ if $m \le 3)$ exists and equals the sum of
the infinite series $T := \sum_{\nu=1}^{\infty} t_\nu = \lim_{\nu \to \infty} T_\nu$. In closing the section, we note that
the convergence of T is established in Proposition 5.1 below.

4 ANALOGUES

In Remark 4.1, we show that the above themes persist for the analogous questions
involving arbitrary monic cubic integral polynomials. Then, in Remark 4.2, we raise
some related questions.

REMARK 4.1 (a) Fix an integer $m \ge 3$. Using polynomials of the form $X^m + aX^2 + bX + c \in \mathbf{Z}[X]$, define a_n^*, b_n^*, c_n^* analogously to the way a_n, b_n, c_n were defined in
terms of polynomials of the form $X^m + aX + b$. Of course, $c_n^* = (2n + 1)^3 \sim 8n^3$.
We next sketch how all the results of Sections 2 and 3 extend naturally in studying
$\overline{\lim}$ and $\underline{\lim}$ for the sequences $\frac{a_n^*}{b_n^*}$ and $\frac{a_n^*}{c_n^*}$.

First, we show that if $m = 3$, then $b_n^* \sim 8n^3$ (that is, the analogue of Proposition
2.1 holds). Consider $f = X^3 + aX^2 + bX + c \in \mathbf{Z}[X]$, with a rational root r. As
in the proof of Proposition 2.1, $f = (X - r)(X^2 + \lambda X + \mu)$ for some $\lambda, \mu \in \mathbf{Z}$ and
so, by equating corresponding coefficients, we have that $a = \lambda - r, b = \mu - r\lambda$, and

$c = -r\mu$. Thus, if $c \neq 0$, one verifies that $b = -\frac{c}{r} - r(r + a)$. The upshot is that for all $a \in \mathbf{Z}$, for all nonzero $c \in \mathbf{Z}$, and for all integral divisors r of c, there exists at most one $b \in \mathbf{Z}$ such that $X^3 + aX^2 + bX + c$ is reducible in $\mathbf{Q}[X]$. It follows that

$$b_n^* \geq (2n+1)^3 - (2n+1) \sum_{1 \leq |c| \leq n} 2\tau(c) - (2n+1)^2.$$

Using Dirichlet's asymptotic formula for the average of the τ function [2, Theorem 6.30] as in the proof of Proposition 2.1, one then shows that $b_n^* \sim 8n^3$.

The analogue of Section 3 holds for a_n^* as well (with no restriction on $m \geq 3$). In particular, the error estimate from Lemma 3.2 (c) carries over, albeit tediously, using already identified facts (and the convergence of $\sum_p \frac{1}{p^k}$ for $k \geq 2$). As a consequence, the $\underline{\lim}$'s in the analogue of Proposition 3.3 are bounded above by $S := \sum_p (\frac{1}{p^3} - \frac{1}{p^4})$. Of course, $S < S = \sum_p (\frac{1}{p^2} - \frac{1}{p^3})$, the upper bound from Proposition 3.3. Our interpretation of this inequality is the intuitive observation that the probability that an integral polynomial f satisfies Eisenstein's Criterion decreases as the number of nonzero coefficients of f increases.

The inclusion/exclusion processes of Section 3 carry over naturally to the present context. The analogue of Proposition 3.7 (a) is an inequality of the form

$$8n^3 \mathcal{T}_\nu - \mathcal{E}_\nu n^2 - \mathcal{F}_\nu n - \mathcal{G}_\nu \leq a_n.$$

One then obtains $\mathcal{T}_\nu$ as a lower bound for the $\underline{\lim}$'s in the analogue of Proposition 3.7 (b). Needless to say, $\mathcal{T}_\nu$ is just as effectively computable via explicit formulas as was T_ν in Lemma 3.6.

(b) Before undertaking the present project, we conducted an informal poll by asking several mathematicians what they thought "the probability that Eisenstein's Criterion applies to an arbitrary integral polynomial" might be. The consensus reply, which conformed to our intuition, was "very small, if not zero." Proposition 3.7 not withstanding, we now show how to use the methodology of this paper to prove that the consensus reply is correct, in the sense that the underlying frequency can be made arbitrarily small by considering polynomials with sufficiently many nonzero coefficients.

Let $\epsilon > 0$. Consider a positive integer $m \geq 3$ (m is to be chosen more carefully in a moment). Using polynomials of the form

$$X^m + v_1 X^{m-1} + v_2 X^{m-2} + \cdots + v_{m-1} X + v_m \in \mathbf{Z}[X],$$

define a_n^{**}, c_n^{**} analogously to the way a_n, c_n and a_n^*, c_n^* were defined above. Carrying out the analogue of the error estimate argument from Lemma 3.2 (c), we see (shades of S from (a)!) that

$$0 \leq \underline{\lim} \frac{a_n^{**}}{c_n^{**}} \leq \overline{\lim} \frac{a_n^{**}}{c_n^{**}} \leq \mathcal{S}_m := \sum_p \left(\frac{1}{p^m} - \frac{1}{p^{m+1}} \right).$$

A coarse application of the integral test, for the function defined by $f(x) := x^{-m}$, gives that

$$\mathcal{S}_m \le \frac{1}{2^m} + \frac{1}{(m-1)(2^{m-1})} \le \frac{1}{2^{m-1}} < \epsilon$$

provided only that $m > \max(3, 1 + \log_2(\frac{1}{\epsilon}))$. In particular, for all $\epsilon > 0$, there exists a positive integer M such that for all $m > M$, there exists a positive integer N such that the approximating probability $\frac{a_n^{**}}{c_n^{**}} < \epsilon$ for all $n > N$. This completes the proof of the "consensus reply." □

REMARK 4.2 (a) As Remark 4.1 reveals, the methodology of this paper adapts to any analogous Eisenstein-theoretic study where the ambient integral polynomials have any specified number of possibly nonzero coefficients. While analogues of a_n are well in hand, we suggest that the most pressing problem is to study b_n for $m \ge 4$. Specifically, we ask whether, in the asymptotic sense of Proposition 2.1, "most" m-th degree monic integral polynomials are irreducible in $\mathbf{Q}[X]$.

(b) As is shown by the the example $f = X^3 + X - 1$ (see [1, page 125]), some polynomials $f \in \mathbf{Z}[X]$ which are irreducible in $\mathbf{Q}[X]$ cannot have their irreducibility confirmed via Eisenstein's Criterion even if f is permitted a linear change of variable. It would be interesting to know how rare such polynomials f are. Accordingly, we propose a study in the spirit of this paper in which a_n and a_n^* are replaced by counting monic integral polynomials (with coefficients in $[-n, n]$) which become Eisensteinable after some linear change of variables.

(c) Finally, we raise the question of adapting the probabilistic concerns of this paper to the study of irreducibility criteria other than Eisenstein's, in particular to those discussed in [3] which feature Newton polygons.

5 ON EMPIRICAL EVIDENCE

At the close of Section 3, we mentioned our conjecture that the limiting probabilities $\lim_{n\to\infty} \frac{a_n}{c_n}$ and (if $m \le 3$) $\lim_{n\to\infty} \frac{a_n}{b_n}$ are given by the sum of the infinite series $T = \sum_{\nu=1}^{\infty} t_\nu$. Because of the complexity of the inclusion/exclusion processes leading up to the construction of this infinite series, we have deferred until now a proof that T actually converges. In fact, Proposition 5.1 establishes more, namely that T converges absolutely. To prepare for this proof, we first need to introduce additional notation to explicate further the construction of T.

Once again, n is a fixed positive integer, and $p_1, p_2, \ldots, p_\nu, \ldots$ is the enumeration of the positive prime integers in increasing order. If $1 \le i_1 < \ldots < i_k$ are integers, the approximation process which is described in Section 3 replaces $g_{i_1 \ldots i_k}$ with $4n^2 h_{i_1 \ldots i_k}$, where

$$h_{i_1 \ldots i_k} = \sum_{\delta_{i_j} \in \{2,3\}} \frac{\pm 1}{p_{i_1}^{\delta_{i_1}} \ldots p_{i_k}^{\delta_{i_k}}},$$

with the numerator of a term chosen as $+1$ (resp., -1) if the denominator of the term has $\delta_{i_j} = 3$ for an even (resp., odd) number of j's. It then follows from the formula for t_ν^* and the definition of t_ν that

$$t_\nu = \sum_{\substack{1 \le i_1 < \dots < i_k < \nu \\ 0 \le k \le \nu - 1}} (-1)^k h_{i_1 \dots i_k \nu}$$

for each $\nu \ge 1$.

PROPOSITION 5.1   *The infinite series T converges absolutely.*

Proof: Recall that if an infinite series converges absolutely, then so does any rearrangement of that series (see Dirichlet's Theorem [5, Theorem II, page 404]); and that a nonnegative series converges if it is dominated termwise by a convergent nonnegative series. It therefore follows from the above descriptions of t_ν and $h_{i_1 \dots i_k}$ that it suffices to prove that the infinite series $A := \sum_{k=1}^\infty \alpha_k$ converges (absolutely), where

$$\alpha_k := \sum_{\substack{1 \le i_1 < \dots < i_k \\ \delta_{i_j} \in \{2,3\}}} \frac{1}{p_{i_1}^{\delta_{i_1}} \dots p_{i_k}^{\delta_{i_k}}}.$$

We next bound the terms α_k. Consider the (absolutely) convergent infinite series

$$B = \sum_p \frac{1}{p^2} \text{ and } C = \sum_p \frac{1}{p^3}$$

where, as usual, the index of summation p runs over the set of all positive prime integers. For each $k \ge 2$, Merten's Theorem [5, Theorem II, page 406] allows $(B + C)^k$ to be calculated via Cauchy multiplication, since $B + C$ is absolutely convergent. Using the results on rearrangements and dominance mentioned above, one then verifies that $\alpha_k \le (B + C)^k$ for each $k \ge 1$.

It suffices to prove that $B + C < 1$, for A would then be dominated by a convergent nonnegative geometric series. Of course, $C \le B$ since $\frac{1}{p^3} < \frac{1}{p^2}$ for each p, and so it suffices to show that $B < 0.5$. As in the proof of Lemma 3.1, the integral test comes to the rescue. Write $B = \sum_{\nu=1}^\infty \frac{1}{p_\nu^2}$. A calculation finds the partial sum

$$B_7 := \sum_{\nu=1}^7 \frac{1}{p_\nu^2} = \frac{1}{4} + \frac{1}{9} + \frac{1}{25} + \frac{1}{49} + \frac{1}{121} + \frac{1}{169} + \frac{1}{289} < 0.44$$

with $B - B_7 = \sum_{\nu=8}^\infty \frac{1}{p_\nu^2} \le \int_{18}^\infty \frac{dx}{x^2} = \frac{1}{18} < 0.056$, whence $B < 0.496 < 0.5$. $\square$

COROLLARY 5.2 $T \le \varliminf \frac{a_n}{c_n}$. *If* $2 \le m \le 3$, *then* $T \le \varliminf \frac{a_n}{b_n} = \varliminf \frac{a_n}{c_n}$.

Proof: Let $\nu \to \infty$ in the assertions of Proposition 3.7 (b). $\square$

Recall from Proposition 3.3 that the limiting probabilities which are studied in our main theorem are related to the infinite series $S = \sum_p (\frac{1}{p^2} - \frac{1}{p^3})$. The above description of the series T reveals that S is much simpler conceptually than T in the following sense: if $p = p_\nu$, then the term $\frac{1}{p^2} - \frac{1}{p^3}$ of S is just h_ν and, hence, is only one of the summands in the above-displayed description of t_ν.

Table 5.1 The Probability $\frac{a_n}{b_n}$ for $n \le 30, m = 3$.

n	a_n	b_n	$\frac{a_n}{b_n}$
2	6	14	0.4286
3	12	32	0.375
4	16	56	0.2857
5	22	88	0.25
6	48	126	0.3810
7	54	176	0.3068
8	62	228	0.2719
9	70	292	0.2397
10	86	360	0.2389
11	92	440	0.2091
12	126	522	0.2414
13	132	620	0.2129
14	182	720	0.2528
15	232	828	0.2802
16	288	946	0.3044
17	294	1074	0.2737
18	360	1202	0.2995
19	366	1348	0.2715
20	412	1494	0.2758
21	474	1652	0.2869
22	548	1816	0.3018
23	554	1992	0.2781
24	624	2164	0.2884
25	640	2356	0.2716
26	726	2552	0.2845
27	750	2756	0.2721
28	804	2966	0.2711
29	810	3190	0.2539
30	956	3418	0.2797

Recall from Lemma 3.1 that $S \approx 0.28$. Another piece of evidence that the probability $\lim_{n\to\infty} \frac{a_n}{b_n}$ exists and is approximately 0.28 is provided by Table 5.1. This records a_n, b_n, and $\frac{a_n}{b_n}$ (rounded off to four decimal places) for $1 \le n \le 30$.

Compiling the data in Table 5.1 was very time-consuming. For each $n \le 30$, we listed all the polynomials $f = X^3 + aX + b \in \mathbf{Z}[X]$ such that $|a| \le n$ and $|b| \le n$; and then a_n (resp., b_n) counted those f to which Eisenstein's criterion applies (resp., which are irreducible in $\mathbf{Q}[X]$). For instance,

$$a_2 = |\{X^3 + 2, X^3 + 2X + 2, X^3 - 2X + 2, X^3 - 2, X^3 + 2X - 2, X^3 - 2X - 2\}| = 6$$

and $b_2 = |\{X^3 + X + 1, X^3 - X + 1, X^3 + 2X + 1, X^3 + X - 1, X^3 - X - 1,$
$X^3 + 2X - 1, X^3 - X + 2, X^3 - X - 2\}| + a_2 = 14,$

whence $\frac{a_2}{b_2} = \frac{6}{14} \approx 0.4286$. Nevertheless, despite all the effort represented by Table

5.1, we believe that $\lim_{n\to\infty} \frac{a_n}{b_n}$ is *not* approximately 0.28, for we believe that $\lim_{n\to\infty} \frac{a_n}{b_n}$ *is* approximately T. Thus, although Table 5.1 suggests there is *an* answer for the value of $\lim_{n\to\infty} \frac{a_n}{b_n}$, we believe that it points toward the *wrong* answer!

Our interpretation is the following. The relatively low values of n in Table 5.1 point to $\frac{a_n}{b_n} \to 0.28 \approx S = \sum(\frac{1}{p^2} - \frac{1}{p^3})$ because these values of n do not accommodate enough "highly composite" n to illustrate the attenuating effect of the inclusion/exclusion processes entering into the calculation of T. Thus we are moved to close with the following observation, which we intend both as moral and as pun: the strategy of relying on empirical evidence has its limits!

References

[1] D. Dobbs and R. Hanks, *A Modern Course on the Theory of Equations*, 2nd ed., Polygonal Publ. House, Washington, N.J., 1992.

[2] W.J. LeVeque, *Fundamentals of Number Theory*, Addison-Wesley Publ. Co., Reading, Mass., 1977.

[3] J.L. Mott, *Eisenstein-type irreducibility criteria*, Zero-dimensional Commutative Rings (D.F. Anderson and D.E. Dobbs, eds.), Lecture Notes in Pure and Appl. Math., vol. 171, Marcel Dekker, Inc., New York, 1995, pp. 307–329.

[4] I. Niven and H.S. Zuckerman, *An Introduction to the Theory of Numbers*, 4th ed., John Wiley & Sons, New York, 1980.

[5] J.M.H. Olmsted, *Advanced Calculus*, Appleton-Century-Crofts, Inc., New York, 1961.

When Is $D + M$ n-Coherent and an (n, d)-Domain?

DAVID E. DOBBS   Department of Mathematics, University of Tennessee, Knoxville, Tennessee 37996-1300

SALAH-EDDINE KABBAJ   Department of Mathematical Sciences, KFUPM, P.O.Box 849, Dhahran 31261, Saudi Arabia.

NAJIB MAHDOU   Département de Mathématiques et Informatique, Faculté des Sciences et Techniques Fès-Saïss, Université de Fès, Fès, Morocco.

MOHAMED SOBRANI   Département de Mathématiques et Informatique, Faculté des Sciences et Techniques Fès-Saïss, Université de Fès, Fès, Morocco.

1 INTRODUCTION

All rings considered below are commutative with unit, typically (integral) domains, and all modules and ring homomorphisms are unital. As its title suggests, this article contributes to a program which was begun in [7]. That article determined, i. a., when the classical $D + M$ construction (in which the ambient domain $K + M$ is a valuation domain) produces a coherent domain. In [8], to which the present article may be considered a sequel, [8, Theorem 3.6] treated the more general problem of characterizing n-coherence for the classical $D + M$ construction, with a complete answer being given in case D has quotient field K. (All relevant definitions, including that of n-coherence, will be recalled three paragraphs hence. For the moment, recall that 1-coherence is equivalent to coherence [8, page 270].)

It was noted in [3] that many of the themes and techniques in [7] carry over to more general $D + M$ contexts (in which $K + M$ need not be a valuation domain). In this spirit, our main result, Theorem 2.1, studies the transfer of n-coherence between a general $D+M$ construction and the associated ring D, with best results in case the ambient $K + M$ is a Bézout domain. In Theorem 2.8, we return to the classical $D + M$ context, to study the possible transfer of the strong n-coherence property between $D + M$ and D. Moreover, one upshot of Proposition 3.3 is that for coherent domains, strong n-coherence is equivalent to n-coherence.

There is also a markedly homological aspect to this article. For instance, Theorem 3.4 establishes that for a (context more general than a) general $D + M$ construction, $K + M$ is a flat module over $D + M$ if and only if D has quotient field K. (The proofs of many of our results, including Theorem 3.4, depend on resolutions, specifically, finding that the kernels of certain homomorphisms on $(D + M)^{(n)}$ are canonically isomorphic to $M^{(n-1)}$. While this observation is prominent in the homological considerations in [7, proof of Theorem 3], our first use of it occurred in the first-named author's proof of Proposition 4.5 (ii) in "On going-down for simple overrings II", Comm. Algebra 1 (1974), 439–458. This occurrence predates by two years its oft-cited occurrence in [3, Theorem 3].) Theorem 3.4 may be viewed as a companion for the results in [7, Theorem 7 and Corollary 8] on flatness of ideals in the classical $D + M$ construction.

In addition to pursuing resolution-theoretic themes from [7], the homological aspect of this work owes much to the classification of non-Noetherian rings initiated by Costa in [4]. Specifically, in addition to the n-coherence results described above, Theorem 2.1 also studies the transfer of the weak (n, d)-domain property between $D+M$ and D, Theorem 2.8 also studies the transfer of the (n, d)-domain property between $D + M$ and D, and Corollary 3.2 establishes that for coherent domains, the (n, d)- and the weak (n, d)-properties are equivalent.

This paragraph collects background from [8], [4] and [5] on the concepts mentioned above. Following [4] and [8], if n is a nonnegative integer, we say that an R-module E is n-presented if there is an exact sequence

$$F_n \longrightarrow F_{n-1} \longrightarrow \cdots \longrightarrow F_0 \longrightarrow E \longrightarrow 0$$

of R-modules in which each F_i is finitely generated and free; and that $\lambda(E) = \lambda_R(E) = sup\{n : E \text{ is an } n\text{-presented } R\text{-module}\}$. If $n \geq 1$, we say that R is n-coherent if each $(n-1)$-presented ideal of R is n-presented; and that R is strong n-coherent if each n-presented R-module is $(n+1)$-presented. (For other inequivalent usages of "n-coherent", see [8, page 270].) Given nonnegative integers n and d, we say that a ring R is an (n, d)-ring if $pd_R(E) \leq d$ for each n-presented R-module E (as usual, pd denotes projective dimension); and that R is a weak (n, d)-ring if $pd_R(I) \leq d - 1$ for each $(n - 1)$-presented ideal I of R. Since, in case R

is a domain, every finitely generated torsionfree R-module can be embedded in a finitely generated free R-module, R is an (n, d)-domain if and only if $pd_R(E) \leq d-1$ for each $(n - 1)$-presented torsionfree R-module E. We note, for motivation, that the (n, d)- and weak (n, d)-ring concepts are relevant to a sequel to [8] because, i.a., Prüfer domains are the $(1, 1)$-domains and the (possibly weak) $(2, 1)$-domains of $D + M$ type are tractable: cf. [4, Theorem 5.1], [5].

It is convenient to use "local" to refer to (not necessarily Noetherian) rings with a unique maximal ideal. Also, unadorned tensor products $\otimes$ are generally taken over the implicit base ring, not necessarily Z. Finally, note that the riding assumptions and notations for Section 2 are announced at its outset.

2 n-COHERENCE AND THE (n, d)-PROPERTY

Throughout this section, we adopt the following riding assumptions and notations: T is a domain of the form $T = K + M$, where K is a field and M is a nonzero maximal ideal of T; D is a subring of K; the quotient field of D is $k = qf(D) \subseteq K$; $R = D + M$; and $T_0 = k + M$.

THEOREM 2.1 *Let T, T_0 and R be as above. Then:*
1) *R is n-coherent $\Longrightarrow$ D is n-coherent;*
   R is a weak (n, d)-domain $\Longrightarrow$ D is a weak (n, d)-domain.
2) *Suppose that T is a Bézout domain and $[K : k] = \infty$. Then:*
   a) T_0 is a weak $(2, 1)$-domain but not coherent. In particular, T_0 is n-coherent $\forall n \geq 2$.
   b) R is not coherent. Moreover, $\forall n \geq 2$ and $\forall d \geq 1$, we have:
   R is n-coherent $\Longleftrightarrow$ D is n-coherent;
   R is a weak (n, d)-domain $\Longleftrightarrow$ D is a weak (n, d)-domain.
3) *Suppose that T is a Bézout domain, with $1 \neq [K : k] < \infty$, and M is not a principal ideal of T. Then:*
   a) T_0 is a weak $(2, 1)$-domain but not coherent. In particular, T_0 is n-coherent $\forall n \geq 2$.
   b) R is not coherent. Moreover, $\forall n \geq 2$ and $\forall d \geq 1$, we have:
   R is n-coherent $\Longleftrightarrow$ D is n-coherent;
   R is a weak (n, d)-domain $\Longleftrightarrow$ D is a weak (n, d)-domain.
4) *Suppose that T is a Bézout domain, with $1 \neq [K : k] < \infty$, and that D is a local $(n, 1)$-domain, for some $n \geq 2$. Then R is a weak $(n, 1)$-domain; in particular, R is m-coherent, $\forall m \geq n$.*
5) *Suppose that T is a Bézout domain and $k = K$. Then:*
   R is n-coherent $\Longleftrightarrow$ D is n-coherent;

R is a weak (n, d)-domain $\Longleftrightarrow$ *D is a weak (n, d)-domain*.

6) *Suppose that T is a local weak $(n, 1)$-domain for some $n \geq 1$ and $K = k$. Then:*

   R is n-coherent $\Longleftrightarrow$ *D is n-coherent;*

   R is a weak (n, d)-domain $\Longleftrightarrow$ *D is a weak (n, d)-domain*.

Before proving Theorem 2.1, we establish the following six Lemmas.

LEMMA 2.2 *Suppose that T is a Bézout domain. If I is a finitely generated ideal of R, then $I = Wa + Ma$, for some $a \in IT$ and some D-submodule W of K.*

Proof: Let I be a finitely generated ideal of R; without loss of generality, $I \neq 0$. Since T is a Bézout domain and IT is a nonzero finitely generated ideal of T, we have that $IT = Ta$, for some nonzero element $a \in IT$. As $aM = aTM = ITM = IM \subseteq I$, it follows that $M \subseteq (1/a)I$. Also, $(1/a)I \subseteq (1/a)IT = T$, and so $M \subseteq (1/a)I \subseteq T$. Put $W = (1/a)I \cap K$; evidently, W is a D-submodule of K. Moreover, $(1/a)I \cap M = M$, since $M \subseteq (1/a)I$. Hence $(1/a)I = (1/a)I \cap T = ((1/a)I \cap K) + ((1/a)I \cap M) = W + M$, and so $I = Wa + Ma$, as asserted.

LEMMA 2.3 *Let $A \to B$ be an injective flat ring homomorphism and let Q be an ideal of A such that $QB = Q$. Let E be an A-module such that $E \otimes_A B$ is B-flat. Then:*

1) $\lambda_A(E) \geq n \Longleftrightarrow \lambda_B(E \otimes B) \geq n$ *and* $\lambda_{A/Q}(E \otimes A/Q) \geq n$.

2) $pd_A(E) \leq d \Longleftrightarrow pd_B(E \otimes B) \leq d$ *and* $pd_{A/Q}(E \otimes A/Q) \leq d$.

Proof: 1) The assertion for the case $n = 0$ is a well-known result concerning finitely generated modules: cf. [11, Theorem 5.1.1(3)], [9].

Now, using induction on n, suppose the assertion holds for some $n \geq 0$ and let E be an $(n+1)$-presented A-module such that $E \otimes_A B$ is B-flat. We have an exact sequence $0 \to K \to A^m \to E \to 0$, where $\lambda_A(K) \geq n$ and m is some nonnegative integer. By hypothesis, B is a flat A-module; moreover, as in [5, proof of Lemma 1], $Tor_A^1(E, A/Q) = 0$. (Indeed, the exact sequence $0 \to Q \to A \to A/Q \to 0$ yields the exact sequence $0 \to Tor_A^1(E, A/Q) \to E \otimes Q \to E \to E/QE \to 0$. Since we still have $E \otimes_A Q \cong (E \otimes_A B) \otimes_B Q$, $E \otimes_A B \cong (E \otimes_A B) \otimes_B B$, $Q \subseteq B$ and $E \otimes_A B$ B-flat, it follows from [5, diagram (3)] that $Tor_A^1(E, A/Q) = 0$ as claimed). So tensoring over A with B and A/Q respectively, we get the following exact sequences :

(*) $0 \to B \otimes K \to B \otimes A^m (\cong B^m) \to B \otimes E \to 0$ and

         $0 \to A/Q \otimes K \to A/Q \otimes A^m (\cong (A/Q)^m) \to A/Q \otimes E \to 0$

of B- and A/Q-modules, respectively. On the other hand, since $\lambda_A(K) \geq n$ and

$K \otimes_A B$ is B-flat (using (*), since $E \otimes_A B$ is B-flat), the induction assumption applies to the A-module K; thus, $\lambda_B(B \otimes K) \geq n$ and $\lambda_{A/Q}(A/Q \otimes K) \geq n$. Therefore, the exact sequences (*) and [8, Lemma 2.2(b)] allow us to conclude that $\lambda_B(B \otimes E) \geq n+1$ and $\lambda_{A/Q}(A/Q \otimes E) \geq n+1$.

Conversely, let E be any A-module such that $\lambda_B(B \otimes E) \geq n+1$, $\lambda_{A/Q}(A/Q \otimes E) \geq n+1$, and $E \otimes_A B$ is B-flat. For some $m \geq 0$, we have an exact sequence $0 \to K \to A^m \to E \to 0$ of A-modules. The exact sequences (*), in conjunction with [8, Lemma 2.2(c)], yield that $\lambda_B(B \otimes K) \geq n$, $\lambda_{A/Q}(A/Q \otimes K) \geq n$, and $K \otimes_A B$ is B flat (since $E \otimes_A B$ is B-flat). By the induction assumption, it follows that $\lambda_A(K) \geq n$; and the exact sequence $0 \to K \to A^m \to E \to 0$, together with [2, Lemma 2.2(b)], shows that $\lambda_A(E) \geq n+1$.

2) We induct on d. The case $d = 0$ is well known: cf. [11, Theorem 5.1.1(1)], [14]; and the case $d = 1$ follows from the proof of [5, Lemma 1]. Let $d > 1$ and assume that 2) is true for any integer $d' < d$. Let E be an A-module such that $E \otimes_A B$ is B-flat. Suppose that $pd_A(E) \leq d$. Since B is A-flat, we have that $pd_B(E \otimes_A B) \leq d$. Choose an exact sequence of A-modules $0 \to K \to F \to E \to 0$ in which F is free. Hence, $pd_A(K) \leq d-1$. By the induction assumption, $pd_{A/Q}(K \otimes_A A/Q) \leq d-1$. Hence, by (*), $pd_{A/Q}(E \otimes_A A/Q) \leq d$.

Conversely, suppose that $pd_B(E \otimes_A B) \leq d$ and $pd_{A/Q}(E \otimes_A A/Q) \leq d$. As above, $Tor_1^A(E, A/Q) = 0$, so that $pd_{A/Q}(K \otimes_A A/Q) \leq d-1$, where K is the kernel of an epimorphism from a free A-module to E. Reasoning as above, $K \otimes_A B$ is B-flat and $pd_B(K \otimes_A B) \leq d-1$. Then, by the induction assumption, $pd_A(K) \leq d-1$, so that $pd_A(E) \leq d$, and this completes the proof.

LEMMA 2.4 *Consider the pullback*

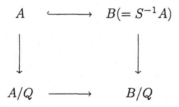

where $B = S^{-1}A$ *for some multiplicative subset* S *of* A, $A \to B$ *is an injective flat ring homomorphism, and* Q *is an ideal of both* A *and* B. *Then:*

1) *Assume that* B *is a local weak* (n, 1)-*domain. Let* I *be any nonzero* (n − 1)-*presented ideal of* A. *Then there exists* $0 \neq x \in B$ *and an ideal* $I' \supseteq Q$ *of* A *such that* $I \otimes A/Q \cong I'/Q$ *as* A/Q-*modules and* $I = xI' \cong I'$ *as* A-*modules.*

2) *Assume that* B *is a local weak* (n, 1)-*domain. Then:*

A/Q is n-coherent $\Longrightarrow$ *A is n-coherent;*

A/Q is a weak (n,d)-domain $\Longrightarrow$ *A is a weak (n,d)-domain.*

3) *Assume that B is a valuation domain and Q, the maximal ideal of B, is a finitely generated ideal of B. Then:*

A is n-coherent $\Longleftrightarrow$ *A/Q is n-coherent;*

A is a weak (n,d)-domain $\Longleftrightarrow$ *A/Q is a weak (n,d)-domain.*

Proof: 1) Let $I = \sum_{i=1}^{m} a_i A$ be any nonzero $(n-1)$-presented ideal of A. We have $I \otimes A/Q \cong I/IQ$. Since $IB(\cong I \otimes_A B)$ is an $(n-1)$-presented B-module and B is a local weak $(n,1)$-domain, IB is a nonzero projective, hence principal ideal of B. Hence there exists $0 \neq x \in B$ such that $IB = xB$; then $IQ = IQB = Q(IB) = xBQ = xQ$. Also, by replacing x with a suitable x', we may assume without loss of generality that $I = xI'$, where I' is an ideal of A. (In detail: $\forall i = 1, \ldots, m$, we have $a_i \in I \subseteq IB = xB$, then $\exists b_i \in A$ and $\exists s_i \in S$ such that $a_i = x(b_i/s_i)$. Thus, for $x' = x/\prod_{j=1}^{m} s_j \in B$, we have $a_i = x'b_i'$, where $b_i' = (\prod_{j=1, j \neq i}^{m} s_j)b_i \in A$ and $I' = \sum_{i=1}^{m} Ab_i'$. Then $I = x'I'$; and $IB = xB = x'B$ since elements of S are units in B.) Therefore, $IQ = xQ$ and $I = xI' \cong I'$ as A-modules, where I' is an ideal of A, so that we have: $I \otimes A/Q \cong I/IQ = xI'/xQ \cong I'/Q$ as A/Q-modules.

2) A/Q **is n-coherent** $\Longrightarrow$ A **is n-coherent:** Let I be any nonzero $(n-1)$-presented ideal of A. Since $I \otimes_A B \cong IB$ is a nonzero projective, hence principal ideal of B, we have that $I \otimes_A B$ is B-flat. Then Lemma 2.3, 1) may be applied to the given pullback and the A-module $E = I$, giving $\lambda_{A/Q}(I \otimes A/Q) \geq n - 1$. Express I via x and I' as in 1). Observe that $I \otimes_A B \cong IB = xB \cong B$ which, in particular, is an n-presented B-module. Now, $I \otimes A/Q \cong I'/Q$ is an $(n-1)$-presented ideal of the n-coherent ring A/Q, so $\lambda_{A/Q}(I \otimes A/Q) = \lambda_{A/Q}(I'/Q) \geq n$. Thus from Lemma 2.3, 1), $\lambda_A(I) \geq n$; and so A is n-coherent.

A/Q **is a weak (n,d)-domain** $\Longrightarrow$ A **is a weak (n,d)-domain:** Argue as above, using both Lemma 2.3, 1) and Lemma 2.3, 2).

3) A **is n-coherent** $\Longleftrightarrow$ A/Q **is n-coherent:** Since any valuation domain is a local weak $(n,1)$-domain $\forall n \geq 1$, then A/Q is n-coherent implies that A is n-coherent by 2). Conversely, let $J = I/Q$ be any nonzero $(n-1)$-presented ideal of A/Q, where I is an ideal of A such that $Q \subset I$. Moreover, IB is a finitely generated ideal of B since Q is. Then $I \otimes_A B \cong IB = xB \cong B$ for some $x \in B$, since B is a valuation domain and IB is a finitely generated ideal of B. In particular, $I \otimes_A B$ is B-flat. We apply Lemma 2.3 to the above pullback and the A-module $E = I$. We have $I \otimes_A B \cong B$ which, in particular, is an $(n-1)$-presented B-module. Moreover, $I \otimes A/Q \cong I/IQ = I/Q(=: J)$. Indeed, since $Q \subset I$, then $\exists b \in I \setminus Q$;

then b is a unit in B, whence $Q = bQ \subseteq IQ \subseteq Q$ and $Q = IQ$. As J is an $(n-1)$-presented A/Q-module, we now see from Lemma 2.3, 1) that $\lambda_A(I) \geq n - 1$. But A is assumed to be n-coherent, and so $\lambda_A(I) \geq n$. Thus from Lemma 2.3, 1), $\lambda_{A/Q}(J) = \lambda_{A/Q}(I \otimes A/Q) \geq n$, and so A/Q is n-coherent.

A **is a weak** (n, d)-**domain** $\Longleftrightarrow A/Q$ **is a weak** (n, d)-**domain:** Argue as above, using both Lemma 2.3, 1) and Lemma 2.3, 2). The proof is complete.

LEMMA 2.5 *Suppose that* T *is a Bézout domain (but not a field). Then each nonzero 1-presented ideal of* T_0 *is isomorphic to* T_0. *Consequently,* T_0 *is a weak* $(2, 1)$-*domain and* n-*coherent* $\forall n \geq 2$, *in each of the following cases:*

1) $[K : k] = \infty$,

2) $1 \neq [K : k] < \infty$ *and* M *is not a principal ideal of* T.

Proof: We first claim that M is not a finitely generated ideal of T_0. Indeed, [3, Lemma 1] shows that if $[K : k] = \infty$, then M is not a finitely generated ideal of T_0. On the other hand, if $1 \neq [K : k] < \infty$ and M is not a principal ideal of T, then M is not a finitely generated ideal of T_0. (Otherwise, M would be finitely generated, hence principal, over T, since T is a Bézout domain.) Thus, the claim has been established.

We shall prove that each nonzero 1-presented ideal I of T_0 is projective, in fact principal, over T_0. Use Lemma 2.2 to write $I = Wa + Ma$, where W is a k-submodule of K and $a \in IT$. Since M is not a finitely generated ideal of T_0, we have $W \neq 0$. Now, $I \otimes k \cong I \otimes T_0/M \cong I/IM = (Wa + Ma)/M(Wa + Ma) = (Wa + Ma)/Ma \cong Wa \cong W$ is a finite dimensional k-vector space, since I is a finitely generated ideal of T_0. Thus, there exists a nonnegative integer p such that $W \cong k^p$. We claim that $p = 1$.

Indeed, if $p \geq 2$, let $e_1, ..., e_p$ be a k-vector space basis of W, and consider the surjective T_0-module homomorphism

$$u : (k + M)^p \longrightarrow W + M (\cong I), \text{ given by}$$

$$(d_1 + m_1, \ldots, d_p + m_p) \mapsto \sum_{i=1}^{p}(d_i + m_i)e_i.$$

(To verify that u is surjective, it suffices to show that $im(u)$ contains each nonzero element $m \in M$. Consider a nonzero element $\alpha = \sum \delta_i e_i \in W$, with each $\delta_i \in k$. A straightforward calculation shows that $u(\delta_1 \alpha^{-1}m, \ldots, \delta_p \alpha^{-1}m) = m$.) Since I is a 1-presented T_0-module, $ker(u)$ is a finitely generated T_0-module. On the other hand, we have

$$ker(u) = \{(d_1 + m_1, \ldots, d_p + m_p) : \sum_{i=1}^{p}(d_i + m_i)e_i = 0\} = [\text{since } K \cap M = 0 \text{ and}$$

$$\{e_i\} \text{ is linearly independent over } k] = \{(m_1, \ldots, m_p) : \sum_{i=1}^{p} m_i e_i = 0\}.$$

Since $Me_i = M$ for each i, it follows that $ker(u) \cong M^{p-1}$. As $p - 1 \geq 1$ and $ker(u)$ is finitely generated over T_0, so is M, a contradiction. This proves the claim that $p = 1$, and so $W \cong k$. Hence $I = Wa + Ma \cong W + M \cong k + M = T_0$ as a T_0-module. In fact, we have also proved that each nonzero $(n-1)$-presented ideal of T_0 is isomorphic to T_0, hence infinitely presented, $\forall n \geq 2$, that is T_0 is n-coherent $\forall n \geq 2$, to complete the proof.

LEMMA 2.6 *Suppose that T is a Bézout domain (but not a field) such that*

$1 \neq [K : k] < \infty$. *Suppose also that D is a local $(n, 1)$-domain (but not a field) for some $n \geq 2$. Then each nonzero $(n-1)$-presented ideal of R is isomorphic to R. Consequently, R is a weak $(n, 1)$-domain and is m-coherent $\forall m \geq n$.*

Proof: Let I be any nonzero $(n-1)$-presented ideal of R. Use Lemma 2.2 to write $I = Wa + Ma$, where W is a D-submodule of K and $a \in IT$. Now, $M(\cong Ma)$ is not a finitely generated ideal of R (by [3, Lemma 1] since D is not a field), and so $W \neq 0$. Since R is D- flat, we have $\lambda_R(W \otimes_D R) = \lambda_R(WR) = \lambda_R(W(D+M)) = \lambda_R(W + M) = \lambda_R(I) \geq n - 1$; therefore, $\lambda_D(W) \geq n - 1$ since R is a faithfully flat D-module. On the other hand, since $W \subseteq K \cong k^r$ where $r = [K : k] < \infty$, there exists $0 \neq \delta \in D$ such that $W \cong \delta W \subseteq D^r$ since $k = qf(D)$. So, there exists a nonnegative integer m such that $W \cong D^m$ as a D-module (since D is a local $(n, 1)$-domain). We claim that $m = 1$.

Indeed, if $m \geq 2$, let $e_1, \ldots, e_m$ be a basis for W as a D-module. Consider the R-module homomorphism

$$v : (D + M)^m \to W + M (\cong I), \text{ given by}$$

$$(d_1 + m_1, \ldots, d_m + m_m) \mapsto \sum_{i=1}^{m} (d_i + m_i)e_i.$$

As in the proof of Lemma 2.5, v is surjective and $ker(v) \cong M^{m-1}$. Hence, since I is an $(n-1)$-presented R-module, $ker(v)$ is an $(n-2)$-presented R-module; in particular, $ker(v) \cong M^{m-1}$ is a finitely generated R-module. Thus, M is a finitely generated ideal of R, a contradiction. This proves the claim that $m = 1$. Therefore, $W \cong D$ and $I = Wa + Ma \cong W + M \cong D + M = R$ as R-modules, to complete the proof.

LEMMA 2.7 *Let n, d be nonnegative integers. If R is a weak (n, d)-domain, then so is D.*

Proof: Mimic the end of the proof of [5, Lemma 2], with Lemma 2.3 replacing the role of [5, Lemma 1].

Proof of Theorem 2.1: 1) Since R is faithfully flat over D, the first assertion follows from [8, Theorem 2.12, page 274]; and the second assertion is the conclusion of Lemma 2.7.

2) Assume that T is a Bézout domain and $[K : k] = \infty$.

a) T_0 is a weak $(2, 1)$-domain and n-coherent $\forall n \geq 2$ by Lemma 2..5, 1). On the other hand, T_0 is not coherent by [3, Theorem 3] since $[K : k] = \infty$.

b) R is not coherent by [3, Theorem 3]. Let $n \geq 2$ and $d \geq 1$.

R is n-**coherent** $\iff D$ is n-**coherent**: If R is n-coherent, then D is n-coherent by 1). Conversely, assume that D is n-coherent, and let I be any nonzero $(n - 1)$-presented ideal of R. By Lemma 2.2, write $I = Wa + Ma$, where W is a D-submodule of K and $a \in IT$. Since $Ma \cong M$ is not a finitely generated ideal of R by [3, Lemma 1], we have $W \neq 0$. Since R is D-flat, $\lambda_R(W \otimes_D R) = \lambda_R(WR) = \lambda_R(W(D + M)) = \lambda_R(W + M) = \lambda_R(I) \geq n - 1$; therefore, $\lambda_D(W) \geq n - 1$, since R is a faithfully flat D-module. Moreover, since T_0 is R-flat, $I \otimes T_0 \cong IT_0 = kWa + Ma$ is an $(n - 1)$-presented ideal of T_0. Thus, by Lemma 2.5, 1), since $n \geq 2$, IT_0 is isomorphic to T_0. Also, by the proof of Lemma 2.5, 1), we can identify $kW(\cong W \otimes_D k) \cong k$. Since W is finitely generated over D, there exists $0 \neq \delta \in D$ such that $W \cong \delta W \subseteq D$. But D is n-coherent, so $\lambda_D(W) = \lambda_D(\delta W) \geq n$ (since δW is an $(n - 1)$-presented ideal of D). Therefore, since R is D-flat, $\lambda_R(I) = \lambda_R(W + M) = \lambda_R(W \otimes_D R) \geq n$, and so R is n-coherent.

R is a weak (n, d)-**domain** $\iff D$ is a weak (n, d)-**domain**: If R is a weak (n, d)-domain, then D is a weak (n, d)-domain by 1). Conversely, assume that D is a weak (n, d)-domain, and let J be any nonzero $(n - 1)$-presented ideal of R. By Lemma 2.2, write $J = Wa + Ma$, where W is a D-submodule of K and $a \in JT$. Since $Ma \cong M$ is not a finitely generated ideal of R by [3, Lemma 1], $W \neq 0$. As in the above argument, we have $\lambda_D(W) \geq n - 1$ and there exists $0 \neq \delta \in D$ such that $W \cong \delta W \subseteq D$. Since D is a weak (n, d)-domain, $pd_D(W) = pd_D(\delta W) \leq d - 1$. Therefore, $pd_R(J) = pd_R(W + M) = pd_R(W \otimes_D R) \leq d - 1$ (the inequality holding since R is a flat D-module). Thus, R is a weak (n, d)-domain.

3) Argue as for 2).

4) This is a restatement of Lemma 2.6.

5) Assume that T is a Bézout domain and $K = k$.

R is n-**coherent** $\iff D$ is n-**coherent**: By 1), it remains to show that if D is n-coherent, then R is n-coherent. Without loss of generality, $R \neq T$, and so D is not a field. Let I be a nonzero $(n - 1)$-presented ideal of R. Write $I = Wa + Ma$, where W is a D-submodule of K and $a \in IT$. Since $Ma \cong M$ is not a finitely generated ideal of R by [3, Lemma 1], $W \neq 0$. We have $I \otimes T \cong IT$ (since $T = (D \setminus \{0\})^{-1} R$ is R-flat) $= Ta \cong T$ is T-flat, and Lemma 2.3 may be applied to $A = R, B = T, E = I$, and the pullback

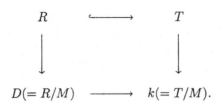

Now, $I \otimes T \cong Ta$ which is, in particular, an n-presented ideal of T. Also, $I \otimes R/M =$
$I/IM = (Wa + Ma)/(Wa + Ma)M = (Wa + Ma)/Ma \cong Wa \cong W$ is an $(n-1)$-
presented D-module by Lemma 2.3, 1) and so there exists $0 \neq \delta \in D$ such that
$\delta W \subseteq D$. Then we have $\lambda_{R/M}(I \otimes R/M) = \lambda_D(W) = \lambda_D(\delta W) \geq n$ since D is
n-coherent and δW is an $(n-1)$-presented ideal of D. Thus, by Lemma 2.3, 1),
$\lambda_R(I) \geq n$, and so R is n-coherent.

R **is a weak** (n, d)**-domain** $\Longleftrightarrow$ D **is a weak** (n, d)**-domain:** Argue as
above, using both Lemma 2.3, 1) and Lemma 2.3, 2).

6) Since $K = k$, we have that $T = S^{-1}R$, with $S = D \setminus \{0\}$. The assertions
now follow by combining 1) and Lemma 2.4, 2).

THEOREM 2.8 *Suppose that T is a valuation domain. Suppose also that one of
the following three conditions holds:*
(a) $[K : k] = \infty$; (b) $1 < [K : k] < \infty$ and $M = M^2$; (c) $K = k$.
Let n and d be nonnegative integers such that $n \geq 2$. Then:
1) *R is an (n, d)-domain* $\Longleftrightarrow$ *D is an (n, d)-domain.*
2) *R is strong n-coherent* $\Longleftrightarrow$ *D is strong n-coherent.*

We need the following lemma before proving Theorem 2.8:

LEMMA 2.9 *Suppose that T is an $(n_0, 1)$-domain for some $n_0 \geq 1$ and that $k = K$.
Let n and d be nonnegative integers such that $n \geq n_0$. Then:*
1) *R is an (n, d)-domain* $\Longleftrightarrow$ *D is an (n, d)-domain.*
2) *R is strong n-coherent* $\Longleftrightarrow$ *D is strong n-coherent.*

Proof: 1) $\Longleftarrow$) Using a criterion mentioned in the introduction, it suffices to show
that if E is an $(n-1)$-presented torsionfree R-module, then $pd_R(E) \leq d-1$. Since
$k = K$, T is R-flat. Thus, $E \otimes_R T$ is an $(n-1)$-presented torsionfree T-module
(since R is a domain, E embeds in some free R-module F, hence $E \otimes_R T$ embeds in

the free T-module $F \otimes_R T$); hence, it is T-projective since T is an $(n_0, 1)$-domain and $n \geq n_0$. Now $E \otimes_R D$ is $(n-1)$-presented by Lemma 2.3, 1) and is a torsionfree D-module as in the proof of [5, Lemma 2]. Since D is an (n, d)-domain, $pd_D(E \otimes_R D) \leq d - 1$. It follows from Lemma 2.3, 2) that $pd_R(E) \leq d - 1$.

$\Longrightarrow$) Assume that R is an (n, d)-domain. Let E be an $(n-1)$-presented torsionfree D-module. Replacing [5, Lemma 1] by our Lemma 2.3, we may mimic the end of the proof of [5, Lemma 2] to show that $pd_D(E) \leq d - 1$.

2) Argue as above, using Lemma 2.3, 1). This achieves the proof.

Proof of Theorem 2.8: If c) holds, Lemma 2.9 gives the conclusion. Next, suppose that a) or b) hold. Then [4, Corollary 5.2] shows that R is a $(2, 1)$-domain. Replacing T by T_0, Lemma 2.9 once more gives the result.

3   FURTHER RESULTS

In [8, page 277], the question was raised whether strong n-coherence is equivalent to n-coherence for $n \geq 2$. An affirmative answer is given in Proposition 3.3 for rings satisfying certain properties P_n, Q_n (defined below). It is shown in Proposition 3.1 that the P_n, Q_n conditions also imply the equivalence of the (n, d)-domain and the weak (n, d)-domain conditions. Since coherence implies P_n, Proposition 3.1 may be viewed as a companion of the result of [5, Proposition 2] that for a coherent ring, one has equivalence of the $(1, d)$-ring and the weak (n, d)-ring conditions. Finally, the section concludes, in the spirit of [7], by characterizing when $K + M$ is $(D + M)$-flat.

We next focus the setting for (3.1) - (3.3). Let R be a domain with quotient field Q, and M be a torsionfree R-module. As usual, $rank(M)$ denotes the Q-vector space dimension of $Q \otimes_R M$. An R-submodule M' of M is said to be *pure* (in M) if M/M' is a torsionfree R-module.

Let n be a positive integer. We say that R satisfies P_n if, for every $(n-1)$-presented torsionfree R-module M, there exists $f \in Hom(M, R^{rank(M)-1})$ such that $f(M)$ is n-presented. This is equivalent to saying that every nonzero $(n-1)$-presented torsionfree R-module M has a proper $(n-1)$-presented pure submodule. Observe that if R is a coherent domain, then R satisfies P_n, $\forall n$.

We say that R satisfies Q_n if, for every $(n-1)$-presented torsionfree R-module M, there exists a projective submodule M' of $R^{rank(M)-1}$ such that $M + M'$ is a projective R-module.

PROPOSITION 3.1 *Let* n, d *be positive integers. Let* R *be a domain which satisfies* P_n *or* Q_n. *Then the following conditions are equivalent:*

  a) R *is a weak* (n, d)-*domain;*

  b) R *is an* (n, d)-*domain.*

Proof: The implication b) $\Rightarrow$ a) holds even without the hypothesis of P_n or Q_n. Conversely, assume a), and let M be a nonzero $(n-1)$-presented torsionfree R-module. We have to show that $pd_R(M) \leq d-1$. We proceed by induction on $p = rank(M)$. If $p = 1$, then M is finitely generated over R and embeds canonically in the quotient field of R, whence M is isomorphic to an ideal of R and the assertion follows from a). We now proceed to the induction step, with $p > 1$.

  Suppose first that R satisfies P_n, so that M has a proper $(n-1)$-presented pure submodule M'. As $rank(M')$, $rank(M/M') \leq rank(M) - 1 = p - 1$, it follows from the induction assumption that $pd_R(M')$, $pd_R(M/M') \leq d-1$, whence $pd_R(M) \leq d-1$ as desired.

  Suppose next that R satisfies Q_n, so that $R^{rank(M)-1}$ has a projective submodule M' such that $M + M'$ is projective. Then $pd_R(M) \leq pd_R(M \cap M')$. Note that if $M \cap M' \neq 0$, then $M \cap M'$ satisfies the induction assumption. Thus, in all cases, $pd_R(M \cap M') \leq d-1$, whence $pd_R(M) \leq d-1$, completing the proof.

COROLLARY 3.2 *Let R be a coherent domain, and let n, d be positive integers. Then the following conditions are equivalent:*

  a) R *is a weak* (n, d)-*domain;*

  b) R *is an* (n, d)-*domain.*

Proof: Since coherence implies the P_n-property, Proposition 3.1 applies.

  By reasoning as in the proof of Proposition 3.1, one can prove the following result (cf. also [5, Proposition 2]).

PROPOSITION 3.3 *Let n be a positive integer. Let R be a domain which satisfies P_n or Q_n (for instance, let R be coherent). Then the following conditions are equivalent:*

  a) R *is n-coherent;*

  b) R *is strong n-coherent.*

  Finally, we turn to questions involving flatness in the $D + M$ construction. In view of a result [7, Theorem 7] for the classical $D + M$ context in which T is a valuation domain, one might well conjecture that if T is a domain and $k \neq K$, then T is not T_0-flat. This assertion is included in Corollary 3.5 below. First, we show that T is R-flat if and only if $k = K$.

THEOREM 3.4 *Let T be a domain of the form $K + M$, where K is a field and M is a nonzero ideal of T. Let $R = D + M$, where D is a subring of K. Then T is R-flat if and only if $qf(D) = K$.*

Proof: If $qf(D) = K$, then $T = S^{-1}R$ is R-flat, where $S = D \setminus \{0\}$. Conversely, assume that T is R-flat. Let $T_0 = k + M$, where $k = qf(D)$. Since T is R-flat, then $T \otimes_R T_0$ is T_0-flat. Now, $T \otimes_R T_0 = T \otimes_R S^{-1}R \cong S^{-1}T = T$, and so T is T_0- flat. Our aim is to show that $K = k$. Assume, on the contrary, that $K \neq k$. Choose a k-vector space basis $\{e_i : i \in L\}$ of K; well-order $L = \{1, 2, \ldots\}$. Consider the surjective T_0-module homomorphism

$$u : F(= T_0^{(L)}) \longrightarrow T(= K + M), \text{ given by}$$
$$(t_i)_i \mapsto \sum_i t_i e_i. \text{ Put } E = ker(u). \text{ Then}$$

$$E = \{(a_i + m_i)_i \in F : \sum_i (a_i + m_i)e_i = 0\} = [\text{since } K \cap M = 0]$$

$$= \{(m_i)_i \in F : \sum_i m_i e_i = 0\} \subseteq M^{(L)}.$$

Since $T \cong F/E$ is T_0-flat, we have from [13, Theorem 3.55, page 88] that $EI = E \cap FI$ for each ideal I of T_0. Consider $I = Ta$, where $0 \neq a \in M$. We have $EI \subseteq M^{(L)}I = (MI)^{(L)} = (MTa)^{(L)} = (Ma)^{(L)} = (M)^{(L)}a$. On the other hand, $FI = T_0^{(L)}I = (T_0I)^{(L)} = (T_0Ta)^{(L)} = (Ta)^{(L)} = (I)^{(L)}$. Let $m_1 = a$ and $m_2 = -(e_1/e_2)m_1 = -(e_1/e_2)a$. Set $f = (m_1, m_2, 0, 0, \ldots)$. Since $m_1 e_1 + m_2 e_2 = 0$, we have $f \in E$, and so $f \in E \cap FI$. However, $f \notin EI \subseteq M^{(L)}a$, since $m_1 = a \notin Ma$. This contradiction shows that $K = k$, thus completing the proof.

COROLLARY 3.5 *Under the hypothesis of Theorem 3.4, put $k = qf(D)$ and $T_0 = k + M$. Then T is T_0-flat if and only if $k = K$.*

Proof: This is the conclusion of Theorem 3.4 for the special case in which D is a field (for then $D = k$ and $T_0 = R$).

We close by noting that [3, Theorem 5] and [14, Theorem 1.1] lead to a direct proof of the special case of Corollary 3.5 in which T is assumed to be a Prüfer domain.

REFERENCES

1. N. Bourbaki, *Algèbre Commutative*, chapitres 1–4, Masson, Paris, 1985.
2. E. Bastida and R. Gilmer, *Overrings and divisorial ideals of rings of the form D + M*, Michigan Math. J. **20** (1973), 79–95.

3. J. W. Brewer and E. A. Rutter, *D + M constructions with general overrings*, Michigan Math. J. **23** (1976), 33–42.

4. D. L. Costa, *Parameterizing families of non-noetherian rings*, Comm. Algebra **22** (1994), 3997–4011.

5. D. L. Costa and S.E. Kabbaj, *Classes of D + M rings defined by homological conditions*, Comm. Algebra **24** (1996), 891–906.

6. D. E. Dobbs, *On the global dimensions of D + M*, Canad. Math. Bull. **18** (1975), 657–660.

7. D. E. Dobbs and I. J. Papick, *When is D + M coherent?*, Proc. Amer. Math. Soc. **56** (1976), 51–54.

8. D. E. Dobbs, S. E. Kabbaj and N. Mahdou, *n-coherent rings and modules*, Lecture Notes in Pure and Appl. Math., vol. 185, Marcel Dekker, Inc., New York, 1997, pp. 269–281.

9. D. Ferrand, *Descente de la platitude par un homomorphisme fini*, C. R. Acad. Sc. Paris **269** (1969), 946–949.

10. B. Greenberg, *Coherence in cartesian squares*, J. Algebra **50** (1978), 12–25.

11. S. Glaz, *Commutative Coherent Rings*, Lecture Notes in Math., vol. 1371, Springer-Verlag, Berlin, 1989.

12. I. Kaplansky, *Commutative Rings*, rev. ed., Univ. Chicago Press, Chicago, 1974.

13. J. J. Rotman, *An Introduction to Homological Algebra*, Academic Press, New York, 1979.

14. W. V. Vasconcelos, *Conductor, projectivity and injectivity*, Pacific J. Math. **46** (1973), 603–608.

Kaplansky Ideal Transform: A Survey

Marco Fontana   Università degli Studi "Roma Tre", Italy

0 INTRODUCTION

In 1956, M. Nagata [N1] introduced the *ideal transform* $T_R(I) = \cup_n(R : I^n)$ *of an integral domain R with respect to an ideal I* of R (cf. (3.3)). This transform proved very useful in his series of papers on the Fourteenth Problem of Hilbert (cf. [N1], [N2], [N3], [N4] and [N5]).

Hilbert's XIVth Problem
Let k be a field, $x_1, x_2, \ldots, x_n$ algebraically independent elements over k and let L be a subfield of $k(x_1, x_2, \ldots, x_n)$ containing k. Is the ring $k[x_1, x_2, \ldots, x_n] \cap L$ finitely generated over k?

Hilbert's problem was motivated by the following problem of invariant theory:

Hilbert's XIVth Problem (strict form)
Let k be a field, $x_1, x_2, \ldots, x_n$ algebraically independent elements over k and let G be a subgroup of $GL(n, k)$. Is the ring of invariants, $k[x_1, x_2, \ldots, x_n]^G$, subring of the polynomial ring $k[x_1, x_2, \ldots, x_n]$, finitely generated over k?

Positive answers to the Hilbert's XIVth problem were given, in particular cases, by D. Hilbert, E. Fischer, E. Noether and H. Weyl (cf. for instance [N5, Chapter 0]). The next significant contributions were made after Zariski generalized, in 1954, the original form of the problem in the following way:

Zariski's Problem [Z]
Let k be a field and A a finitely generated and integrally closed k-algebra with quotient field K. Let L be a subfield of K containing k. Is $A \cap L$ a finitely generated k-algebra?

Zariski answered this question, in the affirmative, when $\mathrm{tr.deg}_k L \leq 2$ and D. Rees in 1957 [Re] gave a counterexample when $\mathrm{tr.deg}_k L = 3$. Finally, in 1959, Nagata [N2] gave a counterexample to Hilbert's XIVth Problem, when $\mathrm{tr.deg}_k L = 4$.

One of the key steps for a negative solution to this type of problem, made by Nagata [N1], lies in the following result that shows clearly the rôle of the ideal transform:

A finitely generated field extension L of a given field k is called a Zariski field over k, if for each finitely generated and integrally closed k-algebra A with quotient field K and $K \supseteq L$, then $A \cap L$ is a finitely generated k-algebra. Then, L is a Zariski field if and only if, for each finitely generated and integrally closed k-algebra B, with quotient field L, and for each ideal I of B, the ideal transform $T_B(I)$ is finitely generated over B.

Ideal transforms have been proved to be very useful in other contexts of commutative algebra.

Nagata [N7] noted that the ideal transform $T_R(I)$ may be used in the study of the Catenary Chain Conditions.

Brewer [Br] introduced the ideal transform in the study of the overrings of an integral domain. After his work, several authors pursued the investigation of overrings by related means (cf., for instance, Brewer-Gilmer [BrG], Arnold-Brewer [AB], Heinzer-Ohm-Pendleton [HOP], Gilmer-Huckaba [GHu], Hedstrom [He1], [He2], Hays [Hy], Anderson-Bouvier [AnB], Fontana-Popescu [FP3], Fontana-Houston [FH]).

The ideal transform was also studied intensively in the not necessarily integral domain setting and the problem of when $T_R(I)$ is finitely generated, flat or integral over R was investigated by Anderson [A], Brodmann [Bro], Játem [J], Katz [Ka], Kiyek [Ki], McAdam-Ratliff [MR], Matijevic [Ma], Nishimura [Ni], Schenzel [S1], [S2], Zöschinger [Zö].

Ideal transform are also closely tied in with local cohomology and with affineness of the open subspaces of the prime spectrum (cf. Hartshorne [Ha], Serre [Se], Arezzo-Ramella [AR], Ohi [O] and the following Section 4).

However, outside of Noetherian setting, the behaviour of the ideal transform, as defined by Nagata, is not entirely satisfactory. For instance, in the study of the overrings, Brewer and Gilmer [BrG] obtained complete results, when considering ideal transforms with respect to finitely generated ideals, but only partial results (and conjectures) in the general case. Another aspect of the non satisfactory behaviour of the ideal transform, as defined by Nagata, will be examined in this paper when, in Section 4, we will look for a general result on the affineness of the open subspace $D(I) := \{P \in \mathrm{Spec}(R) : P \not\supseteq I\}$.

In the case of a non finitely generated ideal I of an integral domain R with quotient field K, a variant of the notion of ideal transform was introduced by Kaplansky [K2]. We call the *Kaplansky (ideal) transform of R with respect to an ideal I of R* the following overring of R:

$$\Omega_R(I) := \{z \in K : \mathrm{rad}(R :_R zR) \supseteq I\} .$$

Note that $\Omega_R(I)$ is an overring of the Nagata (ideal) transform of R with respect to I, since

$$T_R(I) = \{z \in K : (R :_R zR) \supseteq I^n \text{ for some } n \geq 1\}$$

and, if I is finitely generated, $\Omega_R(I) = T_R(I)$.

A natural and general approach to ideal transforms is to use multiplicative systems of ideals and generalized ring of fractions. This is the point of view that will be used in the present paper.

In Section 1, we review and complete some properties, concerning localizing systems of ideals and flatness, partially contained in [FHP, Chapter 5], in [HOP] and in a paper by Gabelli [Ga] published in this volume. In particular, we focus our attention on spectral localizing systems and to related overring properties (e.g. QQR, GQR, $\mathcal{F}QR$).

The second section is devoted to the study of flatness and finiteness of the overrings of an integral domain, by means of localizing systems. In particular, in this section we introduce and study a new class of saturated multiplicative systems of ideals, defined as follows: given a family of ideals of a domain R, $\mathcal{I} := \{I_\alpha : \alpha \in A\}$, consider $\mathcal{K}(\mathcal{I}) := \{J : \text{rad}(J) \supseteq \prod_{k=1}^{n} I_{\alpha_k}$, where $n \geq 1$ and $\{\alpha_1, \alpha_2, \ldots, \alpha_n\} \subseteq A\}$. The multiplicative system of ideals $\mathcal{K}(\mathcal{I})$, in several relevant cases, is a localizing system. Furthermore, in the Noetherian case, every localizing system is of the type $\mathcal{K}(\mathcal{I})$, for some family of ideals $\mathcal{I}$ (Lemma 2.9). Among other results, following the lines of a paper by E.L. Popescu [Po], we recover, in the integral domain case, a result proved by Schenzel [S3] on when a flat algebra is of finite type (Theorem 2.11).

In Section 3, we introduce the Kaplansky transform $\Omega_R(I)$ of an ideal I in a integral domain R as the ring of fractions of R with respect to the spectral localizing system associated to the open set $D(I) = \{P \in \text{Spec}(R) : P \not\supseteq I\}$. After verifying that this definition is equivalent to the classical one, given above [K2, p. 57], we recover the principal properties of the Kaplansky transform and its links with the Nagata transform. Following some ideas of Hays [Hy] (also developed later by Rhodes [Rh]), we study the Ω-ideal I^Ω (i.e. the unique maximal element in the set of the ideals J of R such that $\Omega(J) = \Omega(I)$). Among other properties, we give in Proposition 3.17 a "topological" interpretation of I^Ω, when considering the Zariski topological space $\text{Loc}(R)$ of the localizations of R [ZS]. From this topological point of view, we are led to consider the following radical ideal of R, associated to each R-submodule E of the quotient field of R:

$$\Omega^-(E) := \bigcap_{z \in E} \text{rad}((R :_R zR)) \ .$$

We show that the operators Ω and Ω^- establish a "sort of duality" between radical ideals and overrings of R (Theorem 3.26). In particular, we reobtain, for the integral domain case, some of the results proved by Rhodes [Rh] in the commutative ring case.

In the last section, we start by recalling a "geometric" interpretation of the Nagata transform. More precisely, if R is a Noetherian domain, $\text{Spec}(T_R(I))$ is canonically homeomorphic to the open subspace $D(I) = \{P \in \text{Spec}(R) : P \not\supseteq I\}$ if and only if $IT_R(I) = T_R(I)$ (Proposition 4.1). If R is not Noetherian, $D(I)$ may be an affine space even if $IT_R(I) \neq T_R(I)$. The main result of this section is a generalization of the characterization of the affineness of the open subspace $D(I)$ in the non (necessarily) Noetherian setting, by using the Kaplansky transform $\Omega_R(I)$ instead of the Nagata transform $T_R(I)$ (Theorem 4.4).

1 SPECTRAL LOCALIZING SYSTEMS

Let R be an integral domain with quotient field K. A *multiplicative system (of ideals) of R* is a set $\mathcal{S}$ of integral ideals of R closed under multiplication. The overring

(1.1) $R_\mathcal{S} := \{x \in K : xI \subseteq R$ for some $I \in \mathcal{S}\} = \cup\{(R : I) : I \in \mathcal{S}\}$

is called the *generalized transform* or the *(generalized) ring of fractions of R with respect to $\mathcal{S}$* (cf. [HOP], [H], [AB], [BS]). Note that, if $\mathcal{S}$ is a multiplicative system of

R, then also

(1.2) $\overline{S} := \{J : J$ an ideal of R such that $J \supseteq I$ for some $I \in S\}$

is a multiplicative system of R, called the *saturation of* S, and, moreover

(1.3) $R_{\overline{S}} = R_S$.

Obviously, a *saturated multiplicative system* is a multiplicative system S such that $S = \overline{S}$. We will concentrate our attention on the case where S is *non-trivial* i.e. $S \neq \emptyset$ and $(0) \notin S$ (or, equivalently, $\overline{S}$ is a non–trivial subset of ideals of R).

Given a multiplicative system of ideals S of R and a fractional ideal I of R, we set

(1.4) $I_S := \cup\{(I : J) : J \in S\}$.

It is easy to see that I_S is a fractional ideal of R_S and

(1.5) $IR_S \subseteq I_S$.

LEMMA 1.1. *Let S be a multiplicative system of ideals of R and $\overline{S}$ the saturation of S. Set $T := R_{\overline{S}}$ and $\mathcal{T} := \{JT : J \in S\}$, then $\mathcal{T}$ is a multiplicative system of ideals of T. Let $\overline{\mathcal{T}}$ be the saturation of $\mathcal{T}$ and set:*

$$\nabla(\overline{S}) := \{P \in \operatorname{Spec}(R) : P \notin \overline{S}\} \ , \quad \nabla(\overline{\mathcal{T}}) := \{Q \in \operatorname{Spec}(T) : Q \notin \overline{\mathcal{T}}\} \ .$$

Then
(1) If I is an ideal of R
$$I_S = R_S \Leftrightarrow I \in \overline{S} \ .$$

(2) The map
$$\nabla(\overline{S}) \longrightarrow \nabla(\overline{\mathcal{T}}) \ , \quad P \mapsto P_S$$

is an order-preserving bijection, with inverse map $Q \mapsto Q \cap R$.
Moreover, $R_P = T_{P_S}$ for each $P \in \nabla(\overline{S})$.

Proof: [Ga, Lemma 1.1] and [AB, Theorem 1.1].

$\square$

We will say that a *multiplicative system of ideals S of R is finitely generated* if, for each $I \in S$, there exists a finitely generated ideal J of R such that $J \subseteq I$ and $J \in S$.

A distinguished class of multiplicative systems of ideals is given by the localizing (or topologizing) system of ideals introduced by Gabriel [Gb] (cf. also [B, Ch. 2 p.157], [P] and [St]). We recall that a *localizing system (of ideals) $\mathcal{F}$ of R* is a set of integral ideals of R verifying the following conditions:

(LS1) $I \in \mathcal{F}$ and $I \subseteq J \Rightarrow J \in \mathcal{F}$;
(LS2) $I \in \mathcal{F}$, J an ideal of R such that $(J :_R iR) \in \mathcal{F}$, for each $i \in I, \Rightarrow J \in \mathcal{F}$.

To avoid uninteresting cases, we will consider in general only non trivial localizing systems, i.e. localizing systems $\mathcal{F}$ such that $\mathcal{F} \neq \emptyset$ and $(0) \notin \mathcal{F}$.

Since a localizing system $\mathcal{F}$ of R is a saturated multiplicative system of ideals of R [FHP, Proposition 5.1.11], we can consider the (generalized) ring of fractions $R_{\mathcal{F}}$ of R with respect to $\mathcal{F}$. It can be shown that an overring of R can be a (generalized) ring of fractions with respect to a multiplicative system, but not with respect to a localizing system (Example 1.18; cf. also [FP3, Theorem 2.5]).

If P is a prime ideal of R, we set

(1.6) $$\mathcal{F}(P) := \{I : I \text{ an ideal of } R, \ I \not\subseteq P\} \ ,$$

then it is easy to see that $\mathcal{F}(P)$ is a localizing system of R and $R_{\mathcal{F}(P)} = R_P$ [FHP, (5.1d)]. More generally, for each nonempty set Δ of prime ideals of R, we can consider

(1.7) $$\mathcal{F}(\Delta) := \cap\{\mathcal{F}(P) : P \in \Delta\} \ .$$

Since the intersection of localizing systems is a localizing system, we have [FHP, Proposition 5.1.4]:

LEMMA 1.2. *For each nonempty set Δ of prime ideals of an integral domain R, $\mathcal{F}(\Delta)$ is a localizing system of R and $R_{\mathcal{F}(\Delta)} = \cap\{R_P : P \in \Delta\}$.*

$\square$

A localizing system $\mathcal{F}$ of R is called a *spectral localizing system of R* if $\mathcal{F} = \mathcal{F}(\Delta)$ for some set Δ of prime ideals of R. Not every localizing system is spectral (Example 1.10).

LEMMA 1.3. *Let Δ be a nonempty subset of $\mathrm{Spec}(R)$. If we denote by*

(1.8) $$\Delta^{\downarrow} := \{Q \in \mathrm{Spec}(R) : Q \subseteq P \text{ for some } P \in \Delta\}$$

the closure under generizations of Δ inside $\mathrm{Spec}(R)$, then for each set Λ, with $\Delta \subseteq \Lambda \subseteq \Delta^{\downarrow}$, we have

(1.9) $$\mathcal{F}(\Lambda) = \mathcal{F}(\Delta) \quad (\text{in particular, } R_{\mathcal{F}(\Lambda)} = R_{\mathcal{F}(\Delta)}) \ .$$

Proof: Since $\Delta \subseteq \Lambda$, then $\mathcal{F}(\Lambda) \subseteq \mathcal{F}(\Delta)$. If $I \in \mathcal{F}(\Delta)$, then $I \not\subseteq P$ for each $P \in \Delta$, thus *a fortiori* $I \not\subseteq Q$ for each $Q \in \Lambda$, because $\Lambda \subseteq \Delta^{\downarrow}$. Therefore, $I \in \mathcal{F}(\Lambda)$.

$\square$

By standard arguments on partially ordered sets [AM, Proposition 6.1], we have:

LEMMA 1.4. *The following conditions are equivalent:*
(i) each nonempty subset Δ of $\mathrm{Spec}(R)$ has a maximal element;
(ii) R satisfies the acc on prime ideals.

$\square$

LEMMA 1.5. *Let R be an integral domain satisfying the acc on prime ideals. If Δ is a nonempty subset of $\mathrm{Spec}(R)$ and if Δ_0 is the (nonempty) subset consisting of the maximal elements of Δ, then*

$$\mathcal{F}(\Delta) = \mathcal{F}(\Delta_0) \ .$$

Proof: Since $\Delta^{\downarrow} = \Delta_0^{\downarrow}$, the conclusion follows from Lemma 1.3.

$\square$

If we start from a nontrivial localizing system $\mathcal{F}$ of R, we can associate to $\mathcal{F}$ the following nonempty subset of $\mathrm{Spec}(R)$:

(1.10) $$\nabla(\mathcal{F}) := \{P \in \mathrm{Spec}(R) : P \notin \mathcal{F}\} \ .$$

It is easy to see that $\nabla(\mathcal{F}) = \nabla(\mathcal{F})^{\downarrow}$ and [FHP, (5.1e)]:

LEMMA 1.6. *For each nontrivial localizing system $\mathcal{F}$ of R,*

$$\mathcal{F} \subseteq \mathcal{F}(\nabla(\mathcal{F})) \ .$$

$\square$

Conversely, if we start from a nonempty subset Δ of $\operatorname{Spec}(R)$, we have:

LEMMA 1.7. *Let $\mathcal{F} = \mathcal{F}(\Delta)$ be a spectral localizing system of an integral domain R, then*
(1) $\nabla(\mathcal{F}) = \Delta^{\downarrow}$;
(2) $\mathcal{F} = \mathcal{F}(\nabla(\mathcal{F}))$.

Proof: (1) If we show that $\Delta \subseteq \nabla(\mathcal{F}) \subseteq \Delta^{\downarrow}$, then the conclusion will follow because $\nabla(\mathcal{F}) = \nabla(\mathcal{F})^{\downarrow}$. Since, for each prime ideal P, $P \notin \mathcal{F}(P)$, then it is clear that

$$P \in \Delta \Rightarrow P \notin \mathcal{F}(\Delta) \Rightarrow P \in \nabla(\mathcal{F}) \ .$$

Moreover, if $Q \in \nabla(\mathcal{F})$ then $Q \notin \mathcal{F} = \cap\{\mathcal{F}(P) : P \in \Delta\}$, hence there exists $P \in \Delta$ such that $Q \notin \mathcal{F}(P)$. Therefore $Q \subseteq P$ for some $P \in \Delta$.
(2) follows from (1) and Lemma 1.3.

$\square$

COROLLARY 1.8. *For each integral domain R, the map*

$$\{\mathcal{F} : \mathcal{F} \text{ is a spectral localizing system of } R\} \rightarrow \{\Delta \subseteq \operatorname{Spec}(R) : \Delta = \Delta^{\downarrow}\}, \ \mathcal{F} \mapsto \nabla(\mathcal{F})$$

is an order–reversing bijection.

$\square$

Given a spectral localizing system $\mathcal{F} = \mathcal{F}(\Delta)$, we say that Δ *is irredundant for $\mathcal{F}$* if $\mathcal{F}(\Delta) \neq \mathcal{F}(\Delta \smallsetminus P)$ for each $P \in \Delta$.

We say that $\mathcal{F}$ is an *irredundant spectral localizing system* if there exists $\Delta \subseteq \operatorname{Spec}(R)$ such that $\mathcal{F} = \mathcal{F}(\Delta)$ and Δ is irredundant for $\mathcal{F}$. A nonempty subset Δ of $\operatorname{Spec}(R)$ is called 0-*dimensional* if, for each pair of distinct prime ideals P and Q in Δ, $P \nsubseteq Q$.

EXAMPLE 1.9. *A spectral localizing system which is not irredundant.*

Let V be a valuation domain, having the following prime spectrum:

$$(0) = P_0 \subset P_1 \subset P_2 \subset \cdots \subset P_n \subset \cdots \subset M = \cup\{P_n : n \geq 0\} \ .$$

Let $\mathcal{F} = \cap\{\mathcal{F}(P_n) : n \geq 0\} = \{M, V\}$. Clearly $\mathcal{F}$ is a spectral localizing system of V and $\nabla(\mathcal{F}) = \operatorname{Spec}(V) \smallsetminus \{M\}$. It is easy to see that $\mathcal{F}$ is not irredundant.

EXAMPLE 1.10. *A localizing system which is not spectral.*

Let V be a n–dimensional valuation domain, with $n \geq 2$. Suppose that the maximal ideal M of V is idempotent. Then $\mathcal{F} = \{M, V\}$ is a non–spectral localizing system of V, since $\mathcal{F}(M) = \{V\}$ and $\mathcal{F}(P) \supsetneq \{M, V\}$ for each prime ideal $P \neq M$.

The following result is due to S. Gabelli [Ga]:

LEMMA 1.11. *Let Δ be a nonempty subset of* Spec(R), *ordered under the set theoretical inclusion. Then Δ is irredundant for $\mathcal{F}(\Delta)$ if and only if Δ is 0–dimensional.*

$\square$

COROLLARY 1.12. *Let $\mathcal{F}$ be a nontrivial localizing system of R. Then $\mathcal{F}$ is an irredundant spectral localizing system if and only if each prime ideal of $\nabla(\mathcal{F})$ is contained in a maximal element (of $\nabla(\mathcal{F})$).*

Proof: By Lemma 1.11, $\mathcal{F}$ is an irredundant spectral localizing system if and only if there exists a 0–dimensional subset Δ of Spec(R) such that $\mathcal{F} = \mathcal{F}(\Delta)$. Since $\Delta^{\downarrow} = \nabla(\mathcal{F})$ (Lemma 1.7), then each $Q \in \nabla(\mathcal{F})$ is contained in some $P \in \Delta \subseteq \nabla(\mathcal{F})$. Therefore Δ necessarily coincides with the set of maximal elements of $\Delta(\mathcal{F})$. Conversely, if ∇_0 is the set of maximal elements of $\nabla(\mathcal{F})$, ∇_0 is obviously a 0–dimensional subset of Spec(R) and $\mathcal{F}(\nabla_0) = \mathcal{F}(\nabla(\mathcal{F})) = \mathcal{F}$ (Lemma 1.7 (2)).

$\square$

COROLLARY 1.13. *If R is an integral domain with* acc *on prime ideals, then each spectral localizing system is irredundant.*

Proof: This is an easy consequence of Corollary 1.12 and Lemma 1.4.

$\square$

A relevant class of irredundant spectral localizing systems is the class of the finitely generated localizing systems. We recall that a *localizing system $\mathcal{F}$* is *finitely generated* if, for each $I \in \mathcal{F}$, there exists a finitely generated ideal $J \in \mathcal{F}$ with $J \subseteq I$. Every finitely generated localizing system is spectral, in fact:

LEMMA 1.14. *Let $\mathcal{F}$ be a localizing system of R. The following statements are equivalent:*
 (i) *$\mathcal{F}$ is a spectral localizing system of R;*
 (ii) *$\mathcal{F} = \cap\{\mathcal{F}_\alpha : \alpha \in A\}$, where $\mathcal{F}_\alpha$ is a finitely generalized localizing system of R, for each $\alpha \in A$;*
 (iii) *for each ideal I of R with $I \notin \mathcal{F}$, there exists a prime ideal P of R such that $I \subseteq P$ and $P \notin \mathcal{F}$.*

Proof: [FHP, Proposition 5.1.7].

$\square$

From the previous lemma we can deduce the following:

COROLLARY 1.15. *If $\mathcal{F}$ is a finitely generated localizing system, then $\mathcal{F}$ is a spectral irredundant localizing system.*

Proof: From Lemma 1.14 ((ii) $\Rightarrow$ (i)), it is obvious that $\mathcal{F}$ is spectral and, from Lemma 1.14 ((ii) $\Rightarrow$ (iii)), we reobtain that $\mathcal{F} = \mathcal{F}(\nabla(\mathcal{F}))$ (cf. also Lemma 1.7). The conclusion will follow from Corollary 1.12, if we show that every chain of prime ideals of $\nabla(\mathcal{F})$ has an upper bound (in $\nabla(\mathcal{F})$). In fact, if $\{P_\lambda : \lambda \in \Lambda\}$ is a chain in $\nabla(\mathcal{F})$ and if $\cup_\lambda P_\lambda \in \mathcal{F}$, then there exists a finitely generated ideal $J \subseteq \cup_\lambda P_\lambda$ with $J \in \mathcal{F}$. It follows easily that $J \subseteq P_{\tilde{\lambda}}$ for some $\tilde{\lambda} \in \Lambda$, and hence $P_{\tilde{\lambda}} \in \mathcal{F}$: a contradiction.

$\square$

We note that not every irredundant spectral localizing system is finitely generated. This fact is a consequence of the following:

LEMMA 1.16. *Let $\mathcal{F}$ be a localizing system of an integral domain R. The following are equivalent:*

(i) *$\mathcal{F}$ is a finitely generated localizing system;*

(ii) *there exists a quasi-compact subspace Δ (i.e., every open covering of Δ has a finite sub-covering) of $\operatorname{Spec}(R)$ such that $\mathcal{F} = \mathcal{F}(\Delta)$;*

(iii) *each prime ideal of $\nabla(\mathcal{F})$ is contained in a prime ideal of $\nabla_0 := \{P \in \nabla(\mathcal{F}) : P \text{ is a maximal element of } \nabla(\mathcal{F})\}$ and ∇_0 is a quasi-compact subspace of $\operatorname{Spec}(R)$.*

Proof: [FHP, Proposition 5.1.8].

$\square$

EXAMPLE 1.17. *An irredundant spectral localizing system which is not finitely generated.*

Let R be a 1-dimensional Prüfer domain with $\operatorname{Max}(R) = \{M_n : n \geq 0\}$, where M_n is principal for each $n \geq 1$ and M_0 is not the radical of a finitely generated ideal (for an explicit example cf. [FHP, Theorems 8.2.1, 8.2.2 and 4.1.6]). Let $\Delta := \operatorname{Max}(R) \setminus \{M_0\}$. Then, clearly $\mathcal{F} := \mathcal{F}(\Delta)$ is an irredundant localizing system of R, since Δ is 0-dimensional (Lemma 1.11). But $\mathcal{F}$ is not finitely generated, since $M_0 \in \mathcal{F}$ because $M_0 \not\subseteq M_n$ for each $n \geq 1$; moreover, if $I \in \mathcal{F}$ and I is finitely generated with $I \subseteq M_0$, then $I \not\subseteq M_n$ for each $n \geq 1$, because $\mathcal{F} = \cap\{\mathcal{F}(M_n) : n \geq 1\}$, hence $M_0 = \operatorname{rad}(I)$ and we reach a contradiction. In a different terminology, we can say that Δ coincides with $\nabla(\mathcal{F}) \setminus \{0\}$ and it is a 0-dimensional non quasi-compact subspace of $\operatorname{Spec}(R)$, because each point $M_n \in \Delta$ is open and closed in the Zariski topology of $\operatorname{Spec}(R)$, for $n \geq 1$.

For each overring T of R, we can consider

$$(1.11) \qquad \Lambda = \Lambda(T) := \{N \cap R : N \in \operatorname{Max}(T)\}$$

which is a quasi-compact subspace of $\operatorname{Spec}(R)$, being the continuous image of a quasi-compact space. Therefore we can consider the finitely generated (irredundant spectral) localizing system $\mathcal{F}(\Lambda)$ of R. Since $R_{N \cap R} \subseteq T_N$, for each $N \in \operatorname{Max}(T)$, it follows easily that:

$$(1.12) \qquad R_{\mathcal{F}(\Lambda)} \subseteq T .$$

It is natural to ask when $R_{\mathcal{F}(\Lambda)} = T$. This is a particular case of the question of when an overring is an intersection of localizations. We introduce some terminology.

A *QR-overring* (respectively: *QQR-overring; GQR-overring; $\mathcal{F}QR$-overring*) T of an integral domain R is an overring such that $T = R_S$ (respectively: $T = \cap\{R_P : P \in Y\}$; $T = R_\mathcal{S}$; $T = R_\mathcal{F}$) for some multiplicative set S of elements of R (respectively: for some subset Y of $\operatorname{Spec}(R)$; for some multiplicative system $\mathcal{S}$ of ideals of R; for some localizing system $\mathcal{F}$ of R). We call a **P**-*domain* an integral domain for which every overring is a **P**-overring, where $\mathbf{P} \in \{QR, QQR, GQR, \mathcal{F}QR\}$. It is rather obvious that:

$$QR\text{–domain} \Rightarrow QQR\text{–domain} \Rightarrow \mathcal{F}QR\text{–domain} \Rightarrow GQR\text{–domain}$$

and it is well known that:

$$\text{Bézout domain} \Rightarrow QR\text{–domain} \Rightarrow \text{Prüfer domain} \Rightarrow QQR\text{–domain}$$

(cf. [G4], [GO], [R], [GH], [M], [Pe] and [D]). Moreover, in the integrally closed case, Prüfer domains, QQR-domains, $\mathcal{F}QR$-domains and GQR-domains coincide (cf. [H, Theorem 2.4], [FP2, Corollary 2.7] and [He3]).

In [FP2, Proposition 2.10], it is proved that each overring of a $\mathcal{F}QR$–domain is still a $\mathcal{F}QR$–domain. The following example, due to Heinzer [H, Example 2.9], shows not only that an overring of a GQR–domain is not necessarily a GQR–domain, as Heinzer proved, but also that there may exist an overring T of an integral domain R such that $T = R_\mathcal{S}$, for some multiplicative system of ideals $\mathcal{S}$ of R, but $T \neq R_\mathcal{F}$ for all localizing systems $\mathcal{F}$ of R.

EXAMPLE 1.18. *A (generalized) ring of fractions $R_\mathcal{S}$ of an integral domain R with respect to a multiplicative system of ideals $\mathcal{S}$ of R, which is not a (generalized) ring of fractions $R_\mathcal{F}$ of R with respect to some localizing system $\mathcal{F}$ of R.*

Let $k \subset K$ be a proper minimal extension of fields and let Y be an indeterminate over K. Pick a countable family of elements $x_1 = Y, x_2, x_3, \ldots$ in $K[[Y]]$ algebraically independent over K and set $F := K(x_1, x_2, x_3, \ldots, x_n, \ldots)$. We can consider

$$V_0 := K[[Y]] \cap F = K + M_0, \quad \text{with } M_0 := YK[[Y]] \cap F$$

which is a 1–dimensional discrete valuation domain with quotient field F. Set

$$K_n := K(x_1, x_2, \ldots, x_n)$$
$$F_{\tilde{n}} := K(\{x_i : i \neq n \ \ i \geq 1\})$$
$$W_n := K[[Y]] \cap F_{\tilde{n}} = V_0 \cap F_{\tilde{n}} \ .$$

Let V_n be the 1–dimensional valuation domain of F, associated to the valuation v_n, obtained extending to F the valuation w_n (associated to W_n) by setting $v_n(x_n) = \pi$, where π is a positive irrational number. It is easy to see that $V_0 \notin \{V_n : n \geq 1\}$ and $V_0 \cap K_n = V_m \cap K_n$ if $m > n$. Let

$$R_0 := k + M_0 \ , \quad \mathcal{S} := \cap \{V_n : n \geq 1\} \ , \quad R := R_0 \cap \mathcal{S} \ , \quad T := V_0 \cap \mathcal{S} \ .$$

It is not difficult to prove that the integral closure of R is T, which is a 1–dimensional Prüfer domain. Furthermore, $M_0 = (R_0 : V_0)$, $V_0 = (M_0 : M_0) = (R_0 : M_0)$ and if $\mathcal{S}_0 := \{M_0^k : k \geq 1\}$ then $(R_0)_{\mathcal{S}_0} = F$, since $\cap_k M_0^k = (0)$. Therefore, V_0 is not a GQR–overring of R_0, since every proper saturated multiplicative system of ideals of R_0 contains $\mathcal{S}_0$. However, Heinzer proved that R is a GQR–domain, even though $R_0 = R_{M_0 \cap R}$ is not GQR–domain. By [FP2, Proposition 2.10], we obtain that R is not a $\mathcal{F}QR$–domain because $R_0 = R_{M_0 \cap R}$ is not a $\mathcal{F}QR$–domain. More explicitly, set $P_0 := M_0 \cap R$, then

Claim. $V_0 = R_\mathcal{S}$, *where $\mathcal{S}$ is the multiplicative system of ideals of R given by $\{P_0^k I :$ I ideal of R with $I \not\subseteq P_0$ and $k \geq 0\}$, but $V_0 \neq R_\mathcal{F}$ for each localizing system $\mathcal{F}$ of R.*

The first part of the claim is proved in [H, p. 147]. Suppose that $V_0 = R_\mathcal{F}$. Note that $P_0 \in \mathcal{F}$, otherwise we would have $R_{P_0} = V_0$ (Lemma 1.1 (2)). Let

$$\mathcal{G} := \{J \text{ is an ideal of } R_0 : J \supseteq IR_0 \text{ for some } I \in \mathcal{F}\} \ .$$

It is easy to prove that $\mathcal{G}$ is a localizing system of R_0. In fact, if H is an ideal of R_0 and $I, I' \in \mathcal{F}$ are such that $(H :_{R_0} iR_0) \supseteq I'R_0$ for each $i \in I$, then $((H \cap R) :_R iR) = (H :_{R_0} iR_0) \cap R \supseteq I'R_0 \cap R \supseteq I'$, hence $H \cap R \in \mathcal{F}$ and thus $H \in \mathcal{G}$. Moreover, $(R_0)_\mathcal{G} \subseteq R_\mathcal{F} = V_0$, because if $xIR_0 \subseteq R_0$ for some $I \in \mathcal{F}$, then $xI \subseteq (xIR_0) \cap R \subseteq R$

and hence $x \in R_{\mathcal{F}}$. On the other hand, since $P_0 \in \mathcal{F}$, we have that $P_0 R_0 = M_0 \in \mathcal{G}$ and thus $(R_0)_{\mathcal{G}} = F$ and this fact leads to a contradiction.

We return to the problem of when the equality holds in (1.12). For this purpose, we consider the following relevant localizing system of R associated to an overring T of R:

$$(1.13) \qquad \mathcal{F}_0 = \mathcal{F}_0(T) := \{I : I \text{ ideal of } R \text{ such that } IT = T\} \,.$$

It is easy to see that $\mathcal{F}_0$ is a finitely generated (spectral irredundant) localizing system of R and, if $\Lambda := \Lambda(T)$, $\mathcal{F}_0 \subseteq \mathcal{F}(\Lambda)$, hence in particular

$$(1.14) \qquad R_{\mathcal{F}_0} \subseteq R_{\mathcal{F}(\Lambda)} \subseteq T \,.$$

LEMMA 1.19. *Let $\mathcal{F}$ be a given localizing system of R. Set $T := R_{\mathcal{F}}$, $\Lambda := \Lambda(T)$, $\nabla := \nabla(\mathcal{F})$ and $\mathcal{F}_0 := \mathcal{F}_0(T)$.*
(1) If $\mathcal{F}$ is a spectral localizing system of R, then

$$\mathcal{F}(\Lambda) \subseteq \mathcal{F} \quad (\text{in particular, } \mathcal{F}_0 \subseteq \mathcal{F}) \,.$$

(2) If $\mathcal{F} = \mathcal{F}_0$ then $\mathcal{F}(\Lambda) = \mathcal{F}$.

Proof: By Lemma 1.1, each $P \in \nabla$ is contained in some $Q := N \cap R \in \Lambda$, where $N \in \text{Max}(T)$. Therefore $\nabla \subseteq \Lambda^{\downarrow}$ and thus, by Lemma 1.3 and 1.7,

$$\mathcal{F}(\Lambda) = \mathcal{F}(\Lambda^{\downarrow}) \subseteq \mathcal{F}(\nabla) = \mathcal{F} \,.$$

Since we have already observed that, in general, $\mathcal{F}_0 \subseteq \mathcal{F}(\Lambda)$, then clearly $\mathcal{F}_0 \subseteq \mathcal{F}$ and, if $\mathcal{F}_0 = \mathcal{F}$, we obviously have $\mathcal{F}(\Lambda) = \mathcal{F}$.

$\square$

From the previous results, we easily deduce a sufficient condition for equality in (1.12):

COROLLARY 1.20. *Let T be an overring of an integral domain R. Set $\Lambda := \Lambda(T)$ and $\mathcal{F}_0 := \mathcal{F}_0(T)$. The following conditions are equivalent:*
(i) $R_{\mathcal{F}_0} = T$;
(ii) T is R–flat;
(iii) $\mathcal{F}_0 = \mathcal{F}(\Lambda)$ and $R_{\mathcal{F}(\Lambda)} = T$.

Proof: The equivalence (i) $\Leftrightarrow$ (ii) is well known (cf. [FHP, Remark 5.1.11 (b)] and also [Ak1] and [AB]).
   (iii) $\Rightarrow$ (i) is trivial.
   (ii) $\Rightarrow$ (iii). By [R, Theorem 2], we deduce that $R_{\mathcal{F}(\Lambda)} = T$. Since we already know that (i) $\Leftrightarrow$ (ii), then $T = R_{\mathcal{F}_0} = R_{\mathcal{F}(\Lambda)}$. The conclusion follows by applying Lemma 1.19 to the (spectral) localizing system $\mathcal{F}_0$.

$\square$

In order to make the references easier, in the next result due to Gabelli [Ga], we will collect some equivalent conditions, each of them will imply, in particular, the equality in (1.12).

PROPOSITION 1.21. *Let $\mathcal{S}$ be a nontrivial multiplicative system of ideals of an integral domain R. Set $T := R_{\mathcal{S}}$, $\Lambda := \Lambda(T)$, $\mathcal{F}_0 := \mathcal{F}_0(T)$, $\nabla := \nabla(\overline{\mathcal{S}})$ and let ∇_0 be the set of maximal elements inside ∇. The following conditions are equivalent:*

(i) *$IT = I_{\mathcal{S}}$, for each ideal I of R;*

(ii) *$JT = T$, for each ideal $J \in \mathcal{S}$;*

(iii) *$\overline{\mathcal{S}} = \mathcal{F}_0$;*

(iv) *$\overline{\mathcal{S}} = \mathcal{F}(\Lambda)$;*

(v) *$\nabla_0^{\downarrow} = \nabla$ and $\mathrm{Max}(T) = \{QT : Q \in \nabla_0\}$;*

(vi) *$\nabla_0 = \Lambda$;*

(vii) *for each ideal H of T, $H = (H \cap R)T = (H \cap R)_{\mathcal{S}}$;*

(viii) *$\mathrm{Spec}(T) = \{PT : P \in \nabla\}$.*

In particular, when these conditions are satisfied T is R-flat.

Proof: In [Ga, Proposition 1.2] the equivalence of all the conditions except (vi) is proved. It is obvious that (v) $\Rightarrow$ (vi) (cf. also Lemma 1.1).

(vi) $\Rightarrow$ (iv). We have already noticed that $I \in \overline{\mathcal{S}}$ if and only if $I_{\mathcal{S}} = T$ (Lemma 1.1 (1)). Let I be an ideal of R with $I \notin \overline{\mathcal{S}}$. Then $I_{\mathcal{S}} \subseteq N$ for some $N \in \mathrm{Max}(T)$, whence $I \subseteq N \cap R$ and thus $I \notin \mathcal{F}(\Lambda) = \mathcal{F}(\nabla_0)$. Conversely, if $I \notin \mathcal{F}(\nabla_0) = \mathcal{F}(\Lambda)$ then $I \subseteq N \cap R$ for some $N \in \mathrm{Max}(T)$. This fact implies that $I_{\mathcal{S}} \subseteq (N \cap R)_{\mathcal{S}}$. Since $N \cap R \in \Lambda = \nabla_0$, $(N \cap R)_{\mathcal{S}}$ is a prime ideal of T (Lemma 1.1 (2)). Therefore $I_{\mathcal{S}} \neq T$, whence $I \notin \overline{\mathcal{S}}$ (Lemma 1.1 (1)). $\square$

EXAMPLE 1.22. *A (non finitely generated) localizing system $\mathcal{F}$ and a prime ideal P in an integral domain R such that $PR_{\mathcal{F}} \neq P_{\mathcal{F}}$.*

Let V, M and $\mathcal{F}$ be as in Example 1.10. Then, in this case, $V_{\mathcal{F}} = (V : M) = V = (M : M) = M_{\mathcal{F}}$ and $MV_{\mathcal{F}} = MV = M$. Note that $M \notin \nabla(\mathcal{F})$ (Lemma 1.1).

COROLLARY 1.23. *Let $\mathcal{F}$ be a localizing system of an integral domain R. Set $T := R_{\mathcal{F}}$, $\Lambda := \Lambda(T)$, $\mathcal{F}_0 := \mathcal{F}_0(T)$. The following conditions are equivalent:*

(i) *$\mathcal{F}_0 = \mathcal{F}$;*

(ii) *$\mathcal{F}(\Lambda) = \mathcal{F}$;*

(iii) *$\mathcal{F}_0 = \mathcal{F}(\Lambda) = \mathcal{F}$.*

Proof: This is an easy consequence of Proposition 1.21. $\square$

We will see later (Example 4.7) that $\mathcal{F}_0$ may be equal to $\mathcal{F}(\Lambda)$, but $\mathcal{F}_0 \subsetneq \mathcal{F}$. The following corollary is also due to Gabelli [Ga, Theorem 1.3]:

COROLLARY 1.24. *Let R be a Prüfer domain and let $\mathcal{S}$ be a multiplicative system of ideals of R. Set $T := R_{\mathcal{S}}$ and $\mathcal{F}_0 := \mathcal{F}_0(T)$. Then:*

$$\mathcal{S} \text{ is finitely generated if and only if } \overline{\mathcal{S}} = \mathcal{F}_0 .$$

Therefore, in the Prüfer case, conditions (i)-(viii) of Proposition 1.21 are equivalent to the following:

(ix) *$\mathcal{S}$ is finitely generated.* $\square$

As a consequence of the previous corollary, we recover the following result proved in [FHP, Theorem 5.1.15]:

COROLLARY 1.25. *Let R be a Prüfer domain. For each overring T of R there exists a unique non trivial finitely generated localizing system $\mathcal{F}$ of R such that $T = R_{\mathcal{F}}$.*

$\square$

2 FLATNESS, FINITENESS AND LOCALIZING SYSTEMS

In some relevant cases, the link between flatness and the finitely generated property for overrings can be studied by using localizing systems. This point of view was developed by E.L. Popescu [Po], in order to extend some results of Schentzel [S3]. We begin by recalling the following characterizations of flat overrings:

LEMMA 2.1. *Let T be an overring of an integral domain R. The statements (i)-(iii) of Corollary 1.20 are equivalent to the following statements:*
(iv) $(R :_R yR)T = T$, *for each $y \in T$;*
 (v) *for each $Q \in \mathrm{Spec}(T)$, $R_{Q \cap R} = T_Q$;*
(vi) $P \in \mathrm{Supp}_R(T/R) \Rightarrow PT = T$.

Proof: The equivalences (ii) $\Leftrightarrow$ (iv) $\Leftrightarrow$ (v) are well known and proved in [R, Theorems 1 and 2].

(v) $\Rightarrow$ (vi). Let $P \in \mathrm{Supp}_R(T/R)$ and assume that $PT \neq T$. Let $\overline{Q} \in \mathrm{Spec}(T)$ such that $\overline{Q} \supseteq PT$ and let $\overline{P} := \overline{Q} \cap R$. By assumption, we deduce that $R_{\overline{P}} = T_{R \smallsetminus \overline{P}} = T_{\overline{Q}}$ and thus $(T/R)_{\overline{P}} = 0$, hence $\overline{P} \notin \mathrm{Supp}_R(T/R)$. This leads to a contradiction since $P \subseteq \overline{P}$ and $(T/R)_P \neq 0$.

(vi) $\Rightarrow$ (v). Assume that there exists a prime ideal Q of T such that $R_P \neq T_Q$, where $P := Q \cap R$. This fact implies that $R_P \neq T_{R \smallsetminus P}$ and thus $P \in \mathrm{Supp}_R(T/R)$. By assumption it follows that $PT = T$ and this is a contradiction, since $PT \subseteq Q$. $\square$

COROLLARY 2.2. *Let T be an overring of an integral domain R. Then*

$$T \text{ is } R\text{--flat} \Rightarrow \mathrm{Supp}_R(T/R) = \{P \in \mathrm{Spec}(R) : PT = T\} \ .$$

Proof: The inclusion $\mathrm{Supp}_R(T/R) \subseteq \{P \in \mathrm{Spec}(R) : PT = T\}$ follows from Lemma 2.1 ((ii) $\Rightarrow$ (vi)). Let $P \in \mathrm{Spec}(R)$ be such that $PT = T$. Assume that $P \notin \mathrm{Supp}_R(T/R)$. Then $R_P = T_{R \smallsetminus P}$ and thus $PR_P = PT_{R \smallsetminus P} \neq T_{R \smallsetminus P}$ and this contradicts the assumption $PT = T$.

$\square$

COROLLARY 2.3. *Let R be a Noetherian domain and T an overring of R.*

$$T \text{ is } R\text{--flat} \Leftrightarrow \mathrm{Ass}_R(T/R) \subseteq \{P \in \mathrm{Spec}(R) : PT = T\} \ .$$

Proof: ($\Rightarrow$) is a consequence of Corollary 2.2 and of the fact that $\mathrm{Ass}_R(T/R) \subseteq \mathrm{Supp}_R(T/R)$, since R is Noetherian [B, Ch. IV § 1 N.3 Corollaire 1].

($\Leftarrow$). In order to prove that T is R–flat, we show that condition (vi) of Lemma 2.1 holds. Let $P \in \mathrm{Supp}_R(T/R)$. Since R is Noetherian, there exists a prime ideal $\overline{P} \subseteq P$ such that $\overline{P} \in \mathrm{Ass}_R(T/R)$ [B, Ch. IV § 1 N.3 Proposition 7]. By assumption, $\overline{P}T = T$ and thus $PT = T$. $\square$

REMARK 2.4. Corollary 2.3 can be generalized outside the Noetherian setting, by using the notion of weak associated prime [B, Ch. IV § 1 Exercise 17], since it is known that if E is a R–module, a prime ideal P of R belongs to $\text{Supp}_R(E)$ if and only if contains a prime ideal $\overline{P}$ inside $\text{Ass}_R^w(E)$, the set of all the weakly associated primes of E.

Mutatis mutandis, it can be shown that *if R is an integral domain and T is an overring of R, then*

$$T \text{ is } R\text{–flat} \Leftrightarrow \text{Ass}_R^w(T/R) \subseteq \{P \in \text{Spec}(R) : PT = T\} \ .$$

COROLLARY 2.5. *If R is a Noetherian domain and T an overring of R, then*

$$T \text{ is } R\text{–flat} \Rightarrow T \text{ is Noetherian} \ .$$

Proof: From the equivalence (ii) $\Leftrightarrow$ (v) of Lemma 2.1, it follows that for each prime ideal Q of T, $(Q \cap R)T = Q$. Since R is Noetherian, $Q \cap R$ is finitely generated. Thus Q is finitely generated, hence T is Noetherian [K1, Theorem 8].

$\square$

We have shown that a flat overring of a Noetherian domain is Noetherian, even if, in general, it is not finitely generated (e.g. the quotient field of an integral Noetherian domain which is not a *G–domain*, i.e. an integral domain R for which (0) is different from its *pseudo–radical* $(0)^* := \cap\{P : P \in \text{Spec}(R), \ P \neq 0\}$ [K1, Theorem 18]). Our next goal is to characterize, in the general setting, the flat overrings that are finitely generated in terms of localizing systems.

Let $\mathcal{I} := \{I_\alpha : \alpha \in A\}$ be a given nonempty family of ideals of a domain R. Set

$$\mathcal{K}(\mathcal{I}) := \{J : J \text{ ideal of } R \text{ such that } \text{rad}(J) \supseteq \prod_{k=1}^{n} I_{\alpha_k},$$

$$\text{where } \{\alpha_1, \ldots, \alpha_n\} \text{ is a finite subset of } A\} \ .$$

LEMMA 2.6. *Let $\mathcal{I} = \{I_\alpha : \alpha \in A\}$ be a nonempty family of ideals of a domain R. Then*

(1) *$\mathcal{K}(\mathcal{I})$ is a saturated multiplicative system of ideals of R.*

(2) *If $\mathcal{I}$ is a finite family or if each ideal in $\mathcal{I}$ is finitely generated, then $\mathcal{K}(\mathcal{I})$ is a localizing system of R.*

(3) *If each ideal in $\mathcal{I}$ is finitely generated, then $\mathcal{K}(\mathcal{I})$ is a localizing system of R and*

$$\mathcal{K}(\mathcal{I}) = \{J : J \text{ ideal of } R \text{ such that } J \supseteq \prod_{k=1}^{n} I_{\alpha_k}^e \text{ where}$$

$$\{\alpha_1, \ldots, \alpha_n\} \text{ is a finite subset of } A \text{ and } e \geq 1\} \ .$$

Proof: (1) is an easy consequence of the fact that $\text{rad}(J'J'') = \text{rad}(J') \cap \text{rad}(J'') \supseteq \text{rad}(J')\text{rad}(J'')$.

(2). It is obvious that $\mathcal{K}(\mathcal{I})$ satisfies condition **(LS1)** of the definition of localizing system. Let H be an ideal of R such that $(H :_R jR) \in \mathcal{K}(\mathcal{I})$, for each $j \in J$, where $J \in \mathcal{K}(\mathcal{I})$, we claim that $H \in \mathcal{K}(\mathcal{I})$.

Case 1: Each ideal in $\mathcal{I}$ is finitely generated.

If $\mathrm{rad}(J) \supseteq \prod_{k=1}^{n} I_{\alpha_k}$ and if $i_{\alpha_k} \in I_{\alpha_k}$, then by the finiteness assumption there exists $N \gg 0$ such that

$$\prod_{k=1}^{n} i_{\alpha_k}^{N} \in J .$$

Therefore, $\mathrm{rad}\left((H :_R (\prod_{k=1}^{n} i_{\alpha_k}^{N}) R)\right) \supseteq \prod_{h=1}^{m} I_{\beta_h}$ for some finite subset $\{\beta_1, \ldots, \beta_m\}$ of A. By the finiteness assumption, we can take the same set $\{\beta_1, \ldots, \beta_m\}$ for all elements $\prod_{k=1}^{n} i_{\alpha_k}^{N}$. If $i_{\beta_h} \in I_{\beta_h}$, then as above, there exists $M \gg 0$ such that

$$\prod_{h=1}^{m} i_{\beta_h}^{M} \prod i_{\alpha_k}^{N} \in H .$$

This fact implies that $\prod_{h=1}^{m} I_{\beta_h} \prod_{k=1}^{n} I_{\alpha_k} \subseteq \mathrm{rad}(H)$ and thus $H \in \mathcal{K}(\mathcal{I})$.

Case 2: $\mathcal{I} = \{I_1, I_2, \ldots, I_t\}$.

If $J \in \mathcal{K}(\mathcal{I})$ then $\mathrm{rad}(J) \supseteq \prod_k I_k$. Let $i_k \in I_k$ and let H be an ideal of R such that $\mathrm{rad}((H :_R \prod_k i_k R)) \supseteq \prod_k I_k$. We can find an integer $N \geq 0$ (depending on $\prod_k i_k$) such that $(\prod_k i_k)^{N} \in (H :_R \prod_k i_k R)$. We conclude that $(\prod_k i_k)^{N+1} \subseteq H$ and thus $\prod_k I_k \subseteq \mathrm{rad}(H)$.

(3) If $\mathrm{rad}(J) \supseteq \prod_{k=1}^{n} I_{\alpha_k}$ and I_{α_k} is a finitely generated ideal of R for each α_k, then there exists an integer $e \gg 0$ such that $J \supseteq (\prod_{k=1}^{n} I_{\alpha_k})^{e}$. Conversely, if $J \supseteq \prod_{k=1}^{n} I_{\alpha_k}^{e}$ then $\mathrm{rad}(J) \supseteq \mathrm{rad}(\prod_{k=1}^{n} I_{\alpha_k}^{e}) = \mathrm{rad}(\prod_{k=1}^{n} I_{\alpha_k}) \supseteq \prod_{k=1}^{n} I_{\alpha_k}$.

$\square$

If $\mathcal{I}$ is the set of all the maximal ideals of R, the ring of fractions of R with respect to the saturated multiplicative system of ideals $\mathcal{K}(\mathcal{I})$ is the *global transform* of R. This ring was introduced in the case of Noetherian rings (not necessarily domains) by Matijevic [Ma] (cf. also [A] and [Zö]).

If $\mathcal{I} = \{I\}$, where I is an ideal of R, we denote simply by $\mathcal{K}(I)$ the localizing system $\mathcal{K}(\{I\})$ of R.

COROLLARY 2.7. *Let I be an ideal of an integral domain R.*
(1) $\mathcal{K}(I) = \{J : J \text{ is an ideal of } R \text{ such that } \mathrm{rad}(I) = \mathrm{rad}(J \cap I)\}$.
(2) *If I is finitely generated, then:*

$$\mathcal{K}(I) = \{J : J \text{ ideal of } R \text{ such that } J \supseteq I^{e} \text{ for some } e \geq 1\} .$$

Proof: (1) follows from the fact that

$$\mathrm{rad}(I) = \mathrm{rad}(I \cap J) = \mathrm{rad}(I) \cap \mathrm{rad}(J) \Leftrightarrow \mathrm{rad}(I) \subseteq \mathrm{rad}(J)$$
$$\Leftrightarrow I \subseteq \mathrm{rad}(J)$$

(2) is a particular case of Lemma 2.6 (3).

$\square$

PROPOSITION 2.8. *Let T be an overring of an integral domain R and let $\mathcal{I} := \{(R :_R yR) : y \in T\}$. Then*

$$T \text{ is } R\text{-flat} \Leftrightarrow \mathcal{F}_0(T) = \mathcal{K}(\mathcal{I}) .$$

Proof: ($\Rightarrow$). Note that if J is an ideal of R then

$$\mathrm{rad}(J)T = T \Leftrightarrow JT = T .$$

Since T is R–flat, by Lemma 2.1 ((ii) $\Rightarrow$ (iv)) it follows that the saturated multiplicative system $\mathcal{K}(\mathcal{I})$ is contained in $\mathcal{F}_0(T)$. The conclusion follows from Proposition 1.21 ((ii) $\Rightarrow$ (iii)).

($\Leftarrow$). Since, for each $y \in T$, $(R :_R yR) \in \mathcal{K}(\mathcal{I})$ then, by assumption, $(R :_R yR) \in \mathcal{F}_0(T)$ and this is equivalent to $(R :_R yR)T = T$. The conclusion follows from Lemma 2.1 ((iv) $\Rightarrow$ (ii)).

□

The following result is due essentially to E.L. Popescu [Po, Theorem 5]:

PROPOSITION 2.9. *Let T be an overring of an integral domain R. Then the following statements are equivalent:*
(i) *T is flat and finitely generated over R;*
(ii) *there exists a (finitely generated) ideal I of R such that $\mathcal{F}_0(T) = \mathcal{K}(I)$ and $T = R_{\mathcal{K}(I)}$.*

Proof: (i) $\Rightarrow$ (ii). Let $T = R[y_1, \ldots, y_n]$ and set

$$I := (R :_R y_1 R)(R :_R y_2 R) \cdots (R :_R y_n R) .$$

Since T is R–flat, then $IT = T$ (Lemma 2.1 ((ii) $\Rightarrow$ (iv))) thus $I \in \mathcal{F}_0(T)$ and, hence, $\mathcal{K}(I) \subseteq \mathcal{F}_0(T)$. Conversely if $JT = T$, then $\sum_{k=1}^{r} j_k t_k = 1$ with $j_k \in J$ and $t_k \in T$. Since each $t_k \in R[y_1, \ldots, y_n]$ then there exists $N \geq 0$ such that $t_k I^N \subseteq R$ for each k, $1 \leq k \leq r$. Therefore

$$I^N = I^N \cdot 1 = I^N \sum_{k=1}^{r} j_k t_k = \sum_{k=1}^{r} j_k t_k I^N \subseteq J$$

and whence $J \in \mathcal{K}(I)$ (Corollary 2.7 (2)).

Furthermore, since we have proved that $\mathcal{K}(I) = \mathcal{F}_0(T)$ and T is R-flat then, by Corollary 1.20, $T = R_{\mathcal{F}_0(T)} = R_{\mathcal{K}(I)}$.

If I is not a finitely generated ideal of R, we can find a finitely generated ideal I_* such that $I_* \subseteq I$ and $I_* \in \mathcal{K}(I) = \mathcal{F}_0(T)$. This fact implies that $\mathcal{K}(I) \subseteq \mathcal{K}(I_*)$. On the other hand, since $I_* \in \mathcal{F}_0(T)$, it is easy to see that $\mathcal{K}(I_*) \subseteq \mathcal{F}_0(T)$. We conclude immediately that $\mathcal{K}(I) = \mathcal{K}(I_*) = \mathcal{F}_0(T)$.

(ii) $\Rightarrow$ (i). By the previous argument, we can assume that I is finitely generated. Let $I = (i_1, \ldots, i_s)$. Since $I \in \mathcal{K}(I)$, then by assumption $IT = T$. Therefore $1 = \sum_{h=1}^{s} i_h y_h$ for $y_h \in T$. We claim that $R[y_1, \ldots, y_s] = T$. We observe that, from the assumption, we have $T = R_{\mathcal{K}(I)} = R_{\mathcal{F}_0(T)}$ and hence T is R-flat (Corollary 1.20). Let $t \in T = R_{\mathcal{K}(I)}$, then $(R :_R tR) \in \mathcal{K}(I)$ and hence there exists $e \geq 1$ such that $I^e \subseteq (R :_R tR)$ (Corollary 2.7 (2)). On the other hand we know that $IR[y_1, \ldots, y_s] = R[y_1, \ldots, y_s]$. Thus also $I^e R[y_1, \ldots, y_s] = R[y_1, \ldots, y_s]$, hence

$$1 = \sum_{k=1}^{r} \iota_k z_k \quad \text{with } \iota_k \in I^e \text{ and } z_k \in R[y_1, \ldots, y_s], \quad r \geq 1 ,$$

and so

$$t = \sum_{k=1}^{r} (\iota_k t) z_k \in R[y_1, \ldots, y_s] ,$$

because $\iota_k \in I^e$ and $I^e t \subseteq R$.

□

In the case of Noetherian domains, we can describe every localizing system in terms of localizing systems of type $\mathcal{K}(\mathcal{I})$, for some family of (prime) ideals $\mathcal{I}$.

LEMMA 2.10. *Let R be a Noetherian domain and $\mathcal{F}$ a localizing system of R. Then, there exists a quasi–compact subspace Δ of $\mathrm{Spec}(R)$ such that, if $\mathcal{I} := \{Q \in \mathrm{Spec}(R) \smallsetminus \Delta\}$, then*

$$\mathcal{F} = \mathcal{F}(\Delta) = \mathcal{K}(\mathcal{I}) \ .$$

Proof: Since every localizing system of a Noetherian domain is finitely generated, then $\mathcal{F} = \mathcal{F}(\Delta)$ where $\Delta := \{P \in \mathrm{Spec}(R) : P \notin \mathcal{F}\}$ (Lemma 1.16 and Lemma 1.7). It is obvious that if $Q \in \mathcal{I} = \mathrm{Spec}(R) \smallsetminus \Delta$ then $Q \in \mathcal{F}(\Delta) = \mathcal{F}$ and thus $\mathcal{K}(\mathcal{I}) \subseteq \mathcal{F} = \mathcal{F}(\Delta)$ (Lemma 2.6 (3)). Conversely, if $J \in \mathcal{F}$ is a proper ideal then the minimal primes of J do not belong to Δ. The conclusion follows since, in a Noetherian ring, every ideal contains a power of its radical, so that there exists $e \geq 1$ and $Q_1, \ldots, Q_t \in \mathcal{I}$ such that $Q_1^e \cdot \ldots \cdot Q_t^e \subseteq J$.

$\square$

COROLLARY 2.11. *Let T be an overring of a Noetherian integral domain. Then the following statements are equivalent:*
(i) *T is R–flat;*
(ii) *$\mathcal{F}_0(T) = \mathcal{K}(\mathrm{Supp}_R(T/R))$.*

Proof: (ii) $\Rightarrow$ (i). Since every $P \in \mathrm{Supp}_R(T/R)$ belongs to $\mathcal{K}(\mathrm{Supp}_R(T/R))$ and, in the Noetherian case, $\mathrm{Ass}_R(T/R) \subseteq \mathrm{Supp}_R(T/R)$ [B, Ch. IV § 1 N.3 Corollaire 1], then every $P \in \mathrm{Ass}_R(T/R)$ belongs to $\mathcal{F}_0(T)$. Therefore T is R–flat by Corollary 2.3.

(i) $\Rightarrow$ (ii) is a consequence of (the proof of) Lemma 2.10, applied to the localizing system $\mathcal{F}_0(T)$, and Corollary 2.2.

$\square$

Now, we are in condition to recover a result proved by Schenzel [S3, Theorem 1] (cf. also [Po, Theorem 8]).

THEOREM 2.12. *Let R be a Noetherian domain and T a flat overring of R. Then, the following statements are equivalent:*
(i) *T is finitely generated over R;*
(ii) *there exists a nonzero element $x \in R$ such that $R \subseteq T \subseteq R_x$.*

Proof: It is obvious that (i) $\Rightarrow$ (ii), since $T = R[x_1/x, \ldots, x_n/x]$ for some $x_1, \ldots, x_n, x \in R$ and $x \neq 0$.

(ii) $\Rightarrow$ (i). We claim that there exists a (finitely generated) ideal I of R such that $\mathcal{F}_0(T) = \mathcal{K}(I)$. We note that $\mathrm{Ass}_R(T/R)$ is finite, since $\mathrm{Ass}_R(T/R) \subseteq \mathrm{Ass}_R(R_x/R) = \mathrm{Ass}_R(R/x^n R) = \mathrm{Ass}_R(R/xR)$, for each $n \geq 1$, and $\mathrm{Ass}_R(R/xR)$ is finite [B, Ch. IV § 1 N.4 Corollaire p. 137]. Let I be the product of all the (finitely many) minimal primes of $\mathrm{Ass}_R(T/R)$. By [B, Ch. IV § 1 N.3 Corollaire 1] I is also the product of all the (finitely many) minimal prime ideals of $\mathrm{Supp}_R(T/R)$. Thus, by Corollary 2.7 (2), $P \in \mathcal{K}(I)$ if and only if P contains a minimal prime ideal of $\mathrm{Supp}_R(T/R)$. Therefore from Lemma 2.6 (3), we deduce that $\mathcal{K}(I) = \mathcal{K}(\mathrm{Supp}_R(T/R))$. By Corollary 2.11, since T is R–flat we have $\mathcal{K}(I) = \mathcal{F}_0(T)$. The conclusion follows immediately from Proposition 2.9.

$\square$

3 THE KAPLANSKY TRANSFORM

The spectral localizing systems of an integral domain R are parameterized by the subsets of $\mathrm{Spec}(R)$ stable under generizations (Corollary 1.8). A relevant class of subsets stable under generizations are the open subspaces of $\mathrm{Spec}(R)$. Let I be a nonzero ideal of R and let

$$D(I) := \{P \in \mathrm{Spec}(R) : P \not\supseteq I\} \,.$$

We can consider

(3.1) $$\mathcal{F} = \mathcal{F}(D(I)) = \cap\{\mathcal{F}(P) : P \in D(I)\} \,.$$

By [K2, Theorem 277] (or [Hy, Theorem 1.7] and [FHP, Proposition 3.2.2])

(3.2) $$R_{\mathcal{F}} = \Omega_R(I) := \{z \in K : \forall a \in I,\ za^n \in R \text{ for some } n \geq 1\} =$$
$$= \{z \in K : \mathrm{rad}(R :_R zR) \supseteq I\}$$

where $\Omega_R(I)$ (or, simply, $\Omega(I)$) is called the *Kaplansky transform of R with respect to I.*

Note that, when $I = R$, $\Omega(R) = R$. In this case $D(I) = \mathrm{Spec}(R)$ and $\mathcal{F}(\mathrm{Spec}(R)) = \{R\}$. If $I = (0)$, then $\Omega((0))$ coincides with the quotient field K of R. On the other hand, $D((0)) = \emptyset$ and we assume that $\mathcal{F}(\emptyset)$ is the trivial localizing system consisting of all the ideals of R, whence $R_{\mathcal{F}(\emptyset)} = K$.

We collect in the following lemma some easy facts concerning $\mathcal{F}(D(I))$ and $\Omega(I)$ (cf. also [Hy]).

LEMMA 3.1. *Let I be an ideal of an integral domain R.*
(a) $\mathcal{F}(D(I)) = \{J : J \text{ an ideal of } R \text{ and } \mathrm{rad}(J) \supseteq \mathrm{rad}(I)\} =$
$= \{J : J \text{ an ideal of } R \text{ and } \mathrm{rad}(J) \supseteq I\}$.
(b) $\Omega(I) = \cup\{(R : J) : \mathrm{rad}(J) \supseteq I\}$.
(c) $\Omega(I) = \Omega(\mathrm{rad}(I))$.
(d) *For each $a \in R$, $a \neq 0$, $\Omega(aR) = R_a$.*
(e) *If J is an ideal of R and $I \subseteq J$, then $\Omega(I) \supseteq \Omega(J)$.*
(f) *If $\{I_\alpha : \alpha \in A\}$ is a nonempty family of ideals of R, then*

$$\Omega(\Sigma_\alpha I_\alpha) = \cap_\alpha \Omega(I_\alpha) \,;$$
$$\Omega(\cap_\alpha I_\alpha) \supseteq \Sigma_\alpha \Omega(I_\alpha) \,.$$

(g) *If $I \neq (0)$, then*
$$\Omega(I) = \cap\{\Omega(aR) : a \in I,\ a \neq 0\} \,.$$

(h) *If J is another ideal of R,*

$$\Omega(I \cap J) = \Omega(IJ) \supseteq \Omega(I)\Omega(J) \supseteq \Omega(I) + \Omega(J) \,.$$

(i) *If J is an ideal of R and $I \subseteq J$ (or, more generally, if $\Omega(J) \subseteq \Omega(I)$) then*

$$\Omega(IJ) = \Omega(I)\Omega(J) = \Omega(I) + \Omega(J) \,.$$

(j) *If J is an invertible ideal of R and I is finitely generated then*

$$\Omega(IJ) = \Omega(I)\Omega(J) \,.$$

(k)

$$\Omega_R(I) = \cap\{R_P : P \not\supseteq I\} =$$
$$= \cap\{\Omega_{R_P}(IR_P) : P \in \mathrm{Spec}(R)\} = \cap\{\Omega_{R_M}(IR_M) : M \in \mathrm{Max}(R)\} \ .$$

(l) *If S is an overring of R, with $R \subseteq S \subseteq \Omega_R(I)$, then*

$$\Omega_S(IS) = \Omega_R(I) \ .$$

Proof: (a) is proved in [FHP, Remark 5.8.5 (a)].

(b) follows from (a) and from (3.2).

(c) is an easy consequence of (b).

(d) follows directly from the definition of $\Omega(aR)$.

(e), (f) and (h) are straightforward, since the fractional ideal $\Omega(I)\Omega(J)$ is the smallest overring of R containing $\Omega(I)$ and $\Omega(J)$.

(g) follows from the first equality of (f).

(i) is a consequence of (h) and of the fact that $\Omega(J) \subseteq \Omega(I)$ implies that $\Omega(I) + \Omega(J) = \Omega(I)$.

(j). From (h), we only need to prove that $\Omega(IJ) \subseteq \Omega(I)\Omega(J)$. Let $z \in \Omega(IJ)$ and let $I = (i_1, i_2, \ldots, i_n)R$ and $J = (j_1, j_2, \ldots, j_m)R$. For each $i_h j_k \in IJ$ there exists an integer $r = r(h, k) \geq 0$ such that $z(i_h j_k)^r \in R$. Henceforth, for a suitable $\bar{r} \geq \sup\{r(h, k) : 1 \leq h \leq n, 1 \leq k \leq m\}$, $z(IJ)^{\bar{r}} = zJ^{\bar{r}}I^{\bar{r}} \subseteq R$, and so $zJ^{\bar{r}} \subseteq \Omega(I)$, i.e. $z \in J^{-\bar{r}}\Omega(I) \subseteq \Omega(J)\Omega(I)$.

(k) follows easily from (3.1), (3.2) and the properties of the localizations, since, by Lemma 1.2,

$$\Omega(I) = R_{\mathcal{F}(D(I))} = \cap\{R_P : P \in \mathrm{Spec}(R), P \not\supseteq I\} \ .$$

(l) is an easy consequence of the definition (cf. also [FHP, Theorem 3.3.2]). □

Note that the inclusions in Lemma 3.1 (f) and (h) may be strict. An example of a Prüfer domain R with two ideals I and J such that $\Omega(IJ) \subsetneq \Omega(I)\Omega(J) = \Omega(I) + \Omega(J)$ is given in [FHP, Example 8.2.4 and Remark 8.2.5].

The Kaplansky transform is intimately related to the ideal transform introduced by Nagata [N1], [N4].

We recall that the *Nagata (ideal) transform of* an integral domain R *with respect to an ideal I of R* is the following overring of R:

(3.3) $$T_R(I) = T(I) := \cup\{(R : I^n) : n \geq 0\} \ .$$

With the terminology, introduced in Section 1, if $\mathfrak{N}(I)$ is the multiplicative system $\{I^n : n \geq 0\}$ of the powers of the ideal I, then $T_R(I)$ is the (generalized) ring of fractions of R with respect to $\mathfrak{N}(I)$:

(3.4) $$T_R(I) = R_{\mathfrak{N}(I)} \ .$$

From the fact that $\mathfrak{N}(I) \subseteq \mathcal{F}(D(I))$, it follows immediately that the Nagata transform is a subring of the Kaplansky transform. More precisely:

LEMMA 3.2. *Let I be an ideal of an integral domain R.*

(1) $\mathcal{F}(D(I)) = \mathcal{K}(I) = \{J : J \text{ ideal of } R \text{ such that } \text{rad}(I) = \text{rad}(J \cap I)\}$.

(2) $J \in \overline{\mathfrak{N}(I)} \Rightarrow \text{rad}(I) = \text{rad}(I \cap J)$.

(3) *If I is a finitely generated ideal, then*

$$\overline{\mathfrak{N}(I)} = \mathcal{F}(D(I)) \ .$$

Proof: (a) follows easily from Corollary 2.7 (1) and Lemma 3.1 (a), since $\text{rad}(I \cap J) = \text{rad}(I) \cap \text{rad}(J)$.

(b) is obvious because, if $I^n \subseteq J$, then $\text{rad}(I^n) = \text{rad}(I) \subseteq \text{rad}(J)$.

(c). By (a) and (b), $\overline{\mathfrak{N}(I)} \subseteq \mathcal{F}(D(I))$. Reciprocally, if $J \in \mathcal{F}(D(I))$ then $I \subseteq \text{rad}(J)$ (Lemma 3.1 (a)). Therefore, since I is finitely generated, we have $I^n \subseteq J$ for some $n \geq 0$.

$\square$

COROLLARY 3.3. *Let I be an ideal of an integral domain R.*

(1) $T(I) \subseteq \Omega(I)$.

(2) *If there exists a finitely generated ideal J of R and an integer $n \geq 1$ such that $I^n \subseteq J \subseteq I$ (in particular, if I is finitely generated), then $T(I) = \Omega(I)$.*

Proof: (1) follows from Lemma 3.2 (1) and (2).

(2). It is easy to see (by Lemma 3.1 (c) and Lemma 3.2 (3)) that

$$\Omega(I) = \Omega(J) = T(J) \subseteq T(I^n) = T(I) \ .$$

The conclusion follows from (1).

$\square$

Given an ideal I of an integral domain R, we set as usual

$$I_v := (R : (R : I)) \ ,$$
$$I_t := \cup\{J_v : J \text{ is finitely generated}, J \subseteq I\} \ .$$

It is obvious that $I \subseteq I_t \subseteq I_v$. The next result is due to Anderson and Bouvier [AnB].

PROPOSITION 3.4. *Let I be an ideal of an integral domain R.*

(1) $T(I) = T(I_v)$.

(2) $\Omega(I) = \Omega(I_t)$.

Proof: (1) is obvious, since $(R : I^n) = (R : I_v^n)$ for each $n \geq 1$.

(2). For each finitely generated ideal J of R, we have

$$\Omega(J_v) \subseteq \Omega(J) = T(J) = T(J_v) \subseteq \Omega(J_v) \ ,$$

hence $\Omega(J_v) = \Omega(J)$. Therefore, by Lemma 3.1 (e) and (f), we have:

$$\Omega(I_t) = \Omega(\cup\{J_v : J \text{ finitely generated}, J \subseteq I\}) =$$
$$= \cap\{\Omega(J_v) : J \text{ finitely generated}, J \subseteq I\} =$$
$$= \cap\{\Omega(J) : J \text{ finitely generated}, J \subseteq I\} \supseteq \Omega(I) \ .$$

The conclusion follows immediately, since in general (Lemma 3.1 (e)) $\Omega(I_t) \subseteq \Omega(I)$.

$\square$

From the previous result and from the fact that, in a Krull domain (respectively, a UFD), for each nonzero ideal I there exists a finitely generated (respectively, principal) ideal J such that $I_v = J_v$ [G3, Corollaries 44.3 and 44.5], we recover in particular a result due to Nagata [N4, Lemma 2.4].

COROLLARY 3.5. *Let I be an ideal in a Krull domain R. Then:*
(1) $T(I) = \Omega(I)$;
(2) *if, moreover, R is a UFD and I is a nonzero ideal, then there exists $f \in I_v$ such that*

$$T(I) = \Omega(I) = R_f \ .$$

Proof: (1). Let J be a finitely generated ideal of R such that $J_v = I_v$. Note that $I_t = J_v = I_v$, since the t-ideals coincide with the v-ideals in a Krull domain [Bo, Lemma 3 (a)]. Therefore:

$$T(I) = T(I_v) = \Omega(J_v) = \Omega(I_t) = \Omega(I) \ .$$

(2) follows from Lemma 3.1 (d) and from the previous statement (1) (and its proof).

$\square$

EXAMPLE 3.6. *An integral domain R for which there exists an ideal I such that $\Omega(I) \neq \Omega(I_v)$.*

Let (V, M) be a 1–dimensional valuation domain with $M = M^2$. In this case, $M_v = V$ [FHP, Corollary 3.1.3], therefore $\Omega(M_v) = \Omega(V) = T(V) = V = T(M_v) = T(M)$. On the other hand, by (Lemma 3.1 (k)), $\Omega(M) = V_{(0)} = qf(V)$, hence $\Omega(M_v) \subsetneqq \Omega(M)$.

EXAMPLE 3.7. *An integral domain R for which there exists an ideal I such that $T(I) \subsetneqq \Omega(I)$. Moreover, $T(I)$ is not a QQR-overring of R.*

Let (V, M) be a 1-dimensional valuation domain such that $M = M^2$ and $V = K + M$ where $K \cong V/M$. Let k be a proper subfield of K and set $R := k + M$. It is easy to see that M is the conductor of $R \subset V$ and, hence $V = (M : M) = (R : M) = T_R(M)$. On the other hand, by (3.2), $\Omega_R(M) = R_{(0)} = qf(R) = qf(V)$, hence $T_R(M) \subsetneqq \Omega_R(M)$. Note also that $\mathrm{Spec}(R) = \{(0), M\}$ and hence $T_R(M)$ is not an intersection of localizations of R.

We have already observed that, given an ideal I of R, there are several ideals J of R such that $\Omega(J) = \Omega(I)$. Set $\mathcal{J} = \mathcal{J}(\Omega_R(I)) := \{J : J$ ideal of R such that $\Omega(J) = \Omega(I)\}$. By Lemma 3.1 (f), if $\{J_\alpha : \alpha \in A\} \subseteq \mathcal{J}$ then $\Sigma_\alpha J_\alpha \in \mathcal{J}$. Therefore, as already observed by Hays [Hy, Theorem 2.3], and by Rhodes [Rh, § 2] in the non integral domain case, in $\mathcal{J}$ there exists a unique maximal element that we denote by I^Ω. We say that an ideal H of R is a Ω-*ideal* if $H = I^\Omega$, for some ideal I of R. It is obvious that $R^\Omega = R$. We will see in a moment that, $(0)^\Omega$ may be larger than (0).

Our next goal is to deepen the study of Ω-ideals.

PROPOSITION 3.8. *Let I and J be two ideals of an integral domain R. Then*
(1) $(I^\Omega)^\Omega = I^\Omega$;
(2) $I^\Omega = (\mathrm{rad}(I))^\Omega = \mathrm{rad}(I^\Omega)$;
(3) $I^\Omega = (I_t)^\Omega = (I^\Omega)_t$;
(4) $I^\Omega = \{x \in R : R_x \supseteq \Omega(I)\}$;
(5) $I \subseteq J \Rightarrow I^\Omega \subseteq J^\Omega$;
(6) $(aR)^\Omega = \mathrm{rad}(aR)$, *for each nonzero element $a \in R$.*

Proof: (1), (2) and (3) are straightforward consequences of the definition of a Ω-ideal, Lemma 3.1 (c) and Proposition 3.4 (2).

(4). If $x \in I^\Omega$, $x \neq 0$, then $R_x = \Omega(xR) \supseteq \Omega(I^\Omega) = \Omega(I)$ (Lemma 3.1 (d) and (e)). If $x = 0$, then assuming that $R_x = K$ (because the saturated multiplicative set containing 0 is R), trivially $R_x = K \supseteq \Omega(I)$.

On the other hand if $x \notin I^\Omega$, then $\Omega(I) = \Omega(I^\Omega) \supsetneq \Omega(xR + I^\Omega) = \Omega(xR) \cap \Omega(I)$ (Lemma 3.1 (f)), whence $R_x = \Omega(xR) \not\supseteq \Omega(I)$.

(5) follows immediately from (4) and Lemma 3.1 (e).

(6). Clearly, by (2), $\operatorname{rad}(aR) \subseteq (aR)^\Omega$. Let $x \in (aR)^\Omega$, $x \neq 0$. Then $R_x = \Omega(xR) \supseteq \Omega(aR) = R_a$, hence $1/a = r/x^n$ for some $r \in R$ and for some integer $n \geq 1$, and so $x^n \in aR$.

$\square$

For each ideal I of R, we call the Ω-*radical* of I the following radical ideal of R:

$$\operatorname{rad}^\Omega(I) := \cap\{P \in \operatorname{Spec}(R) : R_P \not\supseteq \Omega(I)\} \ .$$

It is obvious that $\operatorname{rad}^\Omega(I) = \operatorname{rad}^\Omega(I^\Omega)$. The following result is due to Hays:

THEOREM 3.9. *For each ideal I of R, we have*

$$I^\Omega = \operatorname{rad}^\Omega(I) \ .$$

Proof: From the definition of Kaplansky transform (3.2), for each ideal J of R and for each prime ideal Q of R, we have

(3.9.1) $$R_Q \not\supseteq \Omega(J) \Rightarrow Q \supseteq J \ .$$

We deduce immediately that

$$I^\Omega = \operatorname{rad}(I^\Omega) = \cap\{P \in \operatorname{Spec}(R) : P \supseteq I^\Omega\} \subseteq$$

$$\subseteq \cap\{P \in \operatorname{Spec}(R) : R_P \not\supseteq \Omega(I^\Omega) = \Omega(I)\} = \operatorname{rad}^\Omega(I)$$

and hence

$$\Omega(I) = \Omega(I^\Omega) \supseteq \Omega(\operatorname{rad}^\Omega(I)) \ .$$

The conclusion will follow if we show that $\Omega(I) \subseteq \Omega(\operatorname{rad}^\Omega(I))$. Suppose that Q is a prime ideal of R with $Q \not\supseteq \operatorname{rad}^\Omega(I)$.

Then it is easy to see that $R_Q \supseteq \Omega(I)$. Henceforth

$$\Omega(I) \subseteq \cap\{R_Q : Q \not\supseteq \operatorname{rad}^\Omega(I)\} = \Omega(\operatorname{rad}^\Omega(I)) \ ,$$

as desired.

$\square$

For each prime ideal P of R we denote by P^* the *pseudo–radical of P*, i.e. $P^* := \cap\{Q \in \operatorname{Spec}(R) : Q \supsetneq P\}$ [G1].

COROLLARY 3.10. *Let R be an integral domain.*

(1) $(0)^\Omega = (0)^*$.

(2) *If M is a maximal ideal,*

$$M = M^\Omega \Leftrightarrow R \subsetneq \Omega(M) \Leftrightarrow R_M \not\supseteq \cap\{R_P : P \in \operatorname{Spec}(R) \smallsetminus \{M\}\} \ .$$

Proof: (1). Let P be a prime ideal of R. Note that $R_P \not\supseteq K = \Omega((0))$ if and only if $P \neq (0)$.

(2). Note that $\Omega(M) = \cap\{R_P : P \in \operatorname{Spec}(R) \smallsetminus \{M\}\}$.

$\square$

Theorem 3.9 shows that there exists a link between the Ω-ideals of an integral domain R and the space $\operatorname{Loc}(R) := \{R_P : P \in \operatorname{Spec}(R)\}$, endowed with the Zariski topology [ZS, Ch. VI § 17]. Recall that the Zariski topology on $\operatorname{Loc}(R)$ is defined as follows. For each R-submodule E of K, we can consider

$$\mathcal{L}(E) := \{R_P \in \operatorname{Loc}(R) : R_P \supseteq E\} \ .$$

It is easy to prove the following:

LEMMA 3.11. *Let E_α, E and F be R-submodules of K where α belongs to a nonempty set A.*
(1) $E \subseteq F \Rightarrow \mathcal{L}(F) \subseteq \mathcal{L}(E)$;
(2) $\mathcal{L}(\Sigma_\alpha E_\alpha) = \cap_\alpha \mathcal{L}(E_\alpha)$;
(3) $\mathcal{L}(E) = \cap\{\mathcal{L}(zR) : z \in E\}$;
(4) $\mathcal{L}(\cap_\alpha E_\alpha) \supseteq \cup_\alpha \mathcal{L}(E_\alpha)$.

$\square$

By the previous lemma, it follows in particular that the subsets $\mathcal{L}(E)$, where E is a finitely generated R-submodule of K, form a basis for the open sets in $\mathrm{Loc}(R)$. The induced topology is called *the Zariski topology of* $\mathrm{Loc}(R)$.

It is easy to prove that $\mathrm{Spec}(R)$ and $\mathrm{Loc}(R)$ (with their Zariski topologies) are canonically homeomorphic. More precisely:

LEMMA 3.12. *Let $\varphi : \mathrm{Loc}(R) \to \mathrm{Spec}(R)$ be the canonical bijection defined by $R_P \mapsto PR_P \cap R = P$.*
(1) *For each element $x \in R$, $\varphi^{-1}(D(xR)) = \mathcal{L}(R_x)$.*
(2) *For each element $z \in K$,*

$$\varphi(\mathcal{L}(zR)) = \{P \in \mathrm{Spec}(R) : P \not\supseteq (R :_R zR)\} = D((R :_R zR)).$$

(3) *φ is a homeomorphism.*

$\square$

REMARK 3.13 (a) Let E be a R-submodule of K. If we denote by $R[E]$ the smallest overring of R generated by E, then it is clear that

$$\mathcal{L}(E) = \mathcal{L}(R[E]) .$$

Obviously, if E is a finitely generated R–module then $R[E]$ is a finitely generated R–algebra. Conversely, if S is an overring of R and S is a finitely generated R-algebra, then there exist $y_1, y_2, \ldots, y_n \in S$ such that $S = R[y_1, y_2, \ldots, y_n]$. Let $E := y_1 R + y_2 R + \cdots + y_n R$, then E is a finitely generated R–module and $S = R[E]$.

(b) If x is a nonzero element of R, then $R_x = R[1/x]$ and $(R :_R (1/x)R) = xR$, and hence $\mathcal{L}((1/x)R) = \varphi^{-1}(D(xR))$ (Lemma 3.12 (1)).

(c) If $E = (z_1, \ldots, z_t)R$ is a nonzero finitely generated R-submodule of K then, by Lemma 3.11 (2) and 3.12 (2), the basic open set $\mathcal{L}(E)$ of $\mathrm{Loc}(R)$ is homeomorphic to the open set $D(I_E)$ of $\mathrm{Spec}(R)$, where

$$I_E := \bigcap_{i=1}^{t} (R :_R z_i R) = \left(R :_R \sum_{i=1}^{t} z_i R \right) = E^{-1} \cap R .$$

Since $D(I_E) = D(\mathrm{rad}(I_E))$ and $\mathrm{rad}(I_E)$ is the largest ideal containing I_E with this property, it is natural to associate to the basic open set $\mathcal{L}(E)$ of $\mathrm{Loc}(R)$ the radical ideal $\mathrm{rad}(I_E)$ of R. If $E = zR$, then $\mathrm{rad}(I_E) = \mathrm{rad}((R :_R zR))$. By using Lemma 3.11 (1) and 3.12 ((1) and (3)) it is easy to see that:

$$x \in \mathrm{rad}(I_E) \Leftrightarrow R_x \supseteq E .$$

Motivated by Lemma 3.12 (2) and Remark 3.13 (c), in an integral domain R with quotient field K we introduce, for each element $z \in K$ and for each R-submodule E of K, the following radical ideals of R:

$$\Omega_R^-(zR) = \Omega^-(zR) := \mathrm{rad}((R :_R zR)) \, ,$$

$$\Omega_R^-(E) = \Omega^-(E) := \cap\{\Omega^-(zR) : z \in E\} \, .$$

We note that, if $x \in R$ and $x \neq 0$, then

$$\Omega^-((1/x)R) = \mathrm{rad}(xR) = (xR)^\Omega \, .$$

Moreover, $\Omega^-(0R) = \Omega^-(R) = R$. The following properties are easy consequence of the definitions:

LEMMA 3.14. *Let E and F be two R-submodules of K, $\{E_\alpha : \alpha \in A\}$ a family of R-submodules of K and I an ideal of R. Then:*
(1) $\Omega^-(E) = \{x \in R : R_x \supseteq E\} = \{x \in R : \Omega(xR) \supseteq E\}$.
(2) $E \subseteq F \Rightarrow \Omega^-(F) \subseteq \Omega^-(E)$.
(3) $\Omega^-(I) = R$.
(4) $\Omega^-(K) = (0)^*$.
(5) $\mathrm{rad}(E^{-1} \cap R) \subseteq \Omega^-(E)$ *and, if E is finitely generated, then*

$$\mathrm{rad}(E^{-1} \cap R) = \Omega^-(E) \, .$$

(6) $\Omega^-(\Omega(I)) = I^\Omega$ *and* $\mathrm{rad}((R : \Omega(I))) \subseteq I^\Omega$; *moreover, if $\Omega(I)$ is finitely generated, then* $\mathrm{rad}((R : \Omega(I))) = I^\Omega$.
(7) $\Omega^-(\Sigma_\alpha E_\alpha) = \cap_\alpha \Omega^-(E_\alpha)$.
(8) $\Omega^-(\cap_\alpha E_\alpha) \supseteq \Sigma_\alpha \Omega^-(E_\alpha)$.

□

The following result gives another useful representation of $\Omega^-(E)$:

COROLLARY 3.15. *Let E be a R-submodule of K, then*

$$\Omega^-(E) = \cap\{P \in \mathrm{Spec}(R) : R_P \not\supseteq E\} \, .$$

Proof: For each $z \in K$, from (3.9.2) we have:

$$\Omega^-(zR) = \cap\{P \in \mathrm{Spec}(R) : R_P \not\ni z\}$$

and thus

$$\Omega^-(E) = \cap\{\Omega^-(zR) : z \in E\} = \cap\{P \in \mathrm{Spec}(R) : R_P \not\supseteq E\} \, .$$

□

COROLLARY 3.16. *Let R be an integral domain.*
(1) *For each ideal I of R,*

$$I \subseteq \Omega^-(\Omega(I)) \quad \text{and} \quad \Omega(I) = \Omega(\Omega^-(\Omega(I))) \, .$$

(2) *For each R-submodule E of K,*

$$E \subseteq \Omega(\Omega^-(E)) \quad \text{and} \quad \Omega^-(E) = \Omega^-(\Omega(\Omega^-(E))) \, .$$

Proof: (1) follows from Lemma 3.14 (6).

(2). For each $z \in E$, $zR \subseteq (R : (R :_R zR)) \subseteq \Omega((R :_R zR)) = \Omega(\mathrm{rad}((R :_R zR))) = \Omega(\Omega^-(zR)) \subseteq \Omega(\Omega^-(E))$ and thus $E \subseteq \Omega(\Omega^-(E))$. Moreover, by Lemma 3.14 (2), $\Omega^-(E) \supseteq \Omega^-((\Omega(\Omega^-(E))))$. The conclusion follows, since by (1) for $I := \Omega^-(E)$ we have $\Omega^-(E) \subseteq \Omega^-(\Omega(\Omega^-(E)))$.

□

As a consequence of the previous results we obtain a deeper understanding of the homeomorphism between $\mathrm{Loc}(R)$ and $\mathrm{Spec}(R)$:

PROPOSITION 3.17. *Let R be an integral and $\varphi : \mathrm{Loc}(R) \to \mathrm{Spec}(R)$ the canonical homeomorphism* (Lemma 3.12). *Then*

(1) *For each R-submodule E of K, $\varphi^{-1}(D(\Omega^-(E)))$ coincides with the interior of the subspace $\mathcal{L}(E)$ of $\mathrm{Loc}(R)$, i.e.*

$$\varphi^{-1}(D(\Omega^-(E))) = \mathrm{Int}(\mathcal{L}(E)) .$$

(2) *For each ideal I of R,*

$$\varphi^{-1}(D(I^\Omega)) = \mathrm{Int}(\mathcal{L}(\Omega(I))) .$$

(3) *If E is a finitely generated R-submodule of K,*

$$\varphi^{-1}(D(\Omega^-(E))) = \mathcal{L}(E) .$$

(4) *If I is an ideal of R such that $\Omega(I)$ is a finitely generated R-algebra, then*

$$\varphi^{-1}(D(I^\Omega)) = \mathcal{L}(\Omega(I)) .$$

Proof: (1). Note that $D(\Omega^-(E)) \subseteq \varphi(\mathcal{L}(E))$, since

$$P \not\supseteq \Omega^-(E) \Rightarrow R_P \supseteq E \qquad \text{(Corollary 3.15)} .$$

On the other hand, the closure $\mathrm{cl}(Y)$ of a subspace Y of $\mathrm{Spec}(R)$ coincides with $V(J_Y)$ where $J_Y := \cap\{Q \in Y\}$. Therefore,

$$\mathrm{cl}(\mathrm{Spec}(R) \smallsetminus \varphi(\mathcal{L}(E))) = V(J)$$

where

$$J := \cap \{P \in \mathrm{Spec}(R) \smallsetminus \varphi(\mathcal{L}(E))\} = \cap\{P \in \varphi(\mathrm{Loc}(R) \smallsetminus \mathcal{L}(E))\} =$$
$$= \cap \{P \in \mathrm{Spec}(R) : R_P \not\supseteq E\} = \Omega^-(E) .$$

We deduce that $\mathrm{Int}(\varphi(\mathcal{L}(E))) = D(\Omega^-(E))$.

(2). By Lemma 3.14 (6), this statement is a particular case of (1) for $E = \Omega(I)$.

(3) is a consequence of (1) and of the fact that, in the present situation, $\mathcal{L}(E)$ is an open subspace of $\mathrm{Loc}(R)$.

(4) is a consequence of (2) and (3) (cf. also Remark 3.13 (a)).

$\square$

COROLLARY 3.18. *Let I be an ideal of R such that $\Omega(I)$ is a finitely generated R-algebra. Then*

$$(3.18.1) \qquad R_P \supseteq \cap\{R_x : x \in I\} \Rightarrow R_P \supseteq R_y \text{ for some } y \in I^\Omega .$$

Proof: In general, by Proposition 3.17 (2), for each ideal I of R, we have

$$(3.18.2) \qquad \varphi^{-1}(D(I^\Omega)) \subseteq \mathcal{L}(\Omega(I)) .$$

Moreover, by Lemma 3.1 (g) and Lemma 3.12, equality in (3.18.2) holds if and only if (3.18.1) is verified. The conclusion is a consequence of Proposition 3.17 (4).

$\square$

REMARK 3.19. If I is a nonzero principal ideal of R, $I = xR$, then $\Omega(I) = R_x = R[1/x]$ is a finitely generated R-algebra and hence, by Proposition 3.17 (4), we reobtain that $\varphi^{-1}(D(xR)) = \mathcal{L}(R_x)$ (Lemma 3.12 (1)). It is also easy to see that if $x, y \in R$ are nonzero elements, then

$$\mathcal{L}(R_x) \cap \mathcal{L}(R_y) = \varphi^{-1}(D(xR) \cap D(yR)) = \mathcal{L}(R_{xy}) .$$

Therefore, the family $\{\mathcal{L}(R_x) : x \in R \; x \neq 0\}$ is a basis for the Zariski topology of $\mathrm{Loc}(R)$.

From the previous results we can obtain a characterization of the Ω-ideals of R:

COROLLARY 3.20. *Let R be an integral domain and I an ideal of R. The following statements are equivalent:*

(i) *I is a Ω-ideal;*
(ii) *$I = I^{\Omega}$;*
(iii) *$I = \Omega^-(E)$, for some R-submodule E of K;*
(iv) *$I = \Omega^-(T)$, for some overring T of R.*

Proof: (ii) $\Rightarrow$ (i) $\Rightarrow$ (iv) $\Rightarrow$ (iii) are obvious (cf. also Lemma 3.14 (6)).
   (iii) $\Rightarrow$ (ii). By Corollary 3.16 (2), we have

$$I = \Omega^-(E) = \Omega^-(\Omega(\Omega^-(E))) = \Omega^-(\Omega(I)) = I^{\Omega} .$$

$\square$

We collect in the following proposition some properties of Ω-ideals.

PROPOSITION 3.21. *Let $\{I_{\alpha} : \alpha \in A\}$ be a family of ideals of an integral domain R. Then:*

(1) $(\Sigma_{\alpha} I_{\alpha})^{\Omega} = (\Sigma_{\alpha} I_{\alpha}^{\Omega})^{\Omega}$.
(2) $\cap_{\alpha} I_{\alpha}^{\Omega} = (\cap_{\alpha} I_{\alpha}^{\Omega})^{\Omega}$.
(3) *If each I_{α} is a Ω-ideal, for $\alpha \in A$, then $\cap_{\alpha} I_{\alpha}$ is also a Ω-ideal.*

Proof: (1). From the following easy inclusions (Proposition 3.8 (5)):

$$\Sigma_{\alpha} I_{\alpha} \subseteq \Sigma_{\alpha} I_{\alpha}^{\Omega} \subseteq (\Sigma_{\alpha} I_{\alpha})^{\Omega} \subseteq (\Sigma_{\alpha} I_{\alpha}^{\Omega})^{\Omega}$$

we deduce the statement.
   (2). It is obvious that

$$\cap_{\alpha} I_{\alpha}^{\Omega} \subseteq (\cap_{\alpha} I_{\alpha}^{\Omega})^{\Omega}$$

and, by Proposition 3.8 (1),

$$(\cap_{\alpha} I_{\alpha}^{\Omega})^{\Omega} \subseteq (I_{\alpha}^{\Omega})^{\Omega} = I_{\alpha}^{\Omega} , \quad \text{for each } \alpha \in A .$$

The conclusion follows immediately.
   (3) is a consequence of (2).

$\square$

COROLLARY 3.22. *Let I and J be two nonzero ideals of an integral domain R. Then*

$$(IJ)^{\Omega} = (I \cap J)^{\Omega} .$$

Proof: Recall that $\Omega(IJ) = \Omega(I \cap J)$ (Lemma 3.1 (h)). The conclusion follows from Lemma 3.14 (6).

$\square$

For each overring T of R, we set

$$T_{\Omega} := \Omega(\Omega^-(T)) = \cap\{R_P : P \not\supseteq \Omega^-(T)\} ,$$

where T_{Ω} is an overring of T and $\Omega^-(T_{\Omega}) = \Omega^-(T)$ (Corollary 3.16 (2)). We say that an overring S of R is a Ω-*overring of R* if $S = T_{\Omega}$ for some overring T of R.

PROPOSITION 3.23. *Let R be an integral domain and T on overring of R. The following conditions are equivalent:*
(i) *T is a Ω-overring of R;*
(ii) *$T = T_{\Omega}$;*
(iii) *$T = \Omega(I)$, for some ideal I of R.*

Proof: It is obvious that (iii) $\Rightarrow$ (ii) $\Rightarrow$ (i) (cf. also Corollary 3.16 (1)).
(i) $\Rightarrow$ (iii). If $T = S_{\Omega}$ for some overring S of R, then $T = \Omega(I)$ where $I := \Omega^-(S) = \Omega^-(S_{\Omega}) = \Omega^-(T)$.

$\square$

Let $\{T_{\alpha} : \alpha \in A\}$ be a family of overrings of an integral domain R. We denote by $\vee_{\alpha} T_{\alpha}$ the smallest overring of R containing $\cup_{\alpha} T_{\alpha}$. In the case of a finite family of overrings $\{T_i : 1 \le i \le n\}$, it is obvious that

$$\vee_{i=1}^{n} T_i = \prod_{i=1}^{n} T_i = \left\{ \sum_{k=1}^{r} t_1^{(k)} t_2^{(k)} \cdot \ldots \cdot t_n^{(k)} : t_i^{(k)} \in T_i, \ r \ge 1 \right\} .$$

PROPOSITION 3.24. *Let $\{T_{\alpha} : \alpha \in A\}$ be a family of overrings of R. Then*
(1) *$(\vee_{\alpha} T_{\alpha})_{\Omega} = (\vee_{\alpha} (T_{\alpha})_{\Omega})_{\Omega}$.*
(2) *$\cap_{\alpha} (T_{\alpha})_{\Omega} = (\cap_{\alpha} (T_{\alpha})_{\Omega})_{\Omega}$.*
(3) *If each T_{α} is a Ω-overring of R, then $\cap_{\alpha} T_{\alpha}$ is also a Ω-overring of R.*

Proof: *Mutatis mutandis* (in particular using Corollary 3.16 (2) instead of Corollary 3.16 (1) and the fact that $T \subseteq T_{\Omega}$, $(T_{\Omega})_{\Omega} = T_{\Omega}$ and $S_{\Omega} \subseteq T_{\Omega}$ for each pair of overrings S, T of R with $S \subseteq T$), the proof is analogous to that of Proposition 3.21.

$\square$

COROLLARY 3.25. *Let S and T be two overrings of an integral domain R. Then*

$$(ST)_{\Omega} = (S_{\Omega} T)_{\Omega} = (ST_{\Omega})_{\Omega} = (S_{\Omega} T_{\Omega})_{\Omega} .$$

Proof: We have already observed that $S \vee T = ST$, hence the statement follows easily from Proposition 3.24 (1).

$\square$

From a "global" point of view, some of the results proved above can be restated in the following way:

THEOREM 3.26. *Let R be an integral domain. Set $\mathcal{I}^{\Omega}(R) := \{I$ ideal of $R : I^{\Omega} = I\}$ and $\mathcal{O}_{\Omega}(R) := \{T$ overring of $R : T_{\Omega} = T\}$. The map*

$$\Omega : \mathcal{I}^{\Omega}(R) \longrightarrow \mathcal{O}_{\Omega}(R) , \quad I \mapsto \Omega(I)$$

is an order-reversing bijection, with inverse map

$$\Omega^- : \mathcal{O}_{\Omega}(R) \longrightarrow \mathcal{I}^{\Omega}(R) , \quad T \mapsto \Omega^-(T) .$$

Proof: The statement is a consequence of Proposition 3.8 (5), Lemma 3.14 ((2) and (6)), Corollary 3.20 and Proposition 3.23.

$\square$

If R is not a field, it is not possible that every ideal of R is a Ω-ideal. In fact, if x is a nonzero element of R, and if every ideal of R is a Ω-ideal, then

$$x^2 R = (x^2 R)^\Omega = (\text{rad}(x^2 R))^\Omega = (\text{rad}(xR))^\Omega = (xR)^\Omega = xR$$

and thus x is invertible. It would be interesting to study the integral domains such that every nonzero radical ideal is a Ω-ideal.

On the other hand, it is possible that each overring of R is a Ω-overring. For instance, if V is a finite dimensional valuation domain and if

$$(0) = P_0 \subset P_1 \subset P_2 \subset \cdots \subset P_n = M$$

is the chain of all its prime ideals, then

$$V_{P_k} = \Omega(P_{k+1}) , \qquad \text{for } 0 \leq k \leq n-1 ,$$
$$V = \Omega(uV) , \qquad \text{where } u \in V \smallsetminus M ,$$

and thus every overring of V is a Ω-overring. In this case, every nonzero prime (radical) ideal of V is a Ω-ideal, since $P_{k+1} = \Omega^-(V_{P_k})$ for $0 \leq k \leq n-1$. Note also that the inclusion in Lemma 3.14 (6) may be proper

$$\text{rad}((V : \Omega(P_{k+1}))) = \text{rad}((V : V_{P_k})) = P_k \subsetneqq P^\Omega_{k+1} = P_{k+1} , \quad \text{for } 0 \leq k \leq n-1 .$$

Integral domains with the property that each overring is a Nagata ideal transform were studied in [Br], [BrG], [He1] and [He2].

The integral domains with the property that each overring is a Ω-overring are characterized in [FH]; in particular, a domain of this type is always semilocal with Prüfer integral closure. Note that, if T is an overring of an integral domain R, then:

$$T \subseteq \cap \{R_P : R_P \supseteq T\} \subseteq T_\Omega$$

since if $P \not\supseteq \Omega^-(T)$ then $R_P \supseteq T$ (Corollary 3.15).

PROPOSITION 3.27. *If R is an integral domain and if $S \subseteq T$ are two overrings of R, then*

$$\Omega^-_R(T) = \Omega^-_R(S) \cap \Omega^-_S(T) .$$

Proof: It is obvious, by Lemma 3.14 (1), that

$$\Omega^-_R(T) \subseteq \Omega^-_R(S) \cap \Omega^-_S(T) .$$

Conversely, if $x \in \Omega^-_R(S) \cap \Omega^-_S(T)$ then $R_x \supseteq S$ and $S_x \supseteq T$. On the other hand $R_x = S_x$, since $R_x \supseteq S$ with $x \in R \subseteq S$, thus $R_x \supseteq T$ and hence $x \in \Omega^-_R(T)$.
$$\square$$

The previous result was proved, in a more general setting, by Rhodes [Rh].

4 A GEOMETRIC INTERPRETATION OF THE KAPLANSKY TRANSFORM

An interesting property of the Nagata (ideal) transform is its "geometric" interpretation. For an ideal I in a Noetherian integral domain R, the Nagata transform $T_R(I)$ is the ring of global sections over the open subspace $D(I)$ of $\text{Spec}(R)$. More precisely, from classical results by Chevalley [EGA, I.6.7.1], Nagata [N1] and Hartshorne [Ha] (cf. also Arezzo and Ramella [AR] and Theorem 2.11], the following can be shown:

PROPOSITION 4.1. *Let R be a Noetherian integral domain and I an ideal of R. Set $X := \operatorname{Spec}(R)$, $Y := D(I) = \{P \in X : P \not\supseteq I\}$ and $Z := \operatorname{Spec}(T(I))$. The following statements are equivalent:*
  (i) *Y is an affine open subspace of X;*
  (ii) *the canonical morphism:*

$$(Z, \mathcal{O}_Z) \longrightarrow (Y, \mathcal{O}_Y) , \quad Q \mapsto Q \cap R ,$$

  is a scheme–isomorphism;
(iii) *$IT(I) = T(I)$.*
In particular, when the previous statements hold, $T(I)$ is flat and finitely generated over R.

$\square$

REMARK 4.2. (a) If R is a Noetherian domain, it is easy to see that the image of the canonical map

$$\operatorname{Spec}(T(I)) \longrightarrow \operatorname{Spec}(R) , \quad Q \mapsto Q \cap R$$

contains $D(I)$, and if $T(I)$ is R–flat, then this image coincides with $D(I_v)$ [L, Proposition 4.4]; therefore if $T(I)$ is R-flat then $D(I_v)$ is an affine open subspace of $\operatorname{Spec}(R)$.

Furthermore, it is known that the statements (i)–(iii) of Proposition 4.1 are equivalent to the following:
(iv) *$T(I)$ is R–flat and there exists a divisorial ideal J such that $D(J) = D(I)$;*
 (v) *$T(I)$ is R–flat and $D(I) = D(I_v)$;*
(vi) *$T(I)$ is R–flat and $\operatorname{rad}(I) = \operatorname{rad}(I)_v$.*
[L, Proposition 4.3].

(b) Note that for each nonzero ideal I of R and for each nonzero element $x \in I$, we have:

$$R \subseteq T(I) \subseteq \Omega(I) \subseteq \Omega(xR) = R_x .$$

When R is Noetherian and $T(I) = \Omega(I)$ is R-flat then, by Theorem 2.12, $T(I)$ is also finitely generated over R.

Reciprocally, Schenzel proved that if $R \subseteq T \subseteq R_x$ for some nonzero element x of a Noetherian ring R and if T is R-flat, then there exists an ideal I of R such that $T = T(I)$ and $IT(I) = T(I)$ [S3, Corollary 3].

(c) We will see in Example 4.11 that $T(I)$ may be finitely generated over R without being R-flat. The finiteness of $T(I)$ was studied in several papers (cf., for instance, [Bro], [EHKR] and [Ki]).

If R is not a Noetherian domain, $D(I)$ may be affine with $IT(I) \neq T(I)$ and $\operatorname{Spec}(T(I))$ non isomorphic to $D(I)$. Take, for instance, $R = V$ to be a 2–dimensional valuation domain with idempotent maximal ideal M. Set $I := M$; in this case $D(M) = \operatorname{Spec}(V) \smallsetminus V(M)$ is canonically isomorphic to $\operatorname{Spec}(V_f)$ with $f \in M \smallsetminus P$, where P is the height 1 prime ideal of V. On the other hand, $T(M) = (V : M) = (M : M) = V$, since $M = M^2$, whence $MT(M) = M \neq T(M) = V$. Furthermore, $\operatorname{Spec}(T(M)) = \operatorname{Spec}(V)$ is obviously not isomorphic to $D(M)$.

Our next goal is to generalize Proposition 4.1 to a not necessarily Noetherian context.

LEMMA 4.3. *Let I be an ideal of an integral domain R and let $\Omega(I)$ be the Kaplansky transform of R with respect to I. Set $X := \operatorname{Spec}(R)$ and $Y := D(I)$. The ring of global sections over the open subspace Y of X, $\Gamma(Y, \mathcal{O}_X|_Y)$ coincides with $\Omega(I)$.*

Proof: It follows immediately from (3.1) and (3.2) that

$$\Omega(I) = R_{\mathcal{F}(D(I))} = \cap \{R_P : P \not\supseteq I\} = \Gamma(Y, \mathcal{O}_X|_Y) .$$

$\square$

THEOREM 4.4. *Let R be an integral domain and I an ideal of R. Set $X := \operatorname{Spec}(R)$, $Y := D(I)$ and $W := \operatorname{Spec}(\Omega(I))$.*

The following statements are equivalent:

(i) *Y is an affine open subspace of X;*

(ii) *the canonical morphism*

$$(W, \mathcal{O}_W) \longrightarrow (Y, \mathcal{O}_Y) , \qquad Q \mapsto Q \cap R ,$$

is a scheme-isomorphism;

(iii) *$I\Omega(I) = \Omega(I)$;*

(iv) *$\Omega(I)$ is R-flat and, for each $P \in \operatorname{Spec}(R)$ with $P \supseteq I$, $R_P \not\supseteq \Omega(I)$;*

(v) *$\Omega(I)$ is R-flat and, for each $P \in \operatorname{Spec}(R)$ with $P \supseteq I$, $P\Omega(I) = \Omega(I)$;*

(vi) *$\mathcal{F}_0(\Omega(I)) = \mathcal{F}(D(I))$.*

In particular, if the previous conditions are verified then $\Omega(I)$ is a finitely generated R-algebra.

Proof: Let $Y' := D_W(I\Omega(I))$. We need the following fact proved in [FHP, Theorem 3.3.2],

(4.4.1)   *the canonical map $Y' \to Y$, $Q \mapsto Q \cap R$, is a bijection*

and $R_{Q \cap R} = \Omega(I)_Q$, for each $Q \in Y'$.

(i) $\Leftrightarrow$ (ii). By Lemma 4.3, Y is an affine subspace of X if and only if the canonical map $Y \to \operatorname{Spec}(\Gamma(Y, \mathcal{O}_X|_Y)) = W$ [EGA, I.2.3.2] defines a scheme isomorphism.

(ii) $\Leftrightarrow$ (iii). It is a straightforward consequence of (4.4.1), since $I\Omega(I) = \Omega(I)$ if and only if $Y' = W$.

(iii) $\Rightarrow$ (iv). The flatness of $\Omega(I)$ follows immediately from [R, Theorem 2] and (4.4.1). Moreover if, for some prime ideal P of R with $P \supseteq I$, it happens that $R_P \supseteq \Omega(I)$, then $PR_P \cap \Omega(I)$ is a prime ideal of $\Omega(I)$ containing $I\Omega(I)$. This contradicts (iii).

(iv) $\Rightarrow$ (iii). If there exists a maximal ideal Q of $\Omega(I)$ such that $Q \supseteq I\Omega(I)$ then $Q \cap R \supseteq I$ and the hypothesis then yields $\Omega(I) \not\subseteq R_{Q \cap R}$. On the other hand, by the flatness of $\Omega(I)$ over R, we have that $R_{Q \cap R} = \Omega(I)_Q$ and this leads to a contradiction.

(iv) $\Leftrightarrow$ (v). Under the assumption that $\Omega(I)$ is R-flat, we claim that, for each prime ideal P of R with $P \supseteq I$,

(4.4.2) $\Omega(I) \not\subseteq R_P \Leftrightarrow P\Omega(I) = \Omega(I) .$

In fact, if $P\Omega(I) \subsetneq \Omega(I)$ then there exists a maximal ideal Q of $\Omega(I)$ with $Q \supseteq P\Omega(I)$, so that $\Omega(I)_Q = R_{Q \cap R} \subseteq R_P$, and $\Omega(I) \subseteq R_P$. Conversely, if $\Omega(I) \subseteq R_P$ then $P\Omega(I) \subseteq PR_P \cap \Omega(I) \subsetneq \Omega(I)$.

(vi) $\Rightarrow$ (iii). It is sufficient to note that $I \in \mathcal{F}(D(I))$.

(iii) $\Rightarrow$ (vi). Denote simply by $\mathcal{F}_0$ the localizing system $\mathcal{F}_0(\Omega(I)) = \{J : J \text{ ideal of } R$ such that $J\Omega(I) = \Omega(I)\}$ and by $\mathcal{K}$ the localizing system $\mathcal{F}(D(I))$. Since we know already that (iii) implies that $\Omega(I)$ is R-flat, then we have $R_\mathcal{K} = \Omega(I) = R_{\mathcal{F}_0}$ and $\mathcal{F}_0 \subseteq \mathcal{K}$ (Lemma 1.19 (1), Corollary 1.20 and (3.2)). Moreover, if $J \in \mathcal{K}$ then $I \subseteq \text{rad}(I) = \text{rad}(J \cap I) \subseteq \text{rad}(J)$ (Lemma 3.2 (1)) and thus $\text{rad}(J) \in \mathcal{F}_0$, because $I \in \mathcal{F}_0$ by assumption. On the other hand, it is easy to see that $\text{rad}(J)\Omega(I) = \Omega(I)$ if and only if $J\Omega(I) = \Omega(I)$, whence $\text{rad}(J) \in \mathcal{F}_0$ is equivalent to $J \in \mathcal{F}_0$.

The last statement follows from Proposition 2.9 ((ii) $\Rightarrow$ (i)) (cf. also [FHP, Proposition 3.3.5]).

$\square$

COROLLARY 4.5. *Let I be an ideal in an integral domain R such that $I\Omega(I) = \Omega(I)$, and write*

$$1 = \sum_{h=1}^{s} i_h y_h \quad \text{with } i_h \in I \text{ and } y_h \in \Omega(I) .$$

Set $I_* := \sum_{h=1}^{s} i_h R$ and $E := \sum_{h=1}^{s} y_h R$. Then
(1) $\Omega(I) = \Omega(I_*) = T(I_*)$.
(2) $I^\Omega = \text{rad}(I)$.
(3) $I^\Omega = I_*^\Omega = \text{rad}(I_*) = \text{rad}(I_*)_t = \text{rad}(E^{-1} \cap R)$.
(4) $D(I)$ is a quasi-compact affine (open) subspace of $\text{Spec}(R)$.

Proof: (1). As in the proof of Proposition 2.8, we have that $\Omega(I) = R[E]$ and $\mathcal{K}(I) = \mathcal{K}(I_*)$ and thus $\Omega(I) = \Omega(I_*)$ (Lemma 3.2 (1), (3.1) and (3.2)). Since I_* is finitely generated $T(I_*) = \Omega(I_*)$ (Corollary 3.3 (2)).

(2). From Theorem 4.4 ((iii) $\Rightarrow$ (iv)) we deduce that $\text{rad}(I) = \text{rad}^\Omega(I) = I^\Omega$.

(3). Since $\Omega(I) = \Omega(I_*)$, $I^\Omega = I_*^\Omega$ and, by (2), $I_*^\Omega = \text{rad}(I_*)$. Moreover, by (1), $\Omega(I_*) = R[E]$ thus $I_*^\Omega = \Omega^-(E) = \text{rad}(E^{-1} \cap R)$ (Lemma 3.14 (5)). Since I^Ω is a t-ideal (Proposition 3.8 (3)), then necessarily $\text{rad}(I_*) = \text{rad}(I_*)_t$.

(4). In this situation, by Theorem 4.4 ((iii) $\Rightarrow$ (ii)), $D(I_*) \cong \text{Spec}(\Omega(I_*)) = \text{Spec}(\Omega(I)) \cong D(I)$, whence $\text{rad}(I) = \text{rad}(I_*)$ with I_* finitely generated.

$\square$

COROLLARY 4.6. *If I is an ideal of an integral domain R such that $\text{rad}(I)$ is locally the radical of a nonzero principal ideal (e.g. I is invertible), then $D(I)$ is a quasi-compact affine (open) subspace of $\text{Spec}(R)$.*

Proof: Note that $I\Omega(I) = \Omega(I)$ if (and only if) $IR_P\Omega(IR_P) = \Omega(IR_P)$, for each prime ideal P of R.

As a matter of fact, if $I\Omega(I) \neq \Omega(I)$ then there exists a prime ideal Q of $\Omega(I)$ such that $I\Omega(I) \subseteq Q$. Let $P := Q \cap R$. Then $\Omega_R(I)_Q = \Omega_{R_P}(IR_P)$ because $R_P \subseteq \Omega(I)_Q$ (Lemma 3.1 (1)). By the choice of Q, we have $I\Omega(I)_Q \neq \Omega(I)_Q$. This fact leads to a contradiction since, by assumption, $IR_P\Omega_{R_P}(IR_P) = \Omega_{R_P}(IR_P)$.

The conclusion will follow from Corollary 4.5 if we show that if, given a prime ideal P of R, $\text{rad}(IR_P) = \text{rad}(xR_P)$ for some $x \in R_P$, $x \neq 0$, then $IR_P\Omega_{R_P}(IR_P) = \Omega_{R_P}(IR_P)$. Note that $\Omega_{R_P}(IR_P) = \Omega_{R_P}(\text{rad}(xR_P)) = \Omega_{R_P}(xR_P) = R_P[1/x]$, then $xR_P[1/x] = R_P[1/x]$ and thus $\text{rad}(IR_P)\Omega_{R_P}(IR_P) = \text{rad}(xR_P)R_P[1/x] = R_P[1/x] = \Omega_{R_P}(IR_P)$. Therefore, $IR_P\Omega_{R_P}(IR_P) = \Omega_{R_P}(IR_P)$.

$\square$

The previous result was proved, in the Noetherian setting, by Hartshorne [Ha, § 2 Example 2], making use of cohomological techniques, (cf. also [Se, Theorem 1] and [N1, Theorem 5]).

COROLLARY 4.7. (1) *If I is a nonzero finitely generated ideal in a Prüfer domain, then $D(I)$ is affine.*

(2) *For each nonzero ideal I in a Dedekind domain, $D(I)$ is affine.*

Proof: (1). Note that, in a Prüfer domain, a nonzero finitely generated ideal is invertible [G3, Theorem 22.1].

(2) follows from (1), since a Dedekind domain is a Noetherian Prüfer domain. $\square$

Note that it is not difficult to prove that statement (2) of Corollary 4.7 holds more generally for each *nonzero* ideal in a 1-dimensional domain with the property that each nonzero element lies in only finitely many maximal ideals.

We close with some examples.

EXAMPLE 4.8. Let $\mathcal{F}$ be a localizing system of an integral domain R. Set $\mathcal{F}_0 := \mathcal{F}_0(R_{\mathcal{F}}) = \{I : I \text{ ideal of } R \text{ and } IR_{\mathcal{F}} = R_{\mathcal{F}}\}$, $\Lambda := \Lambda(R_{\mathcal{F}}) = \{N \cap R : N \in \text{Max}(R_{\mathcal{F}})\}$ and $\mathcal{F}(\Lambda) = \cap\{\mathcal{F}(Q) : Q \in \Lambda\}$.

We give an *example of an integral domain R having a localizing system $\mathcal{F}$ such that $\mathcal{F}_0 = \mathcal{F}(\Lambda)$, but $\mathcal{F}_0 \neq \mathcal{F}$.*

Claim. *If I is an ideal of an integral domain R such that $\Omega(I)$ is R-flat, then $\mathcal{F}_0(\Omega(I)) = \mathcal{F}(\Lambda(\Omega(I)))$.*

In fact, by flatness $\Omega(I) = R_{\mathcal{F}_0(\Omega(I))}$ (Corollary 1.20) and, by Lemma 1.19 (2), we conclude that $\mathcal{F}(\Lambda(\Omega(I))) = \mathcal{F}_0(\Omega(I))$.

Let $R = V$ be a valuation domain having an ascending chain of prime ideals of the following type:

$$P_1 \subset P_2 \subset \cdots \subset P_n \subset \cdots \subset P , \quad \text{with } \bigcup_{n \geq 1} P_n = P .$$

In this case, $\Omega(P) = \bigcap_{n \geq 1} V_{P_n} = V_P$ is trivially V-flat, hence $\mathcal{F}_0 = \mathcal{F}_0(\Omega(P)) = \mathcal{F}(\Lambda(\Omega(P))) = \mathcal{F}(P)$. On the other hand, if $\mathcal{K} = \mathcal{K}(P) = \mathcal{F}(D(P))$, then we know that $R_{\mathcal{K}} = \Omega(P)$.

Note that $\mathcal{F}_0 \subsetneq \mathcal{K}$, because if $I \in \mathcal{F}_0$ then $I \supsetneq P$ and thus $I \in \mathcal{K}$; moreover $P \in \mathcal{K}$ but $P \notin \mathcal{F}_0$. Furthermore, $V_{\mathcal{F}_0} = \Omega(P) = V_{\mathcal{K}}$ and thus $\mathcal{K}$ is a nonfinitely generated localizing system (Corollary 1.25).

PROPOSITION 4.9. *Let I be an ideal of an integral domain R. The localizing system $\mathcal{F}(D(I))$ is a finitely generated localizing system of R if and only if $D(I)$ is a quasi-compact (open) subspace of $\text{Spec}(R)$.*

Proof: ($\Rightarrow$). Since $I \in \mathcal{F}(D(I))$ and $\mathcal{F}(D(I))$ is finitely generated, there exists a finitely generated ideal $J \subseteq I$ with $J \in \mathcal{F}(D(I))$. This fact implies that $\text{rad}(J) = \text{rad}(J \cap I) = \text{rad}(I)$ (Corollary 2.7 (1) or Lemma 3.2 (1)) and hence $D(J) = D(I)$ is quasi-compact [EGA I.1.1.4].

($\Leftarrow$). This implication is a particular case of Lemma 1.16. $\square$

EXAMPLE 4.10. *An ideal I of an integral domain R such that $\mathcal{F}(D(I))$ is a finitely generated localizing system of R, but $\mathcal{F}_0(\Omega(I)) \neq \mathcal{F}(D(I))$ (in other words, $D(I)$ is quasi–compact but not affine; cf. Theorem 4.4 and Proposition 4.9).*

Let (R, M) be a local integrally closed Noetherian domain of dimension 2. In this case, since M is finitely generated, $D(M)$ is obviously a quasi-compact open subspace of Spec(R). On the other hand, $\Omega(M) = T(M) = \cap\{R_P : P \in \mathrm{Spec}(R)\ P \neq M\} = R$, because R is a Krull domain. Therefore $M\Omega(M) = MR = M \neq \Omega(M)$ and thus $D(M)$ is not affine.

Note that, in this case, $\Omega(M)$ is trivially finitely generated over R, but $M\Omega(M) \neq \Omega(M)$.

EXAMPLE 4.11. *An ideal I of an integral domain R such that $\Omega(I)$ is a finitely generated proper overring of R, but $\Omega(I)$ is not R-flat.*

Let K be a field and X, Y two indeterminates over K. Let $(V, M = \pi V)$ be a 1-dimensional discrete valuation domain of $F := K(X, Y)$ dominating the local ring $K[X, Y]_{(X,Y)}$ and with residue field K. Let $W := K[X, Y]_{(X-1,Y)}$, let N be the maximal ideal of W and set
$$\overline{R} := V \cap W .$$

It is easy to see that $\overline{R}$ is a semilocal Noetherian integrally closed domain with two maximal ideals $m := M \cap \overline{R}$ and $n := N \cap \overline{R}$ such that $\overline{R}_m = V$, $\overline{R}_n = W$. Let $J(\overline{R}) = M \cap N = \pi N$ be the Jacobson radical of $\overline{R}$ and set

$$R := K + J(\overline{R}) .$$

Then, R is a 2-dimensional local subring of $\overline{R}$ such that $\overline{R}$ is the integral closure of R, $\overline{R}$ is finitely generated over R (hence R is Noetherian) and $(R : \overline{R}) = J(\overline{R})$ [N6, E2.1].

It is easy to see that

$$T_R(J(\overline{R})) = \Omega_R(J(\overline{R})) \supseteq (J(\overline{R}) : J(\overline{R})) \supseteq \overline{R} ,$$

and thus (Lemma 3.1 (e)):

$$\Omega_R(J(\overline{R})) = \Omega_{\overline{R}}(J(\overline{R})) = \Omega_{\overline{R}}(\pi N) = \overline{R}[1/\pi] = \overline{R}_\pi = W .$$

Therefore, $R \subset \overline{R} \subset \Omega_R(J(\overline{R})) = \overline{R}[1/\pi]$ is finitely generated, but $J(\overline{R})\Omega_R(J(\overline{R})) = J(\overline{R})W = N \neq \Omega_R(J(\overline{R})) = W$, whence $\Omega_R(J(\overline{R}))$ is not R-flat (cf. also Theorem 4.4).

ACKNOWLEDGMENT

The author wishes to acknowledge the partial support of a NATO Collaborative Research Grant (N. 970140). The author also wishes to thank Evan Houston for several useful conversations on the topics treated in this paper and the referee for his careful reading and helpful comments.

REFERENCES

[Ak1] T. Akiba, *Remarks on generalized rings of quotients*, Proc. Japan Acad. **40** (1964), 801–806.

[Ak2] T. Akiba, *Remarks on generalized rings of quotients, II*, J. Math. Kyoto Univ. **5** (1965), 39–44.

[A] D.D. Anderson, *Global transforms and Noetherian pairs*, Hiroshima Math. J. **10** (1980), 69–74.

[AnB] D.F. Anderson and A. Bouvier, *Ideal transforms and overrings of a quasilocal integral domain*, Ann. Univ. Ferrara **32** (1986), 15–38.

[AR] D. Arezzo and L. Ramella, *Sur les ouverts affines d'un schéma affine*, Einsegn. Math **25** (1979), 313–323.

[AB] J. Arnold and J. Brewer, *On flat overrings, ideal transforms and generalized transforms of a commutative ring*, J. Algebra **18** (1971), 254–263.

[AM] M. Atiyah and I.G. MacDonald, Introduction to Commutative Algebra, Addison-Wesley, Readings Mass. 1969.

[BO] M. Beattie and M. Orzech, *Prime ideals and finiteness conditions for Gabriel topologies over commutative rings*, Rocky Mount. J. Math. **22** (1992), 423–439.

[B] N. Bourbaki, Algèbre Commutative, Hermann, Paris 1961–1965.

[Bo] A. Bouvier, *Le groupe des classes d'un anneau intègre*, 107ᵉ Congrès National des Sociétes Savantes, Brest 1982, 85–92.

[BB1] W. Brandal and E. Barbut, *Localization of torsion theories*, Pacific J. Math. **107** (1983), 27–37.

[BB2] W. Brandal and E. Barbut, *Torsion theories over commutative rings*, J. Algebra **101** (1986), 136–150.

[BB3] W. Brandal and E. Barbut, Torsion theories over commutative rings, BCS Associates, Moscow 1996.

[Br] J. Brewer, *The ideal transform and overrings over an integral domain*, Math. Z. **107** (1968), 301–306.

[BrG] J. Brewer and R. Gilmer, *Integral domains whose overrings are ideal transforms*, Math. Nach. **51** (1971), 755–767.

[Bro] M.P. Brodmann, *Finiteness of ideal transforms*, J. Algebra **63** (1980), 162–185.

[BS] H. Butts and C. Spaht, *Generalized quotient rings*, Math. Nach. **51** (1971), 181–210.

[C] P.J. Cahen, *Commutative torsion theories*, Trans. Amer. Math. Soc. **184** (1973), 73–85.

[D] E.D. Davis, *Overrings of commutative rings, II: Integrally closed overrings*, Trans. Amer. Math. Soc. **110** (1964), 196–212.

[EHKR] P.M. Eakin Jr., W. Heinzer, D. Katz and L.J. Ratliff Jr., *Note on ideal-transform, Rees rings and Krull rings*, J. Algebra **110** (1987), 407–419.

[FH] M. Fontana and E. Houston, *On integral domains whose overrings are Kaplansky ideal transforms*, in preparation.

[FHP] M. Fontana, J.A. Huckaba and I.J. Papick, Prüfer domains, M. Dekker, New York 1997.

[FP1] M. Fontana and N. Popescu, *Sur une classe d'anneaux qui généralisent les anneaux de Dedekind*, J. Algebra **73** (1995), 44–66.

[FP2] M. Fontana and N. Popescu, *On a class of domains having Prüfer integral closure: the FQR-domains*, Commutative Ring Theory, II, Dekker Lect. Notes Pure Appl. Math. **185** (1997), 303–312.

[FP3] M. Fontana and N. Popescu, *Nagata transform and localizing systems*, submitted.

[Ga] S. Gabelli, *Prüfer (##)-domains and localizing system of ideals*, Commutative Ring Theory, III, Dekker Lect. Notes Pure Appl. Math. (1999).

[Gb] P. Gabriel, *Des catégories abéliennes*, Boll. Soc. Math. France **90** (1962), 323–448.

[G1] R.W. Gilmer, *The pseudo-radical of a commutative ring*, Pacific J. Math. **19** (1966), 275–284.

[G2] R.W. Gilmer, *Overrings of Prüfer domains*, J. Algebra **4** (1966), 331–340.

[G3] R.W. Gilmer, Multiplicative ideal theory, M. Dekker, New York 1972.

[G4] R.W. Gilmer, *Prüfer-like conditions on the set of overrings of an integral domain*, Springer Lect. Notes Math. **311** (1973), 90–112.

[GH] R.W. Gilmer and W. Heinzer, *Intersections of quotient rings of an integral domain*, J. Math. Kyoto Univ. **7** (1967), 133–150.

[GHu] R.W. Gilmer and J. Huckaba, *The transform formula for ideals*, J. Algebra **21** (1972), 191–215.

[GO] R.W. Gilmer and J. Ohm, *Integral domains with quotient overrings*, Math. Ann. **153** (1964), 97–103.

[Go] O. Goldman, *Rings and modules of quotients*, J. Algebra **13** (1969), 10–47.

[EGA] A. Grothendieck and J. Dieudonné, Eléments de Géométrie Algébrique I, Springer, 1971.
A. Grothendieck, Eléments de Géométrie Algebrique II, Publ. Math. I.H.E.S. **8** (1961).

[Ha] R. Hartshorne, *Cohomological dimension of algebraic varietes*, Ann. Math. **88** (1968), 401–450.

[Hy] J. Hays, *The S-transform and the ideal transform*, J. Algebra **57** (1979), 223–229.

[He1] J.R. Hedstrom, *Domains of Krull-type and ideal transforms*, Math. Nach. **69** (1975), 145–148.

[He2] J.R. Hedstrom, *G-domains and property (T)*, Math. Nach. **56** (1973), 125–129.

[He3] J.R. Hedstrom, *Integral closure of generalized quotient rings*, Math. Nach. **69** (1975), 145–148.

[H] W. Heinzer, *Quotient overrings of integral domains*, Mathematika **17** (1970), 139–148.

[HOP] W. Heinzer, J. Ohm and R. Pendleton, *On integral domains of the form $\cap D_P$, P minimal*, J. Reine Angew. Math. **241** (1970), 147–159.

[J] J. Játem, *Idealtransformierte von multiplikativen Systemen von Idealen und deren Endlichkeit*, Beiträge Algebra Geom. **28** (1989), 99–112.

[K1] I. Kaplansky, Commutative rings, Allyn and Bacon, Boston 1970.

[K2] I. Kaplansky, Topics in commutative ring theory, Lecture Notes, University of Chicago 1974.

[Ka] D. Katz, *On the integrity of the ideal transforms*, Houston J. Math. **10** (1984), 415–421.

[KR] D. Katz and L.J. Ratliff Jr., *Two notes on ideal transforms*, Math. Proc. Cambridge Phil. Soc. **102** (1987), 389–397.

[Ki] K.H. Kiyek, *Anwendung von Ideal–Transformationen*, Manuscripta Math. **34** (1981), 327–353.

[L] D. Lazard, *Epimorphismes plats*, Seminaire P. Samuel 1967/68, N. 4.

[MR] S. McAdam and L.J. Ratliff, *Finite transforms of a Noetherian ring*, J. Algebra **101** (1986), 479–489.

[Ma] J.R. Matijevic, *Maximal ideal transform of Noetherian rings*, Proc. Amer. Math. Soc. **54** (1976), 49–52.

[M] J. Mott, *Integral domains with quotient overrings*, Math. Ann. **166** (1966), 229–232.

[N1] M. Nagata, *A treatise on the 14-th problem of Hilbert*, Mem. Coll. Sci. Kyoto, Math. **30** (1956–57), 57–70. Addition and correction to it, 197–200.

[N2] M. Nagata, *On the 14-th problem of Hilbert*, Ann. J. Math. **81** (1959), 766–772.

[N3] M. Nagata, *A theorem on finite generation of a ring*, Nagoya Math. J. **27** (1966), 193–205.

[N4] M. Nagata, *Some sufficient conditions for the fourteenth problem of Hilbert*, Actas Coloquio International sobre Geometria Algebraica, Madrid 1965, 107–121.

[N5] M. Nagata, Lectures on the fourteenth problem of Hilbert, Notes by M.P. Murthy, Tata Institute of Fundamental Research, Bombay 1965.

[N6] M. Nagata, Local rings, Interscience, New York 1962.

[N7] M. Nagata, *Note on a chain condition for prime ideals*, Mem. Coll. Sci. Kyoto Univ. **32** (1959), 85–90.

[NP] C. Năstăsescu and N. Popescu, *On the localization ring of a ring*, J. Algebra **15** (1970), 41–56.

[Ni] J.-I. Nishimura, *On ideal transform of Noetherian rings, I*, J. Math. Kyoto Univ. **19** (1979), 41-46.

[O] T. Ohi, *Affine open subset of an affine scheme*, TRU Math. **11** (1975), 23–25.

[Pe] R. Pendleton, *A characterization of Q–domains*, Bull. Amer. Math. Soc. **72** (1966), 499-500.

[P] N. Popescu, Abelian categories with applications to rings and modules, Academic Press, New York 1973.

[Po] E.L. Popescu, *On a paper of Peter Schenzel*, Rev. Roumaine Math. Pures Appl. **40** (1995), 521–525.

[Ra] L. Ramella, *On the Nagata transform of an ideal*, Rend. Sem. Mat. Univ. Politec. Torino **37** (1979), 149–156.

[Re] D. Rees, *On a problem of Zariski*, Illinois J. Math. **2** (1958), 145–149.

[R] F. Richman, *Generalized quotient rings*, Proc. Amer. Math. Soc. **16** (1965), 794–799.

[Rh] C.P.L. Rhodes, *The S–transform and its dual with applications to Prüfer extensions*, Rend. Sem. Mat. Univ. Padova **95** (1996), 201–216.

[S1] P. Schenzel, *Finiteness of relative Rees rings and asymptotic prime divisors*, Math. Nach. **129** (1986), 123–148.

[S2] P. Schenzel, *Flatness and ideal-transform of finite type*, Commutative Algebra (Salvador, 1988), Springer Lect. Notes Math. **1430** (1990), 88-97.

[S3] P. Schenzel, *When is a flat algebra of finite type?*, Proc. Amer. Math. Soc. **109** (1990), 287–290.

[Se] J.P. Serre, *Sur la cohomologie des variétés algébriques*, J. Math. Pures Appl. **36** (1957), 1–16.

[St] B. Stenström, Rings of quotients, Springer–Verlag, New York 1975.

[Z] O. Zariski, *Interpretations algébrico-géometriques du quatorzième problème de Hilbert*, Bull. Sc. Math. **78** (1954), 155–168.

[ZS] O. Zariski and P. Samuel, Commutative Algebra, Von Nostrand, 1958–1960.

[Zö] H. Zöschinger, *Die globale Transformation eines lokalen Ringes*, J. Algebra **168** (1994), 877–902.

Dipartimento di Matematica
Università degli Studi, Roma Tre
Largo San L. Murialdo, 1
00146 Roma, Italy
e-mail: fontana@mat.uniroma3.it

Polynomial Closure in Essential
Domains and Pullbacks

Marco Fontana *, Università degli Studi "Roma Tre", Italy

Lahoucine Izelgue, Université de Marrakech, Morocco

Salah-Eddine Kabbaj, King Fahd University, Saudi Arabia

Francesca Tartarone, Università degli Studi "La Sapienza", Italy

ABSTRACT. Let D be a domain with quotient field K. Let $E \subseteq K$ be a subset; the ring of D-integer-valued polynomials over E is $\text{Int}(E, D) := \{f \in K[X]; f(E) \subseteq D\}$. The polynomial closure in D of a subset $E \subseteq K$ is the largest subset $F \subseteq K$ containing E such that $\text{Int}(E, D) = \text{Int}(F, D)$, and it is denoted by $cl_D(E)$. We study the polynomial closure of ideals in several classes of domains, including essential domains and domains of strong Krull-type, and we relate it with the t-closure. For domains of Krull-type we also compute the Krull dimension of $\text{Int}(D)$.

INTRODUCTION

Let D be any integral domain with quotient field K. For each subset $E \subseteq K$, $\text{Int}(E, D) := \{f \in K[X]; f(E) \subseteq D\}$ is called the *ring of D-integer-valued polynomials over E*. As usual, when $E = D$, we set $\text{Int}(D) := \text{Int}(D, D)$. When E is "large enough", it may happen that $\text{Int}(E, D) = D$ (for instance, $\text{Int}(S^{-1}\mathbb{Z}, \mathbb{Z}) = \mathbb{Z}$ for each nontrivial multiplicative subset S of $\mathbb{Z}$, [CC2, Corollary I.1.10]). This does not happen if E is a *D-fractional subset* of K, i.e. if there exists $d \in D \backslash (0)$ such that $dE \subseteq D$. Indeed, in this case, $dX \in \text{Int}(E, D)$. It is well known that

* Partially supported by a NATO Collaborative Research Grant No. 970140.

Int$(D) \subseteq$ Int(E, D) if and only if $E \subseteq D$ [CC2, Corollary I.1.7]. Two subsets E and F of K may be distinct while Int$(E, D) =$ Int(F, D). When this happens, we say that E and F are *polynomially D-equivalent*. For instance, $\mathbb{N}$ and $\mathbb{Z}$ are polynomially $\mathbb{Z}$-equivalent [CC2, Corollary I.1.2]. In particular, if Int$(E, D) =$ Int(D) we say that E is a *polynomially dense* subset of D (so, $\mathbb{N}$ is a polynomially dense subset of $\mathbb{Z}$). In [G1] Gilmer characterized the polynomially dense subsets of $\mathbb{Z}$ and, in [C2] and [C4], Cahen studied the polynomial density, with special emphasis to the Noetherian domains case. McQuillan, pursuing Gilmer's work, investigated the polynomially D-equivalent subsets of a Dedekind domain D [Mc]. Among other results, he proved that two fractional ideals I and J of a Dedekind domain D are polynomially equivalent if and only if $I = J$. After noticing that, Cahen introduced the notion of *polynomial closure* (in D) of a subset E of K as follows:

$$cl_D(E) := \{x \in K; f(x) \in D, \text{ for each } f \in \text{Int}(E, D)\},$$

that is, $cl_D(E)$ is the largest subset $F \subseteq K$ such that Int$(E, D) =$ Int(F, D). Obviously, E is said to be *polynomially dense* in D if $cl_D(E) = D$ and *polynomially D-closed* if $cl_D(E) = E$.

In the first section of this paper, we consider *essential domains*, that is, domains D such that

$$(1) \hspace{4cm} D = \cap_{P \in \mathcal{P}} D_P,$$

where $\mathcal{P}$ is a subset of Spec(D) and D_P is a valuation domain. In particular, among these domains, we will focus our attention on the *strong Krull-type domains*, that is the essential domains D such that the intersection (1) is locally finite (i.e., each nonzero of D belongs to finitely many prime ideals $P \in \mathcal{P}$) and the valuation rings D_P are pairwise independent. Examples of strong Krull-type domains are *Krull domains* and *generalized Krull domains* [G1, p. 524]. We prove that if E is a fractional subset of a strong Krull-type domain D, then $cl_D(E) = \cap_{P \in \mathcal{P}} cl_{D_P}(E_P)$, where $cl_{D_P}(E_P)$ denotes the polynomial D_P-closure of $E_P := \{e/s; e \in E, s \in D \backslash P\}$. This yields a generalization of a result proved by Cahen for Krull domains, [C3] or [C4]. Moreover, we study the polynomial closure as a star-operation and we relate it to the t-operation. We find that if D is an essential domain, then $I_t \subseteq cl_D(I)$, for each fractional ideal I of D and, for some distinguished classes of domains, as Krull domains, Prüfer domains in which each nonzero ideal is divisorial, $I_t = cl_D(I)$.

In the second section we compute the Krull dimension of the ring of Int(D), when D is a domain with a locally finite representation. By using this result we show that if D is a domain of Krull-type then dim(Int(D)) = dim($D[X]$), obtaining further evidence for the validity of the conjecture about the Krull dimension of Int(D) stating that dim(Int(D)) $\leq$ dim($D[X]$) for each integral domain D (cf. [Ch], [C1], and [FIKT]).

In Section 3, we study the quotient of the polynomial closure of a subset modulo a divided prime ideal, and we apply this result to some classes of domains defined by making use of pullback constructions.

1. POLYNOMIAL CLOSURE IN STRONG KRULL-TYPE AND NOETHERIAN DOMAINS

We start this section by studying the polynomial closure of the ideals in a *strong Krull-type domain*, that is a domain D having the following representation:

$$(1.0.I) \qquad\qquad D = \cap_{P \in \mathcal{P}} D_P,$$

where $\mathcal{P} \subseteq \mathrm{Spec}(D)$, the intersection $(1.0.I)$ is locally finite and the rings D_P are pairwise independent valuation domains.

We prove for this class of domains some results already proved by Cahen in case of Noetherian and Krull domains [C3, Proposition 1.3], [C4, Proposition 3.2, 3.5, 3.6, Corollaries 3.7 and 3.8].

THEOREM 1.1. *With the notation above, let D be a strong Krull-type domain with quotient field K and let E be any D-fractional subset of K. Then:*
(1) $\mathrm{Int}(E, D)_P = \mathrm{Int}(E, D_P)$, *for each $P \in \mathcal{P}$;*
(2) $cl_D(E) = \cap_{P \in \mathcal{P}} cl_{D_P}(E_P)$;
(3) *if $E \subseteq D$, then E is polynomially dense in D if and only if E is polynomially dense in D_P for each $P \in \mathcal{P}$.*

Proof. (1) The argument of the proof runs parallel with the one used in [C4, Proposition 3.2] (cf. also [CC2, Proposition I.2.8 and IV.2.9]). We wish to prove that $\mathrm{Int}(E, D_{\overline{P}}) \subseteq \mathrm{Int}(E, D)_{\overline{P}}$, for each fixed ideal $\overline{P} \in \mathcal{P}$ (the opposite inclusion holds in general, [CC2, Lemma I.2.4]).

Since E is a fractional D-subset of K, with a standard argument we can easily assume, without loss of generality, that $E \subseteq D$. Let $f \in \mathrm{Int}(E, D_{\overline{P}})$, $f \neq 0$. It is obvious that there exists $d \in D$, $d \neq 0$, such that $df \in D[X]$. Set

$$\mathcal{P}(d) := \{Q \in \mathcal{P}; d \in Q \text{ and } Q \not\subseteq \overline{P}\}.$$

Since the given representation of D is locally finite, then $\mathcal{P}(d)$ is a finite set.

We claim that there exists $a \in D$ such that $a \in dD_Q \setminus \overline{P}$, for each $Q \in \mathcal{P}(d)$, that is, $\mathbf{v}_Q(a) \geq \mathbf{v}_Q(d)$, for each $Q \in \mathcal{P}(d)$, and $\mathbf{v}_{\overline{P}}(a) = 0$, where $\mathbf{v}_P$ is the valuation associated to the ring D_P for each $P \in \mathcal{P}$.

The Approximation Theorem for valuations [B, Ch. VI § 7 n. 2, Corollaire 1, p. 135] states that there exists an element $b \in K$ such that $\mathbf{v}_{\overline{P}}(b) = 0$ and $\mathbf{v}_Q(b) \geq \mathbf{v}_Q(d)$, for each $Q \in \mathcal{P}(d)$. Now, there exist $a, c \in D$, $a \notin \overline{P}$, such that $b = a/c$. For each $P \in \mathcal{P}$, we have that $\mathbf{v}_P(a) = \mathbf{v}_P(b) + \mathbf{v}_P(c)$. Thus, if $Q \in \mathcal{P}(d)$, then $\mathbf{v}_Q(c) \geq 0$ (since $c \in D \subseteq D_Q$) and $\mathbf{v}_Q(a) \geq \mathbf{v}_Q(b) \geq \mathbf{v}_Q(d)$. Moreover, since $\mathbf{v}_{\overline{P}}(b) = \mathbf{v}_{\overline{P}}(c) = 0$, then also $\mathbf{v}_{\overline{P}}(a) = 0$.

Therefore $af \in D_P[X]$ for each $P \in \mathcal{P}(d)$ and $P \in \mathcal{P}$ such that $d \notin P$. As a matter of fact, for each $P \in \mathcal{P} \setminus \mathcal{P}(d)$, with $d \notin P$, we have that d is a unit in D_P. Thus, since $df \in D[X]$, we deduce that $f \in D_P[X]$ and $af \in D_P[X]$. If $Q \in \mathcal{P}(d)$, then $ad^{-1} \in D_Q$ and $af = (ad^{-1})(df) \in D_Q[X]$. In these cases, since $E \subseteq D$ and $af \in D_P[X]$, then $af(E) \subseteq D_P$, that is, $af \in \mathrm{Int}(E, D_P)$. On the other hand, $f \in \mathrm{Int}(E, D_{\overline{P}})$ and $a \in D$, whence $af(E) \subseteq D_{\overline{P}}$. If $d \in P$ and $P \subseteq \overline{P}$, then $D_P \supseteq D_{\overline{P}}$. Thus $f(E) \subseteq D_{\overline{P}} \subseteq D_P$ and $f \in \mathrm{Int}(E, D_P)$.

We conclude that $af(E) \subseteq \cap_{P \in \mathcal{P}} D_P = D$, that is $af \in \mathrm{Int}(E, D)$. Hence $f \in \mathrm{Int}(E, D)_{\overline{P}}$, because $a \in D \setminus \overline{P}$.

(2) follows from (1) and from [C4, Proposition 3.5].

(3) is a straightforward consequence of (2). □

In order to deepen the study of the polynomial closure of fractional ideals, we recall some properties about star-operations.

Let D be an integral domain with quotient field K, let $\mathfrak{F}(D)$ denote the set of nonzero fractional ideals of D and let $\mathfrak{F}_{\mathrm{fg}}(D)$ denote the subset of $\mathfrak{F}(D)$ of nonzero finitely generated fractional ideals of D. A mapping $I \mapsto I^*$ of $\mathfrak{F}(D)$ into $\mathfrak{F}(D)$ is called a *star-operation on D* if the following conditions hold for all $a \in K \setminus \{0\}$ and $I, J \in \mathfrak{F}(D)$:

(*1) $(aD)^* = aD$;

(*2) $(aI)^* = aI^*$;

(*3) $I \subseteq I^*$;

(*4) $I \subseteq J \Rightarrow I^* \subseteq J^*$;

(*5) $I^{**} = I^*$.

A fractional ideal $I \in \mathfrak{F}(D)$ is called a *star-ideal* if $I = I^*$. A star-operation $*$ on D is said to be *of finite character* if, for each $I \in \mathfrak{F}(D)$,

$$I^* = \cup \{J^*; J \subseteq I \text{ and } J \in \mathfrak{F}_{\mathrm{fg}}(D)\}.$$

Given a star-operation $*$, then the function $*_{\mathrm{s}}$ defined as follows:

$$I \mapsto I^{*_{\mathrm{s}}} = \cup \{J^*; J \subseteq I \text{ and } J \in \mathfrak{F}_{\mathrm{fg}}(D)\}$$

is a star-operation of finite character. The star-operation $*_{\mathrm{s}}$ is called the *star-operation of finite character associated to $*$*. It is obvious that:

$$J^* = J^{*_{\mathrm{s}}}, \text{ for each } J \in \mathfrak{F}_{\mathrm{fg}}(D),$$

$$I^{*_{\mathrm{s}}} \subseteq I^*, \text{ for each } I \in \mathfrak{F}(D).$$

The *v-operation*

$$I \mapsto I_v := (D : (D : I))$$

is a star-operation. The *t-operation*

$$I \mapsto I_t := \cup \{J_v; J \subseteq I \text{ and } J \in \mathfrak{F}_{\mathrm{fg}}(D)\}$$

is the star-operation of finite character associated to the *v*-operation (cf. [G2, Sections 32 and 34]).

The following result is implicitly proved by Cahen [C4, Lemma 1.2].

LEMMA 1.2. *Let D be an integral domain, then the polynomial closure*

$$cl_D : \mathfrak{F}(D) \to \mathfrak{F}(D), \quad I \mapsto cl_D(I),$$

is a star-operation. □

COROLLARY 1.3. *Let D be an integral domain. Then, for all I, J in $\mathfrak{F}(D)$ and for each subset $\{I_\alpha; \alpha \in A\}$ of $\mathfrak{F}(D)$, we have:*

(1) $cl_D(\sum_\alpha I_\alpha) = cl_D(\sum_\alpha cl_D(I_\alpha))$, *if* $\sum_\alpha I_\alpha \in \mathfrak{F}(D)$;

(2) $\cap_\alpha cl_D(I_\alpha) = cl_D(\cap_\alpha cl_D(I_\alpha))$, *if* $\cap_\alpha I_\alpha \neq (0)$;

(3) $cl_D(IJ) = cl_D(Icl_D(J)) = cl_D(cl_D(I)J) = cl_D(cl_D(I)cl_D(J))$;

(4) $cl_D(I) \subseteq I_v$;

(5) $cl_D(I_v) = I_v$.

Proof. It is a straightforward consequence of Lemma 1.2 and of [G2, Proposition 32.2 and Theorem 34.1(4)]. $\square$

Note that, from Corollary 1.3(1) and (3), we recover for the fractional ideals some results proved by Cahen for subsets [C4, Lemma 2.4], in particular we obtain that $cl_D(I) + cl_D(J) \subseteq cl_D(I + J)$, $cl_D(I)cl_D(J) \subseteq cl_D(IJ)$.

We will need the following result, that is a consequence of [C4, Proposition 3.5(2)], in order to deepen the relation between the polynomial closure and the star-operations.

LEMMA 1.4. *Let D be an integral domain and let $\mathcal{P}$ be a subset of $\mathrm{Spec}(D)$ such that $D = \cap_{P \in \mathcal{P}} D_P$. For each $I \in \mathfrak{F}(D)$, we have:*

$$\cap_{P \in \mathcal{P}} cl_{D_P}(ID_P) \subseteq cl_D(I). \quad \square$$

It is well known from the theory of star-operations that, if $\{D_\alpha; \alpha \in A\}$ is a collection of overrings of an integral domain D such that $D = \cap_{\alpha \in A} D_\alpha$, and if $*_\alpha$ is a star-operation on D_α, for each $\alpha \in A$, then the mapping

$$I \mapsto I^{*_A} := \cap\{(ID_\alpha)^{*_\alpha}; \alpha \in A\}$$

is a star-operation on D and $(I^{*_A}D_\alpha)^{*_\alpha} = (ID_\alpha)^{*_\alpha}$, for each $\alpha \in A$ [A, Theorem 2]. If $D = \cap_{P \in \mathcal{P}} D_P$, for some subset $\mathcal{P} \subseteq \mathrm{Spec}(D)$, we call the $\mathcal{P}$-*polynomial closure* *of* $I \in \mathfrak{F}(D)$, the following fractional ideal of D:

$$\mathcal{P}\text{-}cl_D(I) := \cap_{P \in \mathcal{P}} cl_{D_P}(ID_P).$$

PROPOSITION 1.5. *Let D be an integral domain such that $D = \cap_{P \in \mathcal{P}} D_P$, for some subset $\mathcal{P}$ of $\mathrm{Spec}(D)$.*

(1) *The mapping:*
$$I \mapsto \mathcal{P}\text{-}cl_D(I)$$

defines a star-operation on D, with $\mathcal{P}\text{-}cl_D(I) \subseteq cl_D(I)$ for each $I \in \mathfrak{F}(D)$.

(2) *Let $I \in \mathfrak{F}(D)$. If $ID_P \neq D_P$ for finitely many $P \in \mathcal{P}$, then*

$$\mathcal{P}\text{-}cl_D(I)D_P = cl_{D_P}(ID_P).$$

(3) *If $cl_{D_P}(F \cap G) = cl_{D_P}(F) \cap cl_{D_P}(G)$, for each $P \in \mathcal{P}$ and for all $F, G \in \mathfrak{F}(D_P)$, then*

$$\mathcal{P}\text{-}cl_D(I \cap J) = \mathcal{P}\text{-}cl_D(I) \cap \mathcal{P}\text{-}cl_D(J), \text{ for all } I, J \in \mathfrak{F}(D).$$

(4) If $cl_{D_P}((F:_{D_P} G)) = (cl_{D_P}(F):_{D_P} cl_{D_P}(G))$, for each $P \in \mathcal{P}$, $F \in \mathfrak{F}(D_P)$ and $G \in \mathfrak{F}_{fg}(D_P)$, then

$$\mathcal{P}\text{-}cl_D((I:_D J)) = (\mathcal{P}\text{-}cl_D(I):_D \mathcal{P}\text{-}cl_D(J)),$$

for each $I \in \mathfrak{F}(D)$ and $J \in \mathfrak{F}_{fg}(D)$.

(5) If $D = \cap_{P \in \mathcal{P}} D_P$ is locally finite and if, for each $P \in \mathcal{P}$, cl_{D_P} is a star-operation on D_P with finite character, then $\mathcal{P}\text{-}cl_D$ is a star-operation on D with finite character.

Proof. These results are a straightforward consequence of Lemma 1.4, of the definition of the $\mathcal{P}$-polynomial closure and of [A, Theorem 2]. $\square$

COROLLARY 1.6. *If $D = \cap_{P \in \mathcal{P}} D_P$ is a strong Krull-type or a Noetherian domain, then, for each $I \in \mathfrak{F}(D)$, we have that*

$$\mathcal{P}\text{-}cl_D(I) = cl_D(I).$$

Proof. If D is strong Krull-type, the thesis is a consequence of Theorem 1.1(2) and the definition of $\mathcal{P}$-polynomial closure. More precisely, if $\mathrm{Int}(ID_P, D_P) = \mathrm{Int}(I, D)_P$, then $cl_D(I) \subseteq cl_{D_P}(ID_P)$ [C4, Proposition 3.5(1)] and hence $cl_D(I) \subseteq \mathcal{P}\text{-}cl_D(I)$. Therefore $cl_D(I) = \mathcal{P}\text{-}cl_D(I)$ by Lemma 1.4.

If D is Noetherian, then $S^{-1}\mathrm{Int}(I, D) = \mathrm{Int}(S^{-1}I, S^{-1}D)$, for each multiplicative set S of D [CC2, Proposition I.2.7]. As shown above for strong Krull-type domains, also in the Noetherian domain case, for $\mathcal{P} = \mathrm{Max}(D)$, we have that:

$$cl_D(I) = \cap_{P \in \mathcal{P}} cl_{D_P}(ID_P),$$

for each $I \in \mathfrak{F}(D)$, that is, $cl_D(I) = \mathcal{P}\text{-}cl_D(I)$, by definition of $\mathcal{P}$-polynomial closure. $\square$

COROLLARY 1.7. *Let D be a Noetherian domain.*

(1) *For each $M \in \mathrm{Max}(D)$ and for each $I \in \mathfrak{F}(D)$, we have:*

$$cl_D(I)D_M = cl_{D_M}(ID_M).$$

(2) *If S is a multiplicative subset of D, for each $I \in \mathfrak{F}(D)$ we have:*

$$S^{-1}cl_D(I) = cl_{S^{-1}D}(S^{-1}I).$$

In particular, if I is polynomially closed in D then $S^{-1}I$ is polynomially closed in $S^{-1}D$.

(3) *For each $M \in \mathrm{Max}(D)$,*

$$cl_D(M) = (MD_M)_v \cap D = M_v.$$

In particular, M is polynomially closed (respectively, polynomially dense) in D if and only if $M = M_v$ (respectively, $M_v = D$) or, equivalently, if and only if MD_M is polynomially closed (respectively, polynomially dense) in D_M.

(4) *For each nonzero ideal I of D there exists a prime ideal P of D such that $I \subseteq P = cl_D(P)$.*

(5) *If* $\dim(D) = 1$, *then every nonzero prime ideal of* D *is polynomially closed.*

Proof. (1) is an easy consequence of Proposition 1.5(2) and Corollary 1.6, since D is Noetherian and $D = \cap_{M \in \text{Max}(D)} D_M$.

(2) Note that $S^{-1}D = \cap\{D_M; M \in \text{Max}(D) \text{ and } M \cap S = \emptyset\}$ is locally finite. Hence the conclusion follows from (1) and from Corollary 1.6, since

$$S^{-1}cl_D(I) = \cap\{cl_D(I)D_M; M \in \text{Max}(D) \text{ and } M \cap S = \emptyset\} =$$

$$= \cap\{cl_{D_M}(ID_M); M \in \text{Max}(D) \text{ and } M \cap S = \emptyset\} =$$

$$= cl_{S^{-1}D}(S^{-1}I).$$

(3) By Corollary 1.6 we have

$$cl_D(M) = \cap_{N \in \text{Max}(D)} cl_{D_N}(MD_N) =$$

$$= cl_{D_M}(MD_M) \cap (\cap\{D_N; N \in \text{Max}(D), N \neq M\}) = cl_{D_M}(MD_M) \cap D.$$

Moreover, by the proof of [C4, Proposition 2.3], we know that if $(R, \mathbf{m})$ is a local Noetherian domain then $cl_R(\mathbf{m}) = \mathbf{m}_v$. Finally, since D is Noetherian, by [G2, Theorem 4.4(4)], we have that $(MD_M)_v = M_v D_M$ and by [G2, Theorem 4.10(3)] we have

$$M_v = (MD_M)_v \cap D.$$

The conclusion is straightforward.

(4) Since D is Noetherian, cl_D is a star-operation on D with finite character (Corollary 1.6). It is well known, in this situation, that each proper star-ideal of D is contained in a maximal proper star-ideal of D and that a maximal proper star-ideal of D is a prime ideal (cf. for example [J]).

(5) follows immediately from (4). □

Note that Corollary 1.7(2) gives a positive answer to Question 3.10 in [C4] and Corollary 1.7(3) generalizes to the nonlocal case [C4, Proposition 2.3]. Note also that Cahen [C4, Example 3.9] has given an example of an ideal I of an integrally closed (non Noetherian) domain D such that $I = cl_D(I)$ and $S^{-1}I \neq cl_{S^{-1}D}(S^{-1}I)$, for some multiplicative set S of D.

The equality in Corollary 1.7(3) does not hold for the nonmaximal ideals, i.e. the inclusion $cl_D(I) \subseteq I_v$ may be a proper inclusion even in the Noetherian local case. In fact, it is enough to consider a local, Noetherian, one-dimensional, analitically irreducible domain D with finite residue field and a nonzero nonmaximal ideal I of D (cf. [C4, Corollary 4.8] or [CC2, Theorem IV.1.15]). For instance, let k be a finite field, $D := k[[X^3, X^4, X^5]]$ and $I := (X^3, X^4)D$. In this case $(D:I) = k[[X]]$, hence $I_v = (X^3, X^4, X^5)D$; but $I = cl_D(I)$ [C4, Corollary 4.8].

We recall some definitions. An *essential domain* is an integral domain D such that $D = \cap_{P \in \mathcal{P}} D_P$, where D_P is a valuation domain for P belonging to a subset $\mathcal{P}$ of $\text{Spec}(D)$. If D is an essential domain with the valuation rings D_P pairwise independent and $D = \cap_{P \in \mathcal{P}} D_P$ is locally finite (i.e. each nonzero element of D belongs to finitely many prime ideals P of $\mathcal{P}$) then D is a domain of strong Krull-type. Obviously, each Prüfer domain is an essential domain and each Krull domain is a domain of strong Krull-type.

A relevant case is when $\mathcal{P}$ is the set $t_m(D)$ of all *t-maximal ideals of D* (i.e. the maximal elements among the integral *t*-ideals of *D*). It is well known that each maximal *t*-ideal is a prime ideal and, for each ideal I of D, $I = \cap_{P \in t_m(D)} I D_P$; in particular $D = \cap_{P \in t_m(D)} D_P$ [Gr, Proposition 4]. A *Prüfer v-multiplication domain* D is an integral domain such that D_P is a valuation domain for each $P \in t_m(D)$. This class of domains was introduced by Griffin [Gr].

In order to study the polynomial closure of fractional ideals in an essential domain, we start by considering the local case, i.e. when $D = V$ is a valuation domain.

PROPOSITION 1.8. *Let V be a valuation domain with maximal ideal M.*

(1) *If M is principal, then, for each nonzero fractional ideal I of V, $I = cl_V(I) = I_v$.*

(2) *If M is not principal, then:*
    a) *$cl_V(M) = M_v = V$;*
    b) *for each nonzero ideal I of V, $cl_V(I) = I_v$; moreover, if $I \neq I_v$, then $cl_V(I) = I_v$ is a principal ideal of V.*

Proof. We recall that, in general, for each integral domain D and for each $I \in \mathfrak{F}(D)$ we have the following inclusions: $I \subseteq cl_D(I) \subseteq I_v$ (Corollary 1.3).

(1) If M is principal, then each nonzero fractional ideal of V is divisorial [G2, Exercise 12, p. 431]. The conclusion follows immediately from the previous tower of inclusions.

(2) If M is not principal, then $\{aM; a \in V, a \neq 0\}$ is the set of all nonzero nondivisorial (integral) ideals of V [G2, Exercise 12, p. 431]. Therefore, in this case, $M \neq M_v$, hence $M_v = V$.

(a) In order to prove that $cl_V(M) = V$, we will show that $\mathrm{Int}(V) = \mathrm{Int}(V, V) = \mathrm{Int}(M, V)$.

Since M is not principal, $V[X] = \mathrm{Int}(V)$ [CC2, Proposition I.3.16]. Let $f := c_0 + c_1 X + \cdots + c_n X^n \in \mathrm{Int}(M, V)$ be a polynomial of degree n. By [CC2, Corollary I.3.3], if $a_0, a_1, \cdots, a_n$ are $n+1$ elements of M and if $d := \prod_{0 \leq i < j \leq n}(a_i - a_j)$, then $df \in V[X]$. Let $\mathbf{v}$ be the valuation associated to V. By using the assumption that M is not finitely generated, we can choose the elements a_i's such that $0 < \mathbf{v}(d) < |\mathbf{v}(c_i)|$, for each c_i such that $\mathbf{v}(c_i) \neq 0$. On the other hand, $df \in V[X]$ hence $\mathbf{v}(dc_i) = \mathbf{v}(d) + \mathbf{v}(c_i) \geq 0$, for each $0 \leq i \leq n$. If $f \notin V[X]$, then $\mathbf{v}(c_i) < 0$ for some i with $0 \leq i \leq n$, hence we have a contradiction. Therefore, we can conclude that $V[X] = \mathrm{Int}(M, V)$ and thus a) holds.

b) It is obvious that, if $I = I_v$, then $cl_V(I) = cl_V(I_v) = I_v$ (Corollary 1.3(5)). If $I \neq I_v$ and $I \subseteq V$, then $I = aM$ for some nonzero element $a \in V$, hence $cl_V(I) = cl_V(aM) = acl_V(M)$ (Lemma 1.2). By point a), we deduce that $cl_V(I) = aV$ is a principal (hence, divisorial) ideal and $cl_V(I) = I_v$. If $I \neq I_v$ and I is a fractional ideal of V, then $bI \subset V$ and $bI \neq bI_v$, for some nonzero element $b \in V$. The conclusion follows easily from the previous argument. $\square$

THEOREM 1.9. *Let $D = \cap_{P \in \mathcal{P}} D_P$ be an essential domain.*

(1) *For each $J \in \mathfrak{F}_{fg}(D)$, we have:*

$$\mathcal{P}\text{-}cl_D(J) = cl_D(J) = J_v.$$

(2) *For each* $I \in \mathfrak{F}(D)$, *we have:*

$$I_t \subseteq \mathcal{P}\text{-}cl_D(I) \subseteq cl_D(I) \subseteq I_v.$$

Proof. (1) If J is finitely generated, then

$$J_v D_P \subseteq (JD_P)_v, \text{ for each prime ideal } P \text{ of } D,$$

[B, Ch. I § 2 n. 11 (11), p. 41]. On the other hand, by Proposition 1.8,

$$J_v \subseteq \cap_{P \in \mathcal{P}} J_v D_P \subseteq \cap_{P \in \mathcal{P}} (JD_P)_v =$$
$$= \cap_{P \in \mathcal{P}} cl_D(JD_P) = \mathcal{P}\text{-}cl_D(J).$$

The conclusion follows by recalling that, in general for each $I \in \mathfrak{F}(D)$, we have:

$$\mathcal{P}\text{-}cl_D(I) \subseteq cl_D(I) \subseteq I_v.$$

(2) Since $I_t := \cup\{J_v; J \subseteq I \text{ and } J \in \mathfrak{F}_{\mathrm{fg}}(D)\}$ then, by Proposition 1.5(1) and by (1), we have:

$$I_t := \cup\{\mathcal{P}\text{-}cl_D(J); J \subseteq I \text{ and } J \in \mathfrak{F}_{\mathrm{fg}}(D)\} \subseteq \mathcal{P}\text{-}cl_D(I). \quad \square$$

COROLLARY 1.10. *Let* $D = \cap_{P \in \mathcal{P}} D_P$ *be an essential domain.*

(1) *For all* $J', J'' \in \mathfrak{F}_{\mathrm{fg}}(D)$:

$$cl_D(J' \cap J'') = cl_D(J') \cap cl_D(J'').$$

(2) *For all* $I', I'' \in \mathfrak{F}(D)$, *then:*

$$\mathcal{P}\text{-}cl_D(I' \cap I'') = \mathcal{P}\text{-}cl_D(I') \cap \mathcal{P}\text{-}cl_D(I'').$$

Proof. (1) follows from (2) and from Theorem 1.9(1).

(2) Since D_P is a valuation domain, for each $P \in \mathcal{P}$, then either $I'D_P \subseteq I''D_P$ or $I''D_P \subset I'D_P$, hence $cl_{D_P}(I'D_P \cap I''D_P) = cl_{D_P}(I'D_P) \cap cl_{D_P}(I''D_P)$. The conclusion follows from Proposition 1.5(3). $\square$

Let $D = \cap_{P \in t_m(D)} D_P$ and let $I \in \mathfrak{F}(D)$. In this case we set $\mathcal{P} = t_m(D)$ and

$$t\text{-}cl_D(I) := \cap_{P \in t_m(D)} cl_{D_P}(ID_P).$$

COROLLARY 1.11. *Let* $D = \cap_{P \in t_m(D)} D_P$ *be a Prüfer v-multiplication domain. Assume that, for each maximal t-ideal P of D, PD_P is a principal ideal. Then, for each* $I \in \mathfrak{F}(D)$, *we have:*

$$I_t = t\text{-}cl_D(I).$$

If, moreover, $D = \cap_{P \in t_m(D)} D_P$ is locally finite and the valuation rings D_P are pairwise independent (i.e. D is an integral domain of strong Krull-type) then, for each $I \in \mathfrak{F}(D)$, *we have*

$$I_t = t\text{-}cl_D(I) = cl_D(I).$$

Proof. In a Prüfer v-multiplication domain D, for each $I \in \mathfrak{F}(D)$, $I_t = \cap_{P \in t_m(D)} ID_P$ (cf. for instance [A, Theorem 6]).

On the other hand, by Proposition 1.8(1),

$$\cap_{P \in t_m(D)} ID_P \subseteq t\text{-}cl_D(I) = \cap_{P \in t_m(D)} cl_{D_P}(ID_P) = \cap_{P \in t_m(D)} ID_P,$$

hence $I_t = t\text{-}cl_D(I)$. The last statement is a consequence of Corollary 1.6. $\square$

COROLLARY 1.12. *Let* $D = \cap_{P \in t_m(D)} D_P$ *be an integral domain of strong Krull-type. Assume that, for each* $P \in t_m(D)$*, there exists a finitely generated ideal* J *of* D *such that* $J \subseteq P$ *and* $J^{-1} = P^{-1}$ *and that each prime t-ideal of* D *is contained in a unique maximal t-ideal. Then, for each* $I \in \mathfrak{F}(D)$*, we have:*

$$I_t = t\text{-}cl_D(I) = cl_D(I) = I_v.$$

Proof. These assumptions characterize the Prüfer v-multiplication domains such that each t-ideal is divisorial [HZ, Theorem 3.1]. The conclusion is a straightforward consequence of Theorem 1.9(2). □

Examples of integral domains satisfying the assumptions of Corollary 1.12 are Krull domains and the Prüfer domains in which each nonzero ideal is divisorial (cf. [H, Theorem 5.1] and [K, 127]).

REMARK 1.13. For each nonzero fractional ideal I of an integral domain D, since I^{-1} is divisorial, $I \subseteq cl_D(I) \subseteq I_v$ and $I^{-1} = I_v^{-1}$, we have:

(1.13.I) $$cl_D(I^{-1}) = cl_D(I)^{-1} = I^{-1}.$$

Since $I_v = (I^{-1})^{-1}$, then the previous identity generalizes the fact that $cl_D(I_v) = I_v$ (Corollary 1.3(5)). From (1.13.I), we deduce that if $D \neq I^{-1}$, then I is not polynomially dense in D. In particular, for a maximal ideal M of D, we obtain that $D \neq M^{-1}$ implies that $M = cl_D(M)$. This statement could be obtained also as a consequence of Corollary 1.3(5), since $M = M_v$ if and only if $D \neq M^{-1}$.

2. KRULL DIMENSION OF INT(D) WHEN D IS OF KRULL-TYPE

This section is devoted to the study of the Krull dimension of the ring $\text{Int}(D)$ for the integral domains D having a locally finite representation. If a domain D has a representation

$$D = \cap_{P \in \mathcal{P}} D_P$$

where $\mathcal{P} \subseteq \text{Spec}(D)$, the ring D_P are valuation domains for each $P \in \mathcal{P}$ and the intersection is locally finite, then D is called a *Krull-type domain*. Strong Krull-type domains studied in Section 1, are a particular case of Krull-type domains.

As a consequence of our main result we prove that, for each domain of Krull-type D, $\dim(\text{Int}(D)) = \dim(D[X])$. This improves the knowledge of the Krull dimension of the ring of integer-valued polynomials giving further evidence for the conjecture stating that $\dim(\text{Int}(D)) \leq \dim(D[X])$, for each integral domain D.

THEOREM 2.1. *Let* D *be an integral domain and* $\mathcal{P}$ *a subset of* $\text{Spec}(D)$*. Assume that* $D = \cap_{P \in \mathcal{P}} D_P$ *is a locally finite representation of* D*. Set* $\mathcal{P}_0 := \{P \in \mathcal{P} \cap \text{Max}(D); \text{Card}(D/P) < \infty\}$ *and* $\mathcal{M} := \text{Max}(D) \setminus \mathcal{P}_0$*. Then*

$$\dim(\text{Int}(D)) = \text{Max}(\{\dim(D_M[X]); M \in \mathcal{M}\}, \{\dim(\text{Int}(D_P)); P \in \mathcal{P}_0\}).$$

Proof. We note that, for each maximal ideal M of D,

(2.1.I) $$D_M = \cap_{P \in \mathcal{P}} (D_P)_{(D \setminus M)} = (\cap_{P \in \mathcal{P}, P \subseteq M} D_P) \cap (\cap_{P \in \mathcal{P}, P \nsubseteq M} (D_P)_{(D \setminus M)})$$

is a locally finite representation of D_M [G2, Proposition 43.5]. Since $D = \cap_{M \in \text{Max}(D)} D_M$, then, by [CC1, Corollaire 3, p. 303],

$$\text{Int}(D) = \cap_{M \in \text{Max}(D)} \text{Int}(D_M).$$

Since $\text{Int}(D) \subseteq \text{Int}(D)_M \subseteq \text{Int}(D_M)$, for each $M \in \text{Max}(D)$, it follows that

(2.1.II) $$\text{Int}(D) = \cap_{M \in \text{Max}(D)} \text{Int}(D)_M.$$

Now, we will show that for each $M \in \mathcal{M}$, $\text{Int}(D_M) = \text{Int}(D)_M = D_M[X]$. As a matter of fact, if $M \in \text{Max}(D) \setminus \mathcal{P}$, for each $P \in \mathcal{P}$, $P \subset M$, D/P is infinite and then $\text{Int}(D_P) = D_P[X]$. If $P \in \mathcal{P}$ with $P \not\subseteq M$, then clearly the maximal ideals of $(D_P)_{(D \setminus M)}$ contract to nonmaximal prime ideals of D. Therefore, $(D_P)_{(D \setminus M)}$ has infinite residue fields and whence $\text{Int}((D_P)_{(D \setminus M)}) = (D_P)_{(D \setminus M)}[X]$ [CC2, Corollary I.3.7]. From (2.1.I) and [CC1, Corollaire 3, p. 303], we deduce that, if $M \in \text{Max}(D) \setminus \mathcal{P}$,

$$\text{Int}(D_M) = \cap_{P \in \mathcal{P}} \text{Int}((D_P)_{(D \setminus M)}) =$$

$$= (\cap_{P \in \mathcal{P}, P \subseteq M} \text{Int}(D_P)) \cap (\cap_{P \in \mathcal{P}, P \not\subseteq M} \text{Int}((D_P)_{(D \setminus M)})) =$$

$$= (\cap_{P \in \mathcal{P}, P \subseteq M} D_P[X]) \cap (\cap_{P \in \mathcal{P}, P \not\subseteq M} (D_P)_{(D \setminus M)}[X]) = D_M[X].$$

It is obvious that if $M \in (\text{Max}(D) \cap \mathcal{P}) \setminus \mathcal{P}_0$ then $\text{Int}(D_M) = D_M[X]$, since D/M is infinite. Since $D_M[X] \subseteq \text{Int}(D)_M \subseteq \text{Int}(D_M)$, we have that $D_M[X] = \text{Int}(D)_M = \text{Int}(D_M)$.
From the previous claim and from (2.1.II), we deduce that:

(2.1.III) $$\text{Int}(D) = (\cap_{M \in \mathcal{M}} D_M[X]) \cap (\cap_{M \in \mathcal{P}_0} \text{Int}(D)_M).$$

We know that there exists $N \in \text{Max}(\text{Int}(D))$ such that $\text{ht}(N) = \dim(\text{Int}(D))$ and $N \cap D$ maximal. (In fact, if $N \cap D = P$ is a nonmaximal prime ideal of D, then $\text{Int}(D)_P = \text{Int}(D_P) = D_P[X]$ and hence $\text{ht}(N) = \dim(D_P[X]) \leq \dim(D[X]) - 1 \leq \dim(\text{Int}(D))$ [CC2, Proposition V.1.6]. Therefore $\dim(\text{Int}(D)) = \dim(D[X]) - 1$. Arguing as in the proof of [CC2, Proposition V.1.6], we can find a chain $\mathcal{C}$ of prime ideals of $D_P[X]$ of length $n = \dim(D_P[X]) = \dim(\text{Int}(D))$,

$$\mathcal{C}: \qquad (0) \subset Q_1 \subset \cdots \subset Q_{n-1} \subset Q_n,$$

such that $(Q_i \cap D_P)[X] \in \mathcal{C}$, for each $i = 1, \cdots, n$. Therefore, $Q_{n-1} = PD_P[X]$. When we contract $\mathcal{C}$ to $\text{Int}(D)$ we get a chain

$$\mathcal{C}': \qquad (0) \subset Q_1' \subset \cdots \subset Q_{n-1}' \subset Q_n',$$

where $Q_{n-1}' = PD_P[X] \cap \text{Int}(D)$. If M is a maximal ideal of D containing P and $a \in D$, then it is easy to see that $Q_{n-1}' \subset P_a \subseteq M_a$, where $Q_a := \{f \in \text{Int}(D); f(a) \in Q\}$ for $Q \in \{P, M\}$ and M_a is a maximal ideal of $\text{Int}(D)$ above M [CC2, Lemma V.1.3]. Therefore we reach a contradiction: $\dim(\text{Int}(D)) \geq n + 1$.

Let $M := (N \cap D) \in \text{Max}(D)$. Since $N \cap (D \setminus M) = \emptyset$, then $N\text{Int}(D)_M$ is a maximal ideal of $\text{Int}(D)_M$. The conclusion follows immediately by examining the two possible cases:

Case 1. $M \in \mathcal{M}$. In this case, $\text{Int}(D)_M = \text{Int}(D_M) = D_M[X]$, hence $\text{ht}(N) = \dim(D_M[X])$.

Case 2. $M \in \mathcal{P}_0$. In this case, $\text{Int}(D)_M = \text{Int}(D_M)$ and $\text{ht}(N) = \dim(\text{Int}(D_M))$. $\square$

COROLLARY 2.2. *With the same notation and hypotheses of Theorem 2.1, if* $\dim(\mathrm{Int}(D_P)) \leq \dim(D_P[X])$ *for each* $P \in \mathcal{P}_0$, *then* $\dim(\mathrm{Int}(D)) \leq \dim(D[X])$. *In particular, the previous inequality holds in the following cases:*

(a) $D_P[X]$ *is a Jaffard domain, for each* $P \in \mathcal{P}_0$ *(e.g. when* D *is a locally Jaffard domain* [ABDFK]);

(b) D_P *is a* $P^n VD$, *with* $n \geq 0$, *for each* $P \in \mathcal{P}_0$ *(e.g. when* D *is a locally PVD domain,* [F] *or* [DF]).

Proof. The first inequality is a straightforward consequence of Theorem 2.1. As concerns the particular cases, we proceed as follows.

(a) We note that, for each $P \in \mathcal{P}_0$, we have:

$$\dim(\mathrm{Int}(D_P)) \leq \dim_v(\mathrm{Int}(D_P)) \leq \dim_v(D_P[X]) = \dim(D_P[X]).$$

(b) follows from [FIKT, Lemma 3.1]. $\square$

COROLLARY 2.3. *If* D *is a domain of Krull-type (e.g. a generalized Krull domain* [G2, p.524]), *then* $\dim(\mathrm{Int}(D)) = \dim(D[X])$.

Proof. In this case, $\mathcal{P} = t_m(D)$, $\mathcal{P}_0 = \{P \in \mathcal{P} \cap \mathrm{Max}(D); \mathrm{Card}(D/P) < \infty\}$ and D_P is a valuation domain, for each $P \in \mathcal{P}$. In particular $\dim(\mathrm{Int}(D_P)) = \dim(D_P)+1 = \dim(D_P[X])$ for each $P \in \mathcal{P}_0$. The conclusion follows from Theorem 2.1. $\square$

3. POLYNOMIAL CLOSURE IN PULLBACK DOMAINS

Let D be any domain and let P be a prime ideal of D with $\mathrm{Card}(D/P) = \infty$. Let I be an ideal of D such that $P \subset I$, thus $I_P = ID_P = D_P$. From [C4, Lemma 3.4] we have

$$\mathrm{Int}(I, D_P) = \mathrm{Int}(I_P, D_P) = \mathrm{Int}(D_P) = D_P[X],$$

hence $\mathrm{Int}(I, D) \subseteq D_P[X]$. Therefore, we can consider the canonical map

$$\varphi : \mathrm{Int}(I, D) \to \mathrm{Int}(I/P, D/P), \quad f \mapsto \overline{f} := f + PD_P[X],$$

where it is easily seen that $\overline{f} \in \mathrm{Int}(I/P, D/P)$.

We start this section with an observation about $\mathrm{Int}(I, D)$ and $\mathrm{Int}(I/P, D/P)$ when P is a divided prime ideal, i.e. $P = PD_P$.

LEMMA 3.1. *Let* D *be an integral domain,* P *a divided prime ideal of* D *with* $\mathrm{Card}(D/P) = \infty$ *and* I *an ideal of* D *with* $P \subseteq I$. *Then,*

(1) *the canonical map* $\varphi : \mathrm{Int}(I, D) \to \mathrm{Int}(I/P, D/P)$ *is a surjective homomorphism;*

(2) $\ker(\varphi) = P[X]$;

(3) $\mathrm{Int}(I, D)/P[X] \cong \mathrm{Int}(I/P, D/P)$.

Proof. (1) Let $g \in (D_P/PD_P)[X]$ such that $g(I/P) \subseteq D/P$. Then it is easy to see that $g = G + PD_P[X] = G + P[X]$, where $G \in D_P[X]$ and $G(i) + P \in D/P$, for each $i \in I$. Therefore $G \in \mathrm{Int}(I, D)$.

(2) It is obvious that $\ker(\varphi) = PD_P[X] \cap \mathrm{Int}(I, D) = P[X]$.

(3) is a straightforward consequence of (1) and (2). $\square$

PROPOSITION 3.2. *Let D, P and I as in* Lemma 3.1. *Then, the canonical homomorphism φ defines the following isomorphism:*

$$cl_D(I)/P \cong cl_{D/P}(I/P).$$

Proof. Let $x \in cl_D(I)$, then $f(x) \in D$ for each $f \in \text{Int}(I, D)$. Hence, $\overline{f}(\overline{x}) = \overline{f(x)} \in D/P$ for each $f \in \text{Int}(I, D)$. Since the map $\varphi : f \mapsto \overline{f}$ is surjective (Lemma 3.1(1)), then $g(\overline{x}) \in D/P$ for each $g \in \text{Int}(I/P, D/P)$, i.e. $\overline{x} \in cl_{D/P}(I/P)$. Therefore $cl_D(I)/P \subseteq cl_{D/P}(I/P)$. Conversely, if $y = x + P \in cl_{D/P}(I/P)$ then for each $g \in \text{Int}(I/P, D/P)$, $g(y) \in D/P$. Since φ is surjective, $g = \overline{f} = f + P[X]$ for some $f \in \text{Int}(I, D)$. By the fact that $g(y) \in D/P$, for each $g \in \text{Int}(I/P, D/P)$, we deduce that $f(x) \in D$, for each $f \in \text{Int}(I, D)$, i.e. $x \in cl_D(I)$. $\square$

COROLLARY 3.3. *Let D be a domain with a divided prime ideal P. Suppose that D/P is a valuation domain V with nonzero principal maximal ideal. Then, each ideal of D containing P is polynomially closed.*

Proof. If I is an ideal of D and $P \subset I$, then from Propositon 3.2 $cl_D(I)/P \cong cl_{D/P}(I/P)$. But $D/P = V$ is a valuation domain with principal maximal ideal and, by Proposition 1.8, $cl_{D/P}(I/P) = I/P$. Therefore $cl_D(I)/P = I/P$ and $cl_D(I) = I$, since they both contain P. $\square$

Relevant examples of divided domains are the pseudo-valuation domains (PVD) or, more generally, the pseudo-valuation domains of type n (P^nVD). We recall that a PVD, D, is defined by a pullback of the following type:

(3.I)
$$
\begin{array}{ccc}
D := \alpha^{-1}(k) & \longrightarrow & k \\
\downarrow & & \downarrow \\
V & \xrightarrow{\ \alpha\ } & V/M
\end{array}
$$

where (V, M) is a valuation domain (called the *valuation overring associated to* D), $\alpha : V \twoheadrightarrow V/M$ is the canonical projection and k is a subfield of the residue field of V. From [HH, Theorem 2.13], every prime ideal of D is divisorial, hence it is polynomially closed. Moreover, if I is any nonprincipal integral ideal of D, then $I_v = IV$ [HH, Corollary 2.14]. It follows immediately that $cl_D(I) \subseteq IV$. Moreover, for a nonvaluation PVD, the t-operation and the v-operation coincide [HZ, Proposition 4.3] so that $cl_D(I) \subseteq I_t$. If $D = V$ is a valuation domain, then it is known that the t-operation and the v-operation coincide if and only if the maximal ideal M of V is principal [HZ, Remark 1.5]; in fact, in this situation, every nonzero ideal of V is divisorial.

In [C4, § 4] the author establishes some relations between the polynomial closure of a fractional subset and its $\mathfrak{A}$-adic closure, where $\mathfrak{A}$ is an ideal of a *Zariski domain* D (i.e. a Noetherian domain, equipped with the $\mathfrak{A}$-adic topology, in which every ideal is $\mathfrak{A}$-adic closed). Next goal is to obtain a link between the polynomial closure and the adic closure for a special class of PVD's.

PROPOSITION 3.4. *Let D be a PVD. Assume that D possesses a height-one prime ideal P such that $P \neq P^2$. Then, the polynomial closure of each D-fractional subset E of K contains the P-adic closure of E.*

Proof. We start by proving that all ideals of D are closed in the P-adic topology. If I is any ideal of D, then its P-adic closure is given by $\overline{I} := \cap_{n \geq 0}(I + P^n)$. If

$I \supseteq P$, it is obvious that $\overline{I} = I$. If $I \subset P$, then $P^2 \subsetneqq \sqrt{I} = P$ and I contains a power of P by [HH, Corollary 2.5]. Therefore, $\overline{I} = I$. Since each ideal of D is closed in the P-adic topology, we can use the same argument of [C4, Theorem 4.1] in order to conclude. $\square$

Recall that a P^nVD, D, is defined by induction on n in the following way. A P^0VD is a PVD and a P^nVD is obtained by a pullback diagram of the following type:

$$
\begin{array}{ccc}
D := \alpha^{-1}(A_{n-1}) & \longrightarrow & A_{n-1} \\
\downarrow & & \downarrow \\
W_{n-1} & \xrightarrow{\ \alpha\ } & F = W_{n-1}/\mathcal{M}
\end{array}
$$

where W_{n-1} is a P^{n-1}VD with maximal ideal $\mathcal{M}$, F is its residue field, $\alpha : W_{n-1} \to F$ is the canonical projection and A_{n-1} is a PVD with quotient field F. For details about P^nVD the reader is referred to [F].

In the next proposition, we will show that also in a P^nVD all prime ideals are polynomially closed.

PROPOSITION 3.5. *Let D be a P^nVD, then all nonzero prime ideals of D are polynomially closed.*

Proof. Since in a P^nVD every prime ideal is divided [F, Theorem 1.9], then if Q is a prime ideal of D, then D_Q is a P^mVD, with $m \leq n$, and $Q = (D : D_Q)$, since $QD_Q = Q$. Then, if $Q \neq 0$, Q is a divisorial ideal, whence it is polynomially closed. $\square$

References

[A]   D.F. Anderson, *Star-operations induced by overrings*, Comm. Algebra **16**, (1988), 2535–2553.

[ABDFK] D.F. Anderson - A. Bouvier - D.E. Dobbs - M. Fontana - S. Kabbaj, *On Jaffard domains*, Expo. Math. **6** (1988), 145–175.

[B]   N. Bourbaki, *Algèbre Commutative*, Hermann, Paris (1961-1965).

[CC1]   P.-J. Cahen - J.L. Chabert, *Coefficients et valeurs d'un polynôme*, Bull. Sci. Math. **95** (1971), 295–304.

[CC2]   P.-J. Cahen - J.L. Chabert, *Integer-Valued Polynomials*, Math. Surveys and Monographs (AMS) **48** (1997).

[C1]   P.-J. Cahen, *Dimension de l'anneau des polynômes à valeurs entières*, Manuscripta Math. **67** (1990), 333–343.

[C2]   P.-J. Cahen, *Parties pleines d'un anneau noethérien*, J. Algebra **157** (1993), 192–212.

[C3]   P.-J. Cahen, *Integer-valued polynomials on a subset*, Proc. Amer. Math. Soc. **117** (1993), 919–929.

[C4]   P.-J. Cahen, *Polynomial closure*, J. Number Theory **61** (1996), 226–247.

[Ch]   J.L. Chabert, *Les idéaux premiers de l'anneau des polynômes à valeurs entières*, J. reine angew. Math. **293/294** (1977), 275–283.

[DF]   D.E. Dobbs - M. Fontana, *Locally pseudo-valuation domains*, Ann. Mat. Pura Appl. **134** (1983), 147–168.

[F]   M. Fontana, *Sur quelques classes d'anneaux divisés*, Rend. Sem. Mat. Fis. Milano **51** (1981), 179–200.

[FIKT] M. Fontana - L. Izelgue - S.E. Kabbaj - F. Tartarone, *On the Krull dimension of domains of integer-valued polynomials*, Expo. Math. **15** (1997), 433–465.

[G1]   R. Gilmer, *Sets that determine integer-valued polynomials*, J. Number Theory **33** (1989), 95–100.

[G2]   R. Gilmer, *Multiplicative Ideal Theory*, Marcel-Dekker, New York (1972).

[Gr]   M. Griffin, *Some Results on v-multiplication Rings*, Canad. J. Math. **10** (1967), 710–722.

[HH]   J.R. Hedstrom - E.G. Houston, *Pseudo-valuation domains*, Pac. J. Math. **75** (1978), 137–147.

[H]   W. Heinzer, *Integral domains in which each nonzero ideal is divisorial*, Mathematika **15** (1968), 164–170.

[HZ]   E.G. Houston - M. Zafrullah, *Integral domains in which each t-ideal is divisorial*, Michigan Math. J. **35** (1988), 291–300.

[J]   P. Jaffard, *Les Systèmes d'Idéaux*, Dunod, Paris (1960).

[K]   W. Krull, *Idealtheorie*, Springer-Verlag, Berlin (1935).

[Mc]   D.L. McQuillan, *On a Theorem of R. Gilmer*, J. Number Theory **39** (1991), 245–250.

DIPARTIMENTO DI MATEMATICA, UNIVERSITÀ DEGLI STUDI "ROMA TRE", 00146 ROMA, ITALY, E-MAIL: FONTANA@MAT.UNIROMA3.IT;

DÉPARTEMENT DE MATHÉMATIQUES, FACULTÉ DES SCIENCES "SEMLALIA", UNIVERSITÉ DE MARRAKECH, MARRAKECH, MOROCCO, E-MAIL: FSSM.MATH@CYBERNET.NET.MA;

DEPARTMENT OF MATHEMATICAL SCIENCES KFUPM, P.O.BOX 849 DHAHRAN 31261, SAUDI ARABIA, E-MAIL: KABBAJ@KFUPM.EDU.SA;

DIPARTIMENTO DI MATEMATICA, UNIVERSITÀ "LA SAPIENZA", 00185 ROMA, ITALY, E-MAIL: TARTARON@MAT.UNIROMA1.IT.

FACULTÉ DES SCIENCES DE SAINT-JÉRÔME , UNIVERSITÉ D'AIX-MARSEILLE III, 13397 MARSEILLE, FRANCE, E-MAIL: FRANCESCA.TARTARONE@VMESA12.U-3MRS.FR.

Polynomial Functions on Finite Commutative Rings

SOPHIE FRISCH   Institut für Mathematik, Technische Universität Graz,
A–8010 Graz, Austria   e–mail: frisch@blah.math.tu–graz.ac.at

Abstract. Every function on a finite residue class ring D/I of a Dedekind domain D is induced by an integer-valued polynomial on D that preserves congruences mod I if and only if I is a power of a prime ideal. If R is a finite commutative local ring with maximal ideal P of nilpotency N satisfying for all a, $b \in R$, if $ab \in P^n$ then $a \in P^k$, $b \in P^j$ with $k + j \geq \min(n, N)$, we determine the number of functions (as well as the number of permutations) on R arising from polynomials in $R[x]$. For a finite commutative local ring whose maximal ideal is of nilpotency 2, we also determine the structure of the semigroup of functions and of the group of permutations induced on R by polynomials in $R[x]$.

INTRODUCTION

Let R be a finite commutative ring with identity. Every polynomial $f \in R[x]$ defines a function on R by substitution of the variable. Not every function $\varphi \colon R \to R$ is induced by a polynomial in $R[x]$, however, unless R is a finite field. (Indeed, if the function with $\varphi(0) = 0$ and $\varphi(r) = 1$ for $r \in R \setminus \{0\}$ is represented by $f \in R[x]$, then $f(x) = a_1 x + \ldots + a_n x^n$ and for every non-zero $r \in R$ we have $1 = f(r) = (a_1 + \ldots + a_n r^{n-1})r$, which shows r to be invertible.)

This prompts the question how many functions on R are representable by polynomials in $R[x]$; and also, in the case that $R = D/I$ is a residue class ring of a domain D with quotient field K, whether every function on R might be induced by a polynomial in $K[x]$? We will address these questions in sections 2 and 1, respectively.

Other related problems are to characterize the functions on R arising from polynomials in $R[x]$ by intrinsic properties of these functions (such as preservation of certain relations), and to determine the structure of the semigroup of polynomial functions on R and that of the group of polynomial permutations of R.

In section 4, we will answer the second question in the special case that R is a local ring whose maximal ideal is of nilpotency 2. Apart from that, the only result I am aware of is Nöbauer's expression of the group of polynomial permutations on $\mathbb{Z}_{p^n}$ as a wreath product $G \wr S_p$, with G a rather inscrutable subgroup (characterized by conditions on the coefficients of the representing polynomials) of the group of polynomial permutations on $\mathbb{Z}_{p^{n-1}}$ [11]. (There is a wealth of literature on the functions induced by polynomials on finite fields, some of it concerning the structure of the subgroup of S_q generated by special polynomials, see e.g. [8] and its references. Methods from the theory of finite fields do not help much with finite rings, however, except when the rings are algebras over a finite field, see [2].)

A characterization of polynomial functions by preservation of relations has been given for $R = \mathbb{Z}_n$ by Kempner [6]. For finite commutative rings in general there is the criterion of Spira [17] that a function is representable by a polynomial if and only if all the iterated divided differences that can be formed by subsets of the arguments and the respective values are in R.

In what follows, all rings are assumed to be commutative with identity, the natural numbers are written as $\mathbb{N} = \{1, 2, 3, \ldots\}$, and the non-negative integers as $\mathbb{N}_0 = \{0, 1, 2, \ldots\}$.

1   FUNCTIONS INDUCED ON RESIDUE CLASS RINGS BY INTEGER–VALUED POLYNOMIALS

In this section we give the answer, for Dedekind rings, to a question asked by Narkiewicz in his "Polynomial Mappings" book [9]. For

$R = \mathbb{Z}$, the 'if' direction has been shown (for several variables, cf. the corollary) by Skolem [16], the 'only if' direction by Rédei and Szele [12, 13].

If D is a domain with quotient field K, a polynomial $f \in K[x]$ is called *integer-valued on D* if $f(d) \in D$ for all $d \in D$. We write $\mathrm{Int}(D)$ for the set of all integer-valued polynomials on D. If I is an ideal of a domain D, we say that a polynomial $f \in \mathrm{Int}(D)$ induces a function $\varphi \colon D/I \to D/I$ if $\varphi(d + I) = f(d) + I$ is well defined, i.e., if $c \equiv d \bmod I$ implies $f(c) \equiv f(d) \bmod I$.

THEOREM 1. *Let R be a Dedekind domain and I an ideal of R of finite index. Every function $\varphi \colon R/I \to R/I$ is induced by a polynomial $f \in \mathrm{Int}(R)$ if and only if I is a power of a prime ideal of R.*

Proof. The case of a finite field or of $I = R = P^0$ is trivial, so we consider R infinite and $I \neq R$. Let P be a prime ideal with $I \subseteq P$. Assume that the characteristic function of $\{0\}$ on R/I is induced by a polynomial $f \in \mathrm{Int}(R)$, then $f(r) \equiv 1 \bmod I$ for $r \in I$ and $f(r) \equiv 0 \bmod I$ for $r \notin I$. We show that I must be a power of P. Suppose otherwise, then $P^n \not\subseteq I$ for all $n \in \mathbb{N}$. Let $c \in R$ and $g \in R[x]$ such that $f(x) = g(x)/c$, and $n = v_P(c)$.

Since g is in $R[x]$, the function $r \mapsto g(r)$ on R preserves congruences mod every ideal of R, in particular mod P^{n+1}. It follows that $r \equiv s$ mod P^{n+1} implies $f(r) \equiv f(s) \bmod P$. Now consider an element $r \in P^{n+1} \setminus I$. On one hand, $f(r) \notin P$, since $f(r) \equiv f(0) \bmod P$ and $f(0) \equiv 1 \bmod I$; on the other hand, since $r \notin I$, we have $f(r) \in I \subseteq P$, a contradiction.

To show that every function on R/P^n (P a prime ideal of finite index) is induced by a polynomial in $\mathrm{Int}(R)$, it suffices to show this for the charcteristic function of $\{0\}$ on the residue class ring. For this, we need only construct a polynomial $f \in \mathrm{Int}(R)$ satisfying $f(r) \in P$ for $r \notin P^n$ and $f(r) \notin P$ for $r \in P^n$; an appropriate power $\tilde{f}(x) = f(x)^m$ will then satisfy $\tilde{f}(r) \in P^n$ for $r \notin P^n$ and $\tilde{f}(r) \equiv 1 \bmod P^n$ for $r \in P^n$.

Let $a_1, \ldots, a_{q^n-1} \in R$ be a system of representatives of the residue classes of P^n other than P^n itself, and let $a_0 \in P^{n-1} \setminus P^n$. Put $h(x) = \prod_{k=0}^{q^n-1} (x - a_k)$ and $\alpha = \sum_{j=1}^{n} \left[\frac{q^n}{q^j} \right] = \frac{q^n-1}{q-1}$, then for all $r \in P^n$

we have $v_P(h(r)) = \alpha - 1$, while $v_P(h(r)) \geq \alpha$ for all $r \in R \setminus P^n$.

Now let $\mathcal{Q} = \{Q \in \mathrm{Spec}(R) \mid Q \neq P; \exists k \, a_k \in Q\}$ and for $Q \in \mathcal{Q}$ define $m_Q = \max\{m \in \mathbb{N} \mid \exists k \, a_k \in Q^m\}$. Pick $c \in R$ such that $c \notin P$ and $c \in Q^{m_Q+1}$ for all $Q \in \mathcal{Q}$, and set $b_k = c^{-1}a_k$ and $g(x) = \prod_{k=0}^{q^n-1}(x - b_k)$.

We now set $f(x) = g(x)/g(0)$ and claim that $f \in \mathrm{Int}(R)$ and that for all $r \in R$, $f(r) \in P$ if and only if $r \notin P^n$. To verify this, we check that for all $Q \in \mathrm{Spec}(R)$ and all $r \in R$, $v_Q(g(r)) \geq v_Q(g(0))$ and that $v_P(g(r)) > v_P(g(0))$ for $r \in R \setminus P^n$, while $v_P(g(r)) = v_P(g(0))$ for $r \in P^n$.

First consider those $Q \in \mathrm{Spec}(R)$ with $v_Q(c) > 0$. We have $v_Q(b_k) < 0$ for all k and therefore $v_Q(g(r)) = \sum_{k=0}^{q^n-1} v_Q(b_k) = v_Q(g(0))$ for all $r \in R$.

Now consider a $Q \in \mathrm{Spec}(R)$ with $v_Q(c) = 0$ and $Q \neq P$, then $v_Q(b_k) = 0$ for all k, and for all $r \in R$ we have $v_Q(g(r)) \geq 0 = v_Q(g(0))$.

Concerning P, we observe that $v_P(r - b_k) = v_P(c^{-1}(cr - a_k)) = v_P(cr - a_k)$, such that $v_P(g(r)) = v_P(h(cr))$. Since $v_P(cr) = v_P(r)$, this implies $v_P(g(r)) \geq \alpha$ for $r \in R \setminus P^n$ and $v_P(g(r)) = \alpha - 1$ for $r \in P^n$. $\square$

If K is the quotient field of a domain D and I an ideal of D, we say that $f \in K[x_1, \ldots, x_m]$ induces a function $\varphi : (D/I)^m \to D/I$ if $\varphi(d_1 + I, \ldots, d_m + I) = f(d_1, \ldots, d_m) + I$ makes sense, i.e., if $f(d_1, \ldots, d_m) \in D$ for all $(d_1, \ldots, d_m) \in D^m$ and $f(d'_1, \ldots, d'_m) \equiv f(d_1, \ldots, d_m) \bmod I$ whenever $d'_i \equiv d_i \bmod I$ for $1 \leq i \leq m$.

COROLLARY. *If R is a Dedekind domain, P a maximal ideal of finite index and $n \in \mathbb{N}$ then every function $f : (R/P^n)^m \to R/P^n$ is induced by a polynomial $f \in K[x_1, \ldots, x_m]$ (K being the quotient field of R).*

Proof. It suffices to have a polynomial $f \in K[x_1, \ldots, x_m]$ that induces the characteristic function of $(0, 0, \ldots, 0) \bmod P^n$. As R/P is a field, there exists a $g \in R[x_1, \ldots, x_m]$ such that $g(r_1, \ldots, r_m) \equiv 1 \bmod P$ if $r_i \in P$ for $1 \leq i \leq m$ and $g(r_1, \ldots, r_m) \equiv 0 \bmod P$ otherwise. By the Theorem, there exists $h \in \mathrm{Int}(R)$ such that $h(r) \in P$ if $r \in P^n$ and $h(r) \notin P$ otherwise. Now $f(x_1, \ldots, x_m) = g(h(x_1), \ldots, h(x_m))$ satisfies $f(r_1, \ldots, r_m) \notin P$ iff $r_i \in P^n$ for $1 \leq i \leq m$, and a suitable power of $g(x) = f(x)^k$ finally satisfies $g(r_1, \ldots, r_m) \equiv 1 \bmod P^n$ if $r_i \in P^n$ for $1 \leq i \leq m$ and $g(r_1, \ldots, r_m) \equiv 0 \bmod P^n$ otherwise, as required. $\square$

Note that the theorem and its proof still hold if we replace Dedekind ring by Krull ring, prime ideal by height 1 prime ideal, and restrict I to ideals with $\operatorname{div}(I) \neq R$.

2 THE NUMBER FORMULAS

For a commutative finite ring R, let us denote by $\mathcal{F}(R)$ the set (or semigroup with respect to composition) of functions on R induced by polynomials in $R[x]$, and by $\mathcal{P}(R)$ the subset (or group) of those polynomial functions on R that are permutations.

When considering the functions induced on a finite commutative ring R by polynomials in $R[x]$, we can restrict ourselves to local rings, since every finite commutative ring is a direct sum of local rings, and addition and multiplication (and therefore evaluation of polynomials in $R[x]$) are performed in each component independently.

For residue class rings of the integers, we know

$$|\mathcal{F}(\mathbb{Z}_{p^n})| = p^{\sum_{k=1}^{n} \beta_p(k)} \quad \text{and} \quad |\mathcal{P}(\mathbb{Z}_{p^n})| = p! p^p (p-1)^p p^{\sum_{k=3}^{n} \beta_p(k)},$$

where p is a prime and $\beta_p(k)$ is the minimal $m \in \mathbb{N}$ such that $p^k \mid m!$ (in other words, the minimal $m \in \mathbb{N}$ such that $\alpha_p(m) \geq k$, with $\alpha_p(m) = \sum_{j \geq 1} \left[\frac{m}{p^j} \right]$).

The most lucid proof, in my opinion, of these two formulas is that by Keller and Olson [5], to whom the second one is due. Kempner's earlier proof [6] of the formula for $|\mathcal{F}(\mathbb{Z}_{p^n})|$ is rather more involved. Singmaster [15] and Wiesenbauer [18] gave proofs for $R = \mathbb{Z}_m$ which do not use reduction to the local ring case. Brawley and Mullen [3] generalized the formulas to Galois rings (rings of the form $\mathbb{Z}[x]/(p^n, f)$, where p is prime and $f \in \mathbb{Z}[x]$ is irreducible over $\mathbb{Z}_p$, see [7]) and Nečaev [10] to finite commutative local principal ideal rings.

We will give a proof along the lines of Keller and Olson of a generalization of the formulas to a class of local rings (the suitable rings defined below) that properly contains the rings considered by Brawley, Mullen and Nečaev.

DEFINITION. Let R be a finite commutative local ring R with maximal ideal P and $N \in \mathbb{N}$ minimal with $P^N = (0)$. We call R

"suitable", if for all a, $b \in R$ and all $n \in \mathbb{N}$,

$$ab \in P^n \Longrightarrow a \in P^k \text{ and } b \in P^j \text{ with } k + j \geq \min(N, n).$$

Note that every finite local ring R with maximal ideal P such that $P^2 = (0)$ is suitable, as well as every finite local ring whose maximal ideal is principal.

We may think of this property as inducing a valuation-like mapping $v \colon R \to H_N$, by $v(r) = k$ if $r \in P^k \setminus P^{k+1}$ and $v(0) = \infty$, where $(H_N, +)$ results from the non-negative integers by identifying all numbers greater or equal N; it is the semigroup with elements $\{0, 1, \ldots, N-1, N=\infty\}$ and $i+j=\min(i+j, N)$, where the operations on the right are just the usual ones on non-negative integers.

DEFINITION. If R is a finite local ring and P its maximal ideal, for $n \geq 0$, let

$$\alpha(n) = \alpha_{(R,P)}(n) = \sum_{j \geq 1} \left[\frac{n}{[R : P^j]} \right]$$

and let $\beta(n) = \beta_{(R,P)}(n)$ be the minimal $m \in \mathbb{N}$ such that $\alpha_{(R,P)}(m) \geq n$. (If R and P are understood, we suppress the subscript (R, P) of α and β.)

REMARK. Note that $\alpha_{(R,P)}(n)$ is finite if and only if $n < |R|$; we will never use $\alpha_{(R,P)}$ outside that range. Also note that, since $[R/P^k : P^j/P^k] = [R : P^j]$ for $j \leq k$, we have $\alpha_{(R,P)}(n) = \alpha_{(R/P^k, P/P^k)}(n)$ in the range where both values are finite, that is for $n < [R : P^k]$.

THEOREM 2. Let R be a suitable finite local ring with maximal ideal P, $q = [R : P]$, and $N \in \mathbb{N}$ minimal, such that $P^N = (0)$. Then

$$|\mathcal{F}(R)| = \prod_{j=0}^{\beta(N)-1} [R : P^{N-\alpha(j)}],$$

where $\alpha(n) = \sum_{j \geq 1} \left[\frac{n}{[R:P^j]} \right]$ and $\beta(n)$ is the minimal $m \in \mathbb{N}$ such that $\alpha(m) \geq n$.

Also, for $N > 1$,

$$|\mathcal{P}(R)| = \frac{q! \, (q-1)^q}{q^{2q}} \, |\mathcal{F}(R)|.$$

If $[P^{k-1} : P^k] = q$ *for* $1 \leq k \leq N$, *the formulas simplify to*

$$|\mathcal{F}(R)| = q^{\sum_{k=1}^{N} \beta_q(k)} \quad \text{and} \quad |\mathcal{P}(R)| = q! q^q (q-1)^q \, q^{\sum_{k=3}^{N} \beta_q(k)},$$

where $\alpha_q(m) = \sum_{j \geq 1} \left[\frac{m}{q^j}\right]$ *and* $\beta_q(k)$ *is the minimal* $m \in \mathbb{N}$ *such that* $\alpha_q(m) \geq k$.

We will prove the expression for $|\mathcal{F}(R)|$ at the end of the next section, and that for $|\mathcal{P}(R)|$ at the end of section 4.

3 A CANONICAL FORM
FOR THE POLYNOMIAL REPRESENTING A FUNCTION

DEFINITION. Let R be a commutative finite local ring with maximal ideal P of nilpotency N. We call a sequence $(a_k)_{k=0}^{\infty}$ of elements in R a P-sequence, if for $0 \leq n \leq N$

$$a_k - a_j \in P^n \iff [R : P^n] \,\big|\, k - j;$$

and if (a_k) is a P-sequence, we call the polynomials

$$\langle x \rangle_0 = 1 \quad \text{and} \quad \langle x \rangle_n = (x - a_0) \ldots (x - a_{n-1}) \quad \text{for } n > 0$$

the "falling factorials" constructed from the sequence (a_k).

A P-sequence (a_k) for R is easy to construct inductively: Let $a_0, \ldots, a_{[R:P]-1}$ be a complete set of residues mod P with $a_0 = 0$. Once a_k has been defined for $k < [R : P^{n-1}]$ (while $n \leq N$), define a_k for $[R : P^{n-1}] \leq k < [R : P^n]$ as follows: let $b_0 = 0$, b_1, $\ldots$, $b_{[P^{n-1}:P^n]-1}$ be a complete set of residues of P^{n-1} mod P^n; then, for $k = j[R : P^{n-1}] + r$ with $0 \leq r < [R : P^{n-1}]$ and $1 \leq j < [P^{n-1} : P^n]$, let $a_k = b_j + a_r$. After $a_0, \ldots, a_{|R|-1}$ have been defined (necessarily a complete enumeration of the elements of R), continue the sequence $|R|$-periodically.

In the following Lemma, we use the convention that $P^{\infty} = (0)$.

LEMMA. *Let R be a suitable finite local ring with maximal ideal P of nilpotency N, and $\langle x \rangle_n$ the falling factorial of degree n constructed from a P-sequence (a_k). Then for all $n \in \mathbb{N}_0$,*

$$\forall r \in R \quad \langle r \rangle_n \in P^{\alpha(n)} \quad \text{and} \quad \text{if } \alpha(n) < N \text{ then } \langle a_n \rangle_n \notin P^{\alpha(n)+1}.$$

Proof. If $n \geq |R|$ (equivalent to $\alpha(n) = \infty$) then, since $a_0, \ldots, a_{|R|-1}$ enumerate all elements of R, $\langle r \rangle_n = 0$ for all r.

If $n < |R|$ then $\alpha(n) = \sum_{k=1}^{N} \left[\frac{n}{[R:P^k]} \right]$, while $\langle r \rangle_n \in P^e$, where $e =$

$$\sum_{k=1}^{N-1} k \left| \{ j \mid 0 \leq j < n; \ r - a_j \in P^k \setminus P^{k+1} \} \right| +$$

$$+ N \left| \{ j \mid 0 \leq j < n; \ r - a_j \in P^N \} \right|$$

$$= \sum_{k=1}^{N} \left| \{ j \mid 0 \leq j < n; \ r - a_j \in P^k \} \right|$$

and (by definition of suitable) $\langle r \rangle_n$ is in no higher power of P if $e < N$.

From the definition of P-sequence, we see that the cardinality of $\{ j \mid 0 \leq j < n; \ r - a_j \in P^k \}$ is either $\left[\frac{n}{[R:P^k]} \right]$ or $\left[\frac{n}{[R:P^k]} \right] + 1$ and the $+1$ doesn't occur for $r = a_n$. $\square$

PROPOSITION 1. *Let R be a suitable finite local ring with maximal ideal P of nilpotency N, (a_k) a P-sequence for R, $\langle x \rangle_k$ the falling factorial of degree k constructed from it and let $0 \leq n \leq N$.*

A polynomial $f \in R[x]$ induces the zero-function on R/P^n if and only if

$$f(x) = \sum_{j \geq 0} c_j \langle x \rangle_j \quad \text{with} \quad c_j \in P^{n-\alpha(j)} \quad \text{for} \quad 0 \leq j < \beta(n).$$

Proof. As $\langle x \rangle_j$ maps R into $P^{\alpha(j)}$, the "if" direction is evident. To show "only if", assume that $f(x) = \sum_{j \geq 0} c_j \langle x \rangle_j$ maps R into P^n. We show $c_j \in P^{n-\alpha(j)}$ for $0 \leq j < \beta(n)$ by induction on j. (There is no condition on the coefficients for $j \geq \beta(n)$, since $\langle x \rangle_j$ already maps R into P^n for those j.)

For $j = 0$, we have $c_0 = f(a_0) \in P^n$. Now assume $c_i \in P^{n-\alpha(i)}$ for $i < j$ and consider $f(a_j)$. Since $\langle x \rangle_i$ maps R into $P^{\alpha(i)}$ and $\langle a_j \rangle_k = 0$ for $k > j$, we have $f(a_j) \equiv c_j \langle a_j \rangle_j \bmod P^n$. Also, $\langle a_j \rangle_j$ is in no higher power of P than $P^{\alpha(j)}$. Therefore $f(a_j) \in P^n$ implies $c_j \in P^{n-\alpha(j)}$. $\square$

COROLLARY 1. *In the situation of the Proposition, for $0 \le j < \beta(n)$, let C_j be a complete set of residues mod $P^{n-\alpha(j)}$. Then every function on R/P^n arising from a polynomial in $R[x]$ arises from a unique polynomial of the form*

$$f(x) = \sum_{j=0}^{\beta(n)-1} c_j \langle x \rangle_j \qquad \text{with} \qquad c_j \in C_j.$$

For $R = \mathbb{Z}_{p^n}$, other canonical forms for the functions representable by polynomials have been given by Dueball [4], Aizenberg, Semion and Tsitkin [1] and Rosenberg [14] (the latter for polynomials in several variables).

COROLLARY 2. *In the situation of the Proposition, if $n > 0$ then for every function induced on the residue classes of P^{n-1} by a polynomial in $R[x]$, there are exactly*

$$\prod_{j=0}^{\beta(n)-1} [P^{n-\alpha(j)-1} : P^{n-\alpha(j)}]$$

different polynomial functions on the residue classes of P^n that reduce to the given function mod P^{n-1}. If $[P^{k-1} : P^k] = q$ for $1 \le k \le N$ then the expression simplifies to $q^{\beta_q(n)}$, where $\beta_q(n)$ is the minimal $m \in \mathbb{N}$ such that $\alpha_q(m) = \sum_{j \ge 1} \left[\frac{n}{q^j} \right] \ge n$.

Proof of the formula for $|\mathcal{F}(R)|$ in Theorem 2:

$$|\mathcal{F}(R)| = \prod_{j=0}^{\beta(N)-1} [R : P^{N-\alpha(j)}]$$

follows immediately from Corollary 1 with $n = N$. In the special case that $[P^{k-1} : P^k] = q$ for $1 \le k \le N$, writing s_k for the number of different functions on R/P^k arising from polynomials in $R[x]$, we see from Corollary 2 that $q^{\beta_q(k)} s_{k-1} = s_k$. Therefore $q^{\sum_{k=1}^{N} \beta(k)} = s_N = |\mathcal{F}(R)|$ in that case. $\square$

4   THE GROUP $\mathcal{P}(R/P^2)$

We want to determine the structure of the group $\mathcal{P}(R/P^2)$ with respect to composition of functions, R being a suitable finite local ring as above. To simplify notation, we consider the group $\mathcal{P}(R)$, where R is a finite local ring with maximal ideal P of nilpotency $N = 2$.

Some notational conventions: We write the group of *invertible elements* of a monoid M as M^*. If M is a monoid and H a monoid acting on a set S then the *wreath product* $M \wr H$ is the monoid defined on the set $H \times M^S$ by the operation

$$(h, (m_s)_{s \in S})(g, (l_s)_{s \in S}) = (hg, (m_{g(s)} l_s)_{s \in S}).$$

If M acts on a set T then the standard action of $M \wr H$ on $S \times T$ is

$$(h, (m_s)_{s \in S})(x, y) = (h(x), m_x(y)).$$

Note that an element $(h, (m_s)_{s \in S})$ is in $(M \wr H)^*$ if and only if $h \in H^*$ and $m_s \in M^*$ for all $s \in S$, and that therefore $(M \wr H)^* \simeq M^* \wr H^*$.

If D is a commutative ring and M a D-module, we write $\mathbb{A}_D(M)$ for the semigroup with respect to compostion of transformations of M of the form $x \mapsto ax + b$ with $a \in D$ and $b \in M$. We have $|\mathbb{A}_D(M)| = |D/\mathrm{Ann}(M) \times M|$.

PROPOSITION 2.   *Let R be a finite local ring with maximal ideal P of nilpotency 2 and $q = [R : P]$. Denote by Q^Q the semigroup of functions from a set of q elements to itself. Then*

$$\mathcal{F}(R) \simeq \mathbb{A}_{R/P}(P) \wr Q^Q \qquad \text{and} \qquad \mathcal{P}(R) \simeq \mathbb{A}^*_{R/P}(P) \wr S_q,$$

and in particular,

$$|\mathcal{F}(R)| = q^q \, |R|^q \qquad \text{and} \qquad |\mathcal{P}(R)| = q! \, (q-1)^q \, |P|^q.$$

Proof. Fix a system of representatives Q of R mod P. We identify R with $Q \times P$ by $r \mapsto (s, t)$ with $s \in Q$, $t \in P$, such that $r = s + t$. Let $f \in R[x]$. We have

$$f(r) = f(s + t) = f(s) + f'(s)t,$$

since this holds mod P^2 by Taylor's Theorem and $P^2 = (0)$ in R. Now let $\varphi(s)$ be the representative in Q of $f(s) + P$, then

$$f(s+t) = \varphi(s) + (f(s) - \varphi(s)) + f'(s)t,$$

with $\varphi(s) \in Q$ and $f(s) - \varphi(s) \in P$. We regard $f'(s)$ as being in R/P. (As it gets multiplied by $t \in P$, only its residue class mod P matters).

If we associate to $f \in R[x]$ the functions $\varphi_f \colon Q \to Q$ and $\psi_f \colon Q \to \mathbb{A}_{R/P}(P)$, where

- $\varphi_f(s)$ is the representative in Q of $f(s) + P$
- $\psi_f(s)$ is the transformation $x \mapsto a_f(s)x + b_f(s)$ on P, where
  - $a_f(s) \in R/P$ is $f'(s)$ mod P,
  - $b_f(s) = f(s) - \varphi(s) \in P$

then φ_f and ψ_f completely determine the function induced by f on R.

Moreover, the function defined on $Q \times P$ by $\varphi \in Q^Q$, $a \in (R/P)^Q$ and $b \in P^Q$ via $(s, t) \mapsto \varphi(s) + a(s)t + b(s)$ determines φ, a and b uniquely, such that for f, $g \in R[x]$ inducing the same function on R we have $\varphi_g = \varphi_f$ and $\psi_g = \psi_f$. Therefore $f \mapsto (\varphi_f, \psi_f)$ depends only on the function induced by $f \in R[x]$ on R and defines a homomorphism from $\mathcal{F}(R)$ to $\mathbb{A}_{R/P}(P) \wr Q^Q$, which takes the action of $\mathcal{F}(R)$ on R (identified with $Q \times P$) to the standard action of $A \wr Q^Q$ arising from the obvious actions of A on P and of Q^Q on Q. We have already seen that this homomorphism is injective.

To check surjectivity, we show that every triple of functions $\varphi \colon Q \to Q$, $b \colon Q \to P$ and $a \colon Q \to R/P$ actually occurs as φ_f, a_f and b_f for some $f \in R[x]$.

Every pair of functions on R/P arises as f mod P and f' mod P for some polynomial $f \in R[x]$, because R/P is a finite field. This takes care of φ_f and a_f. Since the characteristic function of every residue class of P is induced by a polynomial in $R[x]$ (just take a sufficiently high power of a polynomial representing it mod P), we can adjust f to take prescribed values on the $s \in Q$, by adding a P-linear combination of these characteristic functions. This produces a prescribed b_f without disturbing the values of f and f' mod P, since we only add a polynomial in $P[x]$.

If we restrict to polynomials representing permutations or, equivalently, to polynomials for which φ_f is a permutation of Q and $a_f(s) \neq 0 + P$ for all $s \in Q$, we get an isomorphism of $\mathcal{P}(R)$ and $\mathbb{A}^*_{R/P}(P) \wr S_q$, which takes the action of $\mathcal{P}(R)$ on R (identified with $Q \times P$) to the standard action of the wreath product on $Q \times P$ arising from the obvious actions of $\mathbb{A}^*_{R/P}$ on P and of the symmetric group S_q on Q. $\square$

REMARK. We may simplify the expression for $\mathcal{P}(R)$ by noting that $\mathbb{A}_{R/P}(P)$ is isomorphic to the semi-direct product $((R/P)^*, \cdot) \ltimes (P, +)$ with $(R/P)^*$ acting on $(P, +)$ through the scalar mulutiplication of the R/P-vectorspace structure on P.

Proof of the formula for $|\mathcal{P}(R)|$ in Theorem 2: For $n \leq N$, let s_n denote the number of functions on the residue classes of P^n induced by polynomials in $R[x]$ and t_n the number of them that are permutations.

If $n \geq 2$, a polynomial induces a permutation mod P^n if and only if it induces a permutation mod P and its derivative is nowhere zero mod P, cf. [7]. In particular, if $n > 2$, a polynomial induces a permutation mod P^n if and only if it induces one mod P^{n-1}. Together with the fact that every class of polynomial functions mod P^n reducing to the same function mod P^{n-1} contains the same number of elements (Corollary 2 of Proposition 1), this implies that $\frac{t_n}{t_{n-1}} = \frac{s_n}{s_{n-1}}$ for all $n > 2$, and therefore $t_n = \frac{t_2}{s_2} s_n$ for all $n \geq 2$.

From Proposition 2 applied to R/P^2 we get $t_2 = q!(q-1)^q[P : P^2]^q$ and $s_2 = q^q[R : P^2]^q$ and the formula for $|\mathcal{P}(R)|$ follows. $\square$

REFERENCES

[1] N. Aizenberg, I. Semion, and A. Tsitkin, Polynomial representations of logical functions, Automatic Control and Computer Sciences (transl. of Automatika i Vychislitel'naya Tekhnika, Acad. Nauk Latv. SSR (Riga)), 5 (1971), pp. 5–11 (orig. 6–13).

[2] D. A. Ashlock, Permutation polynomials of Abelian group rings over finite fields, J. Pure Appl. Algebra, 86 (1993), pp. 1–5.

[3] J. V. Brawley and G. L. Mullen, Functions and polynomials over Galois rings, J. Number Theory, 41 (1992), pp. 156–166.

[4] F. Dueball, Bestimmung von Polynomen aus ihren Werten mod p^n, Math. Nachr., 3 (1949/50), pp. 71–76.

[5] G. Keller and F. Olson, Counting polynomial functions (mod p^n), Duke Math. J., 35 (1968), pp. 835–838.

[6] A. J. Kempner, Polynomials and their residue systems, Trans. Amer. Math. Soc., 22 (1921), pp. 240–266, 267–288.

[7] B. R. McDonald, Finite Rings with Identity, Dekker, 1974.

[8] G. L. Mullen and H. Niederreiter, The structure of a group of permutation polynomials, J. Austral. Math. Soc. Ser. A, 38 (1985), pp. 164–170.

[9] W. Narkiewicz, Polynomial Mappings, vol. 1600 of Lecture Notes in Mathematics, Springer, 1995.

[10] A. Nechaev, Polynomial transformations of finite commutative local rings of principal ideals, Math. Notes, 27 (1980), pp. 425–432. transl. from Mat. Zametki 27 (1980) 885-897, 989.

[11] W. Nöbauer, Gruppen von Restpolynomidealrestklassen nach Primzahlpotenzen, Monatsh. Math., 59 (1955), pp. 194–202.

[12] L. Rédei and T. Szele, Algebraisch-zahlentheoretische Betrachtungen über Ringe I, Acta Math. (Uppsala), 79 (1947), pp. 291–320.

[13] ——, Algebraisch-zahlentheoretische Betrachtungen über Ringe II, Acta Math. (Uppsala), 82 (1950), pp. 209–241.

[14] I. G. Rosenberg, Polynomial functions over finite rings, Glas. Mat., 10 (1975), pp. 25–33.

[15] D. Singmaster, On polynomial functions (mod m), J. Number Theory, 6 (1974), pp. 345–352.

[16] Th. Skolem, Einige Sätze über Polynome, Avh. Norske Vid. Akad. Oslo, I. Mat.-Naturv. Kl., 4 (1940), pp. 1–16.

[17] R. Spira, Polynomial interpolation over commutative rings, Amer. Math. Monthly, 75 (1968), pp. 638–640.

[18] J. Wiesenbauer, On polynomial functions over residue class rings of $\mathbb{Z}$, in Contributions to general algebra 2 (Proc. of Conf. in Klagenfurt 1982), Hölder-Pichler-Tempsky, Teubner, 1983, pp. 395–398.

Koszul Algebras

R. Fröberg
Matematiska institutionen
Stockholms Universitet
10691 Stockholm
Sweden
email ralff@matematik.su.se

Abstract

A graded k-algebra A is called a Koszul algebra if the minimal graded free A-resolution of k has only linear maps. This article is a survey on results obtained about Koszul algebras since they were introduced by Priddy in 1970. We start with giving several equivalent conditions to being Koszul, and then give lots of examples of Koszul algebras from different fields. We show that the class of Koszul algebras is closed under a number of natural operations. Almost no proofs are given, but ample references to the literature are provided.

1   Introduction

Let k be a field and V a vector space over k with basis $(x_1, \ldots, x_n)$. The tensor algebra (or the noncommutative polynomial ring) $T(V) = \oplus_{i \geq 0}(T(V))_i$ is a graded k-algebra with the monomials $x_{m_1} \cdots x_{m_i}$ as a k-basis for $(T(V))_i$. We will use the notation $k\langle x_1, \ldots, x_n \rangle$ for $T(V)$. With a *graded algebra* we will mean an algebra $A = k\langle x_1, \ldots, x_n \rangle / I$, where I is a two-sided ideal generated by homogeneous elements. The graded algebra A is called *quadratic* if I is generated by elements of degree two. As an example, the usual commutative polynomial ring is quadratic, since $k[x_1, \ldots, x_n] = k\langle x_1, \ldots, x_n \rangle / I$, where I is generated by all commutators $x_i x_j - x_j x_i$. Also algebras $k[x_1, \ldots, x_n]/I$, where I is generated by quadratic forms, are quadratic. The *Hilbert series* $A(z)$ of a graded algebra $A = \oplus A_i$ is the generating function for the k-dimensions of A_i,

$$A(z) = \sum_{i \geq 0} \dim_k A_i \cdot z^i.$$

If $A = k\langle x_1, \ldots, x_n \rangle$ we have $\dim_k A_i = n^i$, so $A(z) = 1 + nz + n^2 z^2 + \cdots = 1/(1 - nz)$, and if $A = k[x_1, \ldots, x_n]$ we have $\dim_k A_i = \binom{i+n-1}{n-1}$, so $A(z) =$

$1/(1-z)^n$. It is well known that if A is a commutative graded algebra, $A = k[x_1, \ldots, x_n]/I$, then $A(z)$ is a rational function, namely $A(z) = p(z)/(1-z)^n$ for some polynomial $p(z) \in \mathbb{Z}[z]$. If A is a graded algebra, then $\oplus_{i>0} A_i$ is a graded maximal ideal which we will denote by A_+.

For any graded algebra A there exists a minimal free graded A-resolution of k

$$\mathbf{F} : \cdots \xrightarrow{\phi_3} A^{b_2} \xrightarrow{\phi_2} A^{b_1} \xrightarrow{\phi_1} A \longrightarrow k.$$

That the resolution is graded means that the nonzero entries of the matrices ϕ_i are homogeneous, that the resolution is minimal means that all nonzero entries have positive degrees. Since $\phi_i \otimes k = 0$ in a minimal resolution we have

$$\mathrm{Tor}_i^A(k, k) \simeq A^{b_i} \otimes k \simeq k^{b_i} \simeq \mathrm{Ext}_A^i(k, k).$$

The *Poincaré series* $P_A(z)$ of A is the generating function for the k-dimensions of $\mathrm{Tor}_i^A(k, k)$,

$$P_A(z) = \sum_{i \geq 0} \dim_k \mathrm{Tor}_i^A(k, k) \cdot z^i = \sum_{i \geq 0} \dim_k \mathrm{Ext}_A^i(k, k) \cdot z^i.$$

If we shift the degrees such that all ϕ_i becomes maps of degree 0, we see that the grading on A induces a grading on $\mathrm{Tor}_i^A(k, k) = \oplus_j (\mathrm{Tor}_i^A(k, k))_j$ and on $\mathrm{Ext}_A^i(k, k) = \oplus_j (\mathrm{Ext}_A^i(k, k))_j$ and we can define a Poincaré series in two variables

$$\mathbf{P}_A(x, y) = \sum_{i,j} \dim_k (\mathrm{Tor}_i^A(k, k))_j \cdot x^i y^j = \sum_{i,j} \dim_k (\mathrm{Ext}_A^i(k, k))_j \cdot x^i y^j.$$

The existence of a minimal resolution gives that $(\mathrm{Tor}_i^A(k, k))_j = (\mathrm{Ext}_A^i(k, k))_j = 0$ if $j < i$.

Koszul algebras were first introduced in [Pr] (under the name homogeneous Koszul algebras). We define a *Koszul algebra* to be a graded algebra such that $(\mathrm{Tor}_i^A(k, k))_j = 0$ if $i \neq j$ or, equivalently, such that $(\mathrm{Ext}_A^i(k, k))_j = 0$ if $i \neq j$. Another way to say this is that the minimal graded A-resolution of k is *linear*, i.e., all nonzero entries of all ϕ_i are of degree one. If $A = k\langle x_1, \ldots, x_n \rangle$ a minimal A-resolution of k looks like

$$0 \longrightarrow A^n \xrightarrow{\phi_1} A \longrightarrow k \longrightarrow 0$$

with $\phi_1 = (x_1 \ \cdots \ x_n)$ and hence it is linear. We get that the Poincaré series of $k\langle x_1, \ldots, x_n \rangle$ equals $1 + nz$, the double Poincaré series equals $1 + nxy$.

If $A = k[x_1, \ldots, x_n]$, the Koszul complex

$$0 \longrightarrow A^{\binom{n}{n}} \xrightarrow{\phi_n} \cdots \longrightarrow A^{\binom{n}{2}} \xrightarrow{\phi_2} A^{\binom{n}{1}} \xrightarrow{\phi_1} A \longrightarrow k \longrightarrow 0$$

is a minimal graded resolution of k with free A-modules. If we denote a basis for $A^{\binom{n}{i}}$ by $\{e_{m_1 \cdots m_i} ; 1 \leq m_1 < \cdots < m_i \leq n\}$, then $\phi_i(e_{m_1 \cdots m_i}) =$

$\sum_{k=1}^{i}(-1)^{k-1}x_{m_k}e_{m_1\cdots\widehat{m_k}\cdots m_i}$ and hence ϕ_i is linear. The Poincaré series of $k[x_1,\ldots,x_n]$ equals $1+\binom{n}{1}z+\cdots+\binom{n}{n}z^n=(1+z)^n$, the double series $(1+xy)^n$.

Thus both $k\langle x_1,\ldots,x_n\rangle$ and $k[x_1,\ldots,x_n]$ are examples of Koszul algebras since k has a linear resolution over these algebras.

If $A=k\langle x_1,\ldots,x_n\rangle/I$, then $\operatorname{Tor}_2^A(k,k)\simeq I/(A_+I+IA_+)$, so $\operatorname{Tor}_2^A(k,k)=(\operatorname{Tor}_2^A(k,k))_2$ if and only if A is quadratic. Hence a Koszul algebra is necessarily quadratic.

The formula above for $\operatorname{Tor}_2^A(k,k)$ is just a special case of a general result in [Go]. There it is shown that

$$\operatorname{Tor}_{2i}^A(k,k)\simeq(A_+I^{i-1}A_+\cap I^i)/(A_+I^i+I^iA_+)$$

and that

$$\operatorname{Tor}_{2i-1}^A(k,k)\simeq(A_+I^{i-1}\cap I^{i-1}A_+)/(A_+I^{i-1}A_++I^i)$$

if $A=k\langle x_1,\ldots,x_n\rangle/I$. This is used in [B1] to get another equivalent condition for an algebra to be Koszul. Let $L(A)$ be the lattice associated to $A=k\langle x_1,\ldots,x_n\rangle/I$, i.e., the lattice generated by $\{A_+^fI^gA_+^h\ ;\ f,g,h\geq 0\}$ under $+$ and $\cap$. Then A is Koszul if and only if A is quadratic and $L(A)$ is *distributive*. It is also shown in [B1] that this is equivalent to that A is quadratic and all lattices $L_j(A)$ generated by $\{A_+^fI^gA_+^h\ ;\ f,g,h\geq 0, f+g+h=j\}$ are distributive.

2 The Koszul Dual

Let $k\langle x_1,\ldots,x_n\rangle=\Lambda$ and Λ/I a quadratic algebra. Let $\Lambda^*=\operatorname{Hom}_k(\Lambda,k)=\oplus_{i\geq 0}\operatorname{Hom}_k(\Lambda_i,k)=\oplus_{i\geq 0}\Lambda_i^*$ with multiplication induced by $\mu\nu(ab)=\mu(a)\nu(b)$, where $\mu\in\Lambda_i^*,\nu\in\Lambda_j^*,a\in\Lambda_i,b\in\Lambda_j$. Let $I_2^\perp=\{\mu\in\Lambda_2^*\ ;\ \mu(I_2)=0\}$ and let $I^\perp$ be the ideal generated by $I_2^\perp$. The *Koszul dual* of A is $A^!=\Lambda^*/I^\perp$. The Koszul dual is a quadratic algebra and $(A^!)^!=A$.

To calculate a presentation of $A^!$ from the presentation of A is just some elementary linear algebra. We give some examples. If $A=k\langle x_1,\ldots,x_n\rangle$, then $A^!=k\langle y_1,\ldots,y_n\rangle/(y_iy_j,1\leq i,j\leq n)$. If $A=k[x_1,\ldots,x_n]$, then $A^!=k\langle y_1,\ldots,y_n\rangle/(y_i^2,1\leq i\leq n,y_iy_j+y_jy_i,1\leq i<j\leq n)$. If $A=k\langle x_1,\ldots,x_n\rangle/I$, where I is generated by monomials of degree two, then $A^!=k\langle y_1,\ldots,y_n\rangle/J$, where J is generated by those monomials y_iy_j such that $x_ix_j\notin I$. If $A=k[x_1,\ldots,x_n]/(f_1,\ldots,f_r)$, where $f_i=\sum_{j\leq k}b_{ijk}x_jx_k$, then $A^!=k\langle y_1,\ldots,y_n\rangle/J$, where $J=(g_1,\ldots,g_s)$, $s=\binom{n}{2}-r$, and $g_i=\sum_{j\leq k}c_{ijk}[y_i,y_j]$ (here $[y_i,y_j]=y_iy_j+y_jy_i$ if $i\neq j$ and $[y_i,y_i]=y_i^2$) and $(c_{ijk})_{jk},i=1,\ldots,s$ is a basis of the solutions to the linear system $\sum_{j\leq k}b_{ijk}X_{jk}=0,i=1,\ldots,r$, cf. [L]. As an example, if

$$A=k[x_1,x_2,x_3]/(x_1^2,x_2x_3,x_1x_3-x_3^2),$$

then

$$A^!=k\langle y_1,y_2,y_3\rangle/(y_2^2,y_1y_2+y_2y_1,y_1y_3+y_3y_1+y_3^2).$$

For any quadratic algebra A there is a natural differential $d : A_i \otimes_k (A_j^!)^* \longrightarrow A_{i+1}^! \otimes_k (A_{j-1}^!)^*$, where $*$ indicates vector space dual, cf. [Pr], [L], or [Ma], namely if $f \in (A^!)_n^*$, then $df \in A_1 \otimes (A^!)_{n-1}^*$ is defined by identifying $A_1 \otimes (A^!)_{n-1}^*$ with $(A_1^! \otimes A_{n-1}^!)^*$ and letting $(df)(x \otimes m) = f(xm)$, where $x \in A_1^!, m \in A_{n-1}^!$ and extend A-linearly. This makes $A \otimes_k A^!$ into a complex U. (If $A = k[x_1, \ldots, x_n]$, then U is the usual Koszul complex.) This "generalized Koszul complex" is exact if and only if A is Koszul.

There is a well known product $\mathrm{Ext}_A^i(k, k) \times \mathrm{Ext}_A^j(k, k) \longrightarrow \mathrm{Ext}^{i+j}(k, k)$, the Yoneda multiplication, making $\mathrm{Ext}_A(k, k)$ into an associative graded algebra. It is shown in [L] (also cf. [Pr]) that $A^!$ is the subalgebra $[\mathrm{Ext}_A^1(k, k)]$ of $\mathrm{Ext}_A(k, k)$ generated by its one-dimensional elements. It is clear that for any graded algebra A we have $\mathrm{Ext}_A^1(k, k) = (\mathrm{Ext}_A^1(k, k))_1$, so $[\mathrm{Ext}_A^1(k, k)] \subseteq \oplus_i (\mathrm{Ext}_A^i(k, k))_i$, but in fact there is equality, cf. [L]. Hence A is a Koszul algebra if and only if $\mathrm{Ext}_A(k, k) = [\mathrm{Ext}_A^1(k, k)]$. In particular we see that if A is a Koszul algebra, then $\mathrm{Ext}_A(k, k)$ is finitely generated.

For any graded algebra A, for each j, the restriction of the minimal A-resolution of k to degree j is a finite exact complex of finite dimensional vector spaces. Using that the alternating sum of the dimensions in this finite dimensional complex equals the alternating sum of its homologies one gets the formula $A(z)\mathbf{P}_A(-1, z) = 1$ for each graded algebra A. If A is a Koszul algebra we have $\mathbf{P}_A(x, y) = P_A(xy)$, so for a Koszul algebra we have the formula $A(z)P_A(-z) = 1$. This formula is in fact *equivalent* to A being Koszul, cf. [L].

We sum up with a theorem

Definition-Theorem 1 *A graded algebra A is Koszul if and only if the following equivalent conditions are satisfied*

i) $\mathrm{Tor}_i^A(k, k) = (\mathrm{Tor}_i^A(k, k))_i$ *for all i.*
ii) $\mathrm{Ext}_A^i(k, k) = (\mathrm{Ext}_A^i(k, k))_i$ *for all i.*
iii) The minimal graded A-resolution of k is linear.
iv) $\mathrm{Ext}_A(k, k) = [\mathrm{Ext}_A^1(k, k)]$.
v) $A(z)P_A(-z) = 1$.
vi) $A(xy)\mathbf{P}_A(-x, y) = 1$.
vii) $P_A(z) = A^!(z)$.
viii) $\mathbf{P}_A(x, y) = A^!(xy)$.
ix) A is quadratic and $L(A)$ is distributive.
x) A is quadratic and $L_j(A)$ are distributive for all j.
xi)-xx) A is quadratic and $A^!$ satisfies any of the above conditions.
xxi) A is quadratic and the complex U is exact.

It follows that a Koszul algebra has a finitely generated Ext-algebra and rational Poincaré series.

There are some further more technical characterizations of Koszul algebras. The algebra A is Koszul if and only if the map $A \longrightarrow A/A_+^2$ is "small", cf. [BF1]. In [L3] there are conditions (for a commutative algebra A) in terms of a "minimal model" for A and of the "homotopy Lie algebra" of A.

3 Examples of Koszul Algebras

We will now give some examples of classes of Koszul algebras. Since Koszul algebras are necessarily quadratic, we will in this section assume that all algebras are of the form $k\langle x_1, \dots, x_n \rangle / I$, where I is generated by homogeneous elements of degree two.

3.1 Commutative Examples

We start with commutative algebras. If A is a complete intersection, i.e., $A = k[x_1, \dots, x_n]/(f_1, \dots, f_r)$ where $(f_1, \dots, f_r)$ is a regular sequence (of forms of degree two), then A is Koszul. The exact sequences

$$0 \longrightarrow k[\mathbf{x}]/(f_1, \dots, f_{i-1}) \xrightarrow{f_i} k[\mathbf{x}]/(f_1, \dots, f_{i-1}) \longrightarrow k[\mathbf{x}]/(f_1, \dots, f_i) \longrightarrow 0$$

easily give that $A(z) = (1 - z^2)^r/(1 - z)^n$. The Poincaré series of a complete intersection was determined in [T], $P_A(z) = (1+z)^n/(1-z^2)^r$. (Tate considered local rings, but the same arguments can be applied.) Since $A(z)P_A(-z) = 1$, A is Koszul.

If $A = k[x_1, \dots, x_n]/I$ and I is generated by an arbitrary set of monomials (of degree two), then A is Koszul, cf. [F1]. A concrete example: If $A = k[x_1, \dots, x_n]/(x_i x_j \, ; \, i \neq j)$ then $A(z) = 1 + nz + nz^2 + \cdots = (1 + (n-1)z)/(1-z)$, so we can conclude that $P_A(z) = (1 + z)/(1 - (n - 1)z)$. There are interesting classes of quadratic monomial ideals coming from combinatorics. If P is a poset on $\{x_1, \dots, x_n\}$, the associated Stanley-Reisner ring $k[P] = k[\mathbf{x}]/(x_i x_j \, ; \, x_i \not\leq x_j, x_j \not\leq x_i)$ is Koszul. When studying general Stanley-Reisner rings $k[\Delta]$ one can sometimes reduce problems to the barycentric subdivision Δ' of Δ, and $k[\Delta']$ is Koszul.

The result above on monomial ideals was extended in [Ko] to algebras $k[\mathbf{x}]/I$, where I is generated by monomials and certain binomials.

If the (finite) $k[\mathbf{x}]$-resolution of I is linear, i.e., if $(\mathrm{Tor}_i^{k[\mathbf{x}]}(A, k))_j = 0$ if $j \neq i + 1$ for all $i \geq 1$ ($\mathrm{Tor}_1^{k[\mathbf{x}]}(A, k) = (\mathrm{Tor}_1^{k[\mathbf{x}]}(A, k))_2$ is always true since A is quadratic), then A is Koszul, cf. [BF1]. If A is CM then $\mathrm{e.dim} A \leq e(A) + \dim A - 1$, where $e(A)$ is the multiplicity or degree of A. If there is equality, A is said to be of *maximal embedding dimension* or *minimal multiplicity*. CM algebras of maximal embedding dimension have linear resolutions. We give some concrete examples. If (x_{ij}) is a $2 \times n$-matrix of indeterminates and I the ideal in $k[x_{ij}]$ generated by all maximal minors in (x_{ij}), then $k[x_{ij}]/I$ has a linear resolution. If (x_{ij}) is a *symmetric* 3×3-matrix of indeterminates and I the ideal of 2×2-minors then $k[x_{ij}]/I$ is another example, cf. [Sc] and [F2] and the references in them. There are more examples of CM rings with linear resolutions in [FLa].

Extremal Gorenstein rings were introduced in [Sc]. These are Koszul algebras (if they are quadratic), cf. [F2]. A concrete example is $k[x_{ij}]/I$ where I is generated by the 4×4-Pfaffians in a skew-symmetric 5×5-matrix (x_{ij}). If I is the ideal of 2×2-minors in a 3×3-matrix (x_{ij}) we get another example of a quadratic extremal Gorenstein ring.

If $f_1, \ldots, f_r$ are "generic" quadratic forms in $k[x_1, \ldots, x_n]$, then we have that $k[x_1, \ldots, x_n]/(f_1, \ldots, f_r)$ is Koszul if and only if $r \leq n$ or $r \geq \binom{n+1}{2} - [n^2/4]$, cf [L2].

If $A = \oplus_{i \geq 0} A_i$ is a graded algebra, the dth *Veronese subalgebra* is $A^{(d)} = \oplus_{i \geq 0} A_{id}$. An element in A_{id} is considered to have degree i. If $\oplus_{i \geq 0} A_i$ and $\oplus_{i \geq 0} B_i$ are graded algebras their *Segre product* is $A \circ B = \oplus_{i \geq 0} A_i \otimes_k B_i$. An element in $A_i \otimes B_i$ has degree i. It is shown in [BM] that, if one starts with a polynomial ring and performs a finite number of Segre products and Veronese subalgebras in any order. the result will be a Koszul algebra. There is a partial generalization to "weighted" Segre products in [C].

A classical theorem in geometry states that any projective variety can be cut out by quadrics (cf. [Mu], but the result is certainly older). This was generalized in [BF1] where it is shown that for any graded algebra A (commutative or not) we have $A^{(d)}$ quadratic if $d >> 0$ (an actual bound is given). In [B3] it is shown that if A is commutative, then $A^{(d)}$ is even Koszul if $d >> 0$. This was sharpened in [ERT]and [BGT], see below.

3.2 Examples from Geometry

There are several examples of coordinate rings of projective varieties that are Koszul. In [Ke1] it is shown that any algebra A with straightening law whose discrete algebra is defined by quadratic monomials is Koszul (called wonderful in [Ke1]). In [VF] it is shown that the coordinate ring of a general curve of genus ≥ 5 is Koszul. There are more examples in [Ke2], [Pa], [Pol], [PP], and [Ra].

3.3 Noncommutative Examples

If I is generated by an arbitrary set of monomials (of degree two) in $k\langle \mathbf{x} \rangle$, then $k\langle \mathbf{x} \rangle/I$ is Koszul, cf. [F1]. A slight generalization of this result was used in [BHV] to calculate the number of walks of certain kinds in a directed graph.

If $(f_1, \ldots, f_r)$ is a sequence of homogeneous elements of degree two in $k[\mathbf{x}]$ then $k[\mathbf{x}]/(f_1, \ldots, f_r)(z) \geq (1 - z^2)^r/(1 - z)^n$ with equality if and only if the sequence is regular. There is a corresponding property in $k\langle x_1, \ldots, x_n \rangle$. If $(f_1, \ldots, f_r)$ is a sequence of homogeneous elements of degree two in $k\langle x_1, \ldots, x_n \rangle$ then $k\langle \mathbf{x} \rangle/(f_1, \ldots, f_r)(z) \geq 1/(1 - nz + rz^2)$, cf. [An2]. If there is equality the sequence is called *strongly free*. That $(f_1, \ldots, f_r)$. $r \geq 1$. is strongly free is equivalent to gl.dim$(k\langle \mathbf{x} \rangle/(f_1, \ldots, f_r)) = 2$. This is shown in [An2]. and then it follows easily that these rings are Koszul.

The results above can also be formulated in a relative situation. If f is a quadratic form in a commutative algebra A. then $A/(f)(z) \geq (1 - z^2)A(z)$ with equality if and only if f is a nonzerodivisor in A. Similarly if f is a quadratic form in any graded algebra A, then $A/(f)(z) \geq A(z)/(1 + z^2 A(z))$, and f is called strongly free in A if there is equality. We will use this below.

4 Gröbner Bases and Koszul Algebras

If $A = k\langle\mathbf{x}\rangle/I$ and I has a (noncommutative) Gröbner basis consisting of elements of degree two, then A is Koszul. This follows from the spectral sequence $\mathrm{Tor}^{k\langle\mathbf{x}\rangle/\mathrm{in}(I)}(k,k) \implies \mathrm{Tor}^{k\langle\mathbf{x}\rangle/I}(k,k)$, where $\mathrm{in}(I)$ denotes the ideal generated by the initial monomials of the elements in the Gröbner basis, cf. [An3]. There is a similar result in the commutative case, cf. [An3], so if I has a quadratic (commutative) Gröbner basis then $k[\mathbf{x}]/I$ is Koszul. Another proof of this fact could be found in [BHV]. As an example, the ideal I of 2×2-minors in a matrix (x_{ij}) has a quadratic Gröbner basis, cf. [St1], so $k[x_{ij}]/I$ is Koszul.

If A is any commutative graded algebra then $A^{(d)}$ has a quadratic Gröbner basis if $d >> 0$, cf. [ERT]. This improves the result in [B3] mentioned above. The result in [ERT] is further generalized to a larger class of algebras (not necessarily generated in degree one) in [BGT].

Let $A = k[x_1, \ldots, x_n]$. The subring of $A^{(d)}$ generated by the monomials $\{x_1^{i_1} \cdots x_n^{i_n} : i_1 + \cdots + i_n = d, 0 \leq i_1 \leq s_1, \ldots, 0 \leq i_n \leq s_n\}$ is called an algebra of *Veronese type*. (If $s_1 = \cdots = s_n = d$ we get $A^{(d)}$.) It is shown in [St1] that the defining ideal of an algebra of Veronese type has a Gröbner basis (in a certain ordering) which is not only quadratic but also squarefree. This has interesting combinatorial consequences, cf. [St1].

5 Operations on Koszul algebras

The class of Koszul algebras is closed under a number of operations. The *coproduct* of two graded algebras A and B over k is the pushout $A \coprod B$ of A and B of $A \longleftarrow k \longrightarrow B$. The *(fibre) product* $A \prod B$ over k is the pullback of $A \longrightarrow k \longleftarrow B$. The following results are proved in [BF1].

Theorem 2 *i)* A *is Koszul if and only if the Veronese subalgebra* $A^{(d)}$ *is Koszul for all* d.

ii) *If* A *and* B *are Koszul then the Segre product* $A \circ B$ *is Koszul.*

iii) $A \coprod B$ *is Koszul if and only if* A *and* B *are both Koszul.*

iv) $A \prod B$ *is Koszul if and only if* A *and* B *are both Koszul.*

v) $A \otimes_k B$ *is Koszul if and only if* A *and* B *are both Koszul.*

vi) *If* f *is strongly free of degree one or two, then* A *is Koszul if and only if* $A/(f)$ *is Koszul.*

vii) *If* f *is a nonzerodivisor of degree one or two in a commutative algebra* A, *then* A *is Koszul if and only if* $A/(f)$ *is Koszul.*

6 More Examples and some Counterexamples

In this section we consider only quadratic algebras.

If $A = k\langle\mathbf{x}\rangle/I$ and I is principal, then A is Koszul, cf. [B2], but there are counterexamples already when I is generated by two elements. The ideals generated by two elements in $k\langle \mathbf{x} \rangle$ are classified up to isomorphisms in [B1], and all possible Hilbert and Poincaré series are determined. There is a small number of exceptions to Koszulness.

On the commutative side, probably the first counterexample to Koszulness (due to C. Lech) is the following. Let $f_1, \ldots, f_5$ be "generic" quadratic forms in $k[x_1, \ldots, x_4]$. (A concrete example is $(f_1, \ldots, f_5) = (x_1^2, x_2^2, x_3^2, x_4^2, x_1 x_2 + x_3 x_4)$.) It is not so hard to see that if $A = k[x_1, \ldots, x_4]/(f_1, \ldots, f_5)$ then $A(z) = 1 + 4z + 5z^2$, so $1/A(-z) = 1 + 4z + 11z^2 + 24z^3 + 41z^4 + 44z^5 - 29z^6 - \cdots$. Since this series has negative coefficients, A is not Koszul.

In [Ro1], J.-E. Roos constructed an example of an algebra A for which $\mathrm{Ext}_A(k, k)$ is not finitely generated, and in [An1] D. Anick gave an example of an algebra with non-rational Poicaré series.

In embedding dimension two all commutative algebras are Koszul. In [BF2] all quadratic ideals in $k[x_1, x_2, x_3]$ generated by three elements and which are not complete intersections are classified up to isomorphisms and their Hilbert and double Poincaré series are determined. There are two exceptions to Koszulness (the ideals $(x_1^2, x_2 x_3, x_1 x_3 + x_2^2)$ and $(x_1 x_2, (x_1 + x_2) x_3, x_1^2 + x_1 x_3 + x_2^2)$) if $\mathrm{char}(k) \neq 3$ and one further (namely $(x_1^2, x_2 x_3, x_1 x_2 + x_1 x_3 + x_2^2)$) if $\mathrm{char}(k) = 3$. It is shown in [BF2] that these are the only counterexamples in embedding dimension three to Koszulness.

J.-E. Roos has made an extensive study of homological properties of quadratic commutative algebras. He has found 83 different double Poincaré series in embedding dimension four and more than 4500 in embedding dimension five. Among the examples in embedding dimension four there are 37 Koszul algebras. In embedding dimension five there are about ten per cent Koszul algebras. There are lots of examples of the homological behaviour of commutative quadratic algebras given in [Ro4], [Ro5], [Ro6], and [Ro7].

If A is commutative and $\dim_k A_2 \leq 2$, then A is Koszul, cf. [B1], and the examples in [BF2] show that there are counterexamples if $\dim_k A_2 = 3$.

There are commutative Koszul algebras which do not have a quadratic Gröbner basis in any ordering, even after any linear change of coordinates. An example from [ERT]: Take three generic forms (of degree two) f_1, f_2, f_3 in $k[x_1, x_2, x_3]$. Since $k[x_1, x_2, x_3]/(f_1, f_2, f_3)$ is a complete intersection it is Koszul. If $I = (f_1, f_2, f_3)$ had a quadratic Gröbner basis, then $\mathrm{in}(I) = (x_1^2, x_2^2, x_3^2)$ since this is the only quadratic monomial ideal with correct Hilbert series. But if $x_1 > x_2 > x_3$, say, and $x_3^2 \in \mathrm{in}(I)$ then $x_3^2 \in I$. But it is easily seen that I does not contain any square. A concrete example is $I = (x_1^2 + x_1 x_2, x_2^2 + x_2 x_3, x_3^2 + x_1 x_3)$, cf. [ERT].

Given the embedding dimension n of A it is natural to ask for a bound $N(n)$ such that if $\mathrm{Tor}_i^A(k, k) = (\mathrm{Tor}_i^A(k, k))_i$ for $i \leq N(n)$ then A is Koszul. That such a bound does not exist for noncommutative algebras follows from a result in [FLo]. As an example, if $A = k\langle x_1, x_2, x_3, x_4 \rangle/(x_1 x_2 - x_1 x_3, x_2 x_3 - x_3 x_2 - \lambda x_3^2, x_2 x_4)$, $\mathrm{char}(k) = 0$, $\lambda^{-1} = l \in \mathbb{N}$, then $\mathrm{Tor}_i^B(k, k) = (\mathrm{Tor}_i^B(k, k))_i$ if $i \leq l+2$ and $\mathrm{Tor}_{l+3}^B(k, k) \neq (\mathrm{Tor}_{l+3}^B(k, k))_{l+3}$, where $B = A^!$. More surprising is perhaps that not even for commutative algebras such a bound exist. An example is given in [Ro2], $k[x_1, \ldots, x_6]/I$, where $I = (x_1^2, x_1 x_2, x_2 x_3, x_3^2, x_3 x_4, x_4^2, x_4 x_5, x_5 x_6, x_6^2,$

$x_1x_3 + \lambda x_3x_6 - x_4x_6, x_3x_6 + x_1x_4 + (\lambda - 2)x_4x_6), \lambda \in \mathbb{N}$. Over this algebra k has a linear resolution up to and including degree λ, but not in degree $\lambda + 1$.

It has been conjectured that the equality $A(z).A^!(-z) = 1$ should imply that A is Koszul. Counterexamples to this is independently given in [Po] (noncommutative algebras) and [Ro3] (commutative algebras). One example in [Ro3] is $k[x_1, \ldots, x_5]/(x_1^2, x_1x_2 + x_3^2, x_3^2 + x_4x_5, x_1x_3, x_2x_3 + x_1x_4, x_3x_4 + x_2x_5, x_3x_5, x_5^2)$.

7 Local Rings

If (A, m) is a local ring and the associated graded algebra $g(A) = \oplus_{i \geq 0} m^i/m^{i+1}$ is a Koszul algebra, then A is called generalized Koszul. For a generalized Koszul algebra it is true that $g(A)(z)P_A(-z) = 1$, cf. [F3] or [HRW]. Concrete examples of generalized Koszul algebras are local CM rings of maximal embedding dimension. Another example is local Gorenstein rings (A, m) of maximal embedding dimension $e(A) + \dim A - 2$, cf. [Sc] and [F2].

8 Semigroup rings

Let S be a subsemigroup of $\mathbb{N}^d$ and let $k[S] = k[\mathbf{x}]/I(S)$ be its semigroup ring. In [HRW] a topological condition for the initial ideal $\text{in}(I(S))$ to be quadratic is given.

In [PRS] it is shown that all normal subsemigroups of $\mathbb{N}^2$ give quadratic initial ideals (in some ordering) and thus are (generalized) Koszul algebras. This result is also proved in [HRW] in another way.

It has been asked if all monomial projective curves with quadratic defining ideal were Koszul, and even if they had a quadratic Gröbner basis. Sturmfels has recently shown that for projective monomial curves in $P^n, n \leq 4$ this is true, and for $n = 5$ there is only one exception (and its symmetric), namely $k[s^{11}, s^8t^3, s^6t^5, s^5t^6, s^4t^7, t^{11}]$, which is not Koszul, cf. [St2]. All Koszul algebras have quadratic Gröbner bases for $n = 5$.

In [RS] projective monomial curves in $P^n, n > 5$ are studied. It is shown that all such Koszul algebras have quadratic Gröbner bases if $n = 6$ and that there is a counterexample to this if $n = 7$. This counterexample is $k[t^{22}, t^{19}s^3, t^{18}s^4, t^{17}s^5, t^{16}s^6, t^{15}s^7, t^{11}s^{11}, s^{22}] \simeq k[x_1, \ldots, x_8]/(x_2^2 - x_1x_5, x_2x_3 - x_1x_6, x_3^2 - x_2x_4, x_3x_4 - x_2x_5, x_2x_6 - x_3x_5, x_4^2 - x_3x_5, x_1x_7 - x_4x_5, x_3x_6 - x_4x_5, x_4x_6 - x_5^2, x_2x_7 - x_6^2, x_1x_8 - x_7^2)$, cf. [RS]. There is also an example of a monomial curve in P^6 for which the Ext-algebra of its coordinate ring is not finitely generated, and an example in P^8 with non-rational Poincaré series.

9 Generalizations

The concept of Koszul algebras has been generalized (starting with [Ma]) to other tensor categories than k-algebras. A good general reference is [BGSo].

We will not discuss this subject, but only give some references: [Ar], [BGSc], [BG], [Be], [GK], [GM], [H], [M-V], [Pl], [PV], [PS], and [Rs].

References

[An1]   D. Anick, *A counterexample to a conjecture of Serre*, Ann. Math. **115** (1982), 1–33.

[An2]   D. Anick, *Non-commutative graded algebras and their Hilbert series*, J. Algebra **78** (1982), 120–140.

[An3]   D. Anick, *On the homology of associative algebras*, Trans. Amer. Math. Soc. **296** (1986), 641–659.

[Ar]   S. M. Arkhipov, *Koszul duality over a complex*, Func. Anal. Appl. **28** (1994), 202–204.

[B1]   J. Backelin, *A distributiveness property of augmented algebras and some related homological results*, Ph D thesis, Stockholm University, 1982.

[B2]   J. Backelin, *La série de Poincaré-Betti d'une algèbre graduée de type fini à une relation est rationelle*, C. R. Acad. Sci. Paris Sér. A **287** (1978), 843–846.

[B3]   J. Backelin, *On the rate of growth of the homologies of Veronese subrings*, in Algebra, Algebraic Topology and their Interactions, J.-E. Roos ed., Lect. Notes Math., **1183**, pp. 79–100, Springer-Verlag, Berlin/New York, 1986.

[BF1]   J. Backelin and R. Fröberg, *Veronese subrings, Koszul algebras and rings with linear resolutions*, Rev. Roum. Pures Appl. **30** (1985), 85–97.

[BF2]   J. Backelin and R. Fröberg, *Poincaré series of short Artinian rings*, J. Algebra **96** (1985), 495–498.

[BM]   S. Barcanescu and N. Manolache, *Betti numbers of Segre-Veronese singularities*, Rev. Roum. Math. Pures Appl. **26** (1981), 549–565.

[BGSc]   A. A. Beilinson, V. A. Ginsburg and V. V. Schechtman, *Koszul duality*, J. Geom. Phys. **5** (1988), 317–350.

[BGSo]   A. Beilinson, V. Ginsburg and W. Soergel, *Koszul duality patterns in representation theory*, J. Amer. Math. Soc. **9** (1996), 473–527.

[Be]   R. Bezrukavnikov, *Koszul DG-algebras arising from configuration spaces*, Geom. Func. Anal. **4** (1994), 119–135.

[BG]   A. Braverman and D. Gaitsgory, *Poincaré-Birkhoff-Witt theorem for quadratic algebras of Koszul type*, J. Algebra **181** (1996), 315–328.

[BGT]   W. Bruns, J. Gubeladze and N. V. Trung, *Normal polytopes, triangulations and Koszul algebras*, J. Reine Ang. Math **485** (1997), 123–160.

[BHV]   W. Bruns, J. Herzog and U. Vetter, *Syzygies and walks*, in Commutative Algebra, A. Simis, N. V. Trung, and G. Valla eds., pp. 36–57, World Scientific, 1994.

[C]   K. Crona, *A new class of Koszul algebras*. C. R. Acad. Sci. Paris Sér. I **323** (1996), 705–710.

[ERT]   D. Eisenbud, A. Reeves and B. Totaro, *Initial ideals, Veronese subrings, and rates of algebras*, Adv. Math. **109** (1994), 168–187.

[F1]   R. Fröberg, *Determination of a class of Poincaré series*, Math. Scand. **37** (1975), 29–39.

[F2]   R. Fröberg, *A study of graded extremal rings and of monomial rings*, Math. Scand. **51** (1982), 22–34.

[F3]   R. Fröberg, *Connections between a local ring and its associated graded ring*, J. Algebra **111** (1987), 300–305.

[FLa]   R. Fröberg and D. Laksov, *Compressed algebras*, in Proc. Conf. on Complete intersections, S. Greco and R. Strano eds., Acireale 1983, pp. 121–151, Lect. Notes in Math. **1092**, Springer-Verlag, Berlin/New York, 1984.

[FLo]   R. Fröberg and C. Löfwall, *On Hilbert series for commutative and noncommutative graded algebras*, J. Pure Appl. Algebra **76** (1991), 33–38.

[Go]   V. E. Govorov, *Dimension and multiplicity of graded algebras* (Russian), Sibirsk. Mat. J. **14** (1973), 1200–1206.

[GK]   V. Ginzburg and M. Kapranov, *Koszul duality for operads*, Duke Math. J. **76** (1994), 203–272.

[GM]   E. Green and R. Martinez-Villa, *Koszul and Yoneda algebras*, in Representation theory of algebras, Cocoyoc, 1994, pp. 247–297, CMS Conf. Proc **18**, Amer. Math. Soc., Providence, RI, 1996.

[H]   P. H. Hai, *Koszul property and Poincaré series of matrix bialgebras of type A_n*, J. Algebra **192** (1997), 734–748.

[HRW]   J. Herzog, V. Reiner and V. Welker, *The Koszul property in affine semigroup rings*, preprint 1997.

[Ke1]   G. Kempf, *Some wonderful rings in algebraic geometry*, J. Algebra
        134 (1990), 222–224.

[Ke2]   G. Kempf, *Projective coordinate ring of Abelian varieties*, in Algebraic
        Analysis, Geometry and Number Theory, J. I. Igusa ed., 225–236, John
        Hopkins Press, Baltimore, 1989.

[Ko]   Y. Kobayashi, *The Hilbert series of some graded algebras and the Po-
        incaré series of some local rings*, Math. Scand. **42** (1978), 19–33.

[L]   C. Löfwall, *On the subalgebra generated by one-dimensional elements
        in the Yoneda Ext-algebra*, in Algebra, Algebraic Topology and their
        Interactions, J.-E. Roos ed., Lect. Notes Math., **1183**, pp. 291–338,
        Springer-Verlag, Berlin/New York, 1986.

[L2]   C. Löfwall, personal communication.

[L3]   C. Löfwall, *Central elements and deformations of local rings*, J. Pure
        Appl. Algebra **91** (1994), 183–192.

[Ma]   Y. I. Manin, *Some remarks on Koszul algebras and quantum groups*,
        Ann. Inst. Fourier **37** (1987), 191–205.

[M-V]   R. Martinez-Villa, *Applications of Koszul algebras: the preprojective
        algebra*, in Representation theory of algebras, Cocoyoc, 1994, pp. 487–
        504, CMS Conf. Proc **18**, Amer. Math. Soc., Providence, RI, 1996.

[Mu]   D. Mumford, *Varieties defined by quadratic equations*, in Questions on
        Algebraic Varieties, E. Marchionna ed., C.I.M.E. III Ciclo, Varenna
        1969, pp. 29–100, Roma, 1970.

[Pa]   G. Pareschi, *Koszul algebras associated to adjunction bundles*, J. Al-
        gebra **157** (1993), 161–169.

[PP]   G. Pareschi and B. P. Purnaprajna, *Canonical ring of a curve is
        Koszul; a simple proof*, Illinois J. Math. **41** (1997), 266–271.

[PRS]   I. Peeva, V. Reiner and B. Sturmfels, *How to shell a monoid*, preprint
        1996.

[PS]   B. Parshall and L. Scott, *Koszul algebras and the Frobenius automorp-
        hism*, Quart. J. Math. Soc. Oxford Ser. 2 **46** (1995), 345–384.

[Pol]   A. Polischuk, *On the Koszul property of the homogeneous coordinate
        ring of a curve*, J. Algebra **178** (1995), 122–135.

[Pl]   P. Polo, *On Cohen-Macaulay posets, Koszul algebras and certain mo-
        dules associated to Schubert varieties*, Bull. London Math. Soc. **27**
        (1995), 425–434.

[Po]   L. Positselski, *The correspondence between Hilbert series of quadratically dual algebras does not imply their having the Koszul property*, Func. Anal. Appl. **29** (1995), 213–217.

[PV]   L. Positselski and A. Vishik, *Koszul duality and Galois cohomology*, Math. Res. Lett. **2** (1995), 771–781.

[Pr]   S. Priddy, *Koszul resolutions*, Trans. Amer. math. Soc. **152** (1970), 39–60.

[Ra]   Ravi, *Coordinate rings of G/P are Koszul*, J. Algebra **177** (1995), 367–371.

[Ro1]   J.-E. Roos *Relations between the Poincaré-Betti series of loop spaces and of local rings*, Lect. Notes Math., **740**, pp. 285–322, Springer-Verlag, Berlin/New York, 1979.

[Ro2]   J.-E. Roos *Commutative non-Koszul algebras having a linear resolution of arbitrary high order. Applications to torsion in loop space homology*, C. R. Acad. Sci. Paris Sér. I **316** (1993), 1123–1128.

[Ro3]   J.-E. Roos *On the characterisation of Koszul algebras. Four counterexamples*, C. R. Acad. Sci. Paris Sér. I **321** (1995), 15–20.

[Ro4]   J.-E. Roos, *On computer-assisted research in homological algebra*, Math. Comp. Sim. **42** (1996), 475–490.

[Ro5]   J.-E. Roos, *A computer-aided study of the graded Lie-algebra of a local commutative noetherian ring*, J. Pure Appl. Algebra **91** (1994), 255–315.

[Ro6]   J.-E. Roos, *A description of the homological behaviour of families of quadratic forms in four variables*, in Syzygies and Geometry, Boston 1995, A. Iarrobino, A. Martsinkovsky and J. Weyman eds., pp. 86–95, Northeastern univ. 1995.

[Ro7]   J.-E. Roos, *Koszul algebras and non Koszul algebras*, in Syzygies and Geometry, Boston 1995, A. Iarrobino, A. Martsinkovsky and J. Weyman eds., pp. 96–99, Northeastern univ. 1995.

[RS]   J.-E. Roos and B. Sturmfels, work in progress.

[Rs]   M. Rosso, *Koszul resolutions and quantum groups*, Nuclear Phys. B Proc. Suppl. **18B** (1990), 269–276.

[Sc]   P. Schenzel, *Über die freien Auflösungen extremaler Cohen-Macaulay Rings*, J. Algebra **64** (1980), 93–101.

[St1]   B. Sturmfels, *Gröbner bases and convex polytopes*, Univ. Lect. Ser. **8**, Amer. Math. Soc., 1996.

[St2]   B. Sturmfels, personal communication.

[T]   J. Tate, *Homology of local and Noetherian rings*, Illinois J. Math. **1**
          (1957), 14–27.

[VF]   A. Vishnik and M. Finkelberg, *The coordinate ring of a general curve
          of genus ≥ 5 is Koszul*, J. Algebra **162** (1993), 535–539.

Primary Decomposition of Ideals

GUANG FU Department of Mathematics, The Florida State University, Tallahassee, Florida 32306

1. INTRODUCTION

Let R be a commutative ring and I an ideal of R. We say that I has **a finite primary decomposition** or I is **decomposable** if I can be represented as a finite intersection $\cap_{i=1}^{n} Q_i$ of primary ideals of R, where Q_i is P_i-primary. If for every i, $\cap\{Q_j \mid 1 \leq j \leq n, \ j \neq i\} \not\subseteq Q_i$ and the ideals P_i are distinct, we say the primary decomposition $\cap_{i=1}^{n} Q_i$ is **reduced**. If every ideal of R is decomposable, we say that R is a **Laskerian** ring. If S is a multiplicative system in a ring R and I an ideal of R, we denote by $S(I)$ the saturation of I with respect to S — that is, the contraction of the extension of I with respect to the quotient ring R_S. If P is a prime ideal of R, we use $S_P(I)$ to denote the saturation of I with respect to the multiplicative system $R - P$. In 1929 Krull proved in [1] that a ring R is Laskerian if and only if the following two conditions hold (The notation (L1) and (L2) is taken from [2, pp. 56–57]):

(L1) For every ideal I of R and for every prime ideal P of R, $S_P(I) = I : x$ for some $x \in R - P$.

(L2) For every chain $S_1 \supseteq S_2 \supseteq ... \supseteq S_n \supseteq ...$ of multiplicative systems in R, the chain $S_1(I) \supseteq S_2(I) \supseteq ... \supseteq S_n(I) \supseteq ...$ stablizes for every ideal I of R.

In the same paper Krull asked the following question: what are necessary and sufficient conditions for a fixed ideal of a commutative ring to be decomposable? This question has remained open to the present. A natural question in this regard is: Suppose that a fixed ideal I of R satisfies the following two conditions:

(LL1) For every prime ideal P of R, $S_P(I) = I : x$ for some $x \in R - P$

(LL2) For every chain $S_1 \supseteq S_2 \supseteq ... \supseteq S_n \supseteq ...$ of multiplicative systems in R, the chain $S_1(I) \supseteq S_2(I) \supseteq ... \supseteq S_n(I) \supseteq ...$ stablizes.

Is I decomposable? We answer this question negatively in Example 3.9 in Section 3. The purpose of this paper is to answer Krull's question where we seek conditions that

351

are stated in terms of the given ideal. In Theorem 3.6 we provide such conditions. In Section 2 we list some known results and prove some basic results about primary decomposition of ideals. Among other results in this section, we prove (Proposition 2.9.) that an ideal I of an arithmetical ring R is decomposable if and only if $|Ass(I)| < \infty$ and each member of $Ass(I)$ is isolated. Therefore every decomposable ideal in an arithmetical ring has a unique reduced primary decomposition. Section 3 is mainly for Theorem 3.6 and some applications. In Section 4 we show (Theorem 4.2) that for a finite dimensional ring R, every finitely generated ideal of R is decomposable if and only if (L1) and (L2) hold for every finitely generated ideal I of R.

We list here some conventions and definitions that will be used throughout this paper. All rings are assumed to be commutative with identity. For definitions and results not given in this paper, the reader is referred to [3]. While there are several different definitions of the associated primes of an ideal (e.g., see [4]), we consistently use the term in the **weak-Bourbaki** sense, whereby P is an associated prime of an ideal I in a ring R if P is a minimal prime ideal of an ideal of the form $I : x$ for some $x \in R$ (a **Bourbaki prime** of I is a prime ideal of the form $I : x$, where $x \in R$). An associated prime P of I is said to be **isolated** if P is a minimal prime ideal of I; a nonminimal associated prime of I is said to be **embedded**. We denote by $Ass(I)$ the set of the associated primes of the ideal I and by $MAss(I)$ the set of the maximal members of $Ass(I)$. If A is a set, we denote by $|A|$ the cardinality of A. If A and B are two sets, we use $A < B$ to mean that A is properly contained in B. If $\mathcal{F}$ is a family of subsets of a ring, we use $\cup \mathcal{F}$ to denote the union of all the members in $\mathcal{F}$. If $A \in \mathcal{F}$ and if $A_n < ... < A_2 < A_1 < A$ is a chain in $\mathcal{F}$ of maximal length strictly descending from A, then we define the **height of A in $\mathcal{F}$**, denoted $ht(A)$, to be n. The height of a minimal member of $\mathcal{F}$ is defined to be zero. The **dimension of $\mathcal{F}$** is denoted $dim \mathcal{F}$ and is defined by $dim \mathcal{F} = \text{Sup} \{ ht(A) \mid A \in \mathcal{F} \}$.

2. BASIC RESULTS

In this section we prove Lemma 2.1, which will be used frequently in the remainder of this paper. We will also give some applications of this lemma and discuss some decomposability relationships among ideals. Recall that if I is a decomposable ideal of a ring T, then $I : J$ is decomposable for each ideal J of T. Moreover, if $f : R \to T$ is a ring homomorphism, then the decomposability of I implies that of $f^{-1}(I)$ if either f is an injection or if f is a surjection. We begin by recording three known results.

LEMMA A. ([6, Ex.17, pp.289])   *Suppose that I is an ideal of a ring R and S is a multiplicative system of R. Then I is contracted from R_S if and only if $S \cap (\cup Ass(I)) = \phi$.*

LEMMA B. ([4, Proposition 1.2])   *Let R' be a quotient ring of R (with respect to* *a multiplicative system), let A be an ideal of R, and let $A' = AR'$, and let P be a prime ideal of R such that $P' = PR' \neq R'$. Then P is an associated prime of A if and only if P' is an associated prime of A'.*

LEMMA C. ([4, Lemma 3.2]) *Let I be an ideal of a ring R. If $\{\sqrt{I : x} \mid x \in R\}$, has a.c.c., then every associated prime of I is of the form $\sqrt{I : x}$.*

Here is our first result. The proof of LEMMA 2.1 is a modification of the proof of [5, Proposition 2.7].

LEMMA 2.1. *Let I be an ideal of a ring R. Then I is decomposable if and only if the following conditions hold:*

(1) $MAss(I)$ has only finitely many members;

(2) For each $P \in MAss(I)$, IR_P is decomposable.

PROOF. If I is decomposable, it is well known that conditions (1) and (2) hold. Conversely, suppose that $MAss(I) = \{M_1, M_2, ..., M_n\}$ and that IR_{M_i} is decomposable for $1 \leq i \leq n$. Let $S_i = R - M_i$ for $1 \leq i \leq n$ and denote by $S_i(I)$ the contraction of the extension of I with respect to R_{S_i}. We will show that $I = \cap_{i=1}^n S_i(I)$ and hence I is decomposable. For this, it suffices to show that $I \supseteq \cap_{i=1}^n S_i(I)$. If $x \in R - I$, then $I : x \subseteq \cup Ass(I)$, so that $I : x \subseteq M_i$ for some i by condition (1). Thus $x \notin S_i(I)$ by LEMMA A, and hence $x \notin \cap_{i=1}^n S_i(I)$. This completes the proof. □

The following corollaries 2.2 and 2.3 are immediate from Lemma 2.1.

COROLLARY 2.2. *Let I be an ideal of a ring R. If $|Ass(I)| < \infty$ and each member of $Ass(I)$ is isolated, then I is decomposable.*

COROLLARY 2.3. *A ring R is Laskerian if and only if the following conditions hold:*

(1) Every ideal of R has only finitely many maximal associated primes.

(2) R_M is Laskerian for each maximal ideal M of R.

COROLLARY 2.4. *Suppose that R is a ring and every nonzero element in R is contained in only finitely many maximal ideals. If R is locally Laskerian, then R is Laskerian.*

PROOF. If R is not a domain, there exist nonzero elements x and y in R such that $xy = 0$. Since every nonzero element in R is contained in only finitely many maximal ideals, this implies that R has only finitely many maximal ideals. Since the localization of R at each maximal ideal is Laskerian, by LEMMA B every ideal of R has only finitely many maximal associated primes. It follows from COROLLARY 2.3 that R is Laskerian. If R is a domain, since R is locally Laskerian, it suffices, by COROLLARY 2.3, to show that every proper ideal of R has only finitely many associated primes. Take a proper ideal I of R. Then I is contained in only finitely many maximal ideals of R. Thus, I must have only finitely many associated primes; otherwise, one of the maximal ideals that contain I, say M, would contain infinitely many associated primes of I, so that by LEMMA B IR_M has infinitely many associated primes in R_M, contradicting the fact that R_M is Laskerian.   □

PROPOSITION 2.5.   *Let R be a ring and J be a finite intersection of maximal ideals of R.*

   (1) If I is an ideal of R, then $I = (I \cap J) : x$ for some $x \in R$.

   (2) If every ideal of R contained in J is decomposable, then R is Laskerian.

   (3) If every ideal of R contained in J has only finitely many associated primes (resp., which are finitely generated), then every ideal of R has only finitely many associated primes (resp., which are finitely generated)

PROOF. (1) If $I \subseteq J = \cap_{i=1}^n M_i$, where each M_i is a maximal ideal of R, the proof is trivial. Hence we may assume that $I \not\subseteq M_1$. Then $I + M_1 = R$, and we may write $1 = a + m$ for some $a \in I$ and $m \in M_1$. Thus $(I \cap M_1) : m = (I : m) \cap (M_1 : m) = I : m = I$; the last equality is true because, if $y \in I : m$, then $ym \in I$, so that $y = ya + ym \in I$. If $n = 1$, we are done. Otherwise, by induction there is an element $r \in R$ with $I \cap M_1 = ((I \cap M_1) \cap (\cap_{i=2}^n M_i)) : r = (I \cap J) : r$. Since $I = (I \cap M_1) : m$, it follows that $I = (I \cap J) : mr$. This completes the proof of part (1).

   (2) and (3) follow immediately from (1).   □

PROPOSITION 2.6.   *Let I be an ideal of a ring R. If $I \cap J$ is decomposable for some ideal $J \not\subseteq \cup Ass(I)$, then I is decomposable.*

PROOF.   Take $j \in J - \cup Ass(I)$. We will show that $I = (I \cap J) : j$. Let $S = R - \cup Ass(I)$. Then $I = S(I)$ by LEMMA A, where $S(I)$ is the contraction of the extension of I with respect to R_S. Since $(I \cap J) : j = (I : j) \cap (J : j) = I : j$ and since $j \in S$ implies that $I : j \subseteq S(I) = I \subseteq I : j$, we have $I = (I \cap J) : j$. Thus, if $I \cap J$ is decomposable, so is I.   □

   The next corollary is immediate from Proposition 2.6.

COROLLARY 2.7.   *Let J be an ideal of a ring R. Suppose that every ideal of R contained in J is decomposable. Then every ideal I of R such that $J \not\subseteq \cup Ass(I)$ is decomposable.*

COROLLARY 2.8.   *Let J be a finitely generated ideal of a coherent ring R. Suppose that every finitely generated ideal of R contained in J is decomposable. Then every finitely generated ideal I of R such that $J \not\subseteq \cup Ass(I)$ is decomposable.*

PROOF.   Since R is coherent and I and J are finitely generated, $I \cap J$ is finitely generated. Hence $I \cap J$ is decomposable by hypothesis, and Proposition 2.6 shows that I is decomposable.   □

   A **chained** ring is one in which any two ideals are comparable under the inclusion relation. A ring R is **arithmetical** if R_M is a chained ring for every maximal ideal

M of *R*.

PROPOSITION 2.9. *Let R be an arithmetical ring. An ideal I of R is decomposable if and only if* $|Ass(I)| < \infty$ *and each member of Ass(I) is isolated. Therefore every decomposable ideal in an arithmetical ring has a unique reduced primary decomposition.*

PROOF. ($\Leftarrow$) This is clear from COROLLARY 2.2.

($\Rightarrow$) Suppose that *I* is decomposable. Then $|Ass(I)| < \infty$. If $P \in Ass(I)$, to see that *P* is isolated we consider the localization R_P. Since *I* is decomposable, IR_P is decomposable in R_P. Because *R* is arithmetical, R_P is a chained ring, hence every decomposable ideal of R_P is a primary ideal. In particular, IR_P is primary. By LEMMA B, PR_P is an associated prime of IR_P, so that PR_P is the unique associated prime of IR_P. It follows from LEMMA B that *P* is a minimal prime ideal of *I*. Therefore, *P* is an isolated prime of *I*. The last part of PROPOSITION 2.9 follows immediately from ([7, Theorem 8, pp. 211]). □

3. MAIN RESULTS

In this section we determine necessary and sufficient conditions, in terms of a given ideal *I* and of ideals that can be defined in terms of *I*, for the ideal *I* to be decomposable. Since an ideal *I* in a ring *R* is decomposable if and only if the zero ideal (0) is decomposable in the ring *R/I*, many results in this section deal with the decomposability of the zero ideal.

DEFINITION 3.1. *Let* (R, M) *be a quasilocal ring and let* $\{M, P_1, P_2, ..., P_n\}$ *be the set of associated primes of the zero ideal. We say that R satisfies the decomposition condition, abbreviated DC, if there exists an M-primary ideal Q with the following property: if* $x \in Q$ *and* $y \in M - \cup_{i=1}^{n} P_i$ *are such that* $xy = 0$, *then* $x = 0$.

THEOREM 3.2. *Let* (R, M) *be a quasilocal ring. Let* $Ass(0) = \{M, P_1, P_2, ..., P_n\}$ *and let* $S = R - \cup_{i=1}^{n} P_i$. *Then* (0) *is decomposable if and only if* $S(0)$ *is decomposable and R satisfies DC.*

PROOF. ($\Leftarrow$) Since *R* satisfies DC, there exists an *M*-primary ideal *Q* such that if $x \in Q$, $y \in M - \cup_{i=1}^{n} P_i$ and $xy = 0$, then $x = 0$. This means that $Q \cap S(0) = (0)$. Hence (0) is decomposable.

($\Rightarrow$) Suppose that (0) is decomposable. It is well known that $S(0)$ is also decomposable. Let $(0) = \cap_{i=0}^{n} Q_i$ be a reduced primary decomposition of (0), where Q_0 is *M*-primary and each Q_i $(i \geq 1)$ is P_i-primary. If $x \in Q_0$ and $y \in M - \cup_{i=1}^{n} P_i$ are such that $xy = 0$, then $x \in S(0) = \cap_{i=1}^{n} Q_i$, so that $x \in \cap_{i=0}^{n} Q_i = (0)$. Therefore, *R* satisfies DC. □

LEMMA 3.3. *Let I be an ideal of a ring R and S a multiplicative system of R. Then* $Ass(S(I)) = \{P \in Ass(I) \mid P \cap S = \phi\}$.

PROOF. If $P \in Ass(I)$ and $P \cap S = \phi$, then $PR_S \in Ass(IR_S)$, and hence $P \in Ass(S(I))$ by LEMMA B.

Conversely, if $P \in Ass(S(I))$, then, since $S(I)$ is contracted from R_S, $P \cap S = \phi$ by LEMMA A. Applying LEMMA B, we have $PR_S \in Ass(S(I)R_S) = Ass(IR_S)$. Therefore, $P \in Ass(I)$ by LEMMA B.   □

THEOREM 3.4. *The zero ideal* (0) *of a ring R is decomposable if and only if the following conditions hold:*

(1) $|Ass(0)| < \infty$;

(2) For each nonminimal member P of $Ass(0)$, R_P *satisfies DC.*

PROOF. ($\Rightarrow$) Suppose that (0) is decomposable in R. Then $|Ass(0)| < \infty$ and by LEMMA 2.1, (0) is decomposable in R_P for each $P \in Ass(0)$. It follows from THEOREM 3.2 that R_P satisfies DC for each nonminimal member of $Ass(0)$.

($\Leftarrow$) We will use induction on $n = dimAss(0)$, the dimension of the set $Ass(0)$. If $n = 0$, then (0) is decomposable by COROLLARY 2.2. Suppose that the result is true for all rings with conditions (1) and (2) and such that $dimAss(0) \leq n$. We consider the case where $dimAss(0) = n + 1$. By LEMMA 2.1 it suffices to prove that for every member M of $MAss(0)$, the zero ideal of R_M is decomposable. By passage to the localization we may assume without loss of generality that R is a quasi-local ring with maximal ideal M, $Ass(0) = \{M, P_1, P_2, ..., P_k\}$, $dimAss(0) = n + 1$, and R_P satisfies DC for each nonminimal member P of $Ass(0)$. Thus, by THEOREM 3.2, we only need to prove that $S(0)$ is decomposable, where $S = R - \cup_{i=1}^k P_i$. Since $|Ass(0)| < \infty$, $M \neq \cup_{i=1}^n P_i$. We then know from LEMMA 3.3 that $Ass(S(0)) = \{P_1, P_2, ..., P_k\}$. To show that $S(0)$ is decomposable, it suffices, by LEMMA 2.1, to show that for every $Q \in MAss(S(0))$, $S(0)R_Q$ is decomposable. Since $S \subseteq R - Q$, $S(0)R_Q = (0)$. Moreover, for each $P \in Ass(S(0)R_Q)$, $(R_Q)_P$ is isomorphic to R_P, and hence $(R_Q)_P$ satisfies DC for each nonminimal member P of $Ass(S(0)R_Q)$. Moreover, $dimAss(S(0)R_Q) \leq dimAss(S(0)) = dimAss(0) - 1 = n$. By the induction hypothesis, $S(0)R_Q$ is decomposable. This completes the proof. □

PROPOSITION 3.5. *Let* (R, M) *be a quasilocal ring and let* $\{M, P_1, P_2, ..., P_n\}$ *be the set of associated primes of the zero ideal. Then R satisfies DC if and only if there exists an M-primary ideal Q such that for every* $x \in Q - \{0\}$, $\sqrt{(0) : x} < M$.

PROOF. ($\Rightarrow$) If R satisfies DC, then there exists an M-primary ideal Q such that for every $x \in Q - \{0\}$ and every $y \in M - \cup_{i=1}^n P_i$, $xy \neq 0$. This shows that $(0) : x \subseteq \cup_{i=1}^n P_i$, and hence $(0) : x \subseteq P_j < M$ for some j. Therefore, $\sqrt{(0) : x} \subseteq P_j < M$.

($\Leftarrow$) To prove this direction we prove the contrapositive of the statement. Suppose that R doesn't satisfy DC. Then for some $x \in Q - \{0\}$ and some $y \in M - \cup_{i=1}^n P_i$, we have $xy = 0$, — that is, $(0) : x \not\subseteq \cup_{i=1}^n P_i$. Since $Ass(0) = \{M, P_1, P_2, ..., P_n\}$, M is the unique minimal prime of $(0) : x$. This implies that $\sqrt{(0) : x} = M$, completing the proof.   □

THEOREM 3.6. *Let I be an ideal of a ring R. Then I is decomposable if and only if the following conditions hold:*

(1) The set $\mathcal{F} = \{\sqrt{I : x} \mid x \in R\}$ is finite;

(2) For each nonminimal member P of $Ass(I)$, there exists a P-primary ideal Q such that for every $x \in Q - ker(\psi_P \varphi)$, $\sqrt{I : x} \subseteq P_x$ for some $P_x \in Ass(I)$ with $P_x < P$, where $\varphi : R \to R/I$ and $\psi_P : R/I \to (R/I)_{(P/I)}$ are canonical homomorphisms.

PROOF. For simplicity, we denote ψ_P by ψ.

($\Rightarrow$) Since I is decomposable, $|Ass(I)| < \infty$; hence $\mathcal{F}$ is finite. To show that (2) holds, we consider the ring $\varphi(R)$ and its localization. In $\varphi(R)$, $Ass(0) = \{\varphi(P) \mid P \in Ass(I)\}$. For each $\varphi(P) \in Ass(0)$, since I is decomposable in R, (0) is decomposable in $\varphi(R)$, and hence (0) is decomposable in $\varphi(R)_{\varphi(P)}$. If P is not a minimal member of $Ass(I)$, $\varphi(P)\varphi(R)_{\varphi(P)}$ is not a minimal member of $Ass(0)$ in $\varphi(R)_{\varphi(P)}$. By THEOREM 3.4, $\varphi(R)_{\varphi(P)}$ satisfies DC. It follows from PROPOSITION 3.5 that there exists a $\varphi(P)\varphi(R)_{\varphi(P)}$-primary ideal Q^* such that for every $z \in Q^* - \{0\}$, $\sqrt{(0) : z} < \varphi(P)\varphi(R)_{\varphi(P)}$. Let Q be the P-primary ideal such that $\varphi(Q)\varphi(R)_{\varphi(P)} = Q^*$. If $x \in Q - ker(\psi\varphi)$, then $\psi\varphi(x) \in Q^* - \{0\}$, so that $\sqrt{(0) : \psi\varphi(x)} < \varphi(P)\varphi(R)_{\varphi(P)}$. This means that $\varphi(P)\varphi(R)_{\varphi(P)}$ is not a minimal prime of $(0) : \psi\varphi(x)$. Thus, there exists $P_x \in Ass(I)$ with $P_x < P$ such that $\varphi(P_x)\varphi(R)_{\varphi(P)}$ is a minimal prime of $(0) : \psi\varphi(x)$, and hence $\sqrt{(0) : \psi\varphi(x)} \subseteq \varphi(P_x)\varphi(R)_{\varphi(P)}$. Now, $\sqrt{(0) : \psi\varphi(x)} = \sqrt{\psi((0) : \varphi(x))} = \psi(\sqrt{(0) : \varphi(x)}) \subseteq \varphi(P_x)\varphi(R)_{\varphi(P)}$, which implies $\sqrt{(0) : \varphi(x)} \subseteq \varphi(P_x)$. It is easy to show that $\varphi(\sqrt{I : x}) \subseteq \sqrt{(0) : \varphi(x)} \subseteq \varphi(P_x)$. Therefore $\sqrt{I : x} \subseteq P_x$.

($\Leftarrow$) Since $\mathcal{F}$ is finite, it follows from LEMMA C that every associated prime of I is of the form $\sqrt{I : x}$, where $x \in R$. Hence $|Ass(I)| \leq |\mathcal{F}| < \infty$. To show that I is decomposable, it suffices to show that (0) is decomposable in $\varphi(R)$, and for this it suffices, by THEOREM 3.4, to show that if $P \in Ass(I)$ and if P is not minimal over I, then $\varphi(R)_{\varphi(P)}$ satisfies DC. By condition (2), there exists a P-primary ideal Q which satisfies the condition described in (2). In order to assume without loss of generality that Q contains I, we want to show that $S_P(Q + I)$ is another P-primary ideal that satisfies the condition described in (2), where $S_P(Q + I)$ is the contraction of the extension $(Q + I)R_P$ with respect to R_P. Since Q is P-primary, it is clear that $(Q + I)R_P$ is primary to the maximal ideal PR_P in R_P, so that $S_P(Q + I)$ is P-primary. If $y \in S_P(Q + I) - ker(\psi_P \varphi)$, then $sy \in Q + I$ for some $s \in R - P$. Since $y \notin ker(\psi_P \varphi)$, we have $sy \notin ker(\psi_P \varphi)$. Write $sy = z + i$, where $z \in Q - ker(\psi_P \varphi)$ and $i \in I$. Then $\sqrt{I : z} \subseteq P_z$ for some $P_z \in Ass(I)$ with

$P_z < P$. Obviously, $\sqrt{I:y} \subseteq \sqrt{I:sy} = \sqrt{I:(z+i)} = \sqrt{I:z} \subseteq P_z$. Hence we can assume from now on that Q contains I. Since $\varphi : R \to R/I$ and $\psi : R/I \to (R/I)_{(P/I)}$ are canonical homomorphisms, $\varphi(Q)\varphi(R)_{\varphi(P)}$ is $\varphi(P)\varphi(R)_{\varphi(P)}$-primary. If $z \in \varphi(Q)\varphi(R)_{\varphi(P)} - \{0\}$, take $x \in Q$ such that $\psi\varphi(x) = z$. Then $x \in Q - ker(\psi\varphi)$, so that $\sqrt{I:x} \subseteq P_x$ for some $P_x \in Ass(I)$ with $P_x < P$. It is easy to show that $(0) : \varphi(x) = \varphi(I:x)$, so that $I:x \subseteq P_x$ implies $(0) : \varphi(x) \subseteq \varphi(P_x)$. Since $\varphi(R)_{\varphi(P)}$ is a quotient ring of $\varphi(R)$, $(0) : \psi\varphi(x) = \psi((0) : \varphi(x)) \subseteq \psi\varphi(P_x) \subseteq \varphi(P_x)\varphi(R)_{\varphi(P)}$. Consequently, $\sqrt{(0) : \psi\varphi(x)} \subseteq \varphi(P_x)\varphi(R)_{\varphi(P)}$ and $\sqrt{(0) : z} < \varphi(P)\varphi(R)_{\varphi(P)}$. This shows that $\varphi(R)_{\varphi(P)}$ satisfies DC, by PROPOSITION 3.5.   □

Let D be a domain with quotient field K. An ideal I of D is said to be **divisorial** if $(I^{-1})^{-1} = I$, where $I^{-1} = \{x \in K \mid xI \subseteq D\}$. A domain D is **Mori** if every ascending chain of divisorial ideals of D is stable. A divisorial ideal I ($\neq R$) is said to be **d-irreducible** if I is not the intersection of two properly larger divisorial ideals.

LEMMA D. ([5, Theorem 2.1]) *Every divisorial ideal in a Mori domain has only finitely many Bourbaki associated primes.*

LEMMA E. ([8, Theorem 1.2]) *Let I be a divisorial ideal of a Mori domain. Then every weak-Bourbaki associated prime of I is a Bourbaki associated prime of I.*

LEMMA F. ([5, Proposition 2.4]) *Let I be a divisorial ideal of a Mori domain R. Then I is a finite intersection of d-irreducible ideals. Moreover, each d-irreducible ideal has the form $(a) : b$ for suitable chosen $a, b \in R$.*

COROLLARY 3.7. *Let D be a Mori domain. Then every divisorial ideal of D is decomposable if and only if for every $a \in D - \{0\}$ and for every nonminimal member P of $Ass(a)$, there exists a P-primary ideal Q such that if $x \in Q$ and if $a|x$ in D_M for all $M \in Ass(a)$ with $M < P$, then $a|x$ in D_P.*

PROOF. ($\Rightarrow$) Suppose that every divisorial ideal of D is decomposable. If $a \in D - \{0\}$, then (a) is decomposable. Take a nonminimal member P of $Ass(a)$. By THEOREM 3.6, there exists a P-primary ideal Q such that if $x \in Q$ and $y \in P - \cup \{M \in Ass(a) \mid M < P\}$ are such that $xy \in (a)$, then $x \in ker(\psi\varphi)$, where $\varphi : R \to R/(a)$ and $\psi : R/(a) \to (R/(a))_{(P/(a))}$ are canonical homomorphisms. Thus, if $x \in Q$ and if $a|x$ in D_M for all $M \in Ass(a)$ with $M < P$, then for each such M, there exists $m \in D_M$ such that $x = am$. Let $m = u/v$, where $u \in D$ and $v \in D - M$. Then $xv = au \in (a)$, which shows that $(a) : x \not\subseteq M$. Since D is a Mori domain, $|Ass(a)| < \infty$ by LEMMA D and LEMMA E. Hence $(a) : x \not\subseteq \cup \{M \in Ass(a) \mid M < P\}$, i.e., there exists $y \in P - \cup \{M \in Ass(a) \mid M < P\}$ such that $xy \in (a)$. Therefore, $x \in ker(\psi\varphi)$. This shows that $xz \in (a)$ for some $z \in D - P$, that is, $a|x$ in D_P.

  ($\Leftarrow$) By LEMMA F, to show that every divisorial ideal of D is decomposable it suffices to show that every principal ideal of D is decomposable. Take $a \in D - \{0\}$. Since D is a Mori domain, $|Ass(a)| < \infty$ by LEMMA D and LEMMA E. We want to show that condition (2) in THEOREM 3.6 holds for (a). If P is a nonminimal member of Ass(a), by hypothesis there exists a P-primary ideal Q such that if $x \in Q$ and if $a|x$ in D_{P_i} for each $P_i \in Ass(a)$ with $P_i < P$, then $a|x$ in D_P. Thus, if $y \in P - \cup \{P_i \in Ass(I) \mid P_i < P\}$ is such that $xy \in (a)$, then $a|x$ in D_{P_i} for each $P_i \in Ass(a)$ with $P_i < P$, so that $a|x$ in D_P. This means that $xv \in (a)$ for some $v \in D - P$, i.e., $x \in ker(\psi\varphi)$, where $\varphi : R \to R/(a)$ and $\psi : R/(a) \to (R/(a))_{(P/(a))}$ are canonical homomorphisms. Therefore, (a) is decomposable. $\square$

Recall (see [9, Theorem 2.3.2, p.45]) that a ring R is **coherent** if $I : x$ is finitely generated for any finitely generated ideal I of R and any $x \in R$; equivalently, R is coherent if $(0) : x$ and $I \cap J$ are finitely generated for any $x \in R$ and any finitely generated ideals I and J of R.

COROLLARY 3.8. *Let R be a coherent ring. Then (0) is decomposable if and only if the following conditions hold:*

(1) The set $\mathcal{F} = \{\sqrt{(0) : x} \mid x \in R\}$ is finite;

(2) If $P = \sqrt{(0) : x}$ is a prime ideal, then there exists a positive integer n such that if $y \in ((0) : x)^n R_P$, then $\sqrt{(0) : y} R_P \neq \sqrt{(0) : x} R_P$.

PROOF. ($\Rightarrow$) Suppose that (0) is decomposable. By THEOREM 3.6, condition (1) holds. Suppose that $P = \sqrt{(0) : x}$ is prime. If P is minimal over (0), then PR_P is the nilradical of R_P. Since R is coherent, so is R_P by [8, Theorem 2.4.2, p. 51]. Thus, $(0) : x$ is finitely generated and hence $((0) : x)^n R_P = (0)$ for some positive integer n. If $y \in ((0) : x)^n R_P$, then $y = 0$ in R_P, so that $(0) : y R_P = R_P > PR_P$. If P is not minimal over (0), then by THEOREM 3.4, R_P satisfies DC. Hence, by PROPOSITION 3.5, there exists a PR_P-primary ideal Q such that if $y \in Q$, then $\sqrt{(0) : y} R_P \neq PR_P$. Take an integer n so that $((0) : x)^n R_P \subseteq Q$. If $y \in ((0) : x)^n R_P$, then $y \in ((0) : x)^n R_P \subseteq Q$, and hence $\sqrt{(0) : y} R_P \neq PR_P = \sqrt{(0) : x} R_P$.

($\Leftarrow$) Suppose that conditions (1) and (2) hold. Take a nonminimal member $P \in Ass(0)$. Then $P = \sqrt{(0) : x}$ for some $x \in R$, by LEMMA C and condition (1). We want to show that R_P satisfies DC. By condition (2), there exists a positive integer n such that if $y \in ((0) : x)^n R_P$, then $\sqrt{(0) : y} R_P \neq \sqrt{(0) : x} R_P$. By PROPOSITION 3.5, R_P satisfies DC. It follows from THEOREM 3.4 that (0) is decomposable. $\square$

As mentioned in the introduction, it is natural to ask whether an ideal I is decomposable if and only if conditions (LL1) and (LL2) are satisfied for the ideal I. While these two conditions are obviously necessary for decomposability of I, the following construction due to P. Eakin will enable us to show that they are not sufficient for decomposability of I.

EXAMPLE 3.9. Let $U, V, W, X_1, X_2, ...$ be algebraically independent over in the field L which properly contains a field K. Put $R = L[U, V, W, \{X_i, X_i/U^i, (V - X_i)/W^i :$

$i = 1, 2, ...\}]$, and $T = R[U^{-1}] \cap R[W^{-1}]$. Then T has a maximal ideal M such that $K + MT_M$ is a Mori domain which contains a divisorial ideal having no primary decomposition. (For details, please see Proposition 5.6 and Example 5.7 in [5].)

To show that there exists an ideal that satisfies (LL1) and (LL2), and that is not decomposable, we will prove the following result:

PROPOSITION 3.10. *Let D be a Mori domain and let I be a divisorial ideal of D. Then*

(1) For every multiplicative system S in D, $S(I) = I : a$ for some $a \in S$.

(2) For every descending chain $S_1 \supseteq S_2 \supseteq ... \supseteq S_n \supseteq ...$ of multiplicative systems in D, there exists an integer n such that $S_n(I) = S_{n+1}(I) = ...$

PROOF. (1) Let $\mathcal{F} = \{I : s \mid s \in S\}$. Since I is a divisorial ideal, so is $I : s$ for every $s \in S$, by [5, Proposition 1.1(i)]. Hence $\mathcal{F}$ satisfies the ascending chain condition and therefore has a maximal element, say $I : a$. We claim that $S(I) = I : a$. Otherwise, $I : b \not\subseteq I : a$ for some $b \in S$. Take $c \in I : b - I : a$. Then $c \in I : ab - I : a$ and $I : a < I : ab$, which contradicts the maximality of $I : a$.

(2) By part (1), $S_i(I) = I : a_i$ for some $a_i \in S$. So $S_1(I) \supseteq S_2(I) \supseteq ... \supseteq S_n(I) \supseteq ...$ is a descending chain of divisorial ideals, and $\cap_{i=1}^{\infty} S_i(I) \supseteq I \neq (0)$. It follows from [5, Proposition 1.1(vii)] that this chain is stable. $\square$

THEOREM 3.11. *Let I be an ideal of a ring R such that $|Ass(I)| < \infty$ and every nonminimal maximal member of $Ass(I)$ is the radical of a principal ideal. Then I is decomposable if and only if for every nonminimal maximal member $M = \sqrt{(m)}$ of $Ass(I)$, $S(I)$ is decomposable and $S(I) = I : m^k$ for some integer $k > 0$, where $S = R - \cup\{P \in Ass(I) \mid P \neq M\}$.*

PROOF. Let $T = (R/I) - \cup\{P/I \mid P \in Ass(I) \text{ and } P \neq M\}$. We first show that $S(I)/I = T(0)$. In fact, if $x \in S(I)$, then $xy \in I$ for some $y \in S$, so $y + I \in T$ and $(x + I)(y + I) = I$. This shows that $x + I \in T(0)$. Conversely, if $x + I \in T(0)$, then $(x + I)(y + I) = I$ for some $y + I \in T$, that is, $xy \in I$ for some $y \in S$. Hence $x \in S(I)$, or $x + I \in S(I)/I$.

Thus, by passage to R/I we may assume that $I = (0)$.

($\Rightarrow$) Suppose that (0) is decomposable. Let $(0) = \cap_{i=0}^n Q_i$ be a reduced primary decomposition of (0) and assume that Q_0 is M-primary. Then $S(0) = \cap_{i=1}^n Q_i$. Since $M = \sqrt{(m)}$, there exists a positive integer k such that $m^k \in Q_0$. Now, since $Q_0 S(0) \subseteq Q_0 \cap S(0) = \cap_{i=0}^n Q_i = (0)$, $m^k S(0) = (0)$, or $S(0) \subseteq (0) : m^k$. Since $M = \sqrt{(m)}$ and M is not contained in any other associated prime of (0), $m^k \in S$. It follows that $S(0) = (0) : m^k$.

($\Leftarrow$) To show that (0) is decomposable, it suffices, by LEMMA 2.1, to show that, for every nonminimal maximal member $M = \sqrt{(m)}$ of $Ass(I)$, $(0)R_M$ is decomposable. Choose an integer $k > 0$ such that $S(0) = (0) : m^k$.

We will first show that $(m^k) \cap S(0) = (0)$. If $x \in (m^k) \cap S(0)$, then $x = m^k r$ for some $r \in R$. Since $m^k r \in S(0)$ and $m^k \in S = R - \cup Ass(S(0))$ by LEMMA 3.3, it follows from LEMMA A that $r \in S(0)$. Thus $x = m^k r \in (m^k) \cdot S(0) = (0)$. This shows that $(m^k) \cap S(0) = (0)$; hence $m^k R_M \cap S(0)R_M = (0)R_M$. Now

$S(0)R_M$ is decomposable since $S(0)$ is decomposable by hypothesis, and $m^k R_M$ is decomposable because $\sqrt{m^k R_M}$ is maximal. Therefore $(0)R_M$ is decomposable. $\square$

LEMMA 3.12. *Let I be an ideal of a ring R such that $|Ass(I)| < \infty$. Suppose that M is a nonminimal maximal member of $Ass(I)$. Write $S = R - \cup\{P \in Ass(I) \mid P \neq M\}$ and $T = R_M - \cup\{PR_M \mid P \in Ass(I) \text{ and } P < M\}$. Then $T(IR_M) = S(I)R_M$.*

PROOF. Since $IR_M \subseteq S(I)R_M$, to show that $T(IR_M) \subseteq S(I)R_M$, it suffices to show that $T(S(I)R_M) \subseteq S(I)R_M$. Since $|Ass(I)| < \infty$ and M is a maximal member of $Ass(I)$, $M \cap S \neq \phi$. By LEMMA 3.3, $Ass(S(I)) = \{P \in Ass(I) \mid P \neq M\}$. Thus, by LEMMA B, $Ass(S(I)R_M) = \{PR_M \mid P \in Ass(I) \text{ and } P < M\}$. It then follows from LEMMA A that $T(S(I)R_M) = S(I)R_M$.

Conversely, if $x \in S(I)R_M$, then there exists $u \in S(I)$ and $y \in R - M$ such that $x = u/y$. Take $v \in S$ so that $uv \in I$. Then $xv = uv/y \in IR_M$. We claim that $v \in T$. Suppose that $v \notin T$. Then, for some $P \in Ass(I)$ with $P < M$, there exists $w \in R - M$ such that $vw \in P$. Since $w \notin P$, $v \in P$, so that $v \notin S$, a contradiction. Hence $v \in T$ and therefore $x \in T(IR_M)$. This shows that $S(I)R_M \subseteq T(IR_M)$. $\square$

PROPOSITION 3.13. *The following conditions are equivalent for a GCD-domain D:*

(1) For every $a \in D$ such that $|Ass(a)| < \infty$, (a) is decomposable.

(2) For every $a \in D$ such that $|Ass(a)| < \infty$ and for any multiplicative system S in D, the saturation $S(a)$ of (a) is a principal ideal of D.

(3) For every $a \in D$ such that $|Ass(a)| < \infty$, the saturation $S(a)$ of (a) is a principal ideal, where S is the complement of the union of any nonempty subfamily of $Ass(a)$.

PROOF. $(1) \Rightarrow (2)$: It is known that for every decomposable ideal I of a ring R and every multiplicative system S of R, $S(I) = I : x$ for some $x \in S$. Since (a) is decomposable, $S(a) = (a) : x$ for any multiplicative system of D and a suitable $x \in S$. Since D is a GCD-domain, $(a) : x$ is principal by [10, Theorem 4.3.2].

$(2) \Rightarrow (3)$: Trivial.

$(3) \Rightarrow (1)$: Let $a \in D$ be such that $|Ass(a)| < \infty$. To show that (a) is decomposable, we use induction on $n = |Ass(a)|$. If $n = 1$, then (a) is primary. Suppose that every principal ideal (x) of D such that $|Ass(x)| \leq n$ is decomposable. We want to show that if $|Ass(a)| = n + 1$, then (a) is decomposable.

If $dim\, Ass(a) = 0$, we are done by COROLLARY 2.2. If $dim\, Ass(a) > 0$, to show that (a) is decomposable, it suffices, by LEMMA 2.1, to show that aD_M is decomposable for every nonminimal maximal member M of $Ass(a)$. Denote by e and c the extension and contraction of ideals with respect to the quotient ring D_N, where $N = D - \cup\{P \in Ass(a) \mid P < M\}$. Then $(a)^{ec}$ is principal by hypothesis and $|Ass((a)^{ec})| \leq n$ by LEMMA 3.3. Hence $(a)^{ec}$ is decomposable by the induction hypothesis. By LEMMA B, $Ass(aD_M) = \{PD_M \mid P \in Ass(a), P \subseteq M\}$. Let $T = \cup\{PD_M \mid P \in Ass(a), P < M\}$ and denote by e' and c' the extension and contraction of ideals with respect to the quotient ring $(D_M)_L$ where $L = D_M - T$.

By LEMMA 3.12, $(aD_M)^{e'c'} = (a)^{ec}D_M$, which is is decomposable and principal. Since $(aD_M)^{e'c'} = \cup\{aD_M : t \mid t \in T\}$, it follows that $(aD_M)^{e'c'} = aD_M : y$ for some $y \in D_M - T$. If y is a unit of D_M, then $(aD_M)^{e'c'} = aD_M$, so that aD_M is decomposable. If y is not a unit of D_M, then $y \in MD_M - T$ and there exists $z \in (aD_M)^{e'c'} - aD_M$ such that $z \in aD_M : y$, or $y \in aD_M : z$. Since $Ass(aD_M) = \{PD_M \mid P \in Ass(a), P \subseteq M\}$, MD_M is the only prime ideal of D_M containing y. Hence MD_M is the radical of (y). It follows from THEOREM 3.11 that aD_M is decomposable. Therefore, (a) is decomposable, completing the proof. □

We note that the assumption that D is a GCD-domain was not used in proving that (3) implies (1) in Proposition 3.13. We remark that if R' is a quotient ring of a ring R with respect to a multiplicative system, then the contraction to R of a principal ideal of R' need not be principal. However, the next result shows that this condition is satisfied if R is a UFD.

COROLLARY 3.14.   *Let D be a UFD and $a \in D$. Then*

(1) The saturation $S(a)$ of (a) is a principal ideal for any multiplicative system S in D.

(2) Let $\mathcal{F}$ be the set of multiplicative systems in D. Then $\{S(a) \mid S \in \mathcal{F}\}$ is a finite set.

PROOF. (1)   Since every principal ideal of a UFD has only finitely many associated primes and each of them is isolated, every principal ideal of a UFD is decomposable by COROLLARY 2.2. It follows from PROPOSITION 3.13 that $S(a)$ is principal.

(2)   For each $S \in \mathcal{F}$, $S(a)$ is principal and $a \in S(a)$. Since D is a UFD, a is contained in only finitely many principal ideals. Hence part (2) follows.   □

PROPOSITION 3.15.   *Let (R, M) be a quasilocal ring and let I be an ideal of R such that $1 < |Ass(I)| < \infty$ and $M \in Ass(I)$. Let $S = R - \cup\{P \in Ass(I) \mid P \neq M\}$.*

(1) If J is an ideal of R such that $I < J < S(I)$, then $Ass(J) = Ass(I)$

(2) If I is decomposable, then so is every ideal lying in between I and $S(I)$.

PROOF.   (1)   Since $J < S(I)$, $JR_S \subseteq S(I)R_S = IR_S \subseteq JR_S$; hence $JR_S = IR_S$. By LEMMA 3.3 and LEMMA B, $Ass(IR_S) = \{PR_S \mid P \in Ass(S(I))$ and $P \cap S = \phi\}$, so $Ass(S(I)) \subseteq Ass(J)$; similarly, $\{P \in Ass(J) \mid P \cap S = \phi\} \subseteq Ass(I)$. To see that $M \in Ass(J)$, we take $x \in S(I) - J$. Then $xy \in I \subseteq J$, or $y \in I : x \subseteq J : x$ for some $y \in S$. Thus $\sqrt{I : x} = M$ and we must have $\sqrt{J : x} = M$. This shows that $Ass(I) \subseteq Ass(J)$.

It remains to show that if $P \in Ass(J)$ is such that $P \cap S \neq \phi$, then $P \in Ass(I)$. Let $T = S - P$. Then T is a multiplicative system of R. Since $T \subseteq R - \cup Ass(S(I))$, $T(S(I)) = S(I)$ by LEMMA A. We claim that $T(J) \neq S(I)$. Otherwise, since $P \in Ass(J)$, $PR_T \in Ass(T(J)R_T) = Ass(S(I)R_T)$, so that $P \in Ass(S(I)) = \{P \in Ass(I) \mid P \cap S = \phi\}$, a contradiction. This shows that $T(J) \neq S(I)$, — that is, $T(J) < S(I)$. Take $x \in S(I) - T(J)$. Then $sx \in I \subseteq J$ for some $s \in S$, and hence $\sqrt{I : x} = M$. Since $x \notin T(J)$, $J : x \subseteq R - T = P \cup \{Q \in Ass(I) \mid Q \cap S = \phi\}$, so that $J : x \subseteq P$. Now, $M = \sqrt{I : x} \subseteq \sqrt{J : x} \subseteq P$ implies that $P = M \in Ass(I)$.

(2) If I is decomposable, then, by THEOREM 3.6, there exists an M-primary ideal Q such that, for any $x \in Q$ and $y \in S$, if $xy \in I$, then $x \in I$. This means $Q \cap S(I) \subseteq I$. Now, if J is an ideal of R such that $I < J < S(I)$, then $S(J) \subseteq S(S(I)) = S(I)$, and hence $Q \cap S(J) \subseteq Q \cap S(I) \subseteq I \subseteq J$. Let $Q' = Q + J$. Then $\sqrt{Q'} = M$, so that Q' is M-primary. Now, $J \subseteq Q' \cap S(J) \subseteq J + J = J$ implies $J = Q' \cap S(J)$. Hence J is decomposable. $\qquad\square$

4. FINITELY GENERATED IDEALS

This section was motivated by our work in considering the following question of D. D. Anderson ([11], Question 1, page 363): If every finitely generated ideal of a ring R is decomposable, is the same true for the polynomial ring $R[x]$ over R? If I is an ideal of R and P is a prime ideal of R, we denote by $S_P(I)$ the saturation of I with respect to R_P. Consider the following statement: Let $\mathcal{C}$ be a class of ideals of R. Then every ideal in $\mathcal{C}$ is decomposable if and only if the following two conditions hold:

(Li) For every ideal I in $\mathcal{C}$ and for every prime ideal P of R, $S_P(I) = I : x$ for some $x \in R - P$.

(Lii) For every chain $S_1 \supseteq S_2 \supseteq ... \supseteq S_n \supseteq ...$ of multiplicative systems in R, the chain $S_1(I) \supseteq S_2(I) \supseteq ... \supseteq S_n(I) \supseteq ...$ stablizes for every ideal I in $\mathcal{C}$.

Krull's result (mentioned in Section 1) is an extreme case of this statement— $\mathcal{C}$ consists of all ideals of R. Example 3.9 shows that this statement is not generally true. We will show that if R has finite dimension, then under weaker conditions, a similar result (Theorem 4.2) is true for the class of finitely generated ideals of R.

LEMMA 4.1. *Let I be an ideal of a ring R with d.c.c. on prime ideals. If, for every descending chain $S_1 \supseteq S_2 \supseteq ... \supseteq S_n \supseteq ...$, where each S_i is the complement of a union of associated primes of I, the chain $S_1(I) \supseteq S_2(I) \supseteq ... \supseteq S_n(I) \supseteq ...$ stablizes, then $|Ass(I)| < \infty$.*

PROOF. Suppose that $|Ass(I)| = \infty$. We first pick a minimal member P_1 of $Ass(I)$. Since $Ass(I)$ satisfies d.c.c. we can pick a minimal member P_2 in $Ass(I) - \{P_1\}$. Assume that we have picked members $P_1, P_2, ..., P_n$ of $Ass(I)$ such that each P_i $(1 < i \leq n)$ is minimal in $Ass(I) - \{P_1, P_2, ..., P_{i-1}\}$. Since $Ass(I)$ satisfies d.c.c. we can pick a minimal member P_{n+1} of $Ass(I) - \{P_1, P_2, ..., P_n\}$. And $P_{n+1} \not\subseteq \cup_{i=1}^{n} P_i$ because P_1 is minimal in $Ass(I)$ and each P_i $(1 < i \leq n)$ is minimal in $Ass(I) - \{P_1, P_2, ..., P_{i-1}\}$. Inductively we can pick an infinite sequence $P_1, P_2, ..., P_n, ...$ of associated primes of I such that $P_n \not\subseteq \cup_{i=1}^{n-1} P_i$ for each $n > 1$. Let $S_n = R - \cup_{i=1}^{n} P_i$ for each n. Then $S_n \subseteq S_{n-1}$ for each n. We will prove that the chain $S_1(I) \supseteq S_2(I) \supseteq ... \supseteq S_n(I) \supseteq ...$ is not stable. Suppose $S_n(I) = S_{n+1}(I)$ for some n. Then $S_{n+1}(S_n(I)) = S_{n+1}(S_{n+1}(I)) = S_{n+1}(I)$, and hence $S_{n+1}(I) = S_{n+1}(S_n(I)) = S_n(S_{n+1}(I))$. By LEMMA A, we must have $S_n \cap (\cup Ass(S_{n+1}(I))) = \phi$. But, by LEMMA 3.3 $Ass(S_{n+1}(I)) \supseteq \{P_i \mid P_i \cap S_{n+1} = $

$\phi\} = \{P_1, P_2, ..., P_n, P_{n+1}\}$. This is a contradiction since $P_{n+1} \not\subseteq \cup_{i=1}^{n} P_i$ implies that $S_n \cap P_{n+1} \neq \phi$. Hence the chain $S_1(I) \supseteq S_2(I) \supseteq ... \supseteq S_n(I) \supseteq ...$ is not stable, which contradicts the hypothesis. Therefore, $|Ass(I)| < \infty$. $\qquad\square$

THEOREM 4.2.   *Let R be a finite-dimensional ring. Every finitely generated ideal of R is decomposable if and only if the following two conditions hold:*

(1) For each finitely generated ideal I of R and for each $P \in Ass(I)$, $S_P(I) = I : x$ for some $x \in R - P$.

(2) For a finitely generated ideal I of R and for every descending chain $S_1 \supseteq S_2 \supseteq ... \supseteq S_n \supseteq ...$, where each S_i is the complement of a union of associated primes of I, the chain $S_1(I) \supseteq S_2(I) \supseteq ... \supseteq S_n(I) \supseteq ...$ stablizes.

PROOF.   If every finitely generated ideal of R is decomposable, then it is well-known that conditions (1) and (2) hold.

Suppose that (1) and (2) hold. We first want to show that conditions (1) and (2) hold in quotient rings of R with respect to any multiplicative system and in R/I, where I is a finitely generated ideal of R.

Let N be a multiplicative system of R. If A is a finitely generated ideal of R_N and $Q \in Ass(A)$, then there is a finitely generated ideal I of R such that $A = IR_N$, and $Q = PR_N$ for some $P \in Ass(I)$ by LEMMA B.

Since condition (1) holds in R, $S_P(I) = I : x$ for some $x \in R - P$. We claim that $S_Q(A) = A : x$. For this, it suffices to show that $xS_Q(A) \subseteq A$. Take $a \in S_Q(A)$. Then there exit $b \in R$, $c \in N$ and $u \in R_N - QR_N$ such that $a = b/c$, and $au \in A = IR_N$. Thus, $a\frac{v}{w} = d/t$ for some $d \in I$, $t \in N$, $v \in R - P$ and $w \in N$. So $btv = cdw \in I$. Since $Q \neq R_N$, we have $N \cap P = \phi$, so that $tv \in R - P$, and hence $b \in S_P(I) = I : x$. This shows that $ax \in IR_N = A$. Therefore, $S_Q(A) = A : x$. Hence condition (1) holds in quotient rings of R.

We next show that condition (1) holds in R/I. If A is a finitely generated ideal in R/I, then there is a finitely generated ideal J of R such that $J/I = A$ since I is finitely generated. It is routine to show that $Ass(A) \subseteq \{P/I \mid P \in Ass(J)\}$. Let $Q = P/I$ be an associated prime of A. Since condition (1) holds in R, $S_P(J) = J : x$ for some $x \in R - P$. We claim that $S_Q(A) = A : X$, where $X = x + I$. If $a \in R$ is such that $a + I \in S_Q(A)$, then $(a + I)Y \in A$ for some $Y \in R/I - Q$. Take $y \in R - P$ such that $Y = y + I$. Then $ay \in J$, that is, $a \in S_P(J) = J : x$. Thus $ax \in J$, and hence $(a + I)(x + I) \in A$. This shows that $S_P(A) \subseteq A : X$. Since we always have $A : X \subseteq S_P(A)$, $S_Q(A) = A : X$.

By LEMMA 4.1 and condition (2) every finitely generated ideal of R has only finitely many associated primes. Since every finitely generated ideal in a quotient ring of R is the extension of a finitely generated ideal in R, it follows from LEMMA B, that condition (2) holds in the quotient rings of R. Let I be a finitely generated ideal of R. If A is a finitely generated ideal of R/I, then there is a finitely generated ideal J of R such that $J/I = A$, and $Ass(A) \subseteq \{P/I \mid P \in Ass(J)\}$. Note that condition (2) holds in R. Applying LEMMA 4.1 we have $|Ass(J)| < \infty$, so that $|Ass(A)| < \infty$. Hence condition (2) holds in R/I.

To show that the ideal I, as described in conditions (1) and (2), is decomposable, we will use induction on the number $d = dim\,R$. When $d = 0$, I has only finitely many associated primes (LEMMA 4.1), each of which is isolated. Hence I is decomposable. Assume that every finitely generated ideal that satisfies conditions (1) and (2) in a ring of dimension less than or equal to n is decomposable. Consider $d = n + 1$. Without loss of generality we may assume that R is a quasilocal ring with maximal ideal $M \in Ass(I)$. If $dim\,Ass(I) = 0$, then I is decomposable. From now on we assume that $dim\,Ass(I) > 0$. Let S be the complement of the union of the minimal primes of I. In R_S, IR_S satisfies conditions (1) and (2), and $dim\,R_S < dim\,R = n + 1$, so IR_S is decomposable by the induction hypothesis. Hence $S(I)$ is also decomposable. Let $\{P_1, P_2, ..., P_k\}$ be the set of minimal primes of I. Denote $N_i = R - P_i$ for $1 \le i \le k$. By condition (1), for each i, $N_i(I) = I : x_i$ for some $x_i \in N_i$. Let $J = (x_1, x_2, ..., x_k)$. Then $J \not\subseteq P_i$ for any i, and hence $J \not\subseteq \cup_{i=1}^k P_i$. Take $x \in J - \cup_{i=1}^k P_i$. Then $x \in S$. Since $S(I) \subseteq N_i(I) = I : x_i$ for all i, $S(I) \subseteq I : x$. This implies that $S(I) = I : x$. We claim that $(x, I) \cap S(I) = I$. For this, it suffices to show that $(x, I) \cap S(I) \subseteq I$. Let $a = ux + vy \in S(I)$, where $y \in I$ and u, v in R. Then $ux^2 + vyx = ax \in I$, so that $ux^2 \in I \subseteq S(I)$. By LEMMA 3.3. and LEMMA A, we have $u \in S(I)$. Thus $ux \in I$ and hence $a \in I$. To show that I is decomposable, we only need to show that (x, I) is decomposable. We consider the ring $R/(x, I)$. Since (x, I) is finitely generated, conditions (1) and (2) hold in $R/(x, I)$. Since every prime ideal of R that contains (x, I) contains a minimal prime ideal of I, $dim\,R/(x, I) < dim\,R$, so the zero ideal of $R/(x, I)$ is decomposable by the induction hypothesis. This shows that (x, I) is decomposable. Therefore, I is decomposable. □

Anderson and Mahaney [12, Theorem 12] showed that if every principal ideal in a ring has a primary decomposition, then an ideal is finitely generated if it is locally principal. Following their idea we obtain the next theorem.

THEOREM 4.3. *Let R be a ring in which every finitely generated ideal is decomposable. Then an ideal of R is finitely generated if and only if it is locally finitely generated.*

PROOF. The necessity direction is obvious. Let I be an ideal of R that is locally finitely generated. Denote $f(I) = \Sigma_{\alpha \in \mathcal{A}}(J_\alpha : I)$, where $\{J_\alpha\}_{\alpha \in \mathcal{A}}$ is the set of all finitely generated ideals contained in I. If M is a maximal ideal of R, then $f(I)R_M = \Sigma(J_\alpha : I)_M$. Since every finitely generated ideal of R is decomposable, $(J_\alpha : I)R_M = J_\alpha R_M : IR_M$ by [12, Lemma 11]. Since I is locally finitely generated, $IR_M = J_\alpha R_M$ for some J_α, so that $f(I)R_M = R_M$ for every maximal ideal M of R. This shows that $f(I) = R$. Thus, the identity of R can be expressed as $1 = \Sigma_{i=1}^n r_i x_i$, where $r_i \in R$ and $x_i \in J_{\alpha_i} : I$ for each i. Let J be the sum of the J_{α_i}s. Then J is a finitely generated ideal contained in I. Clearly, $1 \in J : I$. Hence $I = J$, a finitely generated ideal. □

COROLLARY 4.4. *Let R be a ring in which every finitely generated ideal is decomposable. If R is locally Noetherian, then R is Noetherian.*

PROOF. Since R is locally Noetherian, every prime ideal of R is locally finitely generated, and hence finitely generated by THEOREM 4.3. Therefore R is Noetherian. □

It is known that a ring that is locally coherent may not be coherent. In fact, there are examples [9, p.51] that show that even a locally Noetherian ring may not be coherent. Our next result gives a sufficient condition for a locally coherent ring to be coherent.

COROLLARY 4.5.　*Let R be a ring in which every finitely generated ideal is decomposable. If R is locally coherent, then R is coherent.*

PROOF. Let I be a finitely generated ideal of R and let a be an element of R. To show that $I : a$ is a finitely generated ideal, it suffices, by THEOREM 4.3, to show that $I : a$ is locally finitely generated. If M is a maximal ideal of R, then $(I : a)R_M = IR_M : aR_M$. Since R is locally coherent, $IR_M : aR_M$ is finitely generated, so that $I : a$ is locally finitely generated. This completes the proof. □

ACKNOWLEDGEMENTS

I would like to thank my advisor, Professor Robert Gilmer, for his help, useful suggestions and valuable guidance. I also want to thank the referee for carefully reading and several valuable comments and suggestions.

REFERENCES

1. W. Krull, Über einen Hauptsatz der allgemeinen Idealtheorie, S.-B. Heidelberger Akad. Wiss. (1929), 11-16.
2. M. Atiyah and I. MacDonald, Introduction to Commutative Algebra, Addison-Wesley Publishing Company, 1969.
3. R. Gilmer, Multiplicative Ideal Theory, Queen's Papers Pure Appl. Math., Vol. 90, 1992.
4. W. Heinzer and J Ohm, Locally Noetherian commutative rings, Trans. Amer. Math. Soc., 158(1971), 273-284.
5. E. Houston, T. Lucas and T. Viswanathan, Primary decomposition of divisorial ideals in Mori domains, J. Algebra, 117(1988), 327-342.
6. N. Bourbaki, Commutative Algebra, Addison-Wesley, 1972.
7. O. Zariski and P. Samuel, Commutative Algebra, Vol. 1, Springer Verlag, 1986.
8. P.-J. Cahen, Ascending chain conditions and associated primes. Commutative ring theory (Fès, 1992), 41–46, Lecture Notes in Pure and Appl. Math., 153, Dekker, New York, 1994.
9. S. Glaz, Commutative Coherent Rings, Lecture Notes in Mathematics, 1371. Springer-Verlag, Berlin-New York, 1989.

10. J. Mott, Groups of Divisibility. Lattice-Ordered Groups — Advances and Techniques, Kluwer Academic Publishers, Dordercht/Boston/ London, year 1989, 80–104.

11. D. D. Anderson, Problem session. Zero-dimensional commutative rings (Knoxville, TN, 1994), 363–369, Lecture Notes in Pure and Appl. Math., 171, Dekker, New York, 1995.

12. D. D. Anderson and L. Mahaney, Commutative rings in which every ideal is a product of primary ideals, J. Algebra, 106(1987), 528-535.

Primary Decomposition of Ideals in Polynomial Rings

GUANG FU and ROBERT GILMER Department of Mathematics, The Florida
State University, Tallahassee, Florida 32306

0. INTRODUCTION

Our work in this paper was motivated by a question raised by D. D.
Anderson (see [A1], [A2]). Specifically, Anderson asked whether the
condition that each of its finitely generated ideals admits a finite primary decomposition is inherited by the polynomial ring $R[x]$ from the
coefficient ring R. In exploring the question we encountered general
questions concerning finite primary decomposition that seemed at least
as interesting as the original question, and these answers would be helpful in attacking the original. Perhaps the most fundamental problem in
this regard is that no easily applicable criteria are known for determining whether a fixed ideal of a commutative ring admits a finite primary
decomposition. (Krull raised the problem of determining such criteria
in [Kr, p. 12], and the problem is addressed in [GH2, Section 1] and in
[F].)

We say that an ideal I of a commutative ring R is **decomposable** if
I is a finite intersection of primary ideals of R and we say that R is
finitely Laskerian if each finitely generated ideal of R is decomposable.
In this terminology, Anderson's question asks whether $R[x]$ is finitely
Laskerian if R is finitely Laskerian. While we suspect that the answer
to the question is negative, we lack a counterexample. Our best result
in a positive direction is for the case where R is an arithmetical ring (see
[G2, §18]). If D is a finitely Laskerian integral domain whose integral
closure is a Prüfer domain, we show in Theorem 3.8 that each principal

ideal of $D[x]$ is decomposable.

In Section 1 of the paper we include some general results about decomposable ideals of a ring; we note that the condition that $R[x]$ is finitely Laskerian is strictly stronger than the condition that each principal ideal of $R[x]$ is decomposable.

Section 2 is concerned with primary decomposition of principal ideals of $R[X]$, where X denotes an arbitrary set of indeterminates over R, and Section 3 deals with decomposability of finitely generated ideals of $R[x]$ in the case where R is arithmetical. In Section 4 we consider mainly the case of $D[x]$, where D is a one-dimensional semiquasilocal domain. Such a domain D is Laskerian. In Section 5 we show (Corollary 5.3) that if (R, M) is a zero-dimensional quasilocal ring with $M^2 = (0)$, then $R[x]$ is finitely Laskerian.

Throughout this paper all rings are commutative with identity. For definitions and results not given in this paper, the reader is referred to [G1]. If it is not specified, an **associated prime** P of an ideal I in a ring R is always in the weak-Bourbaki sense (see [B, p. 289], [HO1]) — that is, P is a minimal prime ideal of the ideal $I : x$ for some $x \in R - I$. An associated prime P of I is said to be **isolated** if P is a minimal prime ideal of I; otherwise, P is said to be **embedded**. We denote by $Ass(I)$ the set of associated primes of the ideal I and by $MAss(I)$ the set of the maximal members of $Ass(I)$. If f is an element of a polynomial ring $R[x]$, $C(f)$ denotes the content of f; that is, $C(f)$ is the ideal of R generated by the coefficients of f. If A is a set, we denote by $|A|$ the cardinality of A. If S is a multiplicative system in a ring R and I an ideal of R, we follow [AM, p. 56] and denote by $S(I)$ the saturation of I with respect to S; thus $S(I)$ is the contraction of the extension of I with respect to the quotient ring R_S. If A and B are two sets, we use $A < B$ to mean that A is properly contained in B. If $\mathcal{F}$ is a set of subsets of a ring, we use $\cup \mathcal{F}$ to denote the union of all the members in $\mathcal{F}$. If $A \in \mathcal{F}$ and if $A_n < \ldots < A_2 < A_1 < A$ is a longest chain in $\mathcal{F}$ strictly descending from A, then we define the **height** of A in $\mathcal{F}$ to be $ht(A) = n$. The height of a minimal member of $\mathcal{F}$ is zero. The dimension of $\mathcal{F}$ is defined to be $dim\,\mathcal{F} = \mathrm{Sup}\{ht(A) \mid A \in \mathcal{F}\}$.

1. DECOMPOSABLE IDEALS OF A RING

Let R be a commutative unitary ring. We denote by $\mathcal{D}(R)$ the set of ideals of R that are decomposable. Recall that R is said to be **Laskerian** if $\mathcal{D}(R)$ is the set of all ideals of R; as stated in the introduction, we use the term finitely Laskerian to describe a ring each of whose finitely generated ideals is decomposable. Theorem 1.1 lists some fundamental properties of the set $\mathcal{D}(R)$.

Theorem 1.1. *Let I be an ideal of a ring R and let X be a nonempty set of indeterminates over R.*

(1) $\mathcal{D}(R)$ is closed under finite intersection.

(2) If R is a subring of a ring S and if $I \in \mathcal{D}(S)$, then $I \cap R \in \mathcal{D}(R)$.

(3) If $I \in \mathcal{D}(R)$, then $I : J \in \mathcal{D}(R)$ for each ideal J of R.

(4) $I \in \mathcal{D}(R)$ iff $IR[X] \in \mathcal{D}(R[X])$.

(5) If $\phi : R \to S$ is a surjective homomorphism and if $I \in \mathcal{D}(S)$, then $\phi^{-1}(I)$ is decomposable.

(6) If N is a multiplicative system in a ring R and if $I \in \mathcal{D}(R)$, then $IR_N \in \mathcal{D}(R_N)$ and $S_N(I) \in \mathcal{D}(R)$.

(7) If $Ass(I)$ is finite, if $MAss(I) = \{P_j\}_{j=1}^{k}$ and if $S_j(I)$ is the saturation of I with respect to the multiplicative system $R - P_i$, then $I = \cap_{j=1}^{k} S_j(I)$. Hence I is decomposable if each $S_j(I)$ is decomposable. In particular, I is decomposable if (and only if) IR_{P_i} is decomposable for each i.

Proof. With the possible exception of (7), all parts of Theorem 1.1 are standard, and hence their proofs are omitted. For completeness we include a proof of (7). To do so, we need only show that $\cap_{j=1}^{k} S_j(I)$ is a subset of I. Thus, take $r \in \cap_{j=1}^{k} S_j(I)$. By definition of $S_j(I)$ this means $I : r \nsubseteq P_j$, and since $Ass(I)$ is finite, it follows from the definition of $Ass(I)$ that $r \in I$. □

Corollary 1.2. *Let X be a set of indeterminates over a ring R. If each principal ideal of $R[X]$ is decomposable, then each principal ideal of R is decomposable; if $R[X]$ is finitely Laskerian, then so is R.*

Proof. (1.2) follows at once from part (4) of Theorem 1.1. □

Part (4) of (1.1) also shows that R is Laskerian if $R[X]$ is Laskerian, but Heinzer and Ohm [HO2] have shown that the converse fails by showing that R is Noetherian if $R[X]$ is Laskerian. Thus if S is a Noetherian ring and if the set X is infinite, then $T = S[X]$ is not Laskerian. On the other hand, each finitely generated ideal I of T is extended from $S[Y]$ for some finite subset Y of X, and hence I is decomposable and T is finitely Laskerian. We proceed to note in Example 1.4 that a ring R need not be finitely Laskerian if each principal ideal of R is decomposable. Our presentation of (1.4) uses an auxiliary result, Proposition 1.3.

Proposition 1.3. *Let (R, M) be a quasilocal ring and let A be the inter-section of all M-primary ideals.*
(1) If I is a finitely generated ideal of R, then $\cap_{i=1}^{\infty} I^i \subseteq A$.
(2) If $a \in A$ is such that aQ is an intersection of primary ideals for some M-primary ideal Q, then $a = 0$.

Proof. (1) Since I is finitely generated, every M-primary ideal contains a power of I, so that every M-primary ideal contains $\cap_{i=1}^{\infty} I^i$. Therefore $\cap_{i=1}^{\infty} I^i \subseteq A$.
(2) Let $aQ = \cap Q_\alpha$, where Q_α is P_α-primary. We show that $a \in aQ$ by showing that $a \in Q_\alpha$ for each α. If $P_\alpha = M$, this follows since $a \in A$, and if $P_\alpha < M$ it follows because $aQ \subseteq Q_\alpha$ and $Q \not\subseteq P_\alpha$. Therefore $a \in aQ$, and since R is quasilocal, this implies that $a = 0$.   □

Example 1.4.   In [N] Mauri Nascimento constructs an example of a quasilocal UFD with a finitely generated maximal ideal M such that $\cap_{i=1}^{\infty} M^i \neq (0)$. By part (2) of Proposition 1.3, aM is not decomposable for every $a \in \cap_{i=1}^{\infty} M^i - (0)$. Hence this is an example of a ring in which every principal ideal is decomposable, but some finitely generated ideal is not decomposable.

Remark 1.5.   In contrast with what we have just observed, there are well-known classes of rings with the property that their members are Laskerian if each of their principal ideals is decomposable. If R is zero-dimensional, for example, it is clear that the conditions (1) R is Laske-rian, (2) the ideal (0) is decomposable, and (3) R is semiquasilocal, are equivalent in R. Similarly, if D is a one-dimensional integral domain, the Laskerian property in D is equivalent to the condition that each principal ideal of D is decomposable, which in turn is equivalent to the condition that each nonzero element of D belongs to only finitely many maximal ideals of D.
    We show in Theorem 1.9 that an arithmetical ring also has the prop-erty that it is Laskerian if each of its principal ideals is decomposable. Recall that R is a **chained ring** if the set of ideals of R is linearly ordered under inclusion and that R is **arithmetical** if R_M is a chained ring for each maximal ideal of R. (See [J] or [G2 , §18] for equivalent forms of the definition of an arithmetical ring.) The arithmetical integral do-mains are, of course, the Prüfer domains. We begin with the case of a chained ring; the straightforward proof of Lemma 1.6 is omitted.

Lemma 1.6. *Suppose $\{Q_a\}_{a \in A}$ is a family of P-primary ideals of a ring R. Let $Q = \cap_{a \in A} Q_a$. If $P = \sqrt{Q}$, then Q is primary.*

Lemma 1.7. *If $\{I_a\}_{a \in A}$ is a family of primary ideals of the chained ring R, then $I = \cap_{a \in A} I_a$ is a primary ideal.*

Proof. Since R is chained ring, $P = \sqrt{I}$ is a prime ideal. If $I = P$, then I is prime, hence primary. If $I < P$, then $I_a \subseteq P$ for some a and $I = \cap \{I_a \mid a \in A$ and I_a is P-primary $\}$. Moreover, since $P = \sqrt{I}$, Lemma 1.6 shows that I is P-primary. □

Proposition 1.8. *If R is a chained ring, the following conditions are equivalent.*
(1) R is Laskerian.
(2) Each finitely generated ideal of R is decomposable.
(3) Each principal ideal of R is decomposable.
(4) Either R is 0-dimensional or else R is a one-dimensional valuation domain.
(5) Each ideal of R is primary.

Proof. The implications $(4) \Rightarrow (5) \Rightarrow (1) \Rightarrow (2) \Rightarrow (3)$ are clear. To show that (3) implies (4), we prove the contrapositive. Thus, if (4) fails then there exist prime ideals P and M of R such that $(0) < P < M < R$. Choose $p \in P - (0)$ and $m \in M - P$. Then $pm \in (pm)$ and $m \notin \sqrt{(pm)}$ since $m \notin P$, which contains $\sqrt{(pm)}$. If (pm) were decomposable, it would be primary by Lemma 1.7. Thus we would have $p \in (pm)$, say $p = pmx$, or $p(1 - mx) = 0$. This implies $p = 0$ since $1 - mx$ is a unit of R. This contradiction to the choice of p shows that (pm) is not decomposable. □

Theorem 1.9. *If R is an arithmetical ring, the following conditions are equivalent.*
(1) R is Laskerian.
(2) Each finitely generated ideal of R is decomposable.
(3) Each principal ideal of R is decomposable.
(4) R is a finite direct sum of 0-dimensional chained rings and one-dimensional Laskerian Prüfer domains.

Proof. The implications $(1) \Rightarrow (2)$ and $(2) \Rightarrow (3)$ are clear, and (4) implies (1) since a finite direct sum of Laskerian rings is again Laskerian. Assume (3) is satisfied. If M is a maximal ideal of R, then R_M is a chained ring in which each principal ideal is decomposable (since each principal ideal of R_M is extended from a principal ideal of R). By Proposition 1.8, $dim(R_M) \leq 1$, and because M is arbitrary, $dim(R) \leq 1$.

Moreover, if P is a prime ideal of R properly contained in each M-primary ideal of R, then P is the only P-primary ideal since R_M is a one-dimensional valuation domain. Let $(0) = \cap_{i=1}^{k} Q_i$ be a reduced primary decomposition of (0) in R, where Q_i is P_i-primary. For $i \neq j$, there is no inclusion relation between Q_i and Q_j since the representation is irredundant. Hence Q_i and Q_j are contained in no common maximal ideal of R — that is, Q_i and Q_j are comaximal. It follows [ZS, Theorem 32, p. 51] that $R = R_1 \oplus \ldots \oplus R_k$, where $R_i \simeq R/Q_i$ for each i. If P_i is maximal, R/Q_i is a 0-dimensional chained ring, and if P_i is not maximal, then $Q_i = P_i$ and R/Q_i is a one-dimensional Prüfer domain. Consider a positive integer j such that R_j is a one-dimensional Prüfer domain. If $x \in R_j - (0)$, then since $xR = xR_j$ is decomposable, the known prime ideal structure of R implies that x belongs to only finitely many maximal ideals of R_j. Therefore (see Remark 1.5) R_j is Laskerian. $\qquad\square$

2. DECOMPOSABILITY OF PRINCIPAL IDEALS

In Section 3 we return to a consideration of finitely generated ideals of polynomial rings over arithmetical rings. In this section we briefly survey the situation in regard to decomposability of principal ideals of polynomial rings.

Let X be a set of indeterminates over a ring R. If $f, g \in R[X]$, it is easy to see that $C(fg) \subseteq C(f)C(g)$. The reverse inclusion need not hold in general, but if $C(f)$ is invertible, the Dedekind-Mertens Lemma (cf. [G1, §28]) shows that $C(fg) = C(f)C(g)$ for each $g \in R[X]$. (For work on the converse of this statement, see [HH1, HH2].) In particular the formula $C(fg) = C(f)C(g) = C(g)$ holds if f has unit content — that is, $C(f) = R$.

Proposition 2.1. *Let X be a set of indeterminates over a ring R and let $f \in R[X]$ have unit content.*
(1) If S is an extension ring of R, then $fR[X]$ is contracted from $S[X]$.
(2) If I is an ideal of R, then $IR[X] \cap fR[X] = IfR[X]$.

Proof. (1): For $g \in S[X]$, let $C^*(g)$ denote the R-submodule of S generated by the coefficients of g. It is known that $C^*(fg) = C^*(g)$ for each $g \in S[X]$ [G1, p. 347]. Thus if $h \in fS[X] \cap R[X]$, then $h = fg$ for some $g \in S[X]$ and $C^*(g) = C^*(h) \subseteq R$. Consequently $g \in R[X]$ and $h \in fR[X]$ as asserted.
(2) The inclusion $IfR[X] \subseteq IR[x] \cap fR[X]$ is clear. If $h = fg \in IR[x]$, then $C(g) = C(h) \subseteq I$. Hence $g \in IR[X]$ and $h \in fIR[X]$. $\qquad\square$

Corollary 2.2. *Let the notation and hypothesis be as in the statement of Proposition 2.1. Let I be an ideal of R.*
(1) If R is embeddable in a ring S such that $fS[X]$ is decomposable in $S[X]$, then $fR[X]$ is decomposable in $R[X]$.
(2) If $fR[X]$ and I are decomposable, then $IfR[X]$ is decomposable.
(3) If $IfR[X]$ is decomposable, then I is decomposable; moreover, if I is invertible, then $fR[X]$ is decomposable.

Proof. Parts (1) and (2) of (2.2) are immediate from Proposition 2.1. To prove (3), suppose $IfR[X]$ is decomposable. Since f has unit content, it is not a zero divisor in $R[X]$ [G1, p. 348]. Thus $IfR[X] : fR[X] = IR[X]$, and the decomposability of I follows from parts (2), (3) and (4) of Theorem 1.1. Similarly, if I is invertible then $IfR[X] : IR[X] = fR[X]$ and $fR[X]$ is decomposable. □

An important case in which the hypothesis of part (1) of (2.2) is satisfied is that in which R is embeddable in a Noetherian ring (for more on this condition, see [GH1]); in particular this holds if R is an integral domain. This special case is of enough significance for us to merit a separate statement.

Corollary 2.3. *Let D be an integral domain and assume that $f \in D[X]$ has unit content.*
(1) $fD[X]$ is decomposable.
(2) If I is an invertible ideal of D, then $IfD[X]$ is decomposable if and only if I is decomposable.

Corollary 2.4. *Let D be an integral domain and assume that $f \in D[X]$ is such that $C(f) = aD$ is principal. Then (f) is decomposable if and only if aD is decomposable.*

Proof. We can write $f = ag$, where $g \in D[X]$ has unit content. Corollary 2.4 then follows from (2.3). □

Recall that a **Bezout ring** is a ring in which each finitely generated ideal is principal and that a **Bezout domain** is an integral domain with the property just described.

Corollary 2.5. *Let D be a Bezout domain. If each principal ideal of D is decomposable, then the same is true for $D[X]$.*

There exists rings R for which a principal ideal of $R[X]$ generated by a polynomial of unit content need not be decomposable. For example, $xR[x]$ is decomposable in $R[x]$ if and only if the zero ideal of R is decomposable.

In the case of an integrally closed domain, Corollaries 2.3 and 2.5 can be generalized. These generalizations depend upon properties of the v-operation and a known result. Recall that if D is an integrally closed domain and if F is a nonzero fractional ideal of D, the v-ideal associated with F is denoted by F_v and is defined to be $(F^{-1})^{-1}$; alternatively F_v is the intersection of the set of principal fractional ideals of D that contain F. For more on the v-operation, see [B, Chap. VII] or [G1, §34]. The known result that we use is the following:

Proposition 2.6. (Cf. [G1, Cor. 34.9]) *If D is an integrally closed domain with quotient field K, then for $f \in D[X]$, $fK[X] \cap D[X] = C(f)^{-1}fD[X]$.*

Proposition 2.7. *Let D be an integrally closed domain, let I be an invertible ideal of D, and let $f \in D[X]$ be such that $C(f)^{-1} = D$.*
(1) (f) is decomposable.
(2) $IfD[X] = I[X] \cap fD[X]$.
(3) $IfD[X]$ is decomposable in $D[X]$ if and only if I is decomposable in D.

Proof. (1): Since $K[X]$ is a UFD, $fK[X]$ is decomposable in $K[X]$. Hence Proposition 2.6 shows that $fD[X]$ is decomposable in $D[X]$.

To prove (2) we need to establish the inclusion $I[X] \cap fD[X] \subseteq IfD[X]$. Thus assume that $g = fh \in I[X]$, where $h \in D[X]$. Since D is integrally closed, it is known [G1, (34.8)] that $(C(fh))_v = (C(f)C(h))_v = (C(f)_vC(h)_v)_v = (D \cdot C(h)_v)_v = (C(h))_v$. Moreover, $fh \in I[X]$ implies that $C(fh) \subseteq I$, and hence $C(fh)_v \subseteq I_v$. Since I is invertible, $I_v = I$, so we conclude that $C(h) \subseteq (C(h))_v \subseteq I$. Therefore $h \in I[X]$ and $g = fh \in fI[X]$ as we wished to show.

(3): Parts (1) and (2) show that $IfD[X]$ is decomposable if I is decomposable, and the converse follows from (3) of Corollary 2.2.   □

Corollary 2.8. *Let D be a GCD-domain. If each principal ideal of D is decomposable, then the same is true in $D[X]$.*

Proof. Take $f \in D[X] - \{0\}$. Since D is a GCD-domain, $(C(f))_v = aD$ is a principal ideal. Hence we can write $f = ag$, where $g \in D[X]$ and

$(C(g))_v = D$. This implies that $D = D^{-1} = (C(g))_v^{-1} = C(g)^{-1}$. Part (3) of (2.7) then shows that $agD[X] = fD[X]$ is decomposable. □

3. POLYNOMIAL RINGS OVER ARITHMETICAL RINGS

In this section we consider primary decomposition of ideals in polynomial rings over arithmetical rings. We show that each finitely generated ideal of $D[x]$ is decomposable if and only if D is Laskerian. We will use x to denote a single indeterminate and X to denote a set of indeterminates. To prove our first result in this section, we use results from [F] and [GH2].

Proposition 3.1. ([F, Proposition 3.13.]) *Let (R, M) be a quasilocal ring and I an ideal of R such that $M \in Ass(I)$ and $|Ass(I)| < \infty$. Let $S = R - \cup\{P \in Ass(I) \mid P \neq M\}$.*
(1) If J is an ideal of R such that $I < J < S(I)$, then $Ass(J) = Ass(I)$
(2) If I is decomposable, then so is every ideal lying between I and $S(I)$.

Proposition 3.2 . ([GH2, (1.1)], [F, Corollary 2.2.]) *Let I be an ideal of a ring R. If $|Ass(I)| < \infty$ and each member of $Ass(I)$ is isolated, then I is decomposable.*

Proposition 3.3 *Let R be a zero-dimensional chained ring with maximal ideal M. Let x be an indeterminate over R. Then*
(1) every ideal of $R[x]$ that is not contained in $M[x]$ is decomposable;
(2) every finitely generated ideal of $R[x]$ is decomposable.

Proof. (1) Since R is zero-dimensional, $R[x]$ is one-dimensional and $M[x]$ is the unique minimal prime of $R[x]$. If I is an ideal of $R[x]$ that is not contained $M[x]$, then I has only isolated associated primes. We claim that I has only finitely many associated primes. To see this, we consider the Noetherian domain $(R/M)[x]$, which is isomorphic to the ring $R[x]/M[x]$. In $(R/M)[x]$, every ideal has only finitely many associated primes, so an intersection of infinitely many prime ideals of $R[x]/M[x]$ is the zero ideal. This implies that an intersection of infinitely many prime ideals of $R[x]$ is equal to $M[x]$. Hence I is contained in only finitely many prime ideals of $R[x]$. This shows that I has only finitely many associated primes, each of which is isolated. Therefore I is decomposable by Proposition 3.2.

(2) We first show that every principal ideal of $R[x]$ is decomposable. Take $f \in R[x]$. Since R is a chained ring, $f = ag$ for some $a \in R$ and

some $g \in R[x]$ of unit content, so part (2) of Corollary 2.2 shows that $fR[x]$ is decomposable if aR and $gR[x]$ are decomposable. The ideal aR is M-primary, hence decomposable, and part (1) shows that $gR[x]$ is also decomposable. Therefore $fR[x]$ is decomposable.

Let $I = (f_1, f_2, ..., f_n)$ be a finitely generated ideal of $R[x]$. We may assume that $I \neq (0)$ and $I < M[x]$. Since R is a chained ring, there is an element $a \in R$ such that $f_i = ag_i$ for each $1 \le i \le n$ and $C(g_i) = R$ for some i. We may assume without loss of generality that $C(g_1) = R$. We first show that $Ass(I) \subseteq Ass(f_1)$. Suppose that P is an associated prime of I. Then P is a minimal prime of $I : g$ for some $g \in R[x] - I$. If $P = M[x]$, then $P \in Ass(f_1)$ since $M[x]$ is the unique minimal prime of $f_1 R[x]$. Assume that $P \neq M[x]$ and take $h \in (I : g) - M[x]$. Then $gh \in (f_1, f_2, ..., f_n) = (ag_1, ag_2, ..., ag_n)$, so that $C(gh) \subseteq aR$. Since $C(gh) = C(g)C(h) = C(g)$, $C(g) \subseteq aR$, which implies that $g = ap$ for some $p \in R[x]$. Write $p = bq$, where $b \in R$ and $q \in R[x]$ with $C(q) = R$. Since $f_1 R[x] : ab \subseteq f_1 R[x] : g \subseteq I : g \subseteq P$ and $g_1 \subseteq f_1 R[x] : ab$, $P \in Ass(f_1)$. This shows that $Ass(I) \subseteq Ass(f_1)$. To show that I is decomposable, it suffices to show that for each $P \in Ass(I)$, $IR[x]_P$ is decomposable in $R[x]_P$. If $P = M[x]$, this is trivial. Assume that $P \neq M[x]$. Let $S = R[x]_P - M[x]R[x]_P$. By Proposition 3.1, to show that $IR[x]_P$ is decomposable in $R[x]_P$, it suffices to show that $IR[x]_P \subseteq S(f_1 R[x]_P)$. Since $C(g_1) = R$, $g_1 \notin M[x]R[x]_P$. Now, $f_i g_1 \in ag_1 R[x] = f_1 R[x]$ implies that $IR[x]_P \subseteq S(f_1 R[x]_P)$. This completes the proof. $\square$

Corollary 3.4. *If R is a 0-dimensional semiquasilocal Bezout ring, then each finitely generated ideal of $R[x]$ is decomposable.*

Proof.   This follows from Proposition 3.3 and the fact that $R = \sum_{i=1}^{n} \oplus R_i$, where R_i is a zero-dimensional chained ring, and hence $R[x] = \sum_{i=1}^{n} \oplus R_i[x]$. $\square$

Remark 3.5. Proposition 3.3 and Theorem 1.1(4) show that each of the following conditions (6) and (7) is equivalent to conditions (1)-(5) of Proposition 1.8.
(6) Each finitely generated ideal of $R[x]$ is decomposable.
(7) Each principal ideal of $R[x]$ is decomposable.

We now turn to extending the equivalent conditions of Theorem 1.9, which are stated for an arithmetical ring, to include conditions (6) and (7) of the preceding remark (see Theorem 3.15). In light of Remark 3.5 and condition (4) of Theorem 1.9, the key result needed for such an extension in that $D[x]$ is finitely Laskerian if D is a one-dimensional

Laskerian Prüfer domain. We prove this result in Theorem 3.14. The proof of our next result, Theorem 3.8, uses two results from [BH]; for reference we record these results as Propositions 3.6 and 3.7.

Proposition 3.6. ([BH, Corollary 8]) *Suppose that P is an associated prime of a regular principal ideal of the polynomial ring $R[x]$ and let $Q = P \cap R$. If Q contains a regular element of R, then $P = Q[x]$ and Q is an associated prime of a regular principal ideal. Thus, if R is an integral domain and P is an associated prime of a nonzero principal ideal in $R[x]$, then $P \cap R = (0)$ or $P = (P \cap R)[x]$ and $P \cap R$ is an associated prime of a principal ideal.*

Proposition 3.7. ([BH, Corollary 9]) *Let D be an integral domain and let X be a set of indeterminates over D. If principal ideals of D have only finitely many associated primes, the same is true of $D[X]$.*

Theorem 3.8. *Let D be a one-dimensional domain whose integral closure is Prüfer. Then every principal ideal of D is decomposable if and only if every principal ideal of the polynomial ring $D[x]$ is decomposable.*

Proof. ($\Leftarrow$) This follows from Corollary 1.2.

($\Rightarrow$) Take $f \in D[x] - \{0\}$. Since the integral closure of D is Prüfer, it is known that $D[x]$ is two-dimensional and that $M[x]$ is a height-one prime of $D[x]$ for each maximal ideal M of D [AG, Prop. 2.7 and Cor. 2.8]. We claim that the principal ideal (f) has only isolated associated primes. Suppose that (f) has an embedded associated prime P. Then P properly contains a minimal prime, say Q, of (f). If $Q \cap D \neq (0)$, then $P \cap D \neq (0)$, so that $P \cap D$ is a maximal ideal of D; if $Q \cap D = (0)$, then, since there does not exist a chain of three distinct prime ideals in $D[x]$ with the same contraction in D, $P \cap D$ is again a maximal ideal of D. In either case we have $P \cap D = M$ for some maximal ideal M of D. But, since P is an associated prime of the principal ideal (f), it follows from Proposition 3.6 that $P = M[x]$, which contradicts the fact that $M[x]$ has height 1. This shows that (f) has only isolated associated primes. Since every principal ideal of D is decomposable, by Proposition 3.7 every principal ideal of $D[x]$ has only finitely many associated primes, which are all isolated. Therefore (f) is decomposable. □

Corollary 3.9. *Let D be a one-dimensional domain. The following conditions are equivalent:*
(1) The integral closure of D is Prüfer.
(2) $\dim \mathrm{Ass}(f) = 0$ for every nonunit $f \in D[x]$.

Proof. (1) $\Rightarrow$ (2): This implication is established in the proof of Theorem 3.8.

(2) $\Rightarrow$ (1): Suppose that the integral closure of D is not Prüfer. Then $dim\, D[x] = 3$. By [AG, Corollary 2.10], some maximal ideal M of D extends to a rank-two prime ideal $M[x]$ in $D[x]$, so that the dimension of $D(x)_{MD(x)}$ is 2. Since $D_M[x]_{(MD_M)[x]} = D(x)_{MD(x)}$, the rank of $(MD_M)[x]$ is 2; hence $(MD_M)[x]$ properly contains a nonzero prime ideal, say P, of $D_M[x]$. Take $g \in P - \{0\}$. Then, since $dim\, D_M[x] = 3$, P is a minimal prime of g, and hence an associated prime of g. Obviously $(MD_M)[x]$ is an associated prime of a nonzero principal ideal, so that $(MD_M)[x]$ is also an associated prime of (g) by [BH, Theorem 3]. Thus, $dim\, Ass(g) \geq 1$. But, since $dim\, Ass(f) = 0$ for every nonunit $f \in D[x]$ by hypothesis, the same is true for the quotient ring $D_M[x]$ of $D[x]$. This is a contradiction, and therefore the integral closure of D is Prüfer.   □

We interrupt our drive toward Theorem 3.15 by showing that in Theorem 3.8 the hypothesis that the integral closure of D is Prüfer is not necessary. That is, if D is a one-dimensional domain such that each principal ideal of $D[x]$ is decomposable, then the integral closure of D need not be a Prüfer domain. For this observation we use Proposition 3.10; note that the proof of (3.10) uses a result, Proposition 4.7, from the next section.

Proposition 3.10. *Let (D, M) be a one-dimensional quasilocal domain such that M is also the maximal ideal of the valuation overring V of D. If X is a set of indeterminates over D, then every principal ideal of the polynomial ring $D[X]$ is decomposable.*

Proof. Take a nonzero polynomial $f \in D[X]$. If $C(f) = D$, then $fD[X]$ is decomposable by Corollary 2.3. Assume that $C(f) \neq D$. Since the content $C(f)$ of f is finitely generated in D, $C(f)V$ is principal in the valuation domain V and generated by a coefficient, say a, of f. We claim that $fD[X] : a^3D \subseteq fD[X] : a^2D$. If $g \in fD[X] : a^3D$, then $a^3g = fh$ for some $h \in D[X]$. Since V is a valuation domain, $a^3C(g)V = C(f)V \cdot C(h)V$, so that $C(h) \subseteq a^2V$. Since $C(f) \neq D$, $a \in M$, the common maximal ideal of D and V. Thus, $C(h) \subseteq a^2V \subseteq aM \subseteq aD$, and hence $h = ah_1$ for some $h_1 \in D[X]$. Since $a^3g = fh = afh_1$, $a^2g = fh_1$, or $g \in (f) : a^2$. This shows that $fD[X] : a^3D \subseteq fD[X] : a^2D$. By Proposition 4.7, $fD[X]$ is decomposable.   □

To obtain the example alluded to in the paragraph before (3.10), let

V be a rank-one valuation domain of the form $K(t) + M$, where K is a field, t is an indeterminate over K, and M is the maximal ideal of V. It is well known from the theory of $(D + M)$-constructions that (D, M) is one-dimensional, quasilocal, and integrally closed, but is not a Prüfer domain.

Proposition 3.11. *Suppose M is a maximal ideal of a ring R and I is an ideal of $R[x]$ such that $M[x] < \sqrt{I}$. Then $\mathrm{Ass}(I)$ is finite and 0-dimensional, so I is decomposable.*

Proof. We claim that $R[x]/I$ is a 0-dimensional semiquasilocal ring; if we prove this claim, the assertions of Proposition 3.11 then follow. To prove the claim, we note that $\sqrt{I}/I$ is the nilradical of $R[x]/I$, and hence it suffices to prove that $(R[x]/I)/(\sqrt{I}/I) \simeq R[x]/\sqrt{I}$ is zero-dimensional and semilocal. Because $M[x] < \sqrt{I}$, this follows from the fact that $R[x]/\sqrt{I}$ is a proper homomorphic image of the PID $R[x]/M[x] \simeq (R/M)[x]$. □

Lemma 3.12. *Let D be a 1-dimensional Prüfer domain in which each nonzero element belongs to only finitely many maximal ideals. If $d \in D - \{0\}$, then D/dD is a 0-dimensional semiquasilocal Bezout ring.*

Proof. By assumption, dD is decomposable. Let $dD = \cap_{i=1}^{n} Q_i$ be a reduced primary decomposition of dD, where Q_i is M_i-primary. Since D is 1-dimensional, the ideals Q_i are comaximal. Hence $D/dD = R_1 \oplus ... \oplus R_n$, where $R_i \simeq D/Q_i \simeq D_{M_i}/Q_i D_{M_i}$ is a 0-dimensional chained ring. Therefore D/dD is a 0-dimensional semiquasilocal Bezout ring. □

Proposition 3.13. *Let R be a ring. If B is a P-primary ideal of R and J is an invertible ideal of R such that $J \not\subseteq P$, then $J \cap B = JB$.*

Proof. It is clear that $JB \subseteq J \cap B$. Since J is invertible and $J \cap B \subseteq J$, $J \cap B = JC$ for some ideal C of R. Thus, $JC = J \cap B \subseteq B \subseteq P$. Noticing that $J \not\subseteq P$ and B is a P-primary, we have $C \subseteq B$. Hence $J \cap B = JC \subseteq JB$. □

Theorem 3.14. *If D is a 1-dimensional Laskerian Prüfer domain, then each finitely generated ideal of $D[x]$ is decomposable.*

Proof. By Theorem 3.8, each principal ideal of $D[x]$ is decomposable. Let I be a proper finitely generated ideal of $D[x]$.

Case 1. $I \cap D \neq (0)$. Take $a \in I \cap D - \{0\}$. $I/aD[x]$ is a finitely generated ideal of the ring $D[x]/aD[x] \simeq (D/aD)[x]$. Now D/aD is a 0-dimensional semiquasilocal Bezout ring by Lemma 3.12, so Corollary 3.4 shows that $I/aD[x]$ is decomposable, and hence I is decomposable by part (5) of Theorem 1.1.

Case 2. $I \cap D = (0)$. In this case $IK[x]$ is a proper ideal of the PID $K[x]$. Let $IK[x] = f(x)K[x]$, where $f(x) \in D[x]$, and let $J = IK[x] \cap D[x]$. Then $I \subseteq J = C(f)^{-1}f(x)D[x]$ (see Proposition 2.6). We claim that J is an invertible ideal of $D[x]$ such that $J \not\subseteq M[x]$ for any maximal ideal M of D. It is clear that J is invertible because D is Prüfer and $C(f)$ is finitely generated. Let $f(x) = a_0 + a_1 x + ... + a_k x^k$, where $a_i \in D$. Since $1 \in C(f)^{-1}C(f)$, $1 = a_0 b_0 + a_1 b_1 + ... + a_k b_k$, where $b_i \in C(f)^{-1}$. Let $g(x) = b_k + b_{k-1}x + ... + b_0 x^k$. Then $f(x)g(x) \in J$ and the coefficient of x^k in $f(x)g(x)$ is 1. Consequently, $J \not\subseteq M[x]$ for any maximal ideal M of D. Because $K[x]$ is a quotient ring of $D[x]$, we have $IK[x] = JK[x]$, and because J is invertible and $I \subseteq J$, there is an ideal A of $D[x]$ such that $I = JA$. Moreover, $IK[x] = JK[x]AK[x] = IK[x]AK[x]$. Since $IK[x]$ is principal, $AK[x] = K[x]$, which shows that $A \cap D \neq (0)$. By Case 1, A is decomposable. Let $A = \cap_{i=1}^m Q_i$ be a reduced primary decomposition of A in $D[x]$. Since J is invertible, $I = JA = \cap_{i=1}^m JQ_i$. To show that I is decomposable, it suffices to prove that each JQ_i is decomposable, and we can do this for any ideal JQ, where Q is P-primary and $M = P \cap D$ is maximal in D. There are two cases to consider, according as (i) $P = M[x]$, or (ii) $P > M[x]$. If (i) occurs, $JQ = J \cap Q$ by Proposition 3.13. Since $J = IK[x] \cap D[x]$ is decomposable, so is JQ. In the second case, $L = Q \cap D$ is M-primary. Thus, $JQ = JQ + JL[x] = JQ + (J \cap L[x]) = J \cap (JQ, L[x])$. Since $\sqrt{(L[x])} = M[x]$ and since $JQ \not\subseteq M[x]$, $\sqrt{(JQ, L[x])} > M[x]$. By Proposition 3.11, $(JQ, L[x])$ is decomposable. Because J is decomposable, JQ is decomposable in this case. This completes the proof of Theorem 3.14. $\square$

Theorem 3.15. *Suppose R is an arithmetical ring. The following conditions are equivalent.*
(1) R is Laskerian.
(2) Each finitely generated ideal of R is decomposable.
(3) Each principal ideal of R is decomposable.
(4) R is a finite direct sum of 0-dimensional chained rings and one-dimensional Laskerian Prüfer domains.
(5) Each principal ideal of $R[x]$ is decomposable.
(6) Each finitely generated ideal of $R[x]$ is decomposable.

Proof. The equivalence of (1)-(4) is the content of Theorem 1.9, and the implications (6)$\Rightarrow$(5) and (5)$\Rightarrow$(3) are clear. To see that (6) follows from (1)-(4), we need only show that finitely generated ideals of $R[x]$

are decomposable in the cases where R is a 0-dimensional chained ring or a one-dimensional Laskerian Prüfer domain. The first of these cases follows from Proposition 3.3, while the second follows from Theorem 3.14. □

4. POLYNOMIAL RINGS OVER LOWER DIMENSIONAL RINGS

This section deals with primary decomposition of ideals in polynomial rings over zero- or one-dimensional semiquasilocal rings. We characterize the decomposibility of ideals in these rings in terms of the ideals and the maximal ideals in the coefficient rings. We first recall two results from [F].

Lemma 4.1. *(1) ([F, Lemma A]) Suppose that I is an ideal of a ring R and S is a multiplicative system of R. Then I is contracted from R_S if and only if $S \cap (\cup Ass(I)) = \phi$.*
(2) ([F, Lemma 3.3.]) Let I be an ideal of a ring R and S a multiplicative system of R. Then $Ass(S(I)) = \{P \in Ass(I) \mid P \cap S = \phi\}$.

Our first result in this section characterizes all decomposable ideals in a polynomial ring over a 0-dimensional semiquasilocal ring using only one multiplicative system.

Theorem 4.2. *Let $(R, M_1, M_2, ..., M_n)$ be a 0-dimensional semiquasilocal ring and let x be an indeterminate over R. Then*
(1) every ideal of $R[x]$ that is not contained in $\cup_{i=1}^{n} M_i[x]$ is decomposable;
(2) an ideal I of $R[x]$ is decomposable if and only if $S(I) = I : f$ for some $f \in S = R[x] - \cup_{i=1}^{n} M_i[x]$.

Proof. (1) Let I be an ideal of $R[x]$ such that $I \not\subseteq \cup_{i=1}^{n} M_i[x]$. First observe that $\{M_1[x], M_2[x], ..., M_n[x]\}$ is the set of minimal prime ideals of $R[x]$. Since $dim\,R = 0$, $dim\,R[x] = 1$. Hence I has only isolated associated primes, each of which is a maximal ideal of $R[x]$. If $\mathcal{P}$ is an infinite set of maximal prime ideals of $R[x]$, then some minimal prime ideal of $R[x]$ is contained in infinitely many members of $\mathcal{P}$. Without loss of generality, we may assume that $M_1[x]$ is contained in infinitely many members of $\mathcal{P}$. As observed in the proof of Proposition 3.3, every ideal of $R[x]/M_1[x]$ has only finitely many associated primes, so that an intersection of infinitely many prime ideals in $R[x]/M_1[x]$ is the zero ideal. Hence the intersection of the members of $\mathcal{P}$ is contained in $M_1[x]$. This implies that I is contained in only finitely many prime ideals of $R[x]$, so that I

has only finitely many associated prime ideals, each of which is isolated. Therefore, I is decomposable by Proposition 3.2.

(2) If $I \not\subseteq \cup_{i=1}^{n} M_i[x]$, then I is decomposable by part (1). And if we take $f \in I - \cup_{i=1}^{n} M_i[x]$, then $S(I) = R[x] = I : f$. From now on, we assume that $I \subseteq \cup_{i=1}^{n} M_i[x]$.

Suppose that I is decomposable. Then it is known that $S(I) = I : f$ for some $f \in S$ (see Exercise 15, page 56 of [AM], or apply Proposition 4.9 and Lemma 4.4 of [AM]). Conversely, if $S(I) = I : f$ for some $f \in S$, we first claim that $(I, f) \cap S(I) = I$. Clearly, $I \subseteq (I, f) \cap S(I)$. Take $g \in (I, f) \cap S(I)$. Then $g = ui + vf$ for some $u, v \in R[x]$, so that $uif + vf^2 = gf \in I$, which implies that $vf^2 \in I \subseteq S(I)$. By part (2) of Lemma 4.1, $Ass(S(I)) \subseteq \{M_1[x], M_2[x], ..., M_n[x]\}$. Since $f \in S$, it follows from part (1) of Lemma 4.1 that $v \in S(I)$—that is, $vf \in I$. This shows that $g \in I$. Therefore, $(I, f) \cap S(I) = I$. Since $(I, f) \not\subseteq \cup_{i=1}^{n} M_i[x]$, (I, f) is decomposable by part (1). Since $Ass(S(I)) \subseteq \{M_1[x], M_2[x], ..., M_n[x]\}$, $S(I)$ is also decomposable. This shows that I is decomposable. $\square$

Proposition 4.3.   *Let $(D, M_1, M_2, ..., M_n)$ be a one-dimensional semi-quasilocal domain.*
(1)   Every ideal I of $D[x]$ such that $I \not\subseteq \cup_{i=1}^{n} M_i[x]$ and $I \cap D \neq (0)$ is decomposable.
(2)   An ideal I of $D[x]$ such that $I \cap D \neq (0)$ is decomposable if and only if $S(I) = I : f$ for some $f \in S = D[x] - \cup_{i=1}^{n} M_i[x]$.

Proof.   (1)   Let I be an ideal of $D[x]$ such that $I \not\subseteq \cup_{i=1}^{n} M_i[x]$ and $I \cap D \neq (0)$. Since $I \not\subseteq \cup_{i=1}^{n} M_i[x]$, none of the $M_i[x]$ is an associated prime of I. Since $I \cap D \neq (0)$, $\sqrt{I} \supseteq \sqrt{(I \cap D)D[x]} \supseteq \cap_{i=1}^{n} M_i[x]$. Thus, if P is an associated prime of I, then P properly contains $M_j[x]$ for some $1 \leq j \leq n$, so that $P \cap D \supseteq M_j$, which implies that $P \cap D = M_j$. Similarly, every prime ideal of $D[x]$ that contains $M_j[x]$ contracts to M_j in D. Since there is no chains of three distinct prime ideals in $D[x]$ that contract to the same prime ideal of D, P is a maximal prime ideal of $D[x]$. Hence every associated prime of I is a maximal prime ideal of $D[x]$. We can show (as in the proof of Theorem 4.2) that an intersection of infinitely many maximal prime ideals of $D[x]$ is contained in one of the $M_i[x]$, so that I is contained in only finitely many maximal prime ideals of $D[x]$, and hence I has only finitely many associated primes, each of which is isolated. Therefore I is decomposable.

(2)   If $I \not\subseteq \cup_{i=1}^{n} M_i[x]$, then, since $I \cap D \neq (0)$, I is decomposable by part (1). And if we take $f \in I - \cup_{i=1}^{n} M_i[x]$, then $S(I) = D[x] = I : f$. From now on, we assume that $I \subseteq \cup_{i=1}^{n} M_i[x]$. Note that $I \cap D \neq (0)$ implies that every minimal prime of I is in the set $\{M_1[x], M_2[x], ..., M_n[x]\}$.

By part (2) of Lemma 4.1, $S(I) \subseteq \{M_1[x], M_2[x], ..., M_n[x]\}$, so that $S(I)$ has only finitely many associated primes and they are isolated. This shows that $S(I)$ is decomposable. Since $S(I) = I : f$ for some $f \in S = D[x] - \cup_{i=1}^{n} M_i[x]$, an argument similar to that in the proof of Theorem 4.2 shows that $(I, f) \cap S(I) = I$. Since $I \cap D \neq (0)$, $(I, f) \cap D \neq (0)$. Hence (I, f) is decomposable by part (1). Therefore I is decomposable. $\quad\Box$

In Proposition 4.3 we considered primary decomposition of ideals I in $D[x]$ such that $I \cap D \neq (0)$. Next we turn to principal ideals (f) in $D[x]$ that may contract to (0) in D. We cite three results from [F].

Lemma 4.4. ([F, Lemma 2.1.]) *Let I be an ideal of a ring R. Then I is decomposable if and only if the following conditions hold:*
(1) $MAss(I)$ has only finitely many members.
(2) For each $P \in MAss(I)$, IR_P is decomposable.

Theorem 4.5. ([F, Theorem 3.11.]) *Let I be an ideal of a ring R such that $|Ass(I)| < \infty$ and every nonminimal maximal member of $Ass(I)$ is the radical of a principal ideal. Then I is decomposable if and only if for every nonminimal maximal member $M = \sqrt{(m)}$ of $Ass(I)$, $S(I)$ is decomposable and $S(I) = I : m^k$ for some integer $k > 0$, where $S = R - \cup\{P \in Ass(I) \mid P \neq M\}$.*

Lemma 4.6. ([F, Lemma 3.12]) *Let I be an ideal of a ring R such that $|Ass(I)| < \infty$. Suppose that M is a nonminimal maximal member of $Ass(I)$. Write $S = R - \cup\{P \in Ass(I) \mid P \neq M\}$ and $T = R_M - \cup\{PR_M \mid P \in Ass(I) \text{ and } P < M\}$. Then $T(IR_M) = S(I)R_M$.*

The next result gives a characterization of a decomposable principal ideal in terms of a chain condition in a polynomial ring over a one-dimensional quasilocal domain.

Proposition 4.7. *Let (D, M) be a one-dimensional quasilocal domain. Suppose that $f \in D[x]$ is such that $\dim Ass(f) \neq 0$ and S is the complement of the union of the isolated associated primes of (f). The following conditions are equivalent:*
(1) (f) is decomposable.
(2) $S(f) = (f) : a$ for some $a \in M$.
(3) For some nonzero nonunit $a \in D$, the chain $(f) : a \subseteq (f) : a^2 \subseteq (f) : a^3 \subseteq ...$ is stable.
(4) For every $d \in D$, the chain $(f) : d \subseteq (f) : d^2 \subseteq (f) : d^3 \subseteq ...$ is stable.

Proof. $(1) \Rightarrow (2)$: We first claim that $dim \, Ass(f) \leq 1$. Suppose that $dim \, Ass(f) > 1$. Then we have a chain $P_1 < P_2 < P_3$ of associated primes of (f). Since there does not exist a chain in $D[x]$ of three distinct prime ideals that are contracted to (0) in D, $P_2 \cap D$ and $P_3 \cap D$ are maximal ideals of D. By Proposition 3.6, P_2 and P_3 are extensions of two distinct maximal ideals of D, which contradicts the fact that $P_2 < P_3$. This proves that $dim \, Ass(f) \leq 1$. By hypothesis, $dim \, Ass(f) \neq 0$, so $dim \, Ass(f) = 1$. Note that the argument above also shows that if P is an embedded associated prime of (f), then $M[x] \subseteq P$. Since $P \cap D \neq (0)$ and $P \in Ass(f)$, it follows from Proposition 3.6 that P is the extension of M in $D[X]$, — that is, $P = M[x]$. This shows that $M[x]$ is the unique embedded associated prime of (f). Take $a \in D - \{0\}$. Since $\sqrt{aD} = M$, $\sqrt{aD[x]} = M[x]$. Since (f) is decomposable, it follows from Theorem 4.5 that $S(f) = (f) : a^k$ for some integer $k > 0$.

$(2) \Rightarrow (3)$: Suppose that $S(f) = (f) : a$ for some $a \in M$. Since $S(f) \neq D[x]$, $a \neq 0$; hence $\sqrt{aD} = M$. Since $\sqrt{aD[x]} = M[x]$, $a \in S$, so that $(f) : a^n \subseteq S(f)$ for every integer $n > 0$. But $(f) : a \subseteq (f) : a^n$, so we have $(f) : a = (f) : a^n$ for all n.

$(3) \Rightarrow (4)$: If $d = 0$, then $D[x] = (f) : d = (f) : d^2 = (f) : d^3 ...$, and similarly, $(f) = (f) : d = (f) : d^2 = (f) : d^3 = ...$ for each unit d of D. Assume that d is a nonzero nonunit of D. We may assume without loss of generality that $(f) : a = (f) : a^2 = (f) : a^3 = ...$ Since D is a one-dimensional quasilocal domain, it is known that $\cap_{i=1}^{\infty}(r^i) = (0)$ for any $r \in M$. Take an integer n such that $(d^n) \subseteq (a)$. For any given integer $i > 0$, take an integer m such that $(a^m) \subseteq (d^{n+i})$. Then $(f) : a \subseteq (f) : d^n \subseteq (f) : d^{n+i} \subseteq (f) : a^m$. This shows that $(f) : d^n = (f) : d^{n+i}$. Hence the chain $(f) : d \subseteq (f) : d^2 \subseteq (f) : d^3 \subseteq ...$ is stable.

$(4) \Rightarrow (1)$: By part (2) of Lemma 4.1, $Ass(S(f))$ consists of the isolated associated primes of (f). Since every principal ideal of D is decomposable, $Ass(f)$ is a finite set by Proposition 3.7. Hence $S(f)$ is decomposable by Proposition 3.2. Applying Theorem 4.5 we only need to show that $S(f) = (f) : a$ for some $a \in D$. Take $b \in D - \{0\}$. Then there exists an integer $n > 0$ such that $(f) : b^n = (f) : b^{n+i}$ for all integer $i > 0$. Let $a = b^n$. We will show that $S(f) = (f) : a$. Clearly $(f) : a \subseteq S(f)$. Conversely, if $g \in S(f) - (f)$, then $gh \in (f)$ for some $h \in S$. Hence $M[x]$ is the only minimal prime ideal of $(f) : g$ because $M[x]$ is the only associated prime of (f) that meets S. This means that $\sqrt{(f) : g} = M[x]$. Choose an integer $k > n$ such that $b^k \in (f) : g$. Then $g \in (f) : b^k = (f) : a$. This shows that $S(f) \subseteq (f) : a$, completing the proof. $\qquad\square$

Proposition 4.8. *Let D be a one-dimensional semiquasilocal domain and let f be a polynomial in $D[x]$. If $dim \, Ass(f) \neq 0$, then (f) is de-*

composable if and only if for every multiplicative system S in $D[x]$, the saturation $S(f)$ of (f) with respect to S is of the form $(f) : s$, where $s \in S$.

Proof.   ($\Rightarrow$):   If (f) is decomposable, it is known that for every multiplicative system S, $S(f) = (f) : s$ for some $s \in S$.

($\Leftarrow$):   The proof of Proposition 4.7 shows that $dim \, Ass(f) < 2$. By hypothesis, $dim \, Ass(f) \neq 0$, so $dim \, Ass(f) = 1$. Since $dim \, D = 1$ and D is semiquasilocal, every ideal of D is decomposable; hence it follows from Proposition 3.7 that $|Ass(f)| < \infty$. By Lemma 4.4, (f) is decomposable if and only if $fD[x]_P$ is decomposable for each embedded associated prime P of (f). Since there does not exist a chain in $D[x]$ of three distinct prime ideals that are contracted to (0) in D, $P \cap D$ is a maximal ideal of D. By Proposition 3.6, $P = (P \cap D)[x]$. Since D is semiquasilocal, $P \cap D$ is the radical of a principal ideal of D; hence $(P \cap D)[x]$ is the radical of a principal ideal of $D[x]$. Let $\{P, P_1, ..., P_n\} = \{Q \in Ass(f) \mid Q \subseteq P\}$. Then $Ass(fD[x]_P) = \{PD[x]_P, P_1 D[x]_P, ..., P_n D[x]_P\}$. Let $S = D[x] - \cup_{i=1}^n P_i$ and let $T = D[x]_P - \cup_{i=1}^n P_i D[x]_P$. Then $S(f)D[x]_P = T(fD[x]_P)$ by Lemma 4.6. By hypothesis, $S(f) = (f) : s$ for some $s \in S$, so $T(fD[x]_P) = fD[x]_P : s$ and $s \in T$. Since $PD[x]_P \neq \cup_{i=1}^n P_i D[x]_P$, $Ass(T(fD[x]_P)) = \{P_1 D[x]_P, ..., P_n D[x]_P\}$ by part (2) of Lemma 4.1. This shows that $T(fD[x]_P)$ has only finitely many associated primes and each of them is isolated. Hence $T(fD[x]_P)$ is decomposable by Proposition 3.2. Now if s is a unit in $D[x]_P$, then $fD[x]_P = T(fD[x]_P)$; hence $fD[x]_P$ is decomposable. If s is not a unit in $D[x]_P$, then there exists $g \in T(fD[x]_P) - fD[x]_P$ such that $g \in fD[x]_P : s$, or $s \in fD[x]_P : g$. Since $Ass(fD[x]_P) = \{PD[x]_P, P_1 D[x]_P, ..., P_n D[x]_P\}$, the only prime ideal of $D[x]_P$ that contains s is $PD[x]_P$; hence $PD[x]_P$ is the radical of sD_P. By Theorem 4.5, $fD[x]_P$ is decomposable. This completes the proof.   □

5. ZERO-DIMENSIONAL QUASILOCAL COEFFICIENT RINGS

A natural setting in which to consider Anderson's question is that in which the coefficient ring (R, M) is zero-dimensional and quasilocal. In this case each proper ideal of R is M-primary. We have noted in Remark 3.5 that each finitely generated ideal of $R[x]$ is decomposable if R is a chained ring, and we show in Corollary 5.3 that the same conclusion is valid if $M^2 = (0)$. We first review the notion of the content ideal of an ideal of a polynomial ring (cf. [OR]).

If I is an ideal of a polynomial ring $R[x]$ in the indeterminate x over the commutative unitary ring R, then the set

$\{r \in R \mid r$ is the coefficient of an element of $I\}$

is an ideal of R called the **content ideal** of I and denoted by $C(I)$. If $\{f_\alpha(x)\}_{\alpha \in A}$ is a generating set for I and if Y_α is the set of coefficients of f_α, then $\cup_{\alpha \in A} Y_\alpha$ generates $C(I)$. In particular $C(I)$ is finitely generated if I is finitely generated. Also, if $f \in R[x]$, then $C(f) = C(fR)$.

Lemma 5.1. *Let x be an indeterminate over the ring R. Let I be a finitely generated ideal of $R[x]$ and let S be the multiplicative system in $R[x]$ consisting of polynomials of unit content. If $R/Ann(C(I))$ is a Noetherian ring, then $S(I) = I : f$ for some $f \in S$.*

Proof.   Let $J = C(I)$. We first observe that $S(I) \subseteq J[x]$. Thus, if $g \in S(I)$, then $gh \in I$ for some $h \in S$, and hence $C(g) = C(gh) \subseteq J$. Consequently, $g \in J[x]$ and $S(I) \subseteq J[x]$. Now $J[x]$ is a finitely generated $R[x]$-module and $R[x]/Ann(J[x]) = R[x]/(Ann(J))[x] \simeq (R/Ann(J))[x]$ is a Noetherian ring . Hence $J[x]$ is a Noetherian $R[x]$-module. If $s \in S$, then $I : s \subseteq S(I) \subseteq J[x]$, so among the ideals $I : s$, $s \in S$, we may choose a maximal element $I : f$. We show that $S(I) = I : f$. Thus, take $g \in S(I)$ and $h \in S$ such that $gh \in I$. Maximality of $I : f$ implies that $I : f = I : fh$, and it is clear that $g \in I : fh$. Hence $g \in I : f$ and $I : f = S(I)$, as we wished to show. $\square$

Theorem 5.2. *If (R, M) is a 0-dimensional quasilocal ring such that $R/Ann(M)$ is Noetherian, then each finitely generated ideal of $R[x]$ is decomposable.*

Proof. Let I be a finitely generated ideal of $R[x]$ and let $A = Ann(M) = (0) : M$. If $I \not\subseteq M[x]$, then I is decomposable by part (1) of Theorem 4.2.

On the other hand, if $I \subseteq M[x]$, then $C(I) \subseteq M$, so that $Ann(M) \subseteq Ann(C(I))$. Since $R/Ann(M)$ is Noetherian, so is $R/Ann(C(I))$. Thus $I : f = S(I)$ for some $f \in S = R[x] - M[x]$, the multiplicative system in $R[x]$ consisting of polynomials of unit content. Hence I is decomposable by part (2) of Theorem 4.2. $\square$

Corollary 5.3. *If (R, M) is a 0-dimensional quasilocal ring such that $M^2 = (0)$, then each finitely generated ideal of $R[x]$ is decomposable.*

In relation to Corollary 5.3, it would be interesting to know whether the conclusion remains valid if M is nilpotent, but $M^2 \neq (0)$. Note that in this case $R/Ann(M)$ need not be Noetherian, so Theorem 5.2 is not directly applicable.

REFERENCES

[A1] D. D. Anderson, Twenty questions in commutative ring theory, Midwest-Great Plains Commutative Algebra Workshop, 1991.

[A2] D. D. Anderson, Some problems in commutative ring theory, Zero-dimensional commutative rings (Knoxville, TN, 1994), Lect. Notes in Pure and Appl. Math., Vol. 171, Dekker, New York, 1995, pp. 363-372.

[AG] J. Arnold and R. Gilmer, The dimension sequence of a commutative ring, Amer. J. Math. 96(1974), 385-408.

[AM] M. Atiyah and I. MacDonald, Introduction to Commutative Algebra, Addison-Wesley, Reading, MA,, 1969.

[B] N. Bourbaki, Elements of Mathematics, Commutative Algebra, Addison-Wesley, Reading, MA, 1972.

[BH] J. Brewer and W. Heinzer, Associated primes of principal ideals, Duke Math. J., 41(1974), 1-7.

[F] G. Fu, Primary decomposition of ideals, Proc. 1997 Fes Conf. (to appear).

[G1] R. Gilmer, Multiplicative Ideal Theory, Queen's Papers Pure Appl. Math., Vol. 90, Queen's Univ., Kingston, Ontario, 1992.

[G2] R. Gilmer, Commutative Semigroup Rings, Univ. Chicago Press, Chicago, IL, 1984.

[GH1] R. Gilmer and W. Heinzer, Ideals contracted from a Noetherian extension ring, J. Pure Appl. Algebra, 24(1982), 123-144.

[GH2] R. Gilmer and W. Heinzer, Primary ideals with finitely generated radical in a commutative ring, manuscripta math., 78(1993), 201-221.

[HH1] W. Heinzer and C. Huneke, The Dedekind-Mertens Lemma and the contents of polynomials, Proc. Amer. Math. Soc. (to appear).

[HH2] W. Heinzer and C. Huneke, Gaussian polynomials and content ideals, Proc. Amer. Math. Soc. 125(1997), 739-745.

[HO1] W. Heinzer and J. Ohm, Locally Noetherian commutative rings, Trans. Amer. Math. Soc. 158(1971), 273-284.

[HO2] W. Heinzer and J. Ohm, On the Noetherian-like rings of E. G. Evans, Proc. Amer. Math. Soc. 34(1972), 73-74.

[Hu] J.A. Huckaba, Commutative Rings with Zero Divisors, Marcel Dekker, New York, 1988.

[J] C. U. Jensen, A remark on arithmetical rings, Proc. Amer. Math. Soc. 15(1964), 951-954.

[K] I. Kaplansky, Commutative Rings, Allyn and Bacon, Boston, 1970.

[Kr] W. Krull, Über einen Hauptsatz der allgemeinen Idealtheorie, S.-B. Heidelberger Akad. Wiss. (1929), 11-16.

[N]   M. C. Nascimento, Intersection of powers of prime ideals in Krull domains, Commun. Algebra 20(1992), 777-782.

[OR] J. Ohm and D. Rush, The finiteness of I when R[X]/I is flat, Trans. Amer. Math. Soc. 171(1972), 377-408.

[ZS] O. Zariski and P. Samuel, Commutative Algebra, Vol. I, Springer Verlag, Berlin-Heidelberg-New York, 1986.

Prüfer (##)-Domains and Localizing Systems of Ideals

STEFANIA GABELLI* Dipartimento di Matematica, Università di Roma "La Sapienza", Piazzale A. Moro 5, 00185 Roma, Italy.

e-mail: gabelli@mat.uniroma1.it

INTRODUCTION

This paper continues the study of Prüfer domains satisfying properties (#) and (##) begun in [9] and [11]. It was inspired by [3, Chapter V] and intends to be a complement to it.

We recall that a domain D is called a (#)-*domain* if $\cap\{D_M ; M \in \Lambda_1\} \neq \cap\{D_M ; M \in \Lambda_2\}$ for any pair of distinct nonempty subsets Λ_1 and Λ_2 of $\mathrm{Max}(D)$, and D is a (##)-*domain* if every overring of D is a (#)-domain. These definitions can be reformulated in terms of localizing systems of ideals. As a matter of fact, for any nonempty subset Λ of $\mathrm{Spec}(D)$, the set of ideals $\mathcal{F}(\Lambda) = \{I ; I \text{ integral ideal of } D \text{ such that } I \not\subseteq P, \text{ for each } P \in \Lambda\}$ is a (spectral) localizing system of ideals of D and $D_{\mathcal{F}(\Lambda)} = \cap\{D_P ; P \in \Lambda\}$. Hence a (#)-domain is a domain D such that $D_{\mathcal{F}(\Lambda_1)} \neq D_{\mathcal{F}(\Lambda_2)}$, for any pair of distinct nonempty subsets Λ_1 and Λ_2 of $\mathrm{Max}(D)$.

* Partially supported by research funds of *Ministero dell'Università e della Ricerca Scientifica e Tecnologica* and *NATO grant* 970140.

Prüfer (##)-domains have very good ideal-theoretic properties [11] and well behave with respect to several "trace properties" [2, 14, 15]. For example, (##)-property and radical trace property coincide in Prüfer domains satisfying the ascending chain condition on prime ideals [14, Theorem 2.7]. An update account in this direction is given in [3, Chapter IV].

In [18] it was introduced the apparently stronger condition (#p) . A (#p)-*domain* is a domain D such that $D_{\mathcal{F}(\Lambda_1)} \neq D_{\mathcal{F}(\Lambda_2)}$ for any couple of distinct nonempty subsets Λ_1 , Λ_2 of Spec(D) with the property that $P + Q = D$ for each $P \in \Lambda_1$ and $Q \in \Lambda_2$ such that $P \neq Q$. It is known that conditions (##) and (#p) coincide in those Prüfer domains with the property that $P \neq P^2$ for all nonzero prime ideals P , hence with the property that each nonzero prime ideal P is branched. Moreover, in this case, each localizing system of ideals of D is finitely generated. Precisely, the following Theorem holds.

THEOREM 0.1. [4, Théorème 2.7] (or [3, Theorem 5.5.4]). *Let D be a Prüfer domain. Then the following are equivalent*:

(i)   *D is a (##)-domain and $P \neq P^2$ for each nonzero prime ideal P* ;

(ii)   *D is a (#p)-domain and $P \neq P^2$ for each nonzero prime ideal P* ;

(iii)   *Each localizing system of ideals of D is finitely generated*;

(iv)   *$D_{\mathcal{F}_1} \neq D_{\mathcal{F}_2}$ for any pair of distinct localizing systems of ideals $\mathcal{F}_1$ and $\mathcal{F}_2$ of D* .

Prüfer domains that satisfy the equivalent conditions of Theorem 0.1 are called *generalized Dedekind domains* [18] and are thoroughly investigated in [3, Chapter V]. More recent contributions to the knowledge of ideal-theoretic properties of this class of domains are given in [6, 7, 16].

In Section 2 of this paper, we characterize Prüfer (##)-domains and Prüfer (##)-domains such that each nonzero prime ideal is branched by properties similar to those of Theorem 0.1. In particular, we show that these classes of domains can be characterized by the property that certain spectral localizing systems of ideals are finitely generated (Theorems 2.4 and 2.6), as it is illustrate by the following picture (where D is a Prüfer domain).

$D_{\mathcal{F}_1} \neq D_{\mathcal{F}_2}$ *for any pair of distinct localizing systems $\mathcal{F}_1$ and $\mathcal{F}_2$*		*D is a (##)-domain and $P \neq P^2$ for each nonzero prime ideal P*		*Each localizing system $\mathcal{F}$ of D is finitely generated*
	$\Leftrightarrow$		$\Leftrightarrow$	
$\Downarrow$		$\Downarrow$		$\Downarrow$
$D_{\mathcal{F}_1} \neq D_{\mathcal{F}_2}$ *for any pair of distinct spectral localizing systems $\mathcal{F}_1$ and $\mathcal{F}_2$*	$\Leftrightarrow$	*D is a (##)-domain and each nonzero prime ideal P is branched*	$\Leftrightarrow$	*Each spectral localizing system $\mathcal{F}$ of D is finitely generated*
$\Downarrow$		$\Downarrow$		$\Downarrow$
$D_{\mathcal{F}_1} \neq D_{\mathcal{F}_2}$ *for any pair of distinct irredundant spectral localizing systems $\mathcal{F}_1$ and $\mathcal{F}_2$*	$\Leftrightarrow$	*D is a (##)-domain*	$\Leftrightarrow$	*Each irredundant spectral localizing system $\mathcal{F}$ of D is finitely generated*
$\Downarrow$		$\Downarrow$		
$D_{\mathcal{H}(\Lambda_1)} \neq D_{\mathcal{H}(\Lambda_2)}$ *for any pair of distinct nonempty subsets Λ_1 and Λ_2 of* Max(D)	$\Leftrightarrow$	*D is a (#)-domain*		

As a consequence of our results, we get that properties (##) and (#$_P$) coincide in any Prüfer domain (Theorem 2.5), thus settling a problem left open in [3].

In Section 1 we preliminarly give several characterizations of finitely generated spectral localizing systems of ideals in valuation and Prüfer domains. Finally, in Section 3, we provide examples that illustrate the connections among finitely generated (spectral) localizing systems of ideals and properties (#) and (##).

1. FINITELY GENERATED LOCALIZING SYSTEMS OF IDEALS

Let D be any commutative ring and $\mathcal{F}$ a multiplicative system of ideals of D, that is a set of nonzero integral ideals of D closed under multiplication. The overring $D_{\mathcal{F}} := \cup\{(D:J) ; J \in \mathcal{F}\}$ of D is called the *generalized ring of fractions* of D with respect to $\mathcal{F}$. We refer to [1, Section 1] and [13, Section 4] for basic properties of multiplicative systems of ideals and generalized quotient rings (see also [3, Section 5.1]).

If I is a fractional ideal of D, set $I_{\mathcal{F}} := \cup\{(I:J) ; J \in \mathcal{F}\}$. Then $I_{\mathcal{F}}$ is a fractional ideal of $D_{\mathcal{F}}$ and $ID_{\mathcal{F}} \subseteq I_{\mathcal{F}}$.

If $\mathcal{F}$ is a multiplicative system of ideals of D, we denote by $\mathrm{Sat}(\mathcal{F})$ the *saturation* of $\mathcal{F}$, that is the set of ideals of D containing some ideal in $\mathcal{F}$. Then $\mathrm{Sat}(\mathcal{F})$ is a multiplicative system of ideals of D and clearly $I_{\mathcal{F}} = I_{\mathrm{Sat}(\mathcal{F})}$ for each ideal I of D. We say that $\mathcal{F}$ is *saturated* if $\mathcal{F} = \mathrm{Sat}(\mathcal{F})$, that is, if $I \in \mathcal{F}$ and J is an ideal of D such that $I \subseteq J$, then $J \in \mathcal{F}$.

A multiplicative system of ideals $\mathcal{F}$ of D is *finitely generated* if, for each $I \in \mathcal{F}$, there exists a finitely generated ideal J such that $J \subseteq I$ and $J \in \mathcal{F}$. It is clear that, if $\mathcal{F}$ is finitely generated, then $\mathrm{Sat}(\mathcal{F})$ is finitely generated.

We say that a saturated multiplicative system of ideals $\mathcal{F}$ of a domain D is a *localizing system* if, whenever $I \in \mathcal{F}$ and J is an ideal of D such that $(J :_D xD) \in \mathcal{F}$ for all $x \in I$, then $J \in \mathcal{F}$. A very general reference for localizing systems of ideals is [17].

If P is a prime ideal of D, we set $\mathcal{F}(P) = \{I ; I$ integral ideal of D such that $I \not\subseteq P\}$. Then $\mathcal{F}(P)$ is a localizing system and $D_{\mathcal{F}(P)} = D_P$. Moreover, if Λ is a nonempty family of prime ideals of D, setting $\mathcal{H}(\Lambda) := \cap\{\mathcal{H}(P)\,;\,P \in \Lambda\}$, we have that $\mathcal{H}(\Lambda)$ is a localizing system and $D_{\mathcal{H}(\Lambda)} = \cap\{D_P\,;\,D \in \Lambda\}$ [3, Proposition 5.1.4]. Always $\mathrm{Sat}(\mathcal{F}) \subseteq \cap\{\mathcal{H}(P)\,;\,P \notin \mathrm{Sat}(\mathcal{F})\}$, hence $D_{\mathcal{F}} \subseteq \cap\{D_P\,;\,P \notin \mathrm{Sat}(\mathcal{F})\}$.

We call $\mathcal{F}$ a *spectral* localizing system if $\mathcal{F} = \mathcal{H}(\Lambda)$ for some nonempty subset Λ of $\mathrm{Spec}(D)$. We also say that $\mathcal{H}(\Lambda)$ is an *irredundant* spectral localizing system of D if $\mathcal{H}(\Lambda) \subsetneq \cap\{\mathcal{H}(P)\,;\,P \in \Lambda\backslash\{Q\}\}$, for each $Q \in \Lambda$, equivalently $\cap\{\mathcal{H}(P)\,;\,P \in \Lambda\backslash\{Q\}\} \not\subseteq \mathcal{H}(Q)$, for each $Q \in \mathcal{H}(\Lambda)$. It is easy to see that $\mathcal{H}(\Lambda)$ is irredundant if and only if $P \not\subseteq Q$, for any pair of distinct prime ideals $P, Q \in \Lambda$. Thus, if $\mathcal{H}(\Lambda)$ is irredundant, $(0) \in \Lambda$ if and only if $\Lambda = \{(0)\}$.

LEMMA 1.1. *Let D be any commutative ring and $\mathcal{F}$ a multiplicative system of ideals of D. Then*:

(a) *If I is an integral ideal of D, $I_{\mathcal{F}} = D_{\mathcal{F}}$ if and only if $I \in \mathrm{Sat}(\mathcal{F})$.*

(b) *The correspondence $P \longrightarrow P_{\mathcal{F}}$ is a one-to-one inclusion preserving correspondence between the set of prime ideals P of D such that $P \notin \mathrm{Sat}(\mathcal{F})$ and the set of prime ideals P' of $D_{\mathcal{F}}$ such that $JD_{\mathcal{F}} \not\subseteq P'$ for any $J \in \mathcal{F}$, whose inverse is defined by $P' \longrightarrow P' \cap D$. In addition $(D_{\mathcal{F}})_{P_{\mathcal{F}}} = D_P$ for each $P \notin \mathrm{Sat}(\mathcal{F})$.*

Proof. (a) is easy to check. (b) is [1, Theorem 1.1].

By Lemma 1.1 (a), we get that if $J \subseteq D$ and $JD_{\mathcal{F}} = D_{\mathcal{F}}$, then $J \in \mathrm{Sat}(\mathcal{F})$. The following Proposition gives several conditions equivalent to the converse.

PROPOSITION 1.2. *Let D be any commutative ring, $\mathcal{F}$ a multiplicative system of ideals of D and $\Lambda(\mathcal{F}) := \{P \in \mathrm{Spec}(D)\,;\,P \notin \mathrm{Sat}(\mathcal{F})$ and P is maximal with respect to this property$\}$. Then the following are equivalent*:

(i) $\mathrm{Spec}(D_{\mathcal{F}}) = \{P_{\mathcal{F}}\,;\,P \notin \mathrm{Sat}(\mathcal{F})\}$;

(ii) $JD_{\mathcal{F}} = D_{\mathcal{F}}$, *for each* $J \in \mathcal{F}$;

(iii) $\mathrm{Sat}(\mathcal{F}) = \{J \subseteq D\,;\,JD_{\mathcal{F}} = D_{\mathcal{F}}\}$;

(iv) $ID_{\mathcal{F}} = I_{\mathcal{F}}$, *for each fractional ideal I of D*;

(v)   *If I' is an ideal of $D_{\mathcal{F}}$ and $I := I' \cap D$, then $I' = ID_{\mathcal{F}} = I_{\mathcal{F}}$;*

(vi)   *The set $\Lambda(\mathcal{F})$ is not empty and $\mathrm{Max}(D_{\mathcal{F}}) = \{QD_{\mathcal{F}}; Q \in \Lambda(\mathcal{F})\}$;*

(vii) $\mathrm{Sat}(\mathcal{F}) = \cap \{\,\mathcal{F}(Q)\,\} ; Q = M' \cap D , M' \in \mathrm{Max}(D_{\mathcal{F}})\}$, *in particular* $\mathrm{Sat}(\mathcal{F})$ *is an irredundant spectral localizing system of D .*

Proof. (i) $\Rightarrow$ (ii). If $J \subseteq D$ and $JD_{\mathcal{F}} \neq D_{\mathcal{F}}$, then $JD_{\mathcal{F}} \subseteq Q$, for some $Q \in \mathrm{Spec}(D_{\mathcal{F}})$. Hence $Q \notin \{P_{\mathcal{F}}, P \notin \mathrm{Sat}(\mathcal{F})\}$ by Lemma 1.1 (b).

(ii) $\Rightarrow$ (iv). Always $ID_{\mathcal{F}} \subseteq I_{\mathcal{F}}$. Conversely, if (ii) holds and $x \in I_{\mathcal{F}}$, then $xJ \subseteq I$ for some $J \in \mathcal{F}$. Thus $xJD_{\mathcal{F}} = xD_{\mathcal{F}} \subseteq ID_{\mathcal{F}}$ and so $I_{\mathcal{F}} \subseteq ID_{\mathcal{F}}$.

(iv) $\Rightarrow$ (iii) by Lemma 1.1 (a).

(iv) $\Rightarrow$ (v). It is enough to observe that we have $ID_{\mathcal{F}} \subseteq I' \subseteq I_{\mathcal{F}}$ (because, if $x \in I'$, then $xJ \subseteq I' \cap D = I$ for some $J \in \mathcal{F}$).

(iii) $\Rightarrow$ (ii) and (v) $\Rightarrow$ (i) are immediate.

(i) $\Rightarrow$ (vi). If M' is a maximal ideal of $D_{\mathcal{F}}$ and $M' = Q_{\mathcal{F}}$, then $Q = M' \cap D \in \Lambda(\mathcal{F})$ and $\mathrm{Max}(D_{\mathcal{F}}) = \{QD_{\mathcal{F}}; Q \in \Lambda(\mathcal{F})\}$ through (iv).

(vi) $\Rightarrow$ (vii). If (vi) holds, then $\Lambda(\mathcal{F}) = \{M' \cap D ; M' \in \mathrm{Max}(D_{\mathcal{F}})\}$ and it is clear that $\mathcal{F} = \mathcal{F}\Lambda(\mathcal{F})$.

(vii) $\Rightarrow$ (iii). If (vii) holds, then $ID_{\mathcal{F}} = D_{\mathcal{F}}$ if and only if $I \not\subseteq M' \cap D$ for each $M' \in \mathrm{Max}(D_{\mathcal{F}})$, if and only if $I \in \mathrm{Sat}(\mathcal{F})$.

It is well known that B is a flat overring of D if and only if there exists a multiplicative system of ideals $\mathcal{F}(B)$ of D which satisfies condition (ii) (hence all conditions) of Proposition 1.2 and such that $B = D_{\mathcal{F}(B)}$ [1, Theorem 1.3] (or [3, Remark 5.1.11 (b)]).

The equivalence (ii) $\Leftrightarrow$ (vi) is proved in [5, Lemme 1.1] under the hypothesis that $\mathcal{F}$ be finitely generated. But this is certainly true when the conditions above are satisfied, because the multiplicative set of ideals $\mathcal{F}(D_{\mathcal{F}})$ $:= \{J \subseteq D ; JD_{\mathcal{F}} = D_{\mathcal{F}}\}$ is finitely generated. We will show next that, the converse holds in the class of Prüfer domains. That is, if D is a Prüfer domain, a multiplicative set of ideals $\mathcal{F}$ is finitely generated if and only if it satisfies the equivalent conditions of the previous Proposition.

Note that, if $\mathcal{F}$ is finitely generated, then always the set $\Lambda(\mathcal{F})$ is not empty and $\mathrm{Sat}(\mathcal{F}) = \mathcal{F}\Lambda(\mathcal{F}))$ [5, Lemme 1.1] but, when D is not Prüfer, a maximal ideal of $D_{\mathcal{F}}$ is not necessarily of the form $QD_{\mathcal{F}}$ with $Q \in \Lambda(\mathcal{F})$. For example, if D is noetherian, then each multiplicative set of ideals $\mathcal{F}$ of D is finitely generated but $D_{\mathcal{F}}$ is not necessarily flat over D. A specific example of a noetherian integrally closed domain D such that an overring of D of type $D_{\mathcal{F}(\Lambda)}$, with Λ a nonempty set of heigth-one primes, is not flat over D is given in [8, pag. 32].

THEOREM 1.3. *Let D be a Prüfer domain and $\mathcal{F}$ a multiplicative system of ideals of D. Then $\mathcal{F}$ is finitely generated if and only if the equivalent conditions of Proposition 1.2 hold.*

Proof. We show that $\mathcal{F}$ is finitely generated if and only if $\mathrm{Sat}(\mathcal{F}) = \{J \subseteq D \; ; \; JD_{\mathcal{F}} = D_{\mathcal{F}}\}$. We have already observed that the multiplicative set $\mathcal{F}(D_{\mathcal{F}})$ $:= \{J \subseteq D \; ; \; JD_{\mathcal{F}} = D_{\mathcal{F}}\}$ of D is finitely generated. Conversely, let $J \in \mathcal{F}$ be a finitely generated ideal. Since J is invertible and $(D{:}J) \subseteq D_{\mathcal{F}}$, then $J(D{:}J)$ $= D \subseteq JD_{\mathcal{F}}$. Whence, if $\mathcal{F}$ is finitely generated, then $ID_{\mathcal{F}} = D_{\mathcal{F}}$ for each ideal $I \in \mathrm{Sat}(\mathcal{F})$ and so $\mathrm{Sat}(\mathcal{F}) \subseteq \mathcal{F}(D_{\mathcal{F}})$. The opposite inclusion follows from Lemma 1.1 (a).

Recalling that D is a Prüfer domain if and only if each overring of D is flat, by the previous Theorem, we obtain the following characterization of Prüfer domains in terms of localing systems.

THEOREM 1.4 [3, Theorem 5.1.15]. *D is a Prüfer domain if and only if, for any overring B of D, there exists one and only one finitely generated localizing system $\mathcal{F}$ of D such that $B = D_{\mathcal{F}}$.*

When $\mathcal{F} = \mathcal{F}(\Lambda)$ is a spectral localizing system and Q is a prime ideal of D, then $Q \notin \mathcal{F}$ if and only if $Q \subseteq P$ for some prime $P \in \Lambda$. Therefore, we immediately obtain the following Proposition.

PROPOSITION 1.5. *Let D be a Prüfer domain, Λ a nonempty subset of $\mathrm{Spec}(D)$ and Λ_0 the set of maximal elements of Λ. Then the following conditions are equivalent*:

(i) $\mathcal{H}(\Lambda)$ *is finitely generated;*

(ii) $\mathcal{H}(\Lambda) = \mathcal{H}(D_{\mathcal{H}(\Lambda)}) := \{I \subseteq D ; ID_{\mathcal{H}(\Lambda)} = D_{\mathcal{H}(\Lambda)}\}$;

(iii)   *The set* Λ_0 *is not empty* , $\mathcal{H}(\Lambda) = \mathcal{H}(\Lambda_0)$ *and* $\mathrm{Max}(D_{\mathcal{H}(\Lambda)}) = \{PD_{\mathcal{H}(\Lambda)} ; P \in \Lambda_0\}$.

In particular, if $\mathcal{H}(\Lambda)$ *is irredundant, then* $\mathcal{H}(\Lambda)$ *is finitely generated if and only if* $\mathrm{Max}(D_{\mathcal{H}(\Lambda)}) = \{PD_{\mathcal{H}(\Lambda)} ; P \in \Lambda\}$.

We can say more in the local case. Let V be a valuation domain. If P is any prime ideal of V, set $\mathcal{F}^*(P) := \{I ; I$ integral ideal of V and $P \subseteq I\}$. In general, the set of ideals $\mathcal{F}^*(P)$ is not multiplicatively closed. However, $\mathcal{F}^*(P)$ is a localizing system if and only if $P = P^2$. In this case $\mathcal{H}(P) \subsetneq \mathcal{F}^*(P)$ (because $P \in \mathcal{F}^*(P) \backslash \mathcal{H}(P)$) and $V_P = V_{\mathcal{F}^*(P)}$ [4, Lemme 1.4] (or [3, Proposition 5.1.12]). We observe that $\mathcal{H}(P)$ is finitely generated for all prime ideals P of V. Indeed an ideal I is in $\mathcal{H}(P)$ if and only if $P \subsetneq I$. Hence, if $x \in I \backslash P$ and $J := xV$, then $J \subseteq I$ and $J \in \mathcal{H}(P)$. By Theorem 1.4, we get that, when $P = P^2$, the localizing system $\mathcal{F}^*(P)$ is never finitely generated.

If Λ is any nonempty subset of $\mathrm{Spec}(V)$, consider the prime ideal $P_0 := \cup\{P ; P \in \Lambda\}$. Hence $P_0 = (0)$ if and only if $\Lambda = \{(0)\}$. If I is an ideal of V, then clearly $P \subseteq I$ for any $P \in \Lambda$ if and only if $P_0 \subseteq I$. Therefore, either $\mathcal{H}(\Lambda) = \mathcal{H}(P_0)$, exactly when $P_0 \in \Lambda$, or $\mathcal{H}(\Lambda) = \mathcal{F}^*(P_0)$ (in this case, necessarily $P_0 = P_0^2 \notin \Lambda$). We remark that, in general, given any localizing system $\mathcal{F}$ of V, there exists a prime ideal Q such that $\mathcal{F} = \mathcal{H}(Q)$ or $\mathcal{F} = \mathcal{F}^*(Q)$ [4, Lemme 1.4] (or [3, Proposition 5.1.12]).

Now for $P \in \mathrm{Spec}(V)$ and $P \neq (0)$, define $\Lambda(P) := \{Q \in \mathrm{Spec}(V) ; Q \subsetneq P\} = \{Q \in \mathrm{Spec}(V) ; Q \notin \mathcal{F}^*(P)\}$. Since $P \neq (0)$, then $\Lambda(P)$ is nonempty. If, as above, $P_0 := \cup\{Q \in \Lambda(P)\}$, then $P_0 \subseteq P$ and clearly $\mathcal{F}^*(P) \subseteq \mathcal{H}(\Lambda(P)) \subseteq \mathcal{H}(P_0)$.

Recall that a prime ideal P of a domain D is *branched* if there exists a P-primary ideal Q of D different from P . Branched ideals of valuation and Prüfer domains are characterized respectively in [10, Theorem 17.3] and [10, Theorem 23.3].

PROPOSITION 1.6. *Let V be a valuation domain, P a nonzero prime ideal of V and $\Lambda(P) := \{Q \in \mathrm{Spec}(V) ; Q \subsetneq P\}$. Then the following are equivalent*:

(i) $\mathcal{F}^*(P) \subsetneq \mathcal{F}(\Lambda(P))$;

(ii)   *There exists a prime ideal $P_0 \subsetneq P$ such that $\mathcal{F}(\Lambda(P)) = \mathcal{F}(P_0)$* ;

(iii) $\Lambda(P)$ *has a maximal element P_0* ;

(iv) $\mathcal{F}(\Lambda(P))$ *is finitely generated*;

(v) P *is branched.*

Proof. (ii) $\Leftrightarrow$ (iii) is clear and (iii) $\Leftrightarrow$ (iv) by Proposition 1.5.

(i) $\Leftrightarrow$ (v). By definition, $I \in \mathcal{F}(\Lambda(P)) \setminus \mathcal{F}^*(P)$ if and only if $I \neq P$ and P is minimal over I. This is equivalent to say that P is branched by [10, Theorem 17.3].

(v) $\Rightarrow$ (iii). Let $P_0 = \cap\{H ; H \neq P$ and H is P-primary$\}$. Then P_0 is a maximal element of $\Lambda(P)$ by [10, Theorem 17.3].

(iii) $\Rightarrow$ (v). Since there are no prime ideals of V properly between P_0 and P, then P is branched again by [10, Theorem 17.3].

We remark that, in order to prove the previous Proposition, we do not need the hypothesis that the set of ideals $\mathcal{F}^*(P)$ be a localizing system (that is $P = P^2$).

COROLLARY 1.7. *Let V be a valuation domain, P a nonzero prime ideal of V such that $P = P^2$ and $\Lambda(P) := \{Q \in \mathrm{Spec}(V) ; Q \subsetneq P\}$. Then the following are equivalent*:

(i) $\mathcal{F}^*(P) = \mathcal{F}(\Lambda(P))$;

(ii) $\mathcal{F}^*(P)$ *is a (redundant) spectral localizing system*;

(iii) $\mathcal{F}(\Lambda(P))$ *is not finitely generated*;

(iv) P *is unbranched.*

Proof. (i) $\Leftrightarrow$ (iii) $\Leftrightarrow$ (iv) follow immediately from Proposition 1.6.

(ii) $\Rightarrow$ (i). Always $\mathcal{F}^*(P) \subseteq \mathcal{F}(\Lambda(P))$. If $\mathcal{F}^*(P) = \mathcal{F}(\Lambda)$ for some $\Lambda \subseteq \mathrm{Spec}(V)$, then $\Lambda \subseteq \Lambda(P)$ and so the equality holds.

(iii) $\Rightarrow$ (ii) Since P is unbranched, then $P = \cup\{Q \in \Lambda(P)\}$ [10, Theorem 17.3]. Hence, via (i), $\mathcal{F}^*(P) = \mathcal{F}(\Lambda(P)) = \mathcal{F}(\Lambda(P)) \cap \mathcal{F}(P)$ is a redundant spectral localizing system.

Recalling the discussion before Proposition 1.6, the following Corollary is immediate.

COROLLARY 1.8. *Let* Λ *be a nonempty subset of* Spec(V) *such that* $P_0 := \cup\{P ; P \in \Lambda\} \neq (0)$. *Then the following are equivalent*:

(i) P_0 *is unbranched*;

(ii) $P_0 \notin \Lambda$;

(iii) $\mathcal{F}(\Lambda) \neq \mathcal{F}(P_0)$;

(iv) $\mathcal{F}(\Lambda) = \mathcal{F}^*(P_0) = \mathcal{F}(\Lambda(P_0))$.

THEOREM 1.9. *Let* V *be a valuation domain. Then the following are equivalent*:

(i)   *For each nonempty subset* Λ *of* Spec(V), *there exists a prime ideal* P_0 *such that* $\mathcal{F}(\Lambda) = \mathcal{F}(P_0)$;

(ii) $\mathcal{F}(\Lambda)$ *is finitely generated, for each nonempty subset* Λ *of* Spec(V) ;

(iii) $\mathcal{F}(\Lambda(P))$ *is finitely generated, for each nonzero prime ideal* P *of* V;

(iv) V *satisfies the ascending chain condition on prime ideals*;

(v)   *Each nonzero prime ideal* P *of* V *is branched*.

Proof. (i) $\Rightarrow$ (ii) $\Rightarrow$ (iii) are trivial and (iii) $\Leftrightarrow$ (v) by Proposition 1.6.

(v) $\Rightarrow$ (i). Let Λ be a nonempty subset of Spec(V) and $P_0 := \cup\{P \in \Lambda\}$. If $\mathcal{F}(\Lambda) \neq \mathcal{F}(P_0)$, then $P_0 \neq (0)$ and P_0 is unbranched by Corollary 1.8.

(i) $\Rightarrow$ (iv). Let Λ be a nonempty subset of Spec(V) and $P_0 := \cup\{P \in \Lambda\}$. Since $\mathcal{F}(P_0) \subseteq \mathcal{F}(\Lambda)$, then $P \subseteq P_0$ for all $P \in \Lambda$ and , since $\mathcal{F}(\Lambda) \subseteq \mathcal{F}(P_0)$, then $P_0 \in \Lambda$.

(iv) $\Rightarrow$ (i). Let Λ be a nonempty subset of Spec(V) and $P_0 := \cup\{P ; P \in \Lambda\}$. If $P_0 = (0)$, then $P_0 \in \Lambda = \{(0)\}$. If $P_0 \neq (0)$, then $P_0 \in \Lambda$ by Corollary 1.8. Thus $\mathcal{F}(\Lambda) = \mathcal{F}(P_0)$.

COROLLARY 1.10. [4, Théorème 2.2] (or [3, Proposition 5.3.8]). *Let V be a valuation domain. Then the following are equivalent:*

(i) *Each localizing system $\mathcal{F}$ of V is finitely generated;*

(ii) *For each localizing system $\mathcal{F}$ of V, there exists a prime ideal P such that $\mathcal{F} = \mathcal{F}(P)$;*

(iii) *$P \neq P^2$ for each nonzero prime ideal P of V .*

By a standard argument, we can globalize the equivalence (iv) ⟺ (v) of Theorem 1.9. A global version of this Theorem in terms of localizing systems of ideals is given in Theorem 2.6 below.

THEOREM 1.11. *Let D be a Prüfer domain. Then the following are equivalent:*

(i) *D satisfies the ascending chain condition on prime ideals;*

(ii) *Each nonempty subset Λ of Spec(D) has maximal elements;*

(iii) *Each nonzero prime ideal P of D is branched.*

Recall that a prime ideal P of a Prüfer domain is branched if and only if it is minimal over a finitely generated ideal [10, Theorem 23.3]. In any domain, the condition that each prime ideal P is the radical of a finitely generated ideal is equivalent to the ascending chain condition on radical ideals [3, Theorem 3.1.11].

2. PRÜFER (##)-DOMAINS

A domain D is called a (#)-*domain* (respectively a (#P)-*domain*) if $D_{\mathcal{F}(\Lambda_1)} \neq D_{\mathcal{F}(\Lambda_2)}$ for any pair of distinct nonempty subsets Λ_1 and Λ_2 of Max(D) (respectively for any pair of distinct nonempty subsets Λ_1 and Λ_2 of Spec(D) with the property that $P + Q = D$ for each $P \in \Lambda_1$ and $Q \in \Lambda_2$ such that $P \neq Q$). D is called a (##)-*domain* (respectively a (##P)-*domain*) if every overring of D is a (#)-domain (respectively a (#P)-*domain*).

PROPOSITION 2.1. [3, Proposition 5.5.1]. *Let* $\mathcal{D}(\#)$, $\mathcal{D}(\#\#)$, $\mathcal{D}(\#_P)$ *and* $\mathcal{D}(\#\#_P)$, *denote the classes of Prüfer domains that satisfy* (#), (##), (#_P) *and* (##_P) *respectively. Then* $\mathcal{D}(\#\#_P) = \mathcal{D}(\#_P) \subseteq \mathcal{D}(\#\#) \subseteq \mathcal{D}(\#)$.

PROPOSITION 2.2. *Let* D *be a Prüfer domain. Then:*

(a) D *is a* (#)-*domain if and only if, for each* $N \in \mathrm{Max}(D)$, *there exists a finitely generated ideal* $J \subseteq N$ *such that* $J \not\subseteq M$ *for any maximal ideal* M $\in \mathrm{Max}(D) \setminus \{N\}$.

(b)   *Let* $N \in \mathrm{Max}(D)$ *and* $\Lambda := \mathrm{Max}(D) \setminus \{N\}$. *If* D *is a* (#)-*domain and* Q *is a prime ideal such that* $Q \subsetneqq N$, *then* $D_{\mathcal{F}(\Lambda)} \cap D_Q \not\subseteq D_N$.

(c) D *is a* (##)-*domain if and only if, for each* $P \in \mathrm{Spec}(D)$, *there exists a finitely generated ideal* $J \subseteq P$ *such that,* $J \not\subseteq M$ *when* $P \not\subseteq M$, *for each maximal ideal* M *of* D .

Proof. (a) is [11, Theorem 1, (a) $\Leftrightarrow$ (b)] (or [3, Theorem 4.1.6, (1) $\Leftrightarrow$ (3)]).

(b). If D is a (#)-domain, then (a) holds and we conclude by [11, Corollary 1].

(c) is [11, Theorem 3] (or [3, Theorem 4.1.7]).

PROPOSITION 2.3. *Let* D *be a Prüfer domain. Then:*

(a)   *If the localizing system* $\mathcal{F}(\Lambda)$ *of* D *is finitely generated, for any nonempty subset* Λ *of* $\mathrm{Max}(D)$, *then* D *is a* (##)-*domain, hence a* (#)-*domain.*

(b) D *is a* (#)-*domain if and only if, whenever* $D_{\mathcal{F}(\Lambda_1)} = D_{\mathcal{F}(\Lambda_2)}$, *with* Λ_1 *and* Λ_2 *nonempty subsets of* $\mathrm{Max}(D)$, *then* $\mathcal{F}(\Lambda_1) = \mathcal{F}(\Lambda_2)$.

Proof. (a) Let $P \in \mathrm{Spec}(D)$ and $\Lambda := \{M \in \mathrm{Max}(D) ; P \not\subseteq M\}$. If Λ is not empty, then $P \in \mathcal{F}(\Lambda)$. Since $\mathcal{F}(\Lambda)$ is finitely generated, then there exists a finitely generated ideal $J \subseteq P$ such that $J \not\subseteq M$ when $P \not\subseteq M$. If Λ is empty, this last condition is trivially verified. Hence D is a (##)-domain by part (c) of Proposition 2.2.

(b) It is enough to observe that, when Λ_1 and Λ_2 are nonempty subsets of $\mathrm{Max}(D)$, then $\mathcal{F}(\Lambda_1) \neq \mathcal{F}(\Lambda_2)$ if $\Lambda_1 \neq \Lambda_2$. As a matter of fact, if $N \in \Lambda_1 \setminus \Lambda_2$, then $N \in \mathcal{F}(\Lambda_2) \setminus \mathcal{F}(\Lambda_1)$.

The converse of Proposition 2.3 (a) holds for (##)-domains but not for (#)-domains (see Theorem 2.4 below and Example (c) in Section 3).

THEOREM 2.4. *Let D be a Prüfer domain. Then the following are equivalent*:

(i) D *is a* (##)-*domain*;

(ii) *Each irredundant spectral localizing system $\mathcal{H}(\Lambda)$ of D is finitely generated*;

(iii) $\mathcal{H}(\Lambda)$ *is finitely generated for each nonempty subset Λ of* $\mathrm{Max}(D)$;

(iv) *If $D_{\mathcal{H}(\Lambda_1)} = D_{\mathcal{H}(\Lambda_2)}$, with $\mathcal{H}(\Lambda_1)$, $\mathcal{H}(\Lambda_2)$ spectral irredundant localizing systems, then $\mathcal{H}(\Lambda_1) = \mathcal{H}(\Lambda_2)$* .

Proof. (ii) $\Rightarrow$ (iii) is trivial and (iii) $\Rightarrow$ (i) by Proposition 2.3 (a).

(i) $\Rightarrow$ (ii). By Proposition 1.5, it is enough to show that, for each irredundant spectral localizing system $\mathcal{H}(\Lambda)$ of D, it results $\mathrm{Max}(D_{\mathcal{H}(\Lambda)}) = \{PD_{\mathcal{H}(\Lambda)} ; P \in \Lambda\}$. Set $D' := D_{\mathcal{H}(\Lambda)}$ and let $\Gamma := \{M ; M = M' \cap D , M' \in \mathrm{Max}(D')$ and $P \subseteq M$ for some $P \in \Lambda\}$, $\mathcal{F} := \mathcal{H}(\Gamma)$. Then $D_{\mathcal{F}} \subseteq D' = \cap\{D_M ; M = M' \cap D , M' \in \mathrm{Max}(D')\}$ (because each $M \in \Gamma$ contains some $P \in \Lambda$) and so $D' = D_{\mathcal{F}} = \cap\{D'_{MD'} ; M \in \Gamma\}$. Since D' is a (#)-domain, then $\mathrm{Max}(D') = \{MD' , M \in \Gamma\}$, in particular each maximal ideal M' of D' contains PD' for some $P \in \Lambda$. Assume that, for a fixed $Q \in \Lambda$, QD' is not maximal and let $ND' \in \mathrm{Max}(D')$ be such that let $Q \subsetneqq N$. We have $\cap\{D'_{M'} ; M' \in \mathrm{Max}(D')$ and $M' \neq ND'\} \cap D'_{QD'} \subseteq D_{\mathcal{H}(\Lambda)} := D' \subseteq D'_{ND'}$ (where the first containement holds because $D'_{M'} = D_M \subseteq D_P$, for some $P \in \Lambda$). This contradicts Proposition 2.2 (b) because D' is a (#)-domain. Thus PD' is maximal as desired.

(ii) $\Rightarrow$ (iv) by Theorem 1.4.

(iv) $\Rightarrow$ (ii) For each irredundant spectral localizing system $\mathcal{H}(\Lambda)$ of D, the localizing system $\mathcal{F} := \mathcal{H}(D_{\mathcal{H}(\Lambda)}) = \cap\{\mathcal{H}(M) ; M = M' \cap D , M' \in \mathrm{Max}(D_{\mathcal{H}(\Lambda)})\}$ is a finitely generated spectral irredundant localizing system of D (Theorem 1.3) and $D_{\mathcal{H}(\Lambda)} = D_{\mathcal{F}}$ (Theorem 1.4). Hence, if (iv) holds $\mathcal{H}(\Lambda) = \mathcal{F}$ is finitely generated.

THEOREM 2.5. *A Prüfer domain satisfies property* (##) *if and only if it satisfies property* (#$_P$) .

Proof. (#$_P$)-property implies (##)-property by Proposition 2.1.

Conversely, let D be a Prüfer domain that satisfies property (##) and let Λ_1, Λ_2 distinct nonempty subsets of Spec(D) with the property that $P + Q = D$ for each $P \in \Lambda_1$ and $Q \in \Lambda_2$ such that $P \neq Q$. Set $\Gamma_i := \{M ; M = M' \cap D , M' \in \text{Max}(D_{\mathcal{F}(\Lambda_i)})$ and $P \subseteq M$ for some $P \in \Lambda_i\}$ and $\mathcal{F}_i := \mathcal{F}(\Gamma_i)$, for $i = 1, 2$. Then $\mathcal{F}_i$ is an irredundant spectral localizing system of D and $D_{\mathcal{F}(\Lambda_i)} = D_{\mathcal{F}_i}$ (because each $M \in \Gamma_i$ contains some $P \in \Lambda_i$). Since, by Theorem 2.4, $\mathcal{F}_i$ is finitely generated, then $\text{Max}(D_{\mathcal{F}(\Lambda_i)}) = \{MD_{\mathcal{F}(\Lambda_i)} ; M \in \Gamma_i\}$ (Proposition 1.5) . It follows that $D_{\mathcal{F}(\Lambda_1)} \neq D_{\mathcal{F}(\Lambda_2)}$, because each $M \in \Gamma_1$, containing some $P \in \Lambda_1$, cannot contain any $Q \in \Lambda_2$ such that $P \neq Q$.

THEOREM 2.6. *Let D be a Prüfer domain. Then the following are equivalent:*

(i)   *D is a (##)-domain satisfying the ascending chain condition on prime ideals;*

(ii)   *D is a (##)-domain and each nonzero prime ideal P of D is branched;*

(iii)   *D satisfies the ascending chain conditions on radical ideals;*

(iv)   *Each prime ideal of D is the radical of a principal ideal;*

(v)   *For each nonempty subset Λ of* Spec(D), *the localizing system $\mathcal{F}(\Lambda)$ is finitely generated ;*

(vi)   *If $D_{\mathcal{F}(\Lambda_1)} = D_{\mathcal{F}(\Lambda_2)}$, with Λ_1 , Λ_2 nonempty subsets of* Spec(D), *then $\mathcal{F}(\Lambda_1) = \mathcal{F}(\Lambda_2)$.*

Proof. (i) $\Leftrightarrow$ (ii) by Theorem 1.11.

(i) $\Leftrightarrow$ (iv) by [3, Theorem 4.2.33] and (iii) $\Leftrightarrow$ (iv) by [3, Theorem 3.1.11].

(v) $\Rightarrow$ (vi) by Theorem 1.4 and (vi) $\Rightarrow$ (v) because the localizing system $\mathcal{F} := \mathcal{F}(D_{\mathcal{F}(\Lambda)}) = \cap\{\mathcal{F}(M) ; M = M' \cap D , M' \in \text{Max}(D_{\mathcal{F}(\Lambda)})\}$ is a finitely generated spectral localizing system (Theorem 1.3) and $D_{\mathcal{F}} = D_{\mathcal{F}(\Lambda)}$ (Theorem 1.4).

(i) $\Rightarrow$ (v). By the ascending chain condition on prime ideals, given any nonempty subset Λ of $\mathrm{Spec}(D)$, the set Λ_0 of maximal elements of Λ is not empty (Theorem 1.11). Since $\mathcal{F}(\Lambda_0)$ is irredundant and clearly $\mathcal{F}(\Lambda) = \mathcal{F}(\Lambda_0)$, then $\mathcal{F}(\Lambda)$ is finitely generated by the (##)-property (Theorem 2.4).

(v) $\Rightarrow$ (ii). D is a (##)-domain by Theorem 2.4. Moreover, if $\mathcal{F}(\Lambda)$ is finitely generated, then Λ has maximal elements (Proposition 1.5). If this holds for any nonempty subset Λ of $\mathrm{Spec}(D)$, then each nonzero prime ideal P of D is branched by Theorem 1.11.

3. REMARKS, COMMENTS AND EXAMPLES

(a) A domain with all the maximal ideals invertible is a (#)-domain [11, Theorem 2] (but not necessarily a (##)-domain, as example (c) below shows).

Conversely, if D is a Prüfer (#)-domain and M is a maximal ideal such that $M \neq M^2$, then M must be invertible. Indeed, in this case, there exists a finitely generated ideal J such that $J \subseteq M$ and $J \not\subseteq N$ for any maximal ideal N different from M (Lemma 2. 2 (a)). It follows that, taking $m \in M \setminus M^2$, then M is the radical of the finitely generated ideal $J + mD$. Whence M is invertible (cf. [7, Proposition 1.2]). Thus a noninvertible maximal ideal of a Prüfer (#)-domain is idempotent.

We reobtain that an almost Dedekind domain which is not Dedekind is not a (#)-domain [9, Theorem 3]. For the same reason, if V is a DVR with finite residue field and k is its quotient field, then the ring of integer valued polynomials $\mathrm{Int}(V) := \{f(X) \in k[X], f(V) \subseteq V\}$ is a (two-dimensional) Prüfer domain which is not a (#)-domain (cf. [5, Example 4.4]). An infinite dimensional example of the same type is the ring E of the entire functions over the complex field (see [3, Section 8.1]).

The Prüfer domain constructed in [12, Section 1] furnishes an example of one-dimensional (#)-domain that is not Dedekind. As a matter of fact, the (#)-property holds in this domain because each M is the radical of a principal ideal. An example of a one-dimensional (#)-domain such that $M = M^2$ for each maximal ideal M is given in [8, Example 1].

(b) Any finitely generated nonzero proper ideal of a Prüfer (##)-domain has a finite number of minimal primes [11, Proposition 3] (or [3, Proposition 4.2.32]) and the converse holds in the one-dimensional case [9, Lemma 6]. A Prüfer domain that is not a (#)-domain, though any finitely generated nonzero proper ideal has finitely many minimal primes, is constructed in [11, Example 3].

However, as in the one-dimensional case, if each (finitely generated) nonzero proper ideal of a Prüfer domain D is contained in only finitely many maximal ideals, then D is a (##)-domain. To prove this, it is enough to show that the localizing system $\mathcal{F}(\Lambda)$ of D is finitely generated for any nonempty subset Λ of $\mathrm{Max}(D)$ (Proposition 2.3). This follows from the fact that any nonzero proper ideal I of D contains a finitely generated ideal J which is contained exactly in the maximal ideals containing I. In fact, let $M_1, \ldots, M_s$ be the maximal ideals containing I and let $x \in I, x \neq 0$. Assume that $M_1, \ldots, M_s, N_1, \ldots, N_t$ are all the maximal ideals of D containing x and let $y_i \in I \setminus N_i, i = 1, \ldots, t$. Then the ideal $J := (x, y_1, \ldots, y_t)$ is contained in I and the only maximal ideals of D containing J are $M_1, \ldots, M_s$.

An example of a two-dimensional Prüfer (##)-domain with a finitely generated ideal contained in infinitely many maximal ideals is given in [11, Example 1].

(c) A Prüfer domain D satisfying the (#)-property but not the (##)-property is constructed in [11, Example 2]. This is a two-dimensional Bezout domain with just one maximal ideal M of height two and such that the only nonmaximal prime ideal P is contained in the union of all the maximal ideals of height one. In this case, any finitely generated ideal J contained in P is principal, hence it is contained in some maximal prime of height one. It follows that, if $\Lambda := \mathrm{Max}(D) \setminus \{M\}$, the localizing system $\mathcal{F}(\Lambda)$ is not finitely generated.

We remark that all the maximal ideals of D are principal. In fact the construction shows that each maximal ideal M of D is locally principal. Hence $M \neq M^2$ and so M is invertible (that is, principal) by part (a).

(d) Any Prüfer domain with finitely many maximal ideals is a (##)-domain and has the ascending chain condition on prime ideals in the finite dimensional case.

If V is a valuation domain with the maximal ideal M idempotent, then it easy to check directly that $\mathcal{F} := \mathcal{F}^*(M) = \{V, M\}$ is a localizing system and that $V = V_{\mathcal{F}(M)} = V_{\mathcal{F}}$, because $(V:M) = V$. Moreover $\mathcal{F}$ is not finitely generated, because M is not invertible. Setting $\Lambda = \Lambda(M) := \mathrm{Spec}(V) \setminus \{M\}$, we have that $\mathcal{F} = \mathcal{F}(\Lambda)$ if and only if M is unbranched. In this case V does not satisfy the ascending chain condition on prime ideals (cf. Corollary 1.7 and Theorem 1.9). An example of a valuation domain V such that M is unbranched and $P \neq P^2$ (hence P is branched) for each nonmaximal nonzero prime ideal P is constructed in [3, Example 8.4.8].

(e) A Prüfer (##)-domain is a generalized Dedekind domain if and only if $P \neq P^2$ for each nonzero prime ideal P (Theorem 0.1), in particular each nonzero P is branched. Any finite dimensional valuation domain with the maximal ideal idempotent furnishes an example of a (##)-domain with all the nonzero prime ideals branched (because of the ascending chain condition on prime ideals) which is not a generalized Dedekind domain. Remark that, in this case, $\mathcal{F}^*(M)$ is a localizing system that is not finitely generated, necessarily non-spectral by Theorem 1.9.

REFERENCES

[1]   J. T. Arnold and J. W. Brewer, On flat overrings, ideal transforms and generalized transforms of a commutative ring, J. Algebra 18 (1971), 254-263.

[2]   M. Fontana, J. A. Huckaba and I. J. Papick, Domains satisfying the trace property, J. Algebra 107 (1987), 169-182

[3]   M. Fontana, J. A. Huckaba and I. J. Papick, "Prüfer domains", Dekker, New York, 1997.

[4]   M. Fontana and N. Popescu, Sur une classe d'anneaux qui généralisent les anneaux de Dedekind, J. Algebra 173 (1995), 44-66.

[5]   M. Fontana and N. Popescu, Sur une classe d'anneaux de Prüfer avec groupe de classes de torsion, Comm. Algebra 23 (1995), 4521-4534.

[8]   R. M. Fossum, "The divisor class group of a Krull domain", Springer-Verlag, Berlin, 1973.

[6]   S. Gabelli, A class of Prüfer domains with nice divisorial ideals, Proceedings of the Second International Conference on Commutative Ring Theory, Fès 1995, Lecture Notes in Pure and Applied Mathematics, n. 185, Dekker, New York, 1997.

[7]   S. Gabelli and N. Popescu, Invertible and divisorial ideals of generalized Dedekind domains, J. Pure Appl. Algebra, to appear.

[9]   R. W. Gilmer, Overrings of Prüfer domains, J. Algebra 4 (1996), 331-340.

[10] R. W. Gilmer, "Multiplicative ideal theory", Dekker, New York, 1972.

[11] R. W. Gilmer and W. J. Heinzer, Overrings of Prüfer domains II, J. Algebra 7 (1967), 281-302.

[12] W. J. Heinzer, Quotient overrings of integral domains, Matematika 17 (1970), 139-148.

[13] W. J. Heinzer, J. Ohm and R. L. Pendleton, On integral domains of the form $\cap D_P$, P minimal, J. Reine Angew. Math., 241 (1970), 147-159.

[14] W. J. Heinzer and I. J. Papick, The radical trace property, J. Algebra 112 (1988), 110-121.

[15] T. G Lucas, The radical trace property and primary ideals, J. Algebra 184 (1996), 1093-1112.

[16] B. Olberding, Torsion-free Modules over Prüfer domains, Ph.D. Thesis, Wesleyan University, Middletown, Connecticut (USA), 1996.

[17] N. Popescu, "Abelian Categories with applications to rings and modules", Academic Press, New York, 1973.

[18] N. Popescu, On a class of Prüfer domains, Rev. Roumaine Math. Pures Appl. 29 (1984), 777-786.

Building Noetherian Domains Inside an Ideal-adic Completion II

WILLIAM HEINZER   Department of Mathematics, Purdue University, West Lafayette IN 47907-1968

CHRISTEL ROTTHAUS   Department of Mathematics, Michigan State University, East Lansing MI 48824-1027

SYLVIA WIEGAND   Department of Mathematics & Statistics, University of Nebraska, Lincoln, NE 68588-0323

SUMMARY

Suppose a is a nonzero nonunit of a Noetherian integral domain R. We exhibit new Noetherian domains between R and the (a)-adic completion R^* of R by extending a construction of Heitmann and Nagata. Suppose $X \subset aR^*$ is a set of elements that is algebraically independent over the fraction field K of R. For each finite subset $Y = \{\tau_1, \ldots, \tau_m\}$ of X there is a natural sequence of nested polynomial rings $U_{Y,n}$ in m variables over R between $R[\tau_1, \ldots, \tau_m]$ and $A_Y := K(\tau_1, \ldots, \tau_m) \cap R^*$. If $U_Y := \cup_{n=0}^{\infty} U_{Y,n}$ is Noetherian, it is not hard to show that A_Y is a localization of U_Y and $R^*[1/a]$

The authors would like to thank the National Science Foundation and the University of Nebraska Research Council for support for this research. In addition they are grateful for the hospitality and cooperation of Michigan State, Nebraska and Purdue, where several work sessions on this research were conducted.

411

is flat over $R[\tau_1, \ldots, \tau_m]$. In [HRW4] we prove, conversely, that if $R^*[1/a]$ is flat over $R[\tau_1, \ldots, \tau_m]$, then U_Y is Noetherian and $A_Y = K(\tau_1, \ldots, \tau_m) \cap R^*$ is a localization of U_Y. Thus the flatness of $R^*[1/a]$ over $R[\tau_1, \ldots, \tau_m]$ implies the intersection domain A_Y is Noetherian and is computable as a directed union of localized polynomial rings.

In this article we extend the results just described and obtain similar information concerning $U := \cup\{U_Y : Y \subseteq X$ and Y is finite $\}$ and $A :=$ $K(X) \cap R^* = \cup\{A_Y : Y \subseteq X$ and Y is finite $\}$. In the case where X is infinite, the ring U is not Noetherian, but A is Noetherian and is equal to a suitable localization of U provided $R^*[1/a]$ is flat over $R[X]$. Thus this flatness condition yields again that A is computable.

1. INTRODUCTION, BACKGROUND AND NOTATION

In this article we continue a project of producing new Noetherian domains between a Noetherian domain R and the ideal-adic completion R^* of R with respect to an ideal of R. As in [HRW1], [HRW2], [HRW3] and [HRW4], a paradigm for our work is a construction of Nagata [N2] from forty years ago. The domains we construct are intersections of R^* with intermediate fields L between the fraction field of R and the total quotient ring of R^*; we have a longstanding curiosity about what domains can arise as $A := L \cap R^*$. Successful application of techniques such as this in the past include [BR1], [BR2], [HRS], [H1], [H2], [H3], [L], [N2], [O1], [O2], [R1], [R2], [R3], and [W]. The approach of Ray Heitmann in [H1] for producing Noetherian extension domains involves an interesting method of ring-theoretically adjoining to R a transcendental (over R) power series in a nonzero nonunit a of R. By generalizing Heitmann's approach (and our previous generalization of it in [HRW3] and [HRW4]), we adjoin an arbitrary set of transcendentals. If a certain flatness condition is satisfied, we prove that this procedure yields a Noetherian ring which is also an intersection domain of the type described above.

Let a be a nonzero nonunit of a Noetherian integral domain R. The (a)-adic completion R^* of R is isomorphic to the ring $R[[x]]/(x - a)$ [N1, (17.5), page 55]. Thus elements of the (a)-adic completion may be regarded as power series in a. Suppose that $X \subset aR^*$ is a set of elements that is algebraically independent over the fraction field of R. For each finite subset $Y = \{\tau_1, \ldots, \tau_s\}$ of X, there is a natural sequence of nested polynomial rings $U_{Y,n}$ in s variables over R starting with $U_{Y,0} := R[\tau_1, \ldots, \tau_s]$, and a corresponding sequence of localizations $B_{Y,n}$ of $U_{Y,n}$, all of which fit between $R[\tau_1, \ldots, \tau_s] = R[Y]$ and $A_Y := K(Y) \cap R^*$. In [HRW4] we establish that, for each Y, flatness of $R^*[1/a]$ over $R[Y]$ is equivalent to $U_Y := \cup_{n=0}^{\infty} U_{Y,n}$ being Noetherian and, furthermore, that this flatness condition implies that $B_Y := \cup_{n=0}^{\infty} B_{Y,n}$ is equal to $A_Y := K(Y) \cap R^*$ and thus A_Y is a localization of U_Y. Therefore in this case the intersection domain A_Y is Noetherian.

In this article we consider also $U := \cup\{U_Y : Y \subseteq X \text{ and } Y \text{ is finite }\}$, $A = K(X) \cap R^*$ and $B := \cup\{B_Y : Y \subseteq X \text{ and } Y \text{ is finite }\}$. If X is infinite, we observe that the ring U is not Noetherian, but we prove in Theorem 2.2 that a necessary and sufficient condition in order that $B = A$ and that both of these rings are Noetherian is that $R^*[1/a]$ is flat over $R[X]$.

1.1 NOTATION

Suppose a is a nonzero nonunit of a Noetherian integral domain R and R^* is the (a)-adic completion of R. Let K be the fraction field of R.

For every $\gamma \in aR^*$ and every $n \in \mathbf{N}$, we define the nth-endpiece γ_n with respect to a of γ to be

$$\gamma_n := \sum_{j=n+1}^{\infty} c_j a^{j-n}, \quad \text{where } \gamma := \sum_{j=1}^{\infty} c_j a^j \text{ with each } c_j \in R. \qquad (1.1.1)$$

Suppose $X \subset aR^*$ is a set of elements that is algebraically independent over K. Each $\tau \in X$ has a representation as a power series in a; we use these representations to define for each finite subset $Y \subseteq X$, say $Y = \{\tau_1, \ldots, \tau_m\}$, a sequence $U_{Y,n}$ of polynomial rings in m variables over R.

For $\tau_s \in Y$, let τ_{sn} denote the nth endpiece of τ_s. That is, if

$$\tau_s := \sum_{j=1}^{\infty} r_{sj} a^j, \text{ where the } r_{sj} \in R, \text{ then} \qquad \tau_{sn} := \sum_{j=n+1}^{\infty} r_{sj} a^{j-n}.$$

With $Y = \{\tau_1, \ldots, \tau_m\}$, we define $Y_n = \{\tau_{1n}, \tau_{2n}, \ldots, \tau_{mn}\}$ and

$$U_{Y,0} = R[Y] = R[\tau_1, \ldots, \tau_m] \quad \text{and} \quad U_{Y,n} = R[Y_n] = R[\tau_{1n}, \ldots, \tau_{mn}].$$

We also define $A := K(X) \cap R^*$ and $U_Y := \cup\{U_{Y,n} : n \in \mathbf{N}\}$, and

$$U := \cup\{U_Y : Y \subseteq X \text{ and } Y \text{ is finite }\}, \quad B_Y := (1 + aU_Y)^{-1} U_Y,$$

$$B := \cup\{B_Y : Y \subseteq X \text{ and } Y \text{ is finite }\} = (1 + aU)^{-1} U.$$

1.2 DISCUSSION   (1) For each finite subset Y of X and each nonnegative integer n, we have a birational inclusion of polynomial rings $U_{Y,n} \subset U_{Y,n+1} \subset U_{Y,n}[1/a]$, because $\tau_{sn} = r_{s,n+1} a + a\tau_{s,n+1}$.

(2) The element a belongs to the Jacobson radical of $(1+aR)^{-1}R$, and, for each Y and n, to the Jacobson radical of the ring $(1 + aU_{Yn})^{-1} U_{Yn}$. Therefore a is also in the Jacobson radical of $(1 + aU_Y)^{-1} U_Y = B_Y$ and of $(1 + aR)^{-1} U = B$.

We show in [HRW4]

(1) For each finite subset $Y = \{\tau_1, \ldots, \tau_s\}$ of X, if $U_Y := \cup_{n=0}^{\infty} U_{Y,n}$ is Noetherian, then A is a localization of U_Y and $R^*[1/a]$ is flat over $R[Y]$.

(2) For each finite subset $Y = \{\tau_1, \ldots, \tau_s\}$ of X, if $R^*[1/a]$ is flat over $R[Y]$, then U_Y is Noetherian and $A_Y := K(\tau_1, \ldots, \tau_s) \cap R^*$ is B_Y, which is a localization of U_Y. Thus the flatness of $R^*[1/a]$ over $R[\tau_1, \ldots, \tau_s]$ implies the intersection domain A_Y is Noetherian and is computable.

In this article we prove an analogous result for A_X in the case where the set X is infinite.

2. LARGE NOETHERIAN EXTENSIONS

2.1 PROPOSITION (cf. [HRW2],[HRW3, (2.2)])   Assume the notation and setting of (1.1), and let U^* and A^* denote the (a)-adic completions of U and A. Then

(1) $a^k U = a^k A \cap U = a^k R^* \cap U$ for each positive integer k.

(2) $U^* = A^* = R^*$, so $R/a^k R = U/a^k U = A/a^k A = R^*/a^k R^*$ for each positive integer k.

(3) If U is Noetherian, then R^* is flat over U and A is the localization of U at the multiplicative system $1 + aU$ of U. (As we show in Theorem 2.2, if X is infinite, then U is not Noetherian.)

(4) If $B := (1 + aU)^{-1}U$ is Noetherian, then R^* is flat over both B and U and $A = B$.

Proof: We have $R \subseteq U \subseteq A \subseteq R^*$. Since R is Noetherian, R^* is flat over R [M1, Theorem 8.8, page 60]. Moreover, $a^k R$ is closed in the (a)-adic topology on R, so we have $a^k R^* \cap R = a^k R$ for each positive integer k [ZS, Theorem 8, page 261]. Furthermore, $A = R^* \cap K(X)$ implies $a^k A = a^k R^* \cap A$. It is clear that $a^k U \subseteq a^k R^* \cap U$; thus for (1) and (2) it suffices to show $a^k R^* \cap U \subseteq a^k U$. Moreover, if $aR^* \cap U = aU$, it follows that $a^k R^* \cap U = a^k R^* \cap aU = a(a^{k-1} R^* \cap U)$, and by induction we see that $a^k R^* \cap U = a^k U$. Thus for the proof of (1) and (2), it suffices to show $aR^* \cap U \subseteq aU$.

Let $g \in aR^* \cap U$. Then there is a finite subset $Y := \{\tau_1, \ldots, \tau_s\}$ of X and a positive integer n with $g \in U_{Yn} = R[\tau_{1n}, \ldots, \tau_{sn}]$, where the τ_{in} are the nth-endpieces of the τ_i. Write $g = r_0 + g_0$ where $g_0 \in (\tau_{1n}, \ldots, \tau_{sn})U_{Yn}$ and $r_0 \in R$. From the definition of τ_{in}, we have $\tau_{in} = a\tau_{i,n+1} + a_{in}a$, where $a_{in} \in R$, for each i with $1 \le i \le s$. Thus $r_0 \in aR^* \cap R = aR$, $\tau_{in}U_{Yn} \subseteq aU_{Y,n+1}$ and $g \in aU$. This completes the proof of (1) and (2).

If U is Noetherian, then $U^* = R^*$ is flat over U. Let S be the multiplicative system $1 + aU$. If U is Noetherian, then B is Noetherian. Moreover, if B is Noetherian, then the (a)-adic completion of B is R^* and R^* is faithfully flat over B [M1, Theorem 8.14, page 62]. Therefore if B is Noetherian, then $B = K(X) \cap R^* = A$. This completes the proof of (2.1).

With the notation and setting of (1.1), the representation of the τ_i as power series in a with coefficients in R is, in general, not unique. However, the argument given in [HRW4,(2.3)], shows that the rings U and B are uniquely determined by the τ_i.

2.2 THEOREM   Assume the setting, hypotheses and notation of (1.1) and also assume the set X is infinite. Then

(1) The ring U is not Noetherian.

(2) If $R^*[1/a]$ is flat over $R[X]$, then B equals A and both are Noetherian.

Proof: Since the localization $U[1/a]$ of U is a polynomial ring $R[1/a][X]$ in infinitely many variables over $R[1/a]$, the ring U is not Noetherian. In order to establish the second assertion of Theorem 2.2, we first observe that by [HRW4], for each finite subset Y of X, the integral domain A_Y is Noetherian. To show A is Noetherian it suffices to show that every finitely generated ideal of A is contracted from R^*. Let I be a finitely generated ideal of A. If I is not contracted from R^* then there exists an element b in IR^* such that $b \in A - I$. Choose a finite subset Y of X so that b and a finite set W of generators of I are contained in A_Y. Then $b \notin I = WA$ and so $b \notin WA_Y$. But since R^* is faithfully flat over A_Y, $WR^* \cap A_Y = WA_Y$, so the assumption that $b \in IR^* = WR^*$ implies that $b \in WR^* \cap A_Y \subseteq I$, a contradiction. We conclude that A is Noetherian.

To show $B = A$, it suffices to show for each finite subset Y of X that $B \cap K(Y) = A \cap K(Y)$, for $B = \cup_{Y \subseteq X}(B \cap K(Y)) = \cup B_Y$ and $A = \cup_{Y \subseteq X}(A \cap K(Y)) = \cup A_Y$. Since the flatness hypothesis of Theorem 2.2 implies each $B_Y = A_Y$, it follows that $B = A$.

The following proposition establishes the existence of infinitely many examples satisfying the flatness hypotheses of Theorem 2.2 with an infinite set X algebraically independent over the fraction field of R. Thus the intersection domain A of these examples is Noetherian.

2.3 PROPOSITION   Suppose k is a field, $z_1, \ldots, z_r, y$ are indeterminates over k and $X \subseteq yk[[y]]$ is a set of algebraically independent elements over $k(y)$. Let $R := k[z_1, \ldots, z_r, y]$. Then $R[X] \to R_y^*$ is flat, where $R^* := k[[y]]$, is the (y)-adic completion of R. Equivalently, $A := k(z_1, \ldots, z_r, y, X) \cap R_y^*$ is Noetherian and equals $(1+yU)^{-1}U$, where $U = \cup k[z_1, \ldots, z_r, y][\{\tau_n | \tau \in X\}]$ and for each τ, τ_n is the nth endpiece of τ.

Proof: We note that $\psi : k[y, X] \longrightarrow k[[y]]_y = k((y))$ is flat because the dimension of $k((y))$ is 0 and the prime ideal (0) of $k((y))$ lies over (0) in $k[y, X]$. We show that $\alpha : R[X] = k[y, X] \otimes_k k[z_1, \ldots, z_r] \longrightarrow R_y^*$ is flat. The flatness of ψ implies flatness of

$$\psi_{\underline{z}} : k[y, X] \otimes_k k[z_1, \ldots, z_r] \longrightarrow k((y)) \otimes_k k[z_1, \ldots, z_r].$$

The (y)-adic completion of $k[[y]] \otimes_k k[z_1, \ldots, z_r]$ is $k[z_1, \ldots, z_r][[y]]$. Let $D = k[z_1, \ldots, z_r][[y]]_{(z_1, \ldots, z_r, y)}$. Then R^* is the (y)-adic completion of D. The map α factors as $\beta \circ \psi_{\underline{z}}$, where β is the canonical map of $k[[y]][z_1, \ldots, z_r]$ to the (y)-adic completion of D. Therefore α is flat.

The following diagram displays the situation concerning possible implications between A being a localization of U and the Noetherian properties for A or B:

$$R^*[1/a] \text{ is flat over R}[X] \quad \longleftrightarrow \quad B := (1 + aU)^{-1}U \text{ Noetherian}$$
$$\downarrow \qquad\qquad\qquad\qquad\qquad\qquad \downarrow$$
$$A \text{ is a localization of } U \qquad\qquad A \text{ is Noetherian}$$

Examples given in [HRW3] and [HRW4] show that each of the remaining possible implications fails in general. Example 2.5 of [HRW4] shows that a right-hand North-pointing arrow and a bottom East-pointing arrow cannot be inserted. Theorem 4.4 of [HRW3] shows that a left North-pointing arrow and a bottom West-pointing arrow are not valid.

REFERENCES

[BR1] M. Brodmann and C. Rotthaus, *Local rings with bad sets of formal prime divisors*, J. Algebra **75** (1982), 386–94.

[BR2] M. Brodmann and C. Rotthaus, *A peculiar unmixed domain*, Proc. AMS. **87** (1993), 596–600.

[HRS] W. Heinzer, C. Rotthaus and J. Sally, *Formal fibers and birational extensions*, Nagoya Math. J. **131** (1993), 1–38.

[HRW1] W. Heinzer, C. Rotthaus and S. Wiegand, *Idealwise algebraic independence for elements of the completion of a local domain*, Illinois J. Math. **41** (1997), 272–308.

[HRW2] W. Heinzer, C. Rotthaus and S. Wiegand, *Intermediate rings between a local domain and its completion*, Illinois J. Math. (to appear).

[HRW3] W. Heinzer, C. Rotthaus and S. Wiegand, *Noetherian rings between a semilocal domain and its completion*, J. Algebra **198** (1997), 627–655.

[HRW4] W. Heinzer, C. Rotthaus and S. Wiegand, *Building Noetherian domains inside an ideal-adic completion*, Proc. of the Orsatti 1997 Conference Padova, Italy (to appear).

[H1] R. Heitmann, *Examples of noncatenary rings*, Trans. Amer. Mat. Soc. **247** (1979), 125–136.

[H2] R. Heitmann, *Characterizations of completions of unique factorization domains*, Trans. Amer. Mat. Soc. **337** (1993), 379–387.

[H3] R. Heitmann, *Completions of local rings with an isolated singularity*, J. Algebra **163** (1994), 538–567.

[L] S. Loepp, *Constructing local generic formal fibers*, J. Algebra **187** (1997), 16–38.

[M1] H. Matsumura, *Commutative Ring Theory*, Cambridge University Press, Cambridge (1986).

[M2] H. Matsumura, *Commutative Algebra*, second edition, Benjamin/Cummings, Reading, MA (1980).

[N1] M. Nagata, *Local Rings*, John Wiley (1962).

[N2] M. Nagata, *An example of a normal local ring which is analytically reducible*, Mem. Coll. Sci., Univ. Kyoto **31** (1958), 83–85.

[O1] T. Ogoma, *Non-catenary pseudo-geometric normal rings*, Japan J. Math. **6** (1980), 147–163.

[O2] T. Ogoma, *Cohen-Macaulay factorial domain is not necessarily Gorenstein*, Mem. Fac. Sci. Kochi Univ. **3** (1982), 65–74.

[R1] C. Rotthaus, *Nicht ausgezeichnete, universell japanische Ringe*, Math. Z. **152** (1977), 107–125.

[R2] C. Rotthaus, *Universell japanische Ringe mit nicht offenen regulärem Ort*, Nagoya Math. J. **74** (1979), 123–135.

[R3] C. Rotthaus, *On rings with low dimensional formal fibres*, J.Pure Appl.Algebra **71** (1991), 287–296.

[V] P. Valabrega, *On two-dimensional regular local rings and a lifting problem*, Annali della Scuola Normale Superiore di Pisa **27** (1973), 1–21.

[W] D. Weston, *On descent in dimension two and non-split Gorenstein modules*, J. Algebra **118** (1988), 263-275.

[ZS] O. Zariski and P. Samuel, *Commutative Algebra* **II**, Van Nostrand, Princeton, NJ (1960).

Trace Properties and Integral Domains

SALAH-EDDINE KABBAJ   Department of Mathematical Sciences, King
Fahd University of Petroleum and Minerals, Dhahran, Saudi Arabia

THOMAS G. LUCAS   University of North Carolina at Charlotte,
Charlotte, North Carolina

ABDESLAM MIMOUNI   Département de Mathématiques et Informatique,
Faculté des Sciences Dhar Al-Mehraz, Université de Fès, Fès, Morocco

INTRODUCTION

Throughout this paper, R will denote an integral domain with quotient
field K. For a pair of fractional ideals I and J of a domain R we let $(J : I)$
denote the set $\{t \in K \,|\, tI \subseteq J\}$. Often, we shall use I^{-1} in place of $(R : I)$.
Recall that the "v" of a fractional ideal I is the set $I_v = (R : (R : I))$
and the "t" of I is the set $I_t = \bigcup J_v$ with the union taken over all finitely
generated fractional ideals contained in I. An ideal I is divisorial if $I = I_v$,
and I is a t-ideal if $I = I_t$.

Let R be an integral domain and let M be an R-module. Then the trace
of M is the ideal generated by the set $\{fm \,|\, f \in Hom(M, R) \text{ and } m \in M\}$.
For a fractional ideal I of R, the trace is simply the product of I and I^{-1}.
We call an ideal of R a *trace ideal* of R if it is the trace of some R-module.
An elementary result due to H. Bass is that if J is a trace ideal of R, then
$JJ^{-1} = J$; i.e., $J^{-1} = (J : J)$ [5, Proposition 7.2]. It follows that J is a
trace ideal if and only if $J^{-1} = (J : J)$. (Such ideals are also referred to as
being "strong"; see, for example, [3].) In 1987, D.D. Anderson, J. Huckaba

and I. Papick proved that if I is a noninvertible ideal of a valuation domain V, then $I(V : I)$ is prime [1, Theorem 2.8]. Later in the same year, M. Fontana, Huckaba and Papick began the study of the "trace property" and "TP domains". A domain R is said to satisfy the *trace property* (or to be a *TP domain*) if for each R-module M, the trace of M is equal to either R or a prime ideal of R [8, page 169]. Among other things, they showed that each valuation domain satisfies the trace property [8, Proposition 2.1], and that if R satisfies the trace property, then it has at most one noninvertible maximal ideal [8, Corollary 2.11]. For Noetherian domains they proved that if R is a Noetherian domain, then it is a TP domain if and only if it is one-dimensional, has at most one noninvertible maximal ideal M, and if such a maximal ideal exists, then M^{-1} equals the integral closure of R (or, equivalently, $M^{-1} = (M : M)$ is a Dedekind domain) [8, Theorem 3.5]. In Section 2 of [10], S. Gabelli showed that by replacing "integral closure" with "complete integral closure", the same list of conditions characterizes the class of Mori domains which satisfy the trace property. Recall that a Mori domain is an integral domain which satisfies the ascending chain condition on divisorial ideals.

In 1988, W. Heinzer and Papick introduced the "radical trace property" declaring that an integral domain R satisfies the *radical trace property* (or is an *RTP domain*) if for each noninvertible ideal I, II^{-1} is a radical ideal. For Noetherian domains, they proved that if R is a Noetherian domain, then it satisfies the radical trace property if and only if R_P is a TP domain for each prime ideal P [12, Proposition 2.1]. Gabelli extended this result to Mori domains [10, Theorem 2.14].

For Prüfer domains there are results concerning the trace property in [6], [8] and [16] and the radical trace property in [12] and [16]. For a Prüfer domain R, Theorem 23 of [16] gives the following equivalent conditions:

(1) R satisfies the radical trace property.

(2) For each primary ideal Q, either Q is invertible or QQ^{-1} is prime.

(3) For each primary ideal Q, if Q^{-1} is a ring, then Q is prime.

(4) Each branched prime is the radical of a finitely generated ideal.

(A prime ideal P is said to be branched if there is a P-primary ideal Q such that $Q \neq P$ [11, page 189].)

In Theorem 10, we will show that the following statement can be added to this list:

(5) For each trace ideal I, $IR_P = PR_P$ for each prime P minimal over I.

Moreover, we will give a new proof for the equivalence of (1)–(3).

According to [16], a domain R is said to satisfy the *trace property for primary ideals* (or to be a *TPP domain*), if for each primary ideal Q, either Q is invertible or QQ^{-1} is prime. By Corollary 8 of [16], R is a TPP domain if and only if for each primary ideal Q, either $QQ^{-1} = \sqrt{Q}$, or Q is invertible and $\sqrt{Q}$ is maximal. Also from [16], R is a *PRIP domain* if for each primary ideal Q, Q^{-1} a ring implies Q is prime. We say that a domain is an *LTP domain* if for each trace ideal I, $IR_P = PR_P$ for each prime P minimal over I. It is known that each RTP domain is a TPP domain [16, Theorem 4] and that there are Noetherian domains which satisfy the radical trace property (and even the trace property) but are not PRIP domains (see, for example, [16, Example 30]). We will show that each TPP domain is an LTP domain and that each PRIP domain is an LTP domain (Corollary 3).

It is easy to see that for one-dimensional domains, each LTP domain is also an RTP domain. Also, it is known that for Mori domains, the radical trace property and the trace property for primary ideals are equivalent. In Theorem 18, we show that if R is a Mori domain, then it is an LTP domain if and only if it is an RTP domain. However, in general, we have been unable to determine whether each TPP domain is an RTP domain, or whether each LTP domain is an TPP domain (or RTP domain).

A field is trivially an RTP domain. While most of the results in this paper are true for fields, the emphasis is on integral domains that are not fields. To avoid having to add the phrase "but not a field" when it would be required, we will simply assume that R is an integral domain which is not a field. We shall also assume that all of the ideals are nonzero.

Notation is standard as in [Gilmer]. In particular, "$\subseteq$" denotes containment and "$\subset$" denotes proper containment.

We shall make use of a number of results concerning consequences of I^{-1} being a ring. We close the Introduction with a theorem where we list several of these results.

THEOREM 0   Let R be an integral domain and let I be an ideal of R such that I^{-1} is a ring. Then

(a) $I^{-1} = I_v^{-1} = (I_v : I_v) = (II^{-1} : II^{-1}) = (II^{-1})^{-1}$ ([14, Proposition 2.2]).

(b) $\sqrt{I}^{-1}$ is a ring ([13, Proposition 2.1]). Moreover, $\sqrt{I}^{-1} = (\sqrt{I} : \sqrt{I})$ ([2, Proposition 3.3]).

(c) P^{-1} is a ring for each prime P minimal over I ([13, Proposition 2.1] and [16, Lemma 13]). Moreover, $P^{-1} = (P : P)$ ([14, Proposition 2.3]).

1   LTP DOMAINS

The first lemma we present is a variation on a result which appears in Fossum's book [9, Lemma 3.7]. (See also, Lemmas 0 and 1 of [16].)

LEMMA 1   Let R be an integral domain and let Q be a primary ideal of R with radical P. If P does not contain QQ^{-1}, then $(R : QQ^{-1}) = (QQ^{-1} : QQ^{-1}) = (Q : Q)$ and so $(R : I) = (Q : Q)$ for each ideal I such that $Q \subset I \subseteq QQ^{-1}$ and $I \not\subseteq P$.

Proof. It is always the case that $(Q : Q) \subseteq (QQ^{-1} : QQ^{-1}) = (R : QQ^{-1})$. Assume P does not contain QQ^{-1} and let I be an ideal such that $Q \subset I \subseteq QQ^{-1}$ and $I \not\subseteq P$. Since I contains Q and is contained in QQ^{-1}, $(QQ^{-1} : QQ^{-1}) = (R : QQ^{-1}) \subseteq (R : I) \subseteq (R : Q)$. Obviously, $QI(R : I) \subseteq Q$. Since Q is P-primary and I is not contained in P, $Q(R : I) \subseteq Q$. Hence $(R : I) \subseteq (Q : Q)$ and it follows that $(R : I) = (Q : Q) = (QQ^{-1} : QQ^{-1}) = (R : QQ^{-1})$. ♦

Our first use of Lemma 1 is to establish a characterization of LTP domains in terms of primary ideals.

THEOREM 2   The following are equivalent for a domain R.
  (1) R is an LTP domain.
  (2) For each noninvertible primary ideal Q, $Q(R : Q)R_P = PR_P$ where $P = \sqrt{Q}$.
  (3) If a primary ideal is also a trace ideal, then it is prime.

Proof. $((1) \Rightarrow (2))$ Assume R is an LTP domain and let Q be a noninvertible P-primary ideal of R. Since R is an LTP domain and QQ^{-1} is a trace ideal, it suffices to show that P contains QQ^{-1}. By way of contradiction assume there is an element $t \in QQ^{-1} \backslash P$ and set $I = t^2 R + Q$. Then from Lemma 1, we have $(R : I) = (Q : Q)$. Let $J = I(R : I)$. Then J is also contained in QQ^{-1}. Hence we have $(J : J) = (R : J) = (R : I) = (Q : Q)$.

Let N be a prime minimal over J. Then $JR_N = NR_N$ since R is an LTP domain. In particular, $t \in JR_N$. It follows that there are elements $a \in (R : I) = (J : J)$, $q \in Q$ and $s \in R \backslash N$ such that $st = at^2 + q$. Hence, $q = t(s - at)$. But $a^2 t^2$ is in J since $(R : I) = (J : J)$. Thus $at \in N$ and so $s - at$ is not in P. As neither t nor $s - at$ is in P, we have a contradiction. Hence we must have $QQ^{-1} \subseteq P$.

  $((2) \Rightarrow (3))$ Obvious.

  $((3) \Rightarrow (1))$ Assume that if an ideal is both a primary ideal and a trace ideal, then it is prime. Let I be a trace ideal of R and let P be a minimal prime of I. Then $Q = IR_P \cap R$ is a P-primary ideal which is also a trace ideal. It follows that $Q = P$ and $IR_P = PR_P$. ◆

COROLLARY 3   Let R be an integral domain. If R is an RTP domain, a TPP domain or a PRIP domain, then R is an LTP domain.

Proof. By Theorem 2 it suffices to show that for RTP domains, TPP domains and PRIP domains, if an ideal is both primary and a trace ideal, then it is prime. Let Q be a trace ideal which is also a primary ideal of R. Then obviously Q^{-1} is a ring. Hence if R is a PRIP domain, then Q is prime. Also, if R is either an RTP domain or a TPP domain, then we have $QQ^{-1} = Q$ is prime.◆

Statement (3) in Theorem 2 is very close to the definition of a PRIP domain. To see that the two are not equivalent consider the ring $R = F[[X^3, X^4, X^5]]$ where F is a field. The ideal $Q = (X^3, X^4)$ is primary but not prime and $Q^{-1} = F[[X]]$ is a ring. Thus R is not a PRIP domain. However, note that $QQ^{-1} = (X^3, X^4, X^5)$ is the maximal ideal of R and $(QQ^{-1})^{-1} = F[[X]]$. That R is an LTP domain now follows from [8, Theorem 3.5] and Corollary 3. At this time we do not know whether each LTP domain is a TPP domain and/or whether each TPP domain is an RTP domain. However in Theorem 10, we prove that if R is a Prüfer domain, then each LTP domain is also a PRIP domain, a TPP domain and an RTP domain.

If R is an RTP domain (a TPP domain), then for each prime ideal P, both R_P and R/P are RTP domains (TPP domains) [16, Theorems 3 and 9]. Next, we establish an analogous result for LTP domains.

THEOREM 4   Let P be a prime ideal of a domain R and let $D = R/P$. If R is an LTP domain, then R_P and D are LTP domains.

Proof. Assume R is an LTP domain.

We first show that D is an LTP domain. Let $\overline{I}$ be a trace ideal of D. Since $\overline{(R : I)} \subseteq (D : \overline{I})$ and $\overline{(I : I)} \subseteq (\overline{I} : \overline{I})$, $(I : I) = (R : I)$. Thus for each prime N minimal over I, $IR_N = NR_N$. It follows that $\overline{I}D_N = \overline{N}D_N$. Hence, D is an LTP domain.

To show R_P is an LTP domain, let IR_P be a trace ideal of R_P. Then $B = IR_P \cap R$ is a trace ideal of R. Hence for each prime N minimal over B, $BR_N = NR_N$. The result follows from the fact that $IR_N = BR_N$ for each $N \subseteq P$. ◆

Our next result collects other useful information concerning the prime ideals of an LTP domain.

THEOREM 5   Let R be an LTP domain. Then

(a) Each maximal ideal is a t-ideal.

(b) Each nonmaximal prime ideal is a divisorial trace ideal.

(c) Each maximal ideal is either idempotent or divisorial.

Proof. To prove (a), it suffices to show that if R is an LTP domain, then for each finitely generated ideal I, $(R:I) \neq R$. By way of contradiction, let I be a finitely generated ideal of R for which $I^{-1} = R$. Then we also have $(I^2)^{-1} = R$. Obviously, both I and I^2 are trace ideals of R. While it may be that $IR_P = PR_P$ for some prime P, the same cannot be true for I^2. Hence if R is an LTP domain, $I^{-1} \neq R$ for each finitely generated ideal I.

For the proof of (b), first note that by statement (2) of Theorem 2, $PP^{-1}R_P = PR_P$. Hence we must have $PP^{-1} = P$.

Since $PP^{-1} = P$, we also have $P_v^{-1} = P^{-1} = (P:P) = (P_v:P_v)$ [14, Proposition 2.2]. Lemma 1 no longer applies, but in its place we simply note that each ideal between P_v and P has inverse equal to $(R:P)$. Starting with an ideal $I = r^2 + P$ for some $r \in P_v \setminus P$, we can repeat the proof given for $(1) \Rightarrow (2)$ in Theorem 2 to show that we must have $P = P_v$.

For (c), let M be a maximal ideal which is not idempotent. Since R is an LTP domain, $M \subseteq M^2(R:M^2)$. But $(R:M^2) = ((R:M):M)$. As M is not idempotent, we cannot have $(R:M) = R$. Hence M is divisorial. ♦

For a TPP domain R, it is known that if R and $(I:I)$ satisfy INC for each trace ideal I, then R is an RTP domain [16, Lemma 33]. We wish to show that the same occurs for LTP domains. Before proving this result, we present a pair of useful lemmas and then prove that if I is a trace ideal of an LTP domain R such that R and $(I:I)$ satisfy INC, then I is a radical ideal of R.

LEMMA 6   Let I be a trace ideal of an integral domain R and let J' be an ideal of $(I:I)$.

(a) If J' contains I, then $J' \cap R$ is a trace ideal of R.

(b) If J' is a trace ideal of $(I:I)$, then IJ' is a trace ideal of R.

Proof. For (a), assume J' contains I and set $J = J' \cap R$. Since $I \subseteq J'$, $J^{-1} \subseteq I^{-1} = (I:I)$. Hence $JJ^{-1} \subseteq J(I:I) \cap R \subseteq J' \cap R = J$.

To prove (b), assume J' is a trace ideal of $(I:I) = (R:I)$. Then $IJ'(R:IJ') = IJ'((R:I):J') = IJ'$; i.e., IJ' is a trace ideal of R. ♦

LEMMA 7   Let I be a trace ideal of an LTP domain R and let $P' \subseteq N'$ be a pair of prime ideals of $(I:I)$ which contain I. Then $P' \cap R = N' \cap R$.

Proof. Set $T = (I : I)$ and let Q' be a primary ideal of T that contains I. Then $Q = Q' \cap R$ is a primary ideal of R which is also a trace ideal by Lemma 6. Since R is an LTP domain, Q must be prime. If $P \neq N$, then there is an element $r \in N \backslash P$. Without loss of generality, we may assume N' is minimal over $J' = r^2 T + P'$ and that $Q' = J' T_{N'} \cap T$. As the corresponding ideal $Q = Q' \cap R$ is a prime ideal of R, we must have $Q = N$. But as in the proof of Theorem 2, Q contains r^2 but not r. Hence it must be that $P = N$. ♦

THEOREM 8   Let I be a trace ideal of an LTP domain R. If the pair R and $(I : I)$ satisfy INC, then I is a radical ideal of R.

Proof. Set $T = (I : I)$ and assume R and $(I : I)$ satisfy INC. Let $r \in \sqrt{I}$ and let P' be a prime of T that is minimal over I. By Lemma 7, if N' is a maximal ideal of T that contains P', then $P' \cap R = N' \cap R$. But since R and T satisfy INC, we must then have that $P' = N'$; i.e., each prime of T that is minimal over I is also a maximal ideal of T. Let $J = \{t \in T \,|\, tr \in I\}$. Let $Q' = IT_{P'} \cap T$ and $Q = Q' \cap R$. By Lemma 6, Q is a trace ideal of R. But it is also a primary ideal of R, so Q must be prime. In particular, Q' must contain r. Hence P' cannot contain J. Since J obviously contains I, we must have $J = T$ and, therefore, $I = \sqrt{I}$. ♦

COROLLARY 9   Let R be an LTP domain. If the pair R and $(I : I)$ satisfy INC for each trace ideal I, then R is an RTP domain.

We are now in a position to show that if R is simultaneously a Prüfer domain and an LTP domain, then it is also an RTP domain and a PRIP domain.

THEOREM 10   Let R be a Prüfer domain. Then the following are equivalent

(1) R is an RTP domain.
(2) R is a TPP domain.
(3) R is an LTP domain.
(4) R is a PRIP domain.

Proof. For the equivalence of (1)–(4), Corollary 3 handles the implications of (1)⇒(3), (2)⇒(3) and (4)⇒(3). Furthermore, as each RTP domain is also a TPP domain, all we need prove is that if R is a Prüfer LTP domain, then it is also an RTP domain and a PRIP domain.

Assume R is an LTP domain. Since R is a Prüfer domain, if T is an overring of R, then the primes of T are all extended from primes of R 11, Theorem 26.2].. Hence the pair R and T satisfy INC. That R is an RTP domain now follows from Corollary 9.

Let Q be a primary ideal of R. Since R is Prüfer, if Q^{-1} is a ring, then $Q^{-1} = (Q : Q)$ by Lemma 4.4 of [8]; i.e., Q is a trace ideal. Hence, Q^{-1} a ring implies Q is prime and, therefore, R is a PRIP domain. ♦

In [8], it was noted that if R is an almost Dedekind domain which is not Dedekind, then R is not a TP domain since it contains a maximal ideal M for which $(R : M) = R$. As R_M is a discrete rank one valuation domain, $M^2 \neq M$ yet $(R : M^2) = ((R : M) : M) = (R : M) = R$ so $M^2(R : M^2) = M^2 \neq M$. This same proof shows that R is not an LTP domain. A different way to establish this result is to use Theorem 5 and the fact that the only divisorial maximal ideals of a Prüfer domain are the invertible ones (see, for example, [14, Corollary 3.4]).

COROLLARY 11   Let R be an almost Dedekind domain. Then the following are equivalent

(1) R is a TP domain.
(2) R is an RTP domain.
(3) R is a TPP domain.
(4) R is an LTP domain.
(5) R is Dedekind.

Another corollary to Theorem 5 concerns Prüfer v-multiplication domains. (A domain R is a Prüfer v-multiplication domain (or PVMD for short) if R_P is a valuation domain for each maximal t-ideal P.) For a Prüfer domain, each maximal ideal is also a maximal t-ideal since each finitely generated ideal is invertible. Thus an integral domain is a Prüfer domain if and only if it is a PVMD where each maximal ideal is a maximal t-ideal.

COROLLARY 12   Let R be a PVMD. If R is an LTP domain, then it is a Prüfer domain and also an RTP domain.

Heinzer and Papick proved that the only Krull domains which satisfy the radical trace property are the Dedekind domains [12, page 112]. Since each Krull domain is a PVMD, Corollary 12 gives a different proof of their result.

COROLLARY 13   Let R be an almost Krull domain. Then the following are equivalent

(1) R is a TP domain.
(2) R is an RTP domain.
(3) R is a TPP domain.
(4) R is an LTP domain.
(5) R is a Dedekind domain.
(6) R is a PRIP domain.

Proof. It suffices to show (4) implies (5). Assume R is an LTP domain. Since R is an almost Krull domain, R_P is a Krull domain for each prime ideal P. By Theorem 4, each R_P is also an LTP domain. Hence from Corollary 12, each R_P is a Dedekind domain. It follows that R is an almost Dedekind domain. From Corollary 11, we have that R is Dedekind.◆

In [15], J. Lipman considered ideals of one-dimensional semi-local Macaulay rings. He defined an open ideal I of such a ring to be "stable" if $\bigcup(I^n : I^n) = (I : I)$. Building on Lipman's work, J. Sally and W. Vasconcelos developed a more general notion of stability by declaring an ideal to be stable if it was projective over its endomorphism ring [19, page 323]. For a nonzero ideal I of an integral domain (or just an ideal which contains an element which is not a zero divisor of a ring), their condition is equivalent to saving that I is invertible as an ideal of $(I : I)$ (since for such an ideal, being projective is equivalent to being invertible). In general, an ideal I of a domain R can be such that $\bigcup(I^n : I^n) = (I : I)$ without being stable in the sense of Sally and Vasconcelos. For example, this will be true for an ideal whose inverse is equal to R. But, if an ideal I is stable in the sense of Sally and Vasconcelos, then it will be true that $(I^n : I^n) = (I : I)$ for each positive integer n. Hence, I will be stable in the sense of Lipman. As in [1], we say that an ideal I is *L-stable* (for Lipman-stable) if $\bigcup(I^n : I^n) = (I : I)$ and *SV-stable* (for Sally-Vasconcelos-stable) if I is invertible as an ideal of $(I : I)$.

Heinzer and Papick showed that if R is an RPT domain and I is an integrally closed ideal of R, then I is L-stable [12, Remark 2.13a]. They also observed that in an RTP domain, each ideal J is such that $JJ^{-1} = J^n J^{-n}$ (where J^{-n} denotes the inverse of J^n) [12, Remark 2.13b]. Our next result considers the radical ideals of an RTP domain. In [13], E. Houston and the three authors of this paper proved that if a radical ideal I can be realized as an intersection of divisorial radical ideals which are also trace ideals, then I is a trace ideal [13, Proposition 3.15]. We shall make use of this result in the proof below.

THEOREM 14   Let R be an RTP domain. Then each radical ideal of R is L-stable.

Proof. Let I be a radical ideal of R. We first consider the two opposite cases of I being invertible and I being a trace ideal. Next we show that IR_M is L-stable for each maximal ideal. This will complete the proof since $\bigcap(BR_M : BR_M) = (B : B)$ for each ideal B of R.

If I is invertible, then so is each power of I. Hence I is L-stable since $(I^n : I^n) = R$ for each positive integer n.

If I is a trace ideal of R, then $I^n I^{-n} = II^{-1} = I$ [12, Remark 2.13b].

Hence $(I^n : I^n) \subseteq (I^n I^{-n} : I^n I^{-n}) = (I : I)$ and it follows that I is L-stable.

Let M be a maximal ideal of R. Then R_M is an RTP domain by Theorem 4. If M does not contain I, then $IR_M = R_M$, so IR_M is trivially L-stable. If M not only contains I, but is also minimal over I, then $MR_M = IR_M$. As MR_M is either invertible or a trace ideal of R_M, IR_M is L-stable. If M contains I but is not minimal over I, then each of the minimal primes of IR_M is a divisorial trace ideal of R_M [Theorem 5]. That IR_M is a trace ideal of R_M now follows from [13, Proposition 3.15]. Hence, we again have that IR_M is L- stable. ♦

Two of the questions raised in [16] concerning RTP domains were whether $(I : I)$ will always be an RTP domain when R is an RTP, and whether INC would always hold between R and $(I : I)$ when R is an RTP domain and I is a trace ideal. Our next example shows that the answer to the first of these questions is NO. Then we prove that the answer to both questions is YES when we restrict to trace ideals which are SV-stable.

EXAMPLE 15   Let V be the power series ring $F(X,Y)[[Z]]$ where F is a field and let $R = F + ZV$. Then V is a valuation domain with maximal ideal $M = ZF(X,Y)[[Z]]$ and R is pseudo-valuation domain. By [16, Theorem 31], R is an RTP domain. Let I be the ideal $Z(F[X,Y] + M)$. Then it is clear that $(I : I) = F[X,Y] + M$. There are a number of ways to verify that $(I : I)$ is not an RTP domain. For example: (a) $(I : I)/M = F[X,Y]$ is a Krull domain which is not an RTP domain since it is not Dedekind (Theorem 4 and [12, page 112]); or (b) the maximal ideal $N = (X,Y)(I : I)$ is neither idempotent nor divisorial (Theorem 5); or (c) the ideal $P = X(I : I)$ is a principal prime ideal which is not maximal (Theorem 5).

The ideal I in the example above is not a trace ideal of R. Thus this example leaves open the possibility that $(I : I)$ may be an RTP domain when I is a trace ideal of R. In our next result we show that if I is SV-stable, then not only will $(I : I)$ be an RTP domain, but the pair R and $(I : I)$ will satisfy INC.

THEOREM 16   Let I be a trace ideal of an RTP domain R. If I is SV-stable, then

(a) $(I : I)$ is an RTP domain.
(b) Each ideal of $(I : I)$ that contains I is invertible as an ideal of $(I : I)$.
(c) Each prime of $(I : I)$ that contains I is minimal over I.
(d) The pair R and $(I : I)$ satisfy INC.

Proof. Assume I is SV-stable. To simplify notation, set $T = (I : I)$. Hence $I(T : I) = T$.

We will first show that T is an RTP domain. To this end let J be a trace ideal of T and let $\sqrt[T]{J}$ denote the radical of J in T. Then $\sqrt[T]{J}$ is a trace ideal of T [13, Proposition 2.1]. By Lemma 6, both IJ and $I\sqrt[T]{J}$ are trace ideals of R. Since R is an RTP domain, both are radical ideals. Hence we must have $IJ = I\sqrt[T]{J}$. That $J = \sqrt[T]{J}$ now follows from the assumption that I is an invertible ideal of $T = (I : I)$. Thus $(I : I)$ is an RTP domain.

For part (b), let B be an ideal of $(I : I)$ which contains I. Then $J = B(T : B)$ will be a trace ideal of T that contains I. By part (a), IJ is then a radical ideal of R. It follows that $IJ = I \cap J = I = IT$ since $I \subset J$. As I is an invertible ideal of T, $J = T$; i.e., B is an invertible ideal of T.

By (b), each prime of T that contains I is invertible. That each of these primes must then be maximal ideals of T follows from Lemma 1 (see also 11, Theorem 7.6]). Therefore each such prime is also minimal over I. For a pair of distinct primes $P' \subset N'$ of T where P' does not contain I, then $P = P' \cap R$ and $N = N' \cap R$ will be distinct primes of R (no matter whether I is invertible or not) [7, Theorem 1.4]. It follows that the pair R and $(I : I)$ satisfy INC. $\blacklozenge$

We have not been able to prove an analogous result for either TPP domains or LTP domains. The best we have been able to do is prove that statements (b) and (c) will hold for a prime P' if $P' \cap R$ is minimal over I.

THEOREM 17 Let R be an LTP domain and let I be a trace ideal R which is invertible as an ideal of $(I : I)$. Let P' be a prime of $(I : I)$ which contains I and let $P = P' \cap R$. Then

 (a) P' survives in $(IR_P : IR_P)$.
 (b) If P is minimal over I, then P' is both maximal and invertible as an ideal of $(I : I)$.

Proof. Set $T = (I : I)$.

Since $P = P' \cap R$, $R_P \subseteq T_{P'}$. Since I is invertible as an ideal of T, $T_{P'} = (IT_{P'} : IT_{P'}) = (IT_{P'} : I) \supseteq (IR_P : I) = (IR_P : IR_P)$. Hence P' survives in $(IR_P : IR_P)$.

Assume P is minimal over I. If P' is not invertible as an ideal of T, then there will be a prime N' which contains P' and is a trace ideal of T. Also $N' \cap R = P$ by Lemma 7. So, without loss of generality, we may assume that P' is a trace ideal of T. Hence IP' is a trace ideal of R and P is minimal over IP'. As R is an LTP domain, we have $IP'R_P = PR_P = IR_P$.

Set $T_{(P)} = (IR_P : IR_P)$. Obviously, IR_P is an invertible ideal of $T_{(P)}$. Hence $P'T_{(P)} = P'[IR_P(T_{(P)} : IR_P)] = (P'IR_P)(T_{(P)} : IR_P) = IR_P(T_{(P)} : IR_P) = T_{(P)}$. This contradicts the fact that P' survives in $(IR_P : IR_P)$. Thus it must be that P' is invertible. $\blacklozenge$

2   MORI DOMAINS

If I is an ideal of a Mori domain R, then $I_t = I_v = A_v$ for some finitely generated ideal A contained in I [17, Théorème 1]. This property of a Mori domain makes dealing with the various trace properties much easier. For one thing it guarantees that if R is an LTP domain, then not only is each maximal ideal divisorial, but also that each is the v of a finitely generated ideal. We begin this section by showing that each Mori LTP domain is also a Mori RTP domain. We also give a characterization of Mori LTP domains in terms of SV-stability.

THEOREM 18   Let R be a Mori domain which is not a field. Then the following are equivalent

   (1) R is an RTP domain.
   (2) R is a TPP domain.
   (3) R is an LTP domain.
   (4) For each maximal ideal M and each M-primary ideal Q, M is SV-stable and QQ^{-1} contains M.
   (5) For each maximal ideal M, M is SV-stable and each maximal ideal of $(M : M)$ that contains M is invertible as an ideal of $(M : M)$.
   (6) For each nonzero radical ideal I, I is SV-stable and each maximal ideal of $(I : I)$ that contains I is invertible as an ideal of $(I : I)$.

Proof. Obviously, (6) implies (5). By [16, Theorem 4] and Corollary 3, it suffices to show that (3) implies (4), (4) implies (5), (5) implies (1), and (1) implies (6).

[(3)$\Rightarrow$(4)] Assume R is an LTP domain and let M be a maximal ideal of R. For each M-primary ideal Q, either Q is invertible or QQ^{-1} is M-primary. Thus since R is an LTP domain, QQ^{-1} must contain M. Also from our assumption that R is an LTP domain, each nonmaximal prime ideal is a divisorial trace ideal [Theorem 5]. Obviously, every invertible ideal is SV-stable, so we need only consider the case where M is not invertible as an ideal of R. From (the proof of) Theorem 5, if A is a finitely generated ideal of R, then $A^{-1} = R$ only if $A = R$. But as R is also a Mori domain, $M^{-1} = A^{-1}$ for some finitely generated ideal $A \subseteq M$ [17, Théorème 1]. It follows that $M = M_v = A_v$. Set $T = (R : M)$. As M is not invertible, T is a ring equal to $(M : M)$. Since R is an LTP domain, each nonmaximal prime of R is divisorial. Thus no such prime can contain A. Also, no other maximal ideal can contain A, since each maximal ideal is divisorial. Hence M is minimal over A, and, therefore A is M-primary. By Theorem 2 we have $M \subseteq A(R : A) = A(R : M) = AT \subseteq MT = M$. So M is a finitely generated ideal of T. As M is not invertible as an ideal of R, neither is M^2. Thus, again by Theorem 2, $M \subseteq M^2(R : M^2) = M[M((R : M) : M)] =$

$M[M(T : M)] \subseteq MT = M$. In particular, we have $M[M(T : M)] = M$. As M is finitely generated as an ideal of T, Nakayama's Lemma implies that $M(T : M) = T$; i.e., M is SV-stable.

[(4)$\Rightarrow$(5)] Assume that for each maximal ideal M and each M-primary ideal Q, M is SV-stable and QQ^{-1} contains M. Let M be a fixed maximal ideal of R. As in the proof of (3) implies (4), there is nothing to prove if M is invertible as an ideal of R. Hence we assume that M is not invertible as an ideal of R. Set $T = (M : M)$ and let N be a maximal ideal of T that contains M. Set $B = N(T : N)$. As B is a trace ideal of T, BM is a trace ideal of R by Lemma 6. Since N contains M, BM is M-primary. Hence, by Theorem 2, $BM = M = MT$. As M is invertible as an ideal of T, we have $B = T$.

[(5)$\Rightarrow$(1)] Assume that for each maximal ideal M, M is SV-stable and each maximal ideal of $(M : M)$ that contains M is invertible as an ideal of $(M : M)$. Let M be a maximal ideal of R and let N be a maximal ideal of $(M : M)$ that contains M. Since R is a Mori domain, so is $(M : M)$ [18, page 11], [3, Corollary 11]. It follows that N has height one [4, Theorem 2.5], and, therefore, M has height one. Hence R must be one-dimensional.

Let Q be an M-primary ideal of R. Since the radical trace property and the trace property for primary ideals are known to be equivalent for one-dimensional domains, we need only show that $Q(R : Q)$ contains M. In $(M : M)$ each maximal ideal that contains Q also contains M and, therefore, each such ideal is invertible and minimal over Q. It follows that no maximal ideal of T can contain $Q(T : Q)$ [13, Proposition 2.1]. Hence $Q(T : Q) = T$. Hence, $M = MT = QM(T : Q) \subseteq Q(R : Q)$, and, therefore, R is an RTP domain.

[(1)$\Rightarrow$(6)] Assume R is an RTP domain and let I be a nonzero radical ideal of R. By Proposition 2.1 of [13], there is nothing to prove if I is invertible. Hence we assume that I is not invertible. From the argument above, R is one-dimensional and each maximal ideal is divisorial (or see [10, Section 2]). It follows that I is divisorial. Since R is a Mori domain, only finitely divisorial prime ideals can contain I. In this case that means that I is a finite intersection of maximal ideals. Let $\{M_1, M_2, \ldots, M_n\}$ denote the set of invertible maximal ideals that contain I and let $\{N_1, N_2, \ldots, N_m\}$ denote the set of noninvertible maximal ideals that contain I. Set $A = \bigcap M_k$ and $B = \bigcap N_k$. Then A and B are comaximal with A an invertible ideal of R [13, Proposition 2.1] and B a trace ideal of R [13, Proposition 3.15]. Hence, $I = A \cap B = AB$. Since A is invertible, we have $(I : I) = (AB : AB) = (B : B)$. Thus to show that I is SV-stable, it suffices to show that B is SV-stable. Set $T = (I : I)$. As B is a trace ideal of R, we also have $T = (R : B)$. Since B is divisorial, $B_v = C_v$ for some finitely generated ideal $C \subseteq B$. Moreover $\sqrt{C} = B$. As in the proof of (3) implies (4), $B = BT = CT$. Let J be a trace ideal of T that contains B. Then by Lemma 6, JB is a trace ideal of R. It follows that $JB = J \cap B = B$ since R is an RTP domain. But since B is finitely generated as an ideal of T, $JB = BT$ is possible only if $J = T$. Hence each ideal of T that contains B is invertible as an ideal of T. In particular, B is invertible as an ideal of T, as is each maximal ideal of T that contains B.

If M' is a maximal ideal of T that contains I but not B, then $M' \cap R = M_k$ for some k [7, Theorem 1.4]. Moreover M' is invertible as an ideal of T since each of the M_js are invertible ideals of R [7, Theorem 1.4]. ◆

By combining Theorems 16 and 18, we have the following.

COROLLARY 19   Let R be a Mori RTP domain. Then for each nonzero radical ideal I, $(I : I)$ is an RTP domain and the pair R and $(I : I)$ satisfy INC.

One of the classic examples of a Mori domain which is not Noetherian is the ring $R = F + X F[X, Y]$ where F is a field. If this ring is localized at its maximal ideal $M = X F[X, Y]$, the resulting ring is a two-dimensional quasi-local Mori domain whose maximal ideal is SV-stable. As R_M is two-dimensional, it cannot be an RTP domain. The corresponding power series ring $F + X F[[X, Y]]$ is also a two-dimensional Mori domain. Unlike R, $F + X F[[X, Y]]$ is quasi-local, but like R, the maximal ideal $X F[[X, Y]]$ is SV-stable. We will do more with this ring in Example 21, but first we give an example of a local one-dimensional Noetherian domain where the maximal ideal is SV-stable yet the ring is not an RTP domain.

EXAMPLE 20   Let $R = F[[X^3, X^5, X^7]]$ where F is a field. Then
   (a) R is a local one-dimensional Noetherian domain with maximal ideal $M = (X^3, X^5, X^7)R$.
   (b) $(R : M) = (M : M) = F[[X^2, X^3]]$.
   (c) $M(M : M) = X^3 F[[X^2, X^3]]$ is invertible as an ideal of $(M : M)$.
   (d) R is not an RTP domain. For example, $Q = (X^5, X^6, X^7)R$ is an M-primary ideal for which $(Q : Q) = (R : Q) = F[[X]]$.

For the ring $R = F + X F[[X, Y]]$, the ideal $P = YX F[[X, Y]]$ is a height one prime ideal. Since Y^{-1} is in P^{-1} and X^{-1} is not, $PP^{-1} = X F[[X, Y]]$, the maximal ideal of R. In our next example we show that even though R is not an RTP domain, the maximal ideal $X F[[X, Y]]$ is the trace of each primary ideal whose radical has height one.

EXAMPLE 21   Let $R = F + X F[[X, Y]]$ and let $M = X F[[X, Y]]$. Then
   (a) M is SV-stable and the ideal $M(X, Y)F[[X, Y]]$ is an M-primary trace ideal.
   (b) Each height one prime of R has the form fM for some irreducible $f \in (X, Y)F[[X, Y]] \setminus X F[[X, Y]]$.

(c) If Q is a primary ideal of R whose radical has height one, then $Q = f^n M$ for some irreducible $f \in (X, Y)F[[X, Y]] \setminus X F[[X, Y]]$.

(d) If Q is a primary ideal of R whose radical has height one, then $QQ^{-1} = M$.

Proof. Since M is not invertible as an ideal of R, $(R : M) = (M; M)$ is a ring. Specifically, $(R : M) = F[[X, Y]]$. Obviously, M is invertible as an ideal of $(M : M)$. As $N = (X, Y)F[[X, Y]]$ is a trace ideal of $F[[X, Y]]$, MN is a trace ideal of R by Lemma 6.

Let P be a height one prime ideal of R. Then there is a unique prime ideal P' of $F[[X, Y]]$ which contracts to P, namely $P' = \{g \in F[[X, Y]] \mid gM \subseteq P\}$. Since $F[[X, Y]]$ is a local UFD, P' is a principal prime of $F[[X, Y]]$. Thus $P' = fF[[X, Y]]$ for some irreducible $f \in F[[X, Y]]$. As P has height one, f is not a multiple of X and it follows that $P = MP' = XP' = XfF[[X, Y]]$.

Continuing with the notation above, let Q be a P-primary ideal. Since P does not contain M, $R_P = F[[X, Y]]_{P'}$ is a discrete rank one valuation domain. Hence $QR_P = f^n R_P$ for some integer n. By contracting this ideal to R we see that $Q = Xf^n F[[X, Y]] = f^n M$. Since Q is a principal multiple of M, $Q^{-1} = (1/f^n)M^{-1}$ and from this it follows that $QQ^{-1} = M$.♦

REFERENCES

1. D.D. Anderson, J. Huckaba, I. Papick, A note on stable domains, Houston J. Math., 13: 13–17, 1987.
2. D.F. Anderson, When the dual of an ideal is a ring. Houston J. Math., 9: 325–332, 1983.
3. V. Barucci, Strongly divisorial ideals and complete integral closure of an integral domain. J. Algebra, 99: 132–142, 1986.
4. V. Barucci, S. Gabelli, How far is a Mori domain from being a Krull domain? J. Pure Appl. Algebra, 15: 101–112, 1987.
5. H. Bass, On the ubiquity of Gorenstein rings. Math. Z., 82: 8–28, 1963.
6. P.-J. Cahen, T. Lucas, The special trace property. In: P.-J. Cahen, M. Fontana, E. Houston and S.-E. Kabbaj, ed. Commutative Ring Theory: Proceedings of the II International Conference. New York: Marcel Dekker, 1987, pp. 161–172.
7. M. Fontana, Topologically defined classes of commutative rings. Ann. Math. Pura Appl., 123: 331–355, 1973.
8. M. Fontana, J. Huckaba, I. Papick, Domains satisfying the trace property. J. Algebra, 107: 169–182, 1987.
9. R. Fossum, The Divisor Class Group of a Krull Domain. New York: Springer-Verlag, 1973.

10. S. Gabelli, Domains with the radical trace property and their complete integral closure. Comm. Algebra, 20: 829–845, 1992.

11. R. Gilmer, Multiplicative Ideal Theory. Kingston, Ontario: Queen's Papers in Pure and Applied Mathematics, 1992.

12. W. Heinzer, I. Papick, The radical trace property. J. Algebra, 112: 110–121, 1988.

13. E. Houston, S.-E. Kabbaj, T. Lucas, A. Mimouni, When is the dual of an ideal a ring? J. Algebra, to appear.

14. J. Huckaba and I. Papick, When the dual of an ideal is a ring. Manuscripta Math., 37: 67–85, 1982.

15. J. Lipman, Stable ideals and Arf rings. Amer. J. Math., 93: 649–685, 1971.

16. T. Lucas, The radical trace property and primary ideals. J. Algebra, 184: 1093–1112, 1996.

17. J. Querré, Sur une propriété des anneaux de Krull. Bull. Sci. Math., 2:341–354, 1971.

18. N. Raillard (Dessagnes), Sur les anneaux de Mori. Thèse, Univ. Pierre et Marie Curie, Paris IV, Paris, 1976.

19. J. Sally, W. Vasconcelos, Stable rings. J. Pure Appl. Algebra, 4: 319–336, 1974.

Pullbacks and Coherent-Like Properties

ABDESLAM MIMOUNI [1] [2] Department of Mathematics, Faculty of Sciences "Dhar Al-Mehraz", University of Fez. Fez-Morocco

INTRODUCTION.

Let I be a nonzero ideal of a domain T, $\varphi: T \to T/I$ the natural projection and D a domain contained in T/I. Let $R = \varphi^{-1}(D)$ be the domain arising from the following pullback of canonical homomorphisms.

$$
\begin{array}{ccc}
R & \longrightarrow & D \\
\downarrow & & \downarrow \\
T & \longrightarrow & T/I
\end{array}
$$

We explicitly assume that $R \subset T$ and we shall refer to this as a diagram of type (Δ). If $I = P$ is a prime ideal of T, we use $\chi(P)$ to denote the residue field of T_P and $qf(D)$ the quotient field of D. The case where $T = V$ is a valuation domain is of particular

[1] Partially supported by the ICTP, Trieste, Italy

[2] The author would like to express his sincere thanks to Professor Evan Houston for his helpful suggestions

interest; in this case we use $Z(V/I)$ to denote the set of zerodivisors of V modulo I and we shall say that the diagram is of type (Δ_1). The goal of this paper is to characterize certain coherent-like properties of integral domains in pullback constructions of type (Δ_1). In some sense, this work is similar to that of S. Gabelli and E. Houston in which coherence and several other properties are studied in pullbacks of type $(\sqcap)$ in which I is assumed to be a maximal ideal of T. In the first part of this work, we give some results concerning duality and divisoriality in diagrams of type (Δ) and (Δ_1). The main results are as follows: For the diagram (Δ_1), $I^{-1} = (I : I)$ and I is a v-ideal of R (Theorem 1.(1)), and for diagram (Δ), if I is a nonprime ideal of T or a maximal ideal of T, then I is a v-ideal of R. But when $I = P$ is a nonmaximal prime ideal of T, then two cases are possible (Theorem 4). With the aid of these results, we give the promised characterizations.

The ideal I is crucial, and we give a complete characterization of when I is v-finite as an ideal of R in the case of the diagram (Δ_1). We also show that if I is a nonprime principal (hence invertible) ideal of V, then I is v-finite in R; this is in sharp contrast with a result of Gabelli and Houston [9, Proposition 2.3]. With this characterization and under the assumption $qf(D) \subset V/I$, we also prove that : For the diagram (Δ_1), R is v-coherent if and only if D is v-coherent and I is v-finite in R (Theorem 13). Using some of these results, we construct a family of v-coherent domains that are neither Mori nor quasi-coherent domains more general than the family of "$D + M$" constructions. We also give necessary and sufficient conditions for R to be a Mori domain.

Finally, for the other coherent-like properties such as quasi-coherence, finite conductor domains, v-domains, PVMD's, Prüfer domains, ..., the main remark is that : For each of these properties, D has the same property, I must be a prime ideal P of V and $\chi(P) = qf(D)$.

It will often be the case that an ideal is not only an ideal of the ring R but is also an ideal of a larger ring (say) T. In this case, it is understood that inverses and v-operations are taken with respect to the ring R.

We begin by recalling some definitions. For a nonzero ideal I of a domain R :

$I^{-1} = \{x \in qf(R)/xI \subseteq R\}$, and $I_v = (I^{-1})^{-1}$.

$(I : I) = \{x \in qf(R)/xI \subseteq I\}$. Then $(I : I)$ is a ring and $(I : I) \subseteq I^{-1}$. We say that:

1) I is a v-ideal (or divisorial ideal) of R if $I = I_v$.

2) I is v-finite if $I^{-1} = J^{-1}$ for some finitely generated ideal J of R.

3) R is a v-coherent domain if I^{-1} is a v-finite divisorial (fractional) ideal for each finitely generated (fractional) ideal I of R.

4) R is a v-domain if each finitely generated ideal I of R is v-invertible, that is $(II^{-1})_v = R$.

5) R is quasi-coherent if I^{-1} is finitely generated for each finitely generated ideal I of R.

6) R is a finite conductor domain (F.C) if $Ra \cap Rb$ is finitely generated for each a, b in R.

7) R is a Mori domain if R satisfies the ascending chain conditions on divisorial ideals, equivalentely, if for each ideal I of R, we have $I^{-1} = J^{-1}$ for some finitely generated ideal J of R with $J \subseteq I$.

We use "$\subset$" for strict inclusion.

THEOREM 1. For thc diagram (Δ_1) :

1) $I^{-1} = (I : I)$ and I is a divisorial ideal of R.

2) The following statements are equivalent:

   i) Each ideal of R is comparable to I.

   ii) $Z(V/I) \cap D = 0$.

   iii) $qf(D) \subseteq I^{-1}/I$. (We note that, by 1), I is an ideal of I^{-1})

3) Assume that $qf(D) \subseteq V/I$. Then for each nonzero (fractional) ideal A of D, $\varphi^{-1}(A)$ is a nonzero (fractional) ideal of R and :

   a) A is a finitely generated ideal of D if and only if $\varphi^{-1}(A)$ is a finitely generated ideal of R.

b) $\varphi^{-1}(D : A) = (R : \varphi^{-1}(A))$ and $\varphi^{-1}(A : A) = (\varphi^{-1}(A) : \varphi^{-1}(A))$.

4) Let J be an ideal of R such that $J \subseteq I$, then :

a) If $J_v = I$, then $J^{-1} = I^{-1} = (I : I)$ is a ring.

b) If $J_v \subset I$, consider the following statements :

 i) $(V : JV)$ is a ring, i.e. JV is a nonprincipal prime ideal of V.

 ii) J^{-1} is a ring.

 iii) $(R : JV)$ is a ring and $(R : JV) = J^{-1}$. Then $i) \Longrightarrow ii)$ and $ii) \Longleftrightarrow iii)$. If in addition $Rad_V(JV) \subset Rad_V(I)$, then the three statements are equivalent. (where $Rad_V(.)$ denotes the radical in V).

Proof. 1) Suppose that there is $x \in I^{-1}$ and $a \in I$ such that $xa \notin I$. Since $xa \in R$ and I is a prime ideal of R, then $(xa)^2 \notin I$. So $I \subset (xa)^2 V$ and $(xa)^{-2}I \subset V$. Since $a \in I$, then $x^{-2}a^{-1} = (xa)^{-2}a \in V$. So $x^{-2} = x^{-2}a^{-1}a \in I$. Hence $x^{-1} = x^{-2}x \in II^{-1} \subseteq R$, and therefore $x^{-1} \in I$. Hence $1 = xx^{-1} \in II^{-1}$. So $R = II^{-1}$. Now, $R \subset V \subseteq (I : I) \subseteq (II^{-1} : II^{-1}) = R$, which contradicts our assumption that $R \subset V$. Hence $II^{-1} \subseteq I$. It follows that $I^{-1} = (I : I)$.

To show that I is a v-ideal of R, suppose that $I \subset I_v$. Let $x \in I_v$ such that $x \notin I$. Then $x^2 \notin I$. So $I \subset x^2 V$. Hence $x^{-1}I \subseteq xV$. Since $V \subseteq (I : I) = I^{-1} = (I_v)^{-1} = (I_v : I_v)$ [12, Proposition 2.2], then I_v is an ideal of V. So $xV \subseteq I_v$. Then $x^{-1}I \subseteq xV \subseteq I_v \subseteq R$, and $x^{-1} \in I^{-1} = (I_v)^{-1}$. So $1 = xx^{-1} \in I_v(I_v)^{-1} = I_v$. Hence $I_v = R$, and $R \subset V \subseteq (I : I) = I^{-1} = (I_v)^{-1} = R$, which is a contradiction. Hence $I = I_v$.

2) $i) \Longrightarrow ii)$. If $Z(V/I) \cap D \neq 0$, then let $0 \neq d \in Z(V/I) \cap D$. Write $d = \varphi(x)$ for some $x \in R$. Since $x \notin I$, by $i)$, $I \subset xR$. Hence $x^{-1}I \subseteq R$, and $x^{-1} \in I^{-1} = (I : I)$. Since $d \in Z(V/I)$, then there is $z \in V$, $z \notin I$ such that $\varphi(xz) = \varphi(x).\varphi(z) = d.\varphi(z) = 0$. Hence $xz \in I$ and therfore $z = xz.x^{-1} \in II^{-1} = I$, a contradiction.

 $ii) \Longrightarrow iii)$. Let $\psi : I^{-1} \to I^{-1}/I$ denote the canonical homomorphism, and let $0 \neq d = \varphi(x)$, $x \in R$. Since $x \notin I$, then $I \subset xV$, and $x^{-1}I \subseteq V$. Now, for each $a \in I$, $x^{-1}a \in V$ and $0 = \varphi(a) = \varphi(x).\varphi(x^{-1}a) = d.\varphi(x^{-1}a)$. By $ii)$, $\varphi(x^{-1}a) = 0$. So $x^{-1}a \in I$ and therefore $x^{-1} \in (I : I) = I^{-1}$. Hence $1 = \psi(x.x^{-1}) = \psi(x).\psi(x^{-1}) = \varphi(x).\psi(x^{-1}) = d.\psi(x^{-1})$ (since $x \in R$, then $\varphi(x) = \psi(x)$). So $d^{-1} = \psi(x^{-1}) \in I^{-1}/I$. Hence $qf(D) \subseteq I^{-1}/I$.

$iii) \implies i)$. Let J be an ideal of R. If $J \nsubseteq I$, then let $x \in J$ such that $x \notin I$. Write $\varphi(x) = d \in D \setminus \{0\}$. By $iii)$, $d^{-1} = \psi(z)$ for some $z \in I^{-1}$. So $1 = d.\psi(z) = \varphi(x).\psi(z) = \psi(x).\psi(z) = \psi(xz)$. Hence $1 - xz \in I$. Now, for each $a \in I$, if $xa^{-1} \in I^{-1}$, then $x = a.xa^{-1} \in II^{-1} = I$, which is impossible. Since I^{-1} is a valuation domain, then $ax^{-1} \in I^{-1}$ and therefore $ax^{-1}(1 - xz) \in II^{-1} = I$. So $ax^{-1} - az \in I$. Since $az \in II^{-1} = I$, then $ax^{-1} \in I$. So $a \in xI \subseteq IJ \subseteq J$ and therefore $I \subseteq J$.

3) Assume that $qf(D) \subseteq V/I$. Let $(0) \neq A$ be an ideal of D. It is easy to see that $\varphi^{-1}(A)$ is an ideal of R (which contains I by 2)).

a) $\implies$) Assume that $A = \sum_{i=1}^{i=n} d_i D$ where $0 \neq d_i = \varphi(x_i)$, $x_i \in R$ and $x_i \notin I$. We claim that $\varphi^{-1}(A) = \sum_{i=1}^{i=n} x_i R$. Let $x \in \varphi^{-1}(A)$, then $\varphi(x) \in A$. So $\varphi(x) = \sum_{i=1}^{i=n} d_i \alpha_i$ where $\alpha_i = \varphi(\beta_i)$ for some $\beta_i \in R$. hence $\varphi(x) = \sum_{i=1}^{i=n} \varphi(x_i)\varphi(\beta_i) = \varphi(\sum_{i=1}^{i=n} x_i \beta_i)$. So $x - \sum_{i=1}^{i=n} x_i \beta_i \in I$. Hence $x \in \sum_{i=1}^{i=n} x_i R + I = \sum_{i=1}^{i=n} x_i R$ (since $I \subseteq \sum_{i=1}^{i=n} x_i R$, by the assertions 2)). So $\varphi^{-1}(A) \subseteq \sum_{i=1}^{i=n} x_i R$. It follows that $\varphi^{-1}(A) = \sum_{i=1}^{i=n} x_i R$.

$\impliedby$). This holds in general.

b) Follows from [10, Proposition 6].

4) Let J be an ideal of R such that $J \subseteq I$. Then $J_v \subseteq I_v = I$.

a) If $J_v = I$, then $J^{-1} = (J_v)^{-1} = I^{-1} = (I : I)$ is a ring (by 1)).

b) Assume that $J_v \subset I$. Then $J \subseteq JV \subseteq I$. So JV is an ideal of R and V, and it is clear that $J^{-1} = (R : J) \subseteq (V : JV)$.

$i) \implies ii)$. Since $(V : JV)$ is a ring, then JV is a nonprincipal prime ideal of V, [12, Proposition 3.5]). Assume that $JV = Q$. Then Q is also a prime ideal of R, and $(Q : Q) \subseteq (R : Q) \subseteq (V : Q) = (Q : Q) = V_Q$. Hence $(Q : Q) = (R : Q) = (V : Q) = V_Q$. Since $J \subseteq$

$JV = Q$, then $(R : Q) \subseteq (R : J) \subseteq (V : JV) = (V : Q) = (R : Q)$. Hence $J^{-1} = (V : JV)$.

$ii) \Longrightarrow iii)$. Since $J \subseteq JV$, then $(R : JV) \subseteq J^{-1}$. Since I is a prime ideal of R, $J \subseteq I$, and J^{-1} is a ring, then, by [11, Proposition 2.2], $J^{-1} = (I : J)$. Now, let $x \in J^{-1}$. Then $xJ \subseteq I$. So $xJV \subseteq IV = I \subseteq R$. Hence $x \in (R : JV)$. It follows that $J^{-1} = (R : JV)$.

$iii) \Longrightarrow ii)$. Trivial.

Assume that $Rad_V(JV) \subset Rad_V(I)$. We wish to show that $ii) \Longrightarrow i)$. The remainder of the proof is divided into two steps. In the first step, we show that $(V : J_v) = J^{-1}$. The second step then uses this and the fact that $J^{-1} \subset (V : JV)$ to derive a contradiction.

Let $Q_0 = Rad_V(JV)$ and $P_0 = Rad_V(I)$. Since $Q_0 \subset P_0$, then $Q_0 \subset I$. Since J^{-1} is a ring and $J \subset I$, then $V \subseteq (I : I) = I^{-1} \subseteq J^{-1} = (J_v : J_v)$. So J_v is an ideal of R, V and I^{-1}. Write $J^{-1} = V_{Q_1}$ and $I^{-1} = V_P$ for some prime ideals Q_1 and P of V. Since $J_v \subset I$, then $V_P = I^{-1} \subset J^{-1} = V_{Q_1}$. Hence $Q_1 \subset P$. Now, $JV \subseteq Q_0$, then $JV_{Q_0} \subseteq Q_0 V_{Q_0} = Q_0 \subset I$. So $V_{Q_0} \subseteq (I : J) = J^{-1} = V_{Q_1}$. Hence $Q_1 \subseteq Q_0 \subset I$.

$\underline{step1}$. Let $x \in J^{-1} = (J_v)^{-1}$. Then $xJ_v \subseteq R \subseteq V$. So $x \in (V : J_v)$ and $J^{-1} \subseteq (V : J_v)$. Conversely, let $x \in (V : J_v)$. Then $xJ_v \subseteq V$ which implies that $xJ_v = xJ_v V_{Q_1} \subseteq V_{Q_1}$. So xJ_v is an ideal of V_{Q_1} (since J_v is an ideal of $(J_v : J_v) = (J_v)^{-1} = J^{-1} = V_{Q_1}$). Hence either $xJ_v = V_{Q_1}$, or $xJ_v \subseteq Q_1$.

If $xJ_v = V_{Q_1}$, then $J_v = x^{-1}V_{Q_1}$. Hence, $J^{-1} = (J_v)^{-1} = (R : x^{-1}V_{Q_1}) = x(R : V_{Q_1}) = xQ_1$ [Note that $(R : V_{Q_1}) = Q_1$, since $Q_1 V_{Q_1} = Q_1 \subseteq Q_0 \subset I \subset R$, $(R : V_{Q_1})$ is an ideal of both R and V_{Q_1}, and Q_1 is the maximal ideal of V_{Q_1}]. On the other hand, $J^{-1} = V_{Q_1}$. Then $xQ_1 = V_{Q_1}$. So $Q_1 = x^{-1}V_{Q_1} = J_v$. Hence $J_v = Q_1$ is a nonmaximal prime ideal of V. By [12, Corollary 3.6], $(V : J_v) = (J_v : J_v) = J^{-1}$. Hence $x \in J^{-1}$.

If $xJ_v \subseteq Q_1$, then $x \in (Q_1 : J_v) \subseteq (I : J_v) \subseteq (J_v)^{-1} = J^{-1}$. It follows that $J^{-1} = (V : J_v)$.

$\underline{step2}$. Suppose that $J^{-1} \subset (V : JV)$. Let $x \in (V : JV)$ such that $x \notin J^{-1}$. Then $xJV \subseteq V$ implies that $xJI^{-1} \subseteq I^{-1}$. Hence $I = I_v \subseteq (R : xJI^{-1}) = x^{-1}(R : JI^{-1}) = x^{-1}((R : I^{-1}) : J) = x^{-1}(I_v : J) = x^{-1}(I : J) \subseteq x^{-1}J^{-1} = (xJ)^{-1}$. Hence $xJ_v = (xJ)_v \subseteq I^{-1} = V_P$. So, either $xJ_v = V_P$, or $xJ_v \subseteq PV_P = P$ [since J_v is an ideal of $I^{-1} = V_P$]. If $xJ_v = V_P$, then $J_v = x^{-1}V_P = x^{-1}I^{-1}$. So $J^{-1} = (J_v : J_v) = (x^{-1}I^{-1} : x^{-1}I^{-1}) = I^{-1}$ (since I^{-1} is a ring). Hence $J_v = I$, a contradiction. So $xJ_v \subseteq P$. Hence $x \in (P : J_v) \subseteq (V : J_v) = J^{-1}$, a contradiction. It follows that $(V : JV) = J^{-1}$ is a ring.

EXAMPLE 2.

The following examples illustrate the necessity of some of the hypotheses in Theorem 1. In the example we use $Z(V, I)$ to denote the set of all $x \in V$ such that $xy \in I$ for some $y \notin I$.

1) A local domain T with a prime ideal P and a domain D contained in T/P such that $(P : P) \subset P^{-1}$ (So P^{-1} is not a ring).

Let k be a field, and let X and Y be indeterminates over k. Set $T = k[[X, Y]]$, $P = YT$ and $D = k[[X^2, X^3]]$. Then it is easy to see that $X^2 k[[X]] \subseteq D$ and $T = k[[X]] + P$. Let $R = D + P$. Then for each $f \in T$, write $f = f_0 + Yf_1$ where $f_0 \in k[[X]]$ and $f_1 \in T$. Then $\frac{X^2}{Y}Yf = X^2 f = X^2 f_0 + YX^2 f_1 \in X^2 k[[X]] + P \subseteq D + P = R$. Hence $\frac{X^2}{Y}P \subseteq R$. So $\frac{X^2}{Y} \in P^{-1}$. But $\frac{X^2}{Y} \notin T = (P : P)$.

2) A pullback diagram of type (Δ_1) and an ideal J such that I and J are not comparable.

Let $V = k[[X]] + Yk((X))[[Y]]$, $M_1 = Yk((X))[[Y]]$ and $M = Xk[[X]] + M_1 = XV$ the maximal ideal of V. Let $I = YXV = YM$. Since $Y \notin I$, then $M = Z(V, I)$. So $Z(V/I) = \varphi(Z(V, I)) = \varphi(M))$. Let $D = k[[X]]$, then $M \cap D \neq (0)$ (since $X \in M \cap D$). Since $D \cap I = (0)$, we identify D as a subdomain of V/I.

Let J be the ideal of R given by $J = Xk[[X]] + YM^2 + M_1^2$. It is easy to see that :

1) J is a proper ideal of R.

2) $J \not\subseteq I$ since $X \in J$ and $X \notin I$.

3) $I \not\subseteq J$ since $YX \in I$ and $YX \notin J$.

3) Conditions i) and ii) of Theorem 1.(4)(b) are not (in general) equivalent.

Let V and M_1 be as in Example 2), and set $D = k$, $V_1 = k((X))[[Y]]$ and $J = M_1^2$. Then J is an ideal of both R and V, and $J \subseteq I \subseteq M_1$. By [10, Corollary 3], $(V : M_1) = (M_1 : M_1) = V_1$. Hence $(V : J) = (V : M_1^2) = ((V : M_1) : M_1) = (V_1 : M_1) = \frac{1}{Y}V_1$. Hence $(V : J)$ is not a ring. Now, we claim that $J^{-1} = V_1$. Since J is an ideal of V_1, then $V_1 \subseteq (J : J) \subseteq J^{-1} \subseteq (V : J) = \frac{1}{Y}V_1$. Let $f \in J^{-1}$. Write $f = \frac{1}{Y}f_1$ for some $f_1 \in V_1$. Write $f_1 = g_0 + Yg_1$ where $g_0 \in k((X))$ and $g_1 \in V_1$. Since $XY^2 \in J$, then $XYf_1 = XY^2f \in R$. So $XYg_0 + XY^2g_1 \in R$. Since $XY^2g_1 \in J \subseteq R$, then $XYg_0 \in R$. Hence $g_0 \in k[[X]]$. Now, for each $n \geq 1$, $\frac{Y^2}{X^n} \in J$, then $\frac{Y}{X^n}f_1 = \frac{Y^2}{X^n}f \in R$. So $\frac{Y}{X^n}g_0 + \frac{Y^2}{X^n}g_1 \in R$. Since $\frac{Y^2}{X^n}g_1 \in J \subseteq R$, then $\frac{Y^2}{X^n}g_0 \in R$. So $g_0 \in X^n k[[X]]$. Hence $g_0 = 0$ and $f_1 = Yg_1$. So $f = \frac{1}{Y}f_1 = g_1 \in V_1$ and therefore $J^{-1} = V_1$.

THEOREM 3. For the diagram (Δ):

1) If each minimal prime over I is a maximal ideal of T, then $I^{-1} = (I : I)$ and I is a divisorial ideal of R.

2) Let P be a (nonzero) prime ideal of R, then :

   a) If $I \not\subseteq P$, let Q be the unique prime ideal of T that contracts to P. Then P^{-1} is a ring if and only if $(T : Q)$ is a ring.

   b) If $I \subset P$, then $P = \varphi^{-1}(q)$ for some nonzero prime ideal q of D. In this case, if $qf(D) \subseteq T/I$, then $P^{-1} = \varphi^{-1}(D : q)$ and $(P : P) = \varphi^{-1}(q : q)$.

Proof. 1) Let $x \in I^{-1}$ and $a \in I$. Suppose that there is a minimal prime M over I such that $xa \notin M$. Since M is a maximal ideal of T, then there is $\alpha \in T$ and $m \in M$ such that $1 = m + ax\alpha$. Since $\alpha \in T$, $a \in I$ and I is an ideal of T, then $a\alpha \in I$. So $xa\alpha \in II^{-1} \subseteq R$. Hence $m = 1 - ax\alpha \in R$. Since M is minimal over

I, then $MT_M = \sqrt{I}T_M$. So, there is $t \in T$, $t \notin M$ and a nonzero integer r such that $(tm)^r \in I$. Now, $1 = (m + ax\alpha)^r = z + m^r$ for some $z \in II^{-1} \subseteq R$. So $t^r = t^r z + t^r m^r \in II^{-1}$. Since I is a prime ideal of R, $t^r \in R$, $m \in R$, $t^r m^r = (tm)^r \in I$ and $t \notin M$, then $m \in I$. Hence $1 = m + ax\alpha \in II^{-1}$. So $R = II^{-1}$. Hence $T \subseteq (I : I) \subseteq (II^{-1} : II^{-1}) = R$, a contradiction. So, for each minimal prime M over I, $xa \in M$.

We prove $xa \in I$ by cheking locally. Let N be a maximal ideal of T, then:

If $I \not\subseteq N$, then $xa \in R \subseteq T_N = IT_N$.

If $I \subseteq N$, then N is minimal over I. So $xa \in N \subseteq NT_N = \sqrt{I}T_N$. Hence, there is $\mu \in T$, $\mu \notin N$ and a nonozero integer r such that $(xa\mu)^r \in I$. Since I is an ideal of T, then $a\mu \in I$. So $xa\mu \in II^{-1} \subseteq R$. Hence $xa\mu \in I$. So $xa = \frac{xa\mu}{\mu} \in IT_N$. It follows that $I^{-1} = (I : I)$.

We now show that I is a v-ideal of R. First, we note that, $T \subseteq (I : I) = I^{-1} = (I_v)^{-1} = (I_v : I_v)$, so that I_v is an ideal of T. Let $x \in I_v$. Suppose that there is a minimal prime M over I such that $x \notin M$. Since M is a maximal ideal of T, then there is $\alpha \in T$ and $m \in M$ such that $1 = m + x\alpha$. Since $\alpha \in T$, $x \in I_v$ and I_v is an ideal of T, then $x\alpha \in I_v$. So, $m = 1 - x\alpha \in R$. As in the first part of the proof, there is $t \in T$, $t \notin M$, $z \in I_v$ and a nonzero integer r such that $(tm)^r \in I$ and $1 = z + m^r$. So $t^r = t^r z + t^r m^r \in I_v$. Since $t \notin M$, then $m \in I$. So $1 = z + m^r \in I_v$. Hence $I_v = R$, which is absurd (since $R \subset T \subseteq (I : I) = I^{-1}$). So, for each M minimal over I, $x \in M$.

We prove that $x \in I$ by cheking locally. Let Let N be a maximal ideal of T, then:

If $I \not\subseteq N$, then $x \in R \subseteq T_N = IT_N$.

If $I \subseteq N$, then N is minimal over I. So $x \in N \subseteq NT_N = \sqrt{I}T_N$. Hence, there is $\mu \in T$, $\mu \notin N$ and a nonozero integer r such that $(x\mu)^r \in I$. Since $x\mu \in I_v \subseteq R$, then $x\mu \in I$. So $x = \frac{x\mu}{\mu} \in IT_N$. Hence $x \in I$ and $I = I_v$.

2) Let P be a prime ideal of R such that $I \not\subseteq P$ and Q the unique

prime ideal of T that contracts to P

a) $\Longrightarrow$). Assume that P^{-1} is a ring. Then $P^{-1} = (P : P)$. Let $x \in (T : Q)$. Then $xP \subseteq xQ \subseteq T$. So $xIP \subseteq I \subseteq R$. Hence $xI \subseteq P^{-1} = (P : P)$. So $xIP \subseteq P$. Since $IQ \subseteq I \cap Q \subseteq Q \cap R = P$, we have $xI^2Q \subseteq xIP \subseteq P \subseteq Q$. Since $I^2 \not\subseteq Q$, $xQ \subseteq T$ and Q is a prime ideal of T, then $xQ \subseteq Q$. Hence $x \in (Q : Q)$ and therefore $(T : Q) = (Q : Q)$.

$\Longleftarrow$). Assume that $(T : Q) = (Q : Q)$. Let $x \in P^{-1}$. Since $IQ \subseteq P$, then $xIQ \subseteq xP \subseteq R \subseteq T$. So $xI \subseteq (T : Q) = (Q : Q)$. Hence $xIQ \subseteq Q$. On the other hand, $IP \subseteq IQ$ implies that $xIP \subseteq xIQ \subseteq Q$. Since $xIP \subseteq xP \subseteq R$, then $xIP \subseteq Q \cap R = P$. Now, $I \not\subseteq P$ and P is a prime ideal of R, so that $xP \subseteq P$. Hence $x \in (P : P)$ and $P^{-1} = (P : P)$.

b) By [10, Proposition 6].

THEOREM 4. For the diagram (Δ) :

1) If $T/I \not\subseteq qf(D)$, then $I = J^{-1}$ for some finitely generated fractional ideal J of R with $J \subseteq T$, so I is a v-ideal of R.

2) If $T/I \subseteq qf(D)$, then $I = P$ is a prime ideal of T and $\chi(P) = qf(D)$. If P is not a v-ideal of R, then $(T : P) = P^{-1} = (P : P)$, P is a prime ideal of $(T : P)$ and for each overring B of R, P is not a maximal ideal of B.

Proof. 1) Let $\varphi(x) \in T/I$ such that $\varphi(x) \notin qf(D)$ for some $x \in T$. Let $J = R + xR$. Then J is a finitely generated fractional ideal of R, $J \subseteq T$ and $J^{-1} = R \cap x^{-1}R$. Let $z \in J^{-1}$. Then $z \in R$ and $z = x^{-1}y$ for some $y \in R$. So $zx = y$ and $\varphi(z)\varphi(x) = \varphi(zx) = \varphi(y)$. If $\varphi(z) \neq 0$, then $\varphi(x) = \frac{\varphi(y)}{\varphi(z)} \in qf(D)$, contrary to hypothesis. Hence $\varphi(z) = 0$ and $z \in I$. So $J^{-1} \subseteq I$. Conversely, let $y \in I$. Then $xy \in I \subseteq R$. So $y \in R \cap x^{-1}R = J^{-1}$. Hence $I = J^{-1}$ and therefore I is a v-ideal of R.

2) If $T/I \subseteq qf(D)$, then $I = P$ is a prime ideal of T and $\chi(P) = qf(D)$. Assume that P is not a v-ideal of R. Let $x \in P_v$ with $x \notin P$.

Then $x \in R$ and $xP^{-1} \subseteq R$. So $xPP^{-1} \subseteq P$. Hence $PP^{-1} \subseteq P$ (since $x \notin P$, $PP^{-1} \subseteq R$ and P is prime in R). So $P^{-1} = (P : P)$. Let $x \in (T : P)$. Then $xP \subseteq T$ implies $xP_v = (xP)_v \subseteq T_v$ where $T_v = (R : (R : T))$. Since $P \subseteq (R : T)$, then $T_v \subseteq P^{-1}$. Then $xP_v \subseteq T_v \subseteq P^{-1}$ implies $xP_vP \subseteq PP^{-1} = P$. Since $xP \subseteq T$, $P_v \not\subseteq P$ and P is a prime ideal of T, then $xP \subseteq P$. So $x \in (P : P)$ and $(T : P) = (P : P) = P^{-1}$.

Suppose that there is an overring B of R such that P is a maximal ideal of B. Since $B \subseteq (P : P) = P^{-1} = (P_v : P_v)$, then P_v is an ideal of B. By maximality of P in B, $P_v = B$. So $B = P_v \subseteq R \subseteq B$ and therefore $B = R = P_v$. So $T \subseteq (P : P) = P^{-1} = R$, a contradiction. Now, P is a prime ideal of $(T : P)$ follows from [12, Proposition 2.4].

Finally, from Theorem 4, if I is a nonprime ideal of T or a maximal ideal of T, then it is divisorial as an ideal of R. But if $I = P$ is a nonmaximal prime ideal of T, then the two cases are possible. We illustrate these cases in the following example .

EXAMPLE 5.

1) A local domain T with a prime ideal P which is not a v-ideal of R.

Let k be a field, and let X, Y and Z be indeterminates over k. Set $T = k[[X, Y, Z]]$ and $P = (Y, Z)T$. Since T is a Krull domain and $ht_T P = 2$, then $(T : P) = (P : P) = T$. Let $D = k[[X^2, X^3]]$ and $R = D + P$. Then $P^{-1} = (P : P) = (T : P) = T$ (since $(P : P) \subseteq P^{-1} \subseteq (T : P)$). So $P_v = (R : T)$. Since $X^2 k[[X]] \subseteq D$, then $P \subset X^2 k[[X]] + P \subseteq (R : T) = P_v$. (In fact, $X^2 k[[X]] + P = P_v$).

2) A domain T with a prime ideal P which is a v-ideal of R and P^{-1} is not a ring (So $P^{-1} \neq (P : P)$).

Let $T = k[X, Y]$, $P = YT$, $D = k[X^2, X^3]$ and $R = D + P$. Then, it is easy to see that $\frac{X^2}{Y} \in P^{-1}$ and $\frac{X^2}{Y} \notin T = (P : P)$. Since $T \subset$

$\frac{1}{Y}T=(T:P)$, then P is a v-ideal of R by Theorem 4(2).

Now, using some of the results above (especially Theorem 1 and Theorem 4), we are ready to characterize some coherent-like properties in pullback constructions of type (Δ_1).

COROLLARY 6. For the diagram (Δ), if R is a v-domain, then $I = P$ is a prime ideal of T and $\chi(P) = qf(D)$.

Proof. If $T/I \nsubseteq qf(D)$, by Theorem 4, there is a finitely generated fractional ideal J of R such that $J \subseteq T$ and $J^{-1} = I$. So $JJ^{-1} \subseteq I$ and therefore $(JJ^{-1})_v \subseteq I_v = I$ (Theorem 4). Since R is a v-domain, then $(JJ^{-1})_v = R$, a contradiction. Hence $T/I \subseteq qf(D)$ and therefore $I = P$ is a prime ideal of T with $\chi(P) = qf(D)$.

COROLLARY 7. For the diagram (Δ_1), R is a v-domain if and only if D is a v-domain, $I = P$ is a prime ideal of T and $\chi(P) = qf(D)$.

Proof. If R is a v-domain, then by corollary 6, $I = P$ is a prime ideal of T and $\chi(P) = qf(D)$. Note that P is the maximal ideal of V_P.
    The fact that D is a v-domain now follows from [9, Theorem 4.15]. The converse also follows from [9, Theorem 4.15].

COROLLARY 8. For the diagram (Δ_1), R is a $PVMD$ (respectively Prüfer), if and only if D is a $PVMD$ (respectively Prüfer), $I = P$ is a prime ideal of T and $\chi(P) = qf(D)$.

Proof. This follows from Corollary 6 and [7, Theorem 4.1].

THEOREM 9. For the diagram (Δ_1), R is a Mori domain if and only if I^{-1} is a DVR and $D = k$ is a field.

Proof. $\Longrightarrow$) Assume that R is a Mori domain. By Theorem 1.(1), $I^{-1} = (I : I) = V_P$ for some prime ideal P of V. By [14, Page 11] or [13, Theorem 11.(e)], I^{-1} is a Mori Domain (since it is a ring). Hence I^{-1} is a DVR. Let $P_0 = \sqrt{I}$. Then $P_0 \subseteq P$. Since I^{-1} is a DVR, then $ht_V P = 1$. So $P = P_0$. Let $0 \neq d \in D$, write $d = \varphi(x)(= \psi(x))$ for some $x \in R$ and $x \notin I$. Since $I = P \cap R$, $x \notin P$. So $x^{-1} \in V_P = I^{-1}$. Hence $d^{-1} = (\varphi(x))^{-1} = (\psi(x))^{-1} = \psi(x^{-1}) \in I^{-1}/I$ and therefore $qf(D) \subseteq I^{-1}/I$. By Theorem 1.(2), each ideal of R is comparable to I. Suppose that there is $0 \neq d \in D$ such that $d^{-1} \notin D$. write $d = \varphi(x)(= \psi(x))$ for some $x \in R$ and $x \notin I$. Then for each $n \geq 1$, $x^n R$ is a proper ideal of R which is not contained in I. Hence for each $n \geq 1$, $I \subset x^n R$. So, $I \subseteq \bigcap_{n \geq 1} x^n R = (0)$ (since $(x^n R)_{n \geq 1}$ is a decreasing sequence of proper principal (so, divisorial) ideals in the Mori domain R), a contradiction. Hence $D = k$ is a field.

$\Longleftarrow$) Assume that $D = k$ is a field and I^{-1} is a DVR. Since $R \subset V$, then $k \subset V/I$. By Theorem 4, there is a finitely generated ideal A of R such that $I = A^{-1}$. Now, let J be an ideal of R. Since R is a local domain with maximal ideal I, then either $JJ^{-1} = R$ or $JJ^{-1} \subseteq I$.

If $JJ^{-1} = R$, then J is a principal ideal of R. So $J = J_v$.

If $JJ^{-1} \subseteq I$, then $J^{-1} = (I : J)$. Since I^{-1} is a DVR, then $JI^{-1} = bI^{-1}$ for some $b \in J$. Then $(R : JI^{-1}) = (R : bI^{-1}) = b^{-1}(R : I^{-1}) = b^{-1}I_v = b^{-1}I$. On the other hand, $(R : JI^{-1}) = ((R : I^{-1}) : J) = (I_v : J) = (I : J) = J^{-1}$. Hence $b^{-1}I = J^{-1}$. So $J_v = bI^{-1} = bA_v = (bA)_v$ and bA is a finitely generated ideal of R. It follows that R is a Mori domain.

Let $R \subseteq T$ be an extension of domains. We recall that, T is said to be t-linked over R if for each ideal I of R such that $I^{-1} = R$, $(T : IT) = T$.

PROPOSITION 10. For the diagram (Δ), assume that $D = k$ is a field. Then :

1) $I^{-1} = (I : I)$ and I is a v-ideal of R.

2) R is not a v-domain.

3) T is t-linked over R.

Proof. 1) Note that, I is a maximal ideal of R. If $I \subset II^{-1}$, then $II^{-1} = R$. So $R \subset T \subseteq (I : I) \subseteq (II^{-1} : II^{-1}) = R$, a contradiction. Hence $I = II^{-1}$ and $(I : I) = I^{-1}$.

If $I \subset I_v$, then $I_v = R$. So $T \subseteq (I : I) = I^{-1} = R$, a contradiction. Hence $I = I_v$.

2) If R is a v-domain, then by corollary 6, $I = P$ is a prime ideal of T and $\chi(P) = k$. Since $k \subseteq T/P \subseteq \chi(P) = k$, then $T/P = k$. So $R = T$, a contradiction.

3) Let J be an ideal of R such that $J^{-1} = R$. We wish to show that $(T : JT) = T$. Let $x \in (T : JT)$. Then $xJT \subseteq T$ whence $xJI \subseteq I$. So $xI \subseteq (I : J) = J^{-1} = R$. Hence $x \in I^{-1} = (I : I)$. Since $J_v = R$ and I is a v-ideal of R (by 1), then $J \not\subseteq I$. Then $I + J = R$. So, there is $a \in I$ and $b \in J$ such that $1 = a + b$. So, $x = xa + xb \in T$ since $xa \in II^{-1} \subseteq T$ and $xb \in J(T : JT) \subseteq T$. Hence $(T : JT) = T$.

THEOREM 11. For the diagram (Δ_1), assume that $D = k$ is a field.

1) R is v-coherent if and only if I is v-finite in R.

2) The following statements are equivalent.

  i) R is a quasi-coherent domain.

  ii) R is a Finite Conductor domain.

  iii) I is a finitely generated ideal of R.

Proof. 1) $\Longrightarrow$) By Theorem 4, there is a finitely generated (fractional) ideal A of R such that $A^{-1} = I$. Since R is v-coherent, then $I = A^{-1}$ is v-finite in R.

  $\Longleftarrow$) Assume that I is v-finite in R. Then $I = I_v = B_v$ for some finitely generated ideal B of R. Let J be a finitely generated ideal of R. Since R is a local domain with maximal ideal I, then either $JJ^{-1} = R$ or $JJ^{-1} \subseteq I$.

If $JJ^{-1} = R$, then $J = bR$ is a principal ideal of R. So $J^{-1} = b^{-1}R$ is v-finite.

If $JJ^{-1} \subseteq I$, then $J^{-1} = (I : J)$. Since J is a finitely generated ideal of R, then $JI^{-1} = bI^{-1}$ for some $b \in J$. Then $(R : JI^{-1}) = (R : bI^{-1}) = b^{-1}I_v = b^{-1}I$. On the other hand, $(R : JI^{-1}) = ((R : I^{-1}) : J) = (I_v : J) = (I : J) = J^{-1}$. Hence $J^{-1} = b^{-1}I = b^{-1}B_v = (b^{-1}B)_v$. So J^{-1} is v-finite since $b^{-1}B$ is a finitely generated (fractional) ideal of R. Hence R is v-coherent.

2) $i) \implies ii)$ Trivial.

$ii) \implies iii)$ By the proof of Theorem 4, $I = R \cap x^{-1}R$ which is finitelely generated by $ii)$.

$iii) \implies i)$ Let J be a nonprincipal finitely generated ideal of R. A proof similar to that of 1) shows that $J^{-1} = b^{-1}I$ for some $b \in J$. Hence R is quasi-coherent.

The following Proposition gives a necessary and sufficient conditions for I to be a v-finite ideal of R.

PROPOSITION 12. For the diagram (Δ_1) :

1) If $V/I \nsubseteq qf(D)$, then I is v-finite in R if and only if I is principal in I^{-1}.

2) If $I = P$ is a prime ideal of V, then P is v-finite in R if and only if P is principal in V_P and $qf(D) \subset \chi(P)$.

Proof. Suppose that I is v-finite. Then there is a finitely generated ideal J of R such that $J_v = I_v = I$. Hence $J \subseteq JI^{-1} \subseteq II^{-1} = I$ and therefore $J_v = (JI^{-1})_v = I$. Write $JI^{-1} = bI^{-1}$ for some $b \in J$. Then $(R : JI^{-1}) = (R : bI^{-1}) = b^{-1}(R : I^{-1}) = b^{-1}I_v = b^{-1}I$. So, $I = (JI^{-1})_v = (R : b^{-1}I) = bI^{-1}$ is a principal ideal of I^{-1}.

1) Assume that $I = aI^{-1}$ for some $a \in I$. Let $x \in V$ such that

$\varphi(x)) \notin qf(D)$ and $J = aR + axR$. Then J is a finitely generated ideal of R and $J \subseteq I$. We claim that $I = J_v$, equivalently, that $I^{-1} = J^{-1}$. It is clear that $J^{-1} = a^{-1}R \cap (ax)^{-1}R \supseteq I^{-1}$. Conversely, let $f \in J^{-1}$. Write $f = \frac{f_1}{a} = \frac{f_2}{ax}$, where f_1 and f_2 are in R. So $xf_1 = f_2$. Hence $\varphi(x)\varphi(f_1) = \varphi(xf_1) = \varphi(f_2)$. If $\varphi(f_1) \neq 0$, then $\varphi(x) = \frac{\varphi(f_2)}{\varphi(f_1)} \in qf(D)$, a contradiction. Hence $\varphi(f_1) = 0$ and therefore $f_1 \in I = aI^{-1}$. Hence $f = \frac{f_1}{a} \in I^{-1}$. Hence $J^{-1} \subseteq I^{-1}$. So, $J^{-1} = I^{-1}$. Hence I is v-finite in R.

2) Assume that $I = P$ is a prime ideal of V.

$\Longrightarrow$) By the first part of the proof, if P is v-finite in R, then P is principal in $P^{-1} = (P : P) = V_P$. Let $J = \sum_{i=1}^{i=n} b_i R$ be a finitely generated ideal of R with $J_v = P_v = P$. Suppose that $qf(D) = \chi(P)$. Let $x \in (V_P : P)$. Then $xJ \subseteq xP \subseteq V_P$. Hence for each $i \in \{1, \ldots, n\}$, there is $t_i \in V$, $t_i \notin P$, such that $t_i x b_i \in V$. Let $t = \prod_{i=1}^{i=n} t_i$. Then $t \in V$, $t \notin P$ and for each $i \in \{1, \ldots, n\}$, $txb_i \in V$. Hence $txJ \subseteq V$. So, $\varphi(txJ) \subseteq V/P \subseteq \chi(P) = qf(D)$. Hence, for each $i \in \{1, \ldots, n\}$, there is $\alpha_i \in R$ and $\beta_i \in R$, $\beta_i \notin P$, such that $\varphi(txb_i) = \frac{\varphi(\alpha_i)}{\varphi(\beta_i)}$. So, $\varphi(\beta_i txb_i) = \varphi(\beta_i)\varphi(txb_i) = \varphi(\alpha_i)$. Hence $\beta_i txb_i - \alpha_i \in P$. Let $\lambda = \prod_{i=1}^{i=n} \beta_i$ and $\lambda_i = \prod_{j \neq i} \beta_j$. Then $\lambda \in R$, $\lambda \notin P$, and for each $i \in \{1, \ldots, n\}$, $\lambda txb_i - \lambda_i \alpha_i \in P$. Since $\lambda_i \alpha_i \in R$, then $\lambda txb_i \in R$. So $\lambda txJ \subseteq R$ and therefore $\lambda tx \in J^{-1} = P^{-1} = (P : P) = V_P$. Hence $x = \frac{\lambda tx}{\lambda t} \in V_P$. So $(V_P : P) = V_P$ is a ring, which is absurd (since P is a principal ideal of V_P). It follows that $qf(D) \subset \chi(P)$.

$\Longleftarrow$). Assume that P is principal in V_P and $qf(D) \subset \chi(P)$. Then $P = aV_P$. Let $x \in V$ such that $\varphi(x) \notin qf(D)$. Consider the ideal $J = aR + axR$. Then J is a finirely generated ideal of R, and, (as in the proof of 1), $J_v = P$. Hence P is v-finite in R.

In [9, Proposition 2.3], Gabelli and Houston show that if one has a diagram of type (Δ) in which $I = M$ is a maximal invertible ideal of T and such that the quotient field of D is T/M, then I is not a v-finite in R. However, in a diagram of type (Δ_1), if I is a nonprime principal ideal of V, then I is a v-finite ideal of R.

THEOREM 13. For the diagram (Δ_1), assume that $qf(D) \subset V/I$. Then R is v-coherent if and only if D is v-coherent and I is v-finite in R.

Proof. $\Longrightarrow$) Since $V/I \nsubseteq qf(D)$, by Theorem 4, there is a finitely generated (fractional) ideal J of R, such that $J^{-1} = I$. Since R is v-coherent, then $I = J^{-1}$ is v-finite.
For the converse and the fact that D is v-coherent, we need the following lemma.

LEMMA 14. In the diagram (Δ_1), let J be a finitely generated ideal of R such that $I \subset J$. Then there is a finitely generated ideal A of D such that $J = \varphi^{-1}(A)$.

Proof. Just set $A = \varphi(J)$.

We wish to prove that D is v-coherent. Let A be a nonzero finitely generated (integral) ideal of D and $J = \varphi^{-1}(A)$. By Theorem 1(2), $I \subseteq J$ (since $qf(D) \subseteq V/I \subseteq I^{-1}/I$). Also by Theorem 1(3), J is a finitely generated ideal of R and $J^{-1} = \varphi^{-1}(D : A)$. So $J_v = (R : J^{-1}) = (\varphi^{-1}(D) : \varphi^{-1}(D : A)) = \varphi^{-1}(A_v)$. Since R is v-coherent, then J^{-1} is v-finite in R. So $J_v = L^{-1}$ for some finitely generated ideal L of R. Since $I \subset J$, then $L \subseteq L_v = J^{-1} \subseteq I^{-1}$.
If $LI^{-1} = I^{-1}$, then $I = II^{-1} = LII^{-1} \subseteq LR \subseteq L$. By Lemma 14, there is a finitely generated ideal B of D such that $L = \varphi^{-1}(B)$. Hence $\varphi^{-1}(A_v) = J_v = L^{-1} = (\varphi^{-1}(B))^{-1} = \varphi^{-1}((D : B))$ and therefore $A_v = \varphi(\varphi^{-1}((A_v))) = \varphi(\varphi^{-1}(D : B)) = (D : B)$. Hence $(D : A)$ is v-finite.
Assume that $LI^{-1} \subset I^{-1}$. Since I^{-1} is a valuation domain, then LI^{-1} and I are comparable. If $LI^{-1} \subseteq I$ then $L \subseteq LI^{-1} \subseteq I$ and $R \subset V \subseteq I^{-1} \subseteq L^{-1} = J_v \subseteq R$, which is a contradiction.

Hence $I \subseteq LI^{-1} = aI^{-1}$ for some $a \in L$ (since L is f.g. in R). So $a^{-1}I = (R : LI^{-1}) \subseteq L^{-1} = J_v$. Hence $I \subseteq aJ_v \subseteq aR \subseteq L$. Now, we conclude by Lemma 14 as in the case where $LI^{-1} = I^{-1}$.

Let J be a finitely generated ideal of R. By Theorem 1.(2), I and J are comparable.

Proof of $\Longleftarrow$).

If $I \subset J$, by lemma 14, $J = \varphi^{-1}(A)$ for some nonzero finitely generated ideal A of D. In this case, we conclude by appealing to Theorem 1.(3) and the v-coherence of D.

Assume that $J \subseteq I$. Then, two cases are possible.

<u>Case1</u>. $JJ^{-1} \subseteq I$. Then $J^{-1} = (I : J)$. Write $JI^{-1} = aI^{-1}$ for some $a \in J$. Then $(R : JI^{-1}) = (R : aI^{-1}) = a^{-1}I$. On the other hand $(R : JI^{-1}) = ((R : I^{-1}) : J) = (I_v : J) = (I : J) = J^{-1}$. Hence $J^{-1} = a^{-1}I$ and therefore J^{-1} is v-finite.

<u>case2</u>. $I \subset JJ^{-1}$. Since J is a finitely generated ideal, then $R_I = (JJ^{-1})R_I = (JR_I)(JR_I)^{-1}$. Hence $JR_I = bR_I$ for some $b \in J$. Write $J = \sum_{i=1}^{i=n} x_i R$. Then, for each $\{i \in 1, \dots n\}$, there is $t_i \in R$, $t_i \notin I$ such that $t_i x_i \in bR$. So there is $t \in R$, $t \notin I$ such that $tx_i \in bR$, for each $\{i \in 1, \dots n\}$. Hence $tb^{-1}J \subseteq R$. By Theorem 1.(2), if $tb^{-1}J \subseteq I$, then $R_I = tR_I = tb^{-1}JR_I \subseteq IR_I$, which is absurd. Hence $I \subset tb^{-1}J$. Since $tb^{-1}J$ is a finitely generated ideal of R, by the first part, $(tb^{-1}J)^{-1} = t^{-1}bJ^{-1}$ is v-finite and therefore J^{-1} is v-finite. It follows that R is v-coherent.

Using the preceding theorem, we can construct examples of v-coherent domains which are neither Mori nor quasi-coherent and which lie outside the class of "$D + M$" constructions.

EXAMPLE 15. Let k be a field, and let X, Y, and Z be indeterminates over k. Let $V = k((X))[[Y]] + Zk((X, Y))[[Z]] = k((X)) + M$. Then V is a valuation domain with maximal ideal $M = YV$ and

residue field $k((X))$. Let $I = YZV$, then I is a nonprime principal ideal of V, $I^{-1} = (I : I) = V$ and $Z(V, \ I) = M$. So
$Z(V/I) = \varphi(Z(V, \ I)) = \varphi(M)$. Let D be any v-coherent domain
(which is not a field) contained in $k((X))$. By Proposition 12.(1), I
is v-finite in R. So R is v-coherent (Theorem 13). Since D is not a
field, then R is not a Mori domain (Theorem 9). Also, R is not a
Finite Conductor domain, since I is not a prime ideal, (Theorm 16
below).

THEOREM 16. For the diagram (Δ_1), assume that $Z(V/I) \cap D = 0$
and D is not a field.

1) R is a Finite Conductor domain if and only if D is a Finite Conductor domain, $I = P$ is a prime ideal of V and $qf(D) = \chi(P)$.

2) R is quasi-coherent if and only if D is quasi-coherent, $I = P$ is a
prime ideal of V and $qf(D) = \chi(P)$.

Proof. 1)$\Longrightarrow$) We first show that I is prime ideal P and that $qf(D) =
\chi(P)$. If $V/I \subseteq qf(D)$, this follows from Theorem 4. Suppose $V/I \nsubseteq
qf(D)$. Then there is an $x \in V$ such that $I = R \cap x^{-1}R$ (Theorem 4).
Since R is an F.C. domain, then $I = R \cap x^{-1}R$ is a finitely generated
ideal of R. So I is a principal ideal of V. Assume that $I = aV$ for
some $a \in V$ then $V = (I : I) \subseteq R^* \subseteq V$ where R^* is the integral
closure of R ([8, Proposition 31.4]). So $V = R^*$ and therefore D
and R are a local domains with maximal ideals q and $Q = \varphi^{-1}(q)$
respectively. In addition $M \cap R = Q$ where M is the maximal ideal
of V. Since D is not a field, then there is a nonzero $d \in D$ such that
$d^{-1} \notin D$. So $d \in q$. Write $d = \varphi(x)$ for some $x \in R$ and $x \notin I$. By
Theorem 1.(2), $I \subset xR$. So $x^{-1}I \subset R$ and $x^{-1} \in I^{-1} = V$. Since
$d \in q$, then $x \in Q \subset M$, a contradiction. Hence $I = P$ is a prime
ideal of V and $qf(D) = \chi(P)$.

Now, since P is the maximal ideal of V_P, we can appeal to [9, Theorem 4.8] to conclude that D is a F.C. domain. The converse also
follows from [9, Theorem 4.8].

2) This follows from 1) and [9, Proposition 4.6].

If the following question has a positive answer, then the condition $Z(V/I) \cap D = (0)$ can be omitted in Theorem 16.

Question: In a diagram of type (Δ_1), If I is finitely generated as an ideal of R, does it necessarily follow that D is a field?

We recall that, a domain is called a DVF-domain if each divisorial ideal is v-finite.

In the diagram (Δ_1), if $I = P$ is a prime ideal of V, then we can consider it as the maximal ideal of V_P. In this situation, a complete characterization of when R is a DVF-domain is given in [9, Theorem 4.20]. Our intention is to examine the case where I is not a prime ideal of V in a particular case of a diagram of type (Δ_1). We shall refer to this as a diagram of type (Δ_2). We define this diagram as follows.

Let V be a valuation domain, I a nonzero ideal of V, D a sub-domain of V such that $P \cap D = 0$, where $P = Z(V, \ I)$. We identify D as a subdomain of V/I and we consider the following diagram.

$$
\begin{array}{ccc}
R = D + I & \longrightarrow & D \\
\downarrow & & \downarrow \\
V_P & \longrightarrow & V_P/P
\end{array}
$$

PROPOSITION 17. For the diagram (Δ_2), R is a DVF-domain if and only if D is a DVF domain and for each v-ideal J of R such that $JJ^{-1} \subseteq I$, J is a principal ideal of I^{-1}.

Proof. As in Theorem 1, we note that each ideal of R is comparable to I and for each nonzero ideal A of D, $(A+I)^{-1} = A^{-1} + I$ (since $qf(D) \subset I^{-1} = (I : I)$).

$\Longrightarrow$). By the above remark, it is easy to see that D is a DVF-domain.

Let J be a v- ideal of R such that $JJ^{-1} \subseteq I$. Then $J^{-1} = (I : J)$. Since R is a DVF-domain, then $J = A_v$ for some finitely generated ideal A of R. So $AI^{-1} = aI^{-1}$ for some $a \in A$. Hence $(I : A) = (I_v : A) = ((R : I^{-1}) : A) = (R : AI^{-1}) = (R : aI^{-1}) = a^{-1}I$. Since $A \subseteq J$ and $A_v = J$, then $A^{-1} = J^{-1} = (I : J) \subseteq (I : A) \subseteq A^{-1}$. So $J^{-1} = (I : A) = a^{-1}I$. Hence $J = J_v = aI^{-1}$.

$\Longleftarrow$) Let J be a v-ideal of R:

If $I \subset J$, then $J = A + I$ for some v-ideal of D. Since D is a DVF-domain, then A is v-finite. So J is also v-finite.

If $J \subseteq I$ and $I \subset JJ^{-1}$, then $JR_S = bR_S$, where $S = D - \{0\}$ and $b \in J$ (since R_S is local with maximal ideal $IR_S = I$). Now, let $W = \{\lambda \in qf(D)/\lambda . b \in J\}$. Then W is a D-submodule of $qf(D)$, $D \subseteq W$ and it is easy to see that $J = b(W + I)$. Since J is a v-ideal of R, then W is a v-fractional ideal of D. In this case, we also have J is v-finite in R since W is v-finite in D.

If $JJ^{-1} \subseteq I$, then J is a principal ideal of I^{-1}. Assume that $J = bI^{-1}$. Since I is not a prime ideal of V, then there is $x \in V$, $y \in V$ such that $xy \in I$ and $x, y \notin I$. Hence $I \subset xR$ and $I \subset yR$. So x^{-1}, $y^{-1} \in I^{-1}$. Since V is a valuation domain, then either $xy^{-1} \in V$ or $yx^{-1} \in V$. If $xy^{-1} \in V$, then let $z \in (R + xy^{-1}R)^{-1} = R \cap x^{-1}yR$. Then $zx \in yR$. So $zx^2 \in xyR \subseteq I$. Since I^{-1} is a ring, then $x^{-2} \in I^{-1}$. Hence $z = zx^2 . x^{-2} \in II^{-1} = I$. Now, let $z \in I$. Then $z . xy^{-1} \in I$. So $z \in R \cap x^{-1}yR = (R + xy^{-1}R)^{-1}$. Hence $I = (R + xy^{-1}R)^{-1}$. If $yx^{-1} \in V$. Then, as in the first (if), $(R + yx^{-1}R)^{-1} = I$. Hence there is $u \in V$ such that $I = R \cap u^{-1}R = (R + uR)^{-1}$. Hence $J = bI^{-1} = b(R + uR)_v = (bR + buR)_v$ is v-finite. (Note that, J is also an ideal of V, so $au \in J$ and therefore $aR + auR \subseteq J$).

COROLLARY 18. For the diagram (Δ_2), If I^{-1} is a DVR, in particular if V is a DVR, then R is a DVF-domain if and only if D is a DVF domain.

Combining Theorem 9 and Corollary 18, we construct examples of *DVF*-domains that are not Mori domains.

References

[1] D. F. Anderson, When the dual of an ideal is a ring. Houston J. Math. 9 (1983), 325-332.

[2] D. F. Anderson and D. Dobbs, Pairs of commutative rings with the same prime ideals, Can. J. Math. 32 (1980), 362-384.

[3] V. Barucci, On a class of Mori domains, Comm. Algebra 11 (1983), 1985-2001.

[4] V. Baucci, D. F. Anderson, and D. Dobbs, Coherent Mori domains and the principal ideal theorem, Comm. Algebra 15 (1987), 1119-1156.

[5] D. Dobbs and I. Papick, When is D+M coherent? Proc. Amer. Math. Soc. 56 (1976), 51-54.

[6] M. Fontana, Topologically defined classes of commutative rings, Ann. Mat. Pura Appl. 123 (1980), 331-355.

[7] M. Fontana and S. Gabelli, On the class group and the local class group of a pullback, J. Algebra, to appear.

[8] R. Gilmer, Multiplicative Ideal Theory, Marcel Dekker, New York, 1972.

[9] S. Gabelli and E. Houston, Coherent-like conditions in pullbacks, Michigan Math. J. 44 (1997) 99-123.

[10] E. Houston, S. Kabbaj, T. Lucas, and A. Mimouni. Duals of ideals in pullback constructions. Lecture Note in Pure and Applied Mathematics 171, Marcel Dekker, New York, 1995, pp. 263-276.

[11] E. Houston, S. Kabbaj, T. Lucas, and A. Mimouni. When is the dual of an ideal a ring ? J. Algebra, to appear.

[12] J. Huckaba and I. Papick, When the dual of an ideal is a ring. Manuscripta Math. 37 (1982), 67-85.

[13] T. Lucas, The radical trace property and primary ideals, J. Algebra, 184 (1999), 1093-112.

[14] N. Raillard (Dessagnes), Sur les anneaux de Mori, Thèse, Univ. Pierre et Marie Curie, Paris IV.

Classification of Plane Cubic Curves

S. B. MULAY Department of Mathematics, University of Tennessee, Knoxville, TN 37996

An equation of the form $F(X, Y, Z) = 0$, where F is a nonzero homogeneous polynomial (with coefficients in a field k), defines a curve in the projective plane (over k) with homogeneous coordinates (X, Y, Z). The degree of $F(X, Y, Z)$ is the *degree* of the curve. The curve is said to be irreducible provided $F(X, Y, Z)$ is an irreducible polynomial. Clearly, for any nonzero constant c the equations $F(X, Y, Z) = 0$ and $cF(X, Y, Z) = 0$ define the same curve. Also, geometric properties of such a curve should be independent of the choice of homogeneous coordinates. Thus, whenever (X', Y', Z') is obtained by a nonsingular homogeneous linear transformation of (X, Y, Z) and $G(X', Y', Z')$ is the corresponding transform of $F(X, Y, Z)$, the curve defined by $G(X', Y', Z') = 0$ is geometrically the same as the curve defined by $F(X, Y, Z) = 0$. In other words, the curves $C : F = 0$ and $D : G = 0$ are *projectively equivalent* if there is a nonsingular homogeneous linear change of variables which transforms F to a constant multiple of G. Note that equivalent curves have equal degrees and an irreducible curve can only be equivalent to another irreducible curve. For curves of a fixed degree, *projective classification* constitutes of a description of the equivalence classes. Since all lines are obviously projectively equivalent, no matter what the ground field k is, there is just one equivalence class. For irreducible conics over an algebraically closed field k the story is the same; up to projective equivalence, there is exactly one irreducible conic. Here, the case of characteristic 2 demands special attention. Projective classification of conics over an arbitrary field k is, in essence, the same as the classification of quadratic forms in 3 variables over k.

The main purpose of this article is to provide an accessible (to a non-expert reader) account of the projective classification of irreducible plane cubic curves over an algebraically closed ground field k of arbitrary characteristic. Along the way, we also provide the real-projective as well as the real-affine classification. To the author's best knowledge no textbook, or any source of that type, contains such a classification (especially in characteristics $2, 3$). In view of this, the present article helps to fill a gap.

The birational classification of nonsingular cubic curves, which is dealt with in many textbooks of Algebraic / Arithmetic Geometry (see [SI], [C]), does not readily yield their projective classification (especially when the ground field k is of characteristic 2, or 3). Since the singular (irreducible) cubic curves are *rational*, and hence birationally equivalent to each other, their projective classification is of greater interest. Nevertheless, except for the case of cuspidal cubic curves over a field of characteristic 3, there is no difference between the birational and the projective classification. In either case the equivalence classes are parametrized (essentially) by a single invariant. The projective equivalence classes of nonsingular plane quartic curves are in one to one correspondence with the moduli space of nonhyperelliptic curves of genus 3. The projective classification of singular (irreducible) plane quartic curves is still not known.

This article is a modified version of the author's notes of his lectures at POSTECH, (South) Korea during the Spring of 1993. The course at POSTECH was based on a selection of topics from [A]. For the rudiments of algebraic-geometry, such as Bezout's Theorem, that are tacitly assumed in this article, the reader is referred to [A].

1 POLARS OF HYPERSURFACES

Consider a nonzero polynomial $F(X)$ which is homogeneous of degree $d > 0$ in the variables $X = (X_1, \cdots, X_n)$, with $n \geq 2$. Let the coefficients of $F(X)$ be in a field k. Let u denote the variable n-tuple $(u_1, \cdots, u_n)$. By $u + tX$ we mean $(u_1 + tX_1, \cdots, u_n + tX_n)$. For $0 \leq r \leq d$ let $\Delta_u^r F$ denote the coefficient of t^r in the expansion of $F(u + tX)$. Then, $\Delta_u^r F$ is homogeneous of degree r in X, homogeneous of degree $(d - r)$ in u. Clearly, $\Delta_u^0 F = F(u)$, whereas $\Delta_u^d F = F(X)$. The following is a list of some basic properties of the Δ's.

1. Let $X = X'M$ where $X' = (X'_1, \cdots, X'_n)$ and M is an invertible $n \times n$ matrix. Also, let $u' = uM^{-1}$. Letting $G(X') = F(X'M)$ we have

$$G(u' + tX') = F(u'M + tX'M).$$

Note that $u'M + tX'M = u + tX$. Hence $\Delta_{u'}^r G = \Delta_u^r F$.

2. Fix an integer r such that $0 \leq r \leq d$. Let $Q(X) = \Delta_u^r F$ and $0 \leq s \leq r$. Then $\Delta_u^s Q$, by definition, is the coefficient of λ^s in $Q(u + \lambda X)$ and therefore it is the coefficient of $\lambda^s t^r$ in $F((t+1)u + t\lambda X)$. Since

$$F((t+1)u + t\lambda X) = \sum_{i=0}^{i=d} (t+1)^{d-i}(t\lambda)^i \Delta_u^i F$$

we have $\Delta_u^s Q = \binom{d-s}{r-s}\Delta_u^s F$. In other words

$$\Delta_u^s(\Delta_u^r F) = \binom{d-s}{r-s}\Delta_u^s F.$$

3. Assume $n \geq 2$. Let $x = (x_1, \cdots, x_{n-1})$, $a = (a_1, \cdots, a_{n-1})$. The polynomial $f(x) = F(x, 1)$ can be thought of as a dehomogenization of $F(X)$. Write

$$f(a + tx) = H_0(x) + \cdots + H_j(x)t^j + \cdots + H_d(x)t^d$$

Then, H_j is homogeneous of degree j in x. Substituting x/z for x (*i.e.* x_i/z for each x_i), in $f(a+tx)$ and then multiplying by z^d we get

$$z^d f(a + t(x/z)) = z^d H_0(x) + \cdots + z^{d-j} H_j(x)t^j + \cdots + H_d(x)t^d.$$

On the other hand

$$z^d f(a + t(x/z)) = z^d F(a + t(x/z), 1) = F(za + tx, z).$$

Now put $z = (t+1)$ in $F(za + tx, z)$ to get

$$F(a + t(a + x), 1 + t) = \sum_{0 \leq j \leq d} (t+1)^{d-j} t^j H_j(x).$$

Observe that the left hand side is $F(v + ty)$ where $v = (a_1, \cdots, a_{n-1}, 1)$, $y = (a_1 + x_1, \cdots, a_{n-1} + x_{n-1}, 1)$. Comparing the coefficients of various powers of t we get

$$(\Delta_v^r F)(y) = \sum_{i=0}^{i=r} \binom{d-i}{r-i} H_i(x).$$

Let the projective hypersurface defined by $F(X) = 0$ be denoted by V. The degree of V is the degree of $F(X)$. If $F(X)$ is irreducible over the algebraic closure $\bar{k}$ of k then V is said to be an irreducible hypersurface ; otherwise, V is said to be reducible. If an irreducible polynomial $G(X) \in \bar{k}[X]$, divides $F(X)$ then the hypersurface defined by $G(X) = 0$ is said to be an irreducible component of V. When $G(X)$ is a multiple factor of $F(X)$, the component is said to be a multiple component.

Let P be a point of V and let $(u_1, \cdots, u_n)$ be its homogeneous coordinates. The locus defined by $\Delta_u^r F = 0$ is called the *polar r-ic to V at P*. The property no. 1 above ensures that the definition of a polar is independent of the choice of homogeneous coordinates X. Thus we may, without any ambiguity, use the symbol $\Delta_P^r V$ to denote the polar r-ic to V at P. In general, the polar 1-ic is the tangent hyperplane to V at P, the polar 2-ic is a quadric (or conic) tangent to V at P etc. If $\Delta_u^r F$ is a nonzero polynomial then $\Delta_P^r V$ is a hypersurface of degree r.

It is possible that the polynomial $\Delta_u^r F$ vanishes identically, in which case the polar r-ic is said to be *improper*. Provided none of the binomial coefficients $\binom{d-s}{r-s}$, where $0 \le s \le r$, is divisible by the characteristic of k, in view of property no. 2, $\Delta_u^r F \equiv 0$ only when $\Delta_u^i F \equiv 0$ for $0 \le i \le r$. The point P is said to be a point of *multiplicity* $(r+1)$, or an $(r+1)$-*fold* point of V, if $\Delta_u^i F \equiv 0$ for $0 \le i \le r$ and provided $\Delta_u^{r+1} F$ does not vanish identically. A 1-fold point is called *nonsingular* or *simple*. Points of multiplicity ≥ 2 are called *singular*; they constitute the *singular locus* of V. Observe that an m-fold point P of V is also an m-fold point of $\Delta_P^{m+i} V$ for $0 \le i \le d-m$. The singular locus of V coincides with V only when V itself is multiple *i.e.* when $F(X) = G(X)^e$ for some polynomial $G(X)$ and some integer $e \ge 2$. Here, we must take into account 'real' points (*i.e.* points with coordinates in k) as well as 'imaginary' points (*i.e.* points with coordinates in an algebraic extension of the field k).

The taylor expansion of the affine equation $f(x) = 0$ at the point P is related to the affine equation of a polar $\Delta_P^r V$ by the property 3. If P is an r-fold point of V then, in the notation of prop. no. 3, we have $H_i(x) \equiv 0$ for $0 \le i < r$ and hence the affine equation of $\Delta_P^r V$, after translating P to the origin, is $H_r(x) = 0$. Since $H_r(x)$ is homogeneous in the variables x, $\Delta_P^r V$ is a cone with a vertex at P. This cone is called the *tangent cone* to V at P.

A *flex* , or a *point of inflection* of V is a simple point where the tangent hyperplane has at least triple contact *i.e.* $\Delta_P^1 V$ is a component of the polar quadric $\Delta_P^2 V$. If $d - 1$ is not divisible by the characteristic of k then $\Delta_P^1 V$ is the tangent hyperplane to the polar quadric at P, and hence, must be a component of the polar quadric in case $\Delta_P^2 V$ is reducible. Therefore, in this case the inflection condition is equivalent to requiring $\Delta_P^2 V$ to be reducible. Letting

$$V : ZW^p + XYZ^{p-1} + X^{p+1} = 0$$

($Y = X$ allowed) where *char* $k = p > 0$, we have $Z = 0$, and $XY = 0$ as the tangent hyperplane and the polar quadric at $(0,0,0,1)$. So, the polar quadric is reducible but there is no inflection.

It is well known that, over a field of characteristic $\neq 2$, the reducibility of a quadratic form

$$Q = \sum_{i \leq j} a_{ij} X_i X_j$$

is equivalent to $rank\ [b_{ij}] \leq 2$ where $b_{ii} = a_{ii}$ and $b_{ij} = \frac{1}{2}a_{ij}$ if $i \neq j$. Letting $Q = 2\Delta_u^2 F$ we clearly have

$$a_{ij} = \begin{cases} \frac{\partial^2 F}{\partial X_i^2}(u) & \text{if } i = j, \\ 2\frac{\partial^2 F}{\partial X_i \partial X_j}(u) & \text{if } i \neq j \end{cases}$$

and hence the matrix $[b_{ij}]$ is seen to be the Hessian matrix $H(F, u)$.

The *hessian* of $F(X)$ is the determinant of the Hessian matrix $H(F, X)$; we denote it by $h(F, X)$. Let M be an $n \times n$ matrix, and let M^t denote its transpose. It is straightforward to verify that

$$H(G, X') = M \cdot H(F, X) \cdot M^t,$$

where $X = X'M$ and $G = F(X'M)$. Consequently,

$$h(G, X') = det(M)^2 h(F, X),$$

and hence, the locus

$$hess(V) : h(F, X) = 0$$

is defined independent of the choice of homogeneous coordinates. We note two important properties.

1. $h(F, X)$ is homogeneous of degree $n(d - 2)$. If there exists an invertible homogeneous linear transformation $X = X'M$ such that $F(X'M)$ does not involve X_n', then $h(F, X)$ is identically zero.

2. Replacing the i-th row of $H(F, X)$ by

$$\sum_1^n X_r \times (\text{the } r\text{-th row}),$$

and then expanding, leads to the identity

$$X_i h(F, X) = (d - 1)(h_{i1} F_1 + \cdots + h_{in} F_n)$$

where F_j denotes the derivative of F with respect to X_j, whereas h_{ij} is the cofactor of the ij-th entry in $H(F, X)$. One implication of the identity is that $hess(V)$ passes through every singular point of V. If $char\ k$ divides $(d - 1)$ then, $h(F, X)$ is identically zero.

2 PLANE CUBICS

In this section, we restrict to $d = 3$ and $n = 3$ *i.e.* the case of projective plane cubic curves. Now, it is convenient to rename the variables $X = (X_1, X_2, X_3)$ as (X, Y, Z). To study the hessian in more detail it is necessary to assume $char\ k \neq 2$. In the following K denotes an overfield of k.

(2.1.1) First, assume that the plane cubic

$$V : F(X, Y, Z) = 0$$

has a nonsingular point P with coordinates in K. Up to a change of coordinates, we may assume P to be the point $(0, 1, 0)$ and $Z = 0$ to be the tangent line at P. Then, F is of the form

$$ZY^2 + (aX^2 + bZX + cZ^2)Y + A(X, Z).$$

Replace Y by $Y - (bX + cZ)/2$ if necessary, to get the simplified form

$$F = ZY^2 + aX^2 Y + G(X, Z),$$

with $a \in K$. Computation of the hessian determinant, which we denote by $h(F)$, yields $u + 2Zh(G)$ where

$$u = -8aY^3 - 4G_{XX} Y^2 + 4a(ZG_{ZZ} + 2XG_{XZ})Y - 4a^2 X G_{ZZ}$$

and $h(G) = G_{XX} G_{ZZ} - G_{XZ}^2$ is the hessian of $G(X, Z)$.

Suppose $h(F) = cF$ for some constant c. Then, we must have $c \in K$, $a = 0$, $-4G_{XX} = cZ$, and $2Zh(G) = cG$. Observe that, if $c = 0$

then $F = ZY^2 + \alpha X^3 + \beta Z^3$ with $6\alpha = 0$. When $c \neq 0$, we have $8G = -cZ(X + \lambda Z)^2$ for some $\lambda \in K$ and hence F is reducible over K.

Suppose that F and $h(F)$ have a nonconstant common factor over K. In view of the above paragraph, we only need to consider the case where $h(F)$ is not identically zero, and the highest common factor is of degree ≤ 2. Since F is necessarily reducible over K in our case, we write $F = (Y + \phi)(ZY + \psi)$. Either $(Y + \phi)$ divides $h(F)$ or $(ZY + \psi)$ divides $h(F)$. If $(ZY + \psi)$ divides $h(F)$ then we must have $a = 0$, and $\psi = \theta Z$ for some linear form θ. Therefore, in any case, an irreducible common factor of F and $h(F)$ must be linear in X, Y, Z (with coefficients in K).

Consider the case where P is also on the curve $hess(V)$ and the line $Z = 0$ is tangent to $hess(V)$ at P. From the expression of $h(F)$ we at once see that $a = 0$, and $G_{XX} = cZ$ for some $c \in K$. Consequently, if *char* $k \neq 3$, then Z divides F.

Thus, for an irreducible F with $h(F) \not\equiv 0$, the intersection multiplicity of V and $hess(V)$ at a nonsingular point of V is either 0 or 1 provided *char* $k \neq 3$. When *char* $k = 3$, it is $0, 1, 3,$ or 9.

(2.1.2) Next, assume that V has a 2-fold point P with coordinates in K. To begin with, assume P to be a node *i.e.* there are two distinct tangents at P. These tangent lines are defined over an extension $K_2 \supset K$ of degree at most 2. As before, up to a transformation in $GL(3, K_2)$, we may assume P to be the point $(0, 1, 0)$ and $Z = 0$, $X = 0$ to be the tangents at P. Then, F takes on the form

$$XZY + aX^3 + (rX + sZ)XZ + bZ^3$$

with a, b, r, s in K_2. Replacing Y by $Y - rX - sZ$ we get the simpler form $F = XZY + aX^3 + bZ^3$. The hessian determinant computation yields

$$h(F) = 2XZY - 6(aX^3 + bZ^3).$$

Clearly, F and $h(F)$ have a common factor if and only if $ab = 0$. Otherwise, since $hess(V)$ also has a node at $(0, 1, 0)$ with $X = 0$, $Z = 0$ as tangents, the intersection multiplicity of V and $hess(V)$ at P is exactly 6.

Finally, suppose P is a cusp of V *i.e.* a 2-fold point with a single tangent. In this case the tangent line must be defined over K; hence, up to a transformation in $GL(3, K)$ we may assume P to be the point $(0, 1, 0)$ and $Z = 0$ to be the tangent line at P. Then, F is of the form

$$Z^2Y + aX^3 + bX^2Z + (rX + sZ)Z^2$$

with $a, b, r, s \in K$. Substituting $Y - rX - sZ$ for Y we obtain

$$F = Z^2Y + (aX + bZ)X^2.$$

The hessian determinant is given by

$$h(F) = -8(3aX + bZ)Z^2.$$

Observe that $h(F)$ vanishes identically only when

$$3a = 0 = b.$$

Provided $h(F)$ is not identically zero, F and $h(F)$ have a common factor if and only if $a = 0$. If $3a \neq 0$, the intersection multiplicity of V and $hess(V)$ at P is 8. If $char\ k = 3$ and $a \neq 0$, the intersection multiplicity is 9.

Bezout's Theorem, together with the above calculation of intersection multiplicities, helps establish the following.

(2.1.3) THEOREM : Let $V : F(X, Y, Z) = 0$ be an irreducible plane cubic defined over a field k with $char\ k \neq 2$.

(i) Assume V is nonsingular. Then, V has at least one flex.
    If $char\ k \neq 3$, then V has 9 distinct points of inflection.

(ii) Assume V is a nodal cubic. If $char\ k = 3$, then V has one flex. Otherwise, it has 3 distinct points of inflection.

(iii) A cuspidal cubic V has exactly 1 flex if $char\ k \neq 3$; if $char\ k = 3$, either it has no flex or each of its simple points is a flex.

(2.1.4) Let V have two flexes P, Q with coordinates in K. Clearly, the line through P, Q is also defined over K. Up to a transformation in $GL(3, K)$, we may assume $P = (0, 1, 0)$, $Q = (0, 0, 1)$, $Z = 0$ to be the tangent at P, and $Y = 0$ to be the tangent at Q. Then,

$$F = ZY^2 + cZ^2Y + aXZY + bX^3$$

with $a, b, c \in K$. Evidently, the third point of intersection of $X = 0$ and V is $R = (0, -c, 1)$; it is also a flex of V.

In conclusion, the line joining any two flexes P, Q of V contains a third flex R. If P, Q have coordinates in K, so does R. It follows that the number of flexes of V, defined over K, is $0, 1, 3,$ or 9.

(2.1.5) The case when K is the field of real numbers, is of special interest. From (2.1.3) it follows that the number of real flexes of V must be odd. Suppose V has three non-collinear real flexes P_1, P_2, P_3. Up to a real change of coordinates, we may assume $Z = 0, Y = 0$ to be the tangents at P_1, P_2 respectively, and also $P_1 = (0, 1, 0)$, $P_2 = (0, 0, 1)$, $P_3 = (1, 1, 1)$. Then, the equation of V is

$$ZY^2 + (-1 - a - b)Z^2Y + aXYZ + bX^3 = 0,$$

with real a, b. Since V is irreducible, $b \neq 0$. It is easy to verify that since P_3 is a flex, and $b \neq 0$, we must have

$$a^2 - (3e)a + (3e^2 + 3e + 3) = 0,$$

where $e = a + b$. The discriminant $(3e)^2 - 4(3e^2 + 3e + 3)$ has to be non-negative, which is possible only when $e = -2$ *i.e.* when $a = -3, b = 1$. But then P_3 would be singular, a contradiction.

Thus, a real cubic curve has either 1 real flex or exactly 3 real flexes.

(2.2) Canonical Forms

Assume V to be irreducible and that V has a point rational over K *i.e.* a point with coordinates in K. Given any such point P we choose our homogeneous coordinates so that P is $(0, 1, 0)$.

Suppose P is a flex of V. Clearly, we can arrange $Z = 0$ to be the tangent at P. If the polar conic is either improper or has singularity at P then *char* $k = 2$, and the equation of V is of the form

$$ZY^2 + sZ^2Y + G(X, Z) = 0$$

where $s \in K$ is zero or nonzero depending on whether the polar conic is improper or is singular at P respectively. If *char* $k \neq 2$ then the polar conic at P is neither improper nor is it singular at P; otherwise, P itself would be singular for V. Hence, we can choose our coordinates so that $ZY = 0$ is the polar conic at P. Then, the equation of V acquires the form $ZY^2 + G(X, Z) = 0$. Observe that, since V is irreducible, the coefficient of X^3 in $G(X, Z)$ is a nonzero constant. Given any $0 \neq \nu \in K$, a scale change of the type $Z \to \sigma Z$; $0 \neq \sigma \in K$, will bring the equation to the form

$$ZY^2 + rZ^2Y + \nu X^3 + bX^2Z + cXZ^2 + dZ^3 = 0.$$

Further, if *char* $k \neq 3$, we can kill the X^2Z term.

Suppose P is a nonsingular point of V but not a flex. Let the tangent line at P meet V in a point $Q \neq P$. Then, Q is rational over K. Assume that Q is nonsingular for V but not a flex. Let T denote the tangent at Q. Let T meet V in a point $R \neq Q$. Then, observe that the line T is necessarily distinct from the line $Z = 0$, and the point R is rational over K. Now, we can arrange $Q = (1, 0, 0)$, $R = (0, 0, 1)$, and T to be the line $Y = 0$. Then, the equation of V has the form

$$ZY^2 + (rX^2 + sXZ + cZ^2)Y + eZ^2X = 0,$$

with $r, s, c, e \in K$, and where $re \neq 0$. Substituting X/e for X we obtain the form

$$ZY^2 + (aX^2 + bXZ + cZ^2)Y + Z^2X = 0,$$

with $e^2a = r$, and $eb = s$. If *char* $k \neq 2$ then the transformation

$$X = 2U(U + cW), Y = U(V - bU - W), Z = W(V - bU - W)$$

is birational. For this transformation, the proper-transform of V is given by an equation of the form

$$WV^2 + G(U, W) = 0,$$

and the exceptional locus is given by

$$U(V - bU - W)(U + cW) = 0.$$

Finally, assume P to be a singular point of V. If P is a node with a tangent defined over K then, arranging $ZX = 0$ to be the tangent cone at P, the equation of V can be put in the form

$$XYZ + G(X, Z) = 0.$$

Let

$$G(X, Z) = aX^3 + bZ^3 + XZ(rX + sZ).$$

Substituting $Y - rX - sZ$ for Y, we get the simpler form

$$XYZ + aX^3 + bZ^3 = 0.$$

Likewise, if P happens to be a cusp then, the equation can be brought to the form

$$YZ^2 + aX^3 + bX^2Z = 0.$$

If *char* $k = 2$ and $a \neq 0$ then substituting $Y + (b^2/a)X$ for Y, and $X + (b/a)Z$ for X, we get

$$YZ^2 + aX^3 = 0.$$

If P is a node with none of the tangents defined over K then, the equation has the form $q(X, Z)Y + G(X, Z) = 0$.

(2.3) **Projective Classification.**

Assume V to be irreducible and K to be algebraically closed. We consider the 3 cases : *char* $k \neq 2, 3$, *char* $k = 3$, and *char* $k = 2$ separately.

(2.3.1) The *char* $k \neq 2, 3$ case.

By (2.1.3), our curve V has at least one flex. Let P be a flex of V. To the geometric data (P, T, L, Q), where T is the tangent line at P, $T \bigcup L$ is the polar conic at P, and $Q \neq P$ is a point of V, we associate some algebraic quantities as follows. Choose homogeneous coordinates X, Y, Z so that $Z = 0$ is the tangent at P, $ZY = 0$ is the polar conic at P, and $X = 0$ is the line containing P, Q. Then $P = (0, 1, 0)$, $Q = (0, e, 1)$ for some $e \in K$, and the equation of V is of the form

$$ZY^2 = aX^3 + bX^2Z + cXZ^2 + dZ^3$$

with $a, b, c, e^2 = d \in K$. Conversely, a coordinate system in which the equation of V is expressible in the above form, corresponds to some tuple (P, T, L, Q) of the above type. Irreducibility of V guarantees $a \neq 0$. Let $t = 4(b^2 - 3ac)$,

$$g_2 = \frac{t}{3a^2}, \quad g_3 = \frac{4(9abc - 27a^2d - 2b^3)}{27a^3}, \quad j = \frac{g_2^3}{27g_3^2 - g_2^3}.$$

Another choice of equations for the lines T, L, PQ merely amounts to a scale change of the sort

$$X \to uX, \ Y \to vY, \ Z \to wZ$$

where u, v, w are nonzero constants. Correspondingly, we see that

$$g_2 \to (w/u)^2 g_2, \ g_3 \to (w/u)^3 g_3, \ j \to j.$$

Since the lines T, L are determined by P, the irredundant part of our geometric data is the ordered pair (P, Q).

Now, we examine the dependence of our functions g_2, g_3 and j on the choice of Q. Let $Q' \neq P$ be another point of V. Then, the line PQ' is necessarily given by $X - uZ = 0$ for some $u \in K$. Had we considered

(P, Q') as our basic geometric data to begin with, we would arrive at the equation

$$ZY^2 = a'X^3 + b'X^2Z + c'XZ^2 + d'Z^3$$

where the right hand side equals

$$a(X + uZ)^3 + b(X + uZ)^2Z + c(X + uZ)Z^2 + dZ^3.$$

We leave it to the reader to verify that g_2, g_3, and j all remain unaffected under the change $a \to a', b \to b', c \to c', d \to d'$. Next, we proceed to show that j does not depend on the choice of the flex P either.

Let P' be another flex of V. In view of the above, we may assume $P' = Q$. Then $d = e^2 \neq 0$ and $c = 2se$ where $s^2 = b$. The tangent at Q is given by $sX - Y + eZ = 0$, and the polar conic at Q has the equation

$$(sX - Y + eZ)(sX + Y + 3eZ) = 0.$$

Let
$$Z' = sX - Y + eZ, \ Y' = sX + Y + 3eZ, \text{ and } X' = X$$

Then, the (P', P)-equation of V is

$$Z'Y'^2 = -8aeX'^3 + 4s^2X'^2Z' - 4sX'Z'^2 + Z'^3$$

Substitute $-8ae, 4s^2, -4s, 1$ for a, b, c, d respectively, in the expressions of g_2, g_3 and j to obtain g_2', g_3', and j'. Then,

$$(-2e)^2g_2' = g_2, \ (-2e)^3g_3' = g_3, \text{ and } j' = j.$$

Thus, j is seen to be a *projective invariant* of V.

Returning to the equation

$$ZY^2 = aX^3 + bX^2Z + cXZ^2 + dZ^3,$$

by substituting $\sqrt{a/4}Y$ in place of Y and $X - (b/3a)Z$ in place of X; and then simplifying, we obtain the *Weirstrass canonical form*

$$ZY^2 = 4X^3 - g_2XZ^2 - g_3Z^3.$$

Clearly, $\Delta_P^2 V$ meets V at P, and at the points $P_t = (t, 0, 1)$ where t is a root of $f(x) = 4x^3 - g_2x - g_3$. Note that the discriminant of $f(x)$ is, up to a nonzero constant factor, $g_2^3 - 27g_3^2$. Hence, $f(x)$ has multiple roots iff $g_2^3 - 27g_3^2 = 0$. If t is a double root of $f(x)$ then the corresponding point P_t is a node of V, if t is triple root then V has a cusp at P_t.

Conversely, if $f(x)$ has no multiple root then V is nonsingular, and if $f(x)$ has no triple root then V is non-cuspidal.

If $g_2 g_3 (g_2^3 - 27 g_3^2) \neq 0$ then, the transformation

$$X \to tX, \ Y \to t^{3/2}Y, \ \text{where} \ t = \frac{g_3}{(1+j)g_2},$$

leads to the *j-canonical form*

$$ZY^2 = 4X^3 - 27j(1+j)XZ^2 - 27j(1+j)^2 Z^3.$$

The *j*-canonical forms in the exceptional cases are easily seen to be

$$
\begin{aligned}
ZY^2 &= 4X^3 - Z^3 & \text{if } g_2 = 0 \neq g_3, \\
ZY^2 &= 4X^3 - XZ^2 & \text{if } g_2 \neq 0 = g_3, \\
ZY^2 &= 4X^3 & \text{if } g_2 = 0 = g_3. \\
ZY^2 &= 4X^3 + 4X^2 & \text{if } g_2^3 = 27g_3^2 \neq 0
\end{aligned}
$$

Now suppose

$$aX^3 + bX^2 Z + cXZ^2 + dZ^3 = a(X - \alpha Z)(X - \beta Z)(X - \gamma Z).$$

If at least two of the factors are distinct, say $\alpha \neq \beta$, then the substitution

$$Y \to \sqrt{a}Y, \ X \to X + r\alpha Z, \ Z \to rZ, \ \text{with } r(\alpha - \beta) = 1,$$

leads to the *Legendre canonical form*

$$ZY^2 = X(X - Z)(X - \lambda Z) \ \text{with} \ \lambda \in K.$$

It is straightforward to verify that when $\lambda \neq 0, 1$ we have

$$j = \frac{4N(\lambda)}{27D(\lambda)},$$

where

$$N(\lambda) = (1 - \lambda + \lambda^2)^3, \ \text{and} \ D(\lambda) = \lambda^2(1 - \lambda)^2.$$

The polynomial $\psi(u) = D(u)N(\lambda) - N(u)D(\lambda)$ is a sextic in u with roots $\lambda, 1 - \lambda, 1/\lambda, 1/(1 - \lambda), (\lambda - 1)/\lambda$, and $\lambda/(\lambda - 1)$. Hence, the curves given by $ZY^2 = X(X - Z)(X - \mu Z)$, and $ZY^2 = X(X - Z)(X - \lambda Z)$ have the same j invariant if and only if μ is one of the above six functions of λ.

There is a third canonical form :

$$X^3 + Y^3 + Z^3 = 6mXYZ,$$

which can be obtained from the j-canonical form by means of the transformation

$$\begin{aligned}
X &\rightarrow a^2(1 - 2ms)X - a^2sY - a^2sZ, \\
Y &\rightarrow a^3b(Y - Z), \\
Z &\rightarrow 2mX + Y + Z
\end{aligned}$$

where the constants a, b, s, and m satisfy

$$\begin{aligned}
&(8m^3 - 1)^3 j + (4m^4 + 4m)^3 = 0, \\
&(8m^3 - 1)^2 a^2 - 12(m + m^4)(8m^6 - 20m^3 - 1) = 0, \\
&(8m^3 - 1)b^2 - 3 = 0, \text{ and } (8m^3 - 1)s - 3m^2 = 0.
\end{aligned}$$

The advantage of having the equation in this canonical form is that the hessian is now given by

$$(1 + 2m^3)XYZ = m^3(X^3 + Y^3 + Z^3).$$

Hence, the flexes are located at the nine intersection-points of

$$X^3 + Y^3 + Z^3 = 0 \text{ and } XYZ = 0.$$

(2.3.2) The *char $k = 3$* case.

At first, assume V to be non-cuspidal. Then, by (2.1.3), V has at least one flex and the equation of V can be brought to the canonical form

$$ZY^2 = aX^3 + bX^2Z + cXZ^2 + dZ^3, \quad \text{with } a \neq 0.$$

Let D be the discriminant of $ax^3 + bx^2 + cx + d$ and let $g = b^2$. Then, $D = a^2(b^2c^2 - b^3d - ac^3)$, and an argument entirely similar to the one in (2.3.1), demonstrates the quotient $j = g^3/D$ to be a projective invariant of V. Also, the equation defines a singular cubic if and only if $D = 0$.

As an analogue of the Weirstrass canonical form we may take the

$$ZY^2 = X^3 + bX^2Z + cXZ^2$$

as the equation of V. Correspondingly, $D = c^2(b^2 - c)$ and

$$j = \frac{b^6}{(b^2c^2 - c^3)}.$$

Now suppose

$$ZY^2 = X^3 + b'X^2Z + c'XZ^2$$

has the same j value. If $j = 0$ then we must have $b = 0 = b'$, and the substitution

$$X \rightarrow X, \ Y \rightarrow uY, \ Z \rightarrow (1/u^2)Z, \quad \text{with } c = c'u^4,$$

sends $ZY^2 = X^3 + cXZ^2$ to $ZY^2 = X^3 + c'XZ^2$. In the case $j \neq 0$, let
$v = (b'/b)$, $e = c'/(cv^2)$, and $u = b'(1+e)/(1-e)$. Then,
$c(1-e)^2 = b^2(1+e)$, and the transformation

$$X \to X + uZ, \ Y \to Y, \ Z \to vZ$$

sends the (b, c)-cubic to the (b', c')-cubic.

If V is a nodal cubic, its equation can always be brought to the form

$$ZY^2 = X^2(X + Z).$$

The equation of a cuspidal cubic can be transformed to

$$ZY^2 = X^3$$

if and only if it has a flex. For a flex-less cuspidal cubic the canonical
form is
$$ZY^2 + X^2Y + X^3 = 0.$$

In (2.2) we have seen that these two cuspidal cubics are *birationally*
equivalent. Thus, in this case (and only in this case, as we shall soon
see) the birational classification differs from the projective classification.

(2.3.3) The *char* $k = 2$ case.

At first we assume V to be nonsingular and prove that it has a flex.
Let P be a point of V. We may assume that the tangent to V at P meets
V in a point $Q \neq P$. Otherwise, P is a flex and we are done. Similarly,
the tangent at Q may be assumed to meet V in a point $R \neq Q$. Choose
coordinates in such a way that P, Q are on $Z = 0$, points Q, R are on
$Y = 0$ and R, P are on the line $X = 0$. Then, it is easy to see that for
an equation of V we may take

$$ZY^2 + (aX^2 + bXZ + cZ^2)Y + Z^2X = 0.$$

Note that we must have $a \neq 0$. Consider a *generic* line

$$L(r, s) : Y = rX + sZ$$

where r, s are letters. If there are values of r, s in K for which $L(r, s) \bigcap V$
consists of a single point P then, P must be a flex (since V is nonsingular). Substituting $Y = rX + sZ$ in the equation of V we obtain the
cubic

$$(ar)X^3 + (r^2 + br + as)X^2Z + (1 + cr + bs)XZ^2 + (cs + s^2)Z^3.$$

The conditions for this cubic to be a perfect cube are

(i) $r \neq 0$,

(ii) $(ar)(1 + cr + bs) = (r^2 + br + as)^2$ and

(iii) $(ar)^2(cs + s^2) = (r^2 + br + as)^3$.

The s-resultant of the equations (ii) and (iii) has the form

$$a^6 r^3 (r^9 + \cdots + a^3).$$

Evidently, there exist values of r, s in K satisfying the above conditions (i), (ii), (iii) $i.e.$ V has a flex.

Consequently, a nonsingular cubic V has an equation of the form

$$ZY^2 + aXYZ + bZ^2Y = X^3 + cXZ^2 + dZ^3.$$

This is achieved by sending a flex P to $(0,1,0)$, arranging $Z = 0$ to be the inflectional tangent at P etc. We call this a P-$canonical$ $form$. It follows that the polar conic at P is given by

$$Z(aX + bZ) = 0.$$

For a cubic equation of this type let

$$D = a^6 d + a^5 bc + a^4 c^2 + a^3 b^3 + b^4 \ \text{and} \ g = a^4.$$

Our equation defines a singular cubic if and only if $D = 0$. Define $j = g^3/D$. As before, j is a projective invariant of V. To see this, one must first observe that a coordinate change of the type

$$X \to \alpha X, \ Y \to \beta Y + uZ, \ Z \to \gamma Z$$

does not alter j. Thus the quantity j is independent of a choice of the P-canonical form. To verify the independence of j on the choice of a flex P, let Q be another flex of V. Using the observations made above, Q may be assumed to lie on the X-axis $i.e.$ it has coordinates $(e, 0, 1)$. Then, the tangent at Q is given by

$$(c + e^2)X + (b + ae)Y + dZ = 0,$$

and the inflection condition is

$$e^4 + (ab)e^2 + (b^2 + a^2 c)e + (abc + c^2) = 0.$$

If $e = 0$ then, since Q is nonsingular, we must have $b \neq 0$ and the Q-canonical form is obtained by sending

$$X \to X, \ Y \to (cX + Z)/b, \ Z \to Y.$$

Assume $e \neq 0$. Note that

1. $d + e^3 + ce = 0$ (Q is on V),

2. $d \neq 0$, $b + ae \neq 0$ and $d + ae(b + ae) \neq 0$ (nonsingularity).

To obtain a Q-canonical form, employ the transformation

$$X \rightarrow ed^3 Y$$
$$Y \rightarrow d^3 X + d(d + a^2 e^2 + b^2)Z$$
$$Z \rightarrow d^2(b + ae)X + d^3 Y + b(b + ae)^2 Z.$$

We leave it to the reader to verify the invariance of j. The reader may find it advantageous to use a symbolic algebra system like Maple or Mathematica (on computer) for this purpose.

If $j \neq 0$, we may take the Weirstrass canonical form to be

$$ZY^2 + XYZ + bZ^2 Y = X^3 + cXZ^2 + dZ^3.$$

Correspondingly,

$$j = \frac{1}{(d + bc + c^2 + b^3 + b^4)}.$$

Suppose

$$ZY^2 + XYZ + b'Z^2 Y = X^3 + c'XZ^2 + d'Z^3$$

has the same j. Let $v \in K$ satisfy

$$v^2 + v + (b' + b) = 0.$$

Then, the transformation

$$X \rightarrow X + (b' + b)Z$$
$$Y \rightarrow vX + Y + (c' + c + bv + v^3 + v^4)Z$$
$$Z \rightarrow Z$$

maps the (b, c, d)-equation to the (b', c', d')-equation.

In the case $j = 0$, the Weirstrass canonical form is

$$ZY^2 + Z^2 Y = X^3 + cXZ^2 + dZ^3.$$

Again, if $v, w \in K$ satisfy

$$v^4 + v + c = c' \quad \text{and} \quad w^2 + w + cv^2 + v^6 + d = d'$$

then the substitution

$$X \rightarrow X + v^2 Z$$
$$Y \rightarrow vX + Y + wZ$$
$$Z \rightarrow Z$$

maps our equation to

$$ZY^2 + Z^2Y = X^3 + c'XZ^2 + d'Z^3.$$

The canonical equation of a nodal cubic is

$$XYZ + X^3 + Z^3 = 0.$$

Likewise, if V is cuspidal, its equation has the canonical form

$$YZ^2 + X^3 = 0.$$

(2.4) Relative Invariants

Let $F(X, Y, Z)$ denote the general homogeneous cubic

$$X^2(hX + aY + bZ) + X(cY^2 + dZ^2) + YZ(eX + fZ + gY) + \rho.$$

where $\rho = iY^3 + jZ^3$. Also, let

$$(X, Y, Z) = (X', Y', Z')M$$

denote a generic change of coordinates *i.e.* the entries of M are also regarded as independent variables. Say $F(X, Y, Z) = G(X', Y', Z')$ where we write

$$G(X', Y', Z') = h'X'^3 + i'Y'^3 + j'Z'^3 + a'X'^2Y' + \cdots + g'Y'^2Z'.$$

By a relative invariant of F (over the field k) we mean a polynomial $I(a, \cdots, j)$ homogeneous in the variables $a, \cdots, j$ (with coefficients in k) and having the property that

$$I(a', \cdots, j') = m^w I(a, \cdots, j),$$

where m denotes the determinant of M. If $I \neq 0$ then w is necessarily an integer, called the *weight* of I. Observe that w is also the degree of I. Let

$S = 16c^2d^2 - 8cde^2 + e^4 - 16acdf + 24bcef - 8ae^2f + 16a^2f^2 - 16bcdg + 24adeg - 8be^2g - 16abfg + 16b^2g^2 - 48cf^2h + 24efgh - 48dg^2h - 48ad^2i + 24bdei - 48b^2fi + 144dfhi - 48bc^2j + 24acej - 48a^2gj + 144cghj + 144abij - 216ehij,$

and let

$$T = -64c^3d^3+48c^2d^2e^2-12cde^4+e^6+12(8ac^2d^2f-12bc^2def+2acde^2f+$$
$$3bce^3f - ae^4f + 18b^2c^2f^2 + 8a^2cdf^2 - 12abcef^2 + 4a^2e^2f^2) - 64a^3f^3 +$$
$$12(8bc^2d^2g-12acd^2eg+2bcde^2g+3ade^3g-1be^4g+4abcdfg-12b^2cefg)-$$
$$144a^2defg+24abe^2fg+96a^2bf^2g+96b^2cdg+216a^2d^2g^2-144abdeg^2+$$
$$48b^2e^2g^2 + 96ab^2fg^2 - 64b^3g^3 - 576c^2df^2h - 72ce^2f^2h + 288acf^3h +$$
$$720cdefgh+36e^3fgh-144bcf^2gh-144aef^2gh-576cd^2g^2h-72de^2g^2h-$$
$$144adfg^2h-144befg^2h+288bdg^3h+216f^2g^2h^2+288acd^3i-144bcd^2ei-$$
$$72ad^2e^2i + 36bde^3i - 144b^2cdfi - 576a^2d^2fi + 720abdefi - 72b^2e^2fi -$$
$$576ab^2f^2i-144abd^2gi-144b^2degi+288b^3fgi+864cd^2fhi-648de^2fhi+$$
$$864adf^2hi+864bef^2hi+864d^2eghi-1296bdfghi-864f^3h^2i+216b^2d^2i^2-$$
$$864d^3hi^2 + 288bc^3dj - 144ac^2dej - 72bc^2e^2j + 36ace^3j - 144abc^2fj -$$
$$144a^2cefj-576b^2c^2gj-144a^2cdgj+720abcegj-72a^2e^2gj+288a^3fgj-$$
$$576a^2bg^2j+864c^2efhj+864c^2dghj-648ce^2ghj-6^4acfghj+864bcg^2hj+$$
$$864aeg^2hj-864g^3h^2j-1296abcdij+864b^2ceij+864a^2deij-648abe^2ij+$$
$$864a^2bfij+864ab^2gij-1296cdehij+540e^3hij-1296bcfhij-6^4aefhij-$$
$$6^4adghij-1296beghij+3888fgh^2ij-864b^3i^2j+3888bdhi^2j+216a^2c^2j^2-$$
$$864c^3hj^2 - 864a^3ij^2 + 3888achij^2 - 5832h^2i^2j^2.$$

It can be proved that S and T are relative invariants of F of weights 4 and 6 respectively (see [E]). They satisfy the equation

$$16h(F)T = 48S^2F - h(h(F))$$

where $h(*)$ denotes the hessian determinant of $*$.

When *char* $k = 0$, these form a *complete system* of invariants of F i.e. every relative invariant of F is a polynomial in S, T (see [CG], [E]).
Define

$$J(F) = \frac{S^3}{T^2 - S^3}.$$

Then, $J(F)$ is an (absolute) invariant of F. When F is in the canonical form

$$ZY^2 - 4X^3 + g_2XZ^2 + g_3Z^3,$$

we have

$$S = 192g_2 \quad \text{and} \quad T = -13824g_3.$$

Consequently, the j-invariant defined in (2.3.1) coincides with $J(F)$. Observe that if $G(X', Y', Z')$ is also in the canonical form, say

$$Z'^2Y'^2 - 4X'^3 + g_2'X'Z'^2 + g_3'Z'^3,$$

then

$$g_2' = m^4 g_2 \quad \text{and} \quad g_3' = m^6 g_3.$$

(2.5) Real Projective Classification

Assume K to be the field of real numbers. From (2.1.5) it follows that an irreducible real cubic V has at least one real flex. Hence, employing the methods of (2.3.1), we see that the equation of V can be brought to the Weirstrass canonical form by means of a real linear transformation. Let the equation be $ZY^2 = 4X^3 - g_2 X Z^2 - g_3 Z^3$ and assume that $g_2 g_3 (g_2^3 - 27 g_3^2) \neq 0$. Then, the *real j-canonical form* is

$$ZY^2 = 4X^3 - 27j(1+j)XZ^2 - 27j(1+j)^2 Z^3$$

if

$$\frac{g_3}{(1+j)g_2} > 0,$$

and it is

$$ZY^2 = 4X^3 - 27j(1+j)XZ^2 + 27j(1+j)^2 Z^3$$

if

$$\frac{g_3}{(1+j)g_2} < 0.$$

In the case $g_2 g_3 = 0$, the j-canonical form is

$$ZY^2 = 4X^3 + \sigma_2 X Z^2 + \sigma_3 Z^3$$

where σ_i is defined to be $0, 1, -1$ according to whether g_i is zero, positive, or negative, respectively. If

$$g_2^3 = 27 g_3^2 \neq 0,$$

the canonical form is either

$$ZY^2 = 4X^3 + 4X^2 Z \quad \text{or} \quad ZY^2 = 4X^3 - 4X^2 Z$$

depending on whether $g_3 < 0$ or $g_3 > 0$ respectively.

The last paragraph of (2.4) tells us that the signs of g_2 and g_3 are unaltered by a real transformation. Therefore, it is evident that the various real j-canonical forms listed above are inequivalent under real coordinate changes.

(2.6) **Real Affine Classification**

The classification of real affine plane cubics was done by Newton (see [ST]). In fact he described the canonical forms of real cubic polynomials $f(x, y)$ modulo *affine coordinate changes*. By an affine coordinate change we mean a substitution of the type

$$x \to ax + by + c, \quad y \to a'x + b'y + c'$$

where a, b, c, a', b', c' are numbers, and $(ab' - ba') \neq 0$. Newton's classification is as follows : Every irreducible real affine cubic curve has, in suitable affine coordinates, an equation of one of the following forms.

$$
\begin{aligned}
&(i) && y = x^3 \\
&(ii) && y^2 = x^3 + cx + d \\
&(iii) && xy = x^3 + 1 \\
&(iv) && xy^2 + ey = \pm x^3 + bx^2 + cx + d.
\end{aligned}
$$

Here is a sketch of the proof. Let

$$f = H_0 + H_1 + H_2 + H_3$$

where H_i is homogeneous of degree i. Assume $H_3 \neq 0$. Note that H_3 must always have a real linear factor (a cubic polynomial with real coefficients must have a real root !). There are three cases to consider : H_3 is a perfect cube, it is not a perfect cube but has a square factor, and neither of the above. In the first case we can choose our affine coordinates so that $H_3 = x^3$. Then, the y-degree of $f(x, y)$ is at most 2 and by fine tuning the choice of (x, y) we can bring f either to the form (i), or to the form (ii), or to the form (iii). If H_3 is not a perfect cube then, we can choose coordinates so that $H_3 = x(y^2 + cx^2)$ for some real c. A further adjustment in the coordinates brings f to the form (iv).

It is easy to see that an irreducible cubic with a singularity at finite distance, must have an equation either of the form (ii), or of the form (iv). Further, if the cubic has an oval (a compact component) then its equation must be of the form (ii) or (iv).

Type (i) defines a cubic curve having a cusp at infinity and such that the line at infinity is the tangent at the cusp. Type (ii) corresponds to the cubic for which the line at infinity is an inflectional tangent. Type (iii) curve has a node at infinity, and the line at infinity is a nodal tangent. The line at infinity is nontangent to the cubic if and only if the equation is of the type (iv). Thus, the four types are mutually inequivalent with respect to affine linear transformations.

An equation of type (ii) can be transformed to exactly one of the 5 forms :

$$y^2 = x^3, \quad y^2 = x^3 \pm 1, \quad \text{or} \quad y^2 = x^3 \pm x + d.$$

Evidently, the equivalence classes are parametrized by the (free) parameter d. An equation of the form (iv) can be brought to exactly one of the 4 forms :

$$xy^2 = \pm x^3 + bx^2 + cx + 1, \quad \text{or} \quad xy^2 + y = \pm x^3 + bx^2 + cx + d.$$

For the first 2 forms the equivalence classes are parametrized by the (free) parameters (b, c). Similarly, for those having the form

$$xy^2 + y = -x^3 + bx^2 + cx + d,$$

the equivalence classes are parametrized by the (free) parameters (b, c, d). Observe that there is a "curve" of (b, c, d)-values for which the polynomial $xy^2 + y - x^3 - bx^2 - cx - d$ is reducible (over the reals); so, in this case, the equivalence classes are parametrized by (b, c, d) not on this curve.

REFERENCES

[A] S. S. Abhyankar, *Algebraic Geometry For Scientists And Engineers,* American Mathematical Society, Mathematical Surveys and Monographs No. 35, (1990).

[C] J. W. S. Cassels, *Lectures on Elliptic Curves,* London Mathematical Society Student Texts 24, Cambridge University Press, (1991).

[CG] A. Clebsch, P. Gordon, Ueber cubische ternäre Formen, *Mathematische Annalen,* **6** (1873), 436 - 512.

[E] E. B. Elliot, *An Introduction to The Algebra of Quantics,* Oxford University Press, (1913).

[SI] J. H. Silverman, *The Arithmetic of Elliptic Curves,* Springer-Verlag, (1986).

[ST] D. J. Struik, *A Source Book in Mathematics, 1200 - 1800,* Harvard University Press, (1969).

Commutative Monoid Rings with n-Generated Ideals

JAMES S. OKON AND J. PAUL VICKNAIR, Department of Mathematics, California State University, San Bernardino CA, 92407.

1 INTRODUCTION.

Let R be a commutative ring with identity. Let $\mu(I)$ denote the minimal number of elements required to generate the ideal I. If $\mu(I) \leq n$ then we say that I is n-generated, and if every ideal of R is n-generated then we say that R has the n-generator property. The Dilworth number $d(R)$ of a ring R is defined as $\max\{\mu(I) \mid I \text{ an ideal of } R\}$ and the Sperner number of an Artin local ring (R, m) is $sp(R) = \max\{\mu(m^i) \mid i \geq 0\}$. Let A be a commutative ring with identity and S be a commutative semigroup. The group ring associated to A and S, denoted $A[S]$, is the ring of elements of the form $\sum_{s \in G} a_s x^s$, where $\{a_s \mid s \in S\}$ is a family of elements of A which are almost all zero. We refer to [9] for elementary properties of group and semigroup rings. If $A[G]$ has the n-generator property, then $\dim(A[G]) = \dim(A) + \alpha \leq 1$, where α denotes the torsionfree rank of G. If $\alpha = 0$ then G must be a finite group. If $\alpha = 1$ then $G \cong Z \oplus H$, where H is a finite abelian group and Z denotes the group of integers.

In this paper we give a survey of results on the problem of determining when a Noetherian group or semigroup ring has the n-generator property. It is well known that Dedekind domains have the two-generator property and if R has the n-generator property then R has Krull dimension at most 1 [25, Theorem 1.2]. Let S is a commutative cancellative monoid, $G(S)$ the quotient group of S, and A a Noetherian ring. If $A[S]$ has the n-generator property then $1 \geq \dim(A[S]) = \dim(A[G(S)]) = \dim(A) +$ torsion-free rank of $G(S)$. In [6] Bass began the study of one-dimensional rings which have the 2-generator property, and showed that if such a ring R is reduced and has finite integral closure then finitely generated torsionfree R-modules are isomorphic to direct sums of ideals. One-dimensional reduced rings having finite integral closure are now often called Bass rings. According to [6, p.

19] one of the original motivations for studying the relationship of the two-generator property to decompositions of R-modules was to understand the modules over the integral group rings $Z[G]$ where Z is the rational integers and G a finite abelian group. It was pointed out there [6, p. 20] that an integral abelian group ring $Z[G]$ has the two-generator property if and only if G has square-free order. Since then there has been considerable research to determine when one-dimensional group or semigroup ring $A[S]$ has the n-generator property. Some of the known results in this area are summarized in Section 2.

In Section 3, we consider group rings $R[G]$ with $\dim(R) = 0$ and $rank(G) = 0$. If $d(R[G]) \leq n$ then G must be a finite group. Since $R \cong R_1 \oplus R_2 \oplus ... \oplus R_n$ where each R_i is local and $d(R[G]) = \sup(d(R_i[G]))$, it suffices to consider the case where R is an Artin local ring. The Dilworth number of $R[G]$ is known in the case that R is a field (see [28] and [17]). If R is not a field then the Dilworth number of $R[G]$ is not known in general. However, if R is a principal Artin local ring and G is a finite cyclic p-group (p is not a unit in R) the Dilworth number of $R[G]$ was determined in the unramified case (p generates the maximal ideal) in [22] and for the ramified case in [4]. An important property in the study of these rings is that the Dilworth and Sperner numbers are equal. This theorem of Watanabe is proven for Artin local rings (A, m) with $\mu(m) \leq 2$ is proven in [11]. The result is false for $\mu(m) \geq 3$ (see [11] and [19]). When G is not cyclic, group rings with $d(R[G]) \leq n$ have been characterized only for special cases of n (see [1], [2], [3] and [4])

Additional results in this area for Noetherian rings can be found in [5], [14], [15], [20], [18], [26], [24], and [29]. In this survey we will restrict ourselves to the case of a commutative cancellative monoid over a commutative Noetherian ring. The problem of determining which group or semigroup rings have the property that finitely generated ideals are n-generated has been studied for non-Noetherian rings in [21] and for noncommutative rings in [27]. The case where the semigroup need not be cancellative or contain an identity has been studied in [7] for $n = 1$.

2 THE n-GENERATOR PROPERTY FOR ONE-DIMENSIONAL RINGS.

Theorem 1 gives necessary and sufficient conditions for a group ring $R[G]$ to have the two-generator property where G is a finite group and $\dim(R) = 1$. The case $R = Z$ of this theorem, which was given by Bass [6, pp. 19-20], says that $Z[G]$ has the two-generator property if and only if the order of G is square-free. The theorem was proven by Greither in [10, Theorem 3.1] under the hypothesis that R is a reduced ring, the integral closure $\bar{R}$ is finitely generated as an R-module and the order of the group is not a zero divisor in R. This was extended by Rush in [24] to all Macaulay rings in which each maximal ideal has height one. In [20, Theorem 2.1] it was shown that the same conclusion holds without the assumption that R is Macaulay and the order of the group is a regular element in R. In [17] additional results were obtained on the structure of group rings over Artin local rings which allowed the removal of the condition that maximal ideals have height one.

THEOREM 1. *[17, Theorem 4.8]* *Let $R \cong A \oplus B$ be a one-dimensional ring where A is an artinian ring (maybe zero) and B is a one-dimensional ring for which maximal ideals have height one and let G be a finite abelian group of order $m \neq 0$. Then $d(R[G]) \leq 2$ if and only if the following hold:*
1. B_P is a principal valuation domain for each prime ideal P of B which contains m.
2. If $m = 2^t m_1$ with m_1 odd, then m_1 is a regular element of B, $B/m_1 B$ is reduced (maybe zero), and if 2 is a nonunit of B, then $t \leq 1$.
3. B has the two-generator property.
4. $A = A_1 \oplus A_2 \oplus ... \oplus A_s$ where for each i, (A_i, M_i) is a local artinian ring with $d(A_i) \leq 2$ such that:
  (a) if (A_i, M_i) is not a principal ideal ring, then m is a unit of A_i.
  (b) if (A_i, M_i) is a principal ideal ring (maybe a field), p a prime integer which divides m and $p \in M_i$ then:
  (1) If p is odd, then G_p is cyclic; furthermore, if $M_i^2 \neq 0$, then $G_p \cong Z/pZ$ and $pR_i = M_i$.
  (2) If $p = 2$ then $G_p \cong Z/2^j Z \oplus Z/2^k Z$ where: $j \leq 1$ if $M_i = 0$; $j = 0$ and $k = 1$ if $M_i^2 \neq 0$; and $j = 0$ otherwise.

When $rank(G) = 1$ and $\dim(R) = 0$ it is possible to determine the Dilworth number of $R[G]$. In [14], Matsuda shows that for a polynomial ring $d(R[X]) = \ell(R)$, the length of R as an R-module. The proof given here also illustrates techniques similar to those used to prove Theorem 3. Before giving the proof we first introduce some notation. Let (R, M) be an Artin local ring with $\ell(M) = m$ and let u be the smallest nonnegative integer such that $M^{u+1} = 0$. If $m \geq 1$ then there exist $r_1, ..., r_m \in R$ and $0 < i_1 < ... < i_u = m$ such that $r_{i_{v-1}+1} + M^{v+1}, ..., r_{i_v} + M^{v+1}$ is a basis for the R/M-vector space M^v/M^{v+1} for $1 \leq v \leq u$.

THEOREM 2. *[14, Proposition 5.13]. Let (R, m) be an Artin local ring then $d(R[X]) = \ell(R)$.*

PROOF. To see $d(R[X]) \geq \ell(R)$, suppose the ideal $(r_1 X, r_2 X^2, ..., r_m X^m, X^{m+1})$, where the r_i are as defined above, can be generated by m elements, say $f_1, ..., f_m$. Then, for $1 \leq i \leq m$, we have

$$f_i = a_{i1} X + a_{i2} X^2 + ... + a_{im} X^m + ...$$

where $a_{ik} \in (r_1, ..., r_k)$ for $1 \le k \le m$. Since $r_1 X \in (f_1, ..., f_m)$, we may assume that $a_{11} = r_1, a_{21} = a_{31} = ... = a_{m1} = 0$, and $(a_{2h}, ..., a_{mh}) \subseteq (r_2, ..., r_m)$ for $2 \le h \le m$. Since $r_2 X^2 \in (f_1, ..., f_m)$, we may assume $a_{22} = r_2, a_{32} = ... = a_{m2} = 0$, and $(a_{3h}, ..., a_{mh}) \subseteq (r_3, ..., r_m)$ for $3 \le h \le m$. Continuing this operation we have:

$$
\begin{aligned}
f_1 &= r_1 X + a_{12}X^2 + a_{13}X^3 + ... + a_{1m}X^m + ... \\
f_2 &= \qquad\quad\; r_2 X^2 + a_{23}X^3 + ... + a_{2m}X^m + ... \\
f_3 &= \qquad\qquad\qquad\quad r_3 X^3 + ... + a_{3m}X^m + ... \\
&\;\;\vdots \\
f_m &= \qquad\qquad\qquad\qquad\qquad\qquad\quad r_m X^m + ... \; .
\end{aligned}
$$

Then $X^{m+1} \in (f_1, ..., f_m)$ is a contradiction. Thus we have $d(R[X]) \ge \ell(R)$.

For the opposite inequality, consider the ideal $I = (f_1, ..., f_{m+2})$ where I cannot be generated by $m+1$ elements. Assume that $\deg(f_i) = n_i$ and $n_1 \le ... \le n_{m+2}$ and these generators are chosen so that $\sum \deg(f_i)$ is minimal among all such $m+2$ generators of I. Then we may write $f_i = a_{i1}x^{n_i} + a_{i2}x^{n_1-1} + ...$, where $a_{ij} = c_{ij}(0) + c_{ij}(1)r_1 + ... + c_{ij}(m)r_m$ and each $c_{\alpha\beta}(k)$ is zero or a unit in R. Now consider the matrix C consisting of the leading coefficients of the f_i :

$$
\begin{bmatrix}
c_{11}(0) & c_{11}(1)r_1 & c_{11}(2)r_2 & ... & c_{11}(m)r_m \\
c_{21}(0) & c_{21}(1)r_1 & c_{21}(2)r_2 & ... & c_{21}(m)r_m \\
\vdots & & & & \\
c_{m+2,1}(0) & c_{m+2,1}(1)r_1 & c_{m+2,1}(2)r_2 & ... & c_{m+2,1}(m)r_m
\end{bmatrix}
$$

We may replace a row with a unit multiple of itself or replace a row with itself added to a unit multiple of another row. Locate the first non-zero entry, if any, in the first column of C and use it to eliminate the nonzero entries below it. At this point one must rearrange the entries in the rows which have been altered. Indeed, if one considers the sum $\varepsilon_1 r_j + \varepsilon_2 r_j$, then $\varepsilon_1 + \varepsilon_2$ may be a unit of R, zero, or a nonunit of R. For the first two possibilities, no rearranging is needed. However, for the third one, $\varepsilon_1 r_j + \varepsilon_2 r_j = \varepsilon_{j+1}r_{j+1} + ... + \varepsilon_m r_m$ where the ε's are zero or units of R. Note that for all three cases the terms $r_1, ..., r_{j-1}$ are not involved, i.e., the rearrangement of the rows does not affect prior columns. So proceed to the second column of C and repeat the process. Because of the size of C, we will eventually obtain a row of zero's and arrive at a contradiction. Thus $d(R[x]) \le \ell(R)$ and we are done. ∎

It is possible to extend the above theorem to more general monoids. In [15, Theorem 5.17], Matsuda shows that if S is a commutative cancellative subsemigroup of the set of non-negative integers Z_+ with $G(S) = Z$ (so S

contains no invertible elements) then $d(R[S]) = i\ell(R)$ where i is the least positive element of S. In the next two theorems Matsuda's result is generalized by dropping the restriction that S contain no invertible elements.

THEOREM 3. *[18, Theorem 1.2] Assume* (R, M) *is an Artin local ring and let* S *be a submonoid of* $Z_+ \oplus H$ *with quotient group* $G(S) = Z \oplus H$ *where* H *is a finite abelian group. Let* $\tau(S)$ *denote the invertible elements of* S *and let* $\tau(S) = 0 \oplus U$. *Let* i *be the smallest positive integer such that* $(i, h) \in S$ *for some* $h \in H$. *Then* $\ell(R)i\,|H/U| \leq d(R[S]) \leq \ell(R)i\,|H|$.

Using Theorem 3, it is possible to characterize monoid rings $R[S]$ which have the three-generator property. The cases where S does or does not contain all the invertible elements of $G(S)$ are treated separately.

THEOREM 4. *[18, Theorem 2.1] Let* R *be a commutative ring and let* S *be a cancellative monoid whose quotient group* $G(S)$ *has torsionfree rank one, say* $G(S) = Z \oplus H$ *where* H *is a finite group of order* m_0. *If* $H \subset S$ *then* $R[S]$ *has the three-generator property if and only if*
(i) $R \cong R_1 \oplus R_2 \oplus ... \oplus R_s$ *where each* (R_i, M_i) *is an Artin local ring with* $d(R_i) \leq 2$.
(ii) *If* p *is a divisor of* m_0 *and* $p \in M_i$ *then* $\ell(R_i)\,|H_p| \leq 3$. *In particular,* m_0 *is a unit in each* R_i *which is not a field of characteristic 2 or 3.*
(iii) *One of the following holds:*
    (a) $S \cong Z \oplus H$
    (b) $S \cong Z_+ \oplus H$
    (c) $S \cong T \oplus H$ *where* T *is a submonoid of* $Z_+\backslash\{1\}$ *containing 2, each* R_i *is a field, and if any* R_i *has characteristic 2 or 3, then* m_0 *is a unit in* R_i.
    (d) $S \cong T \oplus H$ *where* T *is a submonoid of* $Z_+\backslash\{1, 2\}$ *containing 3, each* R_i *is a field, and if any* R_i *has characteristic 2 or 3, then* m_0 *is a unit in* R_i.

THEOREM 5. *[18, Theorem 2.4] Let* R *be a ring and let* S *be a submonoid of* $Z_+ \oplus H$ *with quotient group* $G(S) = Z \oplus H$ *where* H *is a finite abelian group not contained in* S. *Then* $R[S]$ *has the three-generator property if and only if*
(i) $R \cong R_1 \oplus R_2 \oplus ... \oplus R_s$ *where each* R_i *is a field.*
(ii) $S = S_1 \oplus K$, K *a finite abelian group with either* $|K| = 0$ *or* $|K|$ *is a unit in each of the* R_i.
(iii) *One of the following holds:*
    (a) S_1 *is a submonoid of* $Z_+ \oplus Z/2^u Z$ *with* $u \geq 1$ *containing* $(0, 2)$ *and* $(1, g)$ *for some* $g \in Z/2^u Z$. *If some* R_i *has characteristic 2, then* $u = 1$.
    (b) S_1 *is a submonoid of* $Z_+ \oplus Z/3^u Z$ *with* $u \geq 1$ *containing* $(0, 3)$ *and* $(1, g)$ *for some* $g \in Z/3^u Z$. *If some* R_i *has characteristic 3, then* $u = 1$.

In the case where S is a rank 1 abelian group it is possible to determine the Dilworth number of $R[S]$. Arnold and Matsuda first consider the case that S is a subgroup of Q.

THEOREM 6. *[5, Theorem 11] Let A be a ring and let G be a subgroup of Q, and set $G_0 = G \cap Q_0$. Then $d(A[G]) = d(A[G_0]) = d(A[Z]) = d(A[X])$.*

Theorems 6 and 2 allow us to obtain the Dilworth number of $R[Z \oplus G]$.

COROLLARY 7. *Let (A, m) be an Artin local ring with $p \in m$ a prime integer. Let G be a nontrivial finite abelian p-group. Then $d(A[Z \oplus G]) = \ell(A)|G|$.*

PROOF. We have $d(A[Z \oplus G]) = d(A[G][Z]) = \ell(A[G]) = \ell(A)|G|$. ∎

We conclude this section with a result of McLean which gives bounds on the Dilworth number of a local ring (R, m) in terms of the number of generators of powers of the maximal ideal and the Dilworth numbers of the factor rings R/m^i.

THEOREM 8. *[16, Theorem 1] Let (R, m) be a local ring and n a positive integer. Then the following conditions are equivalent:*
(1) $d(R) \leq n$.
(2) If $I_0, ..., I_n$ are ideals of R, then $I_i \subseteq \sum_{j \neq i} I_j$ for some integer i such that $0 \leq i \leq n$.
(3) If $I, I_0, ..., I_n$ are ideals of R, then $I \cap \sum_{i=0}^{n} I_i = \sum_{i=0}^{n}(I \cap \sum_{j \neq i} I_j)$.
(4) If I is a proper ideal of R, then either $\mu(I) \leq n - 1$ or I is the sum of a power of m and an ideal that can be generated by $n - 1$ elements.
(5) $d(R/m^{n+1}) \leq n$.
    If $n \leq 2$ each of these conditions is equivalent to
(6) $\mu(m^i) \leq n$ for all integers i such that $1 \leq i \leq n$.
    If R is a one-dimensional Cohen-Macaulay ring (with no restrictions on n), then each of the six conditions above is equivalent to
(7) $\mu(m^n) \leq n$.

3 GROUP RINGS R[G] WITH DIM(R)=0 AND RANK(G)=0.

We begin this section by considering the case when the coefficient ring is a field of characteristic $p > 0$ and G is a finite abelian group. We note that if $G = G_p \oplus H$ where p does not divide the order of H and F is a field of characteristic p then $F[H]$ is a sum of fields. Thus $d(F[G]) = d(F[G_p])$. Therefore it is sufficient to consider the case when G is a p-group. We also note that if $G \cong Z/p^{t_1}Z \oplus ... \oplus Z/p^{t_s}Z$ then $F[G] \cong F[X_1, ..., X_s]/\left(X_1^{p^{t_1}}, ..., X_s^{p^{t_s}}\right)$.

Probably the most important tool in determining the Dilworth number of an artinian group ring is the ability to relate the Dilworth number to the Sperner number. In [28], Watanabe uses a combinatorial theorem of deBruijn, Tengbergen and Kruyswijk to show that if $G \cong Z/p^{t_1}Z \oplus ... \oplus Z/p^{t_s}Z$ then $d(F[G]) = sp(F[G]) = \mu(m^d)$ where $d = \lfloor \frac{1}{2}(\sum_{i=1}^{s} p^{t_i} - s) \rfloor$ (here $\lfloor \ \rfloor$ denotes the greatest integer function) and m is the maximal ideal of $F[G]$.

THEOREM 9. *[28, Proposition 2.5] Let A be a monomial complete intersection, i.e., $A = F[X_1, ..., X_s]/\left(X_1^{p^{t_1}}, ..., X_s^{p^{t_s}}\right)$ where F is a field. Then $d(A) = \mu(m^d)$ where $d = \lfloor \frac{1}{2}(\sum p^{t_i} - s) \rfloor$ and $m = (x_1, ..., x_s)$ is the maximal ideal of A.*

The Dilworth number of $F[G]$ is computed explicitly in [17]. In order to state this theorem we first introduce some notation. Let $t_1, ..., t_s, d, p$ be positive integers. Then

$$D(t_1, ..., t_s; p, d) = \binom{d+s-1}{s-1} + \sum_{i=1}^{s}(-1)^i \sum_{k_i} \binom{d-k_i+s-1}{s-1}$$

where $k_1 \in \{p^{t_u} : d \geq p^{t_u}$ where $1 \leq u \leq s\}$, $k_2 \in \{p^{t_u} + p^{t_v} : d \geq p^{t_u} + p^{t_v}$ where $1 \leq u < v \leq s\}$, etc.

THEOREM 10. *[17, Theorem 1.2] Let F be a field of characteristic $p > 0$ and let G a finite abelian group. Then $d(F[G]) = D(t_1, ..., t_s; p, d)$ where $G_p \cong Z/p^{t_1}Z \oplus ... \oplus Z/p^{t_s}Z$ and $d = \lfloor \frac{1}{2}(\sum p^{t_i} - s) \rfloor$.*

When the coefficient ring is not a field, the Dilworth number of an artinian group ring $A[G]$ is known in general only when A is principal and G is cyclic. To determine the Dilworth number of $A[G]$ in this case we first show the Dilworth and Sperner numbers are equal. This is a special case of the following theorem which appears in [11] and is attributed to Watanabe.

THEOREM 11. *[11, Theorem 4.2] Let (A, m, k) be an Artin local ring of embedding dimension at most two, i.e., $\mu(m) \leq 2$. Then we have $d(A) = sp(A)$.*

Examples are also given in [11] to show that this result is not true for $\mu(m) \geq 4$ even if A is Gorenstein. Ikeda conjectures that $d(A) = sp(A)$ for all Gorenstein Artin local rings with $\mu(m) = 3$. Theorem 22 and Theorem 17 show that the rings $(Z/p^i Z)[Z/pZ \oplus Z/pZ]$ for $i > 2, p > 2$ form a class of rings for which this conjecture is false.

With Theorem 11 and a result of Hassani and Kabbaj [2, Lemma 4] it is possible to obtain a bound on the Dilworth number in terms of a specific power of the maximal ideal.

THEOREM 12. *[4, Lemma 3]* Let (A, m) be an Artin local ring with $\mu(m) \leq 2$. Then $d(A) \leq n$ if and only if m^n is n-generated.

PROOF. If m^n is n-generated, $\mu(I) \leq \mu(I + m^{n-1})$ for each ideal I of A by [2, Lemma 4]. Thus $sp(A) \leq n$. Since $\mu(m) \leq 2$, $d(A) = sp(A)$ by Theorem 11. Thus $d(A) \leq n$ and the converse is obvious. ∎

When the coefficient ring is not a field, we need some additional notation. Let (A, m) be an Artin local ring and let $p \in m$ be a prime integer. If $p \in m^k - m^{k+1}$ then we say the *ramification index of* p, denoted $e(p)$, is k. If $p = 0$ in A, then $e(p) = \infty$. For the group ring $R = A[Z/p^j Z]$, set $s^* = \min\{e(p)(j - t) + p^t \mid 0 \leq t \leq j\}$. In the case $p = 0$ in A, we set $e(p)(j - j) = 0$. When $G \cong Z/p^{n_1} Z \oplus \ldots \oplus Z/p^{n_r} Z$ is a p-group then we may use the isomorphism $A[G] \cong A[X_1, ..., X_r]/(1 - X_1^{p^{n_1}}, ..., 1 - X_r^{p^{n_r}})$, where X_i are indeterminates, to represent elements of the group ring. We will use x_i to denote the image of X_i in the factor ring.

The following proposition of Hassani, Kabbaj, Okon and Vicknair gives some upper and lower bounds.

PROPOSITION 13. *[4, Proposition 6]* Let (A, m) be an Artin local ring and let $p \in m$ be a prime integer. Let $R = A[Z/p^j Z]$ and let M be the maximal ideal of R. If $0 < s < \min\{s^*, l(A)\}$ then M^s requires $\ell(A/m^{s+1})$ generators. In particular, $\ell(A/m^{s^*}) \leq sp(R) \leq d(R) \leq \ell(A)$.

Sketch of Proof. In the case $p = 0$ in R, $R \cong A[x]/(x^{p^j})$ and the result is clear in this case. Now assume $p \neq 0$ in R. Writing $M = (m, 1 - x)$, we will show

$$M^s = (m^s, m^{s-1}(1 - x), ..., m^2(1 - x)^{s-2}, m(1 - x)^{s-1}, (1 - x)^s)$$

requires $\ell(A/m^{s+1})$ generators.

Let N be such that $m^N \neq 0$ and $m^{N+1} = 0$ and for $0 \leq i \leq N$ let $\{m_{i1}, ..., m_{in(i)}\}$ be a minimal set of generators for m^i. If M^s can be generated by fewer elements then, since R is local, one of the generators, say $m_{(s-t)1}(1 - x)^t$, can be written in terms of the others. If $t = s$, then we have the contradiction $1 \in m$. Thus we may assume that $t < s$. We may pass to the ring $A/m^{s-t+1}[Z/p^j Z]$ obtaining

$$m_{(s-t)1}(1 - x)^t = \sum_{\alpha=2}^{n(s-t)} f_{t\alpha} m_{(s-t)\alpha}(1 - x)^t$$

$$+ \sum_{n=t+1}^{s} \sum_{\alpha=1}^{n(s-n)} f_{n\alpha} m_{(s-n)\alpha}(1 - x)^n$$

where $f_{t\alpha}, f_{n\alpha} \in R$. The first summand in the above expression can be written $\sum_{i=0}^{p^j-1} b_i x^i$ where $b_i \in J = \left(m_{(s-t)2}, ..., m_{(s-t)n(s-t)}\right) A$. For $t < n \leq s$, let $f_{n\alpha} = \sum_{i=0}^{p^j-1} a_{n\alpha i} x^i$ where $a_{n\alpha i} \in A$. Expanding $f_{n\alpha}(1-x)^n$ for $t < n \leq s$ and equating coefficients we obtain the system of equations:

$$m_{(s-t)1}\binom{t}{0} = b_0 + \sum_{n=t+1}^{s} \sum_{\alpha=1}^{n(s-n)} m_{(s-n)\alpha} \left(\sum_{r=0}^{n} (-1)^r \binom{n}{r} a_{n\alpha p^j - r} \right)$$

$$-m_{(s-t)1}\binom{t}{1} = b_1 + \sum_{n=t+1}^{s} \sum_{\alpha=1}^{n(s-n)} m_{(s-n)\alpha} \left(\sum_{r=0}^{n} (-1)^r \binom{n}{r} a_{n\alpha p^j - r + 1} \right)$$

$$\vdots$$

$$(-1)^t m_{(s-t)1} = b_t + \sum_{n=t+1}^{s} \sum_{\alpha=1}^{n(s-n)} m_{(s-n)\alpha} \left(\sum_{r=0}^{n} (-1)^r \binom{n}{r} a_{n\alpha p^j - r + t} \right)$$

$$0 = b_{t+1} + \sum_{n=t+1}^{s} \sum_{\alpha=1}^{n(s-n)} m_{(s-n)\alpha} \left(\sum_{r=0}^{n} (-1)^r \binom{n}{r} a_{n\alpha p^j - r + t + 1} \right)$$

$$\vdots$$

$$0 = b_{p^j-1} + \sum_{n=t+1}^{s} \sum_{\alpha=1}^{n(s-n)} m_{(s-n)\alpha} \left(\sum_{r=0}^{n} (-1)^r \binom{n}{r} a_{n\alpha p^j - r + p^j - 1} \right)$$

Now multiply the *kth* equation by $\binom{t+k}{k}$ and sum from $k=0$ to $k=p^j-1$. On the left we get:

$$m_{(s-t)1} \sum_{k=0}^{t} (-1)^k \binom{t}{k} \binom{t+k}{k} = (-1)^t m_{(s-t)1}.$$

The sum on the right turns out to be an element of $J + m^{s-n+e(p)(j-[\log_p t])}$. By our choice of s, $n \leq s < e(p)(j - [\log_p t]) + p^{[\log_p t]}$. Thus, we have

$$s - t < e(p)(j - [\log_p t]) + p^{[\log_p t]} - t$$
$$\leq e(p)(j - [\log_p t])$$
$$\leq s - n + e(p)(j - [\log_p t]).$$

Hence, we have a contradiction. Thus M^s requires $\ell(A/m^{s+1})$ generators. □

The next theorem gives a sharper upper bound to the Dilworth number.

PROPOSITION 14. *[4, Remark 1] Let (A, m) be a principal Artin local ring and let $p \in m$ be a prime integer. Let $R = A\left[Z/p^j Z\right]$. Then $d(R) \leq \min\{\ell(A), s^*\}$.*

It is now possible to determine the Dilworth number of $A[G]$ when A is principal and G is a cyclic p-group.

THEOREM 15. *[4, Theorem 1] Let (A, m) be a principal Artin local ring, let $p \in m$ be a prime integer and let $R = A[Z/p^j Z]$. Then $d(R) = sp(R) = \min\{s^*, \ell(A)\}$.*

PROOF. From Propositions 13 and 14 we have
$$\min\{s^*, \ell(A)\} = \ell(A/m^{s^*}) \leq sp(R) \leq d(R) \leq \min\{s^*, \ell(A)\}. \quad \blacksquare$$

The next corollary gives a stronger version of [27, Theorem 5.2] for the case of an abelian group.

COROLLARY 16. *[22, Corollary 2.8] Let F be a field of characteristic p where p is a prime integer. Let $n_1 \leq n_2 \leq ... \leq n_r$ be positive integers. Then $d(F[Z/p^{n_1} Z \oplus ... \oplus Z/p^{n_r} Z]) = p^{\sum_{i=1}^{r-1} n_i}$ if and only if $p^{n_r} > \sum_{i=1}^{r-1} (p^{n_i} - 1)$.*

We will compute the Dilworth and Sperner numbers of $A[Z/pZ \oplus Z/pZ]$ where (A, m) is a principal Artin local ring and $p \in m \backslash m^2$ is a prime. If $G = Z/pZ \oplus Z/pZ$ then the maximal ideal of $A[G]$ is $M = (p, 1 - x_1, 1 - x_2)$. To simplify notation we will let $y_1 = 1 - x_1$ and $y_2 = 1 - x_2$. Therefore we may express each element of $A[G]$ as a sum of monomials in the generators of the maximal ideal $u p^a y_1{}^b y_2{}^c$ where u is zero or a unit of A. The proof of [4, Lemma 3] shows
$$y_1{}^p = upy_1 \text{ and } y_2 = vpy_2$$
where u, v are units in $A[G]$. A monomial in $A[G]$ of the form $u p^a y_1{}^b y_2{}^c$ where u is a unit of A and one of the following holds: (1) $a = 0, c < p, b \geq 1$ (2) $a = b = 0$ and $c \geq 1$ (3) $b = c = 0$ is said to be in *standard form*. We define the *value* of a monomial in standard form $f = u p^a y_1{}^b y_2{}^c$, denoted $v(f)$, to be n if $f \in M^n \backslash M^{n+1}$ and define an order on monomials by first comparing values, then $b + c$ and finally b.

If (A, m) is a Gorenstein ring then the socle of A, denoted $S(A)$, is the annihilator of the maximal ideal m. When A is an Artin Gorenstein ring $S(A)$ is a one-dimensional vector space over A/m. In particular, the last non-zero power of the maximal ideal must be principal.

The next theorem shows that the rings we will be considering in Lemma 18 through Theorem 22 are Gorenstein. Eilenberg and Nakayama proved an earlier noncommutative version of this theorem in the case where $\text{rank}(G) = 0$ in [8].

THEOREM 17. *[13, p.196] Let S be a commutative ring with identity and Ω be the set of prime integers which are nonunits in S. Let G be an abelian group of finite torsion-free rank and F a free subgroup of G such that G/F is torsion. Then $S[G]$ is locally Gorenstein if and only if S is locally Gorenstein and for each $p \in \Omega$, $(G/F)_p$ is finite.*

LEMMA 18. *Let (A, m) be an Artin local ring with $p \in m \setminus m^2$ a prime and i minimal such that $p^i = 0$. Let $G = Z/pZ \oplus Z/pZ$, let y_1, y_2 be as defined above and let M denote the maximal ideal of $A[G]$. Let $f = p^a y_1^b y_2^c$ be a non-zero monomial in standard form. Then*
(a) $S(A[G]) = (y_1^{i\phi(p)} y_2^{\phi(p)}) = M^{(i+1)\phi(p)}$ and $M^{(i+1)\phi(p)+1} = 0$ where ϕ denotes the Euler ϕ-function.
(b) Every element in $A[G]$ can be expressed as a sum of monomials in standard form.
(c) $v(f) = a + b + c$.

PROOF. (a) From [22, Theorem 4.6], $y_1^{i\phi(p)} \neq 0$ and $y_1^{i\phi(p)+1} = 0$. Thus, if $f = y_1^{i\phi(p)} y_2^{\phi(p)}$, $fp = fy_1 = fy_2 = 0$. Therefore $M^{(i+1)\phi(p)} = (y_1^{i\phi(p)} y_2^{\phi(p)}) \neq 0$ and $M^{(i+1)\phi(p)} = 0$.

(b) We first note that $f = y_1^{i\phi(p)} y_2^{\phi(p)}$ is in standard form and if $g = p^a y_1^b y_2^c$ is a monomial in $A[G]$ such that $g > f$ then $g = 0$. Indeed, if $v(g) > v(f)$ then $g = 0$ by part (a). If $v(g) = v(f)$, $b + c > (i+1)\phi(p)$ and $b > i\phi(p)$ then $g = 0$. If $c > \phi(p)$ we may substitute to get $g = u p y_1^b y_2^{c-\phi(p)} = v y_1^{b+\phi(p)} y_2^{c-\phi(p)}$ where u, v are units of A. We continue until the exponent of y_1 is greater than $i\phi(p)$. If $v(f) = v(g)$, $b + c = (i+1)\phi(p)$ and $b > i\phi(p)$ then $g = 0$ as before.

Now assume $g < y_1^{i\phi(p)} y_2^{\phi(p)}$ and that every $h > g$ can be expressed as a sum of elements in standard form. If $b \geq 1$ then $g = u y_1^{b+\phi(p)a} y_2^c$ plus terms of greater value. If $c > \phi(p)$ we may substitute to get $g = v p y_1^{b+a\phi(p)} y_2^{c-\phi(p)} + \ldots = v' y_1^{(a+1)\phi(p)} y_2^{c-\phi(p)} + \ldots$. We repeat until $c \leq \phi(p)$. Thus we may write g in the form $u y_1^b y_2^c + \ldots$ where $c \leq \phi(p)$.

If $b = 0, c \geq 1$ and $a > 0$ we substitute to get $g = u y_2^{c+a\phi(p)} + \ldots$ where u is a unit in A. If $b = c = 0$ then g is already in standard form. Thus every monomial can be expressed as a sum of monomials in standard form and, since every element in $A[G]$ is a sum of monomials, (b) holds.

(c) Let $f = p^a y_1^b y_2^c$ be a non-zero monomial in standard form. Clearly $v(f) \geq a + b + c$. Suppose $f \in M^{a+b+c+1}$. If $a = 0$, $b \geq 1$ and $c \leq \phi(p)$ then $f y_1^{i\phi(p)-b} y_2^{c-\phi(p)} = y_1^{i\phi(p)} y_2^{\phi(p)} \in M^{(i+1)\phi(p)+1} = 0$, a contradiction. If $a = b = 0$ then $f y_1^{\phi(p)} y_2^{i\phi(p)-c} = y_1^{i\phi(p)} y_2^{\phi(p)} \in M^{(i+1)\phi(p)+1} = 0$, a contradiction. Finally if $b = c = 0$ then $f p^{i-1-a} = p^{i-1} \in M^i$. Applying the augmentation map we get $p^{i-1} = 0$. This proves (c). ∎

Since $A[Z/pZ \oplus Z/pZ]$ is an Artin local ring with maximal ideal $M = (p, y_1, y_2)$, every element of $f \in A[Z/pZ \oplus Z/pZ]$ can be written in the form $f = a_0 + a_1v_1 + ... + a_nv_n$ where $v_i = p^{a_i}y_1{}^{b_i}y_2{}^{c_i}$ is in standard form, $v_1 < v_2 < ... < v_n$, v_n is the largest monomial in R and each a_i is zero or a unit in A. If a_kv_k is the first non-zero term of f we say a_kv_k is the *leading monomial* of f, $v(f) = v(v_k)$ and $depth(f) = n + 1 - k$. If $B = \{f_1, ..., f_m\}$ then $depth(B) = \sum depth(f_i)$.

LEMMA 19. *Let (A, m) be an Artin local ring with $p \in m \backslash m^2$ a prime. Let $G = Z/pZ \oplus Z/pZ$ and let y_1, y_2 be as defined above. Let $f = p^a y_1{}^b y_2{}^c$ be a monomial. Then*

    (a) $v(fy_1) \geq v(f) + 1$,
    (b) $v(fy_2) \geq v(f) + 1$,
    (c) $v(fp) \geq v(f) + 1$ and

equality holds if fy_1, fy_2 or fp is in standard form.

PROOF.   It is clear that the inequalities hold in (a), (b) and (c). The second assertion is immediate from the preceding lemma. ∎

Though, in general, order is not always preserved, the next lemma shows that multiplication by monomials preserves order in some cases.

LEMMA 20. *Let (A, m) be an Artin local ring with $p \in m \backslash m^2$ a prime integer. Let $G = Z/pZ \oplus Z/pZ$, and let $f_1 = p^{a_1}y_1{}^{b_1}y_2{}^{c_1}$ and $f_2 = p^{a_2}y_1{}^{b_2}y_2{}^{c_2} \in A[G]$ be monomials in standard form.*

    (a) If $a_1 = 0, c_1 < p$ and $g = y_1{}^v$ then $f_1g < f_2g$.
    (b) If $a_1 = b_1 = 0$ and $g = y_2{}^w$ then $f_1g < f_2g$.
    (c) If $b_1 = c_1 = 0$ and $g = p^u$ then $f_1g < f_2g$.

PROOF.   (b) We may assume $w = 1$. If $v(f_1) < v(f_2)$ then $v(f_1g) = v(f_1) + 1 < v(f_2) + 1 \leq v(f_2g)$ since f_1g is in standard form. So $f_1g < f_2g$. If $v(f_1) = v(f_2)$ and $c_1 < b_2 + c_2$ then $v(f_1g) \leq v(f_2g)$ as above. Further $c_1 + 1 < b_2 + c_2 + 1$ so $f_1g < f_2g$. The last case $v(f_1) = v(f_2), c_1 = b_2 + c_2$ and $b_2 > 0$ is similar.

The proofs for (a) and (c) are similar to that of (b). ∎

PROPOSITION 21. *Let (A, m) be an Artin ring and let $p \in m \backslash m^2$ be prime. Then $d(A[Z/pZ \oplus Z/pZ]) \leq p + 2$.*

PROOF.   Let $\{f_1, ..., f_t\}$ be a set of generators of I of minimal depth. We may assume that if a_kv_k is the leading term of f_i then $a_k = 1$. Let $E = \{(a, b, c) \mid p^a y_1{}^b y_2{}^c$ is the leading term of f_i for some $i\}$. Let $E_x = \{(a, b, c) \in E \mid a = 0, b \geq 1, c < p\}$. If $(0, b_1, c), (0, b_2, c) \in E_x, b_1 \leq b_2$ represent exponents of the leading monomials of f_1, f_2 respectively then, replacing f_2 with

$f_2 - y_1{}^{b_2-b_1} f_1$ yields a generating set of lower depth. Therefore the cardinality of E_x is at most p.

Let $E_y = \{(a,b,c) \in E \mid a = b = 0, \, c \geq 1\}$. If $(0,0,c_1)$, $(0,0,c_2) \in E_y$, $c_1 \leq c_2$ represent exponents of the leading monomials of f_1, f_2, respectively, then, replacing f_2 with $f_2 - y_2{}^{c_2-c_1} f_1$, yields a generating set of lower depth. Thus the cardinality of E_y is at most 1. Finally, let $E_p = \{(a,0,0) \in E\}$. If $(a_1,0,0)$, $(a_2,0,0) \in E_p$, $a_1 \leq a_2$ represent exponents of the leading monomials of f_1, f_2, respectively, then replacing f_2 with $f_2 - p^{a_2-a_1} f_1$ yields a generating set of lower depth. Therefore the cardinality of E_p is at most 1. We now have that the cardinality of $E = E_x \cup E_y \cup E_p$ is at most $p + 2$. ∎

The upper bound given in Proposition 21 does not give the Dilworth number in general. The next theorem shows that when $m^2 \neq 0$ the upper bound $p + 2$ is the Dilworth number of $A[Z/pZ \oplus Z/pZ]$. Part (c) of the next theorem and Theorem 17 show that Theorem 11 is not true for $\mu(m) > 2$ even if A is Gorenstein. Theorem 22 has recently been extended [23] to the case where p is ramified and $G = (Z/pZ)^k$.

THEOREM 22. *[19, Theorem 6] Let* (A, m) *be a principal Artin local ring, let* $p \in m \backslash m^2$ *be a prime integer, and let* $R = A[Z/pZ \oplus Z/pZ]$. *Then*

a) $sp(R) = \begin{cases} p+2 & \text{if } m^p \neq 0 \\ p+1 & \text{if } m \neq 0 \text{ and } m^p = 0 \\ p & \text{if } m = 0 \end{cases}$.

b) $d(R) = \begin{cases} p+2 & \text{if } m^2 \neq 0 \\ p+1 & \text{if } m \neq 0 \text{ and } m^2 = 0 \\ p & \text{if } m = 0 \end{cases}$.

c) *If* $p > 2$, $m^2 \neq 0$ *and* $m^p = 0$ *then* $d(R) > sp(R)$.

We conclude this section by determining which group rings over a principal Artin local ring have the n-generator property with $n \leq p$ and with the 4-generator property with no restrictions on p. The first result appears in [4] and the second in [3]. Recent progress in this area allow us to give simpler proofs of these results. We now revert to the standard notation $M = (m, 1 - x)$ for the maximal ideal of the group ring $A[G]$ where (A, m) is a local ring, $p \in m$ and G is a cyclic p-group..

PROPOSITION 23. *[4, Proposition 13] Let* (A, m) *be an Artin local ring, let* $p \in m$ *be a rational prime, and let* G *be a finite abelian* p-*group. Then* $d(A[G]) \geq d(A) + d((A/m)[G])$ *and* $sp(A[G]) \geq \mu(m) + d((A/m)[G])$.

LEMMA 24. *Let* (A, m) *be a principal Artin local ring which is not a field with* $2 \in m$ *and let* $R = A[Z/2Z \oplus Z/2Z]$. *Then* $d(R) \leq 4$.

PROOF. If $2 \in m \backslash m^2$ then we are done by Theorem 22. So assume $2 \in m^2$. Let $M = (r, M_G)$ be the maximal ideal of R where $M_G = (1 - x, 1 - y)$. Then, since $(1 - x)^2 = (2(1 - x))$, $M^2 = (r^2, (1 - x)(1 - y)) + rM_G$. Note that $\mu(M^2) = 4$. Further,

$$
\begin{aligned}
M^3 &= (r^3, r^2 M_G, r M_G^2, M_G^3) \\
&= (r^3, r^2 M_G, r(1 - x)(1 - y)) = rM^2.
\end{aligned}
$$

Now let I be an ideal of R. Then, by [16, Lemma 2], $\mu(I) \leq \mu(I + M^2)$. So we can assume $I \supseteq M^2$. If $I = M^2$ we are done, so assume $I \supset M^2$. Let $w \in I \backslash M^2$. By [12, Theorem 158], $\mu(M/(w)) \leq \mu(M) - 1 = 2$. Therefore $M = (w, r, 1 - x)$, $M = (w, r, 1 - y)$ or $M = (w, 1 - x, 1 - y)$. If $M = (w, r, 1 - x)$ then, in $R/(w)$, $\overline{M} = (\overline{r}, \overline{(1 - x)})$ and $\overline{M}^2 = (\overline{r}^2, \overline{(1 - x)}^2)$. Thus $d(R/(w)) \leq 2$ by Theorem 12. Therefore $\mu(I/(w)) \leq 2$ and hence $\mu(I) \leq 3$. The case $M = (w, r, 1 - y)$ is similar. So assume $M = (w, 1 - x, 1 - y)$. Then $\overline{2} \in \overline{M}^2 = \left(\overline{2(1 - x)}, \overline{2(1 - y)}, \overline{(1 - x)(1 - y)} \right)$. Therefore

$$
\begin{aligned}
\overline{M}^2 &= \left(\overline{(1 - x)(1 - y)} \right) + \overline{M}^3 \\
&= \left(\overline{(1 - x)(1 - y)} \right) + \overline{M}^4 = \ldots = \left(\overline{(1 - x)(1 - y)} \right) + \overline{M}^n
\end{aligned}
$$

for all $n \geq 3$. Since $\overline{M}^N = 0$ for some N we have $\overline{M}^2 = \left(\overline{(1 - x)(1 - y)} \right)$. Thus $\mu(\overline{M}^2) \leq 2$ and $\mu(I) \leq 3$ as before. ∎

The next proposition determines which group rings have the n-generator property for $n \leq p$. The ramification index $e(p)$ of a prime integer p and the integer s^* are used in it's statement and proof. They were formally defined in the remarks preceding Proposition 13.

PROPOSITION 25. *[4, Proposition 14] Let (A, m) be a principal Artin local ring, let $p \in m$ be a rational prime and let G be a non-trivial finite p-group. Then $A[G]$ has the n-generator property for $n \leq p$ if and only if*
(1) *If A is a field then $G \cong \begin{cases} Z/pZ \oplus Z/p^j Z \text{ or } Z/p^j Z & \text{if } n = p \\ Z/p^j Z & \text{if } n < p \end{cases}$.*
(2) *If A is not a field then G is cyclic and, if $\ell(A) > n$, then $\min\{e(p)j + 1, p^j\} \leq n$. In particular, if $n < p$ then $e(p)j \leq n - 1$.*

PROOF. We first assume $A[G]$ has the n-generator property and A is a field. Note that the $(p-1)st$ power of the maximal ideal of $A[Z/pZ \oplus Z/pZ \oplus Z/pZ]$ requires more than p generators. Thus G has at most two cyclic summands. If $G \cong Z/p^2 Z \oplus Z/p^j Z$ with $j \geq 2$ then $d(A[G]) = p^2$ by Corollary 16. Therefore (1) holds.

Now assume $A[G]$ has the n-generator property and A is not a field. Since $d\left((A/m^2)[Z/pZ \oplus Z/pZ]\right) \geq d(a) + d(A/m[Z/pZ \oplus Z/pZ]) \geq 1 + p$, G is cyclic. So assume $G = Z/p^j Z$. By Theorem 15, $\min\{s^*, \ell(A)\} \leq n$. If $\ell(A) > p$ then we must have $\min\{e(p)(j-s) + p^s \mid 0 \leq s \leq j\} = s^* \leq n \leq p$. Therefore $s^* = e(p)j + 1$ or p^j. So we must have $\min\{e(p)j + 1, p^j\} \leq n$.

The converse is immediate from Corollary 16 when A is a field and Theorem 15 when A is not a field. ∎

The above proposition shows that for $n \leq p$ there is a general technique for characterizing group rings $A[G]$ with the n-generator property when A is a principal Artin ring and G is a finite abelian p-group. However, for $p < n$ there are currently only ad-hoc techniques. Indeed, the characterization has been given only in the cases $n = 2, 3, 4$. The next theorem, due to Hassani and Kabbaj, gives the technical cases of $p = 2, 3$ to complete the characterization when $n = 4$.

THEOREM 26. *[2, Theorem]* *Let (A, m) be a principal Artin local ring which is not a field, let $p = 2$ or 3 belong to m and let G be a non-trivial finite p-group.*

(a) If $p = 2$ then $R = A[G]$ has the 4-generator property if and only if

$$G_2 \cong \begin{cases} Z/2Z \oplus Z/2^i Z \text{ or } Z/2^i Z, i \geq 1 & \text{if } m^2 = 0. \\ Z/2Z \oplus Z/2Z \text{ or } Z/2^i Z, i \geq 1 & \text{if } m^2 \neq 0 \text{ and } m^4 = 0. \\ Z/2^i Z \text{ with } 1 \leq i \leq 3 & \text{if } m^4 \neq 0 \text{ and } e(p) = 1. \\ Z/2^i Z \text{ with } 1 \leq i \leq 2 & \text{if } m^4 \neq 0 \text{ and } e(p) \geq 2. \end{cases}$$

(b) If $p = 3$ then $R = A[G]$ has the 4-generator property if and only if

$$G_3 \cong \begin{cases} Z/3Z \oplus Z/3Z \text{ or } Z/3^i Z, i \geq 1 & \text{if } m^2 = 0 \text{ and } e(3) = 1. \\ Z/3^i Z \text{ with } i \geq 1 & \text{if } m^4 = 0 \text{ and } m^2 \neq 0. \\ Z/3^i Z \text{ with } 1 \leq i \leq 3 & \text{if } m^4 \neq 0 \text{ and } e(3) = 1. \\ Z/3Z & \text{if } m^4 \neq 0 \text{ and } e(3) \geq 2. \end{cases}$$

PROOF. (a) We first note that $d(A[Z/2Z \oplus Z/2Z \oplus Z/2Z]) > 4$. For this let $M = (r, 1-x, 1-y, 1-z)$ be the maximal ideal of $A[Z/2Z \oplus Z/2Z \oplus Z/2Z]$ using the identification $A[Z/2Z \oplus Z/2Z \oplus Z/2Z] \cong A[x, y, z]/(1-x^2, 1-y^2, 1-z^2)$. Since $(1-x)^2 = 2(1-x)$, $(1-y)^2 = 2(1-y)$ and $(1-z)^2 = 2(1-z)$, $M^2 = (r^2, r(1-x), r(1-y), r(1-z), (1-x)(1-y), (1-x)(1-z), (1-y)(1-z))$. If $r(1-x)$ can be expressed in terms of the other generators, then apply the map sending $y \to 1$ and $z \to 1$ to get the contradiction $r \in (r^2)$. A similar argument shows $r(1-y)$ and $r(1-z)$ cannot be expressed in terms of the other generators. If $(1-x)(1-y)$ is redundant we apply the map sending $z \to 1$ to get $1 \in (r)$, a contradiction. Thus $d(A[Z/2Z \oplus Z/2Z \oplus Z/2Z]) \geq 6$. Therefore G has at most 2 cyclic summands.

Since $d(A[Z/4Z \oplus Z/4Z]) \geq 1 + d((A/m)[Z/4Z \oplus Z/4Z]) = 5$, G is a homomorphic image of $Z/2Z \oplus Z/2^i Z$.

If $m^2 \neq 0$ we show that $d(A[Z/2Z][Z/4Z]) > 4$. Using the notation of Theorem 13, $s^* = \min\{e(p)2 + 1, e(p) + 2, 4\} \geq 3$. Let $M = (r, 1 - x)$ be the maximal ideal of $A[Z/2Z]$. Then, as above, $\{r^2, r(1-x)\}$ is a minimal generating set for M^2. Thus, by Theorem 13, $d(A[Z/2Z][Z/4Z]) \geq \ell(A[Z/2Z]/M^3) \geq 5$. Therefore $G \cong Z/2Z \oplus Z/2Z$ or cyclic.

If $m^4 \neq 0$ and $e(p) = 1$ then $s^* = \min\{4+1, 3+2, 2+4, 8\}$ for the ring $A[Z/16Z]$. Thus $d(A[Z/16Z]) = \min\{\ell(A), 5\} \geq 5$. Thus G is a homomorphic image of $Z/8Z$. If $e(p) \geq 2$ then $d(A[Z/8Z]) = \min\{\ell(A), \min\{e(p)3+1, e(p)2+2, e(p)+4, 8\}\} \geq 5$. Therefore G is a homomorphic image of $Z/4Z$.

For the converse, $d(A[Z/2Z + Z/2^iZ]) \leq \ell(A[Z/2Z]) \leq 4$ if $m^2 = 0$. If $m^2 \neq 0$ then $d(A[Z/2Z \oplus Z/2Z]) = 4$ by Lemma 24.

If G is cyclic and $m^4 = 0$ then $d(A[Z/2^iZ]) \leq \ell(A) \leq 4$. If $m^4 \neq 0$ and $e(p) = 1$ then $s^* = \min\{3+1, 2+2, 1+4, 8\} = 4$ for the ring $A[Z/8Z]$. Thus $d(A[Z/2^iZ]) \leq 4$ for $1 \leq i \leq 3$. Finally, if $e(p) \geq 2$ then $s^* = \min\{e(p)2+1, e(p)+2, 4\} \leq 4$ for the ring $A[Z/4Z]$. Thus $d(A[Z/2^iZ]) \leq 4$ for $1 \leq i \leq 2$.

This completes the proof of (a).

(b) We note that $d((Z/3Z)[Z/3/Z \oplus Z/3Z \oplus Z/3Z]) = 7$ by Theorem 10. Therefore G has at most two cyclic summands. If $R = A[Z/3Z][Z/9Z]$ then $s^* = \min\{e(3)2+1, e(p)+3, 9\} \geq 3$. Therefore $d(R) \geq \ell(A[Z/3Z]/(m, 1-x)^3) \geq 5$. So if G is not cyclic, $G \cong Z/3Z \oplus Z/3Z$. If $e(p) \geq 2$ then for the ring $A[Z/3Z][Z/3Z]$, $s^* = \min\{e(p)+1, 3\} \geq 3$. As before $d(A[Z/3Z \oplus Z/3Z]) \geq 5$. Thus $e(p) = 1$. Finally, if $m^2 \neq 0$, $d(A[Z/3Z \oplus Z/3Z]) > sp(A[Z/3Z \oplus Z/3Z]) \geq 4$ by Proposition 22.

If G is cyclic and $\ell(A) \geq 4$ then

$$d(A[Z/3^iZ]) = \min\{e(p)i + 1, e(p)(i-1) + 3, ..., 3^i\}.$$

If $e(p) \geq 2$ then $i = 1$ and, if $e(p) = 1$, then $i \leq 3$.

The converse is immediate from Theorem 22 and Theorem 15. ■

REFERENCES

[1]   S. Ameziane Hassani, M. Fontana and S. Kabbaj, Group Rings R[G] with 3-generated ideals when R is Artinian, Comm. in Algebra 24(4)(1996), 1253-1280.

[2]   S. Ameziane Hassani, and S. Kabbaj, Group Rings R[G] with 4-generated ideals when R is an Artinian principal ideal ring, Lect. Notes Dekker 185(1997) 1-14.

[3]   S. Ameziane Hassani and S. Kabbaj, Group Rings R[G] with n-generated ideals when R is an Artinian ring with the 2-generator property, manuscript.

[4]   S. Ameziane Hassani, S. Kabbaj, J. Okon and P. Vicknair, The Dilworth number of group rings over an Artin local ring, Comm. in Algebra, to appear.

[5]   J. T. Arnold and R. Matsuda, The n-generator property for semigroup rings, Houston J. Math. 12 (1986), 345-356.

[6]   H. Bass, On the ubiquity of Gorenstein rings, Math. Z. 82 (1963), 8-28.

[7]   F. Decruyenaere, E. Jespers and P. Wauters, On commutative principal ideal semigroup rings, Semigroup Forum 43 (1991) 367-377.

[8]   S. Eilenberg and T. Nakayama, On the dimension of modules and algebras II, Nagoya Math. J. 9 (1955), 1-16.

[9]   R. Gilmer, Commutative Semigroup Rings, University of Chicago Press, Chicago 1984.

[10]   C. Greither, On the two generator problem for ideals of a one-dimensional ring, J. Pure Appl. Algebra 24 (1982), 265-276.

[11]   H. Ikeda, Results on Dilworth and Rees numbers of Artinian local rings, Japan J. Math., 22 (147) (1996), 147-158.

[12]   I. Kaplansky, Commutative rings, University of Chicago Press, Chicago, 1974.

[13]   D. Lantz, Preservation of local properties and chain conditions in commutative group rings, Pacific J. Math. 63 (1) (1976), 193-199.

[14]   R. Matsuda, Torsion free abelian semigroup rings V, Bull. Fac. Sci., Ibaraki Univ. 11(1979), 1-37.

[15]   R. Matsuda, n-Generator property of a polynomial ring, Bull. Fac. Sci., Ibaraki Univ., Series A Math. 16 (1984), 17-23.

[16]   K. McLean, Local rings with bounded ideals, J. Algebra 74 (1982), 328-332.

[17]   J. Okon and P. Vicknair, Group rings with n-generated ideals, Comm. in Algebra, 20(1992), 189-217.

[18]   J. Okon and P. Vicknair, One-dimensional monoid rings with n-generated ideals, Canadian Math. Bull. 36(3) (1993), 344-350.

[19]   J. Okon and P. Vicknair, A Gorenstein ring R with embedding dimension 3 and larger Dilworth number than Sperner number, Canadian Math. Bull., to appear.

[20]   J. Okon, D. Rush and P. Vicknair, Commutative semigroup rings with two-generated ideals, J. London Math. Soc. (2) 45 (1992), 417-432.

[21]   J. Okon, D. Rush and P. Vicknair, Two-generated ideals in non-Noetherian semigroup rings, J. Pure Applied Algebra 111 (1996), 255-276.

[22]   J. Okon, D. Rush and P. Vicknair, Numbers of generators of ideals in zero-dimensional group rings, Comm. in Algebra., 25(3) (1997), 803-831.

[23]   J. Okon, D. Rush and P. Vicknair, The Sperner and Dilworth number of a one-dimensional group ring, manuscript.

[24]   D. Rush, Rings with two-generated ideals, J. Pure Appl. Algebra, 73 (1991), 257-275.

[25]   J. Sally, Numbers of generators of ideals in local rings, Lecture Notes in Pure and Applied Mathematics 35, Marcel Dekker, New York/Basel 1978.

[26]   H. Sekiguchi, The upper bound of the Dilworth number and the Rees number of Noetherian local rings with a Hilbert function, Advances in Math., 124 (1996), 197-206.

[27]   A. Shalev, Dimension subgroups, nilpotency indices, and the number of generators of ideals in p-group algebras, J. Algebra, 129 (1990), 412-438.

[28]   J. Watanabe, The Dilworth number of Artinian rings and finite posets with rank function, Commutative Algebra and Combinatorics, Advanced Studies in Pure Mathematics, Vol. 11, 303-312, Kinokuniya Co./North-Holland, Amterdam 1987.

[29]   J. Watanabe, The Dilworth number of Artin Gorenstein Rings, Advances in Math., 76(1989), 194-199.

About GCD Domains

GABRIEL PICAVET   Laboratoire de Mathématiques pures
Université Blaise Pascal (Clermont II) 63177 Aubière Cedex, France
e-mail: picavet@ucfma.univ-bpclermont.fr

0. INTRODUCTION AND NOTATION

In this paper, we present results on GCD-domains. Most of them are new; but we also improve known results. We use mainly two tools : some topologies defined on the spectrum of a ring and the notion of true grade (or polynomial grade) defined by M. Hochster and refined by D. G. Northcott in his book [19]. In the non-noetherian case, the notion of true grade is the one to consider. If I is an ideal of a ring A and M an A-module, we denote by $\mathrm{Gr}_A(I, M)$ the true grade of I with respect to M. When A is a local ring, the polynomial depth of A is defined to be $\mathrm{Pdp}(A) = \mathrm{Gr}_{A_P}(P \cdot A_P, A_P)$. Our theory is centered on representations of an integral domain A. A subset X of $\mathrm{Spec}(A)$ defines a representation if $A = \cap [A_P; P \in X]$. It appears that representations are linked with topological properties of X.

There are many ways to define associated primes of A-modules. Bourbaki weakly associated primes provide a representation. If we denote by $\mathrm{Specass}(A)$ the set of all prime ideals associated to some A/Aa where a is a nonunit in A, then $\mathrm{Specass}(A)$ defines a representation. D. G. Northcott defined prime ideals attached to a module. If we define $\mathrm{Specatt}(A)$ to be the set of all prime ideals attached to some A/Aa where a is a nonunit in A, then $\mathrm{Specatt}(A)$ defines another representation. Let A be a GCD-domain and $\mathrm{Specess}(A)$ the set of all prime ideals P such that A_P is a valuation ring. P. B. Sheldon proved that $\mathrm{Specess}(A)$ defines a representation. Here are the main results of section 1. If A is an integral domain and P a nonzero prime ideal of A, then P lies in $\mathrm{Specatt}(A)$ if and only if $\mathrm{Gr}_{A_P}(P \cdot A_P) = 1$ while if A is a GCD-domain, P lies in $\mathrm{Specess}(A)$ if and only if $\mathrm{Gr}_A(P) = 1$. It follows that $\mathrm{Specatt}(A) = \mathrm{Specess}(A)$ when A is a GCD-domain. Hence a GCD-domain is an intersection of polynomial depth-one local domains.

In section 2, we use topological tools. There are many topologies defined on a spectrum such as the Zariski topology or the patch topology (in french : topologie constructible). This last topology, defined by A. Grothendieck and J. Dieudonné in the E.G.A (Edition Springer Verlag), is compact. M. Hochster defined the opposite topology on a spectrum. If A is a ring and $X \subset \mathrm{Spec}(A)$, then a prime ideal P belongs to the closure of X if and only if $X \cap \mathrm{V}(I) \neq \emptyset$ for every finitely generated ideal $I \subset P$. In [22] we showed that X is a closed set for the opposite topology if and only if X is the spectral image of a flat morphism $A \to B$. Thus it seems more convenient to call it the flat topology and to denote by $\overline{X}^p$ the closure of X with respect to the flat topology (the f-closure of X). A crucial result is $\overline{\mathrm{Specass}(A)}^p = \mathrm{Specess}(A)$ when A is a GCD-domain. Moreover, when A is a GCD-domain and $X \subset \mathrm{Spec}(A)$ defines a representation of A, we have $\mathrm{Specess}(A) \subset \overline{X}^p$. Assuming that A is a GCD-domain, we get that $\mathrm{Specass}(A)$ is stable under generalizations if and only if every prime ideal of A_P is branched for every $P \in \mathrm{Specass}(A)$ or also, if and only if A_P has a noetherian spectrum whenever $P \in \mathrm{Specass}(A)$. Now, if A is a GCD-domain and $X \subset \mathrm{Spec}(A)$ is a f-closed set defining a representation of finite character, then $\mathrm{Spec}(A)$ is finite and A is a Bézout ring. Our methods allow us to show that a treed GCD-domain (or verifying other properties) is a Prüfer domain.

The reader is referred to Gilmer's book for the theory of $\star$-operations. D.D. Anderson showed that for $X \subset \mathrm{Spec}(A)$ defining a representation of an integral domain, there is an associated $\star$-operation [1]. If A is a GCD-domain and $X \subset \mathrm{Spec}(A)$ defines a representation and its $\star$-operation, then A is said to be an X-Prüfer domain if every nonzero finitely generated ideal is $\star$-invertible. We show that if A is a GCD-domain, then A is an X-Prüfer domain if and only if $\overline{X}^p = \mathrm{Specess}(A)$. Therefore, a GCD-domain is a $\mathrm{Specass}(A)$-Prüfer domain and a $\mathrm{Specess}(A)$-Prüfer domain.

Next we go back to polynomial depth. Let A be a GCD-domain. Then the polynomial depth of A equals one if and only if $\mathrm{Max}(A) \subset \mathrm{Specess}(A)$. Now take a local GCD-domain A with noetherian spectrum and maximal ideal M and set $U = \mathrm{Spec}(A) \setminus \{M\}$. We get that the polynomial depth of A is ≥ 2 and $\mathrm{Pic}(U) = 0$ if and only if A is a parafactorial domain. Moreover, U is affine if and only if $M = \sqrt{Ad}$ for some $d \in A$ and in this case we get $\mathrm{Pic}(U) = 0$. When A is a local noetherian UFD-domain, then A is parafactorial if and only if $\mathrm{Pdp}(A) \geq 2$.

We begin by reviewing some basic facts and some elementary lemmata on GCD-domains.

Let A be a ring and $A[t]$ its polynomial ring where t is an indeterminate. If $f(t) \in A[t]$, then $\mathrm{C}(f(t))$ denotes the content ideal of A, generated by the coefficients of $f(t)$. The ring $A(t)$ is obtained by localizing $A[t]$ at the multiplicative subset S of all polynomials $f(t)$ such that $\mathrm{C}(f(t)) = A$.

If A is any ring, $\mathrm{U}(A)$ the set of all units in A and $a, b \in A$, then $a \approx b$ means that a and b are associates (there is some $u \in \mathrm{U}(A)$ such that $a = ub$).

If A is any ring and I an ideal of A, then $\mathrm{D}(I) = \{P \in \mathrm{Spec}(A) ; I \not\subset P\}$ while $\mathrm{V}(I)$ is its complement in $\mathrm{Spec}(A)$. When $I = Aa$, we set $\mathrm{V}(I) = \mathrm{V}(a)$.

Definition 0.1. If A is a GCD-domain and $I = (a_1, \ldots, a_n)$, we define the gcd-content of I to be $c(I) = A \gcd(a_1, \ldots, a_n)$ where $a_i \neq 0$. Obviously, the definition is independant of the generators of I as well as of the choice of a gcd. Observe that $I \subset c(I)$. If $f(t)$ is a polynomial in $A[t]$, the gcd-content $c(f(t))$ of $f(t)$ is the gcd-content of $C(f(t))$. We denote by $c_{f(t)}$ a generator of $c(f(t))$.

Definition 0.2. Let A be a GCD-domain. A polynomial $f(t) = \sum\limits_{i=1}^{n} a_i t^i$ is said to be primitive if $c(f(t)) = A$, that is to say $\gcd(a_0, \ldots, a_n) \approx 1$. Gauss' Lemma is well known to be true for GCD-domains (the product of two primitive polynomials is a primitive polynomial). For instance, see [15,exercise 8,p.42].

It follows easily that $c_{f(t)} c_{g(t)} \approx c_{f(t)g(t)}$ and $c(f(t)g(t)) = c(f(t))c(g(t))$ for $f(t)$, $g(t)$ in $A[t]$.

We recall that GCD-domains are stable under localizations at multiplicative subsets. If A is a GCD-domain, S a mutiplicative subset and $\frac{a_1}{s_1}, \ldots, \frac{a_n}{s_n} \in A_S \setminus \{0\}$, then we have $\gcd(\frac{a_1}{s_1}, \ldots, \frac{a_n}{s_n}) \approx \frac{\gcd(a_1, \ldots, a_n)}{1}$.

Lemma 0.3. *Let A be a GCD-domain, $0 \neq a \in A$ and $b_1, \ldots, b_n$ nonzero pairwise coprime elements in A; then $\gcd(a, b_1 b_2 \cdots b_n) \approx \prod\limits_{i=1}^{n} \gcd(a, b_i)$.*

Proof. The hypotheses imply that b_n and $b_1 \cdots b_{n-1}$ are coprime. Thus it is enough to give a proof for $n = 2$. Consider two nonzero coprime elements b, c and $d \approx \gcd(b, a)$, $\delta \approx \gcd(c, a)$. Setting $b = db'$ and $c = \delta c'$, then d and δ are coprime. Thus, there is some y such that $a = d\delta y$. Now b' and y are coprime and so are c' and y. Therefore, y and $c'b'$ are coprime. From $\gcd(y, c'b') \approx 1$, we deduce that $d\delta \approx \gcd(yd\delta, d\delta c'b') = \gcd(a, bc)$.

GCD-domains are stable under polynomial extensions. Let K be the quotient field of A. Then an atom of $A[t]$ is either an atom of A or a primitive polynomial $f(t)$ such that $f(t)$ is irreducible in $K[t]$. Denote by $\mathcal{P} = \{p(t)\}$ a set of representatives for non-constant atoms of $A[t]$ (with respect to $\approx$). If $p(t)$ belongs to $\mathcal{P}$, then $P = A[t]p(t)$ is a prime ideal such that $A[t]_P$ is a discrete valuation ring with valuation $v_{p(t)}$. Then a nonzero polynomial $f(t) \in A[t]$ can be written

$$f(t) \approx c_{f(t)} \prod_{p(t) \in \mathcal{P}} p(t)^{v_{p(t)}(f(t))}.$$

From 0.3, we get :

Lemma 0.4. *Let A be a GCD-domain and $f(t)$, $g(t) \in A[t]$. Then we have*

$$\gcd(f(t), g(t)) \approx \gcd(c_{f(t)}, c_{g(t)}) \prod_{p(t) \in \mathcal{P}} p(t)^{\min\{v_{p(t)}(f(t)), v_{p(t)}(g(t))\}}.$$

Therefore, if $f(t)$, $g(t)$ are coprime, so are $c_{f(t)}$ and $c_{g(t)}$.

Lemma 0.5. *Let A be a GCD-domain, a, $b \in A \setminus \{0\}$, $d \approx \gcd(a, b)$ and $a = da'$, $b = db'$ with $\gcd(a', b') \approx 1$. Then we have $(a) : (b) = (a')$. It follows that a and b are coprime if and only if $(a) : (b) = (a)$.*

Proof. Straightforward.

1. REPRESENTATIONS OF GCD-RINGS

Definition 1.1. Let A be an integral domain and $X \subset \text{Spec}(A)$. We say that X defines a representation of A if $A = \cap [A_P \; ; \; P \in X]$.

According to [9, exercise 22, p. 52], X defines a representation of A if and only if $X \subset D((a) : (b)) \Leftrightarrow (b) \subset (a)$ for all $a, b \in A$.

A representation defined by X is said to be of finite character if $X \cap V(a)$ is a finite subset of $\text{Spec}(A)$ for each nonzero element in A.

Many subsets of $\text{Spec}(A)$ are well known to define a representation of A (see examples below).

If M is an A-module, then $Z(M)$ is the set of all elements in A which are zero-divisors of M. Let A be a ring and $a \notin U(A)$. A maximal prime of a is a prime ideal, maximal with respect to the saturated multiplicative subset $A \setminus Z(A/Aa)$ of A. We denote by $\text{Specmax}(A)$ the set of all maximal primes of elements $a \notin U(A)$.

We will have also to consider the following subsets of $\text{Spec}(A)$:
- $\text{Spec}_1(A)$ the set of all prime ideals with height ≤ 1.
- $\text{Specdv}(A)$ the set of all prime ideals P such that A_P is a discrete valuation ring (here, we consider that a field is a discrete valuation ring).
- $\text{Max}(A)$ the set of all maximal ideals.

Examples 1.2. *Let A be an integral domain.*

(1) $\text{Max}(A)$ *defines a representation. Notice that $\text{Max}(A)$ is quasi-compact.*

(2) *If A is a Krull domain, then $\text{Spec}_1(A) = \text{Specdv}(A)$ defines a representation of finite character.*

(3) $\text{Specmax}(A)$ *defines a representation* [15,Theorem 53].

In this section, we are attempting to show that a domain has other interesting representations. We need some definitions.

Definition 1.3. Let A be any ring, M an A-module and P a prime ideal of A.

(1) P is said to be an assassinator of M if there is some $x \in M$ such that $P \in \text{Min}(V(0 : x))$ (see [5, exercise 17, p. 163] or [17]).

(2) P is said to be attached to M if for any finitely generated ideal $I \subset P$ there is some $x \in M$ such that $I \subset 0 : x \subset P$ (see [7], [19,p.178], [13] and [21,p.712]).

(3) The set of all assassinators of M is denoted by $\text{Ass}_A(M)$ and the set of all attached primes to M is denoted by $\text{Att}_A(M)$.

Definition 1.4. Let A be an integral domain, then :

(1) $\text{Specass}(A)$ is the set of all prime ideals P of A such that $P \in \text{Ass}_A(A/Aa)$ for some $a \notin U(A)$. Equivalently, P belongs to $\text{Specass}(A)$ if and only if there is some $a \notin U(A)$ and some $b \in A$ such that $P \in \text{Min}(V((a) : (b)))$.

(2) $\text{Specatt}(A)$ is the set of all prime ideals P of A such that $P \in \text{Att}_A(A/Aa)$ for some $a \notin U(A)$. Equivalently, P belongs to $\text{Specatt}(A)$ if and only if there is some $a \in U(A)$ such that for any finitely generated ideal $I \subset P$ there is some $b \in A$ such that $I \subset (a) : (b) \subset P$ (see [13]).

Notice that $0 \neq a \in P$ for a nonzero prime ideal P in $\text{Specass}(A)$ or $\text{Specatt}(A)$.

Actually, Northcott's original definition of attached primes uses the notion of true grade (or polynomial grade). If P is a prime ideal of A, then P is attached to M if and only if $\text{Gr}_{A_P}(P \cdot A_P; M_P) = 0$. We define the polynomial depth of A_P to be $\text{Pdp}(A_P) = \text{Gr}_{A_P}(P \cdot A_P)$. For more information on the true grade, the interested reader is referred to [19].

Theorem 1.5. *Let A be an integral domain and P a nonzero prime ideal of A. Then P belongs to $\text{Specatt}(A)$ if and only if $\text{Pdp}(A_P) = 1$.*

Proof. Let $0 \neq P$ be an element of $\text{Att}_A(A/Aa)$ where $a \notin \text{U}(A)$. In view of 1.4, a is a nonzero element lying in P. Now a is regular in A_P; we deduce from [19,Theorem 15,p.152] that $\text{Gr}_{A_P}(P \cdot A_P; A_P) = \text{Gr}_{A_P}(P \cdot A_P; (A/Aa)_P) + 1$ so that $\text{Pdp}(A_P) = 1$. Conversely, assume that $\text{Gr}_{A_P}(P \cdot A_P) = \text{Pdp}(A_P) = 1$; then P cannot equal 0. If not, $0 :_{A_P} 0 = A_P \neq 0$ shows that $\text{Gr}_{A_P}(P \cdot A_P) = 0$ [19,Lemma 8, p.149]. Let $a \neq 0$ in P, then a does not lie in $\text{U}(A)$. The above quoted formula shows that $\text{Gr}_{A_P}(P \cdot A_P; (A/Aa)_P) = 0$ whence $P \in \text{Att}_A(A/Aa)$.

Lemma 1.6. *If A is a domain, then $\text{Specass}(A) \subset \text{Specatt}(A)$.*

Proof. According to [21,p.712], we have $\text{Ass}_A(M) \subset \text{Att}_A(M)$ for any A-module M. It follows easily that $\text{Specass}(A) \subset \text{Specatt}(A)$.

Proposition 1.7. *If A is an integral domain, then $\text{Specass}(A)$ and $\text{Specatt}(A)$ define representations of A.*

Proof. The result for $\text{Specass}(A)$ is given by [12,2.0] quoting [26]. Then $\text{Specatt}(A)$ is also a representation of A by 1.6.

Lemma 1.8. *Let A be an integral domain and Q, P prime ideals of A. Then Q_P lies in $\text{Specatt}(A_P)$ (resp. $\text{Specass}(A_P)$) if and only if Q lies in $\text{Specatt}(A)$ (resp. $\text{Specass}(A)$) and $Q \subset P$.*

Proof. Let M be an A-module and P a prime ideal of A. If Q is a prime ideal, then $Q \cdot A_P$ belongs to $\text{Att}_{A_P}(M_P)$ if and only if $Q \subset P$ and $Q \in \text{Att}_A(M)$ [7]. It follows that if $Q \cdot A_P$ belongs to $\text{Specatt}(A_P)$, then $Q \subset P$ and $Q \in \text{Specatt}(A)$. A similar argumentation holds for $\text{Specass}(A)$. The converse is straightforward.

Proposition 1.9. *Let A be an integral domain. Then any prime ideal of A is a union of some elements of $\text{Specatt}(A)$ and any element of $\text{Specatt}(A)$ is a union of some elements of $\text{Specass}(A)$.*

Proof. We showed in [21,5.2] that an element of $\text{Att}_A(M)$ is a union of some elements of $\text{Ass}_A(M)$ for any A-module M. Thus we need only to show that $P \in \text{Spec}(A)$ is a union of elements belonging to $\text{Specatt}(A)$. Thanks to 1.8, we can assume that A is a local ring with maximal ideal P. If $a \in P$, then $\text{Att}_A(A/Aa)$ is not empty by [7,3.4] and a belongs to any element Q of $\text{Att}_A(A/Aa)$. The proof is complete since $Q \subset P$.

Definition 1.10. [24] Let A be a ring and $f(t) = a_0 + a_1 t + \cdots + a_n t^n \in A[t]$ where $a_n \neq 0$ and $n \geq 1$. Then $f(t)$ is said to be a Sharma polynomial when

$$\cap_{i=0}^{n-1} ((a_n) : (a_i)) = (a_n).$$

Lemma 1.11. *Let A be a GCD-domain and nonzero elements $a_0, \ldots, a_n \in A$. Then we have $\gcd(a_0, \ldots, a_n) \approx 1$ if and only if $\cap_{i=0}^{n-1}((a_n) : (a_i)) = (a_n)$. Therefore, in a GCD-domain, a Sharma polynomial is nothing but a primitive polynomial.*

Proof. Set $I = \cap_{i=0}^{n-1}((a_n) : (a_i)) = (a_n) : \overset{n-1}{\underset{i=0}{\Sigma}}(a_i) \supset (a_n)$ and $\gcd(a_0, \ldots, a_{n-1}) \approx \delta$. To begin, if $I = (a_n)$, then $(a_n) : (\delta) = (a_n)$. Let $d \approx \gcd(a_n, \delta) \approx \gcd(a_0, \ldots, a_n)$. Hence a_n and δ are coprime by 0.5 so that $d \approx 1$. Now, if $\gcd(a_0, \ldots, a_n) \approx 1$, then $\gcd(\delta, a_n) \approx 1$. Consider an element $a \in I$ so that $aa_i \in (a_n)$ for each i. Hence, a_n divides $\gcd(aa_0, \ldots, aa_{n-1}) \approx a\delta$ and by Euclid's lemma, we get $a \in (a_n)$ so that $I = (a_n)$.

Proposition 1.12. *Let A be an integral domain with quotient field K and let $f(t) = a_0 + \cdots + a_n t^n \in A[t]$ with $a_n \neq 0$ and $n \geq 1$. The following statements are equivalent :*

(1) *$f(t)$ is a Sharma polynomial.*
(2) *$A :_K C(f(t)) = A$.*
(3) *$\mathrm{Specatt}(A) \subset D(C(f(t)))$.*

Proof. Set $I = C(f(t))$ and assume that (1) holds. If $A :_K I \neq A$, there is some $x \in K \setminus A$ such that $xI \subset A$ and hence there is some $b_i \in A$ such that $xa_i = b_i$ for $i = 0, \ldots, n$. It follows that $x = b_n a_n^{-1}$ whence $b_n a_i = b_i a_n \in (a_n)$. Thus we have $b_n \in \cap((a_n) : (a_i)) = (a_n)$ contradicting $x \notin A$. Thus (1) implies (2). Now assume that (2) holds and let $a \in A$ be such that $aa_i \in (a_n)$ for $i = 0, \ldots, n-1$. Observe that $aa_n^{-1}a_i$ belongs to A for all index i. Therefore, aa_n^{-1} lies in $A :_K I = A$ so that $a \in (a_n)$. Hence (2) implies (1). Suppose that $I \subset P$ for some $P \in \mathrm{Specatt}(A)$. There are $a, b \in A$ such that $I \subset (a) : (b) \subset P$. Set $x = ba^{-1}$. Then x does not lie in A, for if not $b \in (a)$ gives $A = (a) : (b) \subset P$. From $bI \subset (a)$, we get $xI \subset A$. Thus we have shown that $A :_K I \neq A$. Conversely, assume that $A :_K I \neq A$. There is some $x = \frac{b}{a} \notin A$ such that $xI \subset A$. Then there is some $P \in \mathrm{Specatt}(A)$ such that $x \notin A_P$ by 1.7. Suppose that $(a) : (b) \not\subset P$. Then there is some $s \in A \setminus P$ such that $sb \in (a)$. In this case, we get $x \in A_P$, a contradiction. Therefore, we have $I \subset (a) : (b) \subset P$. Consequently, (3) is equivalent to (2).

Definition 1.13. *Let A be an integral domain A. Then $\mathrm{Specess}(A)$ is defined to be the set of all prime ideals P of A such that A_P is a valuation domain (here, a field is a valuation domain). An element of $\mathrm{Specess}(A)$ is also called an essential prime ideal.*

Sheldon proved the following result [25] (actually, Sheldon proved much more and these results will be recovered and completed).

Proposition 1.14. *Let A be a GCD-domain and P a prime ideal of A. Then P is an essential prime ideal if and only if $\gcd(x, y) \in P$ for any $x, y \neq 0$ lying in P. For instance, A_p is essential when p is an atom (actually, A_{A_p} is a discrete valuation domain).*

Theorem 1.15. *Let A be a GCD-domain and P a nonzero prime ideal of A. Then P belongs to $\mathrm{Specess}(A)$ if and only if $\mathrm{Gr}_A(P) = 1$.*

Proof. First observe that $\mathrm{Gr}_A(P) \geq 1$ when P is a nonzero prime ideal, in fact $\mathrm{gr}_A(P) \leq \mathrm{Gr}_A(P)$. Assume that $P \in \mathrm{Specess}(A)$. Then for every nonzero finitely generated ideal $I \subset P$, we have $c(I) \neq A$. According to [19,Theorem 9, p.184], we must have $\mathrm{Gr}_A(I) \leq 1$ (if not, $c(I) = A$). Then $\mathrm{Gr}_A(P) = \sup\{\mathrm{Gr}_A(I); I \subset P$ and I f.g.$\} \leq 1$ by [19,Theorem 11, p.149] so that $\mathrm{Gr}_A(P) = 1$. Conversely, suppose $\mathrm{Gr}_A(P) = 1$ and let $(a_0, \dots, a_n) = I \subset P$, $n \geq 1$, be a nonzero finitely generated ideal. If $c(I) = A$, there is a primitive polynomial $f(t) \in I[t] \subset P[t]$ and, consequently, an atom $p(t) \in P[t]$. Set $R = A[t]$ and $E = R/Rp(t)$. We get $\mathrm{Gr}_R(P[t], E) = \mathrm{Gr}_R(P[t]) - 1 = 0$ by [19,Theorem 15, p.152 and 5.5.3, p.149] whence $\mathrm{Gr}_R(I[t], E) = 0$ by [19,5.5.4]. According to [19,Lemma 8, p.149], we have $0 :_E I[t] \neq \{0\}$. Hence there is some $g(t) \in R \setminus Rp(t)$ such that $g(t)I[t] \subset Rp(t)$. Since $Rp(t)$ is a prime ideal, we get $I[t] \subset Rp(t)$ so that $I = \{0\}$, a contradiction. Thus, we have $c(I) \neq A$ and P lies in $\mathrm{Specess}(A)$. Indeed, take $I = (a, b)$ with $d \approx \gcd(a, b)$ and set $a = da'$, $b = db'$ with $\gcd(a', b') \approx 1$ so that $(a', b') \not\subset P$. From $d(a', b') \subset P$, we get $d \in P$. $\blacksquare$

Theorem 1.16. *If A is a GCD-domain, then $\mathrm{Specess}(A) = \mathrm{Specatt}(A)$. It follows that $\mathrm{Specess}(A[t]) = \mathrm{Specatt}(A[t])$.*

Proof. To begin, we show that a prime ideal $P \in \mathrm{Att}_A(A/Aa)$, $a \notin U(A)$, is stable under the gcd-operation. Consider $x, y \in P \setminus \{0\}$. There is some b in A such that $(x) + (y) \subset (a) : (b) \subset P$. Letting $d \approx \gcd(x, y)$, from $xb, yb \in (a)$ we get $b \gcd(x, y) \approx \gcd(bx, by) \in (a)$ so that $d \in (a) : (b) \subset P$. Therefore, $\gcd(x, y)$ belongs to P. Thus $\mathrm{Specatt}(A) \subset \mathrm{Specess}(A)$ holds by 1.14. Now, let P in $\mathrm{Specess}(A)$. Observe that A_P is a GCD-domain. Clearly, $P \cdot A_P$ lies in $\mathrm{Specess}(A_P)$. According to 1.15, we have $\mathrm{Gr}_{A_P}(P \cdot A_P) = 1$. Thus P belongs to $\mathrm{Specatt}(A)$ by 1.5. It follows that $\mathrm{Specess}(A) \subset \mathrm{Specatt}(A)$. $\blacksquare$

Hence we recover Sheldon's results and more. We summarize this as follows :

Theorem 1.17. *Let A be a GCD-domain.*

(1) *Any prime ideal of A is a union of some essential prime ideals of A.*
(2) *$\mathrm{Pdp}_A(P) = 1$ for any nonzero element of $\mathrm{Specess}(A)$.*
(3) *$\mathrm{Specess}(A)$ defines a representation of A.*

Proposition 1.18. *Let A be a GCD-domain and Q a prime ideal of $A[t]$.*

(1) *If $Q \neq \{0\}$ is such that $Q \cap A = \{0\}$, then $Q = A[t]p(t)$ for some atom $p(t) \in A[t]$.*
(2) *Q belongs to $\mathrm{Specess}(A[t])$ if and only if either $Q = P[t]$ for some $P \in \mathrm{Specess}(A)$ or $Q = A[t]p(t)$ for some atom $p(t) \in A[t]$.*

Proof. Let $Q \neq \{0\}$ be a prime ideal of $A[t]$ such that $Q \cap A = \{0\}$ and $q(t)$ a polynomial of least positive degree in Q. Set $q(t) = ap(t)$ where $a \in A$ and $p(t)$ is primitive so that $p(t)$ is a Sharma polynomial lying in Q. Then (1) follows from [24,Theorem 1]. We show (2). Let $p(t)$ be an atom in $A[t]$ and consider the prime ideal $Q = A[t]p(t)$. Then it is obvious that $Q \in \mathrm{Min}(\mathrm{V}(p(t) : (1)))$ so that $Q \in \mathrm{Specass}(A[t]) \subset \mathrm{Specess}(A[t])$. Hence to prove (2), we need only to consider nonzero prime ideals Q of $A[t]$ such that $Q \cap A = P \neq \{0\}$. If P is a nonzero element of Specess(A), then $\mathrm{Gr}_A(P) = 1$ implies $\mathrm{Gr}_{A[t]}(P[t]) = 1$, whence $P[t]$ lies in Specess($A[t]$). Conversely, let Q be a nonzero element of Specess($A[t]$) such that $P = Q \cap A \neq \{0\}$. Assume that $Q \neq P[t]$ and let $f(t) \in Q \setminus P[t]$. Clearly, we have $c_{f(t)} \notin Q$. Thus, there is some atom $p(t)$ dividing $f(t)$ such that $p(t) \in Q \setminus P[t]$. Setting $R = A[t]$ and $E = R/Rp(t)$, we get $\mathrm{Gr}_R(Q, E) = 0$ by a result used in 1.15, whence $0 :_E Q \neq 0$ by [19,Lemma 8, p.149]. Thus, there is some $g(t) \notin p(t)A[t]$ such that $Qg(t) \subset A[t]p(t)$. It follows that $Q = A[t]p(t)$ lies over $\{0\}$ in A, a contradiction. Therefore, we have $Q = P[t]$ and $\mathrm{Gr}_R(P[t]) = \mathrm{Gr}_A(P) = 1$ yielding $P \in$ Specess(A).

Let A be an integral domain with quotient ring K. We denote by $f(A)$ the set of all nonzero finitely generated fractional ideals. For any fractional ideal I of A, we set $I^{-1} = A :_K I$, $I_v = (I^{-1})^{-1}$ and $I_t = \cup[J_v \,|\, J \subset I \text{ with } J \in f(A)]$. Then a nonzero fractional ideal I of A is said to be a v-ideal (resp. t-ideal) if $I_v = I$ (resp. $I_t = I$). For an outline of these notions and related results, see for instance [14]. The following result is probably well known but does not seem to appear in the literature.

Proposition 1.19. *Let A be an integral domain. Then A is a GCD-domain if and only if I_v is a principal ideal of A for any nonzero finitely generated ideal I of A. In particular, when A is a GCD-domain, we have $I_v = c(I)$ for any nonzero finitely generated ideal I of A.*

Proof. Let $I = (a_1, \ldots, a_n)$ be an ideal of A generated by nonzero elements. Set $d \approx \gcd(a_1, \ldots, a_n)$. It is easy to see that $I^{-1} = Ad^{-1}$. Assuming that $I_v = Ad$, it follows that $c(I) = Ad$.

Corollary 1.20. *Let A be an integral domain. Then A is a GCD-domain if and only if $\mathrm{Pic}(A) = 0$ and every finitely generated v-ideal is invertible.*

Proof. Obviously, 1.19 provides one implication. Indeed, when I is an invertible fractionary ideal, we have $I_v = I$. Assume that $\mathrm{Pic}(A) = 0$ and that I_v is invertible when I is a nonzero finitely generated ideal of A. Then I_v is principal and A is a GCD-domain by 1.19.

Definition 1.21. *Let A be an integral domain. We denote by* Spect(A) *the set of all prime t-ideals together with $\{0\}$.*

The following result is proved by B. G. Kang [14 2.8, 2.9].

Proposition 1.22. *Let A be an integral domain.*

(1) *Each proper integral t-ideal I of A is contained in a maximal proper integral t-ideal M which is a prime ideal.*

(2) *Spect(A) defines a representation of A (actually, only maximal proper integral t-ideals are needed).*

Proposition 1.23. *Let A be an integral domain and $0 \neq P \in$ Spec(A).*

(1) *P is a t-ideal if $P \in$ Specatt(A).*

(2) *If A is a GCD domain, then $\mathrm{Spec}_1(A) \subset$ Specess(A) = Spect(A).*

Proof. Let $P \in \mathrm{Specatt}(A)$ and $I \subset P$ a nonzero finitely generated ideal. There are $a, b \in A$ such that $I \subset (a) : (b) \subset P$. It follows that $I \subset I_v \subset (a) : (b) \subset P$ because $(a) : (b)$ is a v-ideal. Hence, the proof of (1) is complete. The statement (2) seems to be well known but we give a proof. Assume that A is a GCD domain and let P be an essential ideal. We get $I_v = Ad \subset P$ for any finitely generated ideal $I \subset P$. Therefore, P is a t-ideal. Conversely, assume that P is a t-ideal and let $I \subset P$ be a finitely generated ideal. From $c(I) = I_v \subset P$, we deduce that P is stable under the gcd-operation. For $\mathrm{Spec}_1(A) \subset \mathrm{Spect}(A)$, read [11,Introduction].

Remark 1.24. According to [11], if P is a t-prime ideal in an integral domain A, so is $P[t]$ in $A[t]$. But there exist domains $A[t]$ in which there is a nonzero t-prime ideal Q such that $0 \neq (Q \cap A)[t] \neq Q$.

2. TOPOLOGICAL CONSIDERATIONS

If A is a ring, we recall that the generalization of a subset $X \subset \mathrm{Spec}(A)$ is $G(X) = \cap [O ; X \subset O$ and O is open$]$. A prime ideal P lies in $G(X)$ if and only if there is some $Q \in X$ such that $P \subset Q$. Obviously, X is stable under generalizations if and only if $G(X) = X$.

We defined and studied the flat topology on $\mathrm{Spec}(A)$ in [22]. This topology assigns to each subset $X \subset \mathrm{Spec}(A)$ the closure :

$$\overline{X}^p = \cap [O; \ X \subset O \ \text{and} \ O \ \text{is quasi-compact open}]$$

A subset X is closed for the flat topology (f-closed) if and only if there is some flat morphism $f : A \to B$ such that $^a f(\mathrm{Spec}(B)) = X$ or also, if and only if X is quasi-compact and stable under generalizations. If $\mathrm{Spec}(A)$ is noetherian, a subset X of $\mathrm{Spec}(A)$ is f-closed if and only if X is stable under generalizations. Moreover, there is a canonical way to get $\overline{X}^p$. If $A[t]$ is a polynomial ring, then the subset of all polynomials $f(t) \in A[t]$ such that $X \subset D(C(f(t)))$ is a saturated multiplicative subset S of $A[t]$. We set $X(A) = A[t]_S$ and get a flat morphism $A \to X(A)$ with spectral image $\overline{X}^p$. Moreover, we have $\overline{X}^p(A) = X(A)$. When A is an integral domain, there is a factorization $A \to A(t) \to X(A)$ by injective morphisms for any $X \subset \mathrm{Spec}(A)$.

We introduce rings which have already been considered by J. A. Huckaba, I. J. Papick [12] and B. G. Kang [14].

Definition 2.1. If A is an integral domain, then denote by U (resp. V) the set of all polynomials $f(t) \in A[t] \setminus A$ such that $(C(f(t)))^{-1} = A$ (resp. $(C(f(t)))_t = A$). Then U and V are saturated mulplicative subsets of $A[t]$. We set $A\{t\} = A[t]_U$ and $A\langle t \rangle = A[t]_V$.

Obviously, there is a factorization $A \to A(t) \to A\{t\} \to A\langle t \rangle$ by injective morphisms and $A \to A\{t\}$, $A \to A\langle t \rangle$ are flat morphisms while $A \to A(t)$ is faithfully flat.

Moreover, $A\{t\}$ is a Bézout domain when A is either a GCD domain or a Krull domain or an integrally closed coherent domain [12,3.1].

Let X be a subset of $\mathrm{Spec}(A)$. We denote by $X[t]$ the set of all prime ideals $P[t]$ of $A[t]$ such that $P \in X$ and we set $\mathfrak{U}(X) = \cup [P ; P \in X]$.

Theorem 2.2. *Let A be an integral domain.*

(1) *U is the set of all Sharma polynomials in $A[t]$ and $A\{t\} = \operatorname{Specatt}(A)(A)$. Moreover, we have $A[t] \setminus U = \mathfrak{U}(\operatorname{Specass}(A)[t]) = \mathfrak{U}(\operatorname{Specatt}(A)[t])$.*

(2) $A[t] \setminus V = \mathfrak{U}(\operatorname{Spect}(A)[t])$.

(3) *If A is a GCD domain, then we have*

$$A\{t\} = A\langle t\rangle = \operatorname{Specess}(A)(A) = \operatorname{Specass}(A)(A).$$

Proof. To begin with, U is the set of all Sharma polynomials by 1.12. We have $A[t] \setminus U = \mathfrak{U}(\operatorname{Specass}(A)[t])$ by [12,2.0]. Now set $X = \operatorname{Specatt}(A)$ and denote by S the saturated multiplicative subset of all polynomials $f(t)$ such that $X \subset D(C(f(t)))$. If $f(t) = a \in A$, then a is a unit by 1.9 while $f(t) \notin A$ is a Sharma polynomial by 1.12. It follows that $A\{t\} = A[t]_S$. According to [22,IV,Lemme 2], we have $A[t] \setminus S = \mathfrak{U}(X[t])$. This ends the proof of (1), because S and U are saturated. Now (2) is a consequence of [14,2.1]. The first part of (3) follows from (1), (2), 1.16 and 1.23.

Let A be a GCD-domain and T the saturated multiplicative subset of $A[t]$ defined as follows : $f(t)$ belongs to T if and only if $f(t)$ is primitive (T is a saturated multiplicative subset by the formula $c(f(t)g(t)) = c(f(t))c(g(t))$).

Theorem 2.3. *Let A be a GCD-domain.*

(1) $A\{t\}$ *is the localization of $A[t]$ at T.*

(2) *Let $P \neq \{0\}$ be a prime ideal of A. Then $P \in \operatorname{Im}\{\operatorname{Spec}(A\{t\}) \to \operatorname{Spec}(A)\}$ if and only if $c(I) \neq A$ for any finitely generated nonzero ideal $I \subset P$.*

(3) $\operatorname{Specess}(A) = \operatorname{Im}\{\operatorname{Spec}(A\{t\}) \to \operatorname{Spec}(A)\}$.

(4) $\operatorname{Specess}(A)$ *is quasi-compact and stable under generalizations.*

(5) $\overline{\operatorname{Specass}(A)}^{\,p} = \operatorname{Specess}(A)$.

(6) *Any prime ideal of A is a union of some elements of $\operatorname{Specess}(A)$.*

Proof. An appeal to 1.11, 1.12 and 2.2 proves (1). Let $Q \neq \{0\}$ be a prime ideal of $A[t]$. If $Q \cap A = \{0\}$, we know that $Q = A[t]p(t)$ where $p(t)$ is an atom and a primitive polynomial by 1.18. Thus $Q \cap T \neq \emptyset$ shows that Q does not provide a prime ideal of $A\{t\}$. If $P \neq \{0\}$ is a prime ideal of A, then we have $P[t] \cap T \neq \emptyset$ if and only if there is some finitely generated ideal $I \subset P$ such that $c(I) = A$. Now let $P = Q \cap A \neq \{0\}$ such that $Q \cap T = \emptyset$. Then we have $P[t] \cap T = \emptyset$. The above considerations show that (2) holds. We prove (3). Setting $Y = \operatorname{Im}\{\operatorname{Spec}(A\{t\}) \to \operatorname{Spec}(A)\}$, we clearly have $\operatorname{Specess}(A) \subset Y$. Let P be in Y, a, b nonzero elements in P, $d \approx \gcd(a,b)$ and $a = da'$, $b = db'$ where $\gcd(a',b') \approx 1$. Then (a',b') is not included in P so that $d(a',b') \subset P$ implies $d \in P$. Hence P belongs to $\operatorname{Specess}(A)$. We have (4) because $A \to A\{t\}$ is flat. Then we get $\overline{\operatorname{Specass}(A)}^{\,p} = \operatorname{Specess}(A)$ by 2.2, (3) and our earlier facts on flat topology. Thus the proof of (5) is complete. Now (6) follows from 1.9 and 1.16.

Now we examine when $\operatorname{Specass}(A)$ is stable under generalizations. Branched prime ideals are defined in Gilmer's book [9,17.3]. Among other characterizations, a prime ideal P of a valuation ring A is branched if and only if P is the radical of a principal ideal. Let A be a ring and $X \subset \operatorname{Spec}(A)$. Then X is said to be expanded (resp. fathomable) if an ideal (resp. finitely generated ideal) I of A is contained in some element of X whenever $I \subset \mathfrak{U}(X)$. Moreover, X is expanded if and only if X is quasi-compact and fathomable (see [21,2.2]). A ring A is said to be absolutely expanded (resp. fathomable) if every subset X of $\mathrm{Spec}(A)$ is expanded (resp. fathomable). A ring A is absolutely fathomable (resp. absolutely expanded) if and only if for any finitely generated ideal I (resp. prime ideal P) of A there is some $a \in A$ such that $\sqrt{I} = \sqrt{Aa}$ (resp. $P = \sqrt{Aa}$).

Lemma 2.4. *Let A be a valuation ring and P a nonzero prime ideal of A. Then P lies in $\mathrm{Specass}(A)$ if and only if P is branched.*

Proof. If P lies in $\mathrm{Specass}(A)$, then $P \in \mathrm{Min}(V((a) : (b)))$ for some $a, b \in A$. In view of 0.5, we get $P = \sqrt{Aa'}$ and P is branched. Conversely, if P is branched, then $P = \sqrt{Aa}$ so that $P \in \mathrm{Min}(V((a) : (1)))$. Thus P belongs to $\mathrm{Specass}(A)$.

Theorem 2.5. *Let A be a GCD-domain. The following statements are equivalent:*

   (1) *$\mathrm{Specass}(A)$ is stable under generalizations.*
   (2) *Every prime ideal of A_P is branched for any $P \in \mathrm{Specass}(A)$.*
   (3) *A_P is absolutely expanded for any $P \in \mathrm{Specass}(A)$.*
   (4) *$\mathrm{Spec}(A_P)$ is a noetherian space for every $P \in \mathrm{Specass}(A)$.*

If in addition $\mathrm{Spec}(A)$ is a noetherian space, then $\mathrm{Specass}(A) = \mathrm{Specess}(A)$.

Proof. Recall that every A_P is a valuation ring for $P \in \mathrm{Specass}(A) \subset \mathrm{Specess}(A)$ by 1.16 and 1.6. Then (1) $\Rightarrow$ (2) by 2.4 and 1.8. Since a prime ideal P in a valuation ring is branched if and only if $P = \sqrt{Aa}$ for some $a \in A$, we get that (2) $\Leftrightarrow$ (3) by [21,2.1]. Indeed, $P = \sqrt{Aa} \subset \mathfrak{U}(X)$ implies $P \in G(X)$. Now (3) $\Rightarrow$ (4) by [21,4.8]. Now assume that $B = A_P$ has a noetherian spectrum. Then for every ideal I of B, there is some finitely generated ideal J such that $\sqrt{I} = \sqrt{J}$. It follows that $P = \sqrt{Bb}$ for every prime ideal P of B because B is a Bézout ring. Thus every prime ideal of B is branched and (4) $\Rightarrow$ (2). To complete the proof, we show that (2) $\Rightarrow$ (1). Indeed, a branched prime ideal in A_P belongs to $\mathrm{Specass}(A_P)$ and 1.8 gives the conclusion. Now if $\mathrm{Spec}(A)$ is noetherian, so is $\mathrm{Spec}(A_P)$. It follows that $\mathrm{Specass}(A)$ is quasi-compact and stable under generalizations, whence is f-closed. Then $\mathrm{Specass}(A) = \mathrm{Specess}(A)$ is obtained from 2.3, (5).

Lemma 2.6. *Let A be a GCD-domain and let X be a subset of $\mathrm{Spec}(A)$ defining a representation of A. Let $I = (a_1, \ldots, a_n)$ be a nonzero finitely generated ideal of A. Then $X \subset D(I)$ implies $c(I) = A$. This implication is an equivalence when $X = \mathrm{Specess}(A)$ or $X = \mathrm{Specass}(A)$.*

Proof. It suffices to show that $c(I) \neq A \Rightarrow X \cap V(I) \neq \emptyset$. Assume that $c(I) \neq A$ and denote by K the quotient field of A. We set $I^{-1} = A :_K I$. Suppose that $I \not\subset P$ for any $P \in X$. If $\frac{b}{a} \in I^{-1} \subset K$, there is some $s \in I \setminus P$ which yields $s\frac{b}{a} \in A$. Therefore, we get $\frac{b}{a} = \frac{\alpha}{s} \in A_P$ for any $P \in X$. Since X defines a representation of A, we have $\frac{b}{a} \in A$ so that $I :_K A = A$. In view of 1.19, we get $I_v = c(I) = A$, a contradiction. Whence $I \subset P$ for some element in X. Now, if X is $\mathrm{Specass}(A)$ or $\mathrm{Specess}(A)$, assume that $c(I) = A$. We have $X \subset D(I)$ since an element of X is stable under the gcd-operation.

Theorem 2.7. *Let A be a GCD-domain and X a subset of $\mathrm{Spec}(A)$ with $0 \in X$. If X defines a representation of A, we have $\mathrm{Specess}(A) \subset \overline{X}^p$ and $X(A) \subset A\{t\}$.*

Proof. Let P be a nonzero element of $\mathrm{Specess}(A)$ and $I \subset P$ a nonzero finitely generated ideal. By 2.3 we get $\mathrm{c}(I) \neq A$. It follows that $X \cap \mathrm{V}(I) \neq \emptyset$ by 2.6. Thus P lies in $\overline{X}^p$. The second assertion follows since $X \subset \mathrm{D}(\mathrm{C}(f(t)))$ implies primitiveness of $f(t)$ by 2.4.

Remark 2.8. When A is a Krull GCD-domain, the proof of [12,3.1, (c)] shows that $\mathrm{Specass}(A) = \mathrm{Spec}_1(A) = \mathrm{Specdv}(A)$ is stable under generalizations. Therefore, we recover the facts that A_P has a noetherian spectrum (see 2.5) and $\overline{\mathrm{Specass}(A)}^p = \mathrm{Specess}(A)$ since $\mathrm{Specdv}(A) \subset \mathrm{Specess}(A)$.

Recall that an integral domain is said to be a Goldman domain if its quotient field is finitely generated over A. Then an integral domain A is a Goldman domain if and only if $\{0\}$ is open in $\mathrm{Spec}(A)$. Moreover, $A[t]$ is never a Goldman domain.

Proposition 2.9. *Let A be a Goldman domain and let $X \subset \mathrm{Spec}(A)$ be a closed set for the patch topology. If X defines a representation of finite character of A, then X has finitely many elements.*

Proof. Set $X' = X \setminus \{0\}$. If A is a Goldman domain, then X' is patch-closed if X is patch-closed. If X defines a representation of finite character, then $X' \cap \mathrm{V}(a)$ is finite for each nonzero element a. Observe that $\{\mathrm{V}(a)\}_{a \neq 0}$ defines an open cover of X' for the patch topology. By compactness of the patch topology, there is a finite subcover of X'. It follows that X is finite.

Theorem 2.10. *Let A be a Goldman GCD-domain. If there is a closed subset X of $\mathrm{Spec}(A)$ for the flat topology, defining a representation of finite character, then $\mathrm{Specess}(A)$ and $\mathrm{Spec}(A)$ are finite and A is a Bézout domain.*

Proof. A f-closed subset X is patch-closed [22] and $X = \overline{X}^p$. If X defines a representation of finite character, then $\mathrm{Specess}(A) \subset X$ by 2.7 and X is finite by 2.9. In that case, A is a Bézout ring [25,3.6].

Remark 2.11. There are Goldman GCD-domains as well as GCD-domains which are not Goldman domains. For instance, set $A = \mathbb{Z}_{p\mathbb{Z}}$ where p is a prime number. Then A is a Goldman GCD-domain. Now consider the ultraproduct domain $B = A_{\mathcal{U}}^{\mathbb{N}}$ where $\mathcal{U}$ is a nontrivial ultrafilter on $\mathbb{N}$. We showed that B is not a Goldman domain [20,p.77]. It is not difficult to see that an ultraproduct of a GCD-domain is a GCD-domain.

By lack of a best name, we say that a GCD-domain A verifies Bézout's property if $\gcd(a_1, \dots, a_n) \approx 1 \Leftrightarrow (a_1, \dots, a_n) = A$ for any finite set of nonzero elements $\{a_1, \dots, a_n\}$ of A, $n > 1$. Obviously, if A is a GCD-domain, then Bézout's property holds if and only if $Ad = (a_1, \dots a_n)$ for any finite set $a_1, \dots, a_n$ of nonzero elements with $d \approx \gcd(a_1, \dots, a_n)$. This can be translated into : A verifies Bézout's property if and only if $I = \mathrm{c}(I)$ for every nonzero finitely generated ideal I of A. Therefore, a GCD-domain A verifies Bézout's property if and only if A is a Bézout ring. Next we recover some well known results.

Proposition 2.12. *Let A be a GCD-domain. The following statements are equivalent :*

(1) *A verifies Bézout's property (equivalently, A is a Bézout domain).*
(2) *Specess(A) = Spec(A).*
(3) *A is a Prüfer ring.*

Proof. Assume that A verifies (1) and let I be a nonzero finitely generated ideal; we have $I = c(I)$. Therefore, every prime ideal is essential and (2) follows. If (2) holds, let I be a nonzero finitely generated ideal such that $c(I) = A$. Then we have $I \not\subset P$ for every $P \in$ Specess(A) by 2.3 so that $I = A$. Thus (1) holds. Clearly, (2) is equivalent to (3).

Proposition 2.13. *Let A be a GCD-domain such that every prime ideal belongs to* Min$(V(a))$ *for some $a \in A$. Then A is a Prüfer ring. In particular, a GCD-domain with Krull dimension ≤ 1 is a Prüfer ring.*

Proof. Observe that the hypothesis gives Specass(A) = Specess(A) = Spec(A).

Definition 2.14. Let A be an integral domain. Then A is said to be a finite factorization domain (FFD) if every nonzero element of A has a finite number of nonassociate divisors. Then A is a FFD-domain if and only if A is an atomic idf-domain [2].

In general, FFD-domains are not stable under localization at a multiplicative subset. Actually, if A is a GCD-domain, S a multiplicative subset and $a \in A \setminus \{0\}$, it can be proved that the number of classes (with respect to $\approx$) of divisors of a does not increase in A_S.

Proposition 2.15. *Let A be a GCD-domain which is locally FFD. Then we have* Specdv(A) = Specess(A).

Proof. If P is an element of Specess(A), then A_P is a valuation ring. In this case, A_P is a factorial valuation domain, whence a principal ideal domain.

Remark 2.16. Let A be an integral domain. Then A is said to be a generalized GCD-domain (GGCD-domain) if $(a) \cap (b)$ is invertible for each nonzero $a, b \in A$. Then a GGCD-domain is locally a GCD-domain [8,6.2.2]. Morever, coherent regular domains are locally GCD-domains. Also, A is a GGCD-domain if and only if $A(t)$ is a GCD-domain. If A is an integral domain, then we define Specgcd(A) to be the set of all prime ideals of A such that A_P is a GCD-domain. Clearly, Specgcd(A) is stable under generalizations. The next result exhibits a flat embedding of an integral domain in a locally GCD-domain.

Proposition 2.17. *Let A be an integral domain. Assume that $X =$ Specgcd(A) is quasi-compact (for instance, if Spec(A) is noetherian). Then $A \to X(A)$ is an injective flat morphism such that $X(A)$ is a locally GCD-domain. Moreover, if P is a prime ideal of A, there is a prime ideal Q in $X(A)$ lying over P if and only if A_P is a GCD-domain.*

Proof. We can use [22,4.19], because X is closed for the flat topology and $A(t)$ is a GCD-domain if A is a GCD-domain.

Next we look at the quasi-compactness of Specass(A).

Lemma 2.18. *Let A be a ring, $X \subset \mathrm{Spec}(A)$ and set $\mathfrak{U}(X[t]) = A[t] \setminus S_X$ so that $X(A) = A[t]_{S_X}$. The following statements are equivalent :*

(1) *X is quasi-compact.*
(2) *$\mathrm{G}(X) = \overline{X}^p$.*
(3) *$\mathrm{G}(X)$ is quasi-compact.*
(4) *For any prime ideal Q of $A[t]$, there is some $P \in X$ such that $Q \subset P[t]$ whenever $Q \cap S_X = \emptyset$.*

Proof. Assume that X is quasi-compact and let $X \subset \mathrm{D}(I)$ where I is an ideal. Then there is some finitely generated ideal $J \subset I$ such that $X \subset \mathrm{D}(J) \subset \mathrm{D}(I)$. Therefore, $\mathrm{G}(X) = \overline{X}^p$ since we have $\mathrm{G}(X) = \cap[\mathrm{D}(I); X \subset \mathrm{D}(I), I \text{ ideal}]$ and $\overline{X}^p = \cap[\mathrm{D}(J); X \subset \mathrm{D}(J), J \text{ f.g. ideal}]$. Conversely, assume that $\mathrm{G}(X) = \overline{X}^p$ so that $\mathrm{G}(X)$ is quasi-compact. If $X \subset \mathrm{D}(I)$, then $X \subset \mathrm{G}(X) \subset \mathrm{D}(I)$ so that there is some finitely generated ideal $J \subset I$ such that $\mathrm{G}(X) \subset \mathrm{D}(J) \subset \mathrm{D}(I)$ by quasi-compactnes of $\mathrm{G}(X)$. Therefore, (1) and (2) are equivalent. Obviously, (2) implies (3). If (3) holds, then $\mathrm{G}(X)$ is f-closed. Moreover, we have $X \subset \mathrm{G}(X)$ and $\overline{X}^p$ is the f-closure of X. Hence (3) implies (2). Now (1) implies (4) by [22,p.2250]. Assume that (4) holds, we show (2). If R belongs to $\overline{X}^p$, then there is some prime ideal Q in $A[t]$ lying over R such that $Q \cap S_X = \emptyset$. It follows that $Q \subset P[t]$ for some P in X so that $R \subset P$ and consequently, R belongs to $\mathrm{G}(X)$. Thus we have $\mathrm{G}(X) = \overline{X}^p$.

Proposition 2.19. *Let A be an integrally closed domain and X a quasi-compact subset of $\mathrm{Spec}(A)$ such that whenever $Q \subset P[t]$ for some prime ideal $Q \subset A[t]$ and $P \in X$, then Q is extended from a prime ideal of A. Then $X(A)$ is a Bézout domain.*
In particular, if $X = \mathrm{Specess}(A)$ is quasi-compact, then $X(A)$ is a Bézout domain.

Proof. Use [12,3.0] and 2.18, (4) for the first part. The second is a consequence of [9,19.15].

We recall that a ring A is said to be treed if $\mathrm{Spec}(A)$ is treed, that is to say $\mathrm{G}(P)$ is a chain for any prime ideal P.

Proposition 2.20. *Let A be a GCD-domain. In addition, assume that A is absolutely fathomable or treed.*

(1) *A is a Prüfer domain.*
(2) *$\mathrm{Specass}(A)$ is quasi-compact if and only if $\mathrm{Max}(A) \subset \mathrm{Specass}(A)$. In that case, A is a Prüfer domain.*

Proof. Observe that if $\mathrm{Max}(A) \subset \mathrm{Specass}(A)$ or $\mathrm{Specess}(A)$, then $\mathrm{Specess}(A) = \mathrm{Spec}(A)$ because $\mathrm{Specess}(A)$ is stable under generalizations. In this case A is a Prüfer ring. Conversely, let M be a maximal ideal of A. If M does not belong to $\mathrm{Specess}(A)$, pick some $a_P \in M \setminus P$ for each $P \in \mathrm{Specess}(A)$ and denote by I the ideal generated by all the elements a_P. Then we get $\mathrm{Specess}(A) \subset \mathrm{D}(I)$. By quasi-compactness of $\mathrm{Specess}(A)$, there is an ideal $J \subset I \subset M$, generated by finitely many a_P and such that $X \subset \mathrm{D}(J)$. But we have $M = \cup P_\lambda$ for some $P_\lambda \in \mathrm{Specess}(A)$ by 2.3. Since A is absolutely fathomable or treed, $J \subset \cup P_\lambda$ implies $J \subset P_\lambda$

for some $P_\lambda \in \mathrm{Specess}(A)$. This leads to a contradiction, since $\mathrm{Specess}(A) \subset \mathrm{D}(J)$. Therefore, M lies in $\mathrm{Specess}(A)$. A similar argumentation proves that $\mathrm{Max}(A) \subset \mathrm{Specass}(A)$ when $\mathrm{Specass}(A)$ is quasi-compact. Conversely, assume that $\mathrm{Max}(A) \subset \mathrm{Specass}(A)$. It follows that $\mathrm{Spec}(A) = \mathrm{G}(\mathrm{Max}(A)) \subset \mathrm{G}(\mathrm{Specass}(A))$ whence $\mathrm{G}(\mathrm{Specass}(A)) = \mathrm{Spec}(A)$ is quasi-compact. Thus $\mathrm{Specass}(A)$ is quasi-compact by 2.18.

Next we mimic some results of Huckaba and Papick [12]. If A is any ring, we say that $\mathrm{Pdp}(A) = 1$ if $\mathrm{Gr}_A(M) = 1$ for all $M \in \mathrm{Max}(A)$.

Proposition 2.21. *The following statements are equivalent for a GCD-domain A which is not a field.*

(1) $\mathrm{Pdp}(A) = 1$.
(2) $\mathrm{Max}(A) \subset \mathrm{Specess}(A)$.
(3) $A(t) = A\{t\}$.
(4) A is a Prüfer domain.

If in addition A is noetherian, then (1) *is equivalent to* $\mathrm{Max}(A) \subset \mathrm{Specass}(A)$.

Proof. (1) implies (2) by 1.15. We obtain (2) $\Rightarrow$ (3) from 2.2. Indeed, in this case $\mathrm{Specess}(A) = \mathrm{Spec}(A)$ so that $\mathrm{Specess}(A) \subset \mathrm{D}(\mathrm{C}(f(t)))$ implies $\mathrm{C}(f(t)) = A$. Now we show that (3) implies (1). In this case, we get $\mathfrak{U}(\mathrm{Max}(A)[t]) = \mathfrak{U}(\mathrm{Specess}(A)[t])$. The conclusion follows by 2.18 since $\mathrm{Specess}(A)$ is quasi-compact. Indeed, every maximal ideal M belongs to $\mathrm{Specess}(A)$ whence $\mathrm{Gr}_A(M) = 1$ by 1.15. Now, (2) $\Leftrightarrow$ (4) since $\mathrm{Specess}(A)$ is stable under generalizations. The last statement is [12,2.2].

Now we aim to define some $\star$-Prüfer domains in our context. In fact, we will define X-Prüfer domains. We need some definitions and results of D. D. Anderson [1].

Let A be a domain and $X \subset \mathrm{Spec}(A)$ with $\{0\} \in X$. Assume that X defines a representation of A. To recover Anderson's theory, we will deal with $X' = X \setminus \{0\}$. Clearly, X' defines a representation. We denote by $\mathcal{F}(A)$ the set of all nonzero fractionary ideals of A. The reader is referred to Gilmer's book [9] or Anderson's paper [1] for the definition and properties of $\star$-operations on $\mathcal{F}(A)$. Anderson proves the following results [1,Theorem 1]:

• For $I \in \mathcal{F}(A)$, define $I^\star = \cap [I_P \,; P \in X']$.

• The mapping $I \to I^\star$ defines a $\star$-operation on A such that $(I \cap J)^\star = I^\star \cap J^\star$ for $I, J \in \mathcal{F}(A)$.

• $P^\star = P$ for each $P \in X'$ and if I is an integral ideal of A with $I^\star \neq A$, then $I \subset P$ for some $P \in X'$.

This last property is crucial for the following.

Lemma 2.22. *If I is a nonzero integral ideal of A, then*

$$X \subset \mathrm{D}(I) \Leftrightarrow I^\star = A.$$

Proof. Anderson's result shows that $X \subset \mathrm{D}(I)$ implies $I^\star = A$. Conversely, $I^\star = A$ gives $1 \in I_P$ for all $P \in X$ so that $X \subset \mathrm{D}(I)$. If not, $I \subset P$ for some $P \in X$ gives a contradiction with $1 \in I_P$.

Following B. G. Kang [14], we introduce the ring $\star(A) = A[t]_{N_\star}$ where $N_\star$ is the saturated multiplicative subset of all $f(t) \in A[t]$ such that $(C(f(t)))^\star = A$. Then we have $\star(A) = X(A)$ by 2.22.

Next we define $\Gamma(A)$ to be the set of all $\star_s$-maximal ideals of A. Then we have $N_\star = A[t] \setminus \mathfrak{U}(\Gamma(A)[t])$ and $\{M \cdot X(A) \, ; \, M \in \Gamma(A)\} = \operatorname{Max}(X(A))$ [14,2.1].

Definition 2.23. Let A be an integral domain. Then A is said to be a X-Prüfer domain if every nonzero finitely generated integral ideal is X-invertible, that is to say $(II^{-1})^\star = A$.

Theorem 2.24. *Let A be a GCD-domain and $X \subset \operatorname{Spec}(A)$ with $\{0\} \in X$, defining a representation of A. The following statements are equivalent :*

(1) *A is a X-Prüfer domain.*

(2) *A_M is a valuation domain for every $M \in \Gamma(A)$.*

(3) *Every ideal of $X(A)$ is extended from A.*

(4) *$X(A) = A\{t\}$.*

(5) *$\overline{X}^p = \operatorname{Specess}(A)$.*

Proof. Assume that (1) holds and let $J = I_M$ be a nonzero finitely generated ideal of A_M where I is a finitely generated ideal of A. Then I_M is principal by [14,2.7]. Hence A_M is a Bézout local domain whence a valuation ring. If A_M is a valuation ring for all $M \in \Gamma(A)$, then every nonzero finitely generated ideal I is principal in A_M and therefore, I is X-invertible by [14,2.7]. Consequently, (1) is equivalent to (2). Assume that A is a X-Prüfer domain and let $0 \neq f(t) \in A[t]$. Then we get $f(t)X(A) = C(f(t)) \cdot X(A)$ by [14,2.11]. Thus we have (1) $\Rightarrow$ (3). The converse is gotten from [14, 2.12]. Suppose that (2) holds and take $R \in \overline{X}^p$. There is some prime ideal $Q \cdot X(A)$ lying over R and hence there is some $M \in \Gamma(A)$ such that $Q \cdot X(A) \subset M \cdot X(A)$. It follows that $R \subset M$ where $M \in \Gamma(A)$ and thus we get $\overline{X}^p \subset \operatorname{G}(\Gamma(A)) \subset \operatorname{Specess}(A)$. Now, an appeal to 2.7 gives $\overline{X}^p \supset \operatorname{Specess}(A)$ and hence (2) $\Rightarrow$ (5) is proved. Assuming (5), we get $X(A) = A\{t\}$ since $A\{t\} = \operatorname{Specess}(A)(A)$ and $X(A) = \overline{X}^p(A)$. Thus (5) implies (4). Next we show that (4) implies (2). If $M \in \Gamma(A)$, then $M \cdot A\{t\}$ is a prime ideal of $A\{t\}$ so that $M[t] \subset \mathfrak{U}(\operatorname{Specess}(A)[t])$. By quasi-compactness of $\operatorname{Specess}(A)$ and in view of 2.18, there is some $P \in \operatorname{Specess}(A)$ such that $M[t] \subset P[t]$. Whence $M \subset P$ and it follows that A_M is a valuation ring.

Corollary 2.25. *Let A be a GCD-domain and X a subset of $\operatorname{Spec}(A)$ such that $\operatorname{Specass}(A) \subset X \subset \operatorname{Specess}(A)$. Then A is a X-Prüfer domain.*

The following definitions and results on spectral localizing systems can be found in [8,5.1]. Let A be an integral domain with quotient field K and X a subset of $\operatorname{Spec}(A)$. Define A_X to be the set of all elements $x \in K$ such that $X \subset \operatorname{D}(A :_A x)$. Then A_X is an A-subalgebra of K, $P \cdot A_X$ is a prime ideal of A_X for $P \in X$ and $A_X = \cap [A_P \, ; \, P \in X] = \cap [(A_X)_{PA_X} \, ; \, P \in X]$. If $Y = \cap [\operatorname{D}(I \cdot A_X) \, ; \, X \subset \operatorname{D}(I)]$, then the canonical ring morphism $f : A \to A_X$ induces a bijection $Y \to \operatorname{G}(X)$, the inverse φ being given by $\varphi(Q) = Q \cdot A_X$. Moreover, we have ${}^a f^{-1}(\operatorname{G}(X)) = Y = \operatorname{G}(\varphi(X))$, so that Y defines a representation of A_X.

Theorem 2.26. *Let A be a GCD-domain with quotient field K and $X \subset \mathrm{Spec}(A)$. Then there is a multiplicative subset Σ of A such that $A_X = A_\Sigma$. Therefore, A_X is a GCD -domain and Y defines a representation of A_X. If X is quasi-compact, then Y is closed for the flat topology and $\mathrm{Specess}(A_X) \subset Y$.*

Proof. Let $\frac{b}{a}$ be an element of K with $\gcd(b,a) \approx 1$. Then $\frac{b}{a}$ belongs to A_X if and only if $X \subset \mathrm{D}((a) : (b))$. Since a and b are coprime $(a) : (b) = (a)$ by 0.5. Define Σ to be the set of all $a \in A$ such that $X \subset \mathrm{D}(a)$. Then Σ is a saturated multiplicative subset of A and we see by the above that $A_X \subset A_\Sigma$. Conversely, if $\frac{b}{a}$ lies in A_Σ with a such that $X \subset \mathrm{D}(a)$, we get $X \subset \mathrm{D}((a) : (b))$ so that $\frac{b}{a} \in A_X$. Thus the first statement is shown. Assume that X is quasi-compact, then $\mathrm{G}(X)$ is patch-closed and so is $Y = {}^a f^{-1}(\mathrm{G}(X))$. Therefore, Y is quasi-compact and stable under generalizations. To complete the proof, use 2.7.

Remark 2.27. A similar result is proved for UFD-domains, by using different methods [3,2.3].

Lemma 2.28. *Let A be a GCD-domain and $I = (a_1, \ldots, a_p)$ with $a_1, \ldots, a_p \neq 0$, a finitely generated ideal of A. Then $\mathrm{c}(I^n) = \mathrm{c}(I)^n$ for any integer n.*

Proof. We first show that $\gcd(a_1^n, \ldots, a_p^n) \approx \gcd(a_1, \ldots, a_p)^n$. This is obvious for $p = 2$. Then use induction on p. Now set $\mathrm{c}(I^n) = A\delta$, $Ad = \mathrm{c}(I)$ and $a_i = db_i$ with $\gcd(b_1, \ldots, b_p) \approx 1$. We get $\mathrm{c}(I^n) \subset Ad^n$ so that $\delta = ad^n$ for some $a \in A$. Observe that $d^n a$ divides $d^n b_i^n$ so that a divides $\gcd(b_1^n, \ldots, b_p^n) = \gcd(b_1, \ldots, b_p)^n \approx 1$. It follows that $\delta \approx d^n$.

Proposition 2.29. *Let A be a GCD-domain, $U = \mathrm{D}(I)$ a quasi-compact open subset of $\mathrm{Spec}(A)$ where I is a finitely generated ideal and $\Gamma(U)$ its ring of sections.*

(1) $\Gamma(U) = A_\Sigma$ *where Σ is the multiplicative subset of all $a \in A$ such that $U \subset \mathrm{D}(a)$. That is to say $A \setminus \Sigma = \mathfrak{U}(\mathrm{D}(I))$.*
(2) $\Gamma(U) \subset A_d$ *where d is a generator of $\mathrm{c}(I)$.*
(3) $\mathrm{Spec}(\Gamma(U)) = \mathrm{D}(\mathrm{c}(I))$.

Proof. (1) follows from 2.26 and $\Gamma(U) = A_U$. In view of [16,IV.5.2], we have $\Gamma(U) = \cup_{n \geq 0} A :_K I^n$. We show (2). Let $\frac{b}{a} \in A :_K I^n$; from $I^n \subset (a) : (b)$ we deduce that $d^n \in (a) : (b)$ by 2.28 since $(a) : (b)$ is a principal ideal by 0.5. It follows that $\frac{b}{a} \in A_d$, whence $\Gamma(U) \subset A_d$. Next we prove (3). If $Q \in \mathrm{Spec}(\Gamma(U))$ and $P = Q \cap A$, then A_P is isomorphic to $\Gamma(U)_Q$ since $A \to \Gamma(U)$ is a flat epimorphism. Thus we get $A :_K I \subset \Gamma(U) \subset A_P$ and $A :_K I = Ad^{-1}$ by 1.19. It follows that $\mathrm{c}(I) \not\subset P$. Hence we have shown $\mathrm{Spec}(\Gamma(U)) \subset \mathrm{D}(\mathrm{c}(I))$. Conversely, let $P \in \mathrm{D}(\mathrm{c}(I))$ so that $d \notin P$. Consider $\Gamma(U) \subset A_d \subset A_P$ and set $Q = P \cdot A_P \cap \Gamma(U)$. It follows that Q is a prime ideal of $\Gamma(U)$ lying over P and thus the proof that $\mathrm{D}(\mathrm{c}(I)) \subset \mathrm{Spec}(\Gamma(U))$ is completed.

Corollary 2.30. *Let A be a GCD-domain and I a finitely generated ideal of A. The following statements are equivalent :*

(1) $\mathrm{D}(I)$ *is affine.*
(2) $\mathrm{D}(I) = \mathrm{D}(\mathrm{c}(I))$

If in addition A is noetherian, then (1) and (2) are equivalent to $\sqrt{I}$ is a v-ideal.

Proof. If $D(I)$ is affine, $\operatorname{Spec}(\Gamma(U)) = D(I)$ shows that (1) $\Rightarrow$ (2) by 2.29. Then (2) implies (1) is obvious. When A is noetherian see [16,IV.5.3].

These last results allow us to give a characterization of parafactorial GCD-domains [10,IV,21.13]. Let A be a local ring with maximal ideal M and consider the open subset $U = \operatorname{Spec}(A) \setminus \{M\}$. Then A is said to be parafactorial if $A = A_U$ and $\operatorname{Pic}(U) = 0$.

Theorem 2.31. *Let A be a local GCD-domain with maximal ideal M such that $U = \operatorname{Spec}(A)\setminus\{M\}$ is quasi-compact (for instance, if $\operatorname{Spec}(A)$ is noetherian). Then A is parafactorial if and only if $\operatorname{Pdp}(A) \geq 2$ and $\operatorname{Pic}(U) = 0$.*
In particular, if U is affine (that is to say $M = \sqrt{Ad}$ for some $d \in A$), then A is parafactorial if and only if $\operatorname{Pdp}(A) \geq 2$.

Proof. According to 2.7, $A = A_U$ implies $\operatorname{Specess}(A) \subset U$ since U is f-closed. It follows that M does not belong to $\operatorname{Specess}(A)$ whence $\operatorname{Gr}_A(M) \geq 2$ by 1.17. Conversely, $\operatorname{Gr}_A(M) \geq 2$ implies $M \notin \operatorname{Specess}(A)$ so that $\operatorname{Specess}(A) \subset U$. Therefore, we have $A = A_U$. If in addition U is affine, then $\operatorname{Pic}(U) = \operatorname{Pic}(A_\Sigma) = 0$ since A_Σ is a GCD-domain.

Theorem 2.32. *Let A be a local noetherian UFD-domain. Then A is parafactorial if and only if $\operatorname{Pdp}(A) \geq 2$.*

Proof. According to [10,21.6.11], $\operatorname{Pic}(A) \to \operatorname{Pic}(U)$ is surjective while $\operatorname{Pic}(A) = 0$.

Remark 2.33. There exist noetherian parafactorial local domains with Krull-dimension 2 which are not UFD-domains [10,IV,p.318]. Moreover, a local parafactorial GCD-domain A with maximal ideal M is neither treed nor fathomable. To see this, assume there is a GCD-domain with these conditions. The condition $\operatorname{Pdp}(A) \geq 2$ implies that $M = \cup[P; P \in \operatorname{Specess}(A)]$ by 1.17 and $M \notin \operatorname{Specess}(A)$. If $I \subset M$ is a finitely generated ideal, then there is some $P \in \operatorname{Specess}(A)$ such that $I \subset P$. It follows that $M \in \overline{U}^p$ so that $U = \overline{U}^p = \operatorname{Spec}(A)$, a contradiction.

We end with a result beyond the scope of this paper. Mott and Schexnayder quoted a result of Samuel settled for noetherian domains. If A is a GCD-domain and M a projective A-module, then the symetric algebra $S_A(M)$ is a GCD-domain [18,p.394]. They gave an example of a flat module M such that $S_A(M)$ is not a GCD-domain.

Proposition 2.34. *Let A be a GCD-domain with quotient field $K \neq A$. Then $S_A(K) = A + tK[t]$ is a GCD-domain and K is a nonprojective flat A-module.*

Proof. We have $S_A(K) = A + tK[t]$ by [23,2.6]. Moreover, $A + tK[t]$ is a GCD-domain by [4,5.1]. Now observe that the projectiveness of K over A implies that K is a finite A-module, so $K = A$.

REFERENCES

1. D. D. Anderson, *Star-operations induced by overrings*, Comm. Algebra **16** (1988), 2535 – 2553.
2. D. D. Anderson and B. Mullins, *Finite factorizations domains*, Proc. Amer. Math. Soc. **124** (1996), 389 – 396.

3. J. T. Arnold and J. W. Brewer, *On flat overrings, ideal transforms and generalized transforms of a commutative ring*, J. Algebra **18** (1971), 254 - 263.

4. V. Barucci, L. Izelgue and S. Kabbaj, *Some factorization properties of $A + XB[X]$ domains*, Commutative ring theory - Lecture notes in pure and applied mathematics N° 185, Dekker, New York, Basel, Hong Kong, 1997, pp. 69 -78.

5. N. Bourbaki, *Algèbre commutative III-IV*, Hermann, Paris, 1967.

6. J. W. Brewer and W. J. Heinzer, *Associated primes of principal ideals*, Duke Math. J. **41** (1974), 1 -7.

7. P. Dutton, *Prime ideals attached to a module*, Quart. J. Math. Oxford **29** (1978), 403 - 413.

8. M. Fontana, J. A. Huckaba and I. J. Papick, *Prüfer domains*, Dekker, New York, Basel, Hong Kong, 1997.

9. R. Gilmer, *Multiplicative ideal theory*, Dekker, New York, 1992.

10. A. Grothendieck and J. Dieudonné, *Eléments de Géométrie Algébrique N° 32*, I.H.E.S. Presses Universitaires de France, Paris, 1967.

11. E. Houston, *Prime t-ideals in $R[X]$*, Commutative ring theory - Lecture notes in pure and applied mathematics N° 153, Dekker, New York, Basel, Hong Kong, 1994, pp. 163 - 170.

12. J. A. Huckaba and I. J. Papick, *A localization of $R[x]$*, Can. J. Math. **33** (1981), 103 - 115.

13. J. Iroz and D. E. Rush, *Associated primes in non-noetherian rings*, Can. J. Math. **36** (1984), 344 - 360.

14. B. G. Kang, *Prüfer v-multiplication domains and the ring $R[X]_{N_v}$*, J. Algebra **123** (1989), 141 - 170.

15. I. Kaplansky, *Commutative rings*, Allyn and Bacon, Boston, 1970.

16. D. Lazard, *Autour de la platitude*, Bull. Soc. Math. France **97** (1969), 95 - 108.

17. J. Merker, *Idéaux faiblement associés*, Bull. Sci. Math. **93** (1969), 15 - 21.

18. J. L. Mott and M. Schexnayder, *Exact-sequence of semi-value groups*, J. reine angew. Math. **283/284** (1976), 388 - 401.

19. D. G. Northcott, *Finite free resolutions*, Cambridge University Press 71, Cambridge, London, New York, Melbourne, 1976.

20. G. Picavet, *Autour des idéaux premiers de Goldman d'un anneau commutatif*, Ann. Univ. Clermont-Ferrand **39** (1975), 73 - 90.

21. G. Picavet, *Parties sondables d'un spectre et profondeur*, Bollettino U.M.I. **(7) 8-B** (1994), 677 - 730.

22. G. Picavet, *Propriétés et applications de la notion de contenu*, Comm. Algebra **13** (1985), 2231 - 2265.

23. G. Picavet, *About composite rings*, Commutative ring theory - Lecture notes in pure and applied mathematics N° 185, Dekker, New York, Basel, Hong Kong, 1997, pp. 163 - 170.

24. P. K. Sharma, *A note on ideals in polynomial rings*, Arch. Math. **37** (1981), 325 - 329.

25. P. B. Sheldon, *Prime ideals in GCD-domains*, Can. J. Math. **26** (1974), 98 - 107.

26. H. T. Tang, *Gauss' Lemma*, Proc. Amer. Math. Soc. **35** (1972), 372 - 376.

LABORATOIRE DE MATHEMATIQUES PURES, UNIVERSITÉ BLAISE PASCAL (CLERMONT II), 63177 AUBIERE CEDEX, FRANCE

E-mail address: picavet@ucfma.univ-bpclermont.fr

Seminormality and t-Closedness of Algebraic Orders

MARTINE PICAVET-L'HERMITTE   Laboratoire de Mathématiques Pures, Université Blaise Pascal 63177 Aubière Cedex, France
e-mail : picavet@ucfma.univ-bpclermont.fr

1 - Introduction

The ring of integers A of an algebraic number field being integrally closed, A is also t-closed and seminormal. We recall that :

an integral domain A is called **seminormal** if there exists $a \in A$ such that $x = a^2$, $y = a^3$ whenever $x^3 = y^2$ for $(x, y) \in A^2$.

an integral domain A is called **t-closed** if there exists $a \in A$ such that $x = a^2 - ra$, $y = a^3 - ra^2$ whenever $x^3 + rxy - y^2 = 0$ for $(x, y, r) \in A^3$.

We can ask under what conditions a non integrally closed algebraic order is seminormal or t-closed. We have an answer for three cases : D. Dobbs and M. Fontana gave conditions for a quadratic order to be seminormal or GPVD and, if not, they determined its seminormalization [4]. H. Tanimoto gave conditions for the ring $\mathbb{Z}[\sqrt[n]{m}]$, where $m, n \in \mathbb{Z}, n \geq 2$, to be seminormal or quasinormal [15]. Let A be a one-dimensional Noetherian domain with finite integral closure over A; then A is t-closed if and only if A is GPVD if and only if A is quasinormal [7] and [10]. We get in [11] conditions for the ring $\mathbb{Z}[\alpha]$ to be seminormal or t-closed, where α is an algebraic integer.

In Section 3, we determine the seminormalization ^+A and the t-closure ^tA of an algebraic order A : let I be the conductor of A in its integral closure $\bar{A}$ and $P_1, \ldots, P_n$ the maximal ideals of $\bar{A}$ which contain I. Then $^+A = A + \bigcap_{i=1}^{n} P_i$ and

$^tA = \bigcap_{i=1}^{n} (A + P_i)$. This allows to characterize the seminormal and t-closed orders A

with a given conductor in $\bar{A}$.

Section 4 is devoted to an application to algebraic orders in cyclotomic fields. Let $\bar{\mathbb{Z}}$ be the ring of integers of the cyclotomic field generated by a pth root of unity where p is a prime integer and consider a prime integer $q \neq p$ such that $q \not\equiv 1$ (p). Set $I = \bigcap_{i=1}^{r} P_i$ where the P_i are maximal ideals in $\bar{\mathbb{Z}}$ such that $P_i \cap \mathbb{Z} = q\mathbb{Z}$ for each i. There are exactly s^r t-closed orders with conductor I where $f = [\bar{\mathbb{Z}}/P_i : \mathbb{Z}/q\mathbb{Z}]$ and s is the number of proper divisors of f. If $q \equiv 1$ (p) and $r > 1$, there is no t-closed order with conductor I and $\mathbb{Z} + I$ is the only seminormal order with conductor I.

A ring morphism f is said to be **minimal** if f is injective and if any decomposition $f = g \circ h$ where g and h are injective ring morphisms is such that g or h is an isomorphism. Let $A \subset B$ be two orders of algebraic integers with the same quotient field K such that B is finite over A. We showed that any decomposition of $A \to B$ into minimal morphisms is composed of finitely many minimal morphisms, that $A \to B$ may have several decompositions into minimal morphisms with different lengths and that there are three types of minimal morphisms [8] and [9]. Results obtained by D. Dobbs and M. Fontana in [4] allow us to obtain more precise results for a quadratic algebraic number field K in Section 5 : Let K be a quadratic algebraic number field and let $A \subset B$ be two orders of algebraic integers with quotient field K. We show that all the decompositions of $A \to B$ into minimal morphisms have the same length and the types of these minimal morphisms are linked to the decomposition of prime integers in K.

Two examples illustrate the situation in Section 4 and 5.

A generalization of these results can be done without many additions by considering subrings of the integral closure of a Dedekind domain A in a finite algebraic and separable extension of the quotient field of A .

2 - Generalities about seminormal and t-closed orders

First, recall definitions about seminormality, t-closedness and algebraic orders [6], [7], [9], [10] and [16].

Definition 2.1 *An **algebraic order**, related to a finite algebraic extension field K of $\mathbb{Q}$ of degree d is a subring of K and a free $\mathbb{Z}$-module of rank d.*

Properties of algebraic orders

(1) An algebraic order is a one-dimensional Noetherian domain.
(2) If A is an algebraic order which is a free $\mathbb{Z}$-module of rank d, every nonzero ideal of A is a free $\mathbb{Z}$-module of rank d.

Definition 2.2 *We denote by $\bar{\mathbb{Z}}$ the integral closure of an algebraic order A ; then $\bar{\mathbb{Z}}$ is finite over A and is the ring of integers of the quotient field of A. The conductor I of $A \hookrightarrow \bar{\mathbb{Z}}$ is called the **conductor** of A and $I \neq 0$.*

Definition 2.3 *An **order morphism** is a finite injective ring morphism between algebraic orders with the same quotient field.*

Definition and Proposition 2.4

(1) *A ring A is called* ***seminormal*** *if there exists $a \in A$ such that $x = a^2$, $y = a^3$ whenever $x^3 = y^2$ for $(x, y) \in A^2$.*

A ring A is called ***t-closed*** *if there exists $a \in A$ such that $x = a^2 - ra, y = a^3 - ra^2$ whenever $x^3 + rxy - y^2 = 0$ for $(x, y. r) \in A^3$. A t-closed ring is seminormal.*

(2) *Let A be an integral domain with integral closure $\bar{A}$. There exist two A-subalgebras ^+A and ^tA of $\bar{A}$ such that ^+A (resp. ^tA) is the smallest seminormal (resp. t-closed) A-subalgebra of $\bar{A}$.*

The ring ^+A (resp. ^tA) is called the ***seminormalization*** *(resp.* ***t-closure***) *of A. The composite $A \hookrightarrow {}^+A \hookrightarrow {}^tA \hookrightarrow \bar{A}$ is called the* ***canonical decomposition*** *of $A \hookrightarrow \bar{A}$.*

(3) *An injective ring morphism $A \rightarrow B$ is called* ***seminormal*** *(resp.* ***t-closed***) *if an element $b \in B$ is in A whenever $b^2, b^3 \in A$ (resp. whenever there exists $r \in A$ with $b^2 - rb, b^3 - rb^2 \in A$). A t-closed morphism is seminormal.*

(4) *If A is an integral domain with integral closure $\bar{A}$, then A is seminormal (resp. t-closed) if and only if $A \hookrightarrow \bar{A}$ is seminormal (resp. t-closed).*

Proposition 2.5 *Let A be an algebraic order*

(1) *[6, 4.9 and 4.13] A is seminormal (resp. t-closed) if and only if the conductor of A is a radical ideal in $\bar{\mathbb{Z}}$ (resp. the conductor of A is a radical ideal in $\bar{\mathbb{Z}}$ and the map $\mathrm{Spec}\, \bar{\mathbb{Z}} \rightarrow \mathrm{Spec}\, A$ is bijective).*

(2) *[6, 2.6 and 3.5] and [16, 1.1] The seminormalization (resp. t-closure) of A is the largest A-subalgebra C of $\bar{\mathbb{Z}}$ such that $\mathrm{Spec}(C) \rightarrow \mathrm{Spec}(A)$ is bijective and such that the residual extensions of $A \hookrightarrow C$ are isomorphisms (i.e. $A \hookrightarrow C$ is subintegral) (resp. such that the residual extensions of $A \hookrightarrow C$ are isomorphisms (i.e. $A \hookrightarrow C$ is infra-integral)).*

Proposition 2.6 [1, 2.7], [7, 1.14], [10, 2.5] *and* [14, 3.7]

(1) *Let A be an algebraic order. Then A is seminormal (resp. t-closed) if and only if each A-subalgebra of $\bar{\mathbb{Z}}$ is seminormal (resp. t-closed).*

(2) *If $\{A_i\}$ is a family of algebraic orders with the same quotient field, $\cap A_i$ is seminormal (resp. t-closed) if and only if so is A_i for each i.*

Definition 2.7 [9, 1.2, 1.3 and 2.1]

(1) *A ring morphism f is said to be* ***minimal*** *if f is injective and if any decomposition $f = g \circ h$, where g and h are injective ring morphisms, is such that g or h is an isomorphism.*

(2) *The conductor of a minimal order morphism $A \rightarrow B$ is a maximal ideal in A.*

(3) *Let $f : A \rightarrow B$ be an order morphism the conductor of which is a maximal ideal M in A. Then f is a minimal morphism if and only if one of the following conditions is fulfilled :*

    (a) *There exists a maximal ideal P in B such that $P^2 \subset M \subsetneq P$ and $[B/M : A/M] = 2$. Such a morphism is called* ***ramified***.

(b) *There exist two maximal ideals M_1 and M_2 in B lying over M such that $M_1 \cap M_2 = M$ and the morphisms $A/M \to B/M_i$ are isomorphisms for $i = 1, 2$. Such a morphism is called* **decomposed**.

(c) *The ideal M is a maximal ideal in B and $A/M \to B/M$ is a minimal morphism (between fields). Such a morphism is called* **inert**.

(4) *Let $f : A \to B$ be a minimal order morphism. Then*

(a) *f is ramified if and only if $\operatorname{Spec}(B) \to \operatorname{Spec}(A)$ is bijective and the residual extensions of $A \to B$ are isomorphisms.*

(b) *f is decomposed if and only if f is seminormal and the residual extensions of $A \to B$ are isomorphisms.*

(c) *f is inert if and only if f is t-closed.*

Proposition 2.8 [6, 3.18 and Remark after 4.10] *Let $A \to B$ be an order morphism.*
(1) *There exists a unique decomposition of $A \to B$ in $A \to {}^+_B A \to {}^t_B A \to B$, where :*

$A \to {}^+_B A$ *is subintegral : ${}^+_B A$ is called the* **seminormalization** *of A in B and is the smallest A-subalgebra C of B such that $C \to B$ is seminormal*

${}^+_B A \to {}^t_B A$ *is seminormal and infra-integral : ${}^t_B A$ is called the* **t-closure** *of A in B and is the smallest A-subalgebra C of B such that $C \to B$ is t-closed*

${}^t_B A \to B$ *is t-closed.*

Such a decomposition is called the **canonical** *decomposition of $A \to B$.*
(2) *$A \to {}^+_B A$ is composed of ramified morphisms*

${}^+_B A \to {}^t_B A$ *is composed of decomposed morphisms*
${}^t_B A \to B$ *is composed of inert morphisms.*

We end by a lemma useful for the following :

Lemma 2.9 *Let I be the conductor of an algebraic order with quotient field K. Then $\mathbb{Z} + I$ is the smallest algebraic order with conductor I and quotient field K.*

Proof : Let A be an algebraic order with conductor I and with quotient field K. So $\mathbb{Z} + I$ is a subring of A. If d is the rank of the $\mathbb{Z}$-module A, then $\mathbb{Z} + I$ is a free $\mathbb{Z}$-module with rank d and its quotient field is K. The conductor of $\mathbb{Z} + I$ contains the conductor I of A and is contained in I since $\mathbb{Z} + I \subset A$. So we get the proof.

3 - Seminormal or t-closed orders and their conductor

Let us begin to give a new characterization of the seminormalization and the t-closure of an algebraic order.

Theorem 3.1 *Let A be an algebraic order with conductor I. Consider the two primary decompositions : $I = \bigcap_{i=1}^{n} Q_i$ in A and $I = \bigcap_{\substack{i=1 \\ j=1,\ldots,n_i}}^{n} Q_{ij}$ in $\bar{\mathbb{Z}}$, where Q_i (resp. Q_{ij}) are P_i (resp. P_{ij})-primary ideals such that P_{ij} is a maximal ideal in $\bar{\mathbb{Z}}$ lying over P_i in A for $j = 1,\ldots,n_i$. Then : ${}^+A = A + \bigcap_{\substack{i=1 \\ j=1,\ldots,n_i}}^{n} P_{ij} = \bigcap_{i=1}^{n}(A + R_i)$, where*

$$R_i = \bigcap_{j=1}^{n_i} P_{ij} \text{ and } {}^tA = \bigcap_{\substack{i=1 \\ j=1,\ldots,n_i}}^{n} (A + P_{ij}).$$

Proof: Set $A' = \bigcap_{i=1}^{n} (A + R_i)$. As R_i is a radical ideal in $\bar{\mathbb{Z}}$ contained in the conductor

of $A + R_i$, the conductor of $A + R_i$ is itself a radical ideal in $\bar{\mathbb{Z}}$, since $\bar{\mathbb{Z}}$ is a Dedekind domain. By 2.5 (1) we get that $A + R_i$ is seminormal just as A'. So ${}^+A \subset A'$.

On the other hand, the isomorphism $(A + R_i)/R_i \simeq A/P_i$ shows that R_i is a maximal ideal in $A + R_i$ lying over P_i. In fact, R_i is the only prime ideal in $A + R_i$ lying over P_i : if R'_i is another prime ideal in $A + R_i$ lying over P_i, there exists some prime ideal P' in $\bar{\mathbb{Z}}$ lying over R'_i in $A + R_i$ and over P_i in A. Whence P' must be one of the P_{ij}, and $R'_i = R_i$.

Then there is only one prime ideal in A' lying over P_i for each $i = 1, \ldots, n$. So $\mathrm{Spec}(A') \to \mathrm{Spec}(A)$ is bijective (we have only to consider prime ideals in A containing the conductor J of $A \to A'$ but J contains the conductor I of $A \to \bar{\mathbb{Z}}$).

We intend to show that for all prime ideals P in A, we have an isomorphism $A/P \simeq A'/R$, where R is the intersection of all prime ideals in A' lying over P.

If $I \not\subset P$, we have the composed isomorphisms $A_P \to A'_P \to \bar{\mathbb{Z}}_P$, which implies $A/P \simeq A'/R$ since R is the only prime ideal in A' lying over P.

If $I \subset P$, the composite $A \to A' \to A + R_i$ leads to the composite

$$A/P_i \to A'/(A' \cap R_i) \to (A + R_i)/R_i$$

which is an isomorphism as we have seen before. So $A/P_i \simeq A'/(A' \cap R_i)$ for each $i = 1, \ldots, n$.

It follows then that all residual extensions of $A \to A'$ are isomorphisms. By 2.5 (2) we get $A' \subset {}^+A$ which gives the required equality.

Set $K = \bigcap_{\substack{i=1 \\ j=1,\ldots,n_i}}^{n} P_{ij}$. Hence $A + K$ is seminormal since K is a radical ideal

in $\bar{\mathbb{Z}}$. But $K \subset R_i$ leads to $A + K \subset A + R_i$ for each i. So $A + K \subset {}^+A$. The seminormality minimal condition on ${}^+A$ gives then ${}^+A = A + K$.

Consider now $A'' = \bigcap_{\substack{i=1 \\ j=1,\ldots,n_i}}^{n} (A + P_{ij})$. The conductor L of $A + P_{ij}$ contains

P_{ij} and so is either P_{ij} or $\bar{\mathbb{Z}}$. If $L = P_{ij}$ we get that $A + P_{ij}$ is t-closed by [6, 3.14] ; if $L = \bar{\mathbb{Z}}$ then $A + P_{ij} = \bar{\mathbb{Z}}$ is integrally closed and t-closed. So A'' is t-closed by [7, 1.14] and ${}^tA \subset A''$. A similar argument to the one we get for A' gives $(A + P_{ij})/P_{ij} \simeq A/P_i$ which implies isomorphisms between the residual extensions and the result follows.

Now we intend to characterize the seminormal and t-closed orders in a finite algebraic extension of $\mathbb{Q}$.

Proposition 3.2 *Let K be a finite algebraic extension of $\mathbb{Q}$ with $\bar{\mathbb{Z}}$ its ring of integers.*
(1) *The seminormal orders of K are the subrings of $\bar{\mathbb{Z}}$ which contain a ring of the form $\mathbb{Z} + I$ where I is a nonzero radical ideal in $\bar{\mathbb{Z}}$.*
(2) *The t-closed orders of K are the subrings of $\bar{\mathbb{Z}}$ which contain a ring of the form*

$$\bigcap_{i=1}^{n} (\mathbb{Z} + P_i) \text{ where } n \in \mathbb{N}^* \text{ and } P_1, \ldots, P_n \text{ are maximal ideals in } \bar{\mathbb{Z}}.$$

Proof :
(1) By 2.5 (1) the conductor of a seminormal order A is a radical ideal I in $\bar{\mathbb{Z}}$ so $A \supset \mathbb{Z} + I$ by 2.9. Conversely, let I be a radical ideal in $\bar{\mathbb{Z}}$. The conductor of $\mathbb{Z} + I$ contains I, so is a radical ideal in $\bar{\mathbb{Z}}$. By 2.5 (1) $\mathbb{Z} + I$ is seminormal and so are all the subrings of $\bar{\mathbb{Z}}$ which contain $\mathbb{Z} + I$ by 2.6.

(2) As a t-closed order A is seminormal, its conductor I can be written $I = \bigcap_{i=1}^{n} P_i$ where P_i is a maximal ideal in $\bar{\mathbb{Z}}$. Then I is the conductor of $\mathbb{Z} + I$ by 2.9 and $\mathbb{Z}+I \subset A$ so that ${}^t(\mathbb{Z}+I) \subset A$. By 3.1 we get ${}^t(\mathbb{Z}+I) = \bigcap_{i=1}^{n}(\mathbb{Z}+I+P_i) = \bigcap_{i=1}^{n}(\mathbb{Z}+P_i)$ since $I \subset P_i$ for each i and we get the first inclusion.

Conversely if P_i is a maximal ideal in $\bar{\mathbb{Z}}$ for $i = 1, \ldots, n$, then $\mathbb{Z}+P_i$ is a t-closed order with conductor either P_i or $\bar{\mathbb{Z}}$; hence $\bigcap_{i=1}^{n}(\mathbb{Z} + P_i)$ is t-closed and so is any subring of $\bar{\mathbb{Z}}$ containing $\bigcap_{i=1}^{n}(\mathbb{Z} + P_i)$ by 2.6 (2).

We can go further into details in 3.2 by searching for the seminormal or t-closed orders with a given conductor. The following lemma allows to work with simpler assumptions :

Lemma 3.3 *Let A be an algebraic order with conductor I and B a subring of A. Set $I = \bigcap_{\substack{i=1 \\ j=1,\ldots,n_i}}^{n} Q_{ij}$ in $\bar{\mathbb{Z}}$ such that Q_{ij} is a P_{ij}-primary ideal in $\bar{\mathbb{Z}}$ with $P_{ij} \cap B = P_i$ for each j where P_{ij} (resp. P_i) is a maximal ideal in $\bar{\mathbb{Z}}$ (resp. B). For each i, set $R_i = \bigcap_{j=1}^{n_i} Q_{ij}$. There exists a unique family of algebraic orders A_i such that $A = \bigcap_{i=1}^{n} A_i$, where R_i is the conductor of A_i ; then $A_i = A + R_i$.*

Proof : Consider the multiplicative set $S_i = B \setminus P_i$ for a given i. The conductor in B of $A \to \bar{\mathbb{Z}}$ (*i.e.* the annihilator of the B-module $\bar{\mathbb{Z}}/A$) is $I \cap B$. Set $A_i = A + R_i$. As $(R_i)_{S_i} = I_{S_i}$ and $(R_i)_{S_j} = (\bar{\mathbb{Z}})_{S_j}$ for $j \neq i$, we get $(A_i)_{S_i} = A_{S_i}$ and $(A_i)_{S_j} = (\bar{\mathbb{Z}})_{S_j}$ for $j \neq i$. Hence it results that $A_{S_j} = (\bigcap_{i=1}^{n} A_i)_{S_j}$ for each j and $A = \bigcap_{i=1}^{n} A_i$ since $A \subset \bigcap_{i=1}^{n} A_i$ are two B-modules. Moreover the conductor J of A_i contains R_i. The previous equalities obtained by localizing at the maximal ideals of B imply that the conductor of A_i does not meet S_i and meets every S_j for $j \neq i$; we have too $J_{S_i} = I_{S_i} = (R_i)_{S_i}$. It follows then that $J = R_i$.

Consider another intersection $A = \overset{n}{\underset{i=1}{\cap}} A_i'$ where A_i' is an algebraic order with conductor R_i and set $A_i'' = A_i \cap A_i'$ for each i. Localizing at S_i we get :
$(A_i')_{S_i} = A_{S_i} = (A_i)_{S_i} = (A_i'')_{S_i}$ and $(A_i')_{S_j} = (\bar{\mathbb{Z}})_{S_j} = (A_i)_{S_j} = (A_i'')_{S_j}$. So $A_i'' = A_i' = A_i$ since A_i'' is a B-submodule of A_i and A_i'.

Let A be an algebraic order. If K is the quotient field of A then K is a finite separable algebraic extension of $\mathbb{Q}$ of degree m. For a prime integer p, there are n maximal ideals P_i in $\bar{\mathbb{Z}}$ lying over $p\mathbb{Z}$. If $[\bar{\mathbb{Z}}/P_i : \mathbb{Z}/p\mathbb{Z}] = f_i$ there exist integers e_i such that $p\bar{\mathbb{Z}} = \overset{n}{\underset{i=1}{\prod}} P_i^{e_i}$ with $m = \overset{n}{\underset{i=1}{\sum}} f_i e_i$.

We use the above results to give conditions for a radical ideal in the integral closure of a finite algebraic extension of $\mathbb{Q}$ to be the conductor of a seminormal or t-closed order, and, if possible, to find this order.

Proposition 3.4 *Let K be a finite algebraic extension of $\mathbb{Q}$ and $\{P_{ij}\}_{\substack{i=1,\ldots,n \\ j=1,\ldots,n_i}}$ a family of maximal ideals in $\bar{\mathbb{Z}}$ such that $P_{ij} \cap \mathbb{Z} = p_i\mathbb{Z}, j = 1,\ldots,n_i$ where p_i is a prime integer. Consider $I = \overset{n}{\underset{\substack{i=1 \\ j=1,\ldots,n_i}}{\cap}} P_{ij}$ and set $f_{ij} = [\bar{\mathbb{Z}}/P_{ij} : \mathbb{Z}/p_i\mathbb{Z}]$ for each (i,j).*

(1) There exists a seminormal order with conductor I if and only if, for each i, we have $n_i > 1$ or $f_{i1} > 1$ (when $n_i = 1$).
(2) If $n_i > 1$ or $f_{i1} > 1$ for each i, the seminormal orders with conductor I are orders A' such that $A' \supset \mathbb{Z} + I$ and $A' \not\supset \mathbb{Z} + J$ for any ideal $J \underset{\neq}{\supset} I$.

Proof : (1) If there exists an algebraic order with conductor I then $\mathbb{Z} + I$ is an algebraic order with conductor I by 2.9, whence a seminormal order by 2.5 (1). So to prove the existence of a seminormal order with conductor I it is enough to prove that I is the conductor of $\mathbb{Z} + I$. Set $A = \mathbb{Z} + I$ and $P_i = \overset{n_i}{\underset{j=1}{\cap}} P_{ij} \cap A$. Then we have the following isomorphisms :

$$A/P_i = A / \left(\overset{n_i}{\underset{j=1}{\cap}} P_{ij} \cap A \right) \simeq \left(\mathbb{Z} + \overset{n_i}{\underset{j=1}{\cap}} P_{ij} \right) / \overset{n_i}{\underset{j=1}{\cap}} P_{ij} \simeq \mathbb{Z}/p_i\mathbb{Z}.$$

Thus P_i is a maximal ideal in A and P_{ij} lies over P_i for each j. Consider the multiplicative set $S_i = A \setminus P_i$. The conductor J of A contains I. Then $I \neq J$ if and only if there is some i such that $J \not\subset P_i$ if and only if there is some i such that $A_{S_i} = \bar{\mathbb{Z}}_{S_i}$. In that case $\bar{\mathbb{Z}}_{S_i}$ is a local ring : there is only one maximal ideal in $\bar{\mathbb{Z}}$ lying over P_i and $n_i = 1$. This implies $\bar{\mathbb{Z}}_{S_i} = \bar{\mathbb{Z}}_{P_{i1}}$ which gives by considering the residual extensions : $\bar{\mathbb{Z}}/P_{i1} \simeq A/P_i \simeq \mathbb{Z}/p_i\mathbb{Z}$ and $f_{i1} = 1$.

Conversely, assume that there exists some i such that $n_i = 1 = f_{i1}$. There is only one maximal ideal P_{i1} which contain I lying over P_i and $\bar{\mathbb{Z}}/P_{i1} \simeq A/P_i$. We get then $A + P_{i1} = \bar{Z} = Z + P_{i1}$ which implies $\bar{Z}_{S_i} = (Z + P_{i1})_{S_i}$. As $A = Z + I$ is seminormal so is A_{S_i}. Then the conductor of the seminormal morphism $A_{S_i} \to \bar{Z}_{S_i}$ is either P_{iS_i} or $\bar{Z}_{S_i}$ since $I_{S_i} = P_{iS_i}$. There is only one maximal ideal in $\bar{Z}$ lying over P_i so the map $\mathrm{Spec}(\bar{Z}_{S_i}) \to \mathrm{Spec}(A_{S_i})$ is bijective and P_{iS_i} is a maximal ideal in $\bar{Z}_{S_i}$: it follows then that $P_{iS_i} = P_{i1S_i}$ and $\bar{Z}_{S_i} = (A + P_{i1})_{S_i} = A_{S_i}$ so that P_i does not contain the conductor of A. Hence I is not the conductor of A.

(2) If condition (1) is fulfilled, I is indeed the conductor of $Z + I$. If there is an order B with conductor I, we have $Z + I \subset B$. Let B be an order containing $Z + I$ with conductor J. We get then $I \subset J$ and $Z + J \subset B$. It follows that the conductor of B is I if and only if there exists no ideal $J \underset{\neq}{\supset} I$ such that $Z + J \subset B$.

For the t-closedness condition, we first consider a simple case :

Lemma 3.5 *Let K be a finite algebraic extension of $\mathbb{Q}$ and let P be a maximal ideal in $\bar{Z}$ lying over $p\mathbb{Z}$ where p is a prime integer.*
(1) There exists a t-closed order with conductor P if and only if $f = [\bar{Z}/P : \mathbb{Z}/p\mathbb{Z}] > 1$.
(2) If s is the number of proper divisors of f, there are exactly s t-closed orders with conductor P and they are of the form $(\mathbb{Z} + P)[x]$, where x is a representative of a generator over $\mathbb{Z}/p\mathbb{Z}$ of a proper subfield of $\bar{Z}/P$.

Proof : (1) As a t-closed order is seminormal, we remark that the condition (1) of 3.4 must be fulfilled. Moreover a t-closed order with conductor P contains $A = \mathbb{Z} + P$ which is itself a t-closed order with conductor P by the proof of 3.2 (2) if $[\bar{Z}/P : \mathbb{Z}/p\mathbb{Z}] > 1$.

Conversely, if $f = 1$, then $A = \mathbb{Z} + P = \bar{Z}$ and P is not the conductor of A.

(2) Let A' be a t-closed order with conductor P. Then $A \subset A' \underset{\neq}{\subset} \bar{Z}$; we have the field extensions : $\mathbb{Z}/p\mathbb{Z} \simeq A/P \to A'/P \to \bar{Z}/P$ and $\bar{Z}/P$ is a finite field with p^f elements. The subfields of $\bar{Z}/P$ are in bijection with the subgroups of the Galois group of $\bar{Z}/P$ over $\mathbb{Z}/p\mathbb{Z}$, a cyclic group of order f. The number of such proper subgroups and of t-closed orders with conductor P is the number s of proper divisors of f. Moreover a subfield A'/P of $\bar{Z}/P$ has a primitive element $\bar{x} \in \bar{Z}/P$, with $x \in A'$ so that $A' = A[x]$.

We can conclude that there are s t-closed orders with conductor P and they are of the form $(\mathbb{Z} + P)[x]$ where x is a representative in $\bar{Z}$ of a generator over $\mathbb{Z}/p\mathbb{Z}$ of a proper subfield of $\bar{Z}/P$.

Thus we obtain the general case :

Theorem 3.6 *Let K be a finite algebraic extension of $\mathbb{Q}$ and $\{P_{ij}\}_{\substack{i=1,\dots,n \\ j=1,\dots,n_i}}$ a family of maximal ideals in $\bar{Z}$ such that $P_{ij} \cap \mathbb{Z} = p_i\mathbb{Z}$ for each j where p_i is a prime integer.*

Set $f_{ij} = [\bar{Z}/P_{ij} : \mathbb{Z}/p_i\mathbb{Z}]$ for each (i,j) and consider $I = \overset{n}{\underset{\substack{i=1 \\ j=1,\dots,n_i}}{\cap}} P_{ij}$.

(1) I is the conductor of a t-closed order if and only if $f_{ij} > 1$ for each (i,j).

(2) *Assume $f_{ij} > 1$ for each (i,j). An order A is a t-closed order with conductor I if and only if $A \supset \bigcap\limits_{\substack{i=1 \\ j=1,\ldots,n_i}}^{n} (\mathbb{Z} + P_{ij})$ and $A \not\supset \bigcap\limits_{\substack{i=1 \\ j=1,\ldots,n'_i}}^{n'} (\mathbb{Z} + P_{ij})$ for any n' such that $n' < n$ or for any n'_i such that $n'_i < n_i$.*

The t-closed orders with conductor I are of the form $\bigcap\limits_{\substack{i=1 \\ j=1,\ldots,n_i}}^{n} ((\mathbb{Z} + P_{ij})[x_{ij}])$,

where x_{ij} is a representative of a generator over $\mathbb{Z}/p_i\mathbb{Z}$ of a proper subfield of $\bar{\mathbb{Z}}/P_{ij}$.

There are exactly $\prod\limits_{\substack{i=1 \\ j=1,\ldots,n_i}}^{n} s_{ij}$ such t-closed orders, where s_{ij} is the number of proper divisors of f_{ij}.

Proof : (1) As a t-closed order is seminormal, we remark that the condition (1) of 3.4 must be fulfilled. Moreover a t-closed order with conductor I contains $\bigcap\limits_{\substack{i=1 \\ j=1,\ldots,n_i}}^{n} (\mathbb{Z} + P_{ij})$ which is itself a t-closed order with conductor I by the proof of 3.2 (2). So it is enough to find conditions for I to be the conductor of $\bigcap\limits_{\substack{i=1 \\ j=1,\ldots,n_i}}^{n} (\mathbb{Z} + P_{ij})$.

Set $A = \bigcap\limits_{\substack{i=1 \\ j=1,\ldots,n_i}}^{n} (\mathbb{Z} + P_{ij})$ and let J be the conductor of A, a t-closed ring. We have $I \subset J$ and we can assume $J = \bigcap\limits_{\substack{i=1 \\ j=1,\ldots,m_i}}^{m} P_{ij}$ with $m \leq n$ and $m_i \leq n_i$ for each $i \leq m$. So I is the conductor of A if and only if $I = J$.

Assume $J \underset{\neq}{\supset} I$. There exists P_{ij} such that $J \not\subset P_{ij}$. The conductor of $\mathbb{Z} + P_{ij}$ is either P_{ij} or $\bar{\mathbb{Z}}$. But $A \subset \mathbb{Z} + P_{ij}$ implies that the conductor of $\mathbb{Z} + P_{ij}$ contains J and then is $\bar{\mathbb{Z}}$. It follows that $\mathbb{Z} + P_{ij} = \bar{\mathbb{Z}}$ which implies $\bar{\mathbb{Z}}/P_{ij} \simeq \mathbb{Z}/(P_{ij} \cap \mathbb{Z})$ so that $f_{ij} = 1$.

Now, if $f_{ij} = 1$ for some (i,j) then $\mathbb{Z} + P_{ij} = \bar{\mathbb{Z}}$ and $A = \bigcap\limits_{(i',j') \neq (i,j)} (\mathbb{Z} + P_{i'j'})$ implies that $\bigcap\limits_{(i',j') \neq (i,j)} P_{i'j'}$ is contained in the conductor of A and cannot equal I.

To sum up, I is the conductor of A if and only if $f_{ij} > 1$ for each (i,j).

(2) Under the assumptions of (1), a t-closed order B with conductor I contains A. If B contains an intersection of the $\mathbb{Z} + P_{ij}$ in a strictly less number than in A, the conductor of B should be an intersection of the P_{ij} containing strictly I : a contradiction.

Let us determine explicitly the t-closed orders A' with conductor I. With the notations and assumptions of (1), since A is t-closed, $\mathrm{Spec}(\bar{\mathbb{Z}}) \to \mathrm{Spec}(A')$ and $\mathrm{Spec}(\bar{\mathbb{Z}}) \to \mathrm{Spec}(A)$ are bijective by 2.5. So A' is an intersection of t-closed rings $A' + P_{ij}$ by 3.3. By 3.5 we can determine the t-closed orders with conductor P_{ij}. There are s_{ij} t-closed orders with conductor P_{ij} and they are of the form

$(\mathbb{Z} + P_{ij})[x_{ij}]$ where x_{ij} is a representative in $\bar{\mathbb{Z}}$ of a generator over $\mathbb{Z}/p_i\mathbb{Z}$ of a proper subfield of $\bar{\mathbb{Z}}/P_{ij}$. This implies that there are $\prod\limits_{\substack{i=1 \\ j=1,\ldots,n_i}}^{n} s_{ij}$ t-closed orders

with conductor $I = \bigcap\limits_{\substack{i=1 \\ j=1,\ldots,n_i}}^{n} P_{ij}$ and they are of the form $\bigcap\limits_{\substack{i=1 \\ j=1,\ldots,n_i}}^{n} ((\mathbb{Z} + P_{ij})[x_{ij}])$,

where x_{ij} is a representative element of a generator over $\mathbb{Z}/p_i\mathbb{Z}$ of a proper subfield of $\bar{\mathbb{Z}}/P_{ij}$ and s_{ij} is the number of proper divisors of $f_{ij} = [\bar{\mathbb{Z}}/P_{ij} : \mathbb{Z}/p_i\mathbb{Z}]$. The above intersection is written in a unique way by 3.3.

Remark: We did not give here the number and the form of seminormal non t-closed orders with a given conductor. The determination of such order A is much more difficult that in the t-closed case since we cannot use 3.3: in fact, $\mathrm{Spec}(\bar{\mathbb{Z}}) \to \mathrm{Spec}(A)$ is not bijective if A is not t-closed; if the conductor of A is $I = \cap P_{ij}$, where P_{ij} is a maximal ideal in $\bar{\mathbb{Z}}$, we do not know how many P_{ij} lie over the same ideal P_i in A. However we give an example of this situation in the next section.

If the conductor of an order is not known, some special elements may be used to characterize the seminormality.

Proposition 3.7 *Let A be an algebraic order with conductor I.*
(1) Let $I' = I \cap \mathbb{Z}$ be the conductor of A in $\mathbb{Z}$. If $I' = \prod p_i\mathbb{Z}$ where the p_i are distinct prime integers unramified in $\bar{\mathbb{Z}}$, then A is seminormal.
(2) Let d be the discriminant of A. If $d\bar{\mathbb{Z}}$ is a radical ideal in $\bar{\mathbb{Z}}$, then A is seminormal.
(3) If there exists a primitive element x of the quotient field of A such that $A = \mathbb{Z}[x]$, let $f(X)$ be the minimal polynomial of x. If $f'(x)\bar{\mathbb{Z}}$ is a radical ideal in $\bar{\mathbb{Z}}$, then A is seminormal.

Proof: (1) Let $n\mathbb{Z} = I \cap \mathbb{Z}$, with $n = \prod p_i$ where the p_i are distinct unramified integers. Then $p_i\bar{\mathbb{Z}}$ is a radical ideal in $\bar{\mathbb{Z}}$. So is $n\bar{\mathbb{Z}}$ and $n\bar{\mathbb{Z}} \subset I$.
(2) The discriminant of A belongs to I by [2, 3, p.26].
(3) In the same way, if $A = \mathbb{Z}[x]$ where x is a primitive element of the quotient field of A, with minimal polynomial $f(X)$, then $f'(x)$ belongs to the different of A [17, 29, p.303] which is contained in I [17, Remark p.304].

Furthermore, if the conductor I of A contains a radical ideal in $\bar{\mathbb{Z}}$, then I is itself a radical ideal in $\bar{\mathbb{Z}}$ and so A is seminormal in the three cases.

4 - Seminormal and t-closed orders in cyclotomic extensions

We are going to apply the results of Section 3 to algebraic orders of cyclotomic extensions associated to a prime integer. In fact the maximal ideals in the ring of integers of such an extension have a particular behavior.

Recall some properties of these extensions [12, Chapters 10 and 13]:

Let p be a prime integer and consider the field $K = \mathbb{Q}[\zeta]$ where $\zeta \neq 1$ is a pth root of unity. The ring of integers of K is $\bar{\mathbb{Z}} = \mathbb{Z}[\zeta]$. There are two types of decomposition of prime integers in $\bar{\mathbb{Z}}$; set $\xi = 1 - \zeta$:
(1) $\bar{\mathbb{Z}}\xi$ is a maximal ideal in $\bar{\mathbb{Z}}$ such that $p\bar{\mathbb{Z}} = \bar{\mathbb{Z}}\xi^{p-1}$ and $[\bar{\mathbb{Z}}/\xi : \mathbb{Z}/p\mathbb{Z}] = 1$.

(2) Let $q \neq p$ be a prime integer. Then q is unramified in $\bar{\mathbb{Z}}$ and $q\bar{\mathbb{Z}} = P_1 \cdots P_g$ with $P_1, \ldots, P_g$ maximal ideals in $\bar{\mathbb{Z}}$ such that $p - 1 = fg$ where $f = [\bar{\mathbb{Z}}/P_i : \mathbb{Z}/q\mathbb{Z}]$ for each i. In particular $f = 1$ if and only if $q \equiv 1 \ (p)$.

(3) All $\bar{\mathbb{Z}}/P_i$ are isomorphic $\mathbb{F}_q$-algebras as finite fields with q^f elements.

In order to characterize the seminormal or t-closed orders with an arbitrary (radical) conductor, 2.6 and 3.3 allow us to look for the seminormal or t-closed orders with conductor $I = \bigcap\limits_{i=1}^{r} P_i$ where P_i are maximal ideals in $\bar{\mathbb{Z}}$ lying over the same maximal ideal in $\mathbb{Z}$:

Theorem 4.1 *Let* p *be a prime integer,* $\zeta \neq 1$ *a pth root of unity and consider the ring of integers* $\bar{\mathbb{Z}} = \mathbb{Z}[\zeta]$ *of the field* $K = \mathbb{Q}[\zeta]$.

Let $I = \bigcap\limits_{i=1}^{r} P_i$ where P_i are maximal ideals in $\bar{\mathbb{Z}}$ lying over the same maximal ideal $q\mathbb{Z}$ in $\mathbb{Z}$ with $p \neq q$.

(1) *If* $q \equiv 1 \ (p)$ *and* $r > 1$, *there is no t-closed order with conductor* I *and* $\mathbb{Z} + I$ *is the only seminormal order with conductor* I. *If* $q \equiv 1 \ (p)$ *and* $r = 1$, *we have* $\mathbb{Z} + I = \bar{\mathbb{Z}}$.

(2) *If* $q \not\equiv 1 \ (p)$, *there are exactly* s^r *t-closed orders with conductor* I *where* $f = [\bar{\mathbb{Z}}/P_i : \mathbb{Z}/q\mathbb{Z}]$ *and* s *is the number of proper divisors of* f.

$q\bar{\mathbb{Z}}$ *is a maximal ideal in* $\bar{\mathbb{Z}}$ *when* $f = p - 1$ *and all orders containing* $\mathbb{Z} + q\bar{\mathbb{Z}}$ *and distinct from* $\bar{\mathbb{Z}}$ *are t-closed orders with conductor* $q\bar{\mathbb{Z}}$.

Proof : (1) If $q \equiv 1 \ (p)$ then $f = 1 = [\bar{\mathbb{Z}}/P_i : \mathbb{Z}/q\mathbb{Z}]$ for each i. Set $A = \mathbb{Z} + I$, we obtain $A/I \simeq \mathbb{Z}/q\mathbb{Z}$, so A is seminormal if $r > 1$, and non t-closed since all residual extensions of $A \to \bar{\mathbb{Z}}$ are isomorphisms. If $r = 1$ the isomorphism $\bar{\mathbb{Z}}/I \simeq \mathbb{Z}/q\mathbb{Z}$ implies $\bar{\mathbb{Z}} = \mathbb{Z} + I$.

Let A' be an order containing A with conductor J. Then $I \subset J$ so we can assume $J = \bigcap\limits_{i=1}^{r'} P_i$ with $r' \leq r$. By [9, 3.4] a decomposition of $A \to \bar{\mathbb{Z}}$ (resp. $A' \to \bar{\mathbb{Z}}$) into minimal morphisms is composed of $r - 1$ (resp. $r' - 1$) decomposed morphisms. Hence $A = A'$ if and only if $r = r'$. So A is the only order with conductor I.

(2) If $q \not\equiv 1 \ (p)$, we have $q\bar{\mathbb{Z}} = P_1 \cdots P_r \cdots P_g$ where $P_1, \ldots, P_g$ are maximal ideals in $\bar{\mathbb{Z}}$ with $p - 1 = fg$ and $f \neq 1$. By 3.6 (2), as for each i, $[\bar{\mathbb{Z}}/P_i : \mathbb{Z}/q\mathbb{Z}] = f > 1$, there are s^r t-closed orders with conductor I.

If $f = p - 1$, then $q\bar{\mathbb{Z}}$ is a maximal ideal in $\bar{\mathbb{Z}}$ and in $\mathbb{Z} + q\bar{\mathbb{Z}}$. So $q\bar{\mathbb{Z}}$ is a maximal ideal in any order B strictly contained between $\mathbb{Z} + q\bar{\mathbb{Z}}$ and $\bar{\mathbb{Z}}$ and is the conductor of B. As $\mathbb{Z} + q\bar{\mathbb{Z}}$ is t-closed so is B.

As we told after 3.6 we are not able to characterize in an explicit way the seminormal orders with a given conductor. But we can give a method to obtain some seminormal non t-closed orders :

Proposition 4.2 *Let p be a prime integer, $\zeta \neq 1$ a pth root of unity and consider the ring of integers $\bar{\mathbb{Z}} = \mathbb{Z}[\zeta]$ of the field $K = \mathbb{Q}[\zeta]$.*

Let $I = \overset{r}{\underset{i=1}{\cap}} P_i$ where P_i are maximal ideals in $\bar{\mathbb{Z}}$ lying over the same maximal ideal $q\mathbb{Z}$ in $\mathbb{Z}$ with $p \neq q$ and $r > 1$.

Assume that $q \not\equiv 1 \ (p)$. Then $A = \mathbb{Z} + I \to \bar{\mathbb{Z}}$ has a decomposition into minimal morphisms different from those obtained by the canonical decomposition : there exist seminormal orders with conductor I which are not contained in any t-closed order other than $\bar{\mathbb{Z}}$.

Proof: The conductor of A is I by 3.4 (2) since $[\bar{\mathbb{Z}}/P_i : \mathbb{Z}/q\mathbb{Z}] > 1$ for each i. As all $\bar{\mathbb{Z}}/P_i$ are isomorphic $\mathbb{F}_q$-algebras, by [9, 4.1] there exists a subring A' of $\bar{\mathbb{Z}}$ containing A such that $A' \to \bar{\mathbb{Z}}$ is a decomposed morphism with conductor $P'_{r-1} = P_{r-1} \cap P_r$. So A' is a seminormal non t-closed order.

If $r = 2$, then $A \to A'$ is a t-closed morphism so that $A \to A' \to \bar{\mathbb{Z}}$ is not the canonical decomposition. Moreover A' is a seminormal order with conductor I and A' is only contained in the order $\bar{\mathbb{Z}}$.

If $r > 2$, we have $I = \overset{r-1}{\underset{i=1}{\cap}} P'_i$ where $P'_i = P_i \cap A'$. But I is maximal in A and not in A' so I is the conductor of $A \to A'$. Let B be the t-closure of $A \to A'$. Then $B \to A'$ is a t-closed morphism and so is spectrally bijective. If $R_i = P'_i \cap B$ we still have $I = \overset{r-1}{\underset{i=1}{\cap}} R_i$. Moreover for each i, the isomorphisms $A/I \simeq B/R_i$ imply that $B/R_i \to \bar{\mathbb{Z}}/P_i$ is never an isomorphism and so each P_i contains the conductor J of $B \to \bar{\mathbb{Z}}$; as $J \supset I$ then $I = J$. So B is a seminormal non t-closed order with conductor I since A' is not t-closed. The same is true for any order contained between A and B.

This last proof provides a decomposition of $A \to \bar{\mathbb{Z}}$ into minimal morphisms with a minimum number of minimal morphisms, which cannot be gotten in the general case [9, Section 4, Remark] :

Theorem 4.3 *Let p be a prime integer, $\zeta \neq 1$ a pth root of unity and consider the ring of integers $\bar{\mathbb{Z}} = \mathbb{Z}[\zeta]$ of the field $K = \mathbb{Q}[\zeta]$.*

Let $I = \overset{r}{\underset{i=1}{\cap}} P_i$ where P_i are maximal ideals in $\bar{\mathbb{Z}}$ lying over the same maximal ideal $q\mathbb{Z}$ in $\mathbb{Z}$ with $p \neq q$ and $r > 1$.

Assume that $q \not\equiv 1 \ (p)$. The decomposition of $A = \mathbb{Z} + I \to \bar{\mathbb{Z}}$ into minimal morphisms with a minimum number of minimal morphisms is composed of $r - 1$ decomposed morphisms and s inert morphisms where s is the number of prime divisors (with their multiplicity) of $f = [\bar{\mathbb{Z}}/P_i : \mathbb{Z}/q\mathbb{Z}]$.

Proof : In the proof of 4.2 we showed the existence of A' such that P'_{r-1} is the conductor of the decomposed morphism $A' \to \bar{\mathbb{Z}}$. So for each i we get $A'/P'_{r-1} \simeq \bar{\mathbb{Z}}/P_i \simeq A'/P'_i$. Successive applications of [9, 4.1] lead to an order

B containing A with a maximal ideal J such that each P_i lies over J and such that any decomposition of $B \to \bar{Z}$ into minimal morphisms is composed only of decomposed morphisms. Then $\text{Spec}(B) \to \text{Spec}(A)$ is bijective. As $A \to B$ is a seminormal morphism, $A \to B$ is a t-closed morphism with conductor $I = J$ [6, 3.12]. Furthermore $B/I \simeq \bar{Z}/P_i$ for each i. By [9, 3.4] there are exactly s inert morphisms in the decomposition of $A \to B$ into minimal morphisms and $r - 1$ decomposed morphisms in the decomposition of $B \to \bar{Z}$ into minimal morphisms. This is the least decomposition of $A \to \bar{Z}$ into minimal morphisms by [9, 3.5].

Now we look at the prime ideal $\bar{Z}\xi$ in $\bar{Z}$.

Proposition 4.4 *Let p be a prime integer and $\zeta \neq 1$ a pth root of unity. There is no seminormal order with conductor $\bar{Z}(\zeta - 1)$.*

Proof: As $p\bar{Z} = \bar{Z}\xi^{p-1}$ where $\xi = 1 - \zeta$ and $\bar{Z}\xi$ is a maximal ideal in $\bar{Z}$, we get that $[\bar{Z}/\bar{Z}\xi : Z/pZ] = 1$ so that $Z + \bar{Z}\xi = \bar{Z}$. Therefore there is no seminormal order A with conductor $\bar{Z}\xi$.

To conclude we obtain by 3.3 all the seminormal orders with a given conductor I and none of the maximal ideals of $\bar{Z}$ containing I lie over pZ.

We end by an example illustrating these results.

Example: Consider the ring of integers of the cyclotomic extension $K = Q[\zeta]$ where $\zeta \neq 1$ is a 7th root of unity. The decomposition of the prime 2 in $\bar{Z}$ is $2\bar{Z} = P_1 P_2$ where P_i are maximal ideals in $\bar{Z}$ with $[\bar{Z}/P_i : Z/2Z] = 3$ for $i = 1, 2$. We are going to determine the seminormal and t-closed orders with conductor $I = 2\bar{Z}$.

By 4.1 (2) and 3.6 the only t-closed order with conductor I is $(Z + P_1) \cap (Z + P_2)$.

Let us search for seminormal non t-closed orders B with conductor I. So $B \supset A = Z + I$, a seminormal order and $B \not\supset Z + P_i$ for $i = 1, 2$. Let $B \neq A$. As B is not t-closed, $B \to \bar{Z}$ is not t-closed and then not spectrally bijective [6, 4.5]. Then I is a maximal ideal in B as in A and $A \to B$ is a t-closed morphism. So B/I is a field extension of $A/I \simeq Z/2Z$. As $\bar{Z}/P_i$ is a B/I-vector space, the equality $[\bar{Z}/P_i : A/I] = [\bar{Z}/P_i : B/I][B/I : A/I]$ gives $[B/I : A/I] = 3$. Moreover $B/I \simeq \bar{Z}/P_i$ and $B \to \bar{Z}$ is a decomposed morphism. Then $A/I \to B/I$ is a field extension generated by an element $x \in \bar{Z}/I$ the minimal polynomial of which is of degree 3. The only irreducible polynomials with degree 3 in $F_2[X]$ are $\bar{1} + X + X^3$ and $\bar{1} + X^2 + X^3$. It remains to find $x \in \bar{Z}/I$ with minimal polynomial $\bar{1} + X + X^3$ or $\bar{1} + X^2 + X^3$.

If $\bar{\zeta}$ is the residue class of ζ in $\bar{Z}/I$, we get that $\bar{Z}/I$ is a F_2-vector space with basis $\{\bar{1}, \bar{\zeta}, \ldots, \bar{\zeta}^5\}$ since $[\bar{Z}/I : F_2] = 6$ ($\bar{Z}/I \simeq \bar{Z}/P_1 \times \bar{Z}/P_2$). So we can write $x = \sum_{i=0}^{5} \alpha_i \bar{\zeta}^i$ with $\alpha_i \in \{\bar{0}, \bar{1}\}$ and determine α_i such that $\bar{1} + x + x^3 = \bar{0}$ or $\bar{1} + x^2 + x^3 = \bar{0}$.

After tedious calculations we obtain that the roots of $\bar{1} + X + X^3$ in $\bar{Z}/I$ are :
$x_1 = \bar{\zeta} + \bar{\zeta}^2, x_2 = \bar{1} + \bar{\zeta}^2 + \bar{\zeta}^5, x_3 = \bar{\zeta}^3 + \bar{\zeta}^5, x_4 = \bar{\zeta} + \bar{\zeta}^4, x_5 = \bar{\zeta}^2 + \bar{\zeta}^4,$
$x_6 = \bar{1} + \bar{\zeta}^3 + \bar{\zeta}^4, x_7 = \bar{1} + \bar{\zeta} + \bar{\zeta}^2 + \bar{\zeta}^3 + \bar{\zeta}^4, x_8 = \bar{1} + \bar{\zeta} + \bar{\zeta}^2 + \bar{\zeta}^4 + \bar{\zeta}^5$ and

$x_9 = \bar{\zeta}^2 + \bar{\zeta}^3 + \bar{\zeta}^4 + \bar{\zeta}^5$.

The roots of $\bar{1} + X^2 + X^3$ in $\bar{\mathbb{Z}}/I$ are : $y_1 = \bar{1} + \bar{\zeta} + \bar{\zeta}^2, y_2 = \bar{\zeta}^2 + \bar{\zeta}^5, y_3 = \bar{\zeta}^3 + \bar{\zeta}^4$,
$y_4 = \bar{\zeta} + \bar{\zeta}^2 + \bar{\zeta}^3 + \bar{\zeta}^4$, $y_5 = \bar{1} + \bar{\zeta}^2 + \bar{\zeta}^3 + \bar{\zeta}^4 + \bar{\zeta}^5$,
$y_6 = \bar{1} + \bar{\zeta}^3 + \bar{\zeta}^5, y_7 = \bar{\zeta} + \bar{\zeta}^2 + \bar{\zeta}^4 + \bar{\zeta}^5, y_8 = \bar{1} + \bar{\zeta}^2 + \bar{\zeta}^4$ and $y_9 = \bar{1} + \bar{\zeta} + \bar{\zeta}^4$.

But we can remark that : $x_5 = (x_1)^2, x_4 = x_1 + (x_1)^2, y_1 = \bar{1} + x_1, y_3 = (y_2)^2$,
$y_5 = \bar{1} + y_2 + (y_2)^2, x_6 = \bar{1} + (y_2)^2, x_2 = \bar{1} + y_2. x_9 = y_2 + (y_2)^2, x_8 = (x_3)^2$,
$y_4 = x_3 + (x_3)^2, x_7 = \bar{1} + x_3 + (x_3)^2, y_6 = 1 + x_3, x_8 = 1 + y_7, y_8 = 1 + x_5$ and
$y_9 = 1 + x_4$.

This gives the following field extensions of $\mathbb{F}_2$ contained in $\bar{\mathbb{Z}}/I$:

$\mathbb{F}_2[x_1] = \mathbb{F}_2[x_5] = \mathbb{F}_2[y_1] = \mathbb{F}_2[x_4] = \mathbb{F}_2[y_8] = \mathbb{F}_2[y_9] = K_1$
$\mathbb{F}_2[y_2] = \mathbb{F}_2[y_3] = \mathbb{F}_2[y_5] = \mathbb{F}_2[x_6] = \mathbb{F}_2[x_2] = \mathbb{F}_2[x_9] = K_2$
$\mathbb{F}_2[x_3] = \mathbb{F}_2[x_8] = \mathbb{F}_2[y_4] = \mathbb{F}_2[x_7] = \mathbb{F}_2[y_6] = \mathbb{F}_2[y_7] = K_3$.

These three fields are distinct since $y_2, x_3 \notin K_1, x_1, x_3 \notin K_2$ and $y_2, x_1 \notin K_3$.

To sum up, there are four seminormal non t-closed orders with conductor I : the rings $A = \mathbb{Z} + I$, $A_1 = A[\zeta + \zeta^2]$, $A_2 = A[\zeta^3 + \zeta^5]$ and $A_3 = A[\zeta^2 + \zeta^5]$ which give the following diagram :

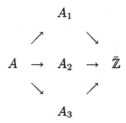

where $A \to A_i$ are inert morphisms and $A_i \to \bar{\mathbb{Z}}$ are decomposed morphisms for $i = 1, 2, 3$. In contrast to what occurs with the canonical decomposition, there does not exist a unique order B such that we get the decomposition $A \to B \to \bar{\mathbb{Z}}$ with $A \to B$ (resp. $B \to \bar{\mathbb{Z}}$) composed of inert (resp. decomposed) morphisms.

5 - Minimal morphisms and quadratic orders

Let d be a square-free integer and consider the quadratic number field $K = \mathbb{Q}(\sqrt{d})$. It is well-known that the ring of integers of K is $\mathbb{Z}[\omega_d]$, where

$$\omega_d = \begin{cases} \dfrac{1 + \sqrt{d}}{2} & \text{if } d \equiv 1 \pmod 4 \\[2mm] \sqrt{d} & \text{if } d \equiv 2, 3 \pmod 4 \end{cases}$$

and $\mathbb{Z}[\omega_d]$ is a free $\mathbb{Z}$-module with basis $\{1, \omega_d\}$.

For a prime integer p we have one of the three decompositions in $\mathbb{Z}[\omega_d]$ (we denote by $\left(\dfrac{d}{p}\right)$ the Legendre symbol of d relative to p) :

• there exists a maximal ideal P in $\mathbb{Z}[\omega_d]$ with $p\mathbb{Z}[\omega_d] = P^2$; then p is **ramified**. This occurs when p is odd and divides d or when $p = 2$ and $d \equiv 2, 3 \pmod 4$.

• there exist two maximal ideals P_1, P_2 in $\mathbb{Z}[\omega_d]$ such that $p\mathbb{Z}[\omega_d] = P_1 \cap P_2$; then p is said to be **decomposed**. This occurs when p is odd, does not divide d and

$$\left(\frac{d}{p}\right) = 1 \text{ or when } p = 2 \text{ and } d \equiv 1 \pmod 8.$$

• $p\mathbb{Z}[\omega_d]$ is a maximal ideal in $\mathbb{Z}[\omega_d]$ and p is said to be **inert**. This occurs when p

is odd, does not divide d and $\left(\dfrac{d}{p}\right) = -1$ or when $p = 2$ and $d \equiv 5 \pmod 8$.

In addition, p ramified implies $\mathbb{Z}[\omega_d]/P \simeq \mathbb{F}_p$, p decomposed implies $\mathbb{Z}[\omega_d]/P_i \simeq \mathbb{F}_p$, for $i = 1, 2$ and p inert implies that $\mathbb{Z}[\omega_d]/p\mathbb{Z}[\omega_d]$ is a quadratic field extension of $\mathbb{F}_p$ (see for example [13]). This is the reason why P. Samuel suggested us to term so the minimal morphisms.

An algebraic order of K is of the form $\mathbb{Z}[n\omega_d]$, where n is an integer such that $n \geq 1$, and is not integrally closed if $n \geq 2$.

We know by [4, 4.1] that for $n, m \in \mathbb{N}^*$ we have $\mathbb{Z}[n\omega_d] \subset \mathbb{Z}[m\omega_d]$ if and only if m divides n and, under this assumption, $\mathbb{Z}[n\omega_d] \to \mathbb{Z}[m\omega_d]$ is an order morphism the conductor of which is $\dfrac{n}{m}\mathbb{Z}[m\omega_d] = \dfrac{n}{m}\mathbb{Z}[m\omega_d] \cap \mathbb{Z}[n\omega_d]$. Moreover, if p is a prime integer, $p\mathbb{Z}[m\omega_d]$ is a maximal ideal of $\mathbb{Z}[pm\omega_d]$ such that $\mathbb{Z}[pm\omega_d]/p\mathbb{Z}[m\omega_d] \to \mathbb{F}_p$ is an isomorphism.

When m divides n, set $n = m \prod p_i^{e_i}$, where the p_i are prime integers with $e_i \geq 1$, $A = \mathbb{Z}[n\omega_d]$ and $B = \mathbb{Z}[m\omega_d]$. So the conductor of $A \to B$ is $I = \prod p_i^{e_i} B = \cap (p_i B)^{e_i}$, where the ideals $(p_i B)^{e_i}$ are pairwise comaximal.

We first study the extensions $\mathbb{Z}[p\omega_d] \hookrightarrow \mathbb{Z}[\omega_d]$ for a prime integer p.

Proposition 5.1 *Let p be a prime integer, then $f : \mathbb{Z}[p\omega_d] \to \mathbb{Z}[\omega_d]$ is a minimal morphism with the same type as p.*

Proof: $p\mathbb{Z}[p\omega_d]$ is a maximal ideal in $\mathbb{Z}[p\omega_d]$ such that $\mathbb{Z}[p\omega_d]/p\mathbb{Z}[\omega_d] \to \mathbb{F}_p$ is a ring isomorphism.

• if p is **ramified**, there exists a maximal ideal P in $\mathbb{Z}[\omega_d]$ such that $p\mathbb{Z}[\omega_d] = P^2 \subsetneq P$. Moreover $\mathbb{Z}[\omega_d]/p\mathbb{Z}[\omega_d]$ is a two-dimensional $\mathbb{F}_p$-vector space;

hence the isomorphism $\mathbb{Z}[p\omega_d]/p\mathbb{Z}[\omega_d] \simeq \mathbb{F}_p$ gives $[\mathbb{Z}[\omega_d]/p\mathbb{Z}[\omega_d] : \mathbb{Z}[p\omega_d]/p\mathbb{Z}[\omega_d]] = 2$ which in turn implies that f is a minimal ramified morphism.

• if p is **decomposed**, there exist two maximal ideals P_1, P_2 in $\mathbb{Z}[\omega_d]$ such that $p\mathbb{Z}[\omega_d] = P_1 \cap P_2$. The isomorphisms $\mathbb{Z}[\omega_d]/P_i \simeq \mathbb{F}_p$, for $i = 1, 2$, yield $\mathbb{Z}[p\omega_d]/p\mathbb{Z}[\omega_d] \simeq \mathbb{Z}[\omega_d]/P_i$, for $i = 1, 2$. So, f is a decomposed morphism.

• if p is **inert**, $p\mathbb{Z}[\omega_d]$ is a maximal ideal in $\mathbb{Z}[\omega_d]$ and we have $[\mathbb{Z}[\omega_d]/p\mathbb{Z}[\omega_d] : \mathbb{Z}[p\omega_d]/p\mathbb{Z}[\omega_d]] = 2$ is a prime integer, so that f is a minimal inert morphism.

Theorem 5.2 *Let m and n be two integers in $\mathbb{N}^*$ such that m divides n. Then $f : \mathbb{Z}[n\omega_d] \to \mathbb{Z}[m\omega_d]$ is a minimal morphism if and only if there exists a prime integer p such that $n = pm$. Under this condition, we have :*
  (a) *if p divides m, then f is a ramified morphism.*
  (b) *if p does not divide m, then f has the same type as p.*

Proof : As m divides n, put $n = m \prod_{i=1}^{k} p_i^{e_i}$, where the p_i are prime integers and $e_i \geq 1$ for each i. The conductor of f is $I = \prod_{i=1}^{k} p_i^{e_i} \mathbb{Z}[m\omega_d]$ and $I \cap \mathbb{Z} = \prod_{i=1}^{k} p_i^{e_i} \mathbb{Z}$.

If f is a minimal morphism, I is a maximal ideal in $\mathbb{Z}[n\omega_d]$, whence $I \cap \mathbb{Z}$ is a maximal ideal in $\mathbb{Z}$. Then $k = 1$ and $e_1 = 1$, *i.e.* $n = pm$, where p is a prime integer.

Now, assume $n = pm$, with p a prime integer. So $p\mathbb{Z}[m\omega_d]$ is a maximal ideal in $\mathbb{Z}[n\omega_d]$ and is the conductor of f. Moreover, $\mathbb{Z}[n\omega_d]/p\mathbb{Z}[m\omega_d]$ is isomorphic to $\mathbb{F}_p$.

(a) If p divides m, set $m = pm'$. Then $P = p\mathbb{Z}[m'\omega_d]$ is a maximal ideal in $\mathbb{Z}[pm'\omega_d] = \mathbb{Z}[m\omega_d]$. So $P^2 = p^2\mathbb{Z}[m'\omega_d] \subset p\mathbb{Z}[m\omega_d] \subsetneq P$, since $pm'\omega_d \in P \setminus p\mathbb{Z}[m\omega_d]$.

The isomorphism of $\mathbb{F}_p$-vector spaces $\mathbb{Z}[m\omega_d]/p\mathbb{Z}[m\omega_d] \simeq \mathbb{F}_p^2$ leads to $[\mathbb{Z}[m\omega_d]/p\mathbb{Z}[m\omega_d] : \mathbb{Z}[n\omega_d]/p\mathbb{Z}[m\omega_d]] = 2$. Thus f is a ramified morphism.

(b) If m and p are coprime, we have the ring isomorphisms :

$$\mathbb{Z}[pm\omega_d]/p\mathbb{Z}[m\omega_d] \simeq \mathbb{F}_p \simeq \mathbb{Z}[p\omega_d]/p\mathbb{Z}[\omega_d].$$

On the other hand, [4, 2.3] and $\omega_d \in \mathbb{Z}[m\omega_d] + p\mathbb{Z}[\omega_d]$ imply :

$$\mathbb{Z}[m\omega_d]/p\mathbb{Z}[m\omega_d] = \mathbb{Z}[m\omega_d]/(p\mathbb{Z}[\omega_d] \cap \mathbb{Z}[m\omega_d])$$

$$\simeq (\mathbb{Z}[m\omega_d] + p\mathbb{Z}[\omega_d])/p\mathbb{Z}[\omega_d] \simeq \mathbb{Z}[\omega_d]/p\mathbb{Z}[\omega_d].$$

Therefore we can identify $\mathbb{Z}[n\omega_d]/p\mathbb{Z}[m\omega_d] \to \mathbb{Z}[m\omega_d]/p\mathbb{Z}[m\omega_d]$ with $\mathbb{Z}[p\omega_d]/p\mathbb{Z}[\omega_d] \to \mathbb{Z}[\omega_d]/p\mathbb{Z}[\omega_d]$ and the result follows from 5.1. Indeed, as $p\mathbb{Z}[m\omega_d]$ is a maximal ideal in $\mathbb{Z}[n\omega_d]$, we have that $f : \mathbb{Z}[n\omega_d] \to \mathbb{Z}[m\omega_d]$ is a minimal order morphism if and only if $\mathbb{Z}[n\omega_d]/p\mathbb{Z}[m\omega_d] \to \mathbb{Z}[m\omega_d]/p\mathbb{Z}[m\omega_d]$ is a minimal ring morphism. Moreover the type of f is given by $\mathbb{Z}[n\omega_d]/p\mathbb{Z}[m\omega_d] \to \mathbb{Z}[m\omega_d]/p\mathbb{Z}[m\omega_d]$ [8, 2.8].

Let m and n be two integers in $\mathbb{N}^*$ such that m divides n, so that $\mathbb{Z}[n\omega_d] \subset \mathbb{Z}[m\omega_d]$. We want to calculate the number of minimal morphisms arising in a decomposition of $\mathbb{Z}[n\omega_d] \to \mathbb{Z}[m\omega_d]$ into minimal morphisms.

In contrast with the behavior of a general order morphism [8, § IV], for quadratic orders the above-mentioned number is independent of the decomposition :

Theorem 5.3 *Let* $f : \mathbb{Z}[n\omega_d] \to \mathbb{Z}[m\omega_d]$ *be an order morphism, with* $n = m \prod_{i=1}^{k} p_i^{e_i}$,

where the p_i *are prime integers and* $e_i > 0$ *for each* i. *Then* f *is decomposed into*

$s = \sum_{i=1}^{k} e_i$ *minimal morphisms. Such a decomposition is of the form :* $f = f_s \circ \cdots \circ f_1$,

where f_j *is a minimal morphism* $\mathbb{Z}[n_j\omega_d] \to \mathbb{Z}[n_{j+1}\omega_d]$ *such that there exists* $i \in \{1, \dots, k\}$ *with* $n_j = p_i n_{j+1}, n_1 = n$ *and* $n_{s+1} = m$.

Proof : By 5.2, a minimal morphism appearing in the decomposition of f into minimal morphisms is of the form $f_j : \mathbb{Z}[n_j\omega_d] \to \mathbb{Z}[n_{j+1}\omega_d]$, where $n_j = p'_j n_{j+1}$, with p'_j a prime integer. Because of $f = f_s \circ \cdots \circ f_1$, we get $n_1 = n$, $n_{s+1} = m$ and

$$n = m \prod_{j=1}^{s} p'_j = m \prod_{i=1}^{k} p_i^{e_i}. \text{ Hence, } \{p_i\}_{i=1,\dots,k} = \{p'_j\}_{j=1,\dots,s} \text{ and } s = \sum_{i=1}^{k} e_i.$$

Remarks :

(1) The evaluation of the number of minimal morphisms appearing in a decomposition of $\mathbb{Z}[n\omega_d] \to \mathbb{Z}[m\omega_d]$ into minimal morphisms is easy (compare with the general case of [8]).

(2) The number of minimal morphisms appearing in a decomposition of $f : \mathbb{Z}[n\omega_d] \to \mathbb{Z}[m\omega_d]$ into minimal morphisms depends only on $\dfrac{n}{m}$ and so is independent of the position of the different types of minimal morphisms in the decomposition of f into minimal morphisms.

Let us consider the order morphism $\mathbb{Z}[n\omega_d] \to \mathbb{Z}[m\omega_d]$ where m divides n; we can write $n = m \prod_{i=1}^{k} p_i^{e_i}$ where the p_i are prime integers with $e_i \geq 1$ for each i. Set $A = \mathbb{Z}[n\omega_d]$ and $B = \mathbb{Z}[m\omega_d]$.

Put $E = \{p_1, \dots, p_k\}$ and define the following partition $E = E_1 \cup E_2 \cup E_3$:
$E_1 = \{p_i \in E \mid p_i \text{ divides } m \text{ or } p \text{ is ramified in } \mathbb{Z}[\omega_d]\}$
$E_2 = \{p_i \in E \mid p_i \text{ does not divide } m \text{ and } p \text{ is decomposed in } \mathbb{Z}[\omega_d]\}$
$E_3 = \{p_i \in E \mid p_i \text{ does not divide } m \text{ and } p \text{ is inert in } \mathbb{Z}[\omega_d]\}$

$$\text{Set } m_1 = \left(\prod_{p_i \in E_1} p_i^{e_i} \right) \left(\prod_{p_i \in E_2 \cup E_3, e_i > 1} p_i^{e_i - 1} \right), m_2 = \prod_{p_i \in E_2} p_i \text{ and } m_3 = \prod_{p_i \in E_3} p_i.$$

So we have $n = m_1 m_2 m_3 m$.

To end, consider $A_1 = \mathbb{Z}[m_2 m_3 m \omega_d]$ and $A_2 = \mathbb{Z}[m_3 m \omega_d]$. Then we get the following theorem :

Theorem 5.4 *With the above notations,* $A \to A_1 \to A_2 \to B$ *is the canonical decomposition of* $A \to B$, *i.e.* $A_1 = {}^+_B A$ *and* $A_2 = {}^t_B A$.

Proof : (1) Set $m' = m_2 m_3 m$; then $n = m_1 m'$, $A_1 = \mathbb{Z}[m'\omega_d]$. For each prime integer p_i dividing m_1, either p_i is ramified or p_i divides m_2, m_3 or m, whence divides m'. By 5.3, any minimal morphism arising in a decomposition of $A \to A_1$ into minimal morphisms is of the form $\mathbb{Z}[p_i n_j \omega_d] \to \mathbb{Z}[n_j \omega_d]$ with p_i dividing m_1 so is a ramified morphism by 5.2.

(2) Set $m'' = m_3 m$; then $m' = m'' \prod\limits_{p_i \in E_2} p_i$. Any $p_i \in E_2$ does not divide $m_3 m$,

is decomposed and has exponent one in $\dfrac{m'}{m''}$. By 5.3, any minimal morphism

arising in the decomposition of $A_1 \to A_2$ into minimal morphisms has the form $\mathbb{Z}[p_i n_j \omega_d] \to \mathbb{Z}[n_j \omega_d]$ with $p_i \in E_2$ such that p_i does not divide n_j; so, by 5.2, this morphism is a decomposed morphism.

(3) In the same way, $A_2 \to B$ is the morphism $\mathbb{Z}[m'' \omega_d] \to \mathbb{Z}[m \omega_d]$, with $m'' = m \prod\limits_{p_i \in E_3} p_i$. A similar argument shows that $A_2 \to B$ is composed only of

inert morphisms.

To sum up, 2.8 gives the canonical decomposition.

Remark : We recover [4, 4.3], which says that $\mathbb{Z}[m \omega_d]$ is the seminormalization of $\mathbb{Z}[n \omega_d]$ in $\mathbb{Z}[m \omega_d]$, when any prime divisor of $\dfrac{n}{m}$ either divides m or is a ramified prime in $\mathbb{Z}[\omega_d]$.

We can now deduce the following corollaries :

Corollary 5.5 *Let m and n be two integers such that m divides n and set $n = m \prod p_i^{e_i}$, where the p_i are prime integers with $e_i \geq 1$ for each i. Then :*

(1) $\mathbb{Z}[n \omega_d] \to \mathbb{Z}[m \omega_d]$ *is a seminormal morphism if and only if, for each i, p_i does not divide m, p_i is decomposed or inert in $\mathbb{Z}[\omega_d]$ and $e_i = 1$.*

(2) $\mathbb{Z}[n \omega_d] \to \mathbb{Z}[m \omega_d]$ *is a t-closed morphism if and only if, for each i, p_i does not divide m, p_i is inert in $\mathbb{Z}[\omega_d]$ and $e_i = 1$.*

Proof : With notations of 5.4, we have :

(1) $\mathbb{Z}[n \omega_d] \to \mathbb{Z}[m \omega_d]$ is seminormal if and only if $\mathbb{Z}[n \omega_d] = A_1$.

(2) $\mathbb{Z}[n \omega_d] \to \mathbb{Z}[m \omega_d]$ is t-closed if and only if $\mathbb{Z}[n \omega_d] = A_2$.

When $m = 1$, we obtain the seminormalization and the t-closure of $\mathbb{Z}[n \omega_d]$ since $\mathbb{Z}[\omega_d]$ is the integral closure of $\mathbb{Z}[n \omega_d]$.

For the seminormalization, D. Dobbs and M. Fontana [4, 4.4 and 4.5], H. Tanimoto [15, 4.4] and A. Ooishi [5, Example 1] have already obtained this result.

For the t-closedness, a characterization was also given in different forms :

(1) H. Tanimoto [15, 5.1] characterized quasinormal quadratic orders and a one-dimensional Noetherian domain A is t-closed if and only if A is quasinormal [7, 4.1].

(2) D. Dobbs and M. Fontana [4, 2.5] characterized the quadratic orders which are GPVD. A GPVD (*i.e.* a globalized pseudo-valuation domain) is an integral domain R with a Prüfer overring T such that the canonical map $\mathrm{Spec}(T) \to \mathrm{Spec}(R)$ is a bijection and there exists a nonzero radical ideal A common to T and R such that each prime ideal of T (resp. R) which contains A is a maximal ideal of T (resp. R).

We have the following equivalence of t-closed rings and GPVD, under suitable assumptions.

Proposition 5.6 *Let A be a one-dimensional integral domain such that its integral closure is a Prüfer domain finite over A. Then A is t-closed if and only if A is a GPVD.*

Proof: Any GPVD is a LPVD (a locally pseudo-valuation domain) [3, Section 3] and A is a LPVD if and only if A is t-closed [10, 3.3].

Conversely, let A be a t-closed domain with integral closure $\bar{A}$. Thus $A \to \bar{A}$ is a finite t-closed morphism and so a seminormal morphism the conductor of which is a nonzero radical ideal [16, 1.3]. Hence A is a GPVD since $\mathrm{Spec}(\bar{A}) \to \mathrm{Spec}(A)$ is bijective by [6, 4.5].

Corollary 5.7 *Let $n = \prod p_i^{e_i}$, where the p_i are prime integers with $e_i \geq 1$ for each i.*

(1) *Set $E_1 = \{p_i \ / \ p_i \text{ is ramified in } \mathbb{Z}[\omega_d]\}$,*
$E_2 = \{p_i \ / \ p_i \text{ is decomposed in } \mathbb{Z}[\omega_d]\}$, $E_3 = \{p_i \ / \ p_i \text{ is inert in } \mathbb{Z}[\omega_d]\}$,

$$m_1 = \left(\prod_{p_i \in E_1} p_i^{e_i}\right)\left(\prod_{p_i \in E_2 \cup E_3, e_i > 1} p_i^{e_i - 1}\right) \ , \ m_2 = \prod_{p_i \in E_2} p_i \text{ and } m_3 = \prod_{p_i \in E_3} p_i. \text{ Then}$$

$\mathbb{Z}[m_2 m_3 \omega_d] = {}^+\mathbb{Z}[n\omega_d]$ and $\mathbb{Z}[m_3 \omega_d] = {}^t\mathbb{Z}[n\omega_d]$.

(2) *$\mathbb{Z}[n\omega_d]$ is seminormal if and only if, for each $i, e_i = 1$ and p_i is decomposed or inert in $\mathbb{Z}[\omega_d]$.*

(3) *$\mathbb{Z}[n\omega_d]$ is t-closed if and only if for each i, p_i is inert in $\mathbb{Z}[\omega_d]$ and $e_i = 1$.*

We end by an example illustrating this last result :

Example : Let $A = \mathbb{Z}[525j] = \mathbb{Z}[525\omega_d]$ where $j = \dfrac{-1 + i\sqrt{3}}{2}$ and where

$\omega_d = \dfrac{1 + i\sqrt{3}}{2} = \dfrac{1 + \sqrt{-3}}{2}$. Thus we have $d = -3 \equiv 1 \pmod 4$. So the integral

closure of A is $\mathbb{Z}[\omega_d] = \mathbb{Z}[j]$. But $525 = 3 \times 5^2 \times 7$.

- 3 is odd and divides d : 3 is ramified

- 5 is odd, does not divide d and $\left(\dfrac{-3}{5}\right) = -1$ so 5 is inert

- 7 is odd, does not divide d and $\left(\dfrac{-3}{7}\right) = 1$ so 7 is decomposed.

Then $m_3 = 5, m_2 = 7$ and ${}^+A = \mathbb{Z}[35j], {}^tA = \mathbb{Z}[5j]$.

REFERENCES

[1] D. F. Anderson, D. E. Dobbs and J. A. Huckaba, *On seminormal overrings*, Comm. Algebra **10** (1982), 1421-1448.

[2] N. Bourbaki, *Algèbre Commutative, Ch. V-VI* (Hermann, Paris, 1964).

[3] D. E. Dobbs and M. Fontana, *Locally pseudo-valuation domains*, Ann. Mat. Pura Appl. **134** (1983), 147-168.

[4] D. E. Dobbs and M. Fontana, *Seminormal rings generated by algebraic integers*, Matematika **34** (1987), 141-154.

[5] A. Oishi, *On seminormal rings (general survey)*, Lecture Notes , RIMS, Kyoto Univ. **374** (1980), 1-17.

[6] G. Picavet et M. Picavet-L'Hermitte, *Morphismes t-clos*, Comm. Algebra **21** (1993), 179-219.

[7] G. Picavet et M. Picavet-L'Hermitte, *Anneaux t-clos*, Comm. Algebra **23** (1995), 2643-2677.

[8] M. Picavet-L'Hermitte, *Minimal order morphisms*, J. Number Theory **45** (1993), 1-27.

[9] M. Picavet-L'Hermitte, *Decomposition of order morphisms into minimal morphisms*, Math. J. Toyama Univ. **19** (1996), 17-45.

[10] M. Picavet-L'Hermitte, *t-closed pairs*, in "Commutative Ring Theory", Lecture notes in pure and appl. math. **185** (1996), Dekker, 401-415.

[11] M. Picavet-L'Hermitte, *When is $\mathbb{Z}[\alpha]$ seminormal or t-closed ?* , to appear in Boll. UMI.

[12] P. Ribenboim, *Algebraic Numbers* (Wiley-Interscience, New York, 1972).

[13] P. Samuel, *Théorie Algébrique des Nombres*, (Hermann, Paris, 1967).

[14] R. G. Swan, *On seminormality*, J. Algebra **67** (1980), 210-229.

[15] H. Tanimoto, *Normality, seminormality and quasinormality of $\mathbb{Z}[\sqrt[n]{m}]$*, Hiroshima Math. J. **17** (1987), 29-40.

[16] C. Traverso, *Seminormality and Picard group*, Ann. Scuola Norm. Sup. Pisa **24** (1970), 585-595.

[17] O. Zariski and P. Samuel, *Commutative Algebra, Vol.I* (Van Nostrand, Princeton, 1960).

Failure of Krull–Schmidt for Direct Sums of Copies of a Module

ROGER WIEGAND   Department of Mathematics, University of Nebraska, Lincoln, NE 68588-0323

In trying to verify finite representation type for rings without the Krull-Schmidt property for direct-sum decompositions, one needs some control over the number of different direct-sum decompositions of a given module. In this note I give a general result of this type. The main point, however, is a series of examples showing dramatic failure of Krull-Schmidt for direct sums of copies of a single module: I construct a one-dimensional local domain R with the following property: Given any positive integer n there are indecomposable finitely generated R-modules M and X (depending on n) such that X is isomorphic to a direct summand of nM (the direct sum of n copies of M), but X is not isomorphic to a direct summand of $(n-1)M$.

1 BACKGROUND AND MOTIVATION

Suppose $(R, \mathbf{m})$ is a local ring (a commutative Noetherian ring with exactly one maximal ideal $\mathbf{m}$). A *maximal Cohen-Macaulay module* (or MCM module) over R is a finitely generated R-module M such that $\mathbf{m}$ contains an

---

This research was partially supported by the National Science Foundation.

M-regular sequence of length $d := \dim(R)$. For one-dimensional domains the MCM modules are exactly the finitely generated torsion-free modules, and for integrally closed domains of dimension two they are the finitely generated reflexive modules. One says that R has finite Cohen-Macaulay type provided there are only finitely many indecomposable maximal Cohen-Macaulay modules up to isomorphism.

For complete (and more generally Henselian) local rings one has the Krull-Schmidt uniqueness theorem for direct-sum decompositions, and a lot is known about Henselian rings of finite representation type and their MCM modules. An excellent exposition of the state of the art can be found in Yoshino's book [Y]. In 1987 F.-O. Schreyer [S] conjectured that the local ring $(R, \mathbf{m})$ has finite Cohen-Macaulay type if and only if its $\mathbf{m}$-adic completion $(\hat{R}, \hat{\mathbf{m}})$ has finite Cohen-Macaulay type. In [W3] I was able to prove the "if" direction in general (that is, finite Cohen-Macaulay type descends from the completion) and "only if" under mild hypotheses. A key component of the proof was a finiteness theorem for direct sums of copies of a single module. In the next section I will state and prove this theorem and use it to prove descent of finite Cohen-Macaulay type.

I spoke about some of this material in the commutative algebra seminar at Purdue University. I am grateful to Lucho Avramov, Bill Heinzer, Craig Huneke and Sylvia Wiegand for many helpful comments. Thanks also to the referee for a careful reading and for several suggestions.

• **Notation** For modules X and Y, we write $X \mid Y$ to indicate that X is isomorphic to a direct summand of Y. The direct sum of n copies of X will be denoted by nX.

2 A FINITENESS THEOREM

The following theorem appears in [W3]:

THEOREM.   Let R be a semilocal (commutative, Noetherian) ring and M a finitely generated R-module. Given a positive integer n, let $S_n(M)$ be the set of isomorphism classes of indecomposable R-modules that can occur as direct summands of the direct sum of n copies of M. Then $\bigcup_{n=1}^{\infty} S_n(M)$ is finite.

While I have not been able to find this result in the literature, it seems likely that experts in representation theory must have been aware of it. Certainly the main ideas in its proof have been used often.

Proof of the Theorem. Let $\mathcal{C}$ be the full subcategory of the category of finitely generated R-modules consisting of direct summands of direct sums of copies of M. Let E be the endomorphism ring of M, acting on the left, and let $\mathbf{P}_E$ denote the category of finitely generated projective right E-modules. The maps $X \to \operatorname{Hom}_R(M, X)$ and $P \to P \otimes_E M$ determine an equivalence of categories between $\mathcal{C}$ and $\mathbf{P}_E$. (This argument goes back to a 1969 paper [D] of Andreas Dress.)

Thus we have translated our problem into a study of projective modules. We shall show that $\mathbf{P}_E$ has only finitely many indecomposables. The first step is to show that E is a module-finite R-algebra. Let F be a free R-module admitting a surjective R-homomorphism $\phi : F \to M$. Applying the functor $\operatorname{Hom}_R(\ , M)$, we get an injection $E \to \operatorname{Hom}_R(F, M)$. Since R is Noetherian and the R-module $\operatorname{Hom}_R(F, M)$ is just a direct sum of copies of M, E is a finitely generated R-module.

Next we show that E is semilocal (but not necessarily commutative). Recall that this means that E/J is Artinian, where J is the Jacobson radical of E. Let L be the Jacobson radical of R. It is easy to see that $LE \subseteq J$, whence E/J is a module-finite R/L-algebra. But R/L is just a direct product of finitely many fields, so clearly E/J is an Artinian ring.

Fuller and Shutters [FS] show that a semilocal ring has only finitely many non-isomorphic indecomposable finitely generated projective right modules, as desired. The proof is so pleasant that it seems worthwhile to give it here. The main combinatorial idea appears in a 1963 paper [J] by Alfredo Jones. Let $X_1 \ldots, X_t$ be the indecomposable right E/J-modules. Given $P \in \mathbf{P}_E$, let $P/PJ \cong n_1 X_1 \oplus \ldots \oplus n_t X_t$. The non-negative integers n_i are uniquely determined (by the Krull-Schmidt theorem for modules of finite length), and we let $\theta(P) = (n_1, \ldots, n_t) \in \mathbf{S} :=$ direct product of t copies of the set of non-negative integers. We give $\mathbf{S}$ the product ordering; thus $(n_1, \ldots, n_t) \leq (m_1, \ldots, m_t)$ if and only if $n_i \leq m_i$ for each i. For $P, Q \in \mathbf{P}_E$ we observe that $P \mid Q$ if and only if $\theta(P) \leq \theta(Q)$. The forward implication is clear. For the converse, note that if $\theta(P) \leq \theta(Q)$ then $P/PJ \mid Q/QJ$. Choose a surjection $\bar{f} : Q/QJ \to P/PJ$. Since Q is projective $\bar{f}$ lifts to an E-homomorphism $Q \to P$, which must be surjective (and hence split) by

Nakayama's lemma. It follows that $P \in \mathbf{P}_E$ is indecomposable if and only if $\theta(P)$ is a minimal element of $\{\theta(Q)|Q \in \mathbf{P}_E\}$. Furthermore, if $\theta(P) = \theta(Q)$ then P and Q must be isomorphic. Therefore it suffices to point out that the set $\mathbf{C}$ of minimal elements of $\{\theta(Q)|Q \in \mathbf{P}_E\}$ is finite since it is a "clutter" (a set of pairwise incomparable elements in a poset). In fact, every clutter in $\mathbf{S}$ is finite. An easy way to see this is to note that if A and B are posets with DCC, both having the property that every clutter is finite, then $A \times B$ (with the product ordering) has the same properties.

Where does the sequence $\mathcal{S}_1(M) \subseteq \ldots \subseteq \mathcal{S}_n(M) \subseteq \mathcal{S}_{n+1}(M) \subseteq \ldots$ stabilize? We will show in the next two sections that no fixed integer n suffices, even when the ring is fixed.

We conclude this section by proving descent of finite Cohen-Macaulay type, using the finiteness theorem. We need one more fact, which follows easily from the results in [G], but we include a direct proof for the convenience of the reader. This proof is the same as that on pages 19 and 20 of [RR], although the result there is stated only for discrete valuation rings.

PROPOSITION.   Let $(R, \mathbf{m})$ be a commutative, Noetherian local ring with $\mathbf{m}$-adic completion $(\hat{R}, \hat{\mathbf{m}})$, and let X and Y be finitely generated R-modules. Then $X \mid Y$ if and only if $\hat{X} \mid \hat{Y}$.

Proof. The "only if" direction is clear; we prove the converse. Choose $\hat{R}$-homomorphisms $\phi : \hat{X} \to \hat{Y}$ and $\psi : \hat{Y} \to \hat{X}$ such that $\psi\phi = 1_{\hat{X}}$. Since $H := \mathrm{Hom}_R(X, Y)$ is a finitely generated R-module (by the argument in the second paragraph of the proof of the theorem above) it follows that $\hat{R} \otimes_R H = \mathrm{Hom}_{\hat{R}}(\hat{X}, \hat{Y})$. But $\hat{R} \otimes_R H = \hat{H}$. Therefore ϕ can be approximated to any order by an element of H. In fact, order 1 will suffice: Choose $f \in \mathrm{Hom}_R(X, Y)$ such that $\hat{f} - \phi \in \hat{\mathbf{m}}\hat{H}$. Similarly, there exists $g \in \mathrm{Hom}_R(Y, X)$ with $\hat{g} - \psi \in \hat{\mathbf{m}}\mathrm{Hom}_{\hat{R}}(\hat{Y}, \hat{X})$. Then the image of $\hat{g}\hat{f} - 1_{\hat{X}}$ is in $\hat{\mathbf{m}}\hat{X}$, and now Nakayama's lemma implies that $\hat{g}\hat{f}$ is surjective, and therefore an isomorphism. It follows that $\hat{g}$ is a split surjection (with splitting map $\hat{f}(\hat{g}\hat{f})^{-1}$). By faithful flatness g is a split surjection.

The proof of descent of finite Cohen-Macaulay type is now easy: Let $Y_1, \ldots, Y_t$ be a complete list of representatives for the isomorphism classes of indecomposable MCM $\hat{R}$-modules. By renumbering we may assume that $i \leq s$ if and only if there exists a MCM R-module, say X_i, such that $Y_i \mid \hat{X}_i$.

Put $X = X_1 \oplus \ldots \oplus X_s$. If, now, Z is a MCM R-module, write $\hat{Z} \cong n_1 Y_1 \oplus \ldots \oplus n_s Y_s$. Then $\hat{Z} \mid n\hat{X}$, where $n = \max\{n_1, \ldots, n_s\}$, and by the proposition $Z \mid nX$. By the finiteness theorem, there are only finitely many indecomposable Z's.

• From now on all rings are assumed to be commutative and Noetherian, and modules are always assumed to be finitely generated.

3 CONSTRUCTION OF THE RING

We will construct our rings and modules via pullbacks, as in [WW]. To make the construction work, all we need is a one-dimensional local domain whose completion is reduced and has two components, each of multiplicity at least 4. For definiteness, let k be a field, and put $A := k[x]/(x^4)$ and $B := k[x]/(x-1)^4$. Let $\widetilde{R}$ be the semilocalization of $k[x]$ at $(x) \cup (x-1)$, let $\pi : \widetilde{R} \to A \times B$ be the obvious surjection, and let $j : k \hookrightarrow A \times B$ be the diagonal embedding. Define R by the following pullback diagram:

$$
\begin{array}{ccc}
R & \longrightarrow & \widetilde{R} \\
\downarrow & & \downarrow{\scriptstyle \pi} \\
k & \xrightarrow{\ j\ } & A \times B
\end{array}
$$

Then R is a one-dimensional local domain, $\widetilde{R}$ is its integral closure, and $\mathbf{m} := \mathrm{Ker}(\pi)$ is the maximal ideal of R and the conductor of R in $\widetilde{R}$. (Proposition 3.1 of [WW] gives all of this information except for the fact that R is local, and this follows from the fact that every element of $R - \mathbf{m}$ is outside $(x) \cup (x-1)$.)

We recall some terminology and results from [W1, §1]. Put $C := A \times B$, and recall that a (k, C)-module is a pair (V, W) consisting of a finitely generated projective C-module W and a k-subspace V of W such that $CV = W$. The (k, C)-modules form an additive category with the Krull–Schmidt uniqueness theorem for direct sum decompositions. If M is a finitely generated torsion-free R-module of rank r, let $\widetilde{R}M = (\widetilde{R} \otimes_R M)/\text{torsion}$. Then $\overline{M} := (M/\mathbf{m}M, \widetilde{R}M/\mathbf{m}M)$ is a (k, C)-module in which the second component $\widetilde{R}M/\mathbf{m}M$ is a *free* C-module. Conversely, given a (k, C) module $E = (V, W)$ with $W \cong C^{(r)}$ we can map $\widetilde{R}^{(r)}$ onto W and form the pullback

$E^{\#}$ of $\widetilde{R}^{(r)}$ and V over W. Then $E^{\#}$ is a finitely generated torsion-free R-module, and $\overline{E^{\#}} \cong E$. Moreover, $E^{\#}$ is unique up to isomorphism. In fact (since R is semilocal) finitely generated torsion-free R-modules M and N are isomorphic if and only if $\overline{M} \cong \overline{N}$.

The *rank* of a (k, C)-module $E = (V, W)$, is the pair (r, s), where $W \cong A^{(r)} \times B^{(s)}$. The failure of Krull-Schmidt for R-modules comes from the fact that only the (k, C)-modules of *constant rank* actually come from R-modules.

If M and N are finitely generated torsion-free R-modules, we note that $M \mid N$ if (and only if) $\overline{M} \mid \overline{N}$. For, if $\overline{M} \oplus E \cong \overline{N}$, then E must have constant rank; and we have $M \oplus E^{\#} \cong N$.

4 CONSTRUCTION OF THE MODULES

Fix an integer $n \geq 1$. A basic pathological construction due to Drozd and Roĭter [DR] (or see [W2, (2.5), (2.6)] for more detail) shows that since A contains elements α and β with $\{1, \alpha, \alpha^2, \beta\}$ linearly independent over k, there is an indecomposable (k, A)-module E of rank n. When viewed as a (k, C)-module, E is still indecomposable, and its rank is $(n, 0)$. Similarly there is an indecomposable (k, C)-module F of rank $(0, n + 1)$. Let $G = (k, B)$, of rank $(0, 1)$. We consider (k, C)-modules of the form $H(a, b, c) := aE \oplus bF \oplus cG$, of rank $(na, (n + 1)b + c)$. Then $H(a, b, c)$ comes from a torsion-free R-module if and only if

$$(\dagger) \qquad\qquad na = (n + 1)b + c, \quad a \geq 0, \quad b \geq 0, \quad c \geq 0.$$

Moreover, since $H(a, b, c) \mid H(a', b', c')$ if and only if $(a, b, c) \leq (a', b', c')$, we see that $H(a, b, c)$ comes from an *indecomposable* torsion-free R-module if and only if (a, b, c) is a *minimal* non-zero solution to $(\dagger)$.

In particular, we have have the indecomposable modules $X := H(1, 0, n)^{\#}$ and $M := H(n, n-1, 1)^{\#}$. Since $H(1, 0, n) \mid H(n^2, n^2 - n, n) = nH(n, n - 1, 1)$, we know that $X \mid nM$. On the other hand $H(1, 0, n)$ is not isomorphic to a direct summand of $H(n^2 - n, (n-1)^2, n-1) = (n-1)H(n, n-1, 1)$, by the Krull-Schmidt theorem for (k, C)-modules; and hence X is not ismorphic to a direct summand of $(n - 1)M$. Thus $\mathcal{S}_{n-1}(M)$ is properly contained in $\mathcal{S}_n(M)$.

REFERENCES

[D] A. Dress, *On the decomposition of modules*, Bull. Amer. Math. Soc. **75** (1969), 984–986.

[DR] Ju. A. Drozd and A. V. Roĭter, *Commutative rings with a finite number of indecomposable integral representations* (Russian), Izv. Akad. Nauk. SSSR, Ser. Mat. **31** (1967), 783–798.

[FS] K. R. Fuller and W. A. Shutters, *Projective modules over noncommutative semilocal rings* Tôhoku Math. J. **27** (1975), 303–311.

[G] R. Guralnick, *Lifting homomorphisms of modules*, Illinois J. Math. **29** (1985), 153–156.

[J] A. Jones, *Groups with a finite number of indecomposable integral representations*, Mich. Math. J. **10** (1963), 257–261.

[RR] I. Reiner and K. W. Roggenkamp, *Integral Representations*, Lecture Notes in Math., vol. 744, Springer-Verlag, Berlin (1979).

[S] F.-O. Schreyer, *Finite and countable CM-representation type*, "Singularities, Representation of Algebras, and Vector Bundles: Proceedings Lambrecht 1985", G.-M. Greuel and G. Trautmann, eds., Lecture Notes in Math., vol. 1273, Springer-Verlag, Berlin (1987), 9–34.

[W1] R. Wiegand, *Cancellation over commutative rings of dimension one and two*, J. Algebra **88** (1984), 438–450.

[W2] R. Wiegand, *Noetherian rings of bounded representation type*, "Commutative Algebra, Proceedings of a Microprogram held June 15–July 2, 1987", Springer, New York (1989), 301–311.

[W3] R. Wiegand, *Local rings of finite Cohen-Macaulay type*, J. Algebra **203** (1998), 156–168.

[WW] R. Wiegand and S. Wiegand, *Stable isomorphism of modules over one-dimensional rings*, J. Algebra **107** (1987), 425–435.

[Y] Y. Yoshino, *Cohen-Macaulay Modules over Cohen-Macaulay Rings*, London Math. Soc. Lect. Notes, vol. 146 (1990).

Index